ENCYCLOPEDIA OF MATERIALS CHARACTERIZATION

MATERIALS CHARACTERIZATION SERIES
Surfaces, Interfaces, Thin Films

Series Editors: C. Richard Brundle and Charles A. Evans, Jr.

Series Titles

Encyclopedia of Materials Characterization, C. Richard Brundle, Charles A. Evans, Jr., and Shaun Wilson

Characterization of Metals and Alloys, Paul H. Holloway and P. N. Vaidyanathan

Characterization of Ceramics, Ronald E. Loehman

Characterization of Polymers, Ned J. Chou, Stephen P. Kowalczyk, Ravi Saraf, and Ho-Ming Tong

Characterization in Silicon Processing, Yale Strausser

Characterization in Compound Semiconductor Processing, Yale Strausser

Characterization of Integrated Circuit Packaging Materials, Thomas M. Moore and Robert G. McKenna

Characterization of Catalytic Materials, Israel E. Wachs

Characterization of Composite Materials, Hatsuo Ishida

Characterization of Optical Materials, Gregory J. Exarhos

Characterization of Tribological Materials, William A. Glaeser

Characterization of Organic Thin Films, Abraham Ulman

ENCYCLOPEDIA OF MATERIALS CHARACTERIZATION

Surfaces, Interfaces, Thin Films

EDITORS

C. Richard Brundle

Charles A. Evans, Jr.

Shaun Wilson

MANAGING EDITOR

Lee E. Fitzpatrick

BUTTERWORTH-HEINEMANN
Boston London Oxford Singapore Sydney Toronto Wellington

MANNING
Greenwich

This book was acquired, developed, and produced by Manning Publications Co.

Recognizing the importance of preserving what has been written, it is the policy of Butterworth-Heinemann and of Manning to have the books they publish printed on acid-free paper, and we exert our best efforts to that end.

Library of Congress Cataloging–in–Publication Data
Brundle, C. R.
Encyclopedia of materials characterization: surfaces, interfaces, thin films/C. Richard Brundle, Charles A. Evans, Jr., Shaun Wilson.
p. cm.—(Materials characterization series)
Includes bibliographical references and index.
ISBN 0–7506-9168-9
1. Surfaces (Technology)—Testing. I. Evans, Charles A. II. Wilson, Shaun. III. Title. IV. Series.
TA418.7.B73 1992 92–14999
620'.44—dc20 CIP

Butterworth-Heinemann
80 Montvale Avenue
Stoneham, MA 02180

Manning Publications Co.
3 Lewis Street
Greenwich, CT 06830

10 9 8 7 6 5 4 3 2 1

Printed in the United States of America

Contents

STRUCTURE DETERMINATION BY DIFFRACTION AND SCATTERING

ELECTRON EMISSION SPECTROSCOPIES

X-RAY EMISSION TECHNIQUES

VISIBLE/UV EMISSION, REFLECTION, AND ABSORPTION

VIBRATIONAL SPECTROSCOPIES AND NMR

ION SCATTERING TECHNIQUES

MASS AND OPTICAL SPECTROSCOPIES

NEUTRON AND NUCLEAR TECHNIQUES

PHYSICAL AND MAGNETIC PROPERTIES

Preface to Series

This *Materials Characterization Series* attempts to address the needs of the practical materials user, with an emphasis on the newer areas of surface, interface, and thin film microcharacterization. The Series is composed of the leading volume, *Encyclopedia of Materials Characterization*, and a set of about 10 subsequent volumes concentrating on characterization of individual materials classes.

In the *Encyclopedia*, 50 brief articles (each 10–18 pages in length) are presented in a standard format designed for ease of reader access, with straightforward technique descriptions and examples of their practical use. In addition to the articles, there are one-page summaries for every technique, introductory summaries to groupings of related techniques, a complete glossary of acronyms, and a tabular comparison of the major features of all 50 techniques.

The 10 volumes in the Series on characterization of particular materials classes include volumes on silicon processing, metals and alloys, catalytic materials, integrated circuit packaging, etc. Characterization is approached from the materials user's point of view. Thus, in general, the format is based on properties, processing steps, materials classification, etc., rather than on a technique. The emphasis of all volumes is on surfaces, interfaces, and thin films, but the emphasis varies depending on the relative importance of these areas for the materials class concerned. Appendixes in each volume reproduce the relevant one-page summaries from the *Encyclopedia* and provide longer summaries for any techniques referred to that are not covered in the *Encyclopedia*.

The concept for the Series came from discussion with Marjan Bace of Manning Publications Company. A gap exists between the way materials characterization is often presented and the needs of a large segment of the audience—the materials user, process engineer, manager, or student. In our experience, when, at the end of talks or courses on analytical techniques, a question is asked on how a particular material (or processing) characterization problem can be addressed the answer often is that the speaker is "an expert on the technique, not the materials aspects, and does not have experience with that particular situation." This Series is an attempt to bridge this gap by approaching characterization problems from the side of the materials user rather than from that of the analytical techniques expert.

We would like to thank Marjan Bace for putting forward the original concept, Shaun Wilson of Charles Evans and Associates and Yale Strausser of Surface Science Laboratories for help in further defining the Series, and the Editors of all the individual volumes for their efforts to produce practical, materials user based volumes.

C.R. Brundle   C.A. Evans, Jr.

Preface

This volume contains 50 articles describing analytical techniques for the characterization of solid materials, with emphasis on surfaces, interfaces, thin films, and microanalytical approaches. It is part of the *Materials Characterization Series*, copublished by Butterworth-Heinemann and Manning. This volume can serve as a stand-alone reference as well as a companion to the other volumes in the Series which deal with individual materials classes. Though authored by professional characterization experts the articles are written to be easily accessible to the materials user, the process engineer, the manager, the student—in short to all those who are not (and probably don't intend to be) experts but who need to understand the potential applications of the techniques to materials problems. Too often, technique descriptions are written for the technique specialist.

With 50 articles, organization of the book was difficult; certain techniques could equally well have appeared in more than one place. The organizational intent of the Editors was to group techniques that have a similar physical basis, or that provide similar types of information. This is not the traditional organization of an encyclopedia, where articles are ordered alphabetically. Such ordering seemed less useful here, in part because many of the techniques have multiple possible acronyms (an *Acronyms Glossary* is provided to help the reader).

The articles follow a standard format for each technique: A clear description of the technique, the range of information it provides, the range of materials to which it is applicable, a few typical examples, and some comparison to other related techniques. Each technique has a "quick reference," one-page summary in Chapter 1, consisting of a descriptive paragraph and a tabular summary.

Some of the techniques included apply more broadly than just to surfaces, interfaces, or thin films; for example X-Ray Diffraction and Infrared Spectroscopy, which have been used for half a century in bulk solid and liquid analysis, respectively. They are included here because they have by now been developed to also apply to surfaces. A few techniques that are applied almost entirely to bulk materials (e.g., Neutron Diffraction) are included because they give complementary information to other methods or because they are referred to significantly in the 10 materials volumes in the Series. Some techniques were left out because they were considered to be too restricted to specific applications or materials.

We wish to thank all the many contributors for their efforts, and their patience and restraint in dealing with the Editors who took a fairly demanding approach to establishing the format, length, style, and content of the articles. We hope the readers will consider our efforts worthwhile. Finally, we would like to thank Lee Fitzpatrick of Manning Publications Co. for her professional help as Managing Editor.

C.R.Brundle   C.A. Evans, Jr.   S.Wilson

Acronyms Glossary

This glossary lists all the acronyms referred to in the encyclopedia together with their meanings. The major technique acronyms are listed alphabetically. Alternatives to these acronyms are listed immediately below each of these entries, if they exist. Related acronyms (variations or subsets of techniques; terminology used within the technique area) are grouped together below the major acronym and indented to the right. Most, but not all, of the techniques listed here are the subject of individual articles in this volume.

AAS	Atomic Absorption Spectroscopy
AA	Atomic Absorption
VPD-AAS	Vapor Phase Decomposition-Atomic Absorption Spectroscopy
GFAA	Graphite Furnace Atomic Absorption
FAA	Flame Atomic Absorption
AES	Auger Electron Spectroscopy
Auger	Auger Electron Spectroscopy
SAM	Scanning Auger Microscopy
SAM	Scanning Auger Microprobe
AED	Auger Electron Diffraction
ADAM	Angular Distribution Auger Microscopy
KE	Kinetic Energy
CMA	Cylindrical Mirror Analyzer
AIS	Atom Inelastic Scattering
BET	Brunauer, Emmett, and Teller equation
BSDF	Bidirectional Scattering Distribution Function
BRDF	Bidirectional Reflective Distribution Function
BTDF	Bidirectional Transmission Distribution Function
CL	Cathodluminescence
CLSM	Confocal Scanning Laser Microscope
EDS	Energy Dispersive (X-Ray) Spectroscopy
EDX	Energy Dispersive X-Ray Spectroscopy
EDAX	Company selling EDX equipment
EELS	Electron Energy Loss Spectroscopy
HREELS	High-Resolution Electron Energy-Loss Spectroscopy
REELS	Reflected Electron Energy-Loss Spectroscopy
REELM	Reflection Electron Energy-Loss Microscopy
LEELS	Low-Energy Electron-Loss Spectroscopy

PEELS	Parallel (Detection) Electron Energy-Loss Spectrscopy
EXELFS	Extended Energy-Loss Fine Structure
EELFS	Electron Energy-Loss Fine Structure
CEELS	Core Electron Energy-Loss Spectroscopy
VEELS	Valence Electron Energy-Loss Spectroscopy
EPMA	Electron Probe Microanalysis
Electron Probe	Electron Probe Microanalysis
ERS	Elastic Recoil Spectrometry
HFS	Hydrogen Forward Scattering
HRS	Hydrogen Recoil Spectrometry
FRS	Forward Recoil Spectrometry
ERDA	Elastic Recoil Detection Analysis
ERD	Elastic Recoil Detection
PRD	Particle Recoil Detection
EXAFS	Extended X-Ray Absorption Fine Structure
SEXAFS	Surface Extended X-Ray Absorption Fine Structure
NEXAFS	Near-Edge X-Ray Absorption Fine Structure
XANES	X-Ray Absorption Near-Edge Structure
XAFS	X-Ray Absorption Fine Structure
FMR	Ferromagnetic Resonance
FTIR	See IR
FT Raman	See Raman
HREELS	See EELS
HRTEM	See TEM
GDMS	Glow Discharge Mass Spectrometry
GDQMS	Glow Discharge Mass Spectrometry using a Quadruple Mass Analyser
Gloquad	Manufacturer name
ICP-MS	Inductively Coupled Plasma Mass Spectrometry
ICP	Inductively Coupled Plasma
LA-ICP-MS	Laser Ablation ICP-MS
ICP-Optical	Inductively Coupled Plasma Optical Emission
ICP	Inductively Coupled Plasma
IETS	Inelastic Electron Tunneling Spectroscopy
IR	Infrared (Spectroscopy)
FTIR	Fourier Transform Infra-Red (Spectroscopy)
GC-FTIR	Gas Chromatography FTIR
TGA-FTIR	Thermo Gravimetric Analysis FTIR
ATR	Attenuated Total Reflection

RA	Reflection Absorption (Spectroscopy)
IRAS	Infrared Reflection Absorption Spectroscopy
ISS	Ion Scattering Spectrometry
LEIS	Low-Energy Ion Scattering
RCE	Resonance Charge Exchange
LEED	Low-Energy Electron Diffraction
LIMS	Laser Ionization Mass Spectrometry
LAMMA	Laser Microprobe Mass Analysis
LAMMS	Laser Microprobe Mass Spectrometry
LIMA	Laser Ionization Mass Analysis
NRMPI	Nonresonant Multi-Photon Ionization
MEISS	Medium-Energy Ion Scattering Spectrometry
MEIS	Medium-Energy Ion Scattering
MOKE	Magneto-Optic Kerr Rotation
SMOKE	Surface Magneto-Optic Kerr Rotation
NAA	Neutron Activation Analysis
INAA	Instrumental Neutron Activation Analysis
NEXAFS	Near Edge X-Ray Absorption Fine Structure
XANES	X-Ray Absorption Near Edge Structure
NIS	Neutron Inelastic Scattering
NMR	Nuclear Magnetic Resonance
MAS	Magic-Angle Spinning
NRA	Nuclear Reaction Analysis
OES	Optical Emission Spectroscopy
PAS	Photoacoustic Spectroscopy
PIXE	Particle Induced X-Ray Emission
HIXE	Hydrogen/Helium Induced X-ray Emission
PL	Photoluminescence
PLE	Photoluminescence Excitation
PR	Photoreflectance
EBER	Electron Beam Electroreflectance
RDS	Reflection Difference Spectroscopy
Raman	Raman Spectroscopy
FT Raman	Fourier Transform Raman Spectroscopy
RS	Raman Scattering
RRS	Resonant Raman Scattering
CARS	Coherent Anti-Stokes Raman Scattering

SERS	Surface Enhanced Raman Spectroscopy
RBS	Rutherford Backscattering Spectrometry
HEIS	High-Energy Ion Scattering
RHEED	Reflected High Energy Electron Diffraction
SREM	Scanning Reflection Electron Microscopy
SALI	Surface Analysis by Laser Ionization
PISIMS	Post-Ionization Secondary Ion Mass Spectrometry
MPNRPI	Multi-Photon Nonresonant Post Ionization
MRRPI	Multiphoton Resonant Post Ionization
RPI	Resonant Post Ionization
MPI	Multi-Photon Ionization
SPI	Single-Photon Ionization
SIRIS	Sputter-Initiated Resonance Ionization Spectroscopy
SARIS	Surface Analysis by Resonant Ionization Spectroscopy
TOFMS	Time-of-Flight Mass Spectrometer
SAM	See AES
SEM	Scanning Electron Microscopy
	Scanning Electron Microprobe
	Secondary Electron Miscroscopy
SE	Secondary Electron
BSE	Backscattered Electron
SEMPA	Secondary Electron Microscopy with Polarization Analysis
SFM	Scanning Force Microscopy
	Scanning Force Microscope
AFM	Atomic Force Microscopy
SPM	Scanning Probe Microscopy
SIMS	Secondary Ion Mass Spectrometry
Dynamic SIMS	Dynamic Secondary Ion Mass Spectrometry
Static SIMS	Static Secondary Ion Mass Spectrometry
Q-SIMS	SIMS using a Quadruple Mass Spectrometer
Magnetic SIMS	SIMS using a Magnetic Sector Mass Spectrometer
Sector SIMS	See Magnetic SIMS
TOF-SIMS	SIMS using Time-of-Flight Mass Spectrometer
PISIMS	Post Ionization SIMS
SNMS	Sputtered Neutrals Mass Spectrometry
	Secondary Neutrals Mass Spectrometry
SNMSd	Direct Bombardment Electron Gas SNMS
SSMS	Spark Source Mass Spectrometry
Spark Source	Spark Source Mass Spectrometry
STEM	See TEM
STM	Scanning Tunneling Microscopy

	Scanning Tunneling Microscope
SPM	Scanning Probe Microscopy
TEAS	Thermal Energy Atom Scattering
TEM	Transmission Electron Microscopy
	Transmission Electron Microscope
CTEM	Conventional Transmission Electron Microscopy
STEM	Scanning Transmission Electron Microscopy
HRTEM	High Resolution Transmission Electron Microscopy
SAD	Selected Area Diffraction
AEM	Analytical Electron Microscopy
CBED	Convergent Beam Electron Diffraction
LTEM	Lorentz Transmission Electron Microscopy
TLC	Thin Layer Chromatography
TSRLM	Tandem Scanning Reflected-Light Microscope
TSM	Tandem Scanning Reflected-Light Microscope
TXRF	See XRF
UPS	Ultraviolet Photoelectron Spectroscopy
	Ultraviolet Photoemission Spectroscopy
MPS	Molecular Photoelectron Spectroscopy
VASE	Variable Angle Spectroscopic Ellipsometry
WDS	Wavelength Dispersive (X-Ray) Spectroscopy
WDX	Wavelength Dispersive X-Ray Spectroscopy
XAS	X-Ray Absorption Spectroscopy
XPS	X-Ray Photoelectron Spectroscopy
	X-Ray Photoemission Spectroscopy
ESCA	Electron Spectroscopy for Chemical Analysis
XPD	X-Ray Photoelectron Diffraction
PHD	Photoelectron Diffraction
KE	Kinetic Energy
XRD	X-RayDiffraction
GIXD	Grazing Incidence X-Ray Diffraction
GIXRD	Grazing Incidence X-Ray Diffraction
DCD	Double Crystal Diffractometer
XRF	X-Ray Fluorescence
XFS	X-Ray Fluorescence Spectroscopy
TXRF	Total Reflection X-Ray Fluorescence
TRXFR	Total Reflection X-Ray Fluorescence
VPD-TXRF	Vapor Phase Decomposition Total X-Ray Fluorescence

Contributors

Mark R. Antonio
BP Research International
Cleveland, OH
Extended X-Ray Absorption Fine Structure

J. E. E. Baglin
IBM Almaden Research Center
San Jose, CA
Elastic Recoil Spectrometry

Scott Baumann
Charles Evans & Associates
Redwood City, CA
Rutherford Backscattering Spectrometry

Christopher H. Becker
SRI International
Menlo Park, CA
Surface Analysis by Laser Ionization

Albert J. Bevolo
Ames Laboratory,
Iowa State University
Ames, IA
Reflected Electron Energy-Loss Spectroscopy

J. B. Bindell
AT&T Bell Laboratories
Allentown, PA
Scanning Electron Microscopy

Filippo Radicati di Brozolo
Charles Evans & Associates
Redwood City, CA
Laser Ionization Mass Spectrometry

C. R. Brundle
IBM Almaden Research Center
San Jose, CA
X-Ray Photoelectron Spectroscopy;
Ultraviolet Photoelectron Spectroscopy

Daniele Cherniak
Rennsselaer Polytechnic Institute
Troy, NY
Nuclear Reaction Analysis

Paul Chu
Charles Evans & Associates
Redwood City, CA
Dynamic Secondary Ion Mass Spectrometry

Carl Colvard
Charles Evans & Assoociates
Redwood City, CA
Photoluminescence

J. Neal Cox
INTEL, Components Research
Santa Clara, CA
Fourier Transform Infrared Spectroscopy

John Gustav Delly
McCrone Research Institute
Chicago, IL
Light Microscopy

Hellmut Eckert
University of California, Santa Barbara
Santa Barbara, CA
Solid State Nuclear Magnetic Resonance

Peter Eichinger
GeMeTec Analysis
Munich
Total Reflection X-Ray Fluorescence

P. Fenter
Rutgers University
Piscataway, NJ
Medium-Energy Ion Scattering with Channeling and Blocking

David E. Fowler
IBM Almaden Research Center
San Jose, CA
Magneto-Optic Kerr Rotation

S. M. Gaspar
University of New Mexico
Albuquerque, NM
Optical Scatterometry

Roy H. Geiss
IBM Almaden Research Center
San Jose, CA
Energy-Dispersive X-Ray Spectroscopy

Torgny Gustafsson
Rutgers University
Piscataway, NJ
Medium-Energy Ion Scattering With Channeling and Blocking

William L. Harrington
Evans East
Plainsboro, NJ
Spark Source Mass Spectrometry

Brent D. Hermsmeier
IBM Almaden Research Center
San Jose, CA
X-Ray Photoelectron and Auger Electron Diffraction

K.C. Hickman
University of New Mexico
Albuquerque, NM
Optical Scatterometry

Tim Z. Hossain
Cornell University
Ithica, NY
Neutron Activation Analysis

Rebecca S. Howland
Park Scientific Instruments
Sunnyvale, CA
Scanning Tunneling Microscopy and Scanning Force Microscopy

John C. Huneke
Charles Evans & Associates
Redwood City, CA
Sputtered Neutral Mass Spectrometry, Glow-Discharge Mass Spectrometry

Ting C. Huang
IBM Almaden Research Center
San Jose, CA
X-Ray Fluorescence

William Katz
Evans Central
Minnetonka, MN
Static Secondary Ion Mass Spectrometry

Michael D. Kirk
Park Scientific Instruments
Sunnyvale, CA
Scanning Tunneling Microscopy and Scanning Force Microscopy

Bruce E. Koel
University of Southern California
Los Angeles, CA

High-Resolution Electron Energy Loss Spectrometry

Max G. Lagally
University of Wisconsin,
Madison, WI

Low-Energy Electron Diffraction

W. A. Lanford
State University of New York,
Albany, NY

Nuclear Reaction Analysis

Charles E. Lyman
Lehigh University
Bethlehem, PA

Scanning Transmission Electron Microscopy

Susan MacKay
Perkin Elmer
Eden Prairie, MN

Surface Analysis by Laser Ionization

John R. McNeil
University of New Mexico
Albuquerque, NM

Optical Scatterometry

Ronald G. Musket
Lawrence Livermore National Laboratory
Livermore, CA

Particle-Induced X-Ray Emission

S. S. H. Naqvi
University of New Mexico
Albuquerque, NM

Optical Scatterometry

Dale E. Newbury
National Institutes of Science and Technology
Gaithersburg, MD

Electron Probe X-Ray Microanalysis

David Norman
SERC Daresbury Laboratory
Daresbury, Cheshire

Surface Extended X-Ray Absorption Fine Structure, Near Edge X-Ray Absorption Fine Structure

John W. Olesik
Ohio State University
Columbus, OH

Inductively Coupled Plasma–Optical Emission Spectroscopy

Fred H. Pollak
Brooklyn College, CUNY
New York, NY

Modulation Spectroscopy

Thomas P. Russell
IBM Almaden Research Center
San Jose, CA

Neutron Reflectivity

Donald E. Savage
University of Wisconsin
Madison, WI

Reflection High-Energy Electron Diffraction

Kurt E. Sickafus
Los Alamos National Laboratory
Los Alamos, NM

Transmission Electron Microscopy

Paul G. Snyder
University of Nebraska
Lincoln, NE

Variable Angle Spectroscopic Ellipsometry

Gene Sparrow
Advanced R&D
St. Paul, MN

Ion Scattering Spectroscopy

Fred A. Stevie
AT&T Bell Laboratories
Allentown, PA

Surface Roughness: Measurement, Formation by Sputtering, Impact on Depth Profiling

Yale E. Strausser
Surface Science Laboratories
Mountainview, CA

Auger Electron Spectroscopy

Barry J. Streusand
Applied Analytical
Austin, TX

Inductively Coupled Plasma Mass Spectrometry

Raymond G. Teller
BP Research International
Cleveland, OH

Neutron Diffraction

Michael F. Toney
IBM Almaden Research Center
San Jose, CA

X-Ray Diffraction

Wojciech Vieth
Charles Evans & Assoociates
Redwood City, CA

Glow-Discharge Mass Spectrometry

William B. White
Pennsylvania State University
University Park, PA

Raman Spectroscopy

S.R. Wilson
University of New Mexico
Albuquerque, NM

Optical Scatterometry

John A. Woollam
University of Nebraska
Lincoln, NE

Variable Angle Spectroscopic Ellipsometry

Ben G. Yacobi
University of California at Los Angeles
Los Angeles, CA

Cathodoluminescence

David J. C. Yates
Consultant
Poway, CA

Physical and Chemical Adsorption for the Measurement of Solid Surface Areas

Nestor J. Zaluzec
Argonne National Laboratory
Argonne, IL

Electron Energy-Loss Spectroscopy in the Transmission Electron Microscope

1

INTRODUCTION AND SUMMARIES

1.0 INTRODUCTION

Though a wide range of analytical techniques is covered in this volume there are certain traits common to many of them. Most involve either electrons, photons, or ions as a probe beam striking the material to be analyzed. The beam interacts with the material in some way, and in some of the techniques the changes induced in the beam (energy, intensity, and angular distribution) are monitored after the interaction, and analytical information is derived from the observation of these changes. In other techniques the information used for analysis comes from electrons, photons, or ions that are ejected from the sample under the stimulation of the probe beam. In many situations several connected processes may be going on more or less simultaneously, with a particular analytical technique picking out only one aspect, e.g., the extent of absorption of incident light, or the kinetic energy distribution of ejected electrons.

The range of information provided by the techniques discussed here is also wide, but again there are common themes. What types of information are provided by these techniques? Elemental composition is perhaps the most basic information, followed by chemical state information, phase identification, and the determination of structure (atomic sites, bond lengths, and angles). One might need to know how these vary as a function of depth into the material, or spatially across the material, and many techniques specialize in addressing these questions down to very fine dimensions. For surfaces, interfaces, and thin films there is often very little material at all to analyze, hence the presence of many microanalytical methods in this volume. Within this field (microanalysis) it is often necessary to identify trace components down to extremely low concentration (parts per trillion in some cases) and a number of techniques specialize in this aspect. In other cases a high degree of accuracy in measuring the presence of major components might be the issue. Usually the techniques that are good for trace identification are not the same ones used to accurately quantify major components. Most complete analyses require the use of multiple techniques, the selection of which depends on the nature of the sample and the desired information.

This first chapter contains one-page summaries of each of the 50 techniques covered in the following chapters. All summaries have the same format to allow easy comparison and quick access to the information. Further comparative information is provided in the introductions to the chapters. Finally, a table is provided at the end of this introduction, in which many of the important parameters describing the capabilities for all 50 techniques are listed.

The subtitle of this Series is "Surfaces, Interfaces, and Thin Films." The definition of a "surface" or of a "thin film" varies considerably from person to person and with application area. The academic discipline of "Surface Science" is largely concerned with chemistry and physics at the atomic monolayer level, whereas the "surface region" in an engineering or applications sense can be much more extensive. The same is true for interfaces between materials. The practical consideration in distinguishing "surface" from "bulk" or "thin" from "thick" is usually connected to the property of interest in the application. Thus, for a catalytic reaction the presence of half a monolayer of extraneous sulfur atoms in the top atomic layer of the catalyst material might be critical, whereas for a corrosion protection layer (for example, Cr segregation to the surface region in steels) the important region of depth may be several hundred Å. For interfaces the epitaxial relationship between he last atomic layer of a single crystal material and the first layer of the adjoining material may be critical for the electrical properties of a device, whereas diffusion barrier interfaces elsewhere in the same device may be 1000 Å thick. In thin-film technology requirements can range from layers μm thick, which for the majority of analytical techniques discussed in this volume constitute bulk material, to layers as thin as 50 Å or so in thin-film magnetic recording technology. Because of these different perceptions of "thick" and "thin," actual numbers are used whenever discussing the depth an analytical technique examines. Thus in Ion Scattering Spectroscopy the signals used in the analysis are generated from only the top atomic monolayer of material exposed to a vacuum, whereas in X-ray photoemission up to 100 Å is probed, and in X-ray fluorescence the signal can come from integrated depths ranging up to 10 μm. Note that in these three examples, two are quoted as having ranges of depths. For many of the techniques it is impossible to assign unique values because the depth from which a signal originates may depend both on the particular manner in which the technique is used, and on the nature of the material being examined. Performing measurements at grazing angles of incidence of the probe beam, or grazing exit angles for the detected signal, will usually make the technique more surface sensitive. For techniques where X-ray, electron, or high-energy ion scattering is the critical factor in determining the depth analyzed, materials consisting of light elements are always probed more deeply than materials consisting of heavy elements.

Another confusing issue is that of "depth resolution." It is a measurement of the technique's ability to clearly distinguish a property as a function of depth. For example a depth resolution of 20 Å, quoted in an elemental composition analysis, means that the composition at one depth can be distinguished from that at another depth if there is at least 20 Å between them.

A depth profile is a record of the variation of a property (such as composition) as a function of depth. Some of the techniques in this volume have essentially no intrinsic depth profiling capabilities; the signal is representative of the material integrated over a fixed probing depth. Most, however, can vary the depth probed by varying the condition of analysis, or by removing the surface, layer by layer, while collecting data.

By varying the angle of incidence, the X-ray, electron, or ion beam energy, etc. many techniques are capable of acquiring depth profiles. Those profiles are generated by combining several measurements, each representative of a different integrated depth. The higher energy ion scattering techniques (Medium Energy Ion Scattering, MEIS, and Rutherford Backscattering, RBS), however, are unique in that the natural output of the methods is composition as a function of depth. By far the most common way of depth profiling is the destructive method of removing the surface, layer by layer, while also taking data. For the mass spectrometry-based techniques of Chapter 10, removal of surface material is intrinsic to the sputtering and ionization process. Other methods, such as Auger Electron Spectroscopy, AES, or X-Ray Photoemission, XPS, use an ancillary ion beam to remove material while constantly ionizing the newly exposed surface. Under the most favorable conditions depth resolutions of around 20 Å can be achieved this way, but there are many artifacts to be aware of and the depth resolution usually degrades rapidly with depth. Some aspects of sputter depth profiling are touched upon in the article "Surface Roughness" in Chapter 12, but for a more complete discussion of the capabilities and limitations of sputter depth profiling the reader is referred to a paper by D. Marton and J. Fine in *Thin Solid Films*, **185**, 79, 1990 and to other articles cited there.

Compilation of Comparative Information on the Analytical Techniques Discussed in This Volume

Article No.	Technique	Main information: Elemental	Chem. state	Phase	Defects	Structure	Image	Other	Depth probed (typical)	Width probed (typical)	Trace capability (typical)	Types of solid sample (typical)	Vacuum needed ?	Commercial Instrument cost	Usage	Service available
2.1	Light Microscopy				•		•	•	Variable	0.2 µm	—	All	N	1	1	Y
2.2	SEM				•		•		sub µm	10 nm	—	Cond., coated ins.	Y	2	1	Y
2.3	STM				•	•	•		sub Å	1 Å	—	Conductors	N	2	3	Y
2.3	SFM				•	•	•		sub Å	1 nm	—	All	N	2	2	Y
2.4	TEM			•	•	•	•		200 nm*	5 nm	—	All; <200 nm thick	Y	3	2	Y
3.1	EDS	•					•		1 µm	0.5 µm	500 ppm	All; $Z > 5$	Y	2	2	Y
3.2	EELS	•	•				•		20 nm*	1 nm	Few %	All; <30 nm thick	Y	2	3	N
3.3	Cathodo-luminescence		•		•			•	10 nm–µm	1 µm	ppm	All; semicond. usually	Y	1	3	N
3.4	STEM			•	•	•	•		100 nm*	1 nm	—	All; <200 nm thick	Y	3	3	N
3.5	EPMA	•					•		1 µm	0.5 µm	100 ppm	All; flat best	Y	3	2	Y
4.1	XRD			•	•	•			10 µm	mm	3%	Crystalline	N	2	1	Y
4.2	EXAFS	•				•			Bulk*	mm	Few %	All	Y/N	—	3	N
4.3	SEXAFS	•				•			1 nm	mm	Few %	Surface and adsorbate	Y	—	3	N
4.3	NEXAFS	•	•			•			1 nm	mm	Few %	Surface and adsorbate	Y	—	3	N
4.4	XPD	•	•			•			3 nm	150 µm	1%	Single crystal	Y	3	3	N
4.5	LEED				•	•			1 nm	0.1 mm	—	Single crystal	Y	—	2	N
4.6	RHEED				•	•			1 nm	0.02 mm	—	Single crystal	Y	—	2	N
5.1	XPS	•	•						3 nm	150 µm	1%	All	Y	3	1	Y
5.2	UPS		•			•			1 nm	mm	—	All	Y	—	3	N
5.3	AES	•	•				•		2 nm	100 nm	0.1%	All, inorganic usually	Y	3	1	Y

Compilation of Comparative Information on the Analytical Techniques Discussed in This Volume

Article No.	Technique	Main information: Elemental	Chem. state	Phase	Defects	Structure	Image	Other	Depth probed (typical)	Width probed (typical)	Trace capability (typical)	Types of solid sample (typical)	Vacuum needed ?	Commercial Instrument cost	Usage	Service available
5.4	REELS	•	•				•		2 nm	100 nm	—	All	Y	—	3	N
6.1	XRF	•							10 μm	mm	0.1%	All	N	2	1	Y
6.2	TXRF	•					•		3 nm	cm	ppb–ppm	Trace heavy metals	Y	3	3	Y
6.3	PIXE	•					•		Few μm	100 μm	10 ppm	All	Y	3	3	Y
7.1	Photo-luminescence		•		•		•	•	Few μm	Few μm	ppb	All, semicond. usually	N	1	2	N
7.2	Modulation Spectroscopy		•		•		•	•	1 μm	100 μm	ppm	All, semicond. usually	N	2	3	N
7.3	VASE							•	1 μm	cm	—	Flat thin films	N	2	3	Y
8.1	FTIR		•		•			•	Few μm	20 μm	Variable	All	N	2	1	Y
8.2	Raman Scattering		•		•			•	Few μm	1 μm	Variable	All	N	2	2	Y
8.3	HREELS		•			•			2 nm	mm	1%	All; flat cond. best	Y	3	3	N
8.4	NMR		•	•		•			Bulk	—	—	All; not all elements	N	3	3	N
9.1	RBS	•			•	•			To 2 μm	mm	0.01–10%	All	Y/N	3	2	Y
9.2	ERS							•	1 μm	mm	0.01%	H containing	Y	—	3	N
9.3	MEIS	•			•	•			1 nm	mm	0.1%–10%	All; usually single crystal	Y	3	3	N
9.4	ISS	•							3 Å	150 μm	50 ppm–1%	All	Y	—	3	Y
10.1	Dynamic SIMS	•					•		2 nm	1 μm	ppb–ppm	All, mostly semicond.	Y	3	1	Y
10.2	Static SIMS	•	•				•		3 Å	100 μm	Few %	All, mostly polymer	Y	3	2	Y
10.3	SALI	•	•						3 Å	100 nm	ppb–ppm	All, mostly inorg.	Y	3	3	N

Compilation of Comparative Information on the Analytical Techniques Discussed in This Volume

Article No.	Technique	Main information: Elemental	Chem. state	Phase	Defects	Structure	Image	Other	Depth probed (typical)	Width probed (typical)	Trace capability (typical)	Types of solid sample (typical)	Vacuum needed ?	Commercial Instrument cost	Usage	Service available
10.4	SNMS	•							1.5 nm	cm	50 ppm	Flat conductors	Y	2	2	Y
10.5	LIMS	•	•						100 nm	2 μm	1–100 ppm	All	Y	3	2	Y
10.6	SSMS	•							3 μm	cm	0.05 ppm	Sample forms electrode	Y	—	2	Y
10.7	GDMS	•							100 nm	cm	ppt–ppb	Sample forms electrode	Y	3	2	Y
10.8	ICPMS	•							5 μm	mm	ppt	All	Y	2	1	Y
10.9	ICPOES	•							5 μm	mm	ppb	All	Y	1	1	Y
11.1	Neutron Diffraction			•		•			Bulk	—	—	Crystalline	N	—	3	N
11.2	Neutron Reflectivity							•	Up to mm	—	—	Flat polymer films	N	—	3	N
11.3	NAA	•							Bulk	—	ppt–ppm	Trace metals	N	2	3	Y
11.4	NRA	•							10–100 nm	10 μm	10–100 ppm	All; $Z < 21$	Y	—	3	Y
12.2	Optical Scatterometry							•	—	mm	—	Flat smooth films	N	1	3	Y
12.3	MOKE						•	•	30 nm	0.5 μm	—	Magnetic films	N	1	2	N
12.4	Adsorption							•	Outer atoms	—	—	Large surface area	Y	—	2	N

Notes: This table should be used as a "quick reference" guide only.

* Measured in transmission.

Commercial Instrument cost: These are typical costs; large ranges depending on sophistication and accessories: 1 means < \$50k; 2 means \$50–300k; 3 means >\$300k. "—" means no complete commercial instrument.

Usage: Numbers refer to usage for anaylsis of solid materials. 1 means Extensive; 2 means medium; 3 means not common.

Trace capability: Guide only. Often very material/conditions dependent. "—" means not used for trace components.

Light Microscopy 1.2.1

The light microscope uses the visible or near visible portion of the electromagnetic spectrum; light microscopy is the interpretive use of the light microscope. This technique, which is much older than other characterization instruments, can trace its origin to the 17th century. Modern analytical and characterization methods began about 150 years ago when thin sections of rocks and minerals, and the first polished metal and metal-alloy specimens were prepared and viewed with the intention of correlating their structures with their properties. The technique involves, at its very basic level, the simple, direct visual observation of a sample with white-light resolution to 0.2 μm. The morphology, color, opacity, and optical properties are often sufficient to characterize and identify a material.

Range of samples characterized	Almost unlimited for solids and liquid crystals
Destructive	Usually nondestructive; sample preparation may involve material removal
Quantification	Via calibrated eyepiece micrometers and image analysis
Detection limits	To sub-ng
Resolving power	0.2 μm with white light
Imaging capabilities	Yes
Main use	Direct visual observation; preliminary observation for final characterization, or preparative for other instrumentation
Instrument cost	$2,500–$50,000 or more
Size	Pocket to large table

Scanning Electron Microscopy (SEM) 1.2.2

The Scanning Electron Microscope (SEM) is often the first analytical instrument used when a "quick look" at a material is required and the light microscope no longer provides adequate resolution. In the SEM an electron beam is focused into a fine probe and subsequently raster scanned over a small rectangular area. As the beam interacts with the sample it creates various signals (secondary electrons, internal currents, photon emission, etc.), all of which can be appropriately detected. These signals are highly localized to the area directly under the beam. By using these signals to modulate the brightness of a cathode ray tube, which is raster scanned in synchronism with the electron beam, an image is formed on the screen. This image is highly magnified and usually has the "look" of a traditional microscopic image but with a much greater depth of field. With ancillary detectors, the instrument is capable of elemental analysis.

Main use	High magnification imaging and composition (elemental) mapping
Destructive	No, some electron beam damage
Magnification range	10×–300,000×; 5000×–100,000× is the typical operating range
Beam energy range	500 eV–50 keV; typically, 20–30 keV
Sample requirements	Minimal, occasionally must be coated with a conducting film; must be vacuum compatible
Sample size	Less than 0.1mm, up to 10 cm or more
Lateral resolution	1–50 nm in secondary electron mode
Depth sampled	Varies from a few nm to a few µm, depending upon the accelerating voltage and the mode of analysis
Bonding information	No
Depth profiling capabilities	Only indirect
Instrument cost	$100,000–$300,000 is typical
Size	Electronics console 3 ft. × 5 ft.; electron beam column 3 ft. × 3 ft.

Scanning Tunneling Microscopy and Scanning Force Microscopy (STM and SFM) 1.2.3

In Scanning Tunneling Microscopy (STM) or Scanning Force Microscopy (SFM), a solid specimen in air, liquid or vacuum is scanned by a sharp tip located within a few Å of the surface. In STM, a quantum-mechanical tunneling current flows between atoms on the surface and those on the tip. In SFM, also known as Atomic Force Microscopy (AFM), interatomic forces between the atoms on the surface and those on the tip cause the deflection of a microfabricated cantilever. Because the magnitude of the tunneling current or cantilever deflection depends strongly upon the separation between the surface and tip atoms, they can be used to map out surface topography with atomic resolution in all three dimensions. The tunneling current in STM is also a function of local electronic structure so that atomic-scale spectroscopy is possible. Both STM and SFM are unsurpassed as high-resolution, three-dimensional profilometers.

Parameters measured	Surface topography (SFM and STM); local electronic structure (STM)
Destructive	No
Vertical resolution	STM, 0.01 Å; SFM, 0.1 Å
Lateral resolution	STM, atomic; SFM, atomic to 1 nm
Quantification	Yes; three-dimensional
Accuracy	Better than 10% in distance
Imaging/mapping	Yes
Field of view	From atoms to > 250 μm
Sample requirements	STM—solid conductors and semiconductors, conductive coating required for insulators; SFM—solid conductors, semiconductors and insulators
Main uses	Real-space three-dimensional imaging in air, vacuum, or solution with unsurpassed resolution; high-resolution profilometry; imaging of nonconductors (SFM).
Instrument cost	$65,000 (ambient) to $200,000 (ultrahigh vacuum)
Size	Table-top (ambient), 2.27–12 inch bolt-on flange (ultrahigh vacuum)

Transmission Electron Microscopy (TEM) 1.2.4

In Transmission Electron Microscopy (TEM) a thin solid specimen ($\leq$ 200 nm thick) is bombarded in vacuum with a highly-focused, monoenergetic beam of electrons. The beam is of sufficient energy to propagate through the specimen. A series of electromagnetic lenses then magnifies this transmitted electron signal. Diffracted electrons are observed in the form of a diffraction pattern beneath the specimen. This information is used to determine the atomic structure of the material in the sample. Transmitted electrons form images from small regions of sample that contain contrast, due to several scattering mechanisms associated with interactions between electrons and the atomic constituents of the sample. Analysis of transmitted electron images yields information both about atomic structure and about defects present in the material.

Range of elements	TEM does not specifically identify elements measured
Destructive	Yes, during specimen preparation
Chemical bonding information	Sometimes, indirectly from diffraction and image simulation
Quantification	Yes, atomic structures by diffraction; defect characterization by systematic image analysis
Accuracy	Lattice parameters to four significant figures using convergent beam diffraction
Detection limits	One monolayer for relatively high-Z materials
Depth resolution	None, except there are techniques that measure sample thickness
Lateral resolution	Better than 0.2 nm on some instruments
Imaging/mapping	Yes
Sample requirements	Solid conductors and coated insulators. Typically 3-mm diameter, < 200-nm thick in the center
Main uses	Atomic structure and Microstructural analysis of solid materials, providing high lateral resolution
Instrument cost	\$300,000–\$1,500,000
Size	100 ft.^{2} to a major lab

Energy-Dispersive X-Ray Spectroscopy (EDS) 1.3.1

When the atoms in a material are ionized by a high-energy radiation they emit characteristic X rays. EDS is an acronym describing a technique of X-ray spectroscopy that is based on the collection and energy dispersion of characteristic X rays. An EDS system consists of a source of high-energy radiation, usually electrons; a sample; a solid state detector, usually made from lithium-drifted silicon, Si (Li); and signal processing electronics. EDS spectrometers are most frequently attached to electron column instruments. X rays that enter the Si (Li) detector are converted into signals which can be processed by the electronics into an X-ray energy histogram. This X-ray spectrum consists of a series of peaks representative of the type and relative amount of each element in the sample. The number of counts in each peak may be further converted into elemental weight concentration either by comparison with standards or by standardless calculations.

Range of elements	Boron to uranium
Destructive	No
Chemical bonding information	Not readily available
Quantification	Best with standards, although standardless methods are widely used
Accuracy	Nominally 4–5%, relative, for concentrations > 5% wt.
Detection limits	100–200 ppm for isolated peaks in elements with $Z > 11$, 1–2% wt. for low-Z and overlapped peaks
Lateral resolution	.5–1 µm for bulk samples; as small as 1 nm for thin samples in STEM
Depth sampled	0.02 to µm, depending on Z and keV
Imaging/mapping	In SEM, EPMA, and STEM
Sample requirements	Solids, powders, and composites; size limited only by the stage in SEM, EPMA and XRF; liquids in XRF; 3 mm diameter thin foils in TEM
Main use	To add analytical capability to SEM, EPMA and TEM
Cost	$25,000–$100,000, depending on accessories (not including the electron microscope)

Electron Energy-Loss Spectroscopy in the Transmission Electron Microscope (EELS) 1.3.2

In Electron Energy-Loss Spectroscopy (EELS) a nearly monochromatic beam of electrons is directed through an ultrathin specimen, usually in a Transmission (TEM) or Scanning Transmission (STEM) Electron Microscope. As the electron beam propagates through the specimen, it experiences both elastic and inelastic scattering with the constituent atoms, which modifies its energy distribution. Each atomic species in the analyzed volume causes a characteristic change in the energy of the incident beam; the changes are analyzed by means of a electron spectrometer and counted by a suitable detector system. The intensity of the measured signal can be used to determine quantitatively the local specimen concentration, the electronic and chemical structure, and the nearest neighbor atomic spacings.

Range of elements	Lithium to uranium; hydrogen and helium are sometimes possible
Destructive	No
Chemical bonding information	Yes, in the near-edge structure of edge profiles information
Depth profiling	None, the specimen is already thin capabilities
Quantification	Without standards ~±10–20% at.; with standards ~1–2% at.
Detection limits	~10^{-21} g
Depth probed	Thickness of specimen (≤ 2000 Å)
Lateral resolution	1 nm–10 μm, depending on the diameter of the incident electron probe and the thickness of the specimen
Imaging capabilities	Yes
Sample requirements	Solids; specimens must be transparent to electrons and ~100–2000 Å thick
Main use	Light element spectroscopy for concentration, electronic, and chemical structure analysis at ultrahigh lateral resolution in a TEM or STEM
Cost	As an accessory to a TEM or STEM: $50,000–$150,000 (does not include electron microscope cost)

Cathodoluminescence (CL) 1.3.3

In Cathodoluminescence (CL) analysis, electron-beam bombardment of a solid placed in vacuum causes emission of photons (in the ultraviolet, visible, and near-infrared ranges) due to the recombination of electron–hole pairs generated by the incident energetic electrons. The signal provides a means for CL microscopy (i.e., CL images are displayed on a CRT) and spectroscopy (i.e., luminescence spectra from selected areas of the sample are obtained) analysis of luminescent materials using electron probe instruments. CL microscopy can be used for uniformity characterization (e.g., mapping of defects and impurity segregation studies), whereas CL spectroscopy provides information on various electronic properties of materials.

Range of elements	Not element specific
Chemical bonding information	Sometimes
Nondestructive	Yes; caution—in certain cases electron bombardment may ionize or create defects
Detection limits	In favorable cases, dopant concentrations down to 10^{14} atoms/cm^3
Depth profiling	Yes, by varying the range of electron penetration (between about 10 nm and several μm), which depends on the electron-beam energy (1–40 keV).
Lateral resolution	On the order of 1 μm; down to about 0.1 μm in special cases
Imaging/mapping	Yes
Sample requirements	Solid, vacuum compatible
Quantification	Difficult, standards needed
Main use	Nondestructive qualitative and quantitative analysis of impurities and defects, and their distributions in luminescent materials
Instrument cost	\$25,000–\$250,000
Size	Small add-on item to SEM, TEM

Scanning Transmission Electron Microscopy (STEM) 1.3.4

In Scanning Transmission Electron Microscopy (STEM) a solid specimen, 5–500 nm thick, is bombarded in vacuum by a beam (0.3–50 nm in diameter) of monoenergetic electrons. STEM images are formed by scanning this beam in a raster across the specimen and collecting the transmitted or scattered electrons. Compared to the TEM an advantage of the STEM is that many signals may be collected simultaneously: bright- and dark-field images; Convergent Beam Electron Diffraction (CBED) patterns for structure analysis; and energy-dispersive X-Ray Spectrometry (EDS) and Electron Energy-Loss Spectrometry (EELS) signals for compositional analysis. Taken together, these analysis techniques are termed Analytical Electron Microscopy (AEM). STEM provides about 100 times better spatial resolution of analysis than conventional TEM. When electrons scattered into high angles are collected, extremely high-resolution images of atomic planes and even individual heavy atoms may be obtained.

Range of elements	Lithium to uranium
Destructive	Yes, during specimen preparation
Chemical bonding information	Sometimes, from EELS
Quantification	Quantitative compositional analysis from EDS or EELS, and crystal structure analysis from CBED
Accuracy	5–10% relative for EDS and EELS
Detection limits	0.1–3.0% wt. for EDS and EELS
Lateral resolution	Imaging, 0.2–10 nm; EELS, 0.5–10 nm; EDS, 3–30 nm
Imaging/mapping capabilities	Yes, lateral resolution down to < 5 nm
Sample requirements	Solid conductors and coated insulators typically 3 mm in diameter and < 200 nm thick at the analysis point for imaging and EDS, but < 50 nm thick for EELS
Main uses	Microstructural, crystallographic, and compositional analysis; high spatial resolution with good elemental detection and accuracy; unique structural analysis with CBED
Instrument cost	$500,000–$2,000,000
Size	3 m × 4 m × 3 m

Electron Probe X-Ray Microanalysis (EPMA) 1.3.5

Electron Probe X-Ray Microanalysis (EPMA) is an elemental analysis technique based upon bombarding a specimen with a focused beam of energetic electrons (beam energy 5–30 keV) to induce emission of characteristic X rays (energy range 0.1–15 keV). The X rays are measured by Energy-Dispersive (EDS) or Wavelength-Dispersive (WDS) X-ray spectrometers. Quantitative matrix (interelement) correction procedures based upon first principles physical models provide great flexibility in attacking unknown samples of arbitrary composition; the standards suite can be as simple as pure elements or binary compounds. Typical error distributions are such that relative concentration errors lie within ±4% for 95% of cases when the analysis is performed with pure element standards. Spatial distributions of elemental constituents can be visualized qualitatively by X-ray area scans (dot maps) and quantitatively by digital compositional maps.

Range of elements	Beryllium to the actinides
Destructive	No, except for electron beam damage
Chemical bonding	In rare cases: from light-element X-ray peak shifts
Depth profiling	Rarely, by changing incident beam energy
Quantification	Standardless or; pure element standards
Accuracy	±4% relative in 95% of cases; flat, polished samples
Detection limits	WDS, 100 ppm; EDS, 1000 ppm
Sampling depth	Energy and matrix dependent, 100 nm–5 μm
Lateral resolution	Energy and matrix dependent, 100 nm–5 μm
Imaging/mapping	Yes, compositional mapping and SEM imaging
Sample requirements	Solid conductors and insulators; typically, < 2.5 cm in diameter, and < 1 cm thick, polished flat; particles, rough surfaces, and thin films
Major uses	Accurate, nondestructive quantitative analysis of major, minor, and trace constituents of materials
Instrument cost	$300,000–$800,000
Size	3 m × 1.5 m × 2 m high

X-Ray Diffraction (XRD) 1.4.1

In X-Ray Diffraction (XRD) a collimated beam of X rays, with wavelength $\lambda \sim 0.5$–2 Å, is incident on a specimen and is diffracted by the crystalline phases in the specimen according to Bragg's law ($\lambda = 2d \sin\theta$, where d is the spacing between atomic planes in the crystalline phase). The intensity of the diffracted X rays is measured as a function of the diffraction angle 2θ and the specimen's orientation. This diffraction pattern is used to identify the specimen's crystalline phases and to measure its structural properties, including strain (which is measured with great accuracy), epitaxy, and the size and orientation of crystallites (small crystalline regions). XRD can also determine concentration profiles, film thicknesses, and atomic arrangements in amorphous materials and multilayers. It also can characterize defects. To obtain this structural and physical information from thin films, XRD instruments and techniques are designed to maximize the diffracted X-ray intensities, since the diffracting power of thin films is small.

Range of elements	All, but not element specific. Low-Z elements may be difficult to detect
Probing depth	Typically a few μm but material dependent; monolayer sensitivity with synchrotron radiation
Detection Limits	Material dependent, but ~3% in a two phase mixture; with synchrotron radiation can be ~0.1%
Destructive	No, for most materials
Depth profiling	Normally no; but this can be achieved.
Sample requirements	Any material, greater than ~0.5 cm, although smaller with microfocus
Lateral resolution	Normally none; although ~10 μm with microfocus
Main use	Identification of crystalline phases; determination of strain, and crystallite orientation and size; accurate determination of atomic arrangements
Specialized uses	Defect imaging and characterization; atomic arrangements in amorphous materials and multilayers; concentration profiles with depth; film thickness measurements
Instrument cost	\$70,000–\$200,000
Size	Varies with instrument, greater than ~70 $ft.^2$

Extended X-Ray Absorption Fine Structure (EXAFS) 1.4.2

An EXAFS experiment involves the irradiation of a sample with a tunable source of monochromatic X rays from a synchrotron radiation facility. As the X-ray energy is scanned from just below to well above the binding energy of a core-shell electron (e.g., K or L) of a selected element, the X-ray photoabsorption process is monitored. When the energy of the incident X-rays is equal to the electron binding energy, X-ray absorption occurs and a steeply rising absorption edge is observed. For energies greater than the binding energy, oscillations of the absorption with incident X-ray energy (i.e., EXAFS) are observed. EXAFS data are characteristic of the structural distribution of atoms in the immediate vicinity (~5 Å) of the X-ray absorbing element. The frequency of the EXAFS is related to the interatomic distance between the absorbing and neighboring atoms. The amplitude of the EXAFS is related to the number, type, and order of neighboring atoms.

Range of elements	Lithium through uranium
Destructive	No
Bonding information	Yes, interatomic distances, coordination numbers, atom types, and structural disorder; oxidation state by inference
Accuracy	1–2% for interatomic distances; 10–25% for coordination numbers
Detection limits	Surface, monolayer sensitivity; bulk, > 100 ppm
Depth probed	Variable, from Å to μm
Depth profiling	Yes, with glancing incidence angles; electron- and ion-yield detection
Lateral resolution	Not yet developed
Imaging/mapping	Not yet developed
Sample requirements	Virtually any material; solids, liquids, gas
Main use	Local atomic environments of elements in materials
Instrument cost	Laboratory facility, < $300,000; synchrotron beam line, > $1,000,000
Size	Small attachment to synchrotron beam line

Surface Extended X-Ray Absorption Fine Structure and Near Edge X-Ray Absorption Fine Structure (SEXAFS / NEXAFS) 1.4.3

In Surface Extended X-Ray Absorption Fine Structure and Near Edge X-Ray Absorption Fine Structure (SEXAFS/NEXAFS) a solid sample, usually placed in ultrahigh vacuum, is exposed to a tunable beam of X rays from a synchrotron radiation source. A spectrum is collected by varying the photon energy of the X rays and measuring the yield of emitted electrons or fluorescent X rays. Analysis of the wiggles in the observed spectrum (the SEXAFS features) gives information on nearest neighbor bond lengths and coordination numbers for atoms at or near the surface. Features near an absorption edge (NEXAFS) are often characteristic of the local coordination (octahedral, tetrahedral, etc.) or oxidation state. For adsorbed molecules, NEXAFS resonances characterize the type of bonding. On a flat surface, the angular variation of intensity of the resonances gives the orientation of the molecule.

Range of elements	Almost all, from C to U
Destructive	No
Chemical bonding information	Yes, through NEXAFS
Accuracy	In nearest neighbor distance, ±0.01 Å with care
Surface sensitivity	Top few monolayers
Detection limits	0.05 monolayer
Lateral resolution	~0.5 mm
Imaging/mapping	No
Sample requirements	Vacuum-compatible solids
Main use of SEXAFS	Adsorbate–substrate bond lengths
Main use of NEXAFS	Orientation of molecular adsorbates
Instrument cost	$400,000, plus cost of synchrotron
Size	Small attachment to synchrotron beam line

X-Ray Photoelectron and Auger Electron Diffraction (XPD and AED) 1.4.4

In X-Ray Photoelectron Diffraction (XPD) and Auger Electron Diffraction (AED), a single crystal or a textured polycrystalline sample is struck by photons or electrons to produce outgoing electrons that contain surface chemical and structural information. The focus of XPD and AED is structural information, which originates from interference effects as the outbound electrons from a particular atom are scattered by neighboring atoms in the solid. The electron–atom scattering process strongly increases the electron intensity in the forward direction, leading to the simple observation that intensity maxima occur in directions corresponding to rows of atoms. An energy dispersive angle-resolved analyzer is used to map the intensity distribution as a function of angle for elements of interest.

Range of elements	All except H and He
Destructive	XPD no; AED may cause e-beam damage
Element specific	Yes
Chemical state specific	Yes, XPD is better than AED
Accuracy	Bond angles to within 1°; atomic positions to within 0.05 Å
Site symmetry	Yes, and usually quickly
Depth Probed	5–50 Å
Depth profiling	Yes, to 30 Å beneath the surface
Detection limits	0.2 at.%
Lateral resolution	150 Å (AED), 150 μm (XPD)
Imaging/mapping	Yes
Sample requirements	Primarily single crystals, but also textured samples
Main use	To determine adsorption sites and thin-film growth modes in a chemically specific manner
Instrument cost	$300,000–$600,000
Size	4 m × 4 m × 3 m

Low-Energy Electron Diffraction (LEED) 1.4.5

In Low-Energy Electron Diffraction (LEED) a collimated monoenergetic beam of electrons in the energy range 10–1000 eV ($\lambda \approx 0.4$–4.0 Å) is diffracted by a specimen surface. In this energy range, the mean free path of electrons is only a few Å, leading to surface sensitivity. The diffraction pattern can be analyzed for the existence of a clean surface or an ordered overlayer structure. Intensities of diffracted beams can be analyzed to determine the positions of surface atoms relative to each other and to underlying layers. The shapes of diffracted beams in angle can be analyzed to provide information about surface disorder. Various phenomena related to surface crystallography and microstructure can be investigated. This technique requires a vacuum.

Range of elements	All elements, but not element specific
Destructive	No, except in special cases of electron-beam damage
Depth probed	4–20 Å
Detection limits	0.1 monolayer; any ordered phase can be detected; atomic positions to 0.1 Å; step heights to 0.1 Å; surface disorder down to ~10% of surface sites
Resolving power	Maximum resolvable distance for detecting disorder: typically 200 Å; best systems, 5 µm
Lateral resolution	Typical beam sizes, 0.1 mm; best systems, ~10 µm
Imaging capability	Typically, no; with specialized instruments (e.g., low-energy electron microscopy), 150 Å
Sample requirements	Single crystals of conductors and semiconductors; insulators and polycrystalline samples under special circumstances; 0.25 cm^2 or larger, smaller with special effort
Main uses	Analysis of surface crystallography and microstructure; surface cleanliness
Cost	≤$75,000; can be home built cheaply
Size	Generally part of other systems; if self-standing, ~8 m^2

Reflection High-Energy Electron Diffraction (RHEED) 1.4.6

In Reflection High-Energy Electron Diffraction (RHEED), a beam of high-energy electrons (typically 5–50 keV), is accelerated toward the surface of a conducting or semiconducting crystal, which is held at ground potential. The primary beam strikes the sample at a grazing angle (~1–5°) and is subsequently scattered. Some of the electrons scatter elastically. Since their wavelengths are shorter than interatomic separations, these electrons can diffract off ordered rows of atoms on the surface, concentrating scattered electrons into particular directions, that depend on row separations. Beams of scattered electrons whose trajectories intersect a phosphor screen placed opposite the electron gun will excite the phosphor. The light from the phosphor screen is called the RHEED pattern and can be recorded with a photograph, television camera, or by some other method. The symmetry and spacing of the bright features in the RHEED pattern give information on the surface symmetry, lattice constant, and degree of perfection, i.e., the crystal structure.

Range of elements	All, but not chemical specific
Destructive	No, Except for electron-sensitive materials
Depth probed	2–100 Å
Depth profiling	No
Lateral resolution	200 μm × 4 mm, in special cases 0.3 nm × 6 nm
Structural information	Measures surface crystal structure parameters, sensitive to structural defects
Sample requirements	Usually single crystal conductor or semiconductor surfaces
Main use	Monitoring surface structures, especially during thin-film epitaxial growth; can distinguish two- and three-dimensional defects
Instrument cost	$50,000–$200,000
Size	~25 sq. ft., larger if incorporated with an MBE chamber

X-Ray Photoelectron Spectroscopy (XPS) 1.5.1

In X-Ray Photoelectron Spectroscopy (XPS) monoenergetic soft X rays bombard a sample material, causing electrons to be ejected. Identification of the elements present in the sample can be made directly from the kinetic energies of these ejected photoelectrons. On a finer scale it is also possible to identify the chemical state of the elements present from small variations in the determined kinetic energies. The relative concentrations of elements can be determined from the measured photoelectron intensities. For a solid, XPS probes 2–20 atomic layers deep, depending on the material, the energy of the photoelectron concerned, and the angle (with respect to the surface) of the measurement. The particular strengths of XPS are semiquantitative elemental analysis of surfaces without standards, and chemical state analysis, for materials as diverse as biological to metallurgical. XPS also is known as electron spectroscopy for chemical analysis (ESCA).

Range of elements	All except hydrogen and helium
Destructive	No, some beam damage to X-ray sensitive materials
Elemental analysis	Yes, semiquantitative without standards; quantitative with standards. Not a trace element method.
Chemical state information	Yes
Depth probed	5–50 Å
Depth profiling	Yes, over the top 50 Å; greater depths require sputter profiling
Depth resolution	A few to several tens of Å, depending on conditions
Lateral resolution	5 mm to 75 μm; down to 5 μm in special instruments
Sample requirements	All vacuum-compatible materials; flat samples best; size accepted depends on particular instrument
Main uses	Determinations of elemental and chemical state compositions in the top 30 Å
Instrument cost	$200,000–$1,000,000, depending on capabilities
Size	10 ft. × 12 ft.

Ultraviolet Photoelectron Spectroscopy (UPS) 1.5.2

If monoenergetic photons in the 10–100 eV energy range strike a sample material, photoelectrons from the valence levels and low-lying core levels (i.e., having lower binding energy than the photon energy) are ejected. Measurement of the kinetic energy distribution of the ejected electrons is known as Ultraviolet Photoelectron Spectroscopy (UPS). The physics of the technique is the same as XPS, the only differences being that much lower photon energies are used and the primary emphasis is on examining the valence electron levels, rather than core levels. Owing to this emphasis, the primary use, when investigating solid surfaces, is for electronic structure studies in surface physics rather than for materials analysis. There are, however, a number of situations where UPS offers advantages over XPS for materials surface analysis.

Elemental analysis	Not usually, sometimes from available core levels
Destructive	No, some beam damage to radiation-sensitive material
Chemical state information	Yes, but complicated using valence levels; for core levels as for XPS
Depth probed	2–100 Å
Depth profiling	Yes, over the depth probed; deeper profiling requires sputter profiling
Lateral resolution	Generally none (mm size), but photoelectron microscopes with capabilities down to the 1-μm range exist
Sample requirements	Vacuum-compatible material; flat samples best; size accepted depends on instrumentation
Main use	Electronic structure studies of free molecules (gas phase), well-defined solid surfaces, and adsorbates on solid surfaces
Instrument cost	No commercial instruments specifically for UPS; usually an add-on to XPS (incremental cost ~$30,000) or done using a synchrotron facility as the photon source
Size	10 ft. × 10 ft. for a stand-alone system

Auger Electron Spectroscopy (AES) 1.5.3

Auger Electron Spectroscopy (AES) uses a focused electron beam to create secondary electrons near the surface of a solid sample. Some of these (the Auger electrons) have energies characteristic of the elements and, in many cases, of the chemical bonding of the atoms from which they are released. Because of their characteristic energies and the shallow depth from which they escape without energy loss, Auger electrons are able to characterize the elemental composition and, at times, the chemistry of the surfaces of samples. When used in combination with ion sputtering to gradually remove the surface, Auger spectroscopy can similarly characterize the sample in depth. The high spacial resolution of the electron beam and the process allows microanalysis of three-dimensional regions of solid samples. AES has the attributes of high lateral resolution, relatively high sensitivity, standardless semiquantitative analysis, and chemical bonding information in some cases.

Range of elements	All except H and He
Destructive	No, except to electron beam-sensitive materials and during depth profiling
Elemental Analysis	Yes, semiquantitative without standards; quantitative with standards
Absolute sensitivity	100 ppm for most elements, depending on the matrix
Chemical state information?	Yes, in many materials
Depth probed	5–100 Å
Depth profiling	Yes, in combination with ion-beam sputtering
Lateral resolution	300 Å for Auger analysis, even less for imaging
Imaging/mapping	Yes, called Scanning Auger Microscopy, SAM
Sample requirements	Vacuum-compatible materials
Main use	Elemental composition of inorganic materials
Instrument cost	$100,000–$800,000
Size	10 ft. × 15 ft.

Reflected Electron Energy-Loss Spectroscopy (REELS) 1.5.4

In Reflected Electron Energy-Loss Spectroscopy (REELS) a solid specimen, placed in a vacuum, is irradiated with a narrow beam of electrons that are sufficiently energetic to induce electron excitations with atoms or clusters of atoms. Some of the incident electrons reemerge from the sample having lost a specific amount of energy relative to the well-defined energy E_0 of the incident electron. The number, direction k, and energy of the emitted electrons can be measured by an electron energy analyzer. Composition, crystal structure, and chemical bonding information can be obtained about the sample's surface from the intensity and line shape of the emitted electron energy-loss spectra by comparison to standards.

Range of elements	Hydrogen to uranium; no isotopes
Destructive	No
Chemical bond information	Yes; energetics and orientation
Depth profiling	Yes; tilting or ion sputtering
Quantification	Standards required
Accuracy	Few percent to tens of percent
Detection limits	Few tenths of a percent
Probing depth	0.07–3.0 nm
Lateral resolution	100 nm–50 μm; sample independent; not limited by rediffused primaries
Imaging/mapping	Yes, called REELM
Sample requirements	Solids; liquids; vacuum compatible; typically < 2.5 cm-diameter, < 1.5 cm-thickness
Main use	Few-monolayer thin-film analysis, e.g., adsorbate and very thin-film reactions; submicron detection of metal hydrides
Instrument cost	\$0–\$700,000, free on any type of electron-excited Auger spectrometer
Size	None extra over Auger spectrometer

X-Ray Fluorescence (XRF) 1.6.1

In X-Ray Fluorescence (XRF), an X-ray beam is used to irradiate a specimen, and the emitted fluorescent X rays are analyzed with a crystal spectrometer and scintillation or proportional counter. The fluorescent radiation normally is diffracted by a crystal at different angles to separate the X-ray wavelengths and therefore to identify the elements; concentrations are determined from the peak intensities. For thin films XRF intensity–composition–thickness equations derived from first principles are used for the precision determination of composition and thickness. This can be done also for each individual layer of multiple-layer films.

Range of elements	All but low-Z elements: H, He, and Li
Accuracy	±1% for composition, ±3% for thickness
Destructive	No
Depth sampled	Normally in the 10-μm range, but can be a few tens of Å in the total-reflection range
Depth profiling	Normally no, but possible using variable-incidence X rays
Detection limits	Normally 0.1% in concentration.
Sensitivity	10–10^5 Å in thickness can be examined
Lateral resolution	Normally none, but down to 10 μm using a microbeam
Chemical bond information	Normally no, but can be obtained from soft X-ray spectra
Sample requirements	≤5.0 cm in diameter
Main use	Identification of elements; determination of composition and thickness
Instrument cost	\$50,000–\$300,000
Size	5 ft. × 8 ft.

Total Reflection X-Ray Fluorescence Analysis (TXRF) 1.6.2

In Total Reflection X-Ray Fluorescence Analysis (TXRF), the surface of a solid specimen is exposed to an X-ray beam in grazing geometry. The angle of incidence is kept below the critical angle for total reflection, which is determined by the electron density in the specimen surface layer, and is on the order of mrad. With total reflection, only a few nm of the surface layer are penetrated by the X rays, and the surface is excited to emit characteristic X-ray fluorescence radiation. The energy spectrum recorded by the detector contains quantitative information about the elemental composition and, especially, the trace impurity content of the surface, e.g., semiconductor wafers. TXRF requires a specular surface of the specimen with regard to the primary X-ray light.

Range of elements	Sodium to uranium
Destructive	No
Chemical bonding information	Not usually
Depth probed	Typically 1–5 nm
Depth profiling capability	Limited (variation of angle of incidence)
Quantification	Yes
Accuracy	1–20%
Detection limits	10^{10}–10^{14} at/cm^2
Lateral resolution	Limited, typically 10 mm
Sample requirements	Specular surface, typically ≥ 2.5-cm diameter
Main use	Multielement analysis, excellent detection limits for heavy metals; quantitative measurement of heavy-metal trace contamination on silicon wafers
Instrument cost	$300,000–$600,000

Particle-Induced X-Ray Emission (PIXE) 1.6.3

Particle-Induced X-Ray Emission (PIXE) is a quantitative, nondestructive analysis technique that relies on the spectrometry of characteristic X rays emitted during irradiation of a specimen with high-energy ionic particles (~0.3–10 MeV). The process is analogous to the emission of characteristic X rays under electron and photon bombardment of a specimen (see the articles on EDS, EMPA, and XRF). With appropriate corrections, X-ray yields (X rays per particle) can be converted to elemental concentrations. The background X-ray radiation for PIXE is much less than that for electron excitation; thus, the detection limits for trace elements using PIXE is orders of magnitude better. PIXE is best for the analysis of thin samples, surface layers, and samples with limited amounts of materials, while photon bombardment (XRF) is better for bulk analysis and thick specimens. Using wavelength-dispersive detectors, PIXE, EMPA, and XRF can provide identification of the chemical bonding of elements. Although EMPA and EDS require that the specimen be in vacuum, PIXE and XRF can be performed with the specimen in vacuum or at atmospheric pressure.

Range of elements	Lithium to uranium
Chemical bonding information	Yes, when spectral resolution is high
Depth probed	$\leq 10\ \mu m$
Depth profiling	Yes, by varying angle of incidence or particle energy.
Detection limits	Thin, freestanding foil, 0.1–10 ppm; surface layers on thick specimens, 10^{13}–5×10^{15} at/cm^2; Bulk specimens, 1–100 ppm
Accuracy	~2–10%, with standards
Lateral resolution	~5 μm–2 mm
Imaging/mapping	Yes
Sample requirements	Solids, liquids, and gases
Main use	Fast analysis for many elements, in all materials, simultaneously
System cost	~$1,000,000, including small ion accelerator (2-MeV H^+)
System size	~100 sq. ft. floor space

Photoluminescence (PL) 1.7.1

In photoluminescence one measures physical and chemical properties of materials by using photons to induce excited electronic states in the material system and analyzing the optical emission as these states relax. Typically, light is directed onto the sample for excitation, and the emitted luminescence is collected by a lens and passed through an optical spectrometer onto a photodetector. The spectral distribution and time dependence of the emission are related to electronic transition probabilities within the sample, and can be used to provide qualitative and, sometimes, quantitative information about chemical composition, structure (bonding, disorder, interfaces, quantum wells), impurities, kinetic processes, and energy transfer.

Destructiveness	Nondestructive
Depth probed	0.1–3 μm; limited by light penetration depth and carrier diffusion length
Lateral resolution	Down to 1–2 μm
Quantitative abilities	Intensity-based impurity quantification to several percent possible; energy quantification very precise
Sensitivity	Down to parts-per-trillion level, depending on impurity species and host
Imaging/mapping	Yes
Sample requirements	Liquid or solid having optical transitions; probe size 2 μm to a few cm
Main uses	Band gaps of semiconductors; carrier lifetimes; shallow impurity or defect detection; sample quality and structure
Instrument cost	Less than $10,000 to over $200,000
Size	Table top to small room

Modulation Spectroscopy 1.7.2

Modulation spectroscopy is a powerful experimental method for measuring the energy of transitions between the filled and empty electronic states in the bulk (band gaps) or at surfaces of semiconductor materials over a wide range of experimental conditions (temperature, ambients, etc.). By taking the derivative of the reflectance (or transmittance) of a material in an analog manner, it produces a series of sharp, derivative-like spectral features corresponding to the photon energy of the transitions. These energies are sensitive to a number of internal and external parameters such as chemical composition, temperature, strain, and electric and magnetic fields. The line widths of these spectral features are a function of the quality of the material.

Destructiveness	Some methods are nondestructive
Depth probed	For bulk applications 0.1–1 μm; for surface applications one monolayer is possible
Lateral resolution	Down to 100 μm
Image/mapping	Yes
Sensitivity	Alloy composition (e.g., $Ga_{1-x}Al_xAs$) $\Delta x = 0.005$; carrier concentration 10^{15}–10^{19} cm^{-3}
Main uses	Contactless, nondestructive monitoring of band gaps in semiconductors; Wide range of temperatures and ambients (air, ultrahigh vacuum); *in-situ* monitoring of semiconductor growth
Instrument cost	\$30,000–\$100,000
Size	For most methods about 2 × 3 ft.

Variable-Angle Spectroscopic Ellipsometry (VASE) 1.7.3

In Variable-Angle Spectroscopic Ellipsometry (VASE), polarized light strikes a surface and the polarization of the reflected light is analyzed using a second polarizer. The light beam is highly collimated and monochromatic, and is incident on the material at an oblique angle. For each angle of incidence and wavelength, the reflected light intensity is measured as a function of polarization angle, allowing the important ellipsometric parameter to be determined. An optimum set of angle of incidence and wavelength combinations is used to maximize measurement sensitivity and information obtained. Physical quantities derivable from the measured parameter include the optical constants of bulk or filmed media, the thicknesses of films (from 1 to a few hundred nm), and the microstructural composition of a multiconstituent thin film. In general only materials with parallel interfaces, and with structural or chemical inhomogeneities on a scale less than about 1/10 the wavelength of the incident light, can be studied by ellipsometry.

Main use	Film thicknesses, microstructure, and optical properties
Optical range	Near ultraviolet to mid infrared
Sample requirements	Planar materials and interfaces
Destructive	No, operation in any transparent ambient, including vacuum, gases, liquids, and air
Depth probed	Light penetration of the material (tens of nm to µm)
Lateral resolution	mm normally, 100 µm under special conditions
Image/mapping	No
Instrument cost	$50,000–$150,000
Size	0.5 m × 1 m

Fourier Transform Infrared Spectroscopy (FTIR) 1.8.1

The vibrational motions of the chemically bound constituents of matter have frequencies in the infrared regime. The oscillations induced by certain vibrational modes provide a means for matter to couple with an impinging beam of infrared electromagnetic radiation and to exchange energy with it when the frequencies are in resonance. In the infrared experiment, the intensity of a beam of infrared radiation is measured before (I_0) and after (I) it interacts with the sample as a function of light frequency, $\{w_i\}$. A plot of I/I_0 versus frequency is the "infrared spectrum." The identities, surrounding environments, and concentrations of the chemical bonds that are present can be determined.

Information	Vibrational frequencies of chemical bonds
Element Range	All, but not element specific
Destructive	No
Chemical bonding information	Yes, identification of functional groups
Depth profiling	No, not under standard conditions
Depth Probed	Sample dependent, from μm's to 10 nm
Detection limits	Ranges from undetectable to < 10^{13} bonds/cc. Sub-monolayer sometimes
Quantification	Standards usually needed
Reproducibility	0.1% variation over months
Lateral resolution	0.5 cm to 20 μm
Imaging/mapping	Available, but not routinely used
Sample requirements	Solid, liquid, or gas in all forms; vacuum not required
Main use	Qualitative and quantitative determination of chemical species, both trace and bulk, for solids and thin films. Stress, structural inhomogeneity
Instrument cost	$50,000–$150,000 for FTIR; $20,000 or more for non-FT spectrophotometers
Instrument size	Ranges from desktop to (2 × 2 m)

Raman Spectroscopy 1.8.2

Raman spectroscopy is the measurement, as a function of wavenumber, of the inelastic light scattering that results from the excitation of vibrations in molecular and crystalline materials. The excitation source is a single line of a continuous gas laser, which permits optical microscope optics to be used for measurement of samples down to a few μm. Raman spectroscopy is sensitive to molecular and crystal structure; applications include chemical fingerprinting, examination of single grains in ceramics and rocks, single-crystal measurements, speciation of aqueous solutions, identification of compounds in bubbles and fluid inclusions, investigations of structure and strain states in polycrystalline ceramics, glasses, fibers, gels, and thin and thick films.

Information	Vibrational Frequencies of chemical bonds
Element range	All, but not element specific
Destructive	No, unless sample is susceptible to laser damage
Lateral resolution	1 μm with microfocus instruments
Depth profiling	Limited to transparent materials
Depth probed	Few μm to mm, depending on material
Detection limits	1000 Å normally, submonolayer in special cases
Quantitative	With difficulty; usually qualitative only
Imaging	Usually no, although imaging instruments have been built
Sample requirements	Very flexible: liquids, gases, crystals, polycrystalline solids, powders, and thin films
Main use	Identification of unknown compounds in solutions, liquids, and crystalline materials; characterization of structural order, and phase transitions
Instrument cost	$150,000–$250,000
Size	1.5 m × 2.5 m

High-Resolution Electron Energy Loss Spectroscopy (HREELS) 1.8.3

In High-Resolution Electron Energy Loss Spectroscopy (HREELS), a highly monoenergetic beam of low energy (1–10 eV) electrons is focused onto a sample's surface, and the scattered electrons are analyzed with high resolution of the scattering energy and angle. Some of the scattered electrons suffer small characteristic energy losses due to vibrational excitation of surface atoms and molecules. A vibrational spectrum can be obtained by counting the number of electrons versus the electron energy loss relative to the elastically scattered (no energy loss) electron beam. This spectrum is used mainly to identify chemical species (functional groups) in the first layer of the surface. Often this layer contains adsorbed species on a solid.

Information	Molecular vibrational frequencies
Main use	Nondestructive identification of the molecular functional groups present at surfaces
Range of elements	Not element specific
Bonding	Any chemical bonds that have vibrations in the range 50–4000 cm^{-1}
Detection limits	0.1% monolayer for strong vibrational bands
Quantification	Difficult, possible with standards
Depth probed	2 nm
Lateral resolution	1 mm^2
Sample requirements	Single-crystal samples of conductors best; other solid samples are suitable, including polycrystalline metals, polymeric materials, semiconductors, and insulators, ultrahigh vacuum compatible; typically ≥ 5 mm diameter, 1–3 mm thick
Instrument cost	$100,000 plus associated techniques and vacuum system
Size	Attaches to vacuum chamber by 8–14 inch diameter flange.

Solid State Nuclear Magnetic Resonance (NMR) 1.8.4

Solid state Nuclear Magnetic Resonance (NMR) exploits the interaction of nuclear magnetic moments with electromagnetic waves in the radio frequency region. In the experiment, a solid specimen (crystalline or amorphous, aligned or randomly oriented) is placed in a strong external magnetic field (typically 1–14 Tesla) and irradiated with intense radio frequency pulses over a frequency range required to excite a specific atomic nucleus from the ground magnetic (spin) state to another higher state. As the nucleus releases back to its ground state the sample re-emits a radio signal at the excitation frequency, which is detected by electromagnetic induction and Fourier transformed to yield a plot of intensity versus frequency. The spectrum thus obtained identifies the presence of the atom and its relative concentration (with standards) and is a sensitive indicator of structural and chemical bonding properties. It can serve for phase identification as well as for the characterization of local bonding environments in disordered materials.

Elements detected	All elements possessing an isotope with a suitable magnetic dipole moment (about half the elements in the periodic table)
Detection limit	On the order of 10^{18} atoms of the nuclear isotope studied
Surface sensitivity	Not intrinsically surface sensitive: Surface areas $> 10\ m^2/g$ required or desirable for surface studies
Typical sample size	10–500 mg, varies greatly with the nucleus studied; sample length, 0.5–5 cm; width, 0.5–2 cm
Measurement conditions	Usually at ambient temperature and pressure
Sample form	Powder, single crystal, randomly oriented, or aligned film
Main use	Element-selective phase identification and quantification, structural characterization of disordered states
Instrument cost	$200,000–$1,200,000, depending mostly on the field strength desired
Space requirement	300 $ft.^2$

Rutherford Backscattering Spectrometry (RBS) 1.9.1

Rutherford Backscattering Spectrometry (RBS) analysis is performed by bombarding a sample target with a monoenergetic beam of high-energy particles, typically helium, with an energy of a few MeV. A fraction of the incident atoms scatter backwards from heavier atoms in the near-surface region of the target material, and usually are detected with a solid state detector that measures their energy. The energy of a backscattered particle is related to the depth and mass of the target atom, while the number of backscattered particles detected from any given element is proportional to concentration. This relationship is used to generate a quantitative depth profile of the upper 1–2 μm of the sample. Alignment of the ion beam with the crystallographic axes of a sample permits crystal damage and lattice locations of impurities to be quantitatively measured and depth profiled. The primary applications of RBS are the quantitative depth profiling of thin-film structures, crystallinity, dopants, and impurities.

Range of elements	Lithium to uranium
Destructive	~10^{13} He atoms implanted; radiation damage.
Chemical bonding information	No
Quantification	Yes, standardless; accuracy 5-20%
Detection limits	10^{12}–10^{16} atoms/cm^2; 1–10 at.% for low-Z elements; 0–100 ppm for high-Z elements
Lateral resolution	1–4 mm, 1 μm in specialized equipment
Depth profiling	Yes and nondestructive
Depth resolution	2–30 nm
Maximum depth	~2 μm, 20 μm with H^+
Imaging/mapping	Under development
Sample requirements	Solid, vacuum compatible
Main use	Nondestructive depth profiling of thin films, crystal damage information
Instrument cost	$450,000–$1,000,000
Size	2 m × 7 m

Elastic Recoil Spectrometry (ERS) 1.9.2

Energetic recoil ions, $^{1}H^{+}$ and $^{2}H^{+}$, are produced when $^{4}He^{+}$ ions having energies in the MeV range undergo elastic nucleus–nucleus collisions within a hydrogen- or deuterium-containing solid sample. Energy spectrometry of the recoiling ions identifies their mass and depth of origin. The total hydrogen content of a thin layer may be determined directly from the recoil fluence. In combination with Rutherford Backscattering (RBS) analysis of the same sample, elastic recoil spectra provide concentration profiles and complete compositional analysis of near-surface regions of the sample material. ERS requires equipment common to RBS analysis. It is the simplest ion beam technique for hydrogen profiling, since ion backscattering (RBS) from hydrogen is not possible.

Range of elements	Unique selection of ^{1}H, ^{2}H
Destructive	Radiation damage may release H in polymers
Chemical bonding information	None
Quantitation	Absolute atoms/cm^{2} $\pm$ 2% typically
Sensitivity	5×10^{13} atom/cm^{2} or 0.01 at.% (typically)
Depth probed	$\leq$ 1 μm typically
Depth profiles	Yes; concentration profile to $\pm$ 1% relative
Depth resolution	Varies with depth; 300–600 Å at depth 1000 Å in Si
Lateral resolution	1–4 mm typically
Sample requirements	Solid, vacuum compatible, dimensions $\geq$ 5 mm
Main use	Determination of H concentrations in thin films; rapid; matrix-independent
Instrument cost	As for RBS; MeV accelerator ($1,000,000–$1,500,000); services available
Size	Requires laboratory $\geq$ 20 ft. $\times$ 50 ft., depending on instrument

Medium-Energy Ion Scattering with Channeling and Blocking (MEIS) 1.9.3

Medium-Energy Ion Scattering (MEIS) with channeling and blocking is a quantitative, real space, nondestructive technique for studying the composition and structure of surfaces and interfaces buried up to a few atomic layers below the surface. Single-crystal or epitaxial samples are required for the structural determinations. The basic quantities measured are the energy and angular distribution of backscattered ions in the 50–400 keV range. The technique has elemental and depth sensitivity. The ion angular distributions are characterized by minima (dips) in intensity, the positions of which are closely connected to the relative positions of atoms in the surface layer. MEIS is more surface sensitive, and more complex instrumentally than other surface ion spectroscopies, though interpretation is straightforward. The technique is useful for the analysis of all ultrahigh vacuum compatible solids, and in particular metals, semiconductors, and overlayers on such surfaces (submonolayer adsorbate concentrations, thin films of silicides, etc.).

Elements detected	all elements
Elemental sensitivity	Scales as the square of nuclear charge; best for heavy elements ($< 10^{-4}$ monolayer); poor for hydrogen
Chemical sensitivity	None
Depth probed	Typically 4–5 atomic layers, but up to 200 Å in special cases
Depth resolution	Optimally on a monolayer level
Quantification	Absolute technique for elemental concentrations
Lateral resolution	None
Destructive	Not inherently
Sample requirements	Ultrahigh vacuum compatibility; practical size ~1 cm in diameter
Main uses	Determination of structural parameters of surfaces and interfaces; very high resolution depth profiling
Accuracy	> 1% (structural parameters); element dependent (composition)
Cost	$1,000,000–$2,000,000
Size	~ 7 m × 3 m

Ion Scattering Spectroscopy (ISS) 1.9.4

In Ion Scattering Spectroscopy (ISS) a low-energy monoenergetic beam of ions is focused onto a solid surface and the energy of the scattered ions is measured at some fixed angle. The collision of the inert ion beam, usually $^{3}He^{+}$, $^{4}He^{+}$, or $^{20}Ne^{+}$, follows the simple laws of conservation of momentum for a binary elastic collision with an atom in the outer surface of the solid. The energy loss thus identifies the atom struck. Inelastic collisions and ions that penetrate deeper than the first atomic layer normally do not yield a sharp, discrete peak. Neighboring atoms do not affect the signal because the kinetics of the collision are much shorter than bond vibrations. A spectrum is obtained by measuring the number of ions scattered from the surface as a function of their energy by passing the scattered ions through an energy analyzer. The spectrum is normally plotted as a ratio of the number of ions of energy E versus the energy of the primary beam E_0. This can be directly converted to a plot of relative concentration versus atomic number, Z. Extremely detailed information regarding the changes in elemental composition from the outer monolayer to depths of 50 Å or more are routinely obtained by continuously monitoring the spectrum while slowly sputtering away the surface.

Range of elements	All but helium; hydrogen indirectly
Sample requirements	Any solid vacuum-compatible material
Sensitivity	< 0.01 monolayer, 0.5% for C to 50 ppm for heavy metals
Quantitation	Relative; 0.5–20%
Speed	Single spectrum, 0.1 s; nominal 100-Å profile, 30 min
Depth of analysis	Outermost monatomic layer to any sputtered depth
Lateral resolution	150 μm
Imaging	Yes, limited
Sample damage	Only if done with sputter profiling
Main uses	Exclusive detection of outer most monatomic layer and very detailed depth profiles of the top 100 Å
Instrument cost	\$25,000–\$150,000
Size	10 ft. × 10 ft.

Dynamic Secondary Ion Mass Spectrometry (Dynamic SIMS) 1.10.1

In Secondary Ion Mass Spectrometry (SIMS), a solid specimen, placed in a vacuum, is bombarded with a narrow beam of ions, called primary ions, that are sufficiently energetic to cause ejection (sputtering) of atoms and small clusters of atoms from the bombarded region. Some of the atoms and atomic clusters are ejected as ions, called secondary ions. The secondary ions are subsequently accelerated into a mass spectrometer, where they are separated according to their mass-to-charge ratio and counted. The relative quantities of the measured secondary ions are converted to concentrations, by comparison with standards, to reveal the composition and trace impurity content of the specimen as a function of sputtering time (depth).

Range of elements	H to U; all isotopes
Destructive	Yes, material removed during sputtering
Chemical bonding information	In rare cases, from molecular clusters, but see Static SIMS
Quantification	Standards usually needed
Accuracy	2% to factor of 2 for concentrations
Detection limits	10^{12}–10^{16} atoms/cm^3 (ppb–ppm)
Depth probed	2 nm–100 μm (depends on sputter rate and data collection time)
Depth profiling	Yes, by the sputtering process; resolution 2–30 nm
Lateral resolution	50 nm–2 μm; 10 nm in special cases
Imaging/mapping	Yes
Sample requirements	Solid conductors and insulators, typically ≤ 2.5 cm in diameter, ≤ 6 mm thick, vacuum compatible
Main use	Measurement of composition and of trace-level impurities in solid materials a function of depth, excellent detection limits, good depth resolution
Instrument cost	$500,000–$1,500,000
Size	10 ft. × 15 ft.

Static Secondary Ion Mass Spectrometry (Static SIMS) 1.10.2

Static Secondary Ion Mass Spectrometry (SIMS) involves the bombardment of a sample with an energetic (typically 1–10 keV) beam of particles, which may be either ions or neutrals. As a result of the interaction of these primary particles with the sample, species are ejected that have become ionized. These ejected species, known as secondary ions, are the analytical signal in SIMS.

In static SIMS, the use of a low dose of incident particles (typically less than 5×10^{12} atoms/cm^2) is critical to maintain the chemical integrity of the sample surface during analysis. A mass spectrometer sorts the secondary ions with respect to their specific charge-to-mass ratio, thereby providing a mass spectrum composed of fragment ions of the various functional groups or compounds on the sample surface. The interpretation of these characteristic fragmentation patterns results in a chemical analysis of the outer few monolayers. The ability to obtain surface chemical information is the key feature distinguishing static SIMS from dynamic SIMS, which profiles rapidly into the sample, destroying the chemical integrity of the sample.

Range of elements	H to U; all isotopes
Destructive	Yes, if sputtered long enough
Chemical bonding information	Yes
Depth probed	Outer 1 or 2 monolayers
Lateral resolution	Down to ~100 μm
Imaging/mapping	Yes
Quantification	Possible with appropriate standards
Mass range	Typically, up to 1000 amu (quadrupole), or up to 10,000 amu (time of flight)
Sample requirements	Solids, liquids (dispersed or evaporated on a substrate), or powders; must be vacuum compatible
Main use	Surface chemical analysis, particularly organics, polymers
Instrument cost	$500,000–$750,000
Size	4 ft. × 8 ft.

Surface Analysis by Laser Ionization (SALI) 1.10.3

In Surface Analysis by Laser Ionization (SALI), a probe beam such as an ion beam, electron beam, or laser is directed onto a surface to remove a sample of material. An untuned, high-intensity laser beam passes parallel and close to but above the surface. The laser has sufficient intensity to induce a high degree of nonresonant, and hence nonselective, photoionization of the vaporized sample of material within the laser beam. The nonselectively ionized sample is then subjected to mass spectral analysis to determine the nature of the unknown species. SALI spectra accurately reflect the surface composition, and the use of time-of-flight mass spectrometers provides fast, efficient and extremely sensitive analysis.

Range of elements	Hydrogen to Uranium
Destructive	Yes, surface layers removed during analysis
Post ionization approaches	Multiphoton ionization (MPI), single-photon ionization (SPI)
Information	Elemental surface analysis (MPI); molecular surface analysis (SPI)
Detection limit	ppm to ppb
Quantification	~10% using standards
Dynamic range	Depth profile mode ~10^4
Probing depth	2–5 Å (to several μm in profiling mode)
Lateral resolution	down to 60 nm
Mass range	1–10,000 amu or greater
Sample requirements	Solid, vacuum compatible, any shape
Main uses	Quantitative depth profiling, molecular analysis using SPI mode; imaging
Instrument cost	\$600,000–\$1,000,000
Size	Approximately 45 sq. ft.

Sputtered Neutral Mass Spectrometry (SNMS) 1.10.4

Sputtered Neutral Mass Spectrometry (SNMS) is the mass spectrometric analysis of sputtered atoms ejected from a solid surface by energetic ion bombardment. The sputtered atoms are ionized for mass spectrometric analysis by a mechanism separate from the sputtering atomization. As such, SNMS is complementary to Secondary Ion Mass Spectrometry (SIMS), which is the mass spectrometric analysis of sputtered ions, as distinct from sputtered atoms. The forte of SNMS analysis, compared to SIMS, is the accurate measurement of concentration depth profiles through chemically complex thin-film structures, including interfaces, with excellent depth resolution and to trace concentration levels. Generically both SALI and GDMS are specific examples of SNMS. In this article we concentrate on post ionization only by electron impact.

Range of elements	Li to U
Destructive	Yes, surface material sputtered
Chemical bonding information	None
Quantification	Yes, accuracy × 3 without standards; 5–10% with analogous standard; 30% with dissimilar standard
Detection limits	10–100 ppm
Depth probed	15 Å (to many μm when profiling)
Depth profiling	Yes, by sputtering
Lateral resolution	A few mm in direct plasma sputtering; 0.1–10 μm using separate, focused primary ion-beam sputtering
Imaging/mapping	Yes, with separate, focused primary ion-beam
Sample requirements	Solid conducting material, vacuum compatible; flat wafer up to 5-mm diameter; insulator analysis possible
Main use	Complete elemental analysis of complex thin-film structures to several μm depth, with excellent depth resolution
Cost	$200,000–$450,000
Size	2.5 ft. × 5 ft.

Laser Ionization Mass Spectrometry (LIMS) 1.10.5

In Laser Ionization Mass Spectrometry (LIMS, also LAMMA, LAMMS, and LIMA), a vacuum-compatible solid sample is irradiated with short pulses (~10 ns) of ultraviolet laser light. The laser pulse vaporizes a microvolume of material, and a fraction of the vaporized species are ionized and accelerated into a time-of-flight mass spectrometer which measures the signal intensity of the mass-separated ions. The instrument acquires a complete mass spectrum, typically covering the range 0–250 atomic mass units (amu), with each laser pulse. A survey analysis of the material is performed in this way. The relative intensities of the signals can be converted to concentrations with the use of appropriate standards, and quantitative or semi-quantitative analyses are possible with the use of such standards.

Range of elements	Hydrogen to uranium; all isotopes
Destructive	Yes, on a scale of few micrometers depth
Chemical bonding information	Yes, depending on the laser irradiance
Quantification	Standards needed
Detection limits	10^{16}–10^{18} at/cm^3 (ppm to 100 ppm)
Depth probed	variable with material and laser power
Depth profiling	Yes, repeated laser shots sample progressively deeper layers; depth resolution 50–100 nm
Lateral resolution	3–5 μm
Mapping capabilities	No
Sample requirements	Vacuum-compatible solids; must be able to absorb ultraviolet radiation
Main use	Survey capability with ppm detection limits, not affected by surface charging effects; complete elemental coverage; survey microanalysis of contaminated areas, chemical failure analysis
Instrument cost	$400,000
Size	9 ft. × 5 ft.

Spark Source Mass Spectrometry (SSMS) 1.10.6

Spark Source Mass Spectrometry (SSMS) is a method of trace level analysis—less than 1 part per million atomic (ppma)—in which a solid material, in the form of two conducting electrodes, is vaporized and ionized by a high-voltage radio frequency spark in vacuum. The ions produced from the sample electrodes are accelerated into a mass spectrometer, separated according to their mass-to-charge ratio, and collected for qualitative identification and quantitative analysis.

SSMS provides complete elemental surveys for a wide range of sample types and allows the determination of elemental concentrations with detection limits in the range 10–50 parts per billion atomic (ppba).

Range of elements	All elements simultaneously
Destructive	Yes, material is removed from surface
Chemical bonding information	No
Sensitivity	Sub-ppma; 0.01–0.05 ppma typical
Accuracy	Factor of 3, without standards, or factor of 1.2, with standards
Bulk analysis	Yes
Depth probed	1 –5-μm depth
Depth profiling	Yes, but only 1–5 μm resolution
Lateral resolution	None
Sample requirements	Bulk solid: 1/16 in × 1/16 in × 1/2 in; powder: 10–100 mg; thin film: 1 cm^2 × ~5 μm
Sample conductivity	Conductors and semiconductors: direct analysis; insulators (>10^7 $(ohm\text{-}cm)^{-1}$): pulverize and mix with a conductor
Main use	Complete trace elemental survey of solid materials with accuracy to within a factor of 3 without standards
Cost	Used instrumentation only: $10,000–$100,000
Size	9 ft. × 10 ft.

Glow-Discharge Mass Spectrometry (GDMS) 1.10.7

Glow-Discharge Mass Spectrometry is the mass spectrometric analysis of material sputtered into a glow-discharge plasma from a cathode. Atoms sputtered from the sample surface are ionized in the plasma by Penning and electron impact processes, giving ion yields that are matrix-independent and very similar for all elements. Sputtering is rapid (about 1 μm/min) and ion currents are high, yielding sub-ppbw detection limits. Thus GDMS provides accurate concentration measurements, as a function of depth, from major to ultratrace levels over the full periodic table.

Range of elements	Lithium to uranium
Destructive	Yes, surface material sputtered
Chemical bonding information	No
Quantitation	Yes, with standards, 20% accuracy, 5% precision
Detection limits	pptw (GDMS), 10 ppbw (GDQMS)
Depth probed	100 nm to many μm, depending on sputter time
Depth profiling	Yes, by sputtering
Lateral resolution	A few mm
Imaging/mapping	No
Sample requirements	Solid conducting material, vacuum compatible; pin sample ($2 \times 2 \times 20$ mm^3) or flat wafer sample (10–20 mm diameter); insulator analysis possible
Main use	Complete qualitative and quantitative bulk elemental analysis of conducting solids to ultratrace levels
Instrument cost	$200,000–$600,000
Size	6.5 ft. × 6.4 ft. (GDMS) 2.3 ft. × 5.7 ft. (GDQMS)

Inductively Coupled Plasma Mass Spectrometry (ICPMS) 1.10.8

Inductively Coupled Plasma Mass Spectrometry (ICPMS) uses an inductively coupled plasma to generate ions that are subsequently analyzed by a mass spectrometer. The plasma is a highly efficient ion source that gives detection limits below 1 ppb for most elements. The technique allows both fully quantitative and semiquantitative analyses. Samples usually are introduced as liquids but recent developments allow the direct sampling of solids by laser ablation-ICPMS, and gases and vapors using a special torch design. Solids or thin films are, however, more usually digested into solution prior to analysis.

Range of elements	Lithium to uranium, all isotopes; some elements excluded
Destructive	Yes
Chemical bonding information	No
Quantification	Yes, both semiquantitative and quantitative
Accuracy	0.2% isotopic; 5% or better quantitative; and 20% or better semiquantitative
Detection limits	Sub-ppb for most elements
Depth probed	1–10 μm per laser pulse, for solids
Depth profiling	Yes, with, laser ablation
Lateral resolution	20–50 μm for laser ablation
Imaging/mapping capabilities	No, but possible for laser ablation
Sample requirements	Solutions, digestible solids, solids, gases, and vapors
Main use	High-sensitivity elemental and isotopic analysis of high-purity chemicals and water
Instrument cost	$150,000–$750,000
Size	8 ft. × 8 ft.

Inductively Coupled Plasma-Optical Emission Spectroscopy (ICP-OES) 1.10.9

In Inductively Coupled Plasma-Optical Emission Spectroscopy (ICP-OES), a gaseous, solid (as fine particles), or liquid (as an aerosol) sample is directed into the center of a gaseous plasma. The sample is vaporized, atomized, and partially ionized in the plasma. Atoms and ions are excited and emit light at characteristic wavelengths in the ultraviolet or visible region of the spectrum. The emission line intensities are proportional to the concentration of each element in the sample. A grating spectrometer is used for either simultaneous or sequential multielement analysis. The concentration of each element is determined from measured intensities via calibration with standards.

Range of elements	At least 70 elements can be determined
Destructive	Yes
Quantification	Standards (often pure aqueous solutions)
Accuracy	10% or better with simple standards; as good as 0.5% with appropriate techniques
Precision	Typically 0.2–0.5% for solutions or dissolved solids; 3–10% for direct solid analysis
Detection limits	Typically sub-ppb to 100 ppb; tens of pg to ng
Sample requirements	Liquids, directly; solids, following dissolution; solids, surfaces, and thin films with special methods (e.g., laser ablation)
Depth probed	μm scale for solids
Sample size	2–5 mL of solution; μL of solution with special techniques; μg to mg of solid
Main uses	Rapid, quantitative measurement of trace to minor elemental composition of solids and solutions; excellent detection limits, with linear calibration over ~5 orders of magnitude
Instrument cost	\$40,000–\$200,000
Size	4–8 ft. × 4 ft.

Neutron Diffraction

1.11.1

Diffraction is a technique that uses interference of short wavelength particles (such as neutrons or electrons) or photons (X or γ rays) reflected from planes of atoms in crystalline materials to yield three-dimensional structural information at the atomic level. Neutron diffraction, like X-ray diffraction is a nondestructive technique that can be used for atomically resolved structure determination and refinement, phase identification and quantification, residual stress measurements, and average particle-size determination of crystalline materials. The major advantages of neutron diffraction compared to other diffraction techniques, namely the extraordinarily greater penetrating nature of the neutron and its direct interaction with nuclei, lead to its use in measurements under special environments, experiments on materials requiring a depth of penetration greater than about 50 μm, or structure refinements of phases containing atoms of widely varying atomic numbers.

Range of elements	All elements detected approximately equally, except vanadium
Destructive	No
Bonding information	No
Depth probed	Yields bulk information of macro-sized samples (thin films for determining magnetic ordering)
Lateral resolution	None
Quantitation	Can be used to quantify crystalline phases
Structural accuracy	Atomic positions to 10^{-13} m, accuracy of phase quantitation ~1% molar
Imaging capabilities	None to date
Sample requirements	Material must be crystalline at data collection temperatures
Main uses	Atomic structure refinements or determinations and residual stress measurements, all in bulk materials
Instrument cost	Instruments are at government-funded facilities; cost for proprietary experiments \$1000–\$9000 per day

Neutron Reflectivity 1.11.2

In neutron reflectivity, neutrons strike the surface of a specimen at small angles and the percentage of neutrons reflected at the corresponding angle are measured. The angular dependence of the reflectivity is related to the variation in concentration of a labeled component as a function of distance from the surface. Typically the component of interest is labeled with deuterium to provide mass contrast against hydrogen. Use of polarized neutrons permits the determination of the variation in the magnetic moment as a function of depth. In all cases the optical transform of the concentration profiles is obtained experimentally.

Range of elements	All elements and their isotopes
Destructive	No
Quantification	Requires model calculations
Detection limits	Not suited for trace element analysis
Depth profiling	Yes
Penetration depth	mm
Depth resolution	1 nm
Lateral resolution	No
Imaging/mapping	No
Sample requirements	Solids or liquids, typically 5–10 cm in diameter, usually deuterium labeled
Main use	Concentration profiles in organic materials and between interfaces of organic materials
Instrument cost	$300,000, requires access to neutrons

Neutron Activation Analysis (NAA) 1.11.3

In Neutron Activation Analysis (NAA), samples are placed in a neutron field typically available in a research nuclear reactor. Following neutron capture, trace impurities present in the sample become radioactive. Samples are removed from the reactor and analyzed using γ-ray spectroscopy. Gamma rays or high-energy photons (~1 MeV) are given off as a result of the radioactive decay process. The spectrometer measures the energies of the γ rays and "counts" the number of γ rays of each energy emitted from the sample. Each radioisotope of an impurity emits a signature, or characteristic, γ ray. Therefore, the energy of the γ ray identifies the element, while the number of counts provides the concentration. Since neutrons and γ rays are penetrating radiations, only a bulk composition is obtained. Surface analysis can be accomplished by combining NAA with chemical etching techniques.

Elements measured	Two-thirds of the periodic table: transition metals, halogens, lanthanides, and platinum-group metals
Destructive	No, sample rendered radioactive
Chemical bonding	No, nuclear process
Quantification	Yes, with or without standard
Accuracy	5–20%
Detection limits	10^8–10^{14} atoms/cc (ppb-ppt)
Depth probed	Bulk technique
Depth resolution	Few μm (using chemical etching, otherwise none)
Lateral resolution	None
Imaging/mapping	No, limited autoradiography
Sample requirements	Conductors, insulators, or plastics; flexible sample size, down to 0.5 gms material
Main use	Simultaneous quantitative trace impurities analysis; particularly sensitive to gold
Instrument cost	$50,000
Size	Specialized radiation laboratories needed

Nuclear Reaction Analysis (NRA) 1.11.4

In Nuclear Reaction Analysis (NRA), a beam of charged particles with energy from a few hundred keV to several MeV is produced in an accelerator and bombards a sample. Nuclear reactions with low-Z nuclei in the sample are induced by the ion beam. Products of these reactions (typically protons, deuterons, tritons, He, α particles, and γ rays) are detected, producing a spectrum of particle yield versus energy. Depth information is obtained from the spectrum using energy loss rates for incident and product ions traveling through the sample. Particle yields are converted to concentrations with the use of experimental parameters and nuclear reaction cross sections.

Range of elements	Hydrogen to calcium; specific isotopes
Destructive	No, but some materials may be damaged by ion beams
Chemical bonding information	No
Depth profiling	Yes
Quantification	Yes, standards usually unnecessary
Accuracy	A few percent to tens of percent
Detection limits	Varies with specific reaction; typically 10–100 ppm
Depth probed	Several μm
Depth resolution	Varies with specific reaction; typically a few nm to hundreds of nm
Lateral resolution	Down to a few μm with microbeams
Imaging/mapping	Yes, with microbeams
Sample requirements	Solid conductors and insulators
Main use	Quantitative measurement of light elements (particularly hydrogen) in solid materials, without standards; has isotope selectivity
Instrument Cost	Several million dollars for high-energy ion accelerator
Size	Large laboratory for accelerator

Surface Roughness: Measurement, Formation by Sputtering, Impact on Depth Profiling 1.12.1

Surface roughness is commonly measured using mechanical and optical profilers, scanning electron microscopes, and atomic force and scanning tunneling microscopes. Angle-resolved scatterometers can also be applied to this measurement. The analysis surface can be roughened by ion bombardment, and roughness will degrade depth resolution in a depth profile. Rotation of the sample during sputtering can reduce this roughening.

Mechanical Profiler

Depth resolution	0.5 nm
Minimum step	2.5–5 nm
Maximum step	~150 μm
Lateral resolution	0.1–25 μm, depending on stylus radius
Maximum sample size	15-mm thickness, 200-mm diameter
Instrument cost	$30,000–$70,000

Optical Profiler

Depth resolution	0.1 nm
Minimum step	0.3 nm
Maximum step	15 μm
Lateral resolution	0.35–9 μm, depending on optical system
Maximum sample size	125-mm thickness, 100-mm diameter
Instrument cost	$80,000–$100,000

SEM (see SEM article)

Scanning Force Microscope (see STM/SFM article)

Depth resolution	0.01 nm
Lateral resolution	0.1 nm
Instrument cost	$75,000–$150,000

Scanning Tunneling Microscope (see STM/SFM article)

Depth resolution	0.001 μm
Lateral resolution	0.1 nm
Instrument cost	$75,000–$150,000

Optical Scatterometer (see next article)

Depth resolution	0.1 nm (root mean square)
Instrument cost	$50,000–$150,000

Optical Scatterometry 1.12.2

Optical scatterometry involves illuminating a sample with light and measuring the angular distribution of light which is scattered. The technique is useful for characterizing the topology of two general categories of surfaces. First, surfaces that are nominally smooth can be examined to yield the root-mean-squared (rms) roughness and other surface statistics. Second, the shapes of structure (lines) of periodically patterned surfaces can be characterized. The intensity of light diffracted into the various diffraction orders from the periodic structure is indicative of the shape of the lines. If the line shape is influenced by steps involved in processing the sample, the scattering technique can be used to monitor the process. This has been applied to several steps involved in microelectronics processing. Scatterometry is noncontact, nondestructive, fast, and often yields quantitative results. For some applications it can be used *in-situ.*

Parameters measured	Surface topography (rms roughness, rms slope, and power spectrum of structure); scattered light; line shape of periodic structure (width, side wall angle, height, and period)
Destructive	No
Vertical resolution	≥ 0.1 nm
Lateral resolution	$\geq \lambda/2$ for topography characterization; much smaller for periodic structure characterization (λ is the laser wavelength used to illuminate the sample)
Main uses	Topography characterization of nominally smooth surfaces; process control when characterizing periodic structure; can be applied *in situ* in some cases; rapid; amenable to automation
Quantitative	Yes
Mapping capabilities	Yes
Instrument cost	\$10,000–\$200,000 or more
Size	1 ft. × 1 ft. to 4 ft. × 8 ft.

Magneto-optic Kerr Effect (MOKE) — 1.12.3

The Magneto-Optic Kerr Effect (MOKE) is an optical technique to determine the orientation and relative magnitude of the net magnetic moment near the surface of a magnetic sample. It is based on the proportionality between the net magnetization M of a material and a small, but measurable, change in the polarization of visible light that has been reflected from the surface of a magnetic sample. The orientation of the magnetization is determined from the sign of the rotation and the geometry of the setup. MOKE measurements can be made as a function of external magnetic field. This gives a determination of the magnetic hysteresis loop of the material. MOKE measurements can be done at MHz frequencies, as well as under dc conditions, making it suitable for examining magnetic domain dynamics or static domain imaging.

Range of elements	Magnetic materials only; not element specific
Destructive	No
Quantification	Standards are needed to find M
Sensitivity	~1 monolayer of magnetic material
Depth probed	20–40 nm
Lateral resolution	Limited by spatial focus of light, greater than about 0.3 μm
Imaging/mapping capabilities	Yes
Sample requirements	Magnetic material of interest must be within optical penetration depth of the probing light
Main use	Hysteresis loops and magnetic anisotropies of ultrathin ferromagnetic films; dynamic magnetic domain imaging (MHz rates) magneto-optic data recording
Instrument cost	$20,000–$150,000
Size	~ 1 m × ~1 m

Physical and Chemical Adsorption for the Measurement of Solid Surface Areas 1.12.4

Physical adsorption isotherms are measured near the boiling point of a gas (e.g., nitrogen, at 77 K). From these isotherms the amount of gas needed to form a monolayer can be determined. If the area occupied by each adsorbed gas molecule is known, then the surface area can be determined for all finely divided solids, regardless of their chemical composition. In the case of metal surfaces, the area can be measured by the chemisorption of simple molecules like H_2 and CO. Chemisorption isotherms will measure selectively only the metal area. This is especially useful when the metal is dispersed on high area oxide supports. Usually H_2 is adsorbed at 25° C; no adsorption of H_2 occurs on the support under these conditions. At finite pressures (~10 cm Hg), each surface metal atom adsorbs one hydrogen atom, giving an adsorbed monolayer. The spacing of metal atoms is usually known, so that the number of hydrogen atoms gives directly the area of metal at the surface, or the dispersion.

Range of elements	Not element specific
Sample requirements	Vacuum compatible solids, stable to 200°C, any shape
Destructive	No
Chemical bonding information	None
Depth examined	Surface adsorbed layers only
Detection limits	Above about 1 m^2/g
Precision	1% or better
Quantification	Standards are available
Main uses	*Physical adsorption*—surface areas of any stable solids, e.g., oxides used as catalyst supports and carbon black: *Chemisorption*—measurements of particle sizes of metal powders, and of supported metals in catalysts
Instrument cost	Homemade, or up to $25,000
Size	2 ft. × 3 ft.

2

IMAGING TECHNIQUES (MICROSCOPY)

2.0 INTRODUCTION

The four techniques included in this chapter all have microscopy in their names. Their role (but certainly not only their only one) is to provide a magnified image. The objective, at its simplest, is to observe features that are beyond the resolution of the human eye (about 100 μm). Since the eye uses visible wavelength light, only a Light Microscope can do this directly. Reflected or transmitted light from the sample enters the eye after passing through a magnification column. All other microscopy imaging techniques use some other interaction probe and response signal (usually electrons) to provide the *contrast* that produces an image. The response signal image, or map, is then processed in some way to provide an optical equivalent "picture" for us to see. We usually think of images as three dimensional, with the object as "solid." The microscopies have different capabilities, not only in terms of magnification and lateral resolution, but also in their ability to represent *depth*. In the light microscope, topological contrast is provided largely by shadowing in reflection. In Scanning Electron Microscopy, SEM, the topological contrast is there because the efficiency of generating secondary electrons (the signal), which originate from the several top tens of nanometers of material, strongly depends on the angle at which the probe beam strikes the surface. In Scanning Tunneling Microscopy/Scanning Force Microscopy, STM/SFM, the surface is directly profiled by scanning a tip, capable of following topology at atomic-scale resolution, across the surface. In Transmission Electron Microscopy, TEM, which can also achieve atomic-scale *lateral* resolution, *no* depth information is obtained because the technique works by having the probe electron beam transmitted through a sample that is up to 200 nm thick.

If one wants only to better identify regions for further examination by other techniques, the Light Microscope is likely to be the first imaging instrument used. Around for over 150 years, it is capable of handling every type of sample (though different types of microscope are better suited to differing applications), and can easily provide magnification up to 1400×, the useful limit for visible wavelengths. By utilizing polarizers, many other properties, in addition to size and shape, become accessible (e. g., refractive index, crystal system, melting point, etc.). There are enormous collections of data (atlases) to help the observer identify what he or she is seeing and to interpret it. Light microscopes are also the cheapest "modern" instrument and take up the least physical space.

The next instrument likely to be used is the SEM where magnified images of up to 300k× are obtainable, the wavelength of electrons not being nearly so limiting as that of visible light, and lateral features down to a few nm become resolvable. Sample requirements are more stringent, however. They must be vacuum compatible, and must be either conducting or coated with a thin conducting layer. A variety of contrast mechanisms exist, in addition to the topological, enabling the production of maps distinguishing high- and low-Z elements, defects, magnetic domains, and even electrically charged regions in semiconductors. The *depth* from which all this information comes varies from nanometers to micrometers, depending on the primary beam energy used and the particular physical process providing the contrast. Likewise, the lateral resolution in these analytical modes also varies and is always poorer than the topological contrast mode. The cost and size range are about a factor of 5 to 10 greater than for light microscopes.

STMs and SFMs are a new breed of instrument invented in 1981 and 1985, respectively. Their enormous lateral resolution capability (atomic for STM; a little lower for SFM) and vertical resolution capability (0.01 Å for STM, 0.1 Å for SFM) come about because the interactions involved between the scanning tip and the surface are such as to be limited to a few atoms on the tip (down to one) and a few atoms on the surface. Though famous for their use in imaging single atoms or molecules, and moving them under control on clean surfaces in pristine UHV conditions, their practical uses in ambient atmosphere, including liquids, to profile large areas at reduced resolution have gained rapid acceptance in applied science and engineering. Features on the nanometer scale, sometimes not easily seen in SEM, can be observed in STM / SFM. There are however no ancillary analytical modes, such as in SEM. Costs are in the same range as SEMs. Space requirements are reduced.

The final technique in this chapter, TEM, has been a mainstay of materials science for 30 years. It has become ever more powerful, specialized, and expensive. A well-equipped TEM laboratory today has 2 or 3 TEMs with widely different capabilities and the highest resolution / highest electron energy TEMs probably cost over $1 million. Sample preparation in TEM is *critical*, since the sample sizes accepted are usually less than 3 mm in diameter and 200 nm in thickness (so that the electron beam can pass through the sample). This distinguishes TEM from the other techniques for which very little preparation is needed. It is quite common for excellent TEMs to stand idle or fail in their tasks because of inadequacy in the ancillary sample preparation equipment or the lack of qualified manpower there. A complex variety of operation modes exist in TEM, all either variations or combinations of *imaging* and *diffraction* methods. Switching from one mode to another in modern instruments is trivial, but interpretation is *not* trivial for the nonspecialist. The combination of imaging (with lateral magnification up to 1M×) with a variety of contrast modes, plus an atomic resolution mode for crystalline material (phase contrast in HREM), together with small and large area diffraction modes, provide a wealth of characterization information for the expert. This is always summed through a column of atoms (maybe 100), however, with *no* depth information included. Clearly then, TEM is a thin-film technique rather than a surface or interface technique, unless interfaces are viewed in cross section.

2.1 Light Microscopy

JOHN GUSTAV DELLY

Contents

Introduction

The practice of light microscopy goes back about 300 years. The light microscope is a deceptively simple instrument, being essentially an extension of our own eyes. It magnifies small objects, enabling us to directly view structures that are below the resolving power of the human eye (0.1 mm). There is as much difference between materials at the microscopic level as there is at the macroscopic level, and the practice of microscopy involves learning the microscopic characteristics of materials. These direct visual methods were applied first to plants and animals, and then, in the mid 1800s, to inorganic forms, such as thin sections of rocks and minerals, and polished metal specimens. Since then, the light microscope has been used to view virtually all materials, regardless of nature or origin.

Basic Principles

In the biomedical fields, the ability of the microscopist is limited only by his or her capacity to remember the thousands of distinguishing characteristics of various tissues; as an aid, atlases of tissue structures have been prepared over the years. Likewise, in materials characterization, atlases and textbooks have been prepared to aid the analytical microscopist. In addition, the analytical microscopist typically has a collection of reference standards for direct comparison to the sample under study. Atlases may be specific to a narrow subfield, or may be quite general and universal. There are microscopical atlases for the identification of metals and alloys,[1] rocks and ores,[2] paper fibers, animal feeds, pollens, foods, woods, animal hairs, synthetic fibers, vegetable drugs, and insect fragments, as well as universal atlases that include everything, regardless of nature or origin,[3, 4] and, finally, atlases of the latest composites.

The familiar light microscope used by biomedical scientists is not suitable for the study of materials. Biomedical workers rely almost solely on morphological characteristics of cells and tissues. In the materials sciences, too many things look alike; however, their structures may be quite different internally and, if crystalline, quite specific. Ordinary white light cannot be used to study such materials principally because the light vibrates in all directions and consists of a range of wavelengths, resulting in a composite of information—which is analytically useless. The instrument of choice for the study of materials is the polarized light microscope. By placing a polarizer in the light's path before the sample, light is made to vibrate in one direction only, which enables the microscopist to isolate specific properties of materials in specific orientations. For example, with ordinary white light, one can determine only morphology (shape) and size; if a polarizer is added, the additional properties of pleochroism (change in color or hue relative to orientation of polarized light) and refractive indices may be determined. By the addition of a second polarizer above the specimen, still other properties may be determined; namely, birefringence (the numerical difference between the principal refractive indices), the sign of elongation (location of the high and low refractive indices in an elongated specimen), and the extinction angle (the angle between the vibration direction of light inside the specimen and some prominent crystal face). Some of these may be determined by simply adding polarizers to an ordinary microscope, but true, quantitative polarized light microscopy and conoscopy (observations and measurements made at the objective back focal plane) can be performed only by using polarizing microscopes with their many graduated adjustments.

Some of the characteristics of materials that may be determined with the polarized light microscope include

- Morphology
- Size
- Transparency or opacity
- Color (reflected and transmitted)
- Refractive indices
- Dispersion of refractive indices

- Pleochroism
- Dispersion staining colors
- Crystal system
- Birefringence
- Sign of elongation
- Optic sign
- Extinction angle
- Fluorescence (ultraviolet, visible, and infrared)
- Melting point
- Polymorphism
- Eutectics
- Degree of crystallinity
- Microhardness.

The modern light microscope is constructed in modular form, and may be configured in many ways depending on the kind of material that is being studied. Transparent materials, whether wholly or partly so, are studied with transmitted light; opaque specimens are studied with an episystem (reflected light; incident light), in which the specimen is illuminated from above. Materials scientists who study all kinds of materials use so-called "universal" microscopes, which may be converted quickly from one kind to another.

Sample Preparation

Sample preparation methods vary widely. The very first procedure for characterizing any material simply is to look at it using a low-power stereomicroscope; often, a material can be characterized or a problem solved at this stage. If examination at this level does not produce an answer, it usually suggests what needs to be done next: go to higher magnification; mount for FTIR, XRD, or EDS; section; isolate contaminants; and so forth.

If the material is particulate, it needs to be mounted in a refractive index liquid for determination of its optical properties. If the sample is a metal, or some other hard material, it may need to be embedded in a polymer matrix and then sawn, ground, polished, and etched[5] before viewing. Polymers may be viewed directly, but usually need to be sectioned. This may involve embedding the sample to support the material and prevent preparation artifacts. Sectioning may be done dry and at room temperature using a hand, rotary, rocking, or sledge microtome (a large bench microtome incorporating a knife that slides horizontally), or it may need to be done at freezing temperatures with a cryomicrotome, which uses glass knives.

If elemental or compound data are required, the material needs to be mounted for the appropriate analytical instrument. For example, if light microscopy shows a sample to be a metal it can be put into solution and its elemental composition determined by classical microchemical tests; in well-equipped microscopy laboratories, some sort of microprobe (for example, electron- or ion-microprobe) is usually available, and as these are nondestructive by comparison, the sample is mounted for them using the low-power stereomicroscope. Individual samples <1 μm are handled freehand by experienced particle handlers under cleanroom conditions. A particle may be mounted on a beryllium substrate for examination by an electron microprobe, using a minimal amount of flexible collodion as an adhesive, or it may be mounted on an aluminum stub for SEM, on the end of a glass fiber for micro-XRD, or on a thin cleavage fragment of sodium chloride ("salt plate") for micro-FTIR. The exact procedures for preparing the instruments and mounting particles for various analyses have been described in detail.[4]

Detection Limits

Many kinds of materials, because of their color by transmitted light and their optical properties, can be detected even when present in sizes below the instrument's resolving power, but cannot be analyzed with confidence. Organized structures like diatom fragments can be identified on sight, even when very small, but an unoriented polymer cannot be characterized by morphology alone. The numerical aperture, which is engraved on each objective and condenser, is a measure of the light-gathering ability of the objective, or light-providing ability of the condenser. Specifically, the numerical aperture *NA* is defined as

$$NA = n \sin \frac{AA}{2} \qquad (1)$$

where *n* is the refractive index of the medium between the cover glass and the objective front lens, and *AA* is the angular aperture of the objective. The maximum theoretical *NA* of a dry system is 1.0; the practical maximum is 0.95. Higher values of *NA* can be obtained only by using oil-immersion objectives and condensers. The oils used for this have a refractive index of 1.515; the practical maximum numerical aperture achieved is 1.4. The significance of the numerical aperture lies in the diffraction theory of microscopical image formation; details on the theoretical and practical limits of the light microscope are readily available.[6]

The theoretical limit to an instrument's resolving power is determined by the wavelength of light used, and the numerical aperture of the system:

$$r = \frac{\lambda}{2NA} \qquad (2)$$

where *r* is the resolving power, λ is the wavelength of light used, and *NA* is the numerical aperture of the system. The wavelength is taken to be 0.55 μm when using white light. The use of ultraviolet microscopy effectively doubles the resolving power, but the lenses must be made of quartz and photographic methods or image converter tubes must be used to image the specimen. The maximum theoretical limit of resolving power is currently about 0.2 μm, using white light and conventional light microscopes. The practical limit to the maximum useful magnification, *MUM*, is 1000 *NA*. In modern microscopes *MUM* = 1400×. Although many instruments easily provide magnifications of 2000–5000×, this is "empty" magnification; i.e., no more detail is revealed beyond that seen at 1400×.

Common Modes of Analysis

Particulate materials are usually analyzed with a polarizing microscope set up for transmitted light. This allows one to determine the shape, size, color, pleochroism, refractive indices, birefringence, sign of elongation, extinction angle, optic sign, and crystal system, to name but a few characteristics. If the sample is colorless, transparent, and isotropic, and is embedded in a matrix with similar properties, it will not be seen, or will be seen only with difficulty, because our eyes are sensitive to amplitude and wavelength differences, but not to phase differences. In this case, the mode must be changed to phase contrast. This technique, introduced by Zernike in the 1930s, converts phase differences into amplitude differences. Normarski differential interference contrast is another mode that may be set up. Both modes are qualitative methods of increasing contrast. Quantitative methods are available via interference microscopy.

Darkfield microscopy is one of the oldest modes of microscopy. Here, axial rays from the condenser are prevented from entering the objective, through the use of an opaque stop placed in the condenser, while peripheral light illuminates the specimen. Thus, the specimen is seen lighted against a dark field.

For studying settled materials in liquids, or for very large opaque specimens, the inverted microscope may be used.

For fluorescence microscopy the light source is changed from an incandescent lamp to a high-pressure mercury vapor burner, which is rich in wavelengths below the visible. Exciter filters placed in the light path isolate various parts of the spectrum. The 365-nm wavelength is commonly used in fluorescence microscopy to characterize a material's primary fluorescence, or to detect a tracer fluorochrome through secondary fluorescence. The 400-nm wavelength region is another commonly used exciter.

Attachment of a hot or cold stage to the ordinary microscope stage allows the specimen to be observed while the temperature is changed slowly, rapidly, or held constant somewhere other than ambient. This technique is used to determine melting and freezing points, but is especially useful for the study of polymorphs, the determination of eutectics, and the preparation of phase diagrams.

Spindle stages and universal stages allow a sample to be placed in any orientation relative to the microscope's optical axis.

Not every sample requires all modes for complete characterization; most samples yield to a few procedures. Let us take as an example some particulate material—this may be a sample of lunar dust fines, a contaminant removed from a failed integrated circuit, a new pharmaceutical or explosive, a corrosion product or wear particle, a fiber from a crime scene, or a pigment from an oil painting—the procedure will be the same. A bit of the sample, or a single particle, is placed on a microscope slide in a suitable mounting medium, and a cover slip is placed on top. The mounting medium is selected from a series of refractive index liquid standards which range from about 1.300 to 1.800—usually something around 1.660 is selected because it provides good contrast with a wide variety of industrial materials. The sample is then placed on the stage of the polarizing microscope and brought into focus. At this point the microscope may be set up for plane-polarized light or slightly uncrossed polarizers—the latter is more useful. Several characteristics will be immediately apparent: the morphology, relative size, and isotropy or anisotropy. If the sample cannot be seen at any orientation between fully crossed polarizers, it is isotropic; it has only one refractive index, and is either amorphous or in the cubic crystal system. If it can be seen, it will display one or more colors in the Newtonian series; this indicates that it has more than one refractive index, or, if it is only spotty, that it has some kind of strain birefringence or internal orientation.

The analyzer is removed and the color of the sample is observed in plane-polarized light. If the sample is colored, the stage is rotated. Colored, anisotropic materials may show pleochroism—a change in color or hue when the orientation with respect to the vibration direction of the polarizer is changed. Any pleochroism should be noted and recorded.

Introducing a monochromatic filter—usually 589 nm—and closing the aperture diaphragm while using a high numerical aperture objective, the focus is changed from best focus position to above best focus. The diffraction halo seen around the particle (Becke line) will move into or away from the particle, thus indicating the relative refractive index. By orienting the specimen and rotating the stage, more than one refractive index may be noted.

With polarizers fully crossed and the specimen rotated to maximum brightness, the sample thickness is determined with the aid of a calibrated eyepiece micrometer, and the polarization (retardation) color is noted. From these the birefringence may be determined mathematically or graphically with the aid of a Michel-Lévy chart.

If the sample is elongated, it is oriented 2 o'clock–8 o'clock, the retardation color is noted, and a compensator is inserted in the slot above the specimen. The retardation colors will go upscale or downscale; i.e., they will be additive or subtractive. This will indicate where the high and low refractive indices are located with respect to the long axis of the sample. This is the sign of elongation, and is said to be positive if the sample is "length slow" (high refractive index parallel to length), or negative if the sample is "length fast" (low refractive index parallel to length).

The elongated sample is next rotated parallel to an eyepiece crosshair, and one notes if the sample goes to extinction; if it does, it has parallel extinction (the vibra-

Figure 1 **Nikon Optiphot-2 polarizing microscope.**

tional directions inside the sample are parallel to the vibrational directions of the polarizer and analyzer). If the sample does not go to extinction, the stage reading is noted and the sample is rotated to extinction (not greater than 45°); the stage reading is again noted, and the difference between the readings is the extinction angle.

If necessary, each refractive index is determined specifically through successive immersion in liquids of various refractive index until one is found where the sample disappears—knowing the refractive index of the liquid, one then knows the refractive index in a particular orientation. There may be one, two, or three principal refractive indices.

The Bertrand lens, an auxiliary lens that is focused on the objective back focal plane, is inserted with the sample between fully crossed polarizers, and the sample is oriented to show the lowest retardation colors. This will yield interference figures, which immediately reveal whether the sample is uniaxial (hexagonal or tetragonal) or biaxial (orthorhombic, monoclinic, or triclinic). Addition of the compensator and proper orientation of the rotating stage will further reveal whether the sample is optically positive or negative.

These operations are performed faster than it takes to describe them, and are usually sufficient to characterize a material. The specific steps to perform each of the above may be found in any textbook on optical crystallography.

Sample Requirements

There are no specific sample requirements; all samples are accommodated.

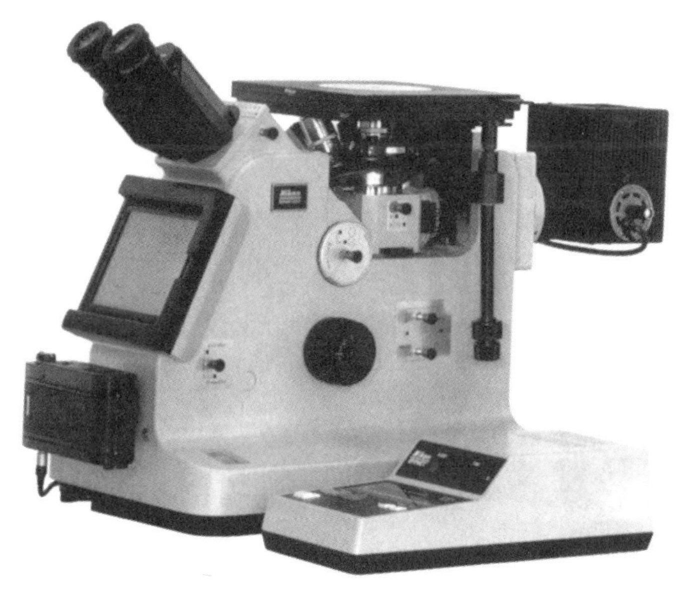

Figure 2 Nikon Epithot inverted metallograph.

Artifacts

Artifacts may be introduced from the environment or through preparative techniques. When assessing individual tiny particles of material, the risk of loss or contamination is high, so that samples of this nature are handled and prepared for examination in a clean bench or a cleanroom (class 100 or better).

Artifacts introduced through sample preparation are common materials; these may be bits of facial tissue, wax, epithelial cells, hair, or dried stain, all inadvertently introduced by the microscopist. Detergent residues on so-called "precleaned" microscope slides and broken glass are common artifacts, as are knife marks and chatter marks from sectioning with a faulty blade, or scratch marks from grinding and polishing.

Quantification

For other quantification, specialized graticules are available, including point counting, grids, concentric circles, and special scales. The latest methods of quantification involve automatic image analysis.

Instrumentation

Figure 1 illustrates a typical, good quality, analytical polarizing microscope. Polarizing microscopes are extraordinarily versatile instruments that enable the trained microscopist to characterize materials rapidly and accurately.

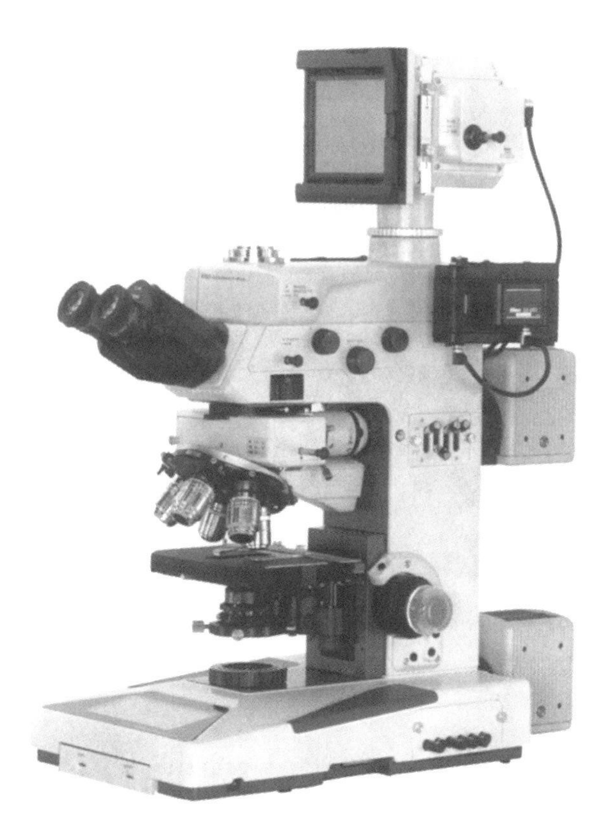

Figure 3 **Nikon Microphot-FXA research microscope for materials science.**

As an example of a more specific application, Figure 2 illustrates a metallograph—a light microscope set up for the characterization of opaque samples. Figure 3 illustrates a research-grade microscope made specifically for materials science, i.e., for optically characterizing all transparent and translucent materials.

Conclusions

The classical polarizing light microscope as developed 150 years ago is still the most versatile, least expensive analytical instrument in the hands of an experienced microscopist. Its limitations in terms of resolving power, depth of field, and contrast have been reduced in the last decade, in which we have witnessed a revolution in its evolution. Video microscopy has increased contrast electronically, and thereby revealed structures never before seen. With computer enhancement, unheard of resolutions are possible. There are daily developments in the X-ray, holographic, acoustic, confocal laser scanning, and scanning tunneling microscopes.[7, 8]

The general utility of the light microscope is also recognized by its incorporation into so many other kinds of analytical instrumentation. Continued development of new composites and materials, together with continued trends in microminiaturization make the simple, classical polarized-light microscope the instrument of choice for any initial analytical duty.

Related Articles in the Encyclopedia

None in this volume.

References

1 ASM Handbook Committee. *Metals Handbook, Volume 7: Atlas of Microstructures.* American Society of Metals, Metals Park, 1972.

2 O. Oelsner. *Atlas of the Most Important Ore Mineral Parageneses Under the Microscope.* Pergamon, London, 1961 (English edition, 1966).

3 A. A. Benedetti-Pichler. *Identification of Materials.* Springer-Verlag, New York, 1964.

4 W. C. McCrone, and J. G. Delly. *The Particle Atlas.* Ann Arbor Science, Ann Arbor, 1973, Volumes 1–4; and S. Palenik., 1979, Volume 5; and J. A. Brown and I. M. Stewart, 1980, Volume 6.

5 G. L. Kehl. *The Principles of Metallographic Laboratory Practice.* McGraw-Hill, New York, 1949.

6 J. G. Delly. *Photography Through The Microscope.* Eastman Kodak Company, Rochester, 1988.

7 *Modern Microscopies.* (P. J. Duke and A. G. Michette, eds.) Plenum, New York, 1990.

8 M. Pluta. *Advanced Light Microscopy*. Elsevier, Amsterdam, 1988.

2.2 SEM

Scanning Electron Microscopy

JEFFREY B. BINDELL

Contents

Introduction

Traditionally, the first instrument that would come to mind for small scale materials characterization would be the optical microscope. The optical microscope offered the scientist a first look at most samples and could be used to routinely document the progress of an investigation. As the sophistication of investigations increased, the optical microscope often has been replaced by instrumentation having superior spatial resolution or depth of focus. However, its use has continued because of the ubiquitous availability of the tool.

For the purpose of a detailed materials characterization, the optical microscope has been supplanted by two more potent instruments: the Transmission Electron Microscope (TEM) and the Scanning Electron Microscope (SEM). Because of its reasonable cost and the wide range of information that it provides in a timely manner, the SEM often replaces the optical microscope as the preferred starting tool for materials studies.

The SEM provides the investigator with a highly magnified image of the surface of a material that is very similar to what one would expect if one could actually "see" the surface visually. This tends to simplify image interpretations considerably, but

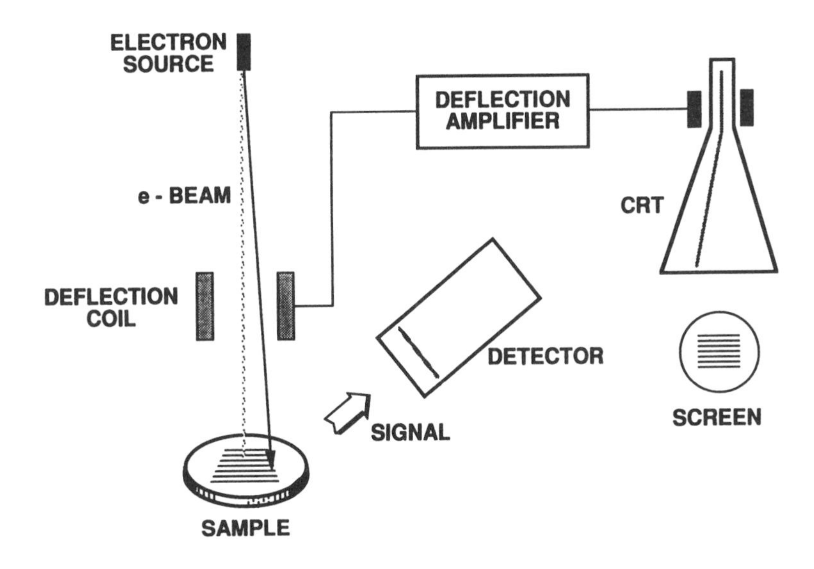

Figure 1 Schematic describing the operation of an SEM.

reliance on intuitive reactions to SEM images can, on occasion, lead to erroneous results. The resolution of the SEM can approach a few nm and it can operate at magnifications that are easily adjusted from about 10×–300,000×.

Not only is topographical information produced in the SEM, but information concerning the composition near surface regions of the material is provided as well. There are also a number of important instruments closely related to the SEM, notably the electron microprobe (EMP) and the scanning Auger microprobe (SAM). Both of these instruments, as well as the TEM, are described in detail elsewhere in this volume.

Physical Basis of Operation

In the SEM, a source of electrons is focused (in vacuum) into a fine probe that is rastered over the surface of the specimen, Figure 1. As the electrons penetrate the surface, a number of interactions occur that can result in the emission of electrons or photons from (or through) the surface. A reasonable fraction of the electrons emitted can be collected by appropriate detectors, and the output can be used to modulate the brightness of a cathode ray tube (CRT) whose x- and y-inputs are driven in synchronism with the x–y voltages rastering the electron beam. In this way an image is produced on the CRT; every point that the beam strikes on the sample is mapped directly onto a corresponding point on the screen. If the amplitude of the saw-tooth voltage applied to the x- and y-deflection amplifiers in the SEM is reduced by some factor while the CRT saw-tooth voltage is kept fixed at the level necessary to produce a full screen display, the magnification, as viewed on the screen, will be increased by the same factor.

The principle images produced in the SEM are of three types: secondary electron images, backscattered electron images, and elemental X-ray maps. Secondary and backscattered electrons are conventionally separated according to their energies. They are produced by different mechanisms. When a high-energy primary electron interacts with an atom, it undergoes either inelastic scattering with atomic electrons or elastic scattering with the atomic nucleus. In an inelastic collision with an electron, some amount of energy is transferred to the other electron. If the energy transfer is very small, the emitted electron will probably not have enough energy to exit the surface. If the energy transferred exceeds the work function of the material, the emitted electron can exit the solid. When the energy of the emitted electron is less than about 50 eV, by convention it is referred to as a secondary electron (SE), or simply a *secondary.* Most of the emitted secondaries are produced within the first few nm of the surface. Secondaries produced much deeper in the material suffer additional inelastic collisions, which lower their energy and trap them in the interior of the solid.

Higher energy electrons are primary electrons that have been scattered without loss of kinetic energy (i.e., elastically) by the nucleus of an atom, although these collisions may occur after the primary electron has already lost some of its energy to inelastic scattering. Backscattered electrons (BSEs) are considered to be the electrons that exit the specimen with an energy greater then 50 eV, including Auger electrons. However most BSEs have energies comparable to the energy of the primary beam. The higher the atomic number of a material, the more likely it is that backscattering will occur. Thus as a beam passes from a low-Z (atomic number) to a high-Z area, the signal due to backscattering, and consequently the image brightness, will increase. There is a built in contrast caused by elemental differences.

One further breaks down the secondary electron contributions into three groups: SEI, SEII and SEIII. SEIs result from the interaction of the incident beam with the sample at the point of enrry. SEIIs are produced by BSE s on exiting the sample. SEIIIs are produced by BSEs which have exited the surface of the sample and further interact with components on the interior of the SEM usually not related to the sample. SEIIs and SEIIIs come from regions far outside that defined by the incident probe and can cause serious degradation of the resolution of the image.

It is usual to define the primary beam current i_0, the BSE current i_{BSE}, the SE current i_{SE}, and the sample current transmitted through the specimen to ground i_{SC}, such that the Kirchoff current law holds:

$$i_0 = i_{BSE} + i_{SE} + i_{SC} \tag{1}$$

These signals can be used to form complementary images. As the beam current is increased, each of these currents will also increase. The backscattered electron yield η and the secondary electron yield δ, which refer to the number of backscattered and secondary electrons emitted per incident electron, respectively, are defined by the relationships:

$$\eta = \frac{i_{BSE}}{i_0} \tag{2}$$

$$\delta = \frac{i_{SE}}{i_0} \tag{3}$$

In most currently available SEMs, the energy of the primary electron beam can range from a few hundred eV up to 30 keV. The values of δ and η will change over this range, however, yielding micrographs that may vary in appearance and information content as the energy of the primary beam is changed. The value of the BSE yield increases with atomic number Z, but its value for a fixed Z remains constant for all beam energies above 5 keV. The SE yield δ decreases slowly with increasing beam energy after reaching a peak at some low voltage, usually around 1 keV. For any fixed voltage, however, δ shows very little variation over the full range of Z. Both the secondary and backscattered electron yields increase with decreasing glancing angle of incidence because more scattering occurs closer to the surface. This is one of the major reasons why the SEM provides excellent topographical contrast in the SE mode; as the surface changes its slope, the number of secondary electrons produced changes as well. With the BSEs this effect is not as prominent, since to fully realize it the BSE detector would have to be repositioned to measure forward scattering.

An additional electron interaction of major importance in the SEM occurs when the primary electron collides with and ejects a core electron from an atom in the solid. The excited atom will decay to its ground state by emitting either a characteristic X-ray photon or an Auger electron (see the article on AES). The X-ray emission signal can be sorted by energy in an energy dispersive X-ray detector (see the article on EDS) or by wavelength with a wavelength spectrometer (see the article on EPMA). These distributions are characteristic of the elements that produced them and the SEM can use these signals to produce elemental images that show the spatial distribution of particular elements in the field of view. The primary electrons can travel considerable distances into a solid before losing enough energy through collisions to be no longer able to excite X-ray emission. This means that a large volume of the sample will produce X-ray emission for any position of the smaller primary beam, and consequently the spatial resolution of this type of image will rarely be better than 0.5 µm.

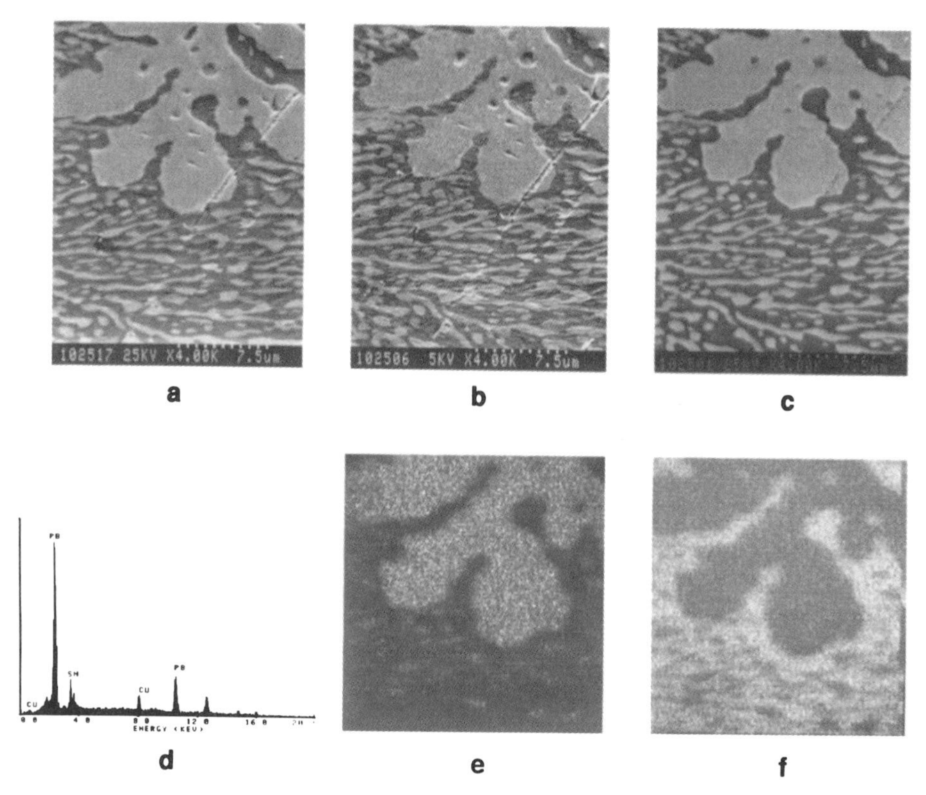

Figure 2 Micrographs of the same region of a specimen in various imaging modes on a high-resolution SEM: (a) and (b) SE micrographs taken at 25 and 5 keV, respectively; (c) backscattered image taken at 25 keV; (d) EDS spectrum taken from the Pb-rich phase of the Pb-Sn solder; (e) and (f) elemental maps of the two elements taken by accepting only signals from the appropriate spectral energy regions.

An illustration of this discussion can be seen in Figure 2, which is a collection of SEM images taken from the surface of a Pb-Sn solder sample contaminated with a low concentration of Cu. Figure 1a, a secondary electron image (SE) taken with a primary energy of 25 keV, distinguishes the two Pb-Sn eutectic phases as brighter regions (almost pure Pb) separated by darker bands corresponding to the Sn-rich phase. The micrograph originally was taken at a magnification of 4000× but care should be exercised when viewing published examples because of the likelihood of photographic enlargement or reduction by the printer. Most SEMs produce a marker directly on the photograph that defines the actual magnification. In the present example, the series of dots at the bottom of the micrograph span a physical distance of 7.5 μm. This can be used as an internally consistent ruler for measurement purposes.

The micrograph also shows the presence of a scratch that goes diagonally across the entire field of view. Note the appearance of depth to this scratch as a result of the variation in secondary electron yield with the local slope of the surface. The spatial resolution of the SEM due to SEIs usually improves with increasing energy of the primary beam because the beam can be focused into a smaller spot. Conversely, at higher energies the increased penetration of the electron beam into the sample will increase the interaction volume, which may cause some degradation of the image resolution due to SEIIs and SEIIIs. This is shown in Figure 2b, which is a SE image taken at only 5 keV. In this case the reduced electron penetration brings out more surface detail in the micrograph.

There are two ways to produce a backscattered electron image. One is to put a grid between the sample and the SE detector with a –50-V bias applied to it. This will repel the SEs since only the BSEs will have sufficient energy to penetrate the electric field of the grid. This type of detector is not very effective for the detection of BSEs because of its small solid angle of collection. A much larger solid angle of collection is obtained by placing the detector immediately above the sample to collect the BSE. Two types of detectors are commonly used here. One type uses partially depleted n-type silicon diodes coated with a layer of gold, which convert the incident BSEs into electron–hole pairs at the rate of 1 pair per 3.8 eV. Using a pair of Si detectors makes it possible to separate atomic number contrast from topographical contrast. The other detector type, the so-called scintillator photo multiplier detector, uses a material that will fluoresce under the bombardment of the high-energy BSEs to produce a light signal that can be further amplified. The photomultiplier detector was used to produce the BSE micrograph in Figure 2c. Since no secondary electrons are present, the surface topography of the scratch is no longer evident and only atomic number contrast appears.

Atomic number contrast can be used to estimate concentrations in binary alloys because the actual BSE signal increases somewhat predictably with the concentration of the heavier element of the pair.

Both energy-dispersive and wavelength-dispersive X-ray detectors can be used for elemental detection in the SEM. The detectors produce an output signal that is proportional to the number of X-ray photons in the area under electron bombardment. With an EDS the output is displayed as a histogram of counts versus X-ray energy. Such a display is shown in Figure 2d. This spectrum was produced by allowing the electron beam to dwell on one of the Pb-rich areas of the sample. The spectrum shows the presence of peaks corresponding to Pb and a small amount of Sn. Since this sample was slightly contaminated with Cu, the small Cu peak at 8 keV is expected. The detectors can be adjusted to pass only a range of pulses corresponding to a single X-ray spectral peak that is characteristic of a particular element. This output can then be used to produce an elemental image or an X-ray map; two X-ray maps using an EDS are shown in Figure 2e for Pb and in Figure 2f for Sn. Note the complementary nature of these images, and how easy it is to iden-

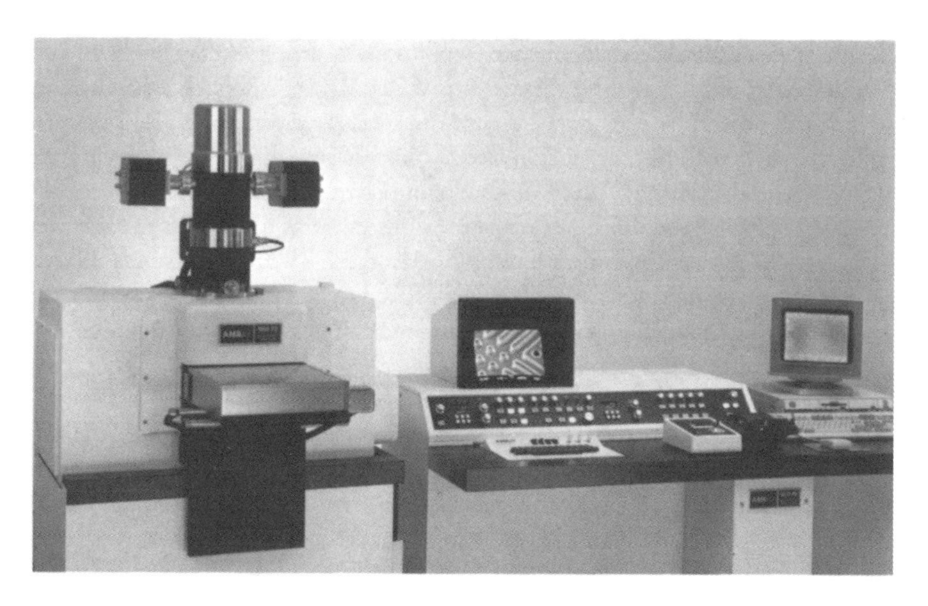

Figure 3 **Photograph of a modern field emission SEM. (Courtesy of AMRAY Inc., Bedford, MA)**

tify portions of the SE or BSE image having specific local compositions. The data usually can be quantified through the use of appropriate elemental standards and well-established computational algorithms.

Instrumentation

Figure 3 shows a photograph of a recent model SEM. The main features of the instrument are the electron column containing the electron source (i.e., the gun), the magnetic focusing lenses, the sample vacuum chamber and stage region (at the bottom of the column) and the electronics console containing the control panel, the electronic power supplies and the scanning modules. A solid state EDS X-ray detector is usually attached to the column and protrudes into the area immediately above the stage; the electronics for the detector are in separate modules, but there has been a recent trend toward integration into the SEM system architecture.

The overall function of the electron gun is to produce a source of electrons emanating from as small a "spot" as possible. The lenses act to demagnify this spot and focus it onto a sample. The gun itself produces electron emission from a small area and then demagnifies it initially before presenting it to the lens stack. The actual emission area might be a few μm in diameter and will be focused eventually into a spot as small as 1 or 2 nm on the specimen.

There are three major types of electron sources: thermionic tungsten, LaB_6, and hot and cold field emission. In the first case, a tungsten filament is heated to allow electrons to be emitted via thermionic emission. Temperatures as high as 3000° C are required to produce a sufficiently bright source. These filaments are easy to work with but have to be replaced frequently because of evaporation. The material LaB_6 has a lower work function than tungsten and thus can be operated at lower temperatures, and it yields a higher source brightness. However, LaB_6 filaments require a much better vacuum then tungsten to achieve good stability and a longer lifetime. The brighter the source, the higher the current density in the spot, which consequently permits more electrons to be focused onto the same area of a specimen. Recently, field emission electron sources have been produced. These tips are very sharp; the strong electric field created at the tip extracts electrons from the source even at low temperatures. Emission can be increased by thermal assistance but the energy width of the emitted electrons may increase somewhat. The sharper the energy profile, the less the effect of chromatic aberrations of the magnetic defocusing lenses. Although they are more difficult to work with, require very high vacuum and occasional cleaning and sharpening via thermal flashing, the enhanced resolution and low voltage applications of field emission tips are making them the source of choice in newer instruments that have the high-vacuum capability necessary to support them.

The beam is defocused by a series of magnetic lenses as shown in Figure 4. Each lens has an associated defining aperture that limits the divergence of the electron beam. The top lenses are called *condenser* lenses, and often are operated as if they were a single lens. By increasing the current through the condenser lens, the focal length is decreased and the divergence increases. The lens therefore passes less beam current on to the next lens in the chain. Increasing the current through the first lens reduces the size of the image produced (thus the term spot size for this control). It also spreads out the beam resulting in beam current control as well. Smaller spot sizes, often given higher dial numbers to correspond with the higher lens currents required for better resolution, are attained with less current (signal) and a smaller signal-to-noise ratio. Very high magnification images therefore are inherently noisy.

The beam next arrives at the final lens–aperture combination. The final lens does the ultimate focusing of the beam onto the surface of the sample. The sample is attached to a specimen stage that provides x- and y-motion, as well as tilt with respect to the beam axis and rotation about an axis normal to the specimen's surface. A final "z" motion allows for adjustment of the distance between the final lens and the sample's surface. This distance is called the *working distance.*

The working distance and the limiting aperture size determine the convergence angle shown in the figure. Typically the convergence angle is a few mrad and it can be decreased by using a smaller final aperture or by increasing the working distance. The smaller the convergence angle, the more variation in the z-direction topography that can be tolerated while still remaining in focus to some prescribed degree. This large depth of focus contributes to the ease of observation of topographical

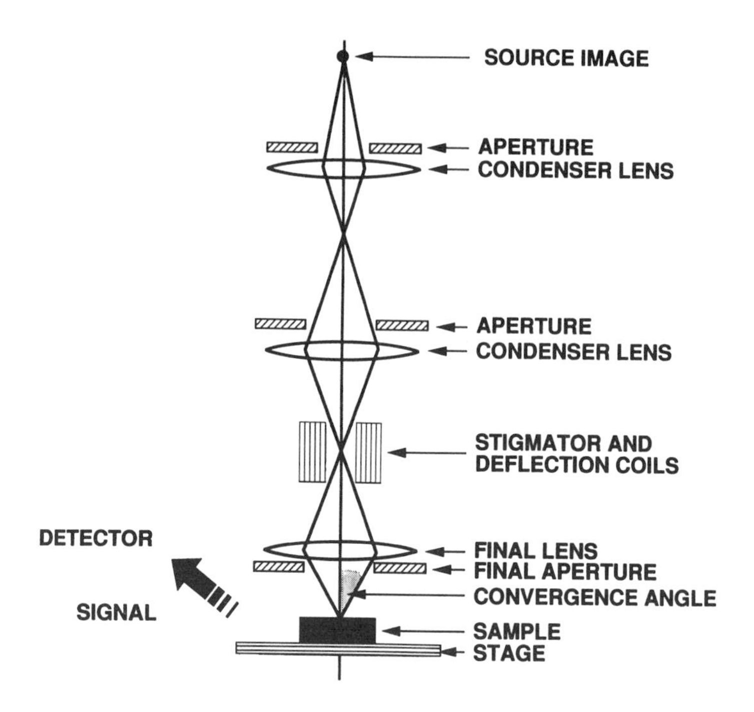

Figure 4 **Schematic of the electron optics constituting the SEM.**

effects. The depth of focus in the SEM is compared in Figure 5 with that of an optical microscope operated at the same magnification for viewing the top of a common machine screw.

Sample Requirements

The use of the SEM requires very little in regard to sample preparation, provided that the specimen is vacuum compatible. If the sample is conducting, the major limitation is whether it will fit onto the stage or, for that matter, into the specimen chamber. For special applications, very large stage–vacuum chamber combinations have been fabricated into which large forensic samples (such as boots or weapons) or 8-in diameter semiconductor wafers can be placed. For the latter case, special final lenses having conical shapes have been developed to allow for observation of large tilted samples at reasonably small working distances.

If the sample is an insulator there are still methods by which it can be studied in the instrument. The simplest approach is to coat it with a thin (10-nm) conducting film of carbon, gold, or some other metal. In following this approach, care must be taken to avoid artifacts and distortions that could be produced by nonuniform coatings or by agglomeration of the coating material. If an X-ray analysis is to be

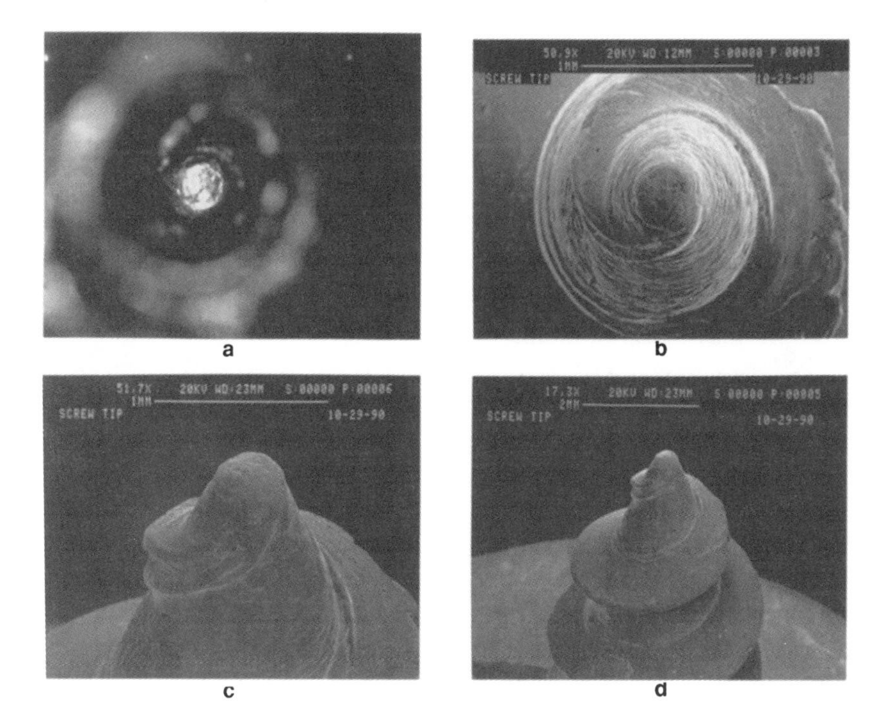

Figure 5 **Micrographs of a machine screw illustrating the great depth of field of the SEM: (a) optical micrograph of the very tip of the screw; (b) and (c) the same area in the SEM and a second image taken at an angle (the latter shows the depth of field quite clearly); (d) lower magnification image.**

made on such a coated surface, care must be taken to exclude or correct for any X-ray peaks generated in the deposited material.

Uncoated insulating samples also can be studied by using low primary beam voltages (< 2.0 keV) if one is willing to compromise image resolution to some extent. If we define the total electron yield as $\sigma = \delta + \eta$, then when $\sigma < 1$ we either must supply or remove electrons from the specimen to avoid charge build-up. Conduction to ground automatically takes care of this problem for conducting samples, but for insulators this does not occur. Consequently, one might expect it to be impossible to study insulating samples in the SEM. The way around this difficulty is suggested in Figure 6 which plots σ as a function of energy of the incident electron. The yield is seen to rise from 0 to some amount greater than 1 and then to decrease back below 1 as the energy increases. The two energy crossovers E_1 and E_2, between which minimal charging occurs, are often quite low, typically less than 2.0 keV.

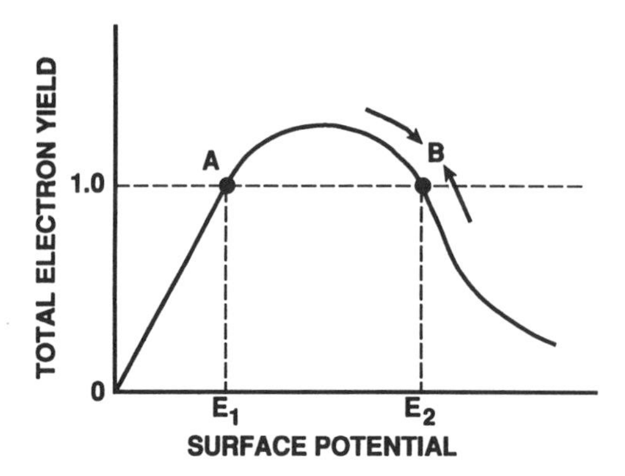

Figure 6 **Total electron yield as a function of the primary electron's energy when it arrives at the surface of the specimen.**

The energy scale in the figure is actually meant to depict the energy of the electron as it arrives at the surface. Because of charging, the electron's energy may be greater or less than the accelerating voltage would suggest. Consider electrons striking the surface with an energy near E_2 as identified by point B in the figure. If the energy is somewhat below that of the crossover point, the total electron yield will be greater than 1 and the surface will be positively charged, thereby attracting incoming electrons and increasing the effective energy of the primary beam. The electron's energy will continue to increase until E_2 is reached. If it overshoots, the yield will drop and some negative charging will begin until, again, a balance is reached at point B. Point B is therefore a stable operating point for the insulator in question, and operating around this point will allow excellent micrographs to be produced. E_1 (point A) does not represent a stable operating condition.

If the sample is a metal that has been coated with a thin oxide layer, a higher accelerating voltage might actually improve the image. The reason for this is that as the high-energy beam passes through the oxide, it can create electron–hole pairs in sufficient numbers to establish local conduction. This effect is often noted while observing semiconductor devices that have been passivated with thin deposited oxide films.

Applications

We have already discussed a number of applications of the SEM to materials characterization: topographical (SE) imaging, Energy-Dispersive X-Ray analysis (EDS) and the use of backscattering measurements to determine the composition of binary alloy systems. We now shall briefly discuss applications that are, in part, specific to certain industries or technologies. Although there is significant literature on applications of the SEM to the biological sciences, such applications will not be covered in this article.

At magnifications above a few thousand, the raster scanning of the beam is very linear, resulting in a constant magnification over the entire image. This is not the case at low magnifications, where significant nonlinearity may be present. Uniform magnification allows the image to be used for very precise size measurements. The SEM therefore can be a very accurate and precise metrology tool. This requires careful calibration using special SEM metrology standards available from the National Institute of Standards and Technology.

The fact that the SE coefficient varies in a known way with the angle that the primary beam makes with the surface allows the approximate determination of the depth (z) variation of the surface morphology from the information collected in a single image. By tilting the specimen slightly (5–8°), stereo pairs can be produced that provide excellent quality three-dimensional images via stereoscopic viewers. Software developments now allow these images to be calculated and displayed in three-dimension-like patterns on a computer screen. Contour maps can be generated in this way. The computer–SEM combination has been very valuable for the analysis of fracture surfaces and in studies of the topography of in-process integrated circuits and devices.

Computers can be used both for image analysis and image processing. In the former case, size distributions of particles or features, and their associated measurement parameters (area, circumference, maximum or minimum diameters, etc.) can be obtained easily because the image information is collected via digital scanning in a way that is directly compatible with the architecture of image analysis computers. In these systems an image is a stored array of 500×500 signal values. Larger arrays are also possible with larger memory capacity computers.

Image processing refers to the manipulation of the images themselves. This allows for mathematical smoothing, differentiation, and even image subtraction. The contrast and brightness in an image can be adjusted in a linear or nonlinear manner and algorithms exist to highlight edges of features or to completely suppress background variations. These methods allow the microscopist to extract the maximum amount of information from a single micrograph. As high-speed PCs and workstations continue to decrease in price while increasing in capacity, these applications will become more commonplace.

Electronics has, in fact, been a very fertile area for SEM application. The energy distribution of the SEs produced by a material in the SEM has been shown to shift linearly with the local potential of the surface. This phenomenon allows the SEM to be used in a noncontact way to measure voltages on the surfaces of semiconductor devices. This is accomplished using energy analysis of the SEs and by directly measuring these energy shifts. The measurements can be made very rapidly so that circuit waveforms at particular internal circuit nodes can be determined accurately.

Very high frequency operation of circuits can be observed by using stroboscopic techniques, blanking the primary beam at very high frequencies and by collecting information only during small portions of the operating cycle. By sliding the viewing window over the entire cycle, a waveform can be extracted with surprising sensitivity and temporal resolution. Recently, special SEM instruments have been designed that can observe high-speed devices under nominal operating conditions. The cost of these complex electron beam test systems can exceed $1,000,000.

Actually, any physical process that can be induced by the presence of an electron beam and that can generate a measurable signal can be used to produce an image in the SEM. For example, when a metal film is deposited on a clean semiconductor surface, a region of the semiconductor next to the metal can be depleted of mobile charge. This depletion region can develop a strong electric field. Imagine the metal to be connected through a current meter to the back side of the semiconductor surface. No current will flow, of course, but we can turn on an electron beam that penetrates the metal and allows the impinging electrons to create electron-hole pairs. The electric field will separate these carriers and sweep them out of the depletion region, thereby generating a current. Among other things, the current will be determined by the perfection or the chemistry of the spot where the beam strikes. By scanning the beam in the SEM mode, an image of the perfection (or chemistry) of the surface can be generated. This is referred to as electron beam-induced conductivity, and it has been used extensively to identify the kinds of defects that can affect semiconductor device operation.

The SEM can also be used to provide crystallographic information. Surfaces that to exhibit grain structure (fracture surfaces, etched, or decorated surfaces) can obviously be characterized as to grain size and shape. Electrons also can be channeled through a crystal lattice and when channeling occurs, fewer backscattered electrons can exit the surface. The channeling patterns so generated can be used to determine lattice parameters and strain.

The X-rays generated when an electron beam strikes a crystal also can be diffracted by the specimen in which they are produced. If a photograph is made of this diffraction pattern (the Kossel pattern) using a special camera, localized crystallographic information can be gleaned.

Further applications abound. Local magnetic fields affect the trajectories of the SEs as well as the BSEs, making the SEM a useful tool for observing the magnetic domains of ferromagnetic materials, magnetic tapes, and disk surfaces. Pulsed electron beams generate both thermal and acoustic signals which can be imaged to provide mechanical property maps of materials. Some semiconductors and oxides produce photons in the ultraviolet, visible, or infrared regions, and these cathodoluminescence signals provide valuable information about the electronic properties of these materials. The application of this method to semiconductor lasers or LED devices is probably self-evident. Even deep-level transient spectroscopy, a method that is particularly difficult to interpret and that provides information about impurities having energy levels within the band gap of semiconductors, has been used to produce images in the SEM.

Conclusions

Every month a new application for the SEM appears in the literature, and there is no reason to assume that this growth will cease. The SEM is one of the more versatile of analytical instruments and it is often the first expensive instrument that a characterization laboratory will purchase.

As time goes on, the ultimate resolution of the SEM operated in these modes will probably level out near 1 nm. The major growth of SEMs now seems to be in the development of specialized instruments. An environmental SEM has been developed that uses differential pumping to permit the observation of specimens at higher pressures. Photographs of the formation of ice crystals have been taken and the instrument has particular application to samples that are not vacuum compatible, such as biological samples.

Other instruments have been described that have application in the electronics field. Special metallurgical hot and cold stages are being produced, and stages capable of large motions with sub-μm accuracy and reproducibility will become common.

Computers will be integrated more and more into commercial SEMs and there is an enormous potential for the growth of computer supported applications. At the same time, related instruments will be developed and extended, such as the scanning ion microscope, which uses liquid-metal ion sources to produce finely focused ion beams that can produce SEs and secondary ions for image generation. The contrast mechanisms that are exhibited in these instruments can provide new insights into materials analysis.

Related Articles in the Encyclopedia

TEM, STEM, EDS, EPMA, Surface Roughness, AES, and CL

References

1 J. I. Goldstein, Dale E. Newbury, P. Echlin, D. C. Joy, C. Fiori, and E. Lifshin. *Scanning Microscopy and X-Ray Microanalysis.* Plenum Press, New York, 1981. An excellent and widely ranging introductory textbook on scanning microscopy and related techniques. Some biological applications are also discussed.

2 D. Newbury, D. C. Joy, P. Echlin, C. E. Fiori, and J. I. Goldstein. *Advanced Scanning Electron Microscopy and X-Ray Microanalysis.* Plenum Press, New York, 1986. A continuation and expansion of Reference 1, *advanced* does not imply a higher level of difficulty.

3 L. Reimer. *Scanning Electron Microscopy.* Springer-Verlag, Berlin, 1985. An advanced text for experts, this is probably the most definitive work in the field.

4 D. B. Holt and D. C. Joy. *SEM Microcharacterization of Semiconductors.* Academic Press, London, 1989. A detailed examination of the applications of the SEM to semiconductor electronics.

5 John C. Russ. *Computer Assisted Microscopy.* Plenum Press, New York, 1990. A highly readable account of the applications of computers to SEMs and other imaging instruments.

2.3 STM and SFM

Scanning Tunneling Microscopy and Scanning Force Microscopy

REBECCA S. HOWLAND AND MICHAEL D. KIRK

Contents

Introduction

Scanning Tunneling Microscopy (STM) and its offspring, Scanning Force Microscopy (SFM), are real-space imaging techniques that can produce topographic images of a surface with atomic resolution in all three dimensions. Almost any solid surface can be studied with STM or SFM: insulators, semiconductors, and conductors, transparent as well as opaque materials. Surfaces can be studied in air, in liquid, or in ultrahigh vacuum, with fields of view from atoms to greater than 250 × 250 μm. With this flexibility in both the operating environment and types of samples that can be studied, STM / SFM is a powerful imaging system.

The scanning tunneling microscope was invented at IBM, Zurich, by Gerd Binnig and Heinrich Rohrer in 1981.[1] In ultrahigh vacuum, they were able to resolve the atomic positions of atoms on the surface of Si (111) that had undergone a 7 × 7 reconstruction (Figure 1). With this historic image they solved the puzzle of the atomic structure of this well studied surface, thereby establishing firmly the credibility and importance of this form of microscopy. For the invention of STM, Binnig and Rohrer earned the Nobel Prize for Physics in 1986.

Figure 1 **Ultrahigh-vacuum STM image of Si (111) showing 7 × 7 reconstruction.**

Since then, STM has been established as an instrument for forefront research in surface physics. Atomic resolution work in ultrahigh vacuum includes studies of metals, semimetals and semiconductors. In particular, ultrahigh-vacuum STM has been used to elucidate the reconstructions that Si, as well as other semiconducting and metallic surfaces undergo when a submonolayer to a few monolayers of metals are adsorbed on the otherwise pristine surface.[2]

Because STM measures a quantum-mechanical tunneling current, the tip must be within a few Å of a conducting surface. Therefore any surface oxide or other contaminant will complicate operation under ambient conditions. Nevertheless, a great deal of work has been done in air, liquid, or at low temperatures on inert surfaces. Studies of adsorbed molecules on these surfaces (for example, liquid crystals on highly oriented, pyrolytic graphite[3]) have shown that STM is capable of even atomic resolution on organic materials.

The inability of STM to study insulators was addressed in 1985 when Binnig, Christoph Gerber and Calvin Quate invented a related instrument, the scanning force microscope.[4] Operation of SFM does not require a conducting surface; thus insulators can be studied without applying a destructive coating. Furthermore, studying surfaces in air is feasible, greatly simplifying sample preparation while reducing the cost and complexity of the microscope.

STM and SFM belong to an expanding family of instruments commonly termed Scanning Probe Microscopes (SPMs). Other common members include the magnetic force microscope, the scanning capacitance microscope, and the scanning acoustic microscope.[5]

Although the first six or seven years of scanning probe microscope history involved mostly atomic imaging, SPMs have evolved into tools complementary to Scanning and Transmission Electron Microscopes (SEMs and TEMs), and optical and stylus profilometers. The change was brought about chiefly by the introduction of the ambient SFM and by improvements in the range of the piezoelectric scanners that move the tip across the sample. With lateral scan ranges on the order of 250 μm, and vertical ranges of about 15 μm, STM and SFM can be used to address larger scale problems in surface science and engineering in addition to atomic-scale research. STM and SFM are commercially available, with several hundred units in place worldwide.

SPMs are simpler to operate than electron microscopes. Because the instruments can operate under ambient conditions, the set-up time can be a matter of minutes. Sample preparation is minimal. SFM does not require a conducting path, so samples can be mounted with double-stick tape. STM can use a sample holder with conducting clips, similar to that used for SEM. An image can be acquired in less than a minute; in fact, "movies" of ten frames per second have been demonstrated.[6]

The three-dimensional, quantitative nature of STM and SFM data permit in-depth statistical analysis of the surface that can include contributions from features 10 nm across or smaller. By contrast, optical and stylus profilometers average over areas a few hundred Å across at best, and more typically a μm. Vertical resolution for SFM / STM is sub-Å, better than that of other profilometers. STM and SFM are excellent high-resolution profilometers.

STM and SFM are free from many of the artifacts that afflict other kinds of profilometers. Optical profilometers can experience complicated phase shifts when materials with different optical properties are encountered. The SFM is sensitive to topography only, independent of the optical properties of the surface. (STM may be sensitive to the optical properties of the material inasmuch as optical properties are related to electronic structure.) The tips of traditional stylus profilometers exert forces that can damage the surfaces of soft materials, whereas the force on SFM tips is many orders of magnitude lower. SFM can image even the tracks left by other stylus profilometers.

In summary, scanning probe microscopes are research tools of increasing importance for atomic-imaging applications in surface science. In addition, SFM and STM are now used in many applications as complementary techniques to SEM, TEM, and optical and stylus profilometry. They meet or exceed the performance of these instruments under most conditions, and have the advantage of operating in an ambient environment with little or no sample preparation. The utility of scanning probe microscopy to the magnetic disk, semiconductor, and the polymer industries is gaining recognition rapidly. Further industrial applications include the analysis of optical components, mechanical parts, biological samples, and other areas where quality control of surfaces is important.

Basic Principles

STM

Scanning tunneling microscopes use an atomically sharp tip, usually made of tungsten or Pt-Ir. When the tip is within a few Å of the sample's surface, and a bias voltage V_t is applied between the sample and the tip, quantum-mechanical tunneling takes place across the gap. This tunneling current I_t depends exponentially on the separation d between the tip and the sample, and linearly on the local density of states. The exponential dependence of the magnitude of I_t upon d means that, in most cases, a single atom on the tip will image the single nearest atom on the sample surface.

The quality of STM images depends critically on the mechanical and electronic structure of the tip. Tungsten tips are sharpened by electrochemical etching, and can be used for a few hours in air, until they oxidize. On the other hand, Pt-Ir tips can be made by stretching a wire and cutting it on an angle with wire cutters. These tips are easy to make and slow to oxidize, but the resulting tip does not have as high an aspect ratio as a tungsten tip. As a result, Pt-Ir tips are not as useful for imaging large structures.

In its most common mode of operation, STM employs a piezoelectric transducer to scan the tip across the sample (Figure 2a). A feedback loop operates on the scanner to maintain a constant separation between the tip and the sample. Monitoring the position of the scanner provides a precise measurement of the tip's position in three dimensions. The precision of the piezoelectric scanning elements, together with the exponential dependence of I_t upon d means that STM is able to provide images of individual atoms.

Because the tunneling current also depends on the local density of states, STM can be used for spatially resolved spectroscopic measurements. When the component atomic species are known, STM can differentiate among them by recording and comparing multiple images taken at different bias voltages. One can ramp the bias voltage between the tip and the sample and record the corresponding change in the tunneling current to measure I versus V or dI/dV versus V at specific sites on the image to learn directly about the electronic properties of the surface. Such measurements give direct information on the local density of electronic states. This technique was pioneered by Hamers, et al., who used tunneling spectroscopy to map the local variations in the bonding structure between Si atoms on a reconstructed surface.[7]

On the other hand, the sensitivity of STM to electronic structure can lead to undesired artifacts when the surface is composed of regions of varying conductivity. For example, an area of lower conductivity will be represented as a dip in the image. If the surface is not well known, separating topographic effects from electronic effects can be difficult.

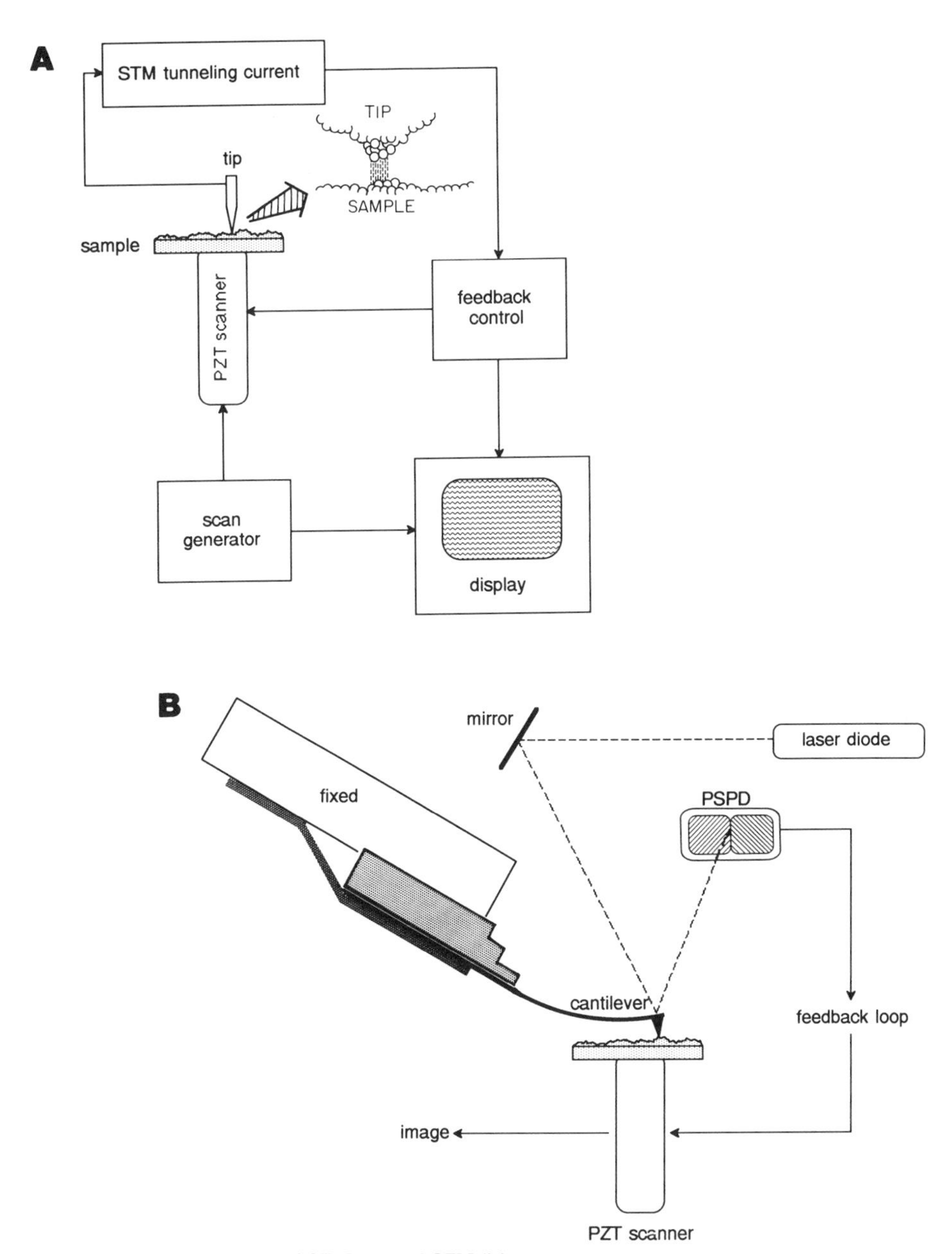

Figure 2 **Schematic of STM (a) and SFM (b).**

SFM

Scanning force microscopes use a sharp tip mounted on a flexible cantilever. When the tip comes within a few Å of the sample's surface, repulsive van der Waals forces

Figure 3 **SEM image of SFM cantilever showing pyramidal tip.**

between the atoms on the tip and those on the sample cause the cantilever to deflect. The magnitude of the deflection depends on the tip-to-sample distance *d*. However, this dependence is a power law, that is not as strong as the exponential dependence of the tunneling current upon *d* employed by STM. Thus several atoms on an SFM tip will interact with several atoms on the surface. Only with an unusually sharp tip and flat sample is the lateral resolution truly atomic; normally the lateral resolution of SFM is about 1nm.

Like STM, SFM employs a piezoelectric transducer to scan the tip across the sample (Figure 2b), and a feedback loop operates on the scanner to maintain a constant separation between the tip and the sample. As with STM, the image is generated by monitoring the position of the scanner in three dimensions.

For SFM, maintaining a constant separation between the tip and the sample means that the deflection of the cantilever must be measured accurately. The first SFM used an STM tip to tunnel to the back of the cantilever to measure its vertical deflection. However, this technique was sensitive to contaminants on the cantilever.[4] Optical methods proved more reliable. The most common method for monitoring the defection is with an optical-lever or beam-bounce detection system.[8] In this scheme, light from a laser diode is reflected from the back of the cantilever into a position-sensitive photodiode. A given cantilever deflection will then correspond to a specific position of the laser beam on the position-sensitive photodiode. Because the position-sensitive photodiode is very sensitive (about 0.1 Å), the vertical resolution of SFM is sub-Å.

Figure 3 shows an SEM micrograph of a typical SFM cantilever. The cantilevers are 100–200 μm long and 0.6 μm thick, microfabricated from low-stress Si_3N_4 with an integrated, pyramidal tip. Despite a minimal tip radius of about 400 Å,

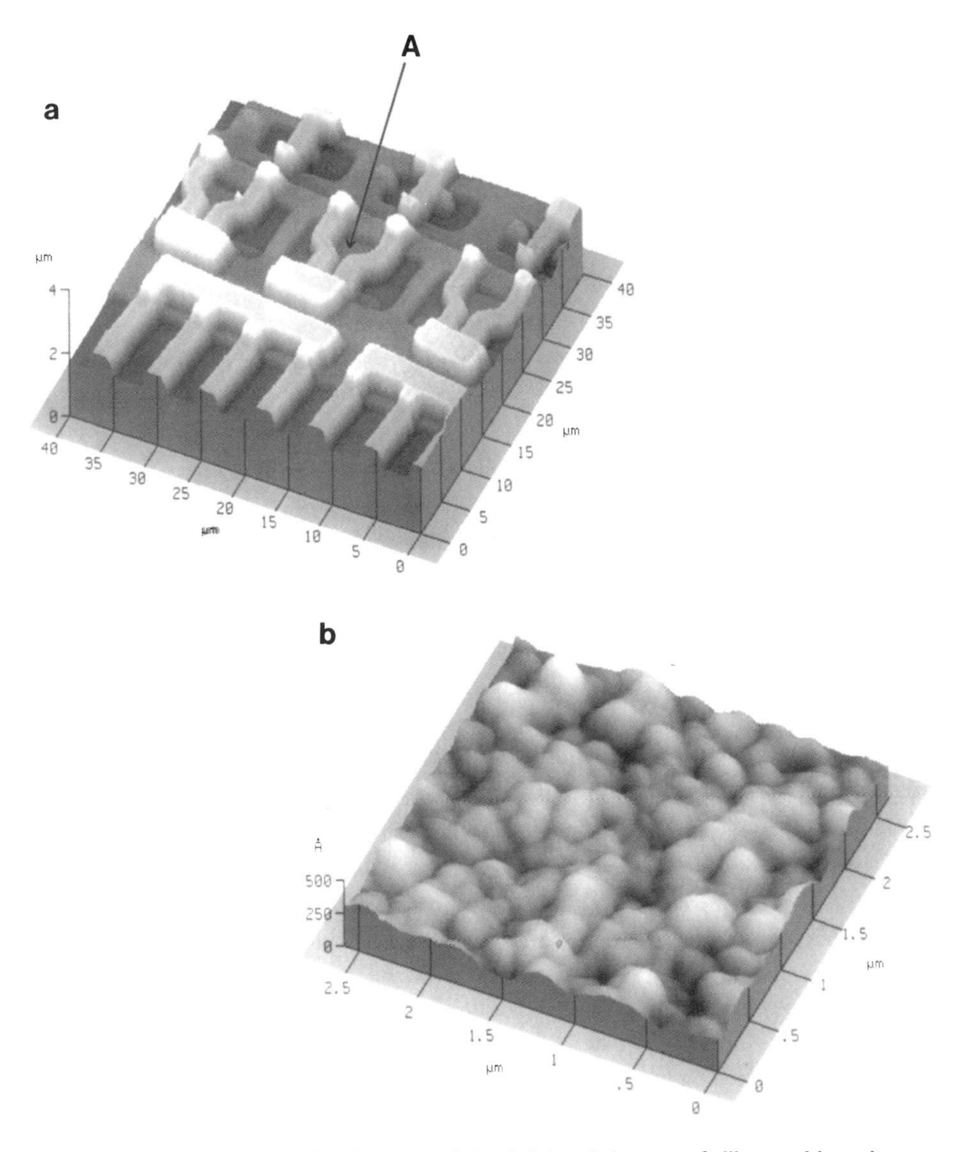

Figure 4 **SFM image of an integrated circuit (a) and close-up of silicon oxide on its surface (b).**

which is needed to achieve high lateral resolution, the pressure exerted on the sample surface is small because of the low force constant of the cantilever (typically 0.2 N / m), and the high sensitivity of the position-sensitive photodiode to cantilever deflection. The back of the cantilever may be coated with gold or another metal to enhance the reflectance of the laser beam into the detector.

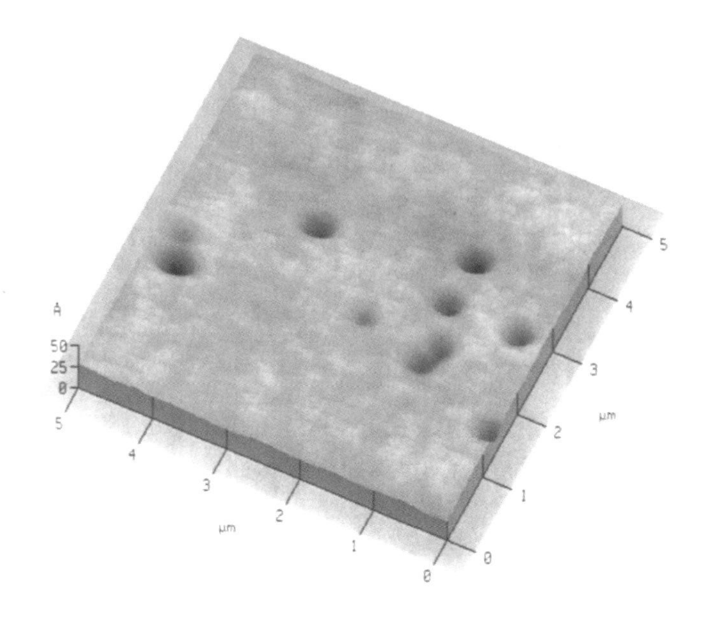

Figure 5 **SFM image of oxidized Si wafer showing pinhole defects 20 Å deep.**

Common Modes of Analysis and Examples

STM and SFM are most commonly used for topographic imaging, three-dimensional profilometry and spectroscopy (STM only).

Topography

Unlike optical or electron microscopes, which rely on shadowing to produce contrast that is related to height, STM and SFM provide topographic information that is truly three-dimensional. The data are digitally stored, allowing the computer to manipulate and display the data as a three-dimensional rendition, viewed from any altitude and azimuth. For example, Figure 4a shows an SFM image of an integrated circuit; Figure 4b is a close-up of the oxide on the surface of the chip in the region marked *A* in Figure 4a. In a similar application, Figure 5 is an SFM image of a Si wafer with pinholes, 20 Å deep. Easily imaged with SFM, these pinholes cannot be detected with SEM.

Profilometry

The three-dimensional, digital nature of SFM and STM data makes the instruments excellent high-resolution profilometers. Like traditional stylus or optical profilometers, scanning probe microscopes provide reliable height information. However, traditional profilometers scan in one dimension only and cannot match SPM's height and lateral resolution.

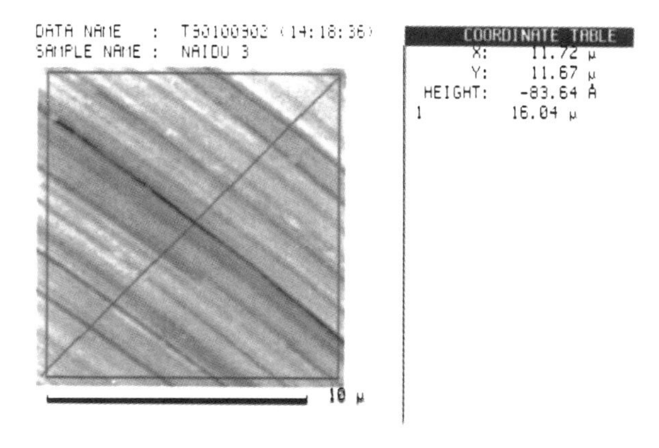

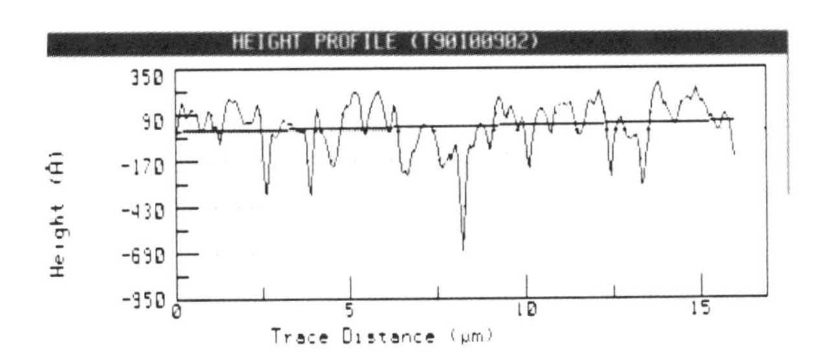

Figure 6 **SFM image of a magnetic storage disk demonstrating roughness analysis.**

In the magnetic storage disk industry, the technology has advanced to the point where surface roughness differences on the order of a few Å have become important. Optical and stylus profilometers, while still preferable for scanning very large distances, cannot measure contributions from small features. Figure 6 is an SFM image of a thin-film storage disk (top), shown top-down, with heights displayed in a linear intensity scale ("gray scale"). Using the mouse, the height profile of any cross section can be displayed and analyzed (bottom). Figure 7 shows a thin-film read–write head. The magnetic poles are recessed about 200 Å; their roughness is comparable to that of the surrounding medium. Note the textural difference between the glass embedding medium and the ceramic. SFM is not affected by differences in optical properties when it scans composite materials.

Profilometry of softer materials, such as polymers, is also possible with SFM, and with STM if the sample is conducting. Low forces on the SFM tip allow imaging of materials whose surfaces are degraded by traditional stylus profilometry. However, when the surface is soft enough that it deforms under pressure from the SFM tip, resolution will be degraded and topography may not be representative of the true

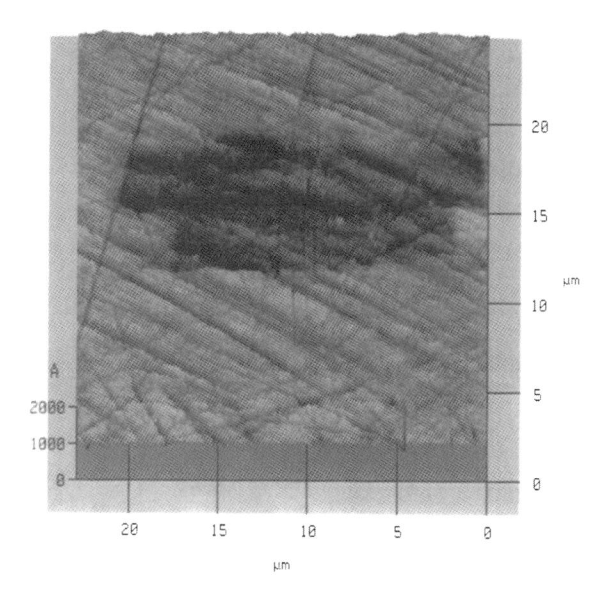

Figure 7 **SFM image of a thin-film read–write head showing magnetic poles (dark rectangles) recessed 200 Å.**

surface. One can investigate the reproducibility of the image by scanning the sample in different directions at various scan rates and image sizes.

Spectroscopy

The preceding topography and profilometry examples have focused on the scanning force microscope. STM also can be used for topographic imaging and profilometry, but the images will be convolutions of the topographic and electronic structure of the surface. A similar effect is seen with SEM, arising from differences in secondary electron coefficients among different materials.

Taking advantage of the sensitivity of the tunneling current to local electronic structure, the STM can be used to measure the spectra of surface-state densities directly. This can be accomplished by measuring the tunneling current as a function of the bias voltage between the tip and sample, or the conductivity, dI/dV, versus the bias voltage, at specific spatial locati ›ns on the surface. Figure 8 is a spectroscopic study of GaAs(110). The image on the left was taken with negative bias voltage on the STM tip, which allows tunneling into unoccupied states, thereby revealing the Ga atoms. Taken simultaneously but with a positive tip bias voltage, the image on the right results from tunneling out occupied states, and shows the positions of the As atoms.

The data above were collected in UHV environment to achieve the most pristine surface. Spectroscopy in air is usually more difficult to interpret due to contamination with oxides and other species, as is the case with all surface-sensitive spectroscopies.

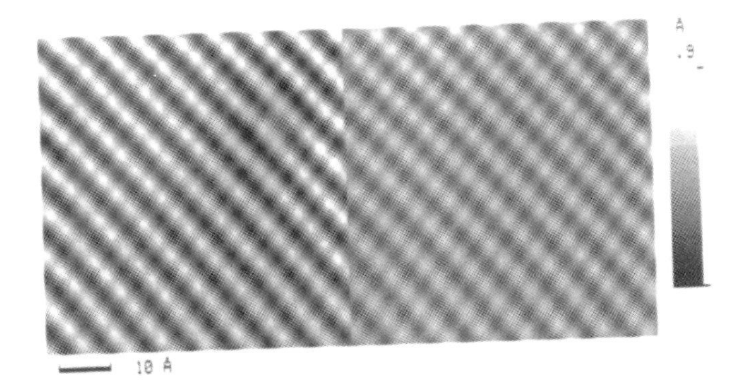

Figure 8 **Spectroscopic study of GaAs(110). With a positive voltage on the STM tip, the left-hand image represents As atoms, while the corresponding negative tip voltage on the right shows Ga atoms. (Courtesy of Y. Yang and J.H. Weaver, University of Minnesota)**

Sample Requirements

For atomic resolution an atomically flat sample is required to avoid tip imaging (see below). STM requires a conducting surface to establish the tunneling current. Doped Si has sufficient conductivity to enable STM imaging, but surfaces of lower conductivity may require a conductive coating. SFM can image surfaces of any conductivity. Both STM and SFM require solid surfaces that are somewhat rigid; otherwise the probes will deform the surfaces while scanning. Such deformation is easily diagnosed by repeatedly scanning the same area and noting changes.

The deformation of soft surfaces can be minimized with SFM by selecting cantilevers having a low force constant or by operating in an aqueous environment. The latter eliminates the viscous force that arises from the thin film of water that coats most surfaces in ambient environments. This viscous force is a large contributor to the total force on the tip. Its elimination means that the operating force in liquid can be reduced to the order of 10^{-9} N.

An example, Figure 9 is an SFM image of a Langmuir-Blodgett film. This film was polymerized with ultraviolet light, giving a periodicity of 200 Å, which is seen in the associated Fourier transform. The low forces exerted by the SFM tip are essential for imaging such soft polymer surfaces.

Poorly cleaned surfaces may not image well. While ordinary dry dust will be brushed aside by the tip and will not affect the image, oily or partially anchored dirt will deflect the SFM tip or interfere with the conductivity in STM. The result is usually a line smeared in the scan direction, exactly as one would expect if the tip began scanning something which moved as it was scanned. If the sample cannot be cleaned, the best procedure is to search for a clean area.

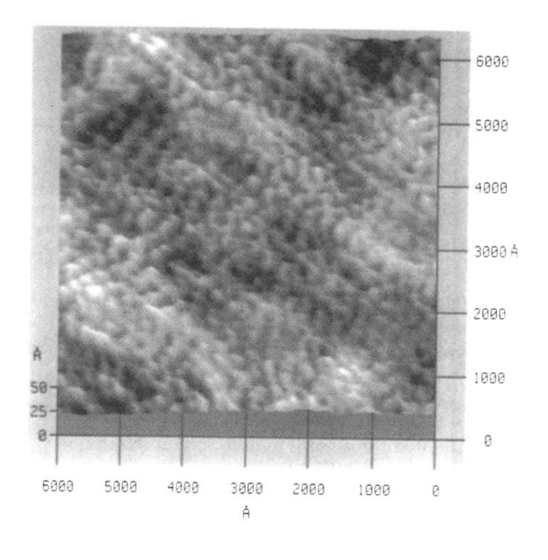

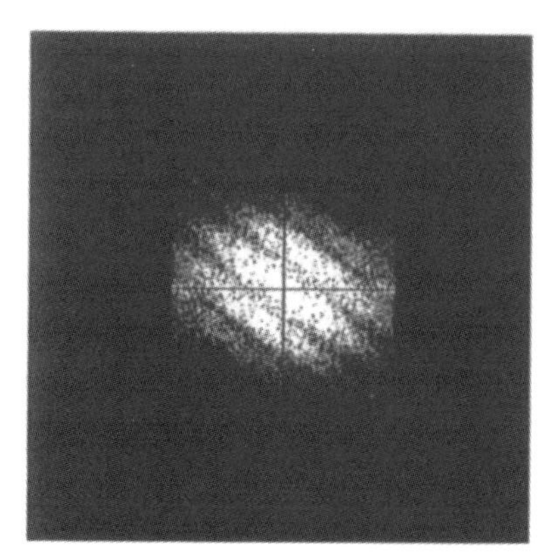

Figure 9 **SFM image of Langmuir-Blodgett film (top) and associated Fourier transform (bottom). (Courtesy of T. Kato, Utsunomiya University)**

Maximum sample sizes that can be accommodated by SFM or STM vary. Current systems can scan a 8-inch Si wafer without cutting it. When industry calls for the capability to scan larger samples, the SPM manufacturers are likely to respond.

Artifacts

The main body of artifacts in STM and SFM arises from a phenomenon known as *tip imaging.*[9] Every data point in a scan represents a convolution of the shape of the tip and the shape of the feature imaged, but as long as the tip is much sharper than the feature, the true edge profile of the feature is represented. However, when the feature is sharper than the tip, the image will be dominated by the edges of the tip. Fortunately, this kind of artifact is usually easy to identify.

Other artifacts that have been mentioned arise from the sensitivity of STM to local electronic structure, and the sensitivity of SFM to the rigidity of the sample's surface. Regions of variable conductivity will be convolved with topographic features in STM, and soft surfaces can deform under the pressure of the SFM tip. The latter can be addressed by operating SFM in the attractive mode, at some sacrifice in the lateral resolution. A limitation of both techniques is their inability to distinguish among atomic species, except in a limited number of circumstances with STM microscopy.

STM

In STM, the tip is formed by an atom or cluster of atoms at the end of a long wire. Because the dependence of the tunneling current upon the tip-to-sample distance is exponential, the closest atom on the tip will image the closest atom on the sample. If two atoms are equidistant from the surface, all of the features in the image will appear doubled. This is an example of multiple tip imaging. The best way to alleviate this problem is to collide the tip gently with the sample, to form a new tip and take another image. Alternatively, a voltage pulse can be applied to change the tip configuration by field emission.

STM tips will last for a day or so in ultrahigh vacuum. Most ultrahigh-vacuum STM systems provide storage for several tips so the chamber does not have to be vented just to change tips. In air, tips will oxidize more rapidly, but changing tips is a simple process.

SFM

At present, all commercial SFM tips are square pyramids, formed by CVD deposition of Si_3N_4 on an etch pit in (100) Si. The etch pit is bounded by (111) faces, which means that the resulting tip has an included angle of about 55°. Therefore the edge profiles of all features with sides steeper than 55° will be dominated by the profile of the tip.

Because many kinds of features have steep sides, tip imaging is a common plague of SFM images. One consolation is that the height of the feature will be reproduced accurately as long as the tip touches bottom between features. Thus the roughness statistics remain fairly accurate. The lateral dimensions, on the other hand, can provide the user with only an upper bound.

Another class of artifacts occurs when scanning vertical or undercut features. As the tip approaches a vertical surface, the side wall may encounter the feature before the end of the tip does. The resulting image will appear to contain a discontinuous shift. Changing the angle of the tip with respect to the sample's surface can minimize the problem. Side wall imaging also occurs in STM, but less frequently since an STM tip has a higher aspect ratio than that of an SFM tip.

Improving the aspect ratio of SFM tips is an area of active research. A major difficulty is that the durability of the tip likely will be compromised as aspect ratios are increased.

Conclusions

Scanning probe microscopy is a forefront technology that is well established for research in surface physics. STM and SFM are now emerging from university laboratories and gaining acceptance in several industrial markets. For topographic analysis and profilometry, the resolution and three-dimensional nature of the data is unequalled by other techniques. The ease of use and nondestructive nature of the imaging are notable.

The main difficulty with STM and SFM techniques is the problem of tip imaging. Neither technique is recommended for obtaining accurate measurements of edge profiles of vertical or undercut surfaces. In addition, SFM tips cannot accurately image the lateral dimensions of features with sides steeper than 55° at present. Obtaining SFM tips with more suitable aspect ratios is an area of active research.

Scanning tunneling and scanning force microscopes are only two members of the family of scanning probe microscopes. Other types of scanning probe microscopes may become widely used in the near future. The magnetic force microscope, for example, may one day be used routinely to study magnetic domains in storage media.

Related Articles in the Encyclopedia

Light Microscopy, SEM, TEM, STEM, and Surface Roughness

References

1. G. Binnig, H. Rohrer, C. Gerber, and E. Weibel. *Phys. Rev. Lett.* **49,** 57, 1982.
2. D. Rugar and P. Hansma. *Physics Today*. October, 23, 1990.
3. J. S. Foster and J. E. Frommer. *Nature.* **333,** 542, 1988.
4. G. Binnig, C. F. Quate, and C. Gerber. *Phys. Rev. Lett.* **54,** 930, 1986.
5. H. K. Wickramsinghe. *Scientific American* . October, 98, 1989.
6. R. Barret and C. F. Quate. To be published.
7. R. J. Hamers, R. M. Tromp, and J. E. Demuth. *Phys. Rev. Lett.* **56,** 1972, 1986.
8. G. Meyer and N. M. Amer. *Appl. Phys. Lett.* **53,** 1045, 1988.
9. S.-I. Park, J. Nogami, and C. F. Quate. *Phys. Rev. B* **36,** 2863, 1987.

2.4 TEM

Transmission Electron Microscopy

KURT E. SICKAFUS

Contents

Introduction

Transmission Electron Microscopy (TEM) has, in three decades time, become a mainstay in the repertoire of characterization techniques for materials scientists. TEM's strong cards are its high lateral spatial resolution (better than 0.2 nm "point-to-point" on some instruments) and its capability to provide both image and diffraction information from a single sample. In addition, the highly energetic beam of electrons used in TEM interacts with sample matter to produce characteristic radiation and particles; these signals often are measured to provide materials characterization using EDS, EELS, EXELFS, backscattered and secondary electron imaging, to name a few possible techniques.

Basic Principles

In TEM, a focused electron beam is incident on a thin (less than 200 nm) sample. The signal in TEM is obtained from both undeflected and deflected electrons that penetrate the sample thickness. A series of magnetic lenses at and below the sample position are responsible for delivering the signal to a detector, usually a fluorescent screen, a film plate, or a video camera. Accompanying this signal transmission is a

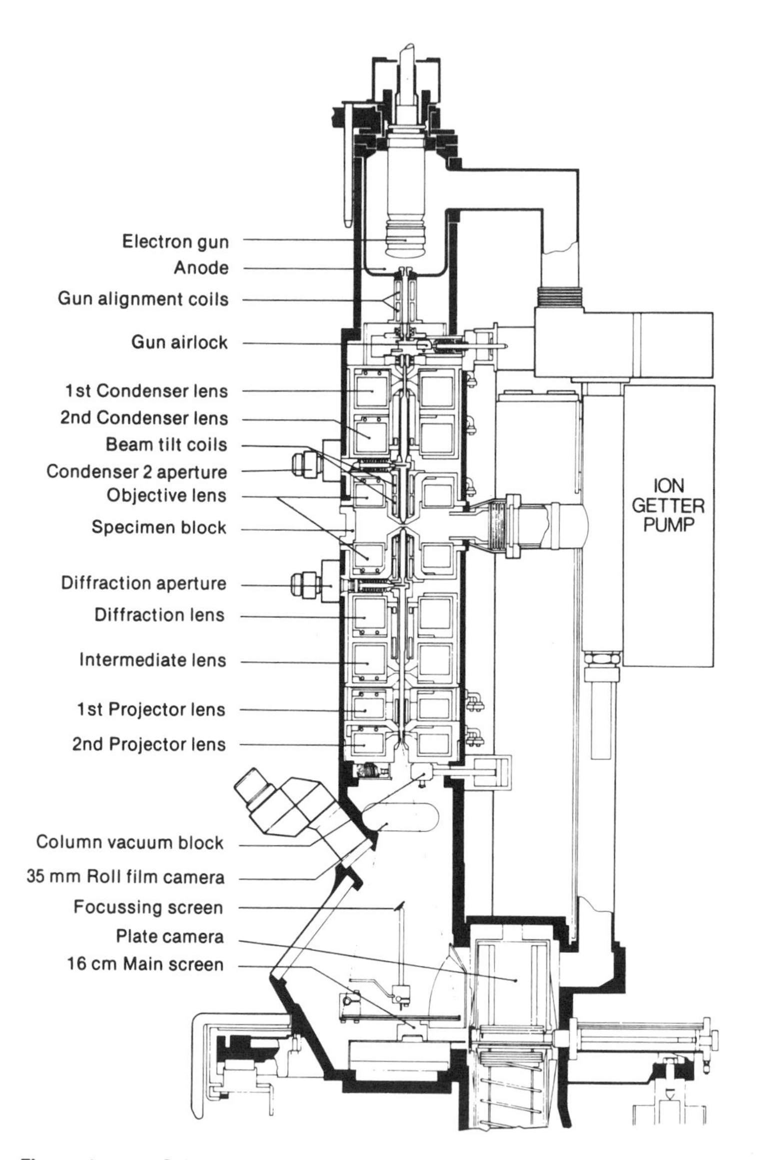

Figure 1a **Schematic diagram of a TEM instrument, showing the location of a thin sample and the principal lenses within a TEM column.**

magnification of the spatial information in the signal by as little as 50 times to as much as a factor of 10^6. This remarkable magnification range is facilitated by the small wavelength of the incident electrons, and is the key to the unique capabilities associated with TEM analysis. A schematic of a TEM instrument, showing the location of a thin sample and the principal lenses within a TEM column, is illustrated in Figure 1a.Figure 1b shows a schematic for the ray paths of both unscattered and scattered electrons beneath the sample.

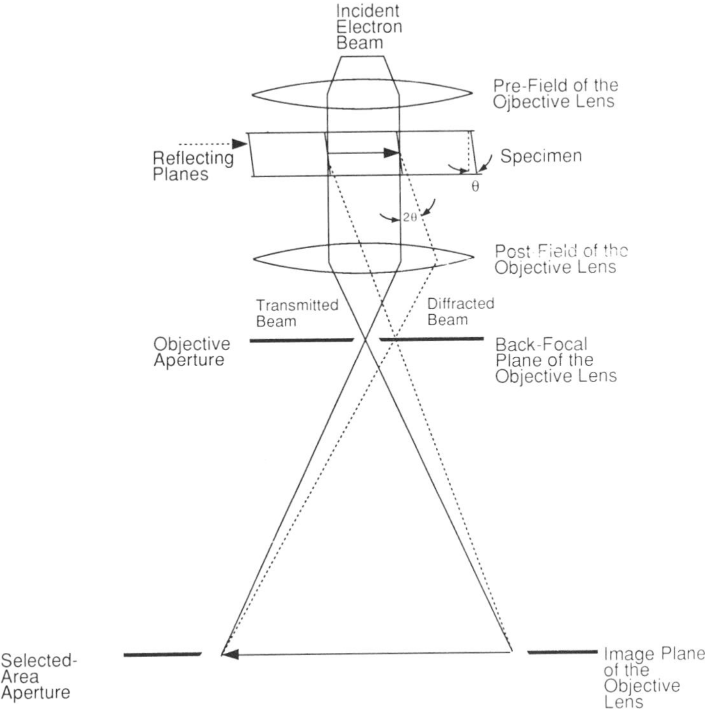

Figure 1b **Schematic representation for the ray paths of both unscattered and scattered electrons beneath the sample.**

Resolution

The high lateral spatial resolution in a TEM instrument is a consequence of several features of the technique. First, in the crudest sense, TEM has high spatial resolution because it uses a highly focused electron beam as a probe. This probe is focused at the specimen to a small spot, often a μm or less in diameter. More importantly, the probe's source is an electron gun designed to emit a highly coherent beam of monoenergetic electrons of exceedingly small wavelength. The wavelength, λ, of 100 keV electrons is only 0.0037 nm, much smaller than that of light, X rays, or neutrons used in other analytical techniques. Having such small wavelengths, since electrons in a TEM probe are in phase as they enter the specimen, their phase relationships upon exiting are correlated with spatial associations between scattering centers (atoms) within the material. Finally, high lateral spatial resolution is maintained via the use of extremely thin samples. In most TEM experiments, samples are thinned usually to less than 200 nm. For most materials this insures relatively few scattering events as each electron traverses the sample. Not only does this limit spreading of the probe, but much of the coherency of the incident source is also retained.

The higher the operating voltage of a TEM instrument, the greater its lateral spatial resolution. The theoretical instrumental point-to-point resolution is proportional[1] to $\lambda^{3/4}$. This suggests that simply going from a conventional TEM instrument operating at 100 kV to one operating at 400 kV should provide nearly a 50% reduction in the minimum resolvable spacing (λ is reduced from 0.0037 to 0.0016 nm in this case). Some commercially available 300 kV and 400 kV instruments, classified as *high-voltage* TEM instruments, have point-to-point resolutions better than 0.2 nm.

High-voltage TEM instruments have the additional advantage of greater electron penetration, because high-energy electrons interact less strongly with matter than low-energy electrons. So, it is possible to work with thicker samples on a high-voltage TEM. Electron penetration is determined by the mean distance between electron scattering events. The fewer the scattering events, either *elastic* (without energy loss) or *inelastic* (involving energy loss), the farther the electron can penetrate into the sample. For an Al sample, for instance, by going from a conventional 100-kV TEM instrument, to a high-voltage 400 kV TEM instrument, one can extend the mean distance between scattering events (both elastic and inelastic) by more than a factor of 2 (from 90 to 200 nm and from 30 to 70 nm, respectively, for elastic and inelastic scattering).[2] This not only allows the user to work with thicker samples but, at a given sample thickness, also reduces deleterious effects due to chromatic aberrations (since inelastic scattering is reduced).

One shortcoming of TEM is its limited depth resolution. Electron scattering information in a TEM image originates from a three-dimensional sample, but is *projected* onto a two-dimensional detector (a fluorescent screen, a film plate, or a CCD array coupled to a TV display). The collapse of the depth scale onto the plane of the detector necessarily implies that structural information along the beam direction is superimposed at the image plane. If two microstructural features are encountered by electrons traversing a sample, the resulting image contrast will be a convolution of scattering contrast from each of the objects. Conversely, to identify overlapping microstructural features in a given sample area, the image contrast from that sample region must be deconvolved.

In some cases, it is possible to obtain limited depth information using TEM. One way is to tilt the specimen to obtain a stereo image pair. Techniques also exist for determining the integrated depth (i.e., specimen thickness) of crystalline samples, e.g., using extinction contours in image mode or using convergent beam diffraction patterns. Alternatively, the width or trace of known defects, inclined to the surface of the foil, can be used to determine thickness from geometrical considerations. Secondary techniques, such as EELS and EDS can in some cases be used to measure thickness, either using plasmon loss peaks in the former case, or by modeling X-ray absorption characteristics in the latter. But no TEM study can escape consideration of the complications associated with depth.

Sensitivity

TEM has no inherent ability to distinguish atomic species. Nonetheless, electron scattering is exceedingly sensitive to the target element. Heavy atoms having large, positively charged nuclei, scatter electrons more effectively and to higher angles of deflection, than do light atoms. Electrons interact primarily with the potential field of an atomic nucleus, and to some extent the electron cloud surrounding the nucleus. The former is similar to the case for neutrons, though the principles of interaction are not related, while the latter is the case for X rays. The scattering of an electron by an atomic nucleus occurs by a Coulombic interaction known as Rutherford scattering. This is equivalent to the elastic scattering (without energy loss) mentioned earlier. The scattering of an electron by the electron cloud of an atom is most often an inelastic interaction (i.e., exhibiting energy loss). Energy loss accompanies scattering in this case because an electron in the incident beam matches the mass of a target electron orbiting an atomic nucleus. Hence, significant electron–electron momentum transfer is possible. A typical example of inelastic scattering in TEM is core-shell ionization of a target atom by an incoming electron. Such an ionization event contributes to the signal that is measured in Electron Energy Loss Spectroscopy (EELS) and is responsible for the characteristic X-Ray Fluorescence that is measured in Energy-Dispersive X-Ray Spectroscopy (EDS) and Wavelength-Dispersive X-Ray Spectroscopy (WDS). The latter two techniques differ only in the use of an energy-dispersive solid state detector versus a wavelength-dispersive crystal spectrometer.

The magnitude of the elastic electron–nucleus interaction scales with the charge on the nucleus, and so with atomic number Z. This property translates into image contrast in an electron micrograph (in the absence of diffraction contrast), to the extent that regions of high-Z appear darker than low-Z matrix material in conventional *bright-field* microscopy. This is illustrated in the bright-field TEM image in Figure 2, where high-Z, polyether sulfone ($-[C_6H_4-SO_2-C_6H_4-O-]_{-n}$) inclusions are seen as dark objects on a lighter background from a low-Z, polystyrene ($-[CH_2-CH(C_6H_5)-]_{-n}$) matrix. [The meaning of bright field is explained later in this article.]

The probability of interaction with a target atom is much greater for electrons than for X rays, with $f_e \sim 10^4 f_X$ (f_e is the electron atomic scattering factor and f_X is X-ray atomic scattering factor; each is a measure of elemental scattering efficiency or equivalently, the elemental sensitivity of the measurement).[3] Unfortunately, with this benefit of elemental sensitivity comes the undesirable feature of multiple scattering. The strong interaction of an incident electron with the potential field of a target atom means that numerous scattering events are possible as the electron traverses the sample. Each scattering event causes an angular deflection of the electron, and often this is accompanied by a loss of energy. The angular deflection upon scattering effectively diminishes the localization of the spatial information in the

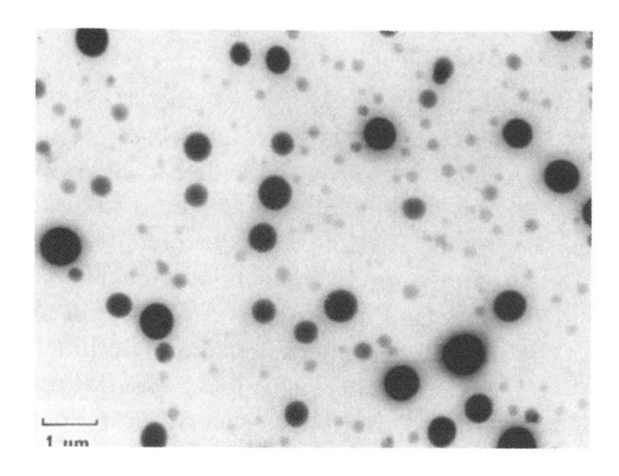

Figure 2 Bright-field TEM image of polyether sulphone inclusions (dark objects; see arrows) in a polystyrene matrix.

TEM signal; the energy losses upon scattering accentuate chromatic aberration effects.

The enormous sensitivity in an electron scattering experiment, in conjunction with the use of a high-brightness electron gun, leads to one of TEM's important features, that of real-time observation. In a conventional TEM, real-time observation is realized by using a W-filament source capable of delivering $\sim 2 \times 10^{19}$ electrons/cm^2-s to the specimen,[4] and a scintillating fluorescent screen to detect the transmitted electrons, viewed through a glass-window flange at the base of the microscope. Recent variations on this theme include the use of better vacuum systems that can accommodate LaB_6 or field-emission gun sources of higher brightness (up to $\sim 6 \times 10^{21}$ electrons/cm^2-s),[4] as well as the use of CCD array–TV displays to enhance detection sensitivity.

TEM Operation

TEM offers two methods of specimen observation, diffraction mode and image mode. In diffraction mode, an electron diffraction pattern is obtained on the fluorescent screen, originating from the sample area illuminated by the electron beam. The diffraction pattern is entirely equivalent to an X-ray diffraction pattern: a single crystal will produce a spot pattern on the screen, a polycrystal will produce a powder or ring pattern (assuming the illuminated area includes a sufficient quantity of crystallites), and a glassy or amorphous material will produce a series of diffuse halos.

The examples in Figure 3 illustrate these possibilities. Figure 3a shows a diffraction pattern from a single crystal Fe thin film, oriented with the [001] crystal axis

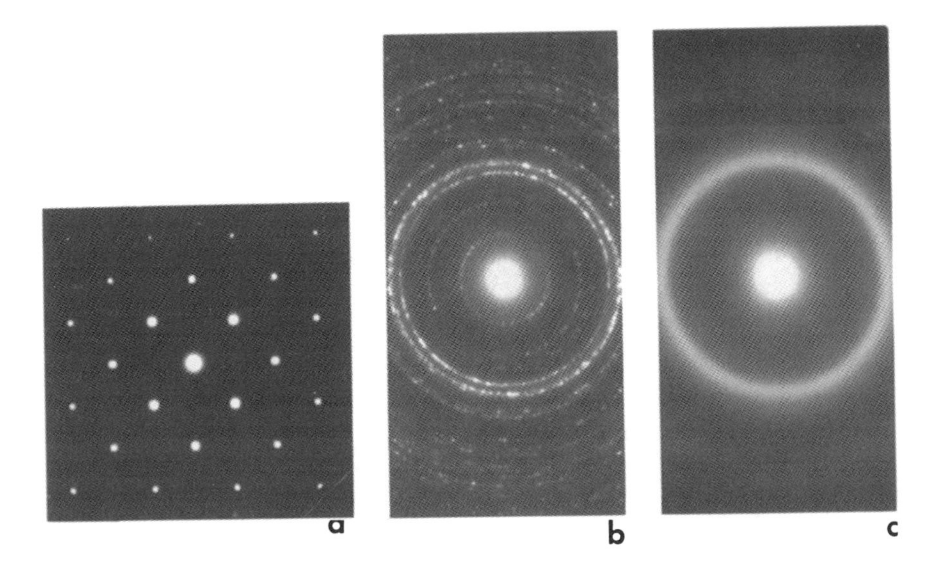

Figure 3 (a) Diffraction pattern from a single crystal Fe thin film, oriented with the [001] crystal axis parallel to the incident electron beam direction. (b) Diffraction pattern from a polycrystalline thin film of Pd_2Si. (c) Diffraction pattern from the same film as in (c), following irradiation of the film with 400-keV Kr^+ ions. See text for discussion (b, c Courtesy of M. Nastasi, Los Alamos National Laboratory)

parallel to the incident electron beam direction. This single crystal produces a characteristic spot pattern. In this case, the four-fold symmetry of the diffraction pattern is indicative of the symmetry of this body-centered cubic lattice. Figure 3b shows a ring pattern from a polycrystalline thin film, Pd_2Si. Figure 3c shows a diffuse halo diffraction pattern from the same film, following irradiation of the film with 400-keV Kr^+ ions. The diffuse halos (the second-order halo here is very faint) are indicative of scattering from an amorphous material, demonstrating a dramatic disordering of Pd_2Si crystal lattice by the Kr^+ ions.

The image mode produces an image of the illuminated sample area, as in Figure 2. The image can contain contrast brought about by several mechanisms: mass contrast, due to spatial separations between distinct atomic constituents; thickness contrast, due to nonuniformity in sample thickness; diffraction contrast, which in the case of crystalline materials results from scattering of the incident electron wave by structural defects; and phase contrast (see discussion later in this article). Alternating between image and diffraction mode on a TEM involves nothing more than the flick of a switch. The reasons for this simplicity are buried in the intricate electron optics technology that makes the practice of TEM possible.

Electron Optics

It is easiest to discuss the electron optics of a TEM instrument by addressing the instrument from top to bottom. Refer again to the schematic in Figure 1a. At the top of the TEM column is an electron source or gun. An electrostatic lens is used to accelerate electrons emitted by the filament to a high potential (typically 100–1,000 kV) and to focus the electrons to a cross-over just above the anode (the diameter of the cross-over image can be from 0.5 to 30 μm, depending on the type of gun[4]). The electrons at the cross-over image of the filament are delivered to the specimen by the next set of lenses on the column, the condensers.

Most modern TEMs use a two-stage condenser lens system that makes it possible to

1. Produce a highly demagnified image of cross-over at the specimen, such that only a very small sample region is illuminated (typically < 1 μm).
2. Focus the beam at "infinity" to produce nearly parallel illumination at the specimen.

The former procedure is the method of choice during operation in the image mode, while the latter condition is desirable for maximizing source coherency in the diffraction mode.

The specimen is immersed in the next lens encountered along the column, the objective lens. The objective lens is a magnetic lens, the design of which is the most crucial of all lenses on the instrument. Instrumental resolution is limited primarily by the spherical aberration of the objective lens.

The magnetic field at the center of the objective lens near the specimen position is large, typically 2–2.5 T (20–25 kG).[4] This places certain restrictions on TEMs applicability to studies of magnetic materials, particularly where high spatial resolution measurements are desired. Nevertheless, low-magnification TEM is often used to study magnetic domain characteristics in magnetic materials, using so-called Lorentz microscopy procedures.[5] In such instances, the objective lens is weakly excited, so that the incident electrons "see" mainly the magnetic field due to the specimen. Changes in this field across domain boundaries produce contrast in the transmitted image.

The final set of magnetic lenses beneath the specimen are jointly referred to as post-specimen lenses. Their primary task is to magnify the signal transferred by the objective lens. Modern instruments typically contain four post-specimen lenses: diffraction, intermediate, projector 1, and projector 2 (in order of appearance below the specimen). They provide a TEM with its tremendous magnification flexibility.

Collectively, the post-specimen lenses serve one of two purposes: they magnify either the diffraction pattern from the sample produced at the back focal plane of the objective lens; or they magnify the image produced at the image plane of the objective lens. These optical planes are illustrated in the electron ray diagram in

Figure 1b. By varying the lenses' strengths so as to alternate between these two object planes, the post-specimen lenses deliver either a magnified diffraction pattern or a magnified image of the specimen to the detector.

The primary remaining considerations regarding the TEM column are the diaphragms or apertures employed at certain positions along the column. The purpose of these apertures is to filter either the source or the transmitted signal. The most important diaphragm is called the objective aperture. This aperture lies in the back focal plane of the objective lens. In this plane the scattered electron waves recombine to form a diffraction pattern. A diffraction pattern corresponds to the angular dispersion of the electron intensity removed from the incident beam by interaction with the specimen. Inserting an aperture in this plane effectively blocks certain scattered waves. The larger the objective aperture, the greater the angular dispersion that is accepted in the transmitted signal. Figure 1b shows an example where the undeflected or transmitted beam is passed by the objective aperture, while the first-order, Bragg-diffracted beam is blocked. Consequently, only intensity in the transmitted beam can contribute to the image formed at the image plane of the objective lens. Use of a small objective aperture while operating in the image mode, which blocks all diffracted beams (as in this example), can serve to enhance significantly image contrast. Use of a large objective aperture, that allows the passage of many diffracted beams, is the *modus operandi* for the technique referred to as high-resolution transmission electron microscopy (HRTEM), discussed later in this article.

Diffraction Mode

A TEM provides the means to obtain a diffraction pattern from a small specimen area. This diffraction pattern is obtained in diffraction mode, where the post-specimen lenses are set to examine the information in the transmitted signal at the back focal plane of the objective lens.

Figure 4 illustrates some of the important aspects of diffraction in TEM. Figure 4a shows a micrograph obtained in image mode of a small region of a Ni_3Al sample illuminated by an electron beam, containing lamellar crystallites with well-defined orientation relationships. Figure 4b shows a selected-area diffraction (SAD) pattern from the same region. In SAD, the condenser lens is defocused to produce parallel illumination at the specimen and a selected-area aperture (see Figure 1b) is used to limit the diffracting volume. Many spots, or reflections, are evident in this pattern, due in part to the special orientation of the sample. The SAD pattern is a superposition of diffraction patterns from crystallites in the illuminated area that possess distinct orientations.

Figures 4c and 4d illustrate what happens when the incident electron probe is focused to illuminate alternately a crystallite in the center of the image (labelled *twin*) (Figure 4c) and another crystallite adjacent to the twin (Figure 4d). This focused-probe technique is sometimes referred to as *micro-diffraction*. Two effects are evident in these micro-diffraction patterns. First, the diffraction patterns consist

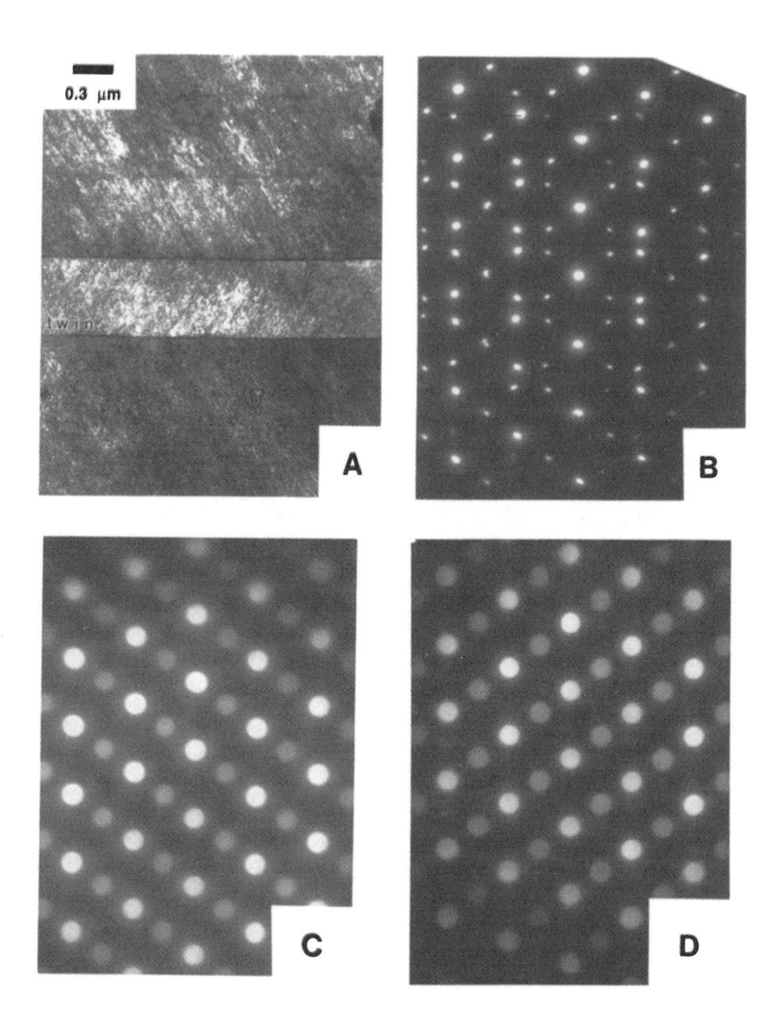

Figure 4 (a) Bright-field image from a small region of a Ni_3Al sample containing oriented crystallites in the center of the illuminated area (one crystallite is labeled *twin* on the micrograph). (b) Selected-area diffraction (SAD) pattern from the same region as in (a). (c) Microdiffraction pattern from the middle region in (a) containing the twin crystallite. (d) Microdiffraction pattern from a crystallite adjacent to the *twin* in (a, c). (Courtesy of G. T. Gray III, Los Alamos National Laboratory)

of "discs" instead of spots. This is a consequence of the use of focused or convergent illumination instead of parallel illumination. Second, the number of reflections in each of these patterns is reduced from that of the SAD pattern in Figure 4b (the reflections are no longer paired). But a superposition of the reflected discs in the microdiffraction patterns can account for all the reflections observed in the SAD pattern. This illustrates the flexibility of a TEM to obtain diffraction information from exceedingly small areas of a sample (in this case, a region of diameter about 0.5 μm or less).

The example in Figure 4 illustrates that the diffraction pattern produced by a crystalline specimen depends on the orientation of the crystal with respect to the incident beam. This is analogous to the way a Laue pattern varies upon changing the orientation of a diffracting crystal relative to an X-ray source.[6] In TEM, this orientation may be varied using the sample manipulation capabilities of a tilting specimen holder. Holders come with a range of tilt capabilities, including single-axis tilt, double-axis tilt, and tilt–rotate stages, with up to ±60° tilting capabilities. But the higher the resolution of the instrument, the more limited the tilting capabilities of a tilt stage (to as low as ±10°). For studies of single crystals or epitaxial thin films, it is important to have access to as much tilt capability as possible.

SAD patterns often are used to determine the Bravais lattice and lattice parameters of crystalline materials. Lattice parameter measurements are made by the same procedures used in X-ray diffraction.[6] Using SAD, each diffracted scattering angle θ is measured in an SAD pattern and an associated atomic interplanar spacing d determined using Bragg's Law, $\lambda = 2d \sin \theta$. Note that at the small electron wavelengths of TEM, typical θ values are small quantities, only 9 mrad for a Au (200) reflection using 100-keV electrons ($\lambda = 0.0037$ nm). By comparison, in a LEED experiment using 150 eV electrons, since $\lambda = 0.1$ nm, a Au (200) reflection would appear at $\theta = 500$ mrad or 30°, using $\lambda = d \sin \theta$; such a large scattering angle is easily observed using the optics of a LEED system, which uses no magnifying lenses for the scattered electrons. Because of the extremely small angle scattering situation in TEM, observation of diffraction patterns is made possible only with the use of magnifying, post-specimen lenses. These lenses greatly magnify the diffraction pattern.

The crystal group or Bravais lattice of an unknown crystalline material can also be obtained using SAD. This is achieved easily with polycrystalline specimens, employing the same powder pattern "indexing" procedures as are used in X-ray diffraction.[6]

Image Mode

In image mode, the post-specimen lenses are set to examine the information in the transmitted signal at the image plane of the objective lens. Here, the scattered electron waves finally recombine, forming an image with recognizable details related to the sample microstructure (or atomic structure).

There are three primary image modes that are used in conventional TEM work, bright-field microscopy, dark-field microscopy, and high-resolution electron microscopy. In practice, the three image modes differ in the way in which an objective diaphragm is used as a filter in the back focal plane.

In bright-field microscopy, a small objective aperture is used to block all diffracted beams and to pass only the transmitted (undiffracted) electron beam. In the absence of any microstructural defects, a bright-field image of a strain-free, single-phase material of uniform thickness, would lack contrast regardless of specimen orientation. Contrast arises in a bright-field image when thickness or compositional variations or structural anomalies are present in the illuminated sample area (or when the sample is strained), so that electrons in some areas are scattered out of the primary beam to a greater extent than in neighboring regions. Regions in which intensity is most effectively removed from the incident beam to become scattered or diffracted intensity appear dark in a bright-field image since this intensity is removed by the objective diaphragm. The images in Figure 2 were obtained using the bright-field imaging procedures.

In multiphase, amorphous or glassy materials, regions containing a phase of high average Z will scatter electrons more efficiently and to higher angles than regions containing a low average Z. The objective aperture in bright-field blocks this scattered intensity, making the high-Z material appear darker (less transmitted intensity) than the low-Z material. This is *mass* contrast, due primarily to incoherent elastic scattering. The scattering is largely incoherent because spatial relationships between scattering centers in these materials are not periodic. A priori there are no well-defined phase relationships between electrons scattered by such materials. Under these circumstances, the transmitted intensity distribution is determined from the principle of the additivity of individual scattered intensities, without consideration for the individual scattered amplitudes.

In crystalline materials, dark contrast regions in bright-field usually originate from areas that are aligned for Bragg diffraction. Here, intensity is removed from the transmitted beam to produce diffracted intensity, that subsequently is blocked by the objective aperture. This is *diffraction* contrast, due to coherent elastic scattering. The scattering is coherent because of the periodic arrangement of scattering centers in crystalline materials. In this case, the transmitted intensity distribution depends on the superposition of the individual scattered amplitudes.

Diffraction contrast is often observed in the vicinity of defects in the lattice. The origins of this contrast are illustrated in Figure 5. Figure 5a shows a thin sample with atomic planes that are close to a Bragg diffraction orientation, but are actually unaligned with respect to an electron beam propagating down the optic axis of the microscope. On the lefthand side of the diagram, the atomic planes are undistorted, as they would be in a perfect crystal. On the righthand side of the diagram, the sample contains an edge dislocation in the middle of the sample thickness. The dislocation lies normal to the page so that it appears in this diagram in cross section. Near the core of the dislocation, the atomic planes are distorted or bent to accommodate the strains associated with the atomic displacements at the dislocation core. See Figure 5a. The result of these local distortions is that some planes near the core adopt a Bragg orientation with respect to the incident beam. This is shown schematically in Figure 5b, where the incident and transmitted electron ray paths are shown for the same sample region. The undistorted crystal on the lefthand side, which is not in

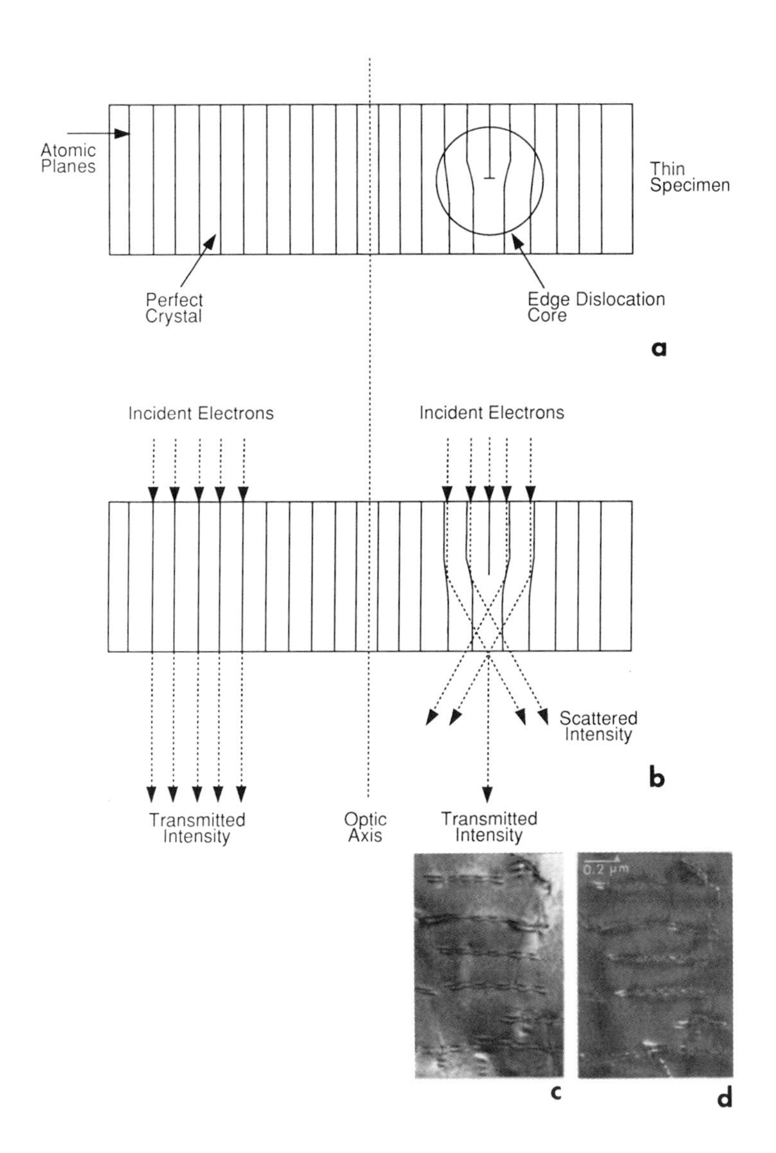

Figure 5 **(a) Schematic of a thin sample with atomic planes that are close to a Bragg diffraction orientation, but which are unaligned with respect to an electron beam propagating down the optic axis of the microscope. The sample contains an edge dislocation in the middle of the sample thickness on the right-hand side of the diagram. (b) Incident and transmitted electron ray paths for electron scattering from the same sample region in (a). (c) Bright-field image of dislocations in shock-deformed Ni_3Al. (d) Dark-field image from the same region as in (c). (c, d courtesy of H. W. Sizek, Los Alamos National Laboratory)**

Bragg alignment, is shown as simply transmitting a similar magnitude of undeflected intensity. The region containing the dislocation, on the other hand, is shown with a deficiency of undeflected, transmitted intensity, because considerable Bragg diffraction occurs near the core of the dislocation. Diffraction is represented by the scattered rays shown in the diagram, which are subsequently blocked by the objective aperture in the bright-field mode. In a bright-field image from this sample, the region containing the edge dislocation would appear dark, surrounded by bright intensity from the neighboring, undistorted crystalline material. In this case, contrast in the image would appear as a dark line across the bright-field image, since this dislocation line lies parallel to the plane of the sample.

This situation is illustrated by the bright-field image in Figure 5c, where a set of dislocations in shock-deformed Ni_3Al is imaged. Each dislocation appears as a dark line on a bright background (each line appears paired in this image because these are dissociated superlattice dislocations). By comparison, Figure 5d is a dark-field image from the same region, which was obtained by placing the objective aperture around a diffracted beam in the SAD pattern instead of the transmitted beam. The same dislocations that were imaged in the bright-field mode in Figure 5c now appear as bright lines on a dark background. The dark background results because the undistorted crystal lattice is not well-aligned for diffraction, so little scattered intensity arises from these regions, to contribute brightness to this dark-field image. But the dislocations appear bright since diffracted intensity from the dislocation cores (that was lost in the bright-field mode) is now captured in the dark-field mode. This is typical of image contrast in the dark-field mode; consequently the name dark-field (i.e., bright objects on a dark background) is applied to this imaging technique. Dark-field microscopy is a powerful technique, but many associated subtleties complicate its practice. A most noteworthy example is the technique of weak-beam dark-field imaging.[5]

The last example of imaging techniques in TEM is high-resolution transmission electron microscopy. High-resolution TEM is made possible by using a large-diameter objective diaphragm that admits not only the transmitted beam, but at least one diffracted beam as well. All of the beams passed by the objective aperture are then made to recombine in the image-forming process, in such a way that their amplitudes and phases are preserved. When viewed at high-magnification, it is possible to see contrast in the image in the form of periodic fringes. These fringes represent direct resolution of the Bragg diffracting planes; the contrast is referred to as *phase* contrast. The fringes that are visible in the high-resolution image originate from those planes that are oriented as Bragg reflecting planes and that possess interplanar spacings greater than the lateral spatial resolution limits of the instrument. The principle here is the same as in the Abbé theory for scattering from gratings in light optics.[7] An example of an HRTEM image is shown in Figure 6. This image is of an epitaxial thin film of $Y_1Ba_2Cu_3O_{7-x}$ grown on $LaAlO_3$ (shown in cross section).

The HRTEM technique has become popular in recent years due to the more common availability of high-voltage TEMs with spatial resolutions in excess of

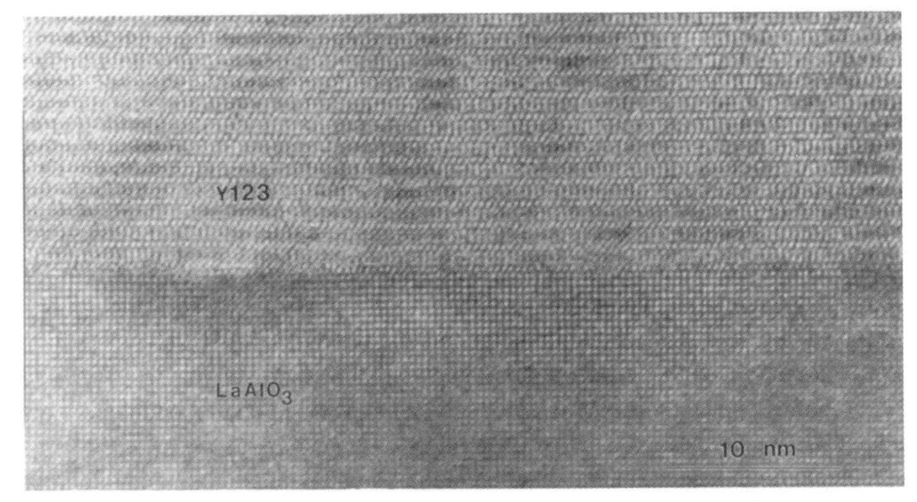

Figure 6 High-resolution transmission electron microscopy image of an epitaxial thin film of $Y_1Ba_2Cu_3O_{7-x}$ grown on $LaAlO_3$, shown in cross section. (Courtesy of T. E. Mitchell, Los Alamos National Laboratory)

0.2 nm. Image simulation techniques are necessary to determine the atomic structure of defects imaged by HRTEM.

Specimen Preparation

Probably the most difficult, yet at the same time, most important aspect of the TEM technique is the preparation of high-quality thin foils for observation. This is an old, ever-expanding, complicated, and intricate field of both science and art. There is no simple way to treat this subject briefly. We will merely mention its importance and list some references for further details. It is important to realize (managers, take notice) that the most labor intensive aspect of TEM is the preparation of a useful sample.

In the early days of TEM, sample preparation was divided into two categories, one for thin films and one for bulk materials. Thin-films, particularly metal layers, were often deposited on substrates and later removed by some sort of technique involving dissolution of the substrate. Bulk materials were cut and polished into thin slabs, which were then either electropolished (metals) or ion-milled (ceramics). The latter technique uses a focused ion beam (typically Ar^+) of high-energy, which sputters the surface of the thinned slab. These techniques produce so-called plan-view thin foils.

Today, there is great interest in a complementary specimen geometry for observation, that of the cross section. Cross sections usually are made of layered materials. The specimens are prepared so as to be viewed along the plane of the layers. The techniques for producing high-quality cross sections are difficult, but rather well established. For additional information on sample preparation, consult Thompson-Russell and Edington[8] and the proceedings of two symposia on TEM sample preparation sponsored by the Materials Research Society.[9, 10]

Conclusions

TEM is an established technique for examining the crystal structure and the microstructure of materials. It is used to study all varieties of solid materials: metals, ceramics, semiconductors, polymers, and composites. With the common availability of high-voltage TEM instruments today, a growing emphasis is being placed on atomic resolution imaging. Future trends include the use of ultrahigh vacuum TEM instruments for surface studies and computerized data acquisition for quantitative image analysis.

Related Articles in the Encyclopedia

STEM, SEM, EDS, EELS

References

1 M. von Heimendahl. *Electron Microscopy of Materials: An Introduction.* Materials Science and Technology Series (A. S. Nowick, ed.) Academic, New York, 1980, Chapter 1. This is an excellent introductory guide to the principles of TEM.

2 L. Reimer. *Transmission Electron Microscopy: Physics of Image Formation and Microanalysis.* Springer-Verlag, Berlin, 1984. This is an advanced but comprehensive source on TEM. Reimer also authored a companion volume on SEM.

3 P. B. Hirsch, A. Howie, R. B. Nicholson, D. W. Pashley, and M. J. Whelan. *Electron Microscopy of Thin Crystals.* Butterworth, Washington, 1965, Chapter 4. This sometimes incomprehensible volume is *the* classic textbook in the field of TEM.

4 R. H. Geiss. Introductory Electron Optics. In: *Introduction to Analytical Electron Microscopy*. (J. J. Hren, J. L. Goldstein, and D. C. Joy, eds.) Plenum, New York, 1979, pp. 43–82.

5 J. W. Edington. *Practical Electron Microscopy in Materials.* van Nostrand Reinhold, New York, 1976. This is an excellent general reference and laboratory handbook for the TEM user.

6 B.D. Cullity. *Elements of X-Ray Diffraction.* Addison-Wesley, Reading, 1956. This is a good general reference concerning X-ray diffraction techniques.

7 E. Hecht and A. Zajac. *Optics.* Addison-Wesley, Reading, 1974, Chapter 14. This entire book is an invaluable reference on the principles of optics.

8 K. C. Thompson-Russell and J. W. Edington. *Electron Microscope Specimen Preparation Techniques in Materials Science. Monographs in Practical Electron Microscopy, No. 5.* Philips Technical Library, Eindhoven & Delaware, 1977.

9 *Specimen Preparation for Transmission Electron Microscopy I.* (J. C. Bravman, R. M. Anderson, and M. L. McDonald, eds.) Volume 115 in MRS symposium proceedings series, 1988.

10 *Specimen Preparation for Transmission Electron Microscopy II.* (R. M. Anderson, ed.) Volume 199 in MRS symposium proceedings series, 1990.

3

ELECTRON BEAM INSTRUMENTS

3.0 INTRODUCTION

Whereas the previous chapter emphasizes imaging using microscopes, this chapter is concerned with analysis (compositional in particular) using fine electron probes, which provide fine spatial resolution. The beams used are either those in the SEM and TEM, discussed in the previous chapter (in which case the analytical techniques described here are used as adjuncts to the imaging capabilities of those instruments), or they involve electron beam columns specially constructed for an analytical mode. The Scanning Transmission Electron Microscope, STEM, and the Electron Microprobe, used for Electron Probe Microanalysis, EPMA, are two examples of the latter that are discussed in this chapter. A third example would be the Auger electron microprobe, used for scanning Auger Electron Spectroscopy, AES, but we choose to discuss this technique in Chapter 5 along with the other major electron spectroscopy methods, since all of them are primarily used to study true surface phenomena (monolayers), which is not generally the case for the techniques in this chapter.

The incoming electron beam interacts with the sample to produce a number of signals that are subsequently detectable and useful for analysis. They are X-ray emission, which can be detected either by Energy Dispersive Spectroscopy, EDS, or by Wavelength Dispersive Spectroscopy, WDS; visible or UV emission, which is known as Cathodoluminescence, CL; and Auger Electron Emission, which is the basis of Auger Electron Spectroscopy discussed in Chapter 5. Finally, the incoming beam itself can lose discrete amounts of energy by inelastic collision, the values of which are determined by an electron energy analyzer. This is the basis of Electron Energy Loss Spectroscopy, EELS. Which of these classes of processes is more dominant, or more useful, depends on a number of factors, including the energy of the electron beam, the nature of the material (high or low Z), and the type of information sought (elemental composition, chemical composition, ultimate in spatial resolution, information limited to the surface, or information throughout the bulk by transmission measurement). A complete perspective for this can be obtained by comparing the articles in this section, plus the AES article, since they interrelate quite strongly. Some brief guidelines are given here.

All the methods, with the exception of CL, provide elemental composition. The most widely used is X-ray emission. If EDS is used the package can be quite inexpensive ($25,000 and up), and can be routinely fitted to SEMs, TEMs, and STEMs. In addition EDS is one of the two detection schemes in EPMA (the other is WDS). Its great advantage is its ability to routinely provide rapid multi element analysis for $Z>11$, with a detection limit of about 200 ppm for elements with resolved peaks. Its major disadvantages are very poor energy resolution, so that peaks are often overlapped; a detector problem that adversely affects detection limits; and the fact that the detector must remain cooled by liquid nitrogen or risk being destroyed. All these shortcomings of the EDS detector can be overcome by using the other detection scheme, WDS. The disadvantages of this scheme are that it is more expensive and cumbersome experimentally and does not have simultaneous multi element detection capability. For these reasons it is not so much used in conjunction with an SEM, TEM, or STEM, but is the heart of the *electron microprobe*, which is designed to combine WDS and EDS in the most effective analytical way.

The spatial resolution of X-ray emission does not usually depend on the diameter of the electron beam, since small beams spread out into a roughly pear-shaped "interaction volume" below the sample surface, and it is from this region that the X-ray signal is generated. This volume varies from a fraction of a micron to several microns depending on the electron beam energy (lower energy, smaller volume), and the material (lower Z, smaller volume). The exceptions are when the beam width is larger than a few microns, in which case it starts to dominate the resolution, or when the sample is very thin (hundreds of angstroms or less) so that the beam passes through before it can spread much. In this case the spatial resolution can be greatly improved toward that of the beam size itself. This is the case for thin samples in a TEM or STEM.

Cathodoluminescence, CL, involves emission in the UV and visible region and as such is not element specific, since the valence/conduction band electrons are involved in the process. It is therefore sensitive to electronic structure effects and is sensitive to defects, dopants, etc., in electronic materials. Its major use is to map out such regions spatially, using a photomultiplier to detect all emitted light without spectral resolution in an SEM or STEM. Spatial resolution and depth resolution capabilities are, in principle, similar to X-ray emission, since the UV/visible emission comes from roughly the same interaction region. In practice lower electron beam energies are sometimes used in CL to improve spatial resolution.

EELS is used in a transmission mode in conjunction with TEMs and STEMs. Samples must be very thin (hundreds of angstroms) and beam energies must be high (100 keV and up) to prevent the single scattered EELS signal from being swamped by a multiple scattering background. A direct consequence of this requirement is that the spatial resolution of transmission EELS is not much worse than the beam size, since a 100-kV electron passing through a sample and scattered only once does not deviate much in direction. Thus, in a STEM with a 2-Å beam size the spatial resolution of EELS for a sample 100 Å thick might be only 3–4 Å! Although the main use of transmission EELS is to provide elemental composition like EDS/WDS it can also provide much information about chemical states and about electronic structure from the line shapes and exact positions of the energy loss peaks. EELS is also used in a reflection mode (REELS) in Auger spectrometers for surface analysis (see Chapter 5).

The STEM instrument itself can produce highly focused high-intensity beams down to 2 Å if a field-emission source is used. Such an instrument provides a higher spatial resolution compositional analysis than any other widely used technique, but to capitalize on this requires very thin samples, as stated above. EELS and EDS are the two composition techniques usually found on a STEM, but CL, and even AES are sometimes incorporated. In addition simultaneous crystallographic information can be provided by diffraction, as in the TEM, but with 100 times better spatial resolution. The combination of diffraction techniques and analysis techniques in a TEM or STEM is termed Analytical Electron Microscopy, AEM. A well-equipped analytical TEM or STEM costs well over $1,000,000.

Electron Probe Microanalysis, EPMA, as performed in an *electron microprobe* combines EDS and WDX to give quantitative compositional analysis in the reflection mode from solid surfaces together with the morphological imaging of SEM. The spatial resolution is restricted by the interaction volume below the surface, varying from about 0.2 μm to 5 μm. Flat samples are needed for the best quantitative accuracy. Compositional mapping over a 100 × 100 micron area can be done in 15 minutes for major components ($Z > 11$), several hours for minor components, and about 10 hours for trace elements.

3.1 EDS

Energy-Dispersive X-Ray Spectroscopy

ROY H. GEISS

Contents

Introduction

With modern detectors and electronics most Energy-Dispersive X-Ray Spectroscopy (EDS) systems can detect X rays from all the elements in the periodic table above beryllium, $Z = 4$, if present in sufficient quantity. The minimum detection limit (MDL) for elements with atomic numbers greater than $Z = 11$ is as low as 0.02% wt., if the peaks are isolated and the spectrum has a total of at least 2.5×10^5 counts. In practice, however, with EDS on an electron microscope, the MDL is about 0.1% wt. because of a high background count and broad peaks. Under conditions in which the peaks are severely overlapped, the MDL may be only 1–2% wt. For elements with $Z < 10$, the MDL is usually around 1–2% wt. under the best conditions, especially in electron-beam instruments.

The accuracy of quantitative analysis has been reported to be better than 2% relative for major concentrations, using well-polished standards having a composition similar to the sample. A more conservative figure of 4–5% relative should be expected for general analysis using pure element standards. For analysis without using standards the accuracy will usually be much worse. The analysis of elements with concentrations less than 5% wt. will typically yield relative accuracies nearer 10%, even with standards. For samples with rough surfaces, such as fracture samples or small particles, the relative accuracy may be as bad as 50%.

Most applications of EDS are in electron column instruments like the scanning electron microscope (SEM), the electron probe microanalyzer (EPMA), and transmission electron microscopes (TEM). TEMs are further classified as conventional transmission (CTEM) or scanning transmission (STEM) instruments, depending on whether scanning is the primary imaging mode. A CTEM equipped with a scanning attachment and an EDS instrument is an Analytical Electron Microscope (AEM). X-ray spectrometers, with X-ray tube generators as sources and Si (Li) detectors have been used for both X-Ray Fluorescence Spectroscopy (XRF) and X-Ray Diffraction (XRD). Portable EDS systems also have been constructed using X-ray tube generators or radioactive sources.

A spectrum can be obtained from almost any sample, as long as it can be put on the specimen stage of the microscope. The choice of accelerating voltage should be determined by the type of sample one is studying, since the X-ray generation volume depends on the electron range in the material. In the study of thin films it is usually desirable to minimize the electron range and use an accelerating voltage E_0 just greater than E_c, the critical ionization voltage for the X-ray line of interest. For bulk samples it is more important to maximize X-ray production regardless of the electron range and, as will be discussed later, the accelerating voltage should be ideally 2–2.5 × E_C. For example, consider the K-shell ionization of copper, for which E_C = 8.98 keV. To analyze a film only a few nm thick on a Si substrate, using the copper Kα, the accelerating voltage should be set near 10 keV. To analyze a bulk sample, more than a few μm thick, an accelerating voltage of 20–25 keV should be used.

With an MDL of 100–200 ppm for most elements, an EDS system is capable of detecting less than a monolayer of metal film on a substrate using Kα lines at moderate accelerating voltages of 5–15 keV. Since many SEMs now have field emission electron guns providing high brightness probes at voltages of 2 keV and less, EDS analysis of even thinner films should be possible, at least in principle, since the electron range and hence, the generated X-ray volume will be very small. In this case, however, since all the X-ray lines will be low energy and in a small energy region, there may be many overlapped peaks that will have to be deconvoluted before quantitative analysis can be attempted. This deconvolution can be tricky, however, since the shape of the background in this energy range is difficult to model. In addition, the shape of the peaks in the low-energy region is often not Gaussian and the peak positions, especially for the K lines from low-Z elements, are often shifted.

Energy-dispersive X-ray spectroscopy has been used for quality control and test analysis in many industries including: computers, semiconductors, metals, cement, paper, and polymers. EDS has been used in medicine in the analysis of blood, tissues, bones, and organs; in pollution control, for asbestos identification; in field studies including ore prospecting, archeology, and oceanography; for identification and forgery detection in the fine arts; and for forensic analysis in law enforcement. With a radioactive source, an EDS system is easily portable and can be used in the field more easily than most other spectroscopy techniques.

The main advantages of EDS are its speed of data collection; the detector's efficiency (both analytical and geometrical); the ease of use; its portability; and the relative ease of interfacing to existing equipment.

The disadvantages are: poor energy resolution of the peaks, (a typical EDS peak is about 100× the natural peak width, limited by the statistics of electron–hole pair production and electronic noise, which often leads to severe peak overlaps); a relatively low peak-to-background ratio in electron-beam instruments due to the high background coming from bremsstrahlung radiation emitted by electrons suffering deceleration on scattering by atoms; and a limit on the input signal rate because of pulse processing requirements.

Principles of X-Ray Production

X-rays are produced as a result of the ionization of an atom by high-energy radiation wherein an inner shell electron is removed. To return the ionized atom to its ground state, an electron from a higher energy outer shell fills the vacant inner shell and, in the process, releases an amount of energy equal to the potential energy difference between the two shells. This excess energy, which is unique for every atomic transition, will be emitted by the atom either as an X-ray photon or will be self-absorbed and emitted as an Auger electron. For example, if the K shell is ionized and the ejected K-shell electron is replaced by an electron from the L_3 shell, the emitted X ray is labeled a characteristic $K\alpha_1$ X ray. (See Figure 2 in the article on electron probe X-ray microanalysis). The hole that exists in the L shell will be filled by an electron from a higher shell, say the M shell, if one exists. This M–L transition may result in the emission of another X ray, labeled in turn according to one of the many M–L transitions possible. The cascade of transitions will continue until the last shell is reached. Thus, in an atom with many shells, many emissions can result from a single primary ionization.

Instrumentation

The heart of the energy-dispersive spectrometer is a diode made from a silicon crystal with lithium atoms diffused, or *drifted*, from one end into the matrix. The lithium atoms are used to compensate the relatively low concentration of grown-in impurity atoms by neutralizing them. In the diffusion process, the central core of the silicon will become intrinsic, but the end away from the lithium will remain p-type and the lithium end will be n-type. The result is a p-i-n diode. (Both lithium-

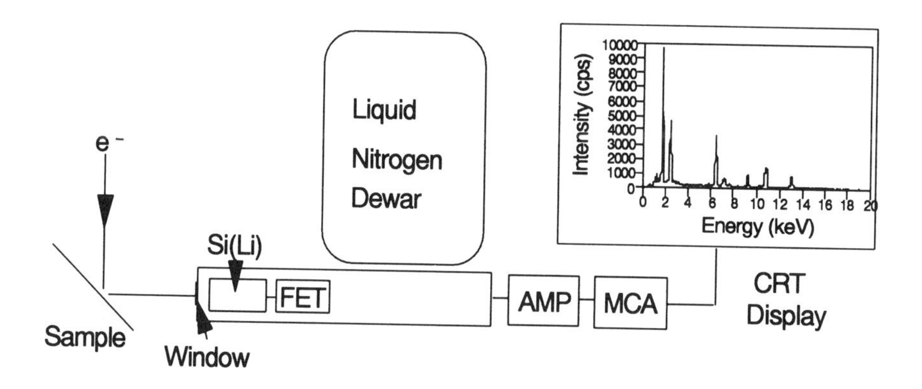

Figure 1 **Schematic of an EDS system on an electron column. The incident electron interacts with the specimen with the emission of X rays. These X rays pass through the window protecting the Si (Li) and are absorbed by the detector crystal. The X-ray energy is transferred to the Si (Li) and processed into a digital signal that is displayed as a histogram of number of photons versus energy.**

drifted and pure intrinsic germanium have been used as detectors, but much less frequently. These will be discussed later). A reverse bias electrical field of 100–1000 volts is applied to thin layers of gold evaporated onto the front and back surfaces of the diode.

When an X-ray photon enters the intrinsic region of the detector through the p-type end, there is a high probability that it will ionize a silicon atom through the photoelectric effect. This results in an X ray or an Auger electron, which in turn produces a number of electron–hole pairs in the Si (Li): one pair per 3.8 eV of energy. For example, a 6.4-keV X ray absorbed by the silicon atoms will produce about 1684 electron–hole pairs or a charge of about 2.7×10^{-13} Coulombs. Both charge carriers move freely through the lattice and are drawn to the detector contacts under the action of the applied bias field to produce a signal at the gate of a specially designed field effect transistor mounted directly behind the detector crystal. The transistor forms the input stage of a low-noise charge-sensitive preamplifier located on the detector housing. The output from the preamplifier is fed to the main amplifier, where the signal is finally amplified to a level that can be processed by the analog-to-digital converter (ADC) of the multichannel analyzer (MCA). See Figure 1. The height of the amplifier output pulse is proportional to the input preamplifier pulse, and hence is proportional to the X-ray energy.

For the amplifier pulse to be recognized in the ADC, it must exceed the lower level set by a discriminator, which is used to prevent noise pulses from jamming the converter. Once the pulse is accepted it is used to charge a capacitor that is discharged through a constant current source attached to an address clock typically

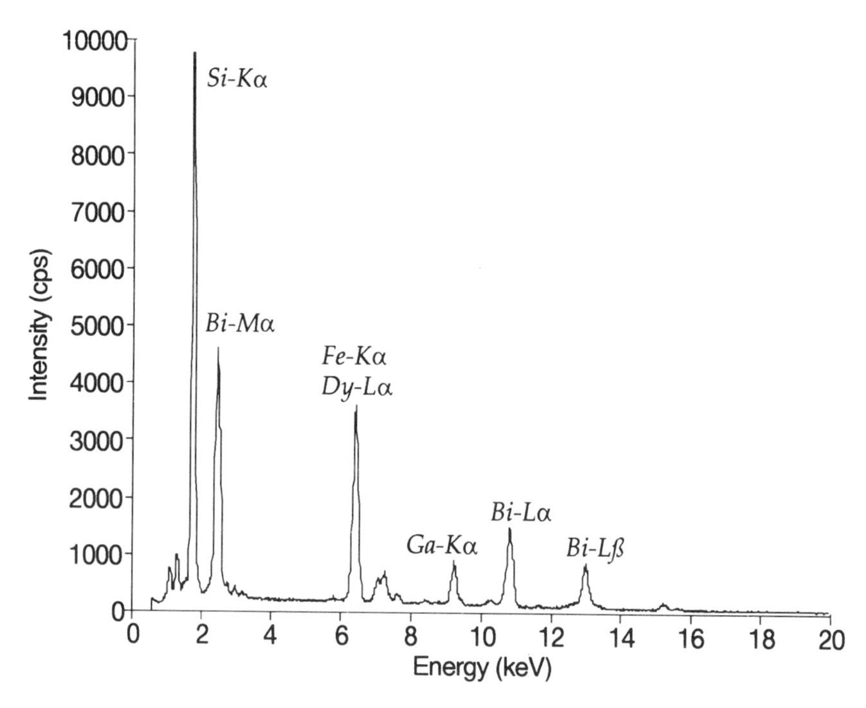

Figure 2 **Standard output of an EDS spectrum. The horizontal axis is the energy scale is and the vertical axis is the number of photons per energy interval. The X-ray identification, element and line, is indicated in the vicinity of the peaks.**

operating at 50 MHz. The time to discharge the capacitor to 0 V is proportional to the pulse amplitude, and hence to the X-ray energy. The 50-MHz clock produces a binary number in one of the 1024 channels typically used by the MCA in accordance with the time of the discharge, and increments the previously collected number in that channel by 1. By an energy calibration of the channels in the MCA, the collection of X-ray pulses may be displayed on a CRT as an energy histogram. A typical spectrum output from a thin alloy film using a TEM is shown in Figure 2.

To partition the incoming X rays into their proper energy channels, it is necessary to measure only single pulses. At high count rates, however, situations often arise where a second pulse reaches the main amplifier during the rise time of the preceding pulse. The two pulses may then combined into a single pulse whose energy is the combined energy of the two individual pulses. This process is known as pulse *pile-up* and the output, which is an artifact, is called a *sum peak.* Pile-up effects can be minimized with the use of electronic pulse rejection using a second pulse amplifier with a much faster response. Operationally, it is usually desirable to collect spectra with a dead time of less than 40%, or at an input count rate of about 5000 cps.

Detectors are maintained under vacuum at liquid nitrogen temperature to reduce electronic noise and to inhibit diffusion of the lithium when the bias voltage is applied. Most have a window covering the entrance to provide vacuum protection. The window in a standard detector is usually made from rolled beryllium foil, a few μm thick. Unfortunately, low-energy, or *soft*, X rays are strongly absorbed by the Be window, limiting the analysis range of these detectors to elements having atomic number $Z > 10$. To reduce the absorption, detectors are built either without windows at all or with windows made of new low-Z materials that can withstand atmospheric pressure. Detectors made with these low-Z windows show a marked increase in sensitivity to X rays from elements having $Z < 10$, compared to the Be window detectors.

Detector sensitivity drops off at the high-energy end (greater than 20 keV), where the Si (Li) crystal, which is typically 3–5 mm thick, becomes transparent to high-energy X rays. This may not be important in an SEM operated at 30 keV, since high-energy X rays will not be excited. But in a TEM operated at 100 keV or higher, the ability to detect hard X rays would be severely limited. In Figure 3, curves are plotted that show the detector's efficiency both at low energy, where the X-ray absorption of a standard Be window is compared to that of a low-Z window, and at high energy, where the high-energy drop-off is shown for the standard thickness Si (Li) crystals used today.

Spectrometers have been made using germanium crystals. The Ge crystals have been either lithium drifted, Ge (Li), or more recently, made from high-purity intrinsic germanium, HPGe. The HPGe crystal has the advantage that it can be allowed to warm to room temperature if it is not being used, saving the hassle of keeping the liquid nitrogen dewar filled. Germanium is much less transparent to high-energy electrons than silicon, because Ge ($Z = 32$) has a higher stopping power than Si ($Z = 14$), and should be able to detect very high energy X rays such as gold $K\alpha$, at 69 keV. This will be a distinct advantage when a TEM, operating at 100 keV and above and therefore capable of exciting high-energy X-ray lines, is used in the study of alloys containing elements that have severe peak overlap in the lower energy lines. However, germanium has some drawbacks; namely, a K-shell absorption edge at 11.1 keV, a complex L-shell absorption edge structure starting at 1.4 keV and a series of escape peaks in the range 2–12 keV. Thus, in the energy range 1–10 keV, the most frequently used range in EDS analysis, the detector is not as well behaved as Si (Li).

A major advantage of the energy-dispersive spectrometer is that it can be positioned very close to the sample and can present a large solid angle for the collection of emitted X rays. The solid angle in a typical EDS configuration is about 10 times greater than that of a WDS. With EDS, more X-ray photons will be collected per incident electron, so that either a smaller probe diameter or lower beam current can be used (which reduces specimen damage). Detectors are usually manufactured with an active area of either 10 mm^2 or 30 mm^2 diameter.

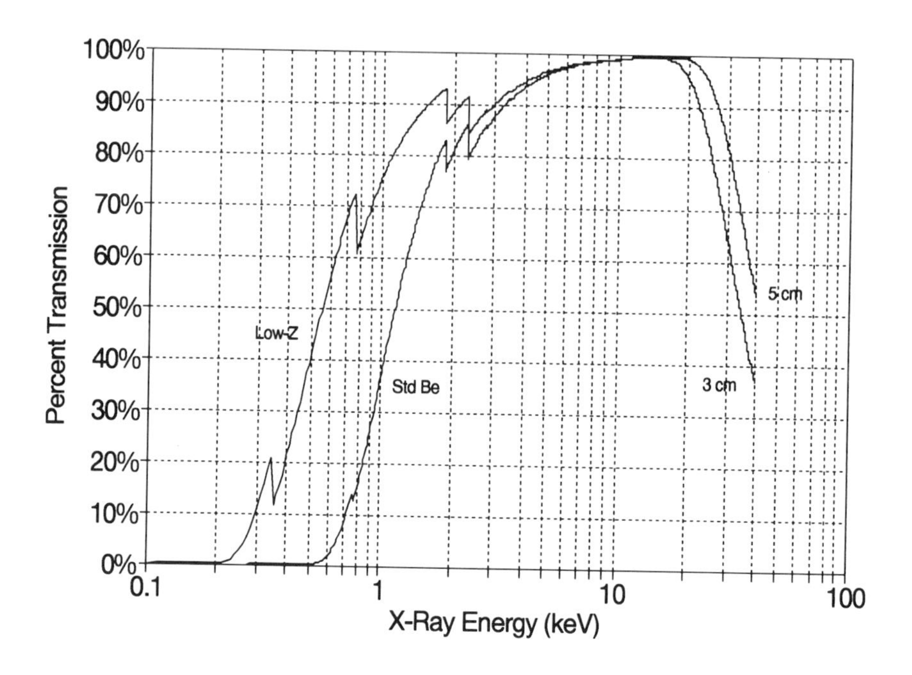

Figure 3 **Curves showing the absorption of the window materials at low energy for a standard Be window and a low-*Z* window. The high-energy region shows the transmission of X rays through 3 mm and 5 mm thick Si (Li) crystals. Most detectors can be represented by a combination of one of the low-energy curves and one of the high-energy curves.**

Since energy-dispersive spectrometers consist mostly of electronic components, they are easy to interface to most instruments. The only limitations are the need for a large liquid nitrogen cryostat to cool the spectrometer and high vacuum for the windowless detectors. Some Si (Li) detectors use mechanical cooling, called *Peltier* cooling, instead of the liquid nitrogen; this eliminates the large cryostat and the nagging requirement to keep it full. Unfortunately, the temperature reached with Peltier refrigerators is not as low as that obtained with liquid nitrogen, and the detector resolution suffers.

Most EDS systems are controlled by minicomputers or microcomputers and are easy to use for the basic operations of spectrum collection and peak identification, even for the computer illiterate. However, the use of advanced analysis techniques, including deconvolution of overlapped peaks, background subtraction, and quantitative analysis will require some extra training, which usually is provided at installation or available at special schools.

Resolution, Peak Overlap, and Minimum Detection

The energy resolution of a solid state detector is defined as the full width at half maximum (FWHM) of the peak obtained at energy E_0 when a monoenergetic beam of X rays having energy E_0 is incident on the detector. Ideally this width would be very small; but due to the statistical nature in the collection process of the electron–hole pairs, there will be some fluctuation in the measured energy of the X rays. For a modern energy-dispersive spectrometer, the FWHM of the peaks ranges from about 70 eV for the C Kα at 0.282 keV to about 150 eV for the Cu Kα at 8.04 keV. Improvements in detector resolution have focused on reducing the electronic noise contribution by improving the FET design and by using nonoptical charge restoration. Noise resolution as low as 40 eV has been achieved, giving a detector resolution of 128 eV FWHM at 5.89 keV.

Peak overlap, which follows from the poor resolution of the detector, is one of the major problems in EDS. For example, severe peak overlap occurs when two peaks of about the same amplitude are separated by less than half of the FWHM of the peaks. In this instance they will merge together and appear as one Gaussian peak. Unfortunately, in X-ray spectroscopy the peaks of the characteristic lines are often closer together then the resolution of the best EDS systems, and significant peak overlap follows. This presents problems not only with the identification of the individual peaks, but also with the determination of the amplitudes of the peaks for quantification. Improvements in detector resolution will obviously provide some relief to the problem, but even for the most optimistic resolution specification, there will still be many instances of peak overlap. An example of this is shown for the $BaTiO_3$ system in Figure 4, where the spectrum from a standard EDS is compared to that obtainable with a standard WDS using a LiF crystal; in the latter case the FWHM of the peaks is only a few eV. Peak overlap presents a particularly difficult problem in the analysis of soft X rays, especially when the K-line X rays from elements with $Z < 10$, the L-line X rays from transition metals or the M-line X rays from the rare earths are present. As previously mentioned, the resolution of most EDS detectors is about 60–90 eV in this region, which ranges from 300 to 1000 eV. To compound the problem, the peak shape is usually not purely Gaussian in this region, making computer deconvolution even more difficult. Here again, the resolution of a wavelength spectrometer will be only 5–10 eV, so that with the proper choice of analyzing crystal, peak overlap will not be a problem.

The minimum detection limit, MDL, of an isolated peak on a uniform background is proportional to the square root of the FWHM. So a 20% reduction in spectrometer resolution will produce about a 10% improvement in MDL. If there is peak overlap, however, then it can be shown that a 20% improvement in resolution can reduce the interference between overlapping peaks by a factor of 3, which gives about a 50% improvement in MDL.

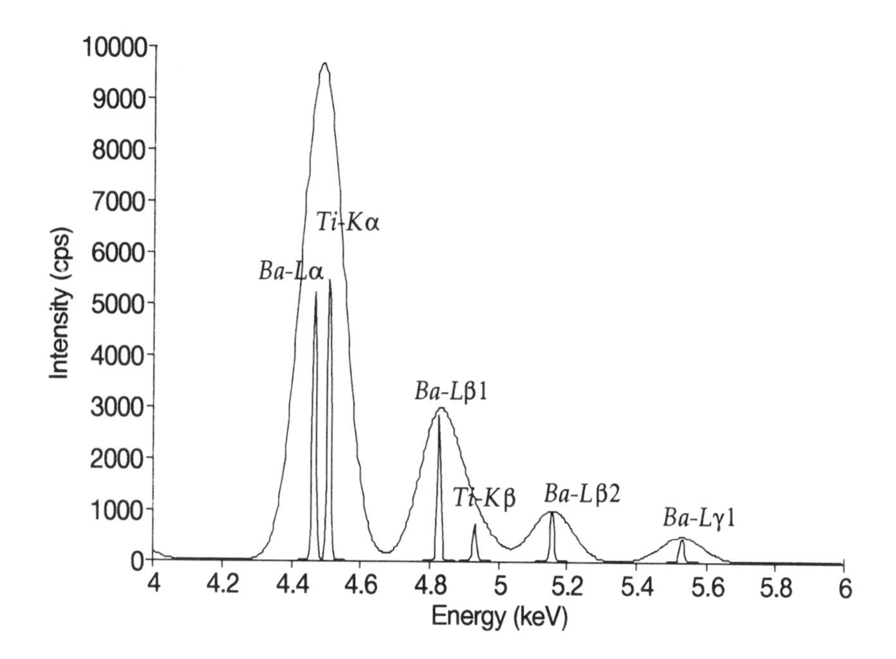

Figure 4 **Superposed EDS and WDS spectra from $BaTiO_3$. The EDS spectrum was obtained with a detector having 135-eV resolution, and shows the strongly overlapped Ba $L\alpha_1$–Ti $K\alpha$ and Ba $L\beta_1$–Ti $K\beta$ peaks. The WDS spectrum from the same material shows the peaks to be completely resolved.**

The MDL, or trace element sensitivity, is inversely proportional to P^2/B, the ratio of the peak counting rate and the background counting rate for a pure element. If data is collected on both EDS and WDS instruments using the low probe current appropriate to EDS (a few nA) the MDL with EDS will be 3–5× better than that obtained with WDS. But if WDS is used at its normally higher probe currents (20–100 nA) the MDL obtained using WDS will be about 10× better! The major obstacle to a lower MDL in EDS, aside from the peak overlap problem, is a much higher background (due to bremsstrahlung radiation) and therefore a much lower P^2/B ratio relative to WDS. In cases where the peaks are not overlapping it should be possible to detect and analyze 0.1–0.2 nm (about one monolayer) of a metal film on a conducting substrate using EDS.

Sample Requirements

Almost any size solid sample can be studied in the SEM or EPMA. The only limitation is the size of the specimen chamber, which is usually at least 10 cm in diameter. In a standard SEM, however, the *x–y* translation is usually limited to 25 mm. In SEMs with an air lock for sample exchange, the maximum sample size may be only 3–5 cm in diameter. A highly polished surface is required for accurate quantitative analysis, since surface roughness will cause undue absorption of the generated X-ray signal, which is difficult to account for in the quantification procedure. Powder samples can be compacted or fixed in epoxy and analyzed in the SEM or EPMA. However, with the high vacuum required in the electron column, liquids are not easily studied. In the TEM, the sample must be transparent to the incident electrons and therefore may be only a few hundred nm thick. The maximum sample diameter is usually 3 mm. In electron column instruments, nonconducting samples usually present a problem due to excess charge build-up. To provide a conducting path to ground they are coated with a thin conducting film of C or Al, which must be accounted for in the final analysis if the absorption by the coating material is significant. For EDS combined with an X-ray tube generator or radioactive source in air, most of these restrictions are relaxed and almost any sample can be analyzed. Coating is not required for nonconducting specimens. As mentioned before, samples with rough surfaces, such as fracture samples, may give a spectrum, but the relative accuracy of the quantitative analysis will not be very good.

Operational Considerations in Electron Microscopes

For inner shell ionization to take place, the energy of the incident radiation must be greater than the ionization energy of the particular atomic shell in question. For electrons, the ionization cross section σ is calculated using a function of the form $(1/U)\ln U$, where $U = E_0/E_c$ is the ratio of the energy of the incident electron E_0 to the critical ionization energy E_c. Qualitative evaluation of this function shows that σ reaches a maximum around $U = 2.5$, but is almost constant for $U > 2$. To a first approximation, therefore, the only requirement is that the accelerating voltage of the electron gun be chosen such that $U > 2$.

In the analysis of trace elements or thin films on substrate using electrons, however, one finds that the MDL, may be increased by choosing E_0 such that U is just greater than 1. The reason for this is that the k factor, which is the ratio of the intensity from the sample to that from the standard, increases as U approaches 1 for thin films. Thus, by maximizing the k factor, the sensitivity is increased. For bulk sample analysis, however, the k factor will usually be a maximum at $U \sim 2.5$.

Another consideration in the determination of the optimum E_0 is the depth of X-ray production in bulk samples, especially if one component strongly absorbs the radiation emitted by another. This is often the case when there is a low-Z element in a high-Z matrix, e.g., C in Fe. Here X rays from carbon generated deep within the sample will be highly absorbed by the Fe and will not exit the sample to be detected. The usual result will be an erroneously low value for the carbon concentration. In these situations the best choice for E_0 will be closer to E_c with $U \approx 1$ rather than a much higher value with $U = 2.5$.

X-ray line E_C (keV)	C Ka 0.283	Si Ka 1.838	Cr Ka 5.988	Cu Ka 8.980	Au La 11.919	K–O Electron range
Matrix element	Range (μm) for E_0 = 20 keV					
Si	4.737	4.653	4.108	3.496	2.743	4.741
Cr	1.760	1.729	1.526	1.299	1.019	1.761
Ag	1.376	1.351	1.193	1.015	0.797	1.377
Au	0.861	0.846	0.747	0.636	0.499	0.862
	Range (μm) for E_0 = 10 keV					
Si	1.486	1.402	0.857	0.245	*	1.490
Cr	0.552	0.521	0.318	0.091	*	0.554
Ag	0.432	0.407	0.249	0.071	*	0.433
Au	0.27	0.255	0.156	0.045	*	0.271

* = no X rays generated.

Table 1 **Range of X-ray generation calculated by the Kanaya-Okayama (K–O) equation for selected X-ray lines in low- to high-density materials for two incident electron energies, E_0. For comparison, the range of electrons with $E_c = 0$, is given in the last column. (Here E_0 is the incident electron energy, E_c is the critical ionization energy, A is the atomic weight, Z is the atomic number, and ρ is the density.) The K–O X-ray range is given by the equation: $R = 0.0276\, A(E_0^{1.67} - E_C^{1.67}) / (Z^{0.889}\rho)$ μm.**

Range and Lateral Resolution of X-Ray Generation

The range over which characteristic X rays are generated depends both on the properties of the sample and the choice of accelerating voltage. X-ray generation ranges calculated for low-, medium- and high-density materials under a few electron-accelerating voltages are given in Table 1 to illustrate this point. For comparison, the electron range is included in the table. As can be seen, the X-ray generation range is always less than the electron range because characteristic X rays are generated only when the electron has energy greater than E_C. But as can be seen in the table, the difference between the X-ray range and the electron range is very significant only when E_C is nearly equal to E_0. Accordingly, the X-ray generation volume is smaller than the electron interaction volume and the volume from which the X rays are measured is even smaller because of X-ray absorption. (Another discussion of this topic can be found in the article on EPMA.)

It also should be recognized that the density of X-ray production is not constant throughout the X-ray generation volume, but is related to the number and length of electron trajectories per unit volume and the average overvoltage along each trajectory. The distribution of the generated X rays as a function of the mass thickness, or depth, in the sample is given by a mathematical expression that has the shape of a Gaussian curve offset at the surface of the sample. This function is often called the $\phi(\rho z)$ function.

The lateral resolution of X-ray analysis in an SEM or EPMA is not generally related to the incident probe size. From the values for the X-ray generation ranges given in Table 1 one can see that the diameter of the X-ray generation volume in thick samples is much greater than a few tens of nm, which would be the nominal diameter of a medium-resolution probe in an SEM. On the other hand, with a thin-foil sample in a TEM, the probe size will be very important in the determining the lateral resolution, since most of the incident electrons pass through the foil with only a few collisions and therefore are not be deflected far from their incident direction.

Modes of Analysis

There are three modes of analysis commonly used: spectrum acquisition; spatial distribution, or dot, mapping of the elements; and elemental line scans.

In the spectrum acquisition mode the probe is either fixed in the spot mode or raster scanned over a small area at high magnification and a complete spectrum acquired. A typical spectrum is shown in Figure 2.

In an electron microscope or microprobe with scanning capability, data from more than 15 elements can be collected simultaneously and used to generate X-ray dot maps displaying the spatial distribution of the elements in the sample. In this technique the brightness of the CRT is modulated as the beam is raster scanned to reflect the X-ray output of the element of interest, which usually is designated by defining a region of interest (ROI) around the peaks in the EDS spectrum at that point. Some analyzers provide the capability to remove the background from the signal before plotting. Multielement maps can be made with a color CRT, using a different color chosen for each element. These maps can be arranged in an overlay with the colors combined to show the presence of various compounds. A long data collection time, sometimes hours, is often required to collect enough data points for a high-resolution dot map.

With the same scanning capability, it is much faster and often more useful to simply scan one line on a sample. The data is again output to a color CRT, but it is presented as the modulation of the y-amplitude, which is determined by the intensity of the X-ray signal production from the ROI of the element of interest. As the probe scans along the line, the CRT plots a graph of the elemental counting rates versus distance. Here again, it is usually possible to plot the data from many elements simultaneously. This mode of display will usually show minor concentration differences much better than the brightness modulation technique used in mapping.

Quantification

The ultimate goal in using an EDS analytical system is to be able to measure the concentrations of all the elements in the sample. To this end, a series of measurements are made in which the peak intensity from each element in the sample is compared to the peak intensity obtained from a reference standard using the same operating conditions. The ratio of these two intensities is the k factor mentioned previously. To convert the measured k factors to the composition, it is usually necessary to make corrections that account for deviations from a linear relationship between the k factor and the composition. These include an atomic number correction accounting for the fraction of electrons backscattered by the sample and the volume of X-ray generation (the Z factor); a correction for the absorption of the generated X rays in the sample (the A factor); and a correction for secondary X-ray fluorescence in the sample (the F factor). Various procedures have been developed to do this; the most common is known as the ZAF correction, with $C(\%) = kZAF$.

In another approach, which was previously mentioned, the mass thickness, or depth distribution of characteristic X-ray generation and the subsequent absorption are calculated using models developed from experimental data into a $\phi(\rho z)$ function. Secondary fluorescence is corrected using the same F factors as in ZAF. The $\phi(\rho z)$ formulation is very flexible and allows for multiple boundary conditions to be included easily. It has been used successfully in the study of thin films on substrates and for multilayer thin films.

Conclusions

EDS is an extremely powerful analytical technique of special value in conjunction with electron column instruments. In a few seconds a qualitative survey of the elements present in almost any sample can be made, and in only a few minutes sufficient data can be collected for quantification. The most frequent application has been to highly polished metal samples, although nonconducting samples may be studied if they can be coated with a thin conducting film. Sample dimensions are usually not a problem in an SEM. Most of the detectors used to date have been of the Si (Li) type, but with improvements in the processing methods of germanium, HPGe detectors may start impacting the market. New cooling methods, e.g., Peltier cooling, are being studied and will be a significant improvement if the ultimate temperature can be lowered to nearer to liquid nitrogen temperatures. Detector resolution is still the most serious problem. It is slowly being reduced through improvements in FET design and new methods of charge restoration. New window materials have made low-Z detectors the standard, opening up the prospects of light element analysis. However, the sensitivity for light elements is generally low and the analytical methods are still being developed because the peak shapes are often not Gaussian and the mass absorption coefficients for the low-Z elements are not very well known.

Overall a "customer" needs to know under what circumstances it is best to use either the electron-beam techniques of EDS and WDS or the X-ray technique of XRF for an analysis problem. If both are equally available, the choice usually resides in whether high spatial resolution is needed, as would be obtained only with electron-beam techniques. If liquids are to be analyzed, the only viable choice is XRF. If one's choice is to use electron-beam methods, the further decision between EDS and WDS is usually one of operator preference. That is, to commence study on a totally new sample most electron-beam operators will run an EDS spectrum first. If there are no serious peak overlap problems, then EDS may be sufficient. If there is peak overlap or if maximum sensitivity is desired, then WDS is usually preferred. Factored into all of this must be the beam sensitivity of the sample, since for WDS analysis the beam current required is 10–100× greater than for EDS. This is of special concern in the analysis of polymer materials.

Related Articles in the Encyclopedia

EPMA, STEM, TEM, SEM, XRF

Bibliography

1 R. Woldseth. *X-Ray Energy Spectrometry.* Kevex Corporation, San Carlos, 1973. A good introduction with emphasis on detectors and electronics. Most of the applications refer to X-ray tube sources. Unfortunately the book is out of print, but many industrial laboratories may have copies.

2 *Quantitative Electron-Probe Microanalysis.* (V. D. Scott and G. Love, eds.) John Wiley & Sons, New York, 1983. Taken from a short course on the electron microprobe for scientists working in the field. A thorough discussion of EDS and WDS is given, including experimental conditions and specimen requirements. The ZAF correction factors are treated extensively, and statistics, computer programs and Monte Carlo methods are explained in detail. Generally, a very useful book.

3 J. I. Goldstein, D. E. Newbury, P. Echlin, D. C. Joy, C. Fiori, and E. Lifshin. *Scanning Electron Microscopy and X-Ray Analysis.* Plenum Press, New York, 1981. Developed from a short course held annually at Lehigh University. The book is concerned with the use and applications of SEM. In the latter context a lengthy discussion of EDS is given. The discussion

on analysis is not as thorough as that in Scott and Love, but serves as an excellent introduction, with all the important topics discussed.

4 D. B.Williams. *Practical Analytical Electron Microscopy in Materials Science.* Verlag Chemie International, Weinheim, 1984. A good monograph discussing the use and applications of AEM, especially at intermediate voltages. The discussion on EDS is an excellent primer for using X-ray analysis on a TEM.

5 *Principles of Analytical Electron Microscopy.* (D. C. Joy, A. D. Romig, and J. I. Goldstein, eds.) Plenum Press, New York, 1986. Another book, more readily available, discussing all aspects of AEM. Approximately one-quarter of the book is devoted to EDS and a discussion of thin-film analysis in the TEM.

3.2 EELS

Electron Energy-Loss Spectroscopy in the Transmission Electron Microscope

NESTOR J. ZALUZEC

Contents

Introduction

Electron Energy-Loss Spectroscopy (EELS) is an analytical methodology which derives its information from the measurement of changes in the energy and angular distribution of an initially nominally monoenergetic beam of electrons that has been scattered during transmission through a thin specimen. This geometry is identical to that which is used routinely in a transmission or Scanning Transmission Electron Microscope (TEM or STEM), and thus in the last decade the technique has been closely associated with these two types of instruments. EELS is an absorption spectroscopy and is similar in many respects to X-Ray Absorption Spectroscopy (XAS). Its characteristic spectral signature, termed the edge profile, is derived from the excitation of discrete inner shell levels to empty states above the Fermi level. Conservation of energy requires that the incident electron beam lose the corresponding amount of energy and intensity expended in exciting each inner shell level, and these are the parameters which are recorded in each experiment. From an analysis of the transmitted electron intensity distribution, the experiment can derive the local elemental concentration of each atomic species present. Additionally, by studying the detailed shape of the spectral profiles measured in EELS, the analyst may derive information about the electronic structure, chemical bonding, and average nearest neighbor distances for each atomic species detected. A related variation of EELS is Reflection Electron Energy-Loss Spectroscopy (REELS). In REELS the energy distribution of electrons scattered from the surface of a specimen is studied. Generally REELS deals with low-energy electrons (< 10 keV), while TEM/STEM-based EELS deals with incident electrons having energies of 100–400 keV. In this article we shall consider only the transmission case. REELS is discussed in Chapter 5.

In principle, EELS can be used to study all the elements in the periodic table; however, the study of hydrogen and helium is successful only in special cases where their signals are not masked by other features in the spectrum. As a matter of experimental practicality, the inner shell excitations studied are those having binding energies less than about 3 keV. Quantitative concentration determinations can be obtained for the elements $3 \leq Z \leq 35$ using a standardless data analysis procedure. In this range of elements, the accuracy varies but can be expected to be ±10–20% at. By using standards the accuracy can be improved to ±1–2% at. Detection limit capabilities have improved over the last decade from $\sim 10^{-18}$ g to $\sim 10^{-21}$ g. These advances have arisen through improved instrumentation and a more complete understanding of the specimen requirements and limitations. The energy resolution of the technique is limited today by the inherent energy spread of the electron source used in the microscope. Conventional thermionic guns typically exhibit an energy spread of 2–3 eV, and LaB_6 a spread of about 1–2 eV; field emission sources operate routinely in the 0.25–1 eV range. In all cases, the sample examined must be extremely thin (typically < 2000 Å) to minimize the adverse effects of multiple inelastic scattering, which can, in the worse cases, obscure all characteristic information.

The uniqueness and desirability of EELS is realized when it is combined with the power of a TEM or STEM to form an Analytical Electron Microscope (AEM). This combination allows the analyst to perform spatially resolved nondestructive analysis with high-resolution imaging (< 3 Å). Thus, not only can the analyst observe the microstructure of interest (see the TEM article) but, by virtue of the focusing ability of the incident beam in the electron microscope, he or she can simultaneously analyze a specific region of interest. Lateral spatial resolutions of regions as small as 10 Å in diameter are achievable with appropriate specimens and probe-forming optics in the electron microscope.

Basic Principles

EELS is a direct result of the Coulombic interaction of a fast nearly monochromatic electron beam with atoms in a solid. As the incident probe propagates through the

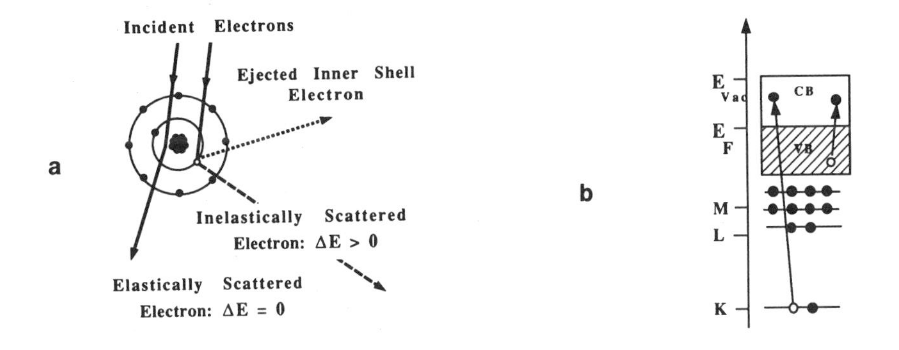

Figure 1 (a) Excitation of inner shells by Coulombic interactions. (b) Energy level diagram illustrating excitation from inner shell and valence band into the conduction band and the creation of a corresponding vacancy.

specimen it experience elastic scattering with the atomic nuclei and inelastic scattering with the outer electron shells (Figure 1a). The inelastic scattering, either with the tightly bound inner shells or with the more loosely bound valence electrons, causes atomic electrons to be excited to higher energy states or, in some cases, to be ejected completely from the solid. This leaves behind a vacancy in the corresponding atomic level (Figure 1b). The complementary analysis techniques of X-ray and Auger spectroscopy (covered in other articles in this book) derive their signals from electron repopulation of the vacancies created by the initial excitation event. After the interaction, the energy distribution of the incident electrons is changed to reflect this energy transfer, the nature and manifestation of which depends upon the specific processes that have occurred. Because EELS is the primary interaction event, all the other analytical signals derived from electron excitation are the result of secondary decay processes. EELS, therefore, yields the highest amount of information per inelastic scattering event of all the electron column-based spectroscopies.

Historically, EELS is one of the oldest spectroscopic techniques based ancillary to the transmission electron microscope. In the early 1940s the principle of atomic level excitation for light element detection capability was demonstrated by using EELS to measure C, N, and O. Unfortunately, at that time the instruments were limited by detection capabilities (film) and extremely poor vacuum levels, which caused severe contamination of the specimens. Twenty-five years later the experimental technique was revived with the advent of modern instrumentation.[1] The basis for quantification and its development as an analytical tool followed in the mid 1970s. Recent reviews can be found in the works by Joy, Maher and Silcox;[2] Colliex;[3] and the excellent books by Raether[4] and Egerton.[5]

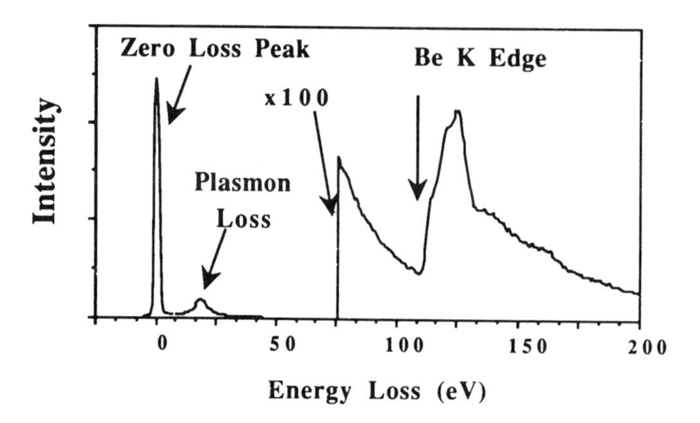

Figure 2 **Example of an energy-loss spectrum, illustrating zero loss, and low-loss valence band excitations and the inner shell edge. The onset at 111 eV identifies the material as beryllium. A scale change of 100× was introduced at 75 eV for display purposes.**

Figure 2 is an experimental energy-loss spectrum measured from a thin specimen of beryllium. At the left, at zero energy loss, is a large, nearly symmetric peak which represents electrons that have passed through the specimen suffering either negligible or no energy losses. These are the elastically scattered and phonon-scattered incident electrons. Following this peak is the distribution of inelastically scattered electrons, which is generally broken up into two energy regimes for simplicity of discussion. The low-loss regime extends (by convention) from about 1 eV to 50 eV, and exhibits a series of broad spectral features related to inelastic scattering with the valence electron structure of the material. In metallic systems these peaks arise due to a collective excitation of the valence electrons, and are termed *plasmon* oscillations or peaks. For most materials these peaks lie in energy range 5–35 eV.

Beyond this energy and extending for thousands of eV one observes a continuously decreasing background superimposed upon which are a series of "edges" resulting from electrons that have lost energy corresponding to the creation of vacancies in the deeper core levels of the atom (K, L_3 L_2, L_1, M_5, and so forth). The edges are generally referred to by the same nomenclature as used in X-ray absorption spectroscopy. The energy needed to ejected electrons amounts to the binding energy of the respective shell (Figure 1b), which is characteristic for each element. By measuring the threshold energy of each edge the analyst can determine the identity of the atom giving rise to the signal, while the net integrated intensity for the edge can be analyzed to obtain the number of atoms producing the signal. This is the basis of quantitative compositional analysis in EELS.

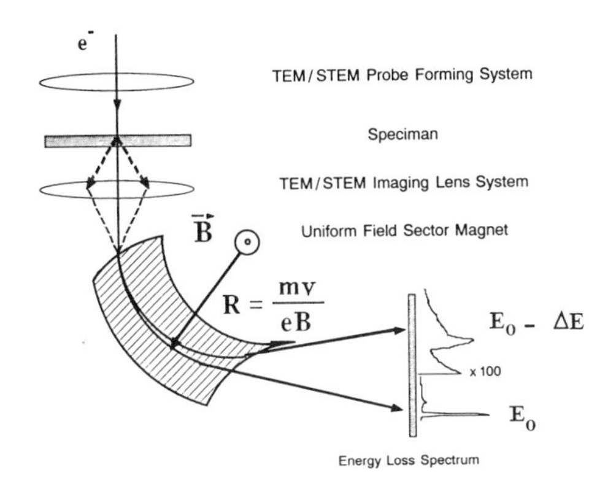

Figure 3 **Schematic representation of EELS analyzer mounted on a TEM / STEM.**

The energy regime most frequently studied by EELS is 0–3 keV. Higher energy losses can be measured; however, a combination of instrumental and specimen-related limitations usually means that these higher loss measurements are more favorable for study by alternative analytical methods, such as X-ray energy-dispersive spectroscopy (see the article on EDS). The practical consequence of this upper energy limit is that for low-Z elements ($1 \leq Z \leq 11$) one studies K-shell excitation; for medium-Z materials ($12 \leq Z \leq 45$), L shells; and for high-Z solids ($19 \leq Z \leq 79$), M, N, and O shells (the latter for $Z > 46$). It is also important to realize that not all possible atomic levels are observed in EELS as edges. The transitions from initial states to final states generally must obey the quantum number selection rules: $\Delta j = 0, \pm 1$, and $\Delta l = \pm 1$. Hence some atomic energy levels, although discrete and well defined, are not discernible by EELS.

Hydrogen and helium are special cases that should be mentioned separately. These elements have absorption edges at ~13 eV and 22 eV, respectively. These values lie in the middle of the low-loss regime, which is dominated by the valence band scattering. Thus, while the physics of inelastic scattering processes dictates that the edges will be present, usually they will be buried in the background of the more intense valence signal. In special cases, for example, when the plasmon losses are well removed, or when the formation of hydrides[6] occurs, presence of hydrogen and helium may be measured by EELS.

The instrumentation used in EELS is generally straightforward. Most commercial apparatus amount to a uniform field magnetic sector spectrometer located at the end of the electron-optical column of the TEM or STEM (Figure 3). Electrons that have traversed the specimen are focused onto the entrance plane of the spectrometer using the microscope lenses. Here the electrons enter a region having a uniform magnetic field aligned perpendicular to their velocity vector, which causes them to be deflected into circular trajectories whose radii vary in proportion to their velocity or energy and inversely with the magnetic field strength ($R = [m_0 v]/eB$). Location of a suitable detector system at the image plane of the spectrometer then allows the analyst to quantitatively measure the velocity–energy distribution. More complex spectrometers that use purely electrostatic or combined electrostatic and electromagnetic systems have been developed; however, these have been noncommercial research instruments and are not used generally for routine studies. More recently, elaborate imaging spectrometers also have been designed by commercial firms and are becoming incorporated into the column of TEM instruments. These newer instruments show promise in future applications, particularly in the case of energy-loss filtered imaging.

Low-Loss Spectroscopy

As we outlined earlier, the low-loss region of the energy-loss spectrum is dominated by the collective excitations of valence band electrons whose energy states lie a few tens of eV below the Fermi level. This area of the spectrum primarily provides information about the dielectric properties of the solid or measurements of valence electron densities. As a fast electron loses energy in transmission through the specimen its interaction—i.e., the intensity of the measured loss spectrum $I(E)$—can be related to the energy-loss probability $P(E, q)$, which in turn can be expressed in terms of the energy-loss function $\mathrm{Im}[-\varepsilon^{-1}(E, q)]$ from dielectric theory.[4] Here q is the momentum vector, and $\varepsilon = (\varepsilon_1 + i\varepsilon_2)$ is the complex dielectric function of the solid.[4] By applying a Kramers–Kronig analysis to the energy-loss function ($\mathrm{Im}\,[-\varepsilon^{-1}(E, q)]$), the real and imaginary parts ($\varepsilon_1, \varepsilon_2$) of the dielectric function can be determined. Using ε_1 and ε_2, one can calculate the optical constants (the refractive index η, the absorption index κ, and the reflectivity R) for the material being examined.[3-5]

In addition to dielectric property determinations, one also can measure valence electron densities from the low-loss spectrum. Using the simple free electron model one can show that the bulk plasmon energy (E_p) is governed by the equation:

$$E_p = \frac{h\omega_p}{2\pi} = \sqrt{\eta \frac{h^2 e^2}{4\pi^2 m \varepsilon_0}} \qquad (1)$$

where e is the electron charge, m is its mass, ε_0 is the vacuum dielectric constant, h is Planck's constant, and η is the valence electron density. From this equation we see that as the valence electron density changes so does the energy of the plasmon-loss peak. Although this can be applied to characterization, it is infrequently done today, as the variation in E_p with composition is small[7] and calibration experiments must be performed using composition standards. A recent application is the use of plasmon losses to characterize hydrides in solids.[6] Figure 4 shows partial EELS spectra from Mg, Ti, Zr, and their hydrides. The shift in the plasmon-loss peaks

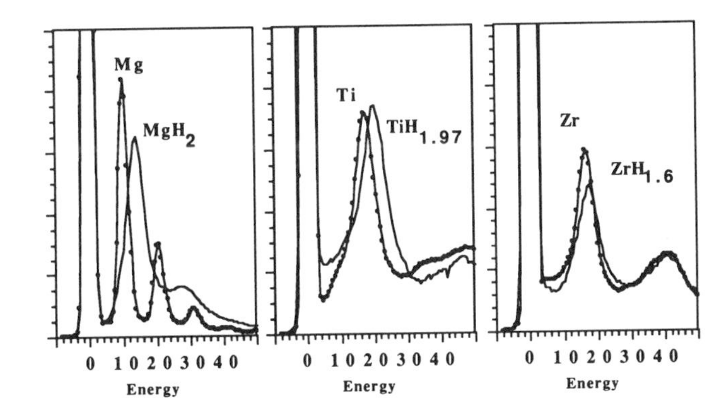

Figure 4 **Experimental low-loss profiles for Mg (10.0), Ti (17.2), Zr(16.6), and their hydrides MgH_2 (14.2), $TiH_{1.97}$ (20.0), and $ZrH_{1.6}$ (18.1). The values in parentheses represent the experimental plasmon-loss peak energies in eV.**

shows that the addition of hydrogen acts to increase the net electron density in these materials.

Inner Shell Spectroscopy

The most prominent spectral feature in EELS is the inner shell edge profile (Figure 2). Unlike EDS, where the characteristic signal profiles are nominally Gaussian-shaped peaks, in EELS the shape varies with the edge type (K, L, M, etc.), the electronic structure, and the chemical bonding. This is illustrated in Figure 5, which compares spectra obtain from a thin specimen of NiO using both windowless EDS and EELS. The difference in spectral profiles are derived from the fact that different mechanisms give rise to the two signals.

In the case of X-ray emission, the energy of the emitted photon corresponds to the energy differences between the initial and final states when a higher energy level electron repopulates the inner shell level, filling the vacancy created by the incident probe (Figure 1b). These levels are well defined and discrete, corresponding to deep core losses. The information derived is therefore mainly representative of the atomic elements present, rather than of the nuances of the chemical bonding or electronic structure. EDS is most frequently used in quantitative compositional measurements, and its poor energy resolution ~100 eV is due to the solid state detectors used to measure the photons and not the intrinsic width of the X-ray lines (about a few eV).

By contrast, in EELS the characteristic edge shapes are derived from the excitation of discrete inner shell levels into states above the Fermi level (Figure 1b) and reflect the empty density of states above E_F for each atomic species. The overall

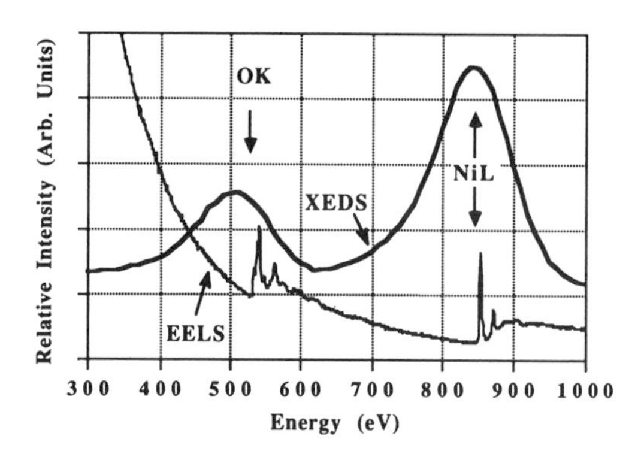

Figure 5 **Comparison of spectral profiles measured from a specimen of NiO using EDS and EELS. Shown are the oxygen K- and nickel L-shell signals. Note the difference in the spectral shape and peak positions, as well as the energy resolution of the two spectroscopies.**

shape of an edge can be approximately described using atomic models, due to the fact that the basic wavefunctions of deep core electrons do not change significantly when atoms condense to form a solid. Thus, the different edge profiles can be sketched as shown in Figure 6. K-shell edges (s $\rightarrow$ p transitions) tend to have a simple hydrogenic-like shape. L-shell edges (p $\rightarrow$ s and p $\rightarrow$ d transitions) vary between somewhat rounded profiles ($11 \leq Z \leq 17$) to nearly hydrogenic-like, with intense "white lines" at the edge onset ($19 \leq Z \leq 28$, and again for $38 \leq Z \leq 46$). In the fourth and fifth periods, these white lines are due to transitions from p to d states. M shells generally tend to be of the delayed-onset variety, due to the existence of an effective centrifugal barrier that is typical of elements with final states having large l quantum numbers. White lines near the M-shell edge onsets are observed when empty d states ($38 \leq Z \leq 46$) or f states ($55 \leq Z \leq 70$) occur, as in the case of the L shells. N and O shells are variable in shape and tend to appear as large, somewhat symmetrically shaped peaks rather than as "edges."

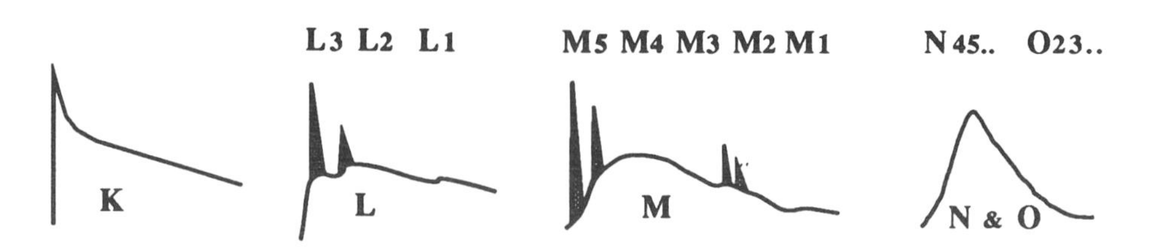

Figure 6 **Schematic illustration of K, L, M, N and O edge shapes; the "white lines" sometimes detected on L and M shells are shown as shaded peaks at the edge onsets. In all sketches the background shape has been omitted for clarity.**

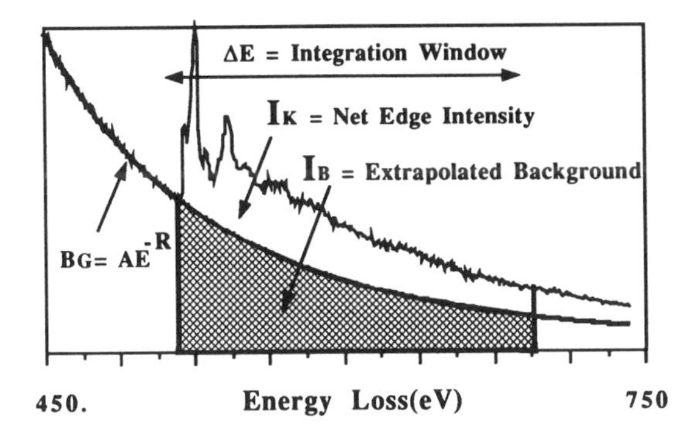

Figure 7 **Details of oxygen K shell in NiO, illustrating NES and EXELFS oscillations and the measurement of the integrated edge intensity I_K used for quantitative concentration determination.**

It is important to note that although specific edge profiles follow these generic shapes somewhat, they can deviate significantly in finer details in the vicinity of edge onsets. This structure arises due to solid state effects, the details of which depend upon the specific state (both electronic and chemical) of the material under scrutiny. Because of this strong variation in edge shape, experimental libraries of edge profiles also have been documented[8, 9] and have proven to be extremely useful supplementary tools. (Calculation of the detailed edge shape requires a significant computational effort and is not currently practical for on-line work.) These solid state effects also give rise to additional applications of EELS in materials research, namely: measurements of the d-band density of states in the transition metal systems,[10] and chemical state determinations[11] using the near-edge structure. The former has been used successfully by several research groups, while the latter application is, as yet, seldom used today in materials science investigations.

A more detailed description of near edge structure requires that one abandon simple atomic models. Instead, one must consider the spectrum to be a measure of the *empty local density of states above the Fermi level* of the elemental species being studied, scaled by the probability that the particular transition will occur. A discussion of such an undertaking is beyond the scope of this article, but EELS derives its capabilities for electronic and chemical bonding determinations from the near-edge structure. Calculation of this structure, which is due to the joint density of states, is involved and the studies of Grunes et al.[12] represent some of the most complete work done to date. The near-edge structure covers only the first few tens of eV beyond the edge onset; however, as we can see intensity oscillations extend for hundreds of eV past the edge threshold. This extended energy-loss fine structure (EXELFS) is analogous to the extended absorption fine structure (EXAFS) visible in X-ray absorption spectroscopy. An example of these undulations can be seen in the weaker oscillations extending beyond the oxygen K edge of Figure 7. The analysis of EXELFS oscillations can be taken virtually from the EXAFS literature and applied to EELS data, and allows the experimentalist to determine the nearest neighbor distances and coordination numbers about individual atomic species.[13]

Quantitative Concentration Measurements

The principles of quantitative concentration measurement in EELS is straightforward and simpler than in EDS. This is due to the fact that EELS is the primary interaction event, while all other electron-column analytical techniques are the result of secondary decay or emission processes. Thus, all other electron microscope-based analytical spectroscopies (EDS, Auger, etc.) must incorporate into their quantitative analysis procedures, corrections terms to account for the variety of competing processes (atomic number effects, X-ray fluorescence yields, radiative partition functions, absorption, etc.) that determine the measured signal. In EELS, the net integrated intensity in the *k*th edge profile for an element corresponds simply to the number of electrons which have lost energy due to the excitation of that particular shell. This is related to the incident electron intensity (I_0) multiplied by the cross section for ionization of the *k*th edges $\sigma_K$times the number of atoms in the analyzed volume (N):

$$I_K = N\sigma_K I_0 \tag{2}$$

Here I_K is the net intensity above background over an integration window of ΔE (Figure 7), while I_0 is the integrated intensity of the zero-loss peak (Figure 2). Generally the background beneath an edge is measured before the edge onset and extrapolated underneath the edge using a simple relationship for the background shape: $BG = AE^{-R}$. Here E is the energy loss, and A and R are fitting parameters determined experimentally from the pre-edge background. From Equation (2), one can express the absolute number of atoms/cm^2 as:

$$N = \frac{I_K}{\sigma_K I_0} \tag{3}$$

Hence by measuring I_K and I_0 and assuming σ_K is known or calculable, the analyst can determine N. Using a hydrogenic model, Egerton[5] has developed a set of FORTRAN subroutines (SigmaK and SigmaL) that are used by the vast majority of analysts for the calculation of K- and L-shell cross sections for the elements lithium through germanium. Leapman et al.[14] have extended the cross section calculations. Using an atomic Hartree-Slater program they have calculated K-, L-, M-, and some N-shell cross sections, however, these calculations are not amenable to use on an entry-level computer and require substantial computational effort.[14] They do, however, extend the method beyond the limits of Egerton's hydrogenic model. Tabular compilations of the cross section are generally not available, nor do they tend to be useful, as parameters used in calculations seldom match the wide range of experimental conditions employed during TEM- or STEM-based analysis.

An alternative approach to the quantitative analysis formalism is the ratio method. Here we consider the ratio of the intensities of any two edges A and B. Using Equation (3) we can show that

$$\frac{N_A}{N_B} = \frac{\sigma_B I_A}{\sigma_A I_B} \tag{4}$$

The elegance of this relationship rests in the fact that all the information one needs to measure the relative concentration ratio of any two elements is simply the ratio of their integrated edge profiles, I_0 having canceled out of the relationship.

This ratio method is generally the most widely used technique for quantitative concentration measurements in EELS. Unfortunately, the assumptions used in deriving this simple relationship are never fully realized. These assumptions are simply that electrons scattering from the specimen are measured over *all* angles and for *all* energy losses. This is physically impossible, since finite angular and energy windows are established or measured in the spectrum. For example, referring to Figure 7, we see that in NiO the Ni L-shell edge is superimposed upon the tail of the oxygen K-shell edge and clearly restricts the integration energy window for oxygen to about 300 eV. Similarly it is impossible in a TEM or STEM to collect all scattered electrons over π sR; an upper limit of about 100 mR is practically attainable. A solution to this problem was devised by Egerton[5] and can be incorporated into Equations (3) and (4) by replacing I_A by $I_A(\Delta E, \beta)$ and σ_A by $\sigma_A(\Delta E, \beta)$, since we measure over a finite energy (ΔE) and angular window (β). The quantity $\sigma_A(\Delta E, \beta)$ is now the partial ionization cross section for the energy and angular windows of ΔE and β, respectively. Using this ratio approach to quantification, accuracies of $\pm$5–10% at. for the same type edges (i.e., both K or L) have been achieved routinely using Egerton's hydrogenic models. When dissimilar edges are analyzed (for example one K and one L shell), the errors increase to $\pm$15–20% at. The major errors here result from the use of the hydrogenic model to approximate all edge shapes.

Although these errors may sound relatively large in terms of accuracy for quantification, it is the simplicity of the hydrogenic model that ultimately gives rise to the problem, and not the principle of EELS quantification. Should it be necessary to achieve greater accuracy, concentration standards can be developed and measured to improve accuracy. In this case, standards are used to accurately determine the experimental ratio $(\sigma_B(\Delta E, \beta)/\sigma_A(\Delta E, \beta)$ by measuring I_A/I_B and knowing the composition N_A/N_B. These σ_B/σ_A values are used when analyzing the unknown specimen, and accuracies to ~1% at. can be obtained in ideal cases. When employing standards, it is essential that the near-edge structure does not vary significantly between the unknown and the standard, since in many cases near-edge structure

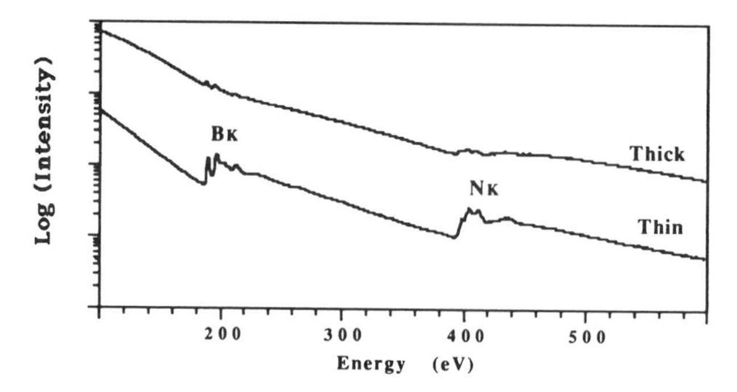

Figure 8 **Illustration of the decrease in the edge / background ratio for the B_K (~188 eV) and N_K (~399 eV) shells in EELS. In the data sets, the upper profile is from the thicker region (~2000 Å) of the BN specimen while the lower is thinner (~200 Å). Note the logarithmic vertical scale.**

contributes substantially to the net integrated edge profiles. This, unfortunately, is usually a difficult situation to realize. As a practical note, standards often are not used due to the fact that they require the analyst to prepare accurate multielement standards in TEM form for each elemental system to be studied and for every set of operating conditions used during the analysis of the unknown.

Limitations and Specimen Requirements

The single most important limitation to the successful application of EELS to problems in materials characterization relates to the specimen, namely, its thickness. Being a transmission technique it is essential that the incident beam penetrate the specimen, interact, and then enter the spectrometer for detection. As the specimen thickness increases, the likelihood of inelastic scattering increases, and hence the EELS signal increases. Unfortunately, the background signal increases at a faster rate than that of the characteristic edges. This results in the edges becoming effectively lost in the background, as illustrated in Figure 8, which shows the decrease in the edge-to-background ratio obtained from different thicknesses of a specimen of boron nitride. As a general rule, if λ is the mean free path for inelastic scattering, the specimen thickness t should not exceed values of $t/\lambda \approx 1$ and preferably should be < 0.5 to minimize the adverse effects of multiple scattering. The mean free path λ is a function of the atomic number and the accelerating voltage. At 100 kV, λ is about 1200 Å for aluminium, decreasing to ~900 Å at nickel, and reaching ~600 Å for gold. Increasing the accelerating voltage of the electron microscope reduces multiple inelastic scattering somewhat; for example, increasing the incident beam voltage from 100 to 300 kV increases λ by a factor of ~1.8, and going from 100 to 1000 kV yields a factor of ~2.5. However, increasing the voltage introduces another set of problems for the experimentalist, that is, electron irradiation (displacement) damage. In this situation, the high-energy electrons have sufficient energy to displace atoms from their normal lattice sites and, in some cases, literally to sputter holes through the specimen.

Conclusions

The combination of EELS with a TEM or STEM yields a powerful tool for the microcharacterization of materials. Its primary applications are in ultrahigh spatial resolution spectroscopy of thin electron-transparent solids. With optimized specimens, EELS can be used to obtain the local elemental composition of a specific region of interest on the specimen, and with more detailed calculations can provide information concerning the electronic or chemical states of the sample. EELS can be applied to any specimen that can be prepared for observation in the TEM or STEM. Future developments will concentrate in the areas of higher speed data acquisition using one- and two-dimensional parallel detectors for combined spectroscopy and parallel imaging, ultrahigh-energy resolution spectroscopy in the 50–100 meV range, and advanced software to make routine the more complex data analyses.

Related Articles in the Encyclopedia

TEM, STEM, EDS, EXAFS, NEXAFS, XPS, and UPS, REELS

References

1 A. V. Crewe, M. Isaacson, and D. E. Johnson. *Rev. Sci. Instr.* **42,** 411, 1971.

2 *Introduction to Analytical Electron Microscopy.* (J. J. Hren, D. C. Joy, and J. I. Goldstein, eds.) Plenum Press, 1979. A good overview of analytical electron microscopy.

3 C. Colliex. In: *Advances in Optical and Electron Microscopy.* (R. Barer and V.E. Cosslett, eds.) Academic Press, 1984, Volume 9. This chapter contains a concise, but detailed, treatment of EELS with significant references to the major studies done.

4 H. Raether. *Springer Tracts in Modern Physics.* **88**, 1980. This book details the wealth of information contained in the low-loss spectrum, and treats the mathematics in considerable detail.

5 R. F. Egerton. *Electron Energy Loss Spectrometry in the Electron Microscope.* Plenum Press, 1986. This is a comprehensive text on the use of EELS in the TEM. It covers instrumentation, theory and practical applications.

6 N. J. Zaluzec, T. Schober, and B. W. Veal. In: *Analytical Electron Microscopy—1982 Proceedings of the Workshop at Vail Colorado.* San Francisco Press, p. 191.

7 D. B. Williams and J. W. Edington. *J. Microsc.* **108,** 113, 1976.

8 N. J. Zaluzec. *Ultramicroscopy.* **9,** 319, 1982; and *J. de Physique. C2* **45** (2), 1984. This work is also available *gratis* from the author, at EM Center, Argonne National Laboratory Materials Science Division-212, Argonne, IL 60439, USA.

9 C. C. Ahn and O. L. Krivanek. *An Atlas of Electron Energy Loss Spectra.* Available from Gatan Inc., Pleasanton, CA 94566, USA.

10 T. I. Morrison, M. B. Brodsky, and N. J. Zaluzec. *Phys. Rev. B.* **32,** (5) 3107, 1985.

11 M. Isaacson. In: Microbeam Analysis in Biology. (C. P. Lechene and R. R. Warner, eds.) Academic Press, New York, 1979, p.53.

12 L. A. Grunes, R. D. Leapman, C. N. Wilker, R. Hoffmann, and A. B. Kunz. *Phys. Rev. B.* **25,** 7157, 1982.

13 M. M. Disko, O. L. Krivanek, and P. Rez. *Phys. Rev. B.* **25,** 4252, 1982.

14 R. D. Leapman, P. Rez, and D. F. Mayers. *J. Chem. Phys.* **72,** 1232, 1980.

3.3 CL

Cathodoluminescence

B.G. YACOBI

Contents

Introduction

Cathodoluminescence (CL), i.e., the emission of light as the result of electron-beam bombardment, was first reported in the middle of the nineteenth century in experiments in evacuated glass tubes. The tubes were found to emit light when an electron beam (cathode ray) struck the glass, and subsequently this phenomenon led to the discovery of the electron. Currently, cathodoluminescence is widely used in cathode-ray tube-based (CRT) instruments (e.g., oscilloscopes, television and computer terminals) and in electron microscope fluorescent screens. With the developments of electron microscopy techniques (see the articles on SEM, STEM and TEM) in the last several decades, CL microscopy and spectroscopy have emerged as powerful tools for the microcharacterization of the electronic properties of luminescent materials, attaining spatial resolutions on the order of 1 μm and less.[1] Major applications of CL analysis techniques include:

1. Uniformity characterization of luminescent materials (e.g., mapping of defects and measurement of their densities, and impurity segregation studies)

2. Obtaining information on a material's electronic band structure (related to the fundamental band gap) and analysis of luminescence centers
3. Measurements of the dopant concentration and of the minority carrier diffusion length and lifetime
4. Microcharacterization of semiconductor devices (e.g., degradation of optoelectronic devices)
5. Analysis of stress distributions in epitaxial layers
6. *In-situ* characterization of dislocation motion in semiconductors
7. Depth-resolved studies of defects in ion-implanted samples and of interface states in heterojunctions.

In CL microscopy, luminescence images or maps of regions of interest are displayed, whereas in CL spectroscopy a luminescence spectrum from a selected region of the sample is obtained. The latter is analogous to a point analysis in X-ray microanalysis (see the article on EPMA). However, unlike X-ray emission, cathodoluminescence does not identify the presence of specific atoms. The lines of characteristic X-rays, which are emitted due to electronic transitions between sharp inner-core levels (see the articles on EDS, EPMA, or XRF), are narrow and are largely unaffected by the environment of the atom in the lattice. In contrast, the CL signal is generated by detecting photons (in the ultraviolet, visible, and near-infrared regions of the spectrum) that are emitted as the result of electronic transitions between the conduction band, or levels due to impurities and defects lying in the fundamental band gap, and the valence band. These transition energies and intensities are affected by a variety of defects, by the surface of the material, and by external perturbations, such as temperature, stress, and electric field. Thus, no universal law can be applied in order to interpret and to quantify lines in the CL spectrum. Despite this limitation, the continuing development of CL is motivated by its attractive features:

1. CL is the only contactless method (in an electron probe instrument) that provides microcharacterization of electronic properties of luminescent materials.
2. A CL system attached to a scanning electron microscope (SEM) provides a powerful means for the uniformity studies of luminescent materials with the spatial resolution of less than 1 μm.
3. The detection limit of impurity concentrations can be as low as 10^{14} atoms/cm^3, which is several orders of magnitude better than that of the X-ray microanalysis mode in the SEM.
4. CL is a powerful tool for the characterization of optical properties of wide band-gap materials, such as diamond, for which optical excitation sources are not readily available.

5. Since the excitation depth can be selected by varying the electron-beam energy, depth-resolved information can be obtained.
6. In optoelectronic materials and devices, it is the luminescence properties that are of practical importance.

CL studies are performed on most luminescent materials, including semiconductors, minerals, phosphors, ceramics, and biological–medical materials.

Basic Principles

The Excitation Process

As the result of the interaction between keV electrons and the solid, the incident electron undergoes a successive series of elastic and inelastic scattering events, with the range of the electron penetration being a function of the electron-beam energy: $R_e = (k/\rho)\, E_b^{\alpha}$, where E_b is the electron-beam energy, k and α depend on the atomic number of the material and on E_b, and ρ is the density of the material. Thus, one can estimate the so-called generation (or excitation) volume in the material. The generation factor, i.e., the number of electron–hole pairs generated per incident beam electron, is given by $G = E_b\,(1-\gamma)/E_i$, where E_i is the ionization energy (i.e., the energy required for the formation of an electron–hole pair), and γ represents the fractional electron-beam energy loss due to the backscattered electrons.

Luminescence Processes

The emission of light in luminescence processes is due to an electronic transition (relaxation) between a higher energy state, E_2, and an empty lower energy state, E_1. (The state E_2 is the excited state caused by the electron-beam excitation process.) The energy, or the wavelength of the emitted photon, $h\nu = hc/\lambda = E_2 - E_1$. The wavelength λ (in µm) of a photon is related to the photon energy E (in eV) by $\lambda \cong 1.2398/E$. In wide band-gap materials luminescence occurs in the visible range (from about 0.4 to 0.7 µm, corresponding to about 3.1 to 1.8 eV). In many cases, luminescence also occurs at longer wavelengths in the near-infrared region.

For a simplified case, one can obtain[1] the rate of CL emission, $L_{CL} = f\eta G I_b/e$, where f is a function containing correction parameters of the CL detection system and that takes into account the fact that not all photons generated in the material are emitted due to optical absorption and internal reflection losses;[1] η is the radiative recombination efficiency (or internal quantum efficiency); I_b is the electron-beam current; and e is the electronic charge. This equation indicates that the rate of CL emission is proportional to η, and from the definition of the latter we conclude that in the observed CL intensity one cannot distinguish between radiative and nonradiative processes in a quantitative manner. One should also note that η depends on various factors, such as temperature, the presence of defects, and the

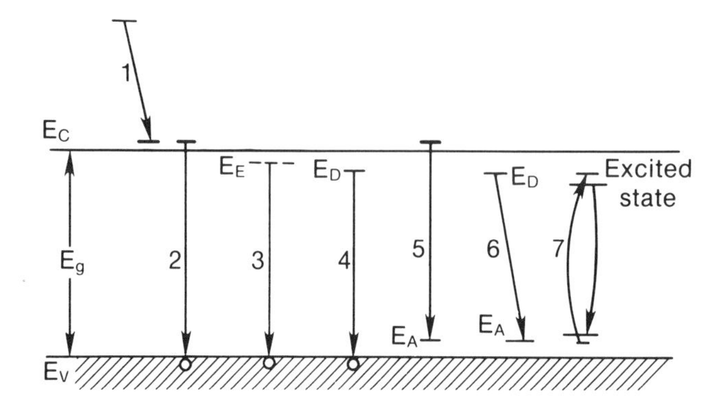

Figure 1 **Schematic diagram of luminescence transitions between the conduction band (E_C), the valence band (E_V), and exciton (E_E), donor (E_D,) and acceptor (E_A) levels in the luminescent material.**

particular dopants and their concentrations. One result of the analysis of the dependence of the CL intensity on the electron-beam energy indicates the existence at the surface of a dead layer, where radiative recombination is absent.[2]

In inorganic solids, luminescence spectra can be categorized as *intrinsic* or *extrinsic.* Intrinsic luminescence, which appears at elevated temperatures as a near Gaussian-shaped band of energies with its peak at a photon energy $h\nu_p \cong E_g$, is due to recombination of electrons and holes across the fundamental energy gap E_g (see Figure 1). Extrinsic luminescence, on the other hand, depends on the presence of impurities and defects. In the analysis of optical properties of inorganic solids it is also important to distinguish between *direct-gap* materials (e.g., GaAs and ZnS) and *indirect-gap* materials (e.g., Si and GaP). This distinction is based on whether the valence band and conduction band extrema occur at the same value of the wave vector **k** in the energy band $E(\mathbf{k})$ diagram of the particular solid. In the former case, no phonon participation is required during the direct electronic transitions. (A phonon is a quantum of lattice vibrations.) In the latter case, phonon participation is required to conserve momentum during the indirect electronic transitions; since this requires an extra particle, the probability of such a process occurring is significantly lower than that of direct transitions. Thus, fundamental emission in indirect-gap materials is relatively weak compared with that due to impurities or defects.

A simplified schematic diagram of transitions that lead to luminescence in materials containing impurities is shown in Figure 1. In process 1 an electron that has been excited well above the conduction band edge dribbles down, reaching thermal equilibrium with the lattice. This may result in phonon-assisted photon emission or, more likely, the emission of phonons only. Process 2 produces intrinsic luminescence due to direct recombination between an electron in the conduction band

and a hole in the valence band, and this results in the emission of a photon of energy $h\nu \cong E_g$. Process 3 is the exciton (a bound electron–hole pair) decay observable at low temperatures; free excitons and excitons bound to an impurity may undergo such transitions. In processes 4, 5, and 6, transitions that start or finish on localized states of impurities (e.g., donors and acceptors) in the gap produce extrinsic luminescence, and these account for most of the processes in many luminescent materials. Shallow donor or acceptor levels can be very close to the conduction and valence bands; to distinguish between the intrinsic band-to-band transitions and those associated with shallow impurity transitions, measurements have to be performed at cryogenic temperatures, where CL spectra are sharpened into lines corresponding to transitions between well-defined energy levels. Process 7 represents the excitation and radiative deexcitation of an impurity with incomplete inner shells, such as a rare earth ion or a transition metal. It should be emphasized that lattice defects, such as dislocations, vacancies, and their complexes with impurity atoms, may also introduce localized levels in the band gap, and their presence may lead to the changes in the recombination rates and mechanisms of excess carriers in luminescence processes.

Spatial Resolution

The spatial resolution of the CL-SEM mode depends mainly on the electron-probe size, the size of the excitation volume, which is related to the electron-beam penetration range in the material (see the articles on SEM and EPMA), and the minority carrier diffusion. The spatial resolution also may be affected by the signal-to-noise ratio, mechanical vibrations, and electromagnetic interference. In practice, the spatial resolution is determined basically by the size of the excitation volume, and will be between about 0.1 and 1 μm[1]

Instrumentation

Two general categories of CL analysis systems are wavelength nondispersive-versus-dispersive, and ambient-versus-cryogenic temperature designs. The first category essentially leads to two basic CL analysis methods, microscopy and spectroscopy. In the former case, an electron microscope (SEM or STEM) is equipped with various CL detecting attachments, and thus CL images or maps of regions of interest can be displayed on the CRT. In the latter case an energy-resolved spectrum corresponding to a selected area of the sample can be obtained. CL detector designs differ in the combination of components used[3]. Although most of these are designed as SEM attachments,[3] several CL collection systems were developed in dedicated STEMs.[4] The collection efficiencies of the CL detector systems vary from several percent for photomultipliers equipped with light guides, to close to 90% for systems incorporating ellipsoidal or parabolic mirrors coupled directly to a monochromator. A relatively simple and inexpensive, but powerful, CL detector using an optical fiber light collection system also has been developed.[5] In these designs, the signal from the photomultiplier can be used to produce micrographs and spectra. When the grating of the monochromator is bypassed, photons of all wavelengths falling on the photomultiplier produce the panchromatic (integral) CL signal. In the dispersive mode, for a constant monochromator setting and a scanning electron-beam condition, monochromatic micrographs can be obtained; and when the monochromator is stepped through the wavelength range of interest and the electron beam is stationary or scans a small area, CL spectra can be derived. The proper choice of a detector is important in CL measurements. In the visible range, photomultipliers are the most efficient detectors. For luminescence in the infrared range, solid state detectors, as well as Fourier transform spectrometry (FTS) can be used. For detailed quantitative analysis, the calibration of the CL detection system for its spectral response characteristics is important in most cases.

Although in many applications noncryogenic CL system designs may be sufficient, for detailed quantitative studies of impurities and defects in various materials it is necessary to use high-efficiency light-collection dispersive systems having the capability of sample cooling, preferably to liquid-helium temperatures. The advantages of sample cooling are to increase the CL intensity, to sharpen the CL spectrum into lines corresponding to transitions between well-defined energy levels that allow the more reliable interpretation of CL spectra, and to reduce the rate of electron bombardment damage in electron-beam sensitive materials.

Another basic approach of CL analysis methods is that of the CL spectroscopy system (having no electron-beam scanning capability), which essentially consists of a high-vacuum chamber with optical ports and a port for an electron gun. Such a system is a relatively simple but powerful tool for the analysis of ion implantation-induced damage, depth distribution of defects, and interfaces in semiconductors.[6]

Optical CL microscopes are instruments that couple electron gun attachments to optical microscopes. Although such systems have a limited spatial resolution, they are used widely in the analysis of minerals.[7]

Quantification

As mentioned above, the interpretation of CL cannot be unified under a simple law, and one of the fundamental difficulties involved in luminescence analysis is the lack of information on the competing nonradiative processes present in the material. In addition, the influence of defects, the surface, and various external perturbations (such as temperature, electric field, and stress) have to be taken into account in quantitative CL analysis. All these make the quantification of CL intensities difficult. Correlations between dopant concentrations and such band-shape parameters as the peak energy and the half-width of the CL emission currently are more reliable as means for the quantitative analysis of the carrier concentration.

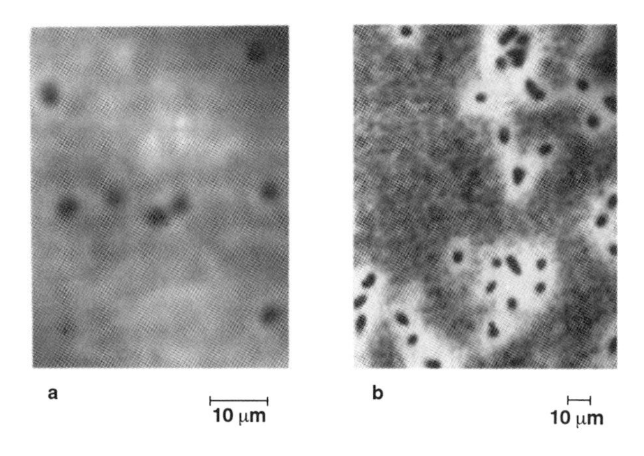

Figure 2 CL micrographs of Te-doped GaAs: dark-dot dislocation contrast (a) in GaAs doped with a Te concentration of 10^{17} cm^{-3}; and dot-and-halo dislocation contrast (b) in GaAs doped with a Te concentration of 10^{18} cm^{-3}.

Nonradiative surface recombination is a loss mechanism of great importance for some materials (e.g., GaAs). This effect, however, can be minimized by increasing the electron-beam energy in order to produce a greater electron penetration range.

A method for quantification of the CL, the so-called MAS corrections, in analogy with the ZAF correction method for X rays (see the article on EPMA), has been proposed[8] to account for the effects of the excess carrier concentration, absorption and surface recombination. In addition, a total internal reflection correction should also be included in the analysis, which leads to the MARS set of corrections. This method can be used for further quantification efforts that also should involve Monte Carlo calculations of the generation of excess carriers.

General Applications and Examples

Major applications of CL microscopy and spectroscopy in the analysis of solids have been listed in the Introduction. Some specific examples of CL applications are outlined below.

An example of the uniformity characterization, as well as of the analysis of the electrically active defects, is shown in Figure 2. These CL micrographs demonstrate two different forms of dislocation contrast (dark-dot and dot-and-halo contrast) for GaAs crystals doped with Te concentrations of 10^{17} cm^{-3} (Figure 2a) and 10^{18} cm^{-3} (Figure 2b). The latter shows variations in the doping concentration around dislocations. This figure also demonstrates that CL microscopy is a valuable tool for determining dislocation distributions and densities in luminescent materi-

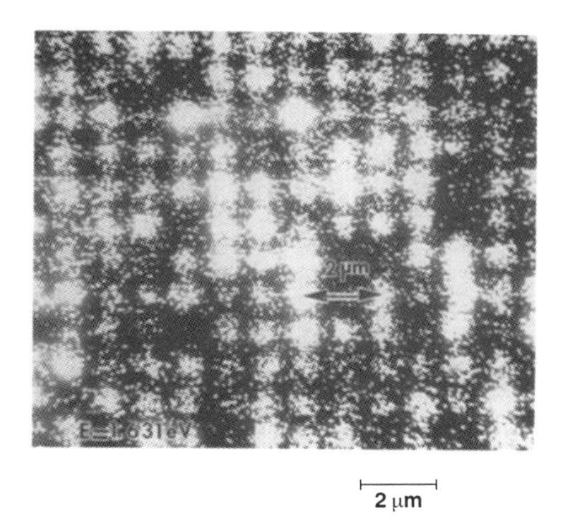

Figure 3 **Monochromatic CL image (recorded at 1.631 eV) of quantum well boxes, which appear as bright spots.[9]**

als. Reliable measurements of dislocation densities up to about 10^6 cm^{-2} can be made with the CL image.

An example of the CL microcharacterization of an array of GaAs/AlGaAs quantum well (QW) boxes[9] is presented in Figure 3, which shows the CL monochromatic image recorded at the energy corresponding to one of the characteristic luminescence lines (i.e., 1.631 eV). In such structures, the carriers are confined by surrounding a smaller band-gap semiconductor layer with wider band-gap layers. Confinement of carrier motion to 0 degrees of freedom will be obtained for the smaller band-gap layer in the form of a box.[9] The monochromatic CL image shows nonuniformities in the luminescence intensity from one box to another, since not all the QW boxes are identical due to variations in the confining potential between them that result from the presence of residual processing-induced damage.[9]

Cathodoluminescence microscopy and spectroscopy techniques are powerful tools for analyzing the spatial uniformity of stresses in mismatched heterostructures,[10] such as GaAs/Si and GaAs/InP. The stresses in such systems are due to the difference in thermal expansion coefficients between the epitaxial layer and the substrate. The presence of stress in the epitaxial layer leads to the modification of the band structure, and thus affects its electronic properties; it also can cause the migration of dislocations, which may lead to the degradation of optoelectronic devices based on such mismatched heterostructures. This application employs low-temperature (preferably liquid-helium) CL microscopy and spectroscopy in conjunction with the known behavior of the optical transitions in the presence of stress to analyze the spatial uniformity of stress in GaAs epitaxial layers. This analysis can reveal,

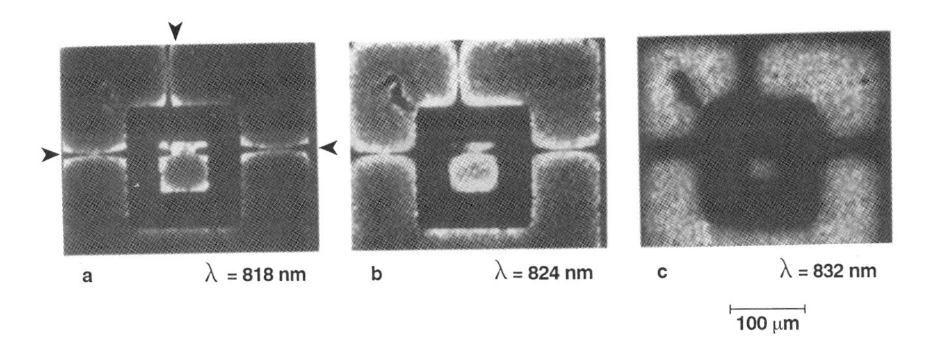

Figure 4 **Monochromatic CL images of the GaAs / Si sample recorded at 818 nm (a), 824 nm (b), and 832 nm (c). Microcracks are indicated by arrows in (a). The sample temperature is about 20 K.**[10]

for example, variations in stress associated with the patterning of GaAs layers grown on mismatched substrates.[10] An example describing stress variations and relief due to patterning in GaAs grown on Si substrates is shown in Figure 4, which presents monochromatic CL images of a GaAs layer at 818, 824, and 832 nm. These images demonstrate that the convex corners and the edges in the patterned regions emit at shorter wavelengths compared to the interiors of these regions. Detailed analysis of the CL spectra in different regions of a GaAs layer indicates strong variations in stress associated with patterning of such layers.[10]

An example of CL depth-resolved analysis of subsurface metal–semiconductor interfaces, using an ultrahigh-vacuum CL system,[6] is shown in Figure 5. This figure presents CL spectra of ultrahigh vacuum-cleaved CdS before and after 50-Å Cu deposition and pulsed laser annealing.[6] The deposition of Cu produces a weak peak at about 1.27 eV, in addition to the CdS band-edge emission at 2.42 eV. Pulsed laser annealing with an energy density of 0.1 J/cm^2 increases the intensity of this peak, which is related to Cu_2S compound formation.[6] This specific example clearly indicates that low-energy CL spectroscopy can be used effectively in the analysis of chemical interactions at buried metal–semiconductor interfaces.

As mentioned earlier, CL is a powerful tool for the characterization of optical properties of wide band-gap materials, such as diamond, for which optical excitation sources are not readily available. In addition, electron-beam excitation of solids may produce much greater carrier generation rates than typical optical excitation. In such cases, CL microscopy and spectroscopy are valuable methods in identifying various impurities, defects, and their complexes, and in providing a powerful means for the analysis of their distribution, with spatial resolution on the order of 1 μm and less.[11]

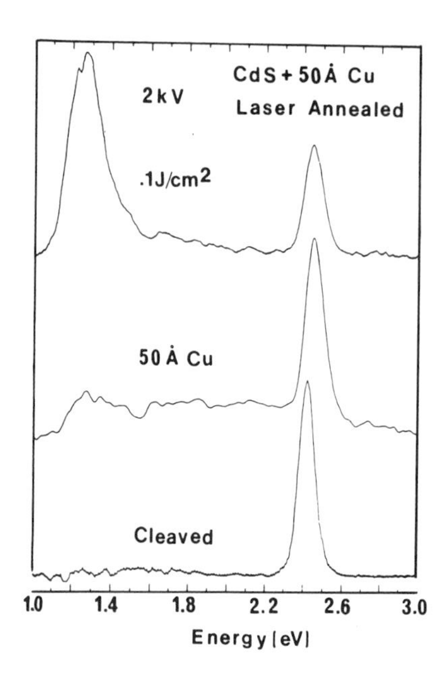

Figure 5 CL spectra of ultrahigh vacuum-cleaved CdS before and after *in situ* deposition of 50 Å of Cu, and after *in situ* laser annealing using an energy density of 0.1 J / cm^2. The electron-beam voltage is 2 kV.[6]

Artifacts

Artifacts in CL analysis of materials may arise from luminescent contaminants, from scintillations caused by backscattered electrons striking components of the CL system, or from incandescence from the electron gun filament reaching the detector. Most of these may be eliminated by chopping the electron beam and detecting the CL signal with a lock-in amplifier. In the CL spectra, ghost peaks may appear due to total internal reflection and self-absorption.[8] Prolonged electron-beam irradiation may cause changes in the CL intensity, localized heating at high beam currents, or may lead to surface contamination of the specimen. To alleviate the latter problem, it is recommended to replace the diffusion pump by a turbomolecular pump, or to use liquid-nitrogen cooled shields around the specimen. Spurious signals also may arise due to charging in nonconducting samples. As a general rule, the electron-beam damage is minimized by maximizing the beam voltage, minimizing the beam current, and using minimum detection times.

It should be noted that during CL observations intensity variations may arise due to sample morphology (e.g., surface roughness), which may lead to nonuniform excitation and to local variations in optical absorption and reflection losses.

Conclusions

In summary, CL can provide contactless and nondestructive analysis of a wide range of electronic properties of a variety of luminescent materials. Spatial resolution of less than 1 μm in the CL-SEM mode and detection limits of impurity concentrations down to 10^{14} at/cm^3 can be attained. CL depth profiling can be performed by varying the range of electron penetration that depends on the electron-beam energy; the excitation depth can be varied from about 10 nm to several μm for electron-beam energies ranging between about 1 keV and 40 keV.

The development of quantitative CL analysis is the most challenging issue. With further development of interpretive theory and with the trend toward the computerization of electron microscopy, quantitative CL analysis should become feasible.

Extensions in wavelength, into both the infrared and the ultraviolet ranges will continue, motivated by increasing interest in narrow band-gap semiconductors and wide band-gap materials.

Applications of CL to the analysis of electron beam-sensitive materials and to depth-resolved analysis of metal–semiconductor interfaces[6] by using low electron-beam energies (on the order of 1 keV) will be extended to other materials and structures.

The continuing development of CL detection systems, cryogenic stages, and signal processing and image analysis methods will further motivate studies of a wide range of luminescent materials, including biological specimens.[12]

Related Articles in the Encyclopedia

EPMA, SEM, STEM, TEM, and PL

References

1 B. G. Yacobi and D. B. Holt. *Cathodoluminescence Microscopy of Inorganic Solids.* Plenum, 1990.

2 D. B. Wittry and D. F. Kyser. *J. Appl. Phys.* **38,** 375, 1967.

3 D. B. Holt. In: *Microscopy of Semiconducting Materials.* IOP, Bristol, 1981, p.165.

4 P. M. Petroff, D. V. Lang, J. L. Strudel, and R. A. Logan. In: *Scanning Electron Microscopy.* SEM Inc., Chicago, 1978, p. 325.

5 M. E. Hoenk and K. J. Vahala. *Rev. Sci. Instr.* **60,** 226, 1989.

6 L. J. Brillson and R. E. Viturro. *Scanning Microscopy.* **2,** 789, 1988.

7 C. E. Barker and T. Wood. In: *Process Mineralogy VI.* The Metallurgical Society of AIME, 1987, p. 159.

8 C. A. Warwick. *Scanning Microscopy* **1,** 51, 1987.

9 P. M. Petroff. In: *Microscopy of Semiconducting Materials.* IOP, Bristol, 1987, p.187.

10 B. G. Yacobi, S. Zemon, C. Jagannath, and P. Sheldon. *J. Cryst. Growth* **95,** 240, 1989.

11 A. T. Collins and S. C. Lawson. *J. Phys. Cond. Matter* **1,** 6929, 1989; L. H. Robins, L. P. Cook, E. N. Farabaugh, and A. Feldman. *Phys. Rev.* **B39,** 13367, 1989.

12 D. B. Holt. In: *Analysis of Organic and Biological Surfaces.* (P. Echlin, ed.) Wiley, New York, 1984, p.301.

3.4 STEM

Scanning Transmission Electron Microscopy

CHARLES E. LYMAN

Contents

Introduction

The Scanning Transmission Electron Microscope (STEM) produces images of the internal microstructure of thin specimens using a high-energy scanning electron beam, in the same way a Scanning Electron Microscope (SEM) produces images of bulk surfaces. The term STEM is also used to describe the group of crystallographic and compositional analysis methods known collectively as Analytical Electron Microscopy (AEM): Convergent Beam Electron Diffraction (CBED), X-ray microanalysis by Energy-Dispersive Spectrometry (EDS), and Electron Energy-Loss Spectrometry (EELS).[1–6] Many STEM images are similar to images from the Transmission Electron Microscope (TEM), and in certain modes the STEM is capable of resolving the atomic lattice of a solid and even single atoms on a thin support.

The STEM is unrivaled in its ability to obtain high-resolution imaging combined with microanalysis from specimens that can be fashioned from almost any solid. Major applications include the analysis of metals, ceramics, electronic devices

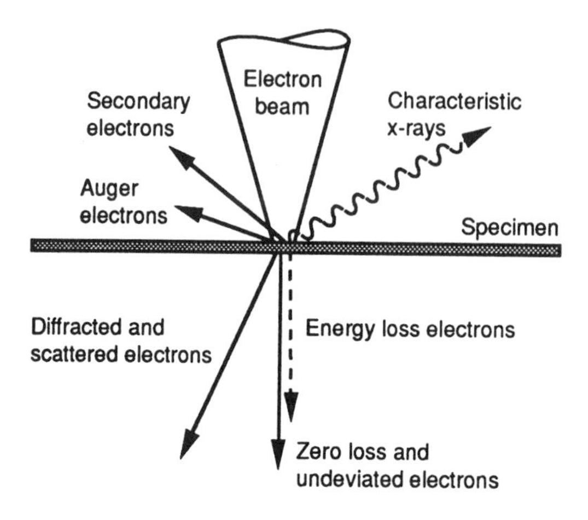

Figure 1 **Signals generated when the focussed electron beam interacts with a thin specimen in a scanning transmission electron microscope (STEM).**

and packaging, joining methods, coatings, composite materials, catalysts, minerals, and biological tissues.

There are three types of instruments that provide STEM imaging and analysis to various degrees: the TEM / STEM, in which a TEM instrument is modified to operate in STEM mode; the SEM/STEM, which is a SEM instrument with STEM imaging capabilities; and dedicated STEM instruments that are built expressly for STEM operation. The STEM modes of TEM / STEM and SEM / STEM instruments provide useful information to supplement the main TEM and SEM modes, but only the dedicated STEM with a field emission electron source can provide the highest resolution and elemental sensitivity.

Analysis capabilities in the STEM vary with the technique used. Crystallographic information may be obtained, including lattice parameters, Bravais lattice types, point groups, and space groups (in some cases), from crystal volumes on the order of 10^{-23} m^3 using CBED. Elemental identification and quantitative microanalysis have been developed for EDS and EELS. Detection limits for each technique are on the order of 0.1 wt % for one element combined with another. The EELS spectrum contains a rich variety of information concerning chemical bonding and dielectric constants in addition to elemental information. Since the STEM provides a through-section analysis (see Figure 1), it is complementary to surface techniques and should be used in conjunction with them. Also analytical signals may be collected as the small STEM electron probe scans across the specimen, providing compositional images in addition to the images typical of the SEM and the TEM. Compositional images showing elemental distributions have been obtained with spatial resolutions in the range 5–50 nm.

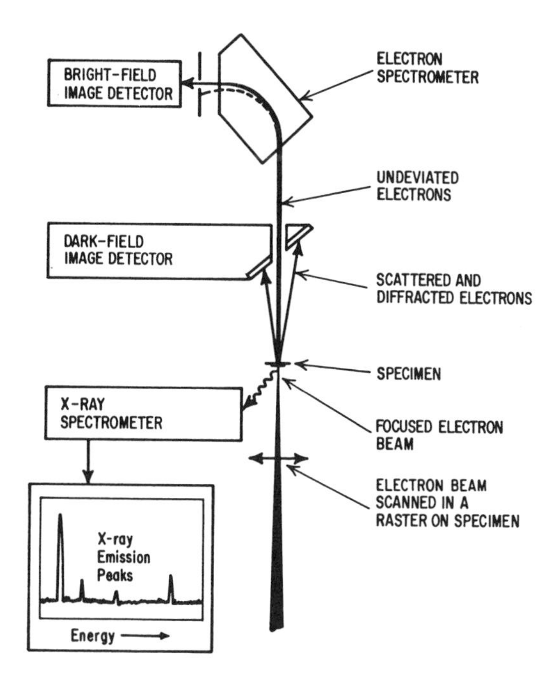

Figure 2 **Schematic of a STEM instrument showing the principal signal detectors. The electron gun and lenses at the bottom of the figure are not shown.**

The development of the STEM is relatively recent compared to the TEM and the SEM. Attempts were made to build a STEM instrument within 15 years after the invention of the electron microscope in 1932. However the modern STEM, which had to await the development of modern electronics and vacuum techniques, was developed by Albert Crewe and his coworkers at the University of Chicago.[7]

Basic Principles

Electron Probe Formation

An electron gun produces and accelerates the electron beam, which is reduced in diameter (demagnified) by one or more electromagnetic electron lenses. Electromagnetic scanning coils move this small electron probe (i.e., the beam) across the specimen in a raster. Electron detectors beyond the specimen collect a signal that is used to modulate the intensity on a cathode-ray tube that is scanned in synchronism with the beam on the specimen. A schematic of the essential components in a dedicated STEM system is shown in Figure 2.

The most important criterion for a STEM instrument is the amount of current in the small electron probe. Generally, 1 nA of probe current is required for high-

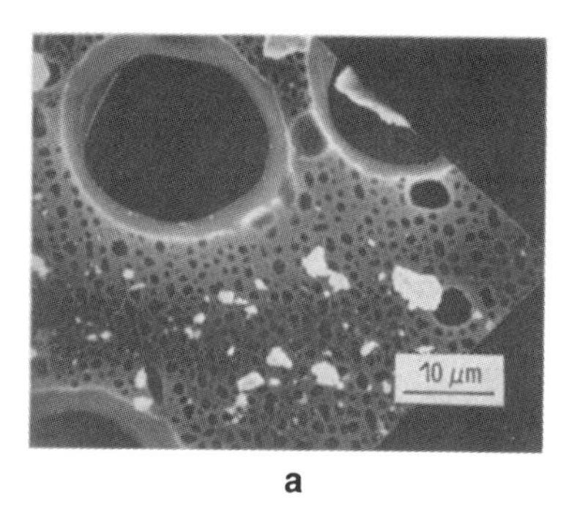

a

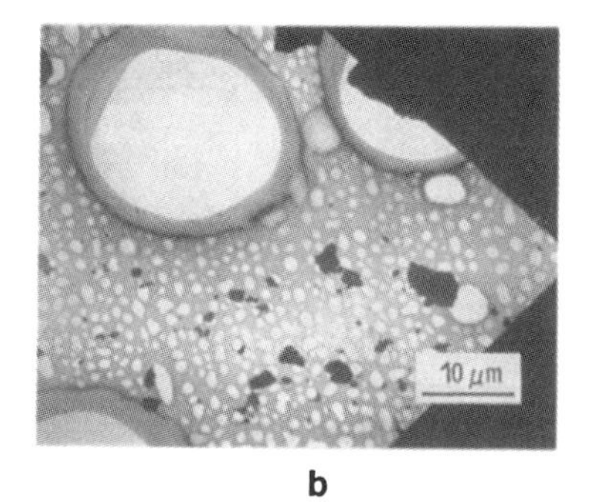

b

Figure 3 **Bright-field (a) and dark-field (b) STEM images of crushed ceramic particles dispersed on a "holey" carbon film supported on an electron microscope grid (shown at the right).**

quality microanalysis. For TEM/STEM and SEM/STEM systems using thermionic electron sources (tungsten wire or LaB_6), electron probes having diameters of 10–30 nm (measured as full width at half maximum) carry about 1 nA, and may be used for imaging and analysis. Smaller probes may be used for imaging, but the current may not be adequate for microanalysis. For the highest spatial resolution and analytical sensitivity, a STEM instrument with a field-emission electron gun must be used to provide 1 nA of current in an electron probe about 1–2 nm in diameter. These systems must use ultrahigh-vacuum technology at least in the electron gun, and preferably throughout the microscope.

Imaging

The scanning electron beam produces a diffraction pattern beyond the specimen similar to that formed in the TEM, and in most cases the image may be interpreted as mass–thickness contrast, diffraction contrast, or phase contrast. Bright-field images are formed by using an aperture to select only the undeviated transmitted beam from the diffraction pattern Figure 2, as in TEM (see the article on TEM). In STEM, the electron signal is collected with either a scintillator–photomultiplier or a semiconductor detector. Bright areas in the image indicate regions of the specimen that suffered little or no interaction with the electron beam (see Figure 3a). Dark-field images may be obtained in two ways: by selecting a single diffracted beam $\mathbf{g}_{hkl}$ to be collected on the detector; or by collecting all of the electrons diffracted or scattered beyond a certain minimum angle on an annular dark-field (ADF) electron detector. The former method gives images similar to the TEM dark-field images used for defect analysis in crystals (see the TEM article). The latter method provides a high-resolution, high-contrast image that is sensitive to specimen thickness and atomic number variations. Bright areas in the dark-field image indicate regions of the specimen that are thick, strongly diffracting, or of high atomic number (see Figure 3b).

The detailed contrast in a STEM image, compared to a TEM image of the same specimen feature, depends on the incident electron beam convergence angle and the electron collection angle at the detector. The theorem of reciprocity states that if appropriate beam angles in TEM and STEM are made equivalent and the sample is inverted, then the STEM and TEM images of a thin specimen will be identical (see Cowley[1]). For example, if the STEM collection angle is reduced to a value typical of the TEM illumination angle, similar phase contrast lattice plane images and structure images may be observed in both STEM and TEM. Often, the STEM collection angle must be enlarged to provide an adequate signal level, which may alter the image contrast. Because of the scanning nature of image generation many other signals, such as secondary electrons and cathodoluminescence (light), also may be used for imaging.

Convergent Beam Electron Diffraction

When the electron beam impinging on the specimen has a high convergence angle (i.e., is in the form of a cone as shown in Figure 1), the electron diffraction pattern becomes an array of disks rather than an array of sharp spots as in TEM. The various manifestations of this type of electron diffraction pattern are known as Convergent Beam Electron Diffraction (CBED). The distance of each diffraction disk from the central beam may be calibrated to yield the interplanar *d*-value for a particular set of *hkl* planes. A diffraction disk containing no contrast detail is produced when a very thin region of a specimen (< 0.1 μm) is under the beam. Other than providing high spatial resolution (on the order of the electron beam size) this pattern of blank disks contains no more information than the typical selected area electron diffraction pattern (see the article on TEM).

Diffraction from thick crystals (0.1–0.5 μm) exhibits intensity variations within the disks caused by dynamical diffraction effects (see Steeds[1]). In this case the symmetry of the intensity variations provides information about the symmetry of the crystal that can be used as a "fingerprint" for phase identification.[8] Because the convergent electron beam senses the three-dimensional aspects of the specimen, CBED patterns from a crystal thicker than about 0.1 μm may be used to determine its point group, and often its space group.[9] If the magnification of the diffraction pattern is made very small and the convergence angle made very large so that the disks overlap, rings of intensity called *higher order Laue zone (HOLZ) rings* may be observed. These rings indicate that diffraction has occurred from other layers of the reciprocal lattice. If the crystal is tilted so that the beam is parallel to the [100], [010], and [001] crystal directions, the crystal lattice parameters along those directions may be determined. Also, fine details of a CBED pattern may yield a relative lattice parameter determination to better than 0.001 nm.

X-Ray Microanalysis

When energetic electrons bombard a solid, characteristic X rays from each element are generated that form the signal used in microanalysis. Characteristic X rays arise from de-excitation of atoms suffering inner-shell electron ionizations, and these X rays allow qualitative elemental identification and quantitative elemental composition determination. The X-ray signal is detected with an EDS placed close to the specimen inside the objective lens of the microscope (see the article on EDS). For materials science specimens, a quantification scheme for specific element pairs is well developed.[10] The ratio of elemental concentrations of two elements in the thin specimen may be determined by multiplying the X-ray intensity ratios of these elements by a sensitivity factor that depends only on the accelerating voltage and the X-ray detector configuration. When the elements in the specimen do not have large differences in their X-ray mass absorption coefficients, or when the specimen is very thin, corrections for X-ray absorption may be negligible and need not be applied to obtain an accuracy of 10–20% relative to the amount of element present. This modest level of analysis is still remarkable when it is realized that it may be obtained from regions of a specimen about 5 nm in diameter. To obtain quantitative results in the 5–10% range, an absorption correction should be applied using an estimate of the specimen thickness from CBED, EELS, or another method.[3–5] Using EDS methods, elemental detection is possible down to a few % wt. for elements having atomic numbers $Z < 11$ and down to 0.1–0.5 wt % for elements having $Z > 11$.

Microanalysis by Electron Energy-Loss Spectrometry

Electrons in the incident beam suffer inelastic collisions with atoms in the specimen; the effect of these collisions may be detected by measuring the energy of the primary electron after it has traversed the specimen. To observe a useful signal above background, very thin specimens (about 10–50 nm thick) must be used. Of the several inelastic events possible, the most useful for elemental analysis is the inner-shell ionization event that leads to characteristic X-ray and Auger electron emission. The shape of the spectrum at these characteristic inner-shell ionization energy losses is similar to the X-ray absorption edge. The signal intensity under the edge and above background can be related to the amount of the element in the specimen.[5] This microanalysis method is somewhat less accurate than EDS X-ray analysis because the ionization cross section, which is needed to convert the collected intensity to chemical composition, is often not well known. Details in the EELS spectrum reveal bonding information and information about the dielectric constant from regions of the specimen as small as 0.5 nm in diameter. Detection limits are similar to EDS, but the method is best applied to the K-edges for light elements from lithium to fluorine.

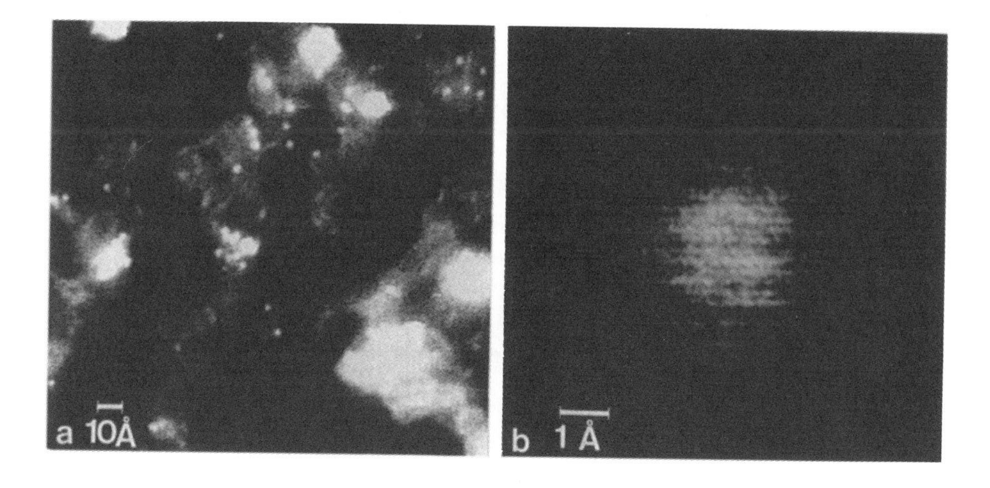

Figure 4 Annular dark-field STEM image of individual gold atoms on a very thin carbon film: (a) individual gold atoms appear as bright spots; and (b) higher magnification image showing a single gold atom. The scan lines are caused by the 0.25-nm electron beam traversing the gold atom about 15 times. (Courtesy of M. Isaacson)

Examples of Common Analysis Modes

The major STEM analysis modes are the imaging, diffraction, and microanalysis modes described above. Indeed, this instrument may be considered a miniature analytical chemistry laboratory inside an electron microscope. Specimens of unknown crystal structure and composition usually require a combination of two or more analysis modes for complete identification.

Conventional Bright-Field Imaging

Bright-field STEM images provide the same morphological and defect analysis typical of TEM images, such as particle sizing, interface analysis, and defect analysis (see Figure 3a). While the contrast may differ from TEM for thin crystalline materials, a dedicated STEM instrument using a field emission gun produces images that are similar enough to use the same image interpretation rationale developed for conventional TEM analysis.

Annular Dark-field Imaging

The annular dark-field detector of the field-emission STEM (see Figure 2) provides a powerful high-resolution imaging mode that is not available in the conventional TEM or TEM/STEM. In this mode, images of individual atoms may be obtained, as shown in Figure 4 (see Isaacson, Ohtsuki, and Utlaut[1]). Some annular dark-field

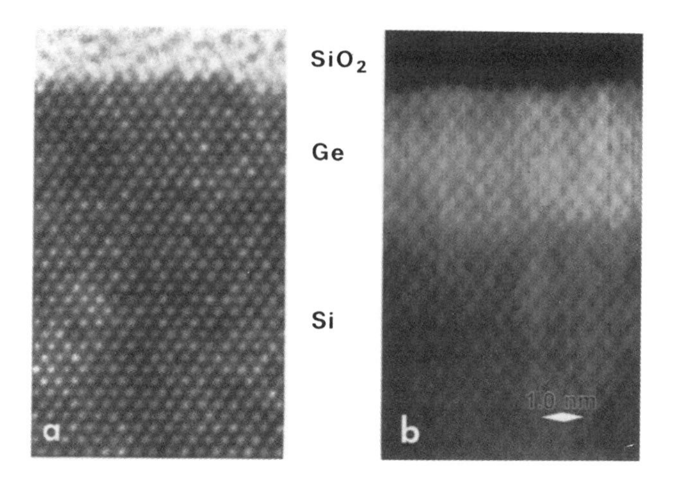

Figure 5 **Images of a thin region of an epitaxial film of Ge on Si grown by oxidation of Ge-implanted Si: (a) conventional TEM phase contrast image with no compositional information; and (b) high-angle dark-field STEM image showing atomically sharp interface between Si and Ge. (Courtesy of S.J. Pennycook)**

detectors have been modified to collect only the electrons that scatter into angles greater than 80 milliradians By collecting only the high-angle Rutherford-scattered electrons, images may be obtained that contain compositional information, as well as atomic-level image detail[11] (see Figure 5).

Point Group and Space Group Determination

Crystallographic structure determination is generally considered the realm of X-ray diffraction. However, three-dimensional sampling of the crystal by CBED allows first-principle determinations of point groups (crystal classes) for crystals as small as 100 nm, which could never be accomplished by X-ray methods. Figure 6 shows CBED patterns of a phase at the triple point of aluminum nitride grains containing Al, N, and O. These patterns helped to determine the point group and space group of this aluminum oxynitride spinel phase as m3m and Fd$\bar{3}$m, respectively.[12]

Microanalysis

Microanalysis of specimen regions in the nm-size range is one of the strongest reasons to use STEM. Figure 7 shows X-ray and EEL spectra taken simultaneously from the same area of a thin catalyst specimen. In each case the energy position of the peak or edge provides element identification, whereas the intensity above background for each peak and edge allows quantitative assessment of the composition. The statistics for these data are atypical owing to an intentionally short acquisition time of a few seconds. Figure 8a shows the statistics for a more typical X-ray spec-

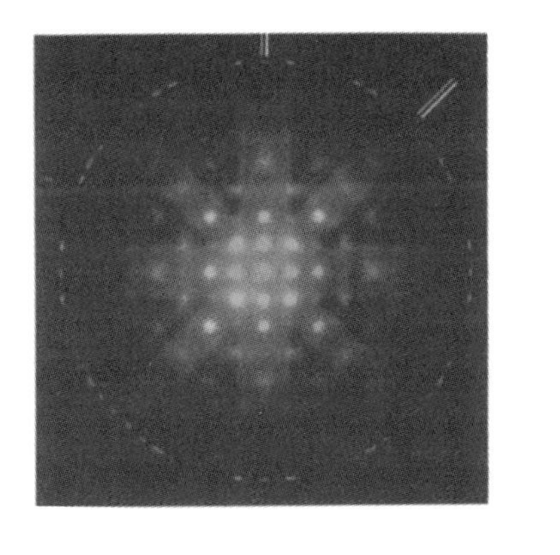

a

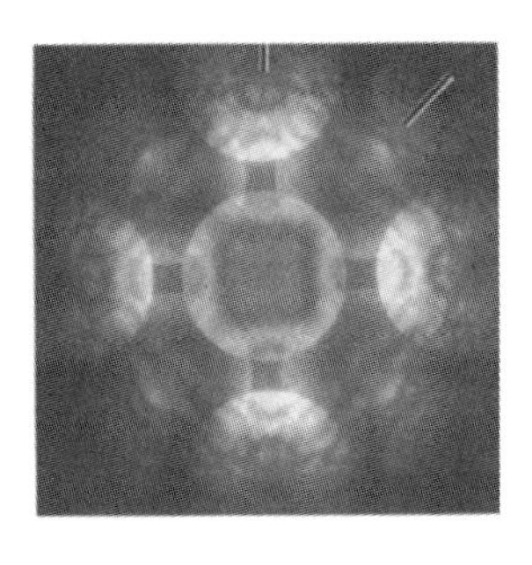

b

Figure 6 **CBED patterns of aluminum oxynitride spinel along the [001] direction. Symmetries in the patterns contributed to the determination of the point group and space group: (a) whole pattern showing 1st Laue zone ring; and (b) 0th order Laue zone. Both patterns show a fourfold rotation axis and two mirror planes parallel to the axis. (Courtesy of V. P. Dravid)**

trum collected over 100 s. Note the low-energy carbon and oxygen peaks collected with a windowless X-ray detector.

Phase Identification

Knowledge of the elements in an unknown phase, as determined by EDS or EELS, usually does not permit identification of the compound. However, this elemental information may be used in combination with interplanar spacing information (*d*-values) from one or more CBED patterns (or even the SAD patterns mentioned in the article on TEM) to render a positive identification. The task of searching all inorganic compounds to make a positive identification is now easier because of a new index to the JCPDS-ICDD Powder Diffraction File and the NIST Crystal Data File called the *Elemental and Lattice Spacing Index.*[13] The elements determined from EDS or EELS analysis are used for the primary search, which places all possible compounds on one or two pages for confirmation by matching to the ten largest interplanar spacings listed for each entry.

Compositional Imaging

Elemental distributions in a thin specimen may be obtained at high resolution from any properly prepared solid specimen using either EDS or EELS signals. These images are sometimes called elemental maps. Elemental images usually collected digitally by setting a region of interest in the spectrum for each element and storing the counts collected in these windows as a function of electron beam position (stor-

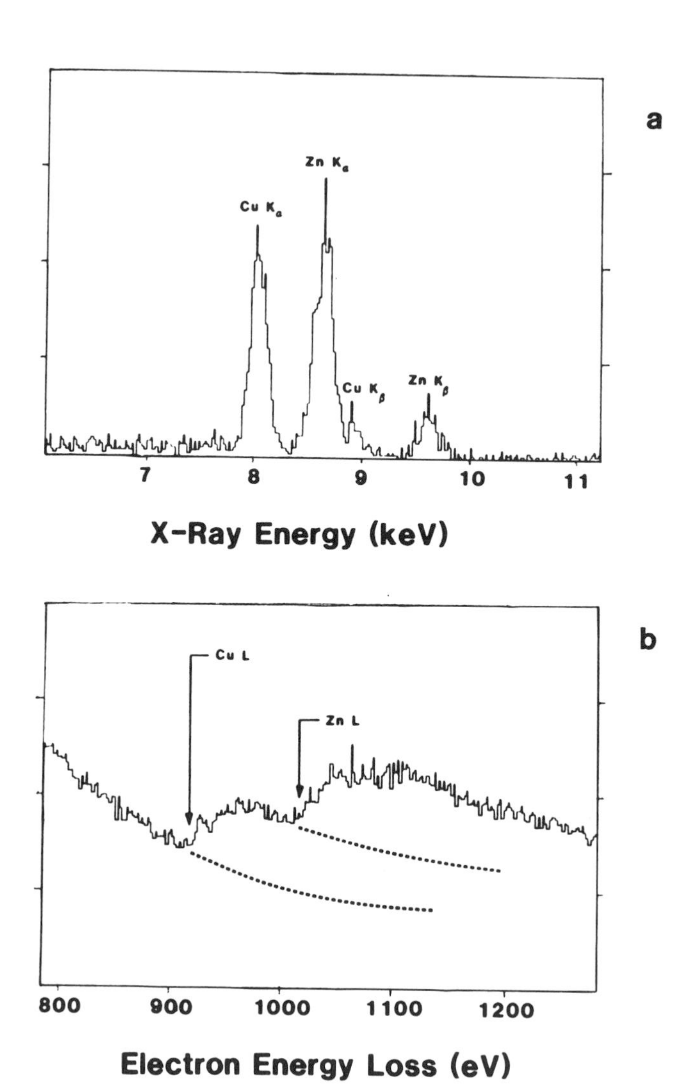

Figure 7 Microanalysis of a CuO/ZnO methanol synthesis catalyst with a field-emission STEM: (a) EDS data showing Cu and Zn K-lines; and (b) EELS data showing Cu and Zn L-edges with dotted lines indicating background levels. Spectra were taken simultaneously from a 2-nm diameter area. Signal intensities above background show that approximately the same relative amounts of Cu and Zn were measured by each method.

age at each image pixel), as shown in Figure 8b. An image of the X-ray background signal was collected separately to determine that the distribution of sulfur in Figure 8 is real.[14] Since the electron probe size is governed by the need for a large probe current (about 1 nA), image magnification, pixel density, and counting time

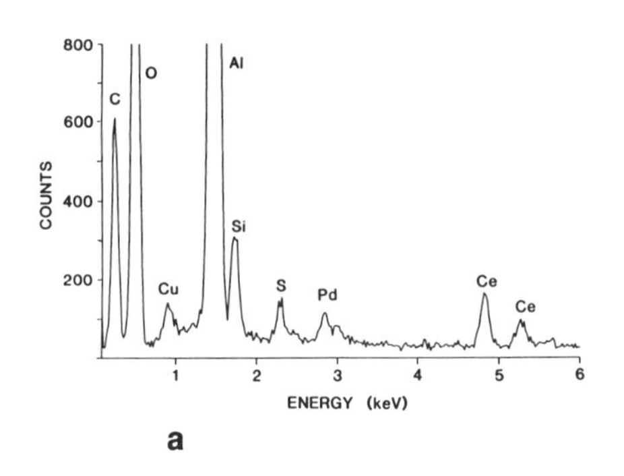

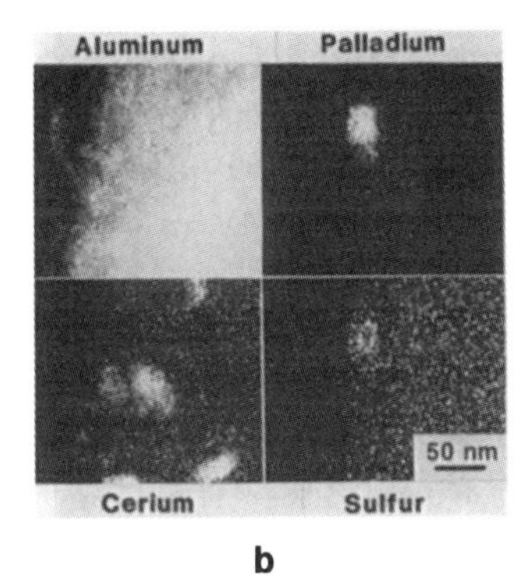

Figure 8 X-ray elemental imaging in a field-emission STEM: (a) EDS data of Pd / Ce / alumina catalyst particle poisoned with SO_2; and (b) 128 × 128 digital STEM images formed using X-ray counts collected at each image pixel for aluminum, palladium, cerium, and sulfur. (Courtesy of North-Holland Publishers)[14]

per pixel must be adjusted for the type of STEM instrument used. For example, if the pixel density is 128 × 128 with a counting time of 100 ms per pixel, a compositional image may be obtained in 28 min. However, for a thermionic source in a TEM–STEM system, where the counting time per pixel must be increased to 3 s per pixel, the total frame time for a similar image is about 14 hours. When elements are present in high concentration (>20 wt %), the resolution of the X-ray elemental image can be < 5 nm. Elemental images using EELS signals can be of even higher resolution, although more computation is necessary to subtract the background.

Sample Requirements

Specimens suitable for imaging in TEM are usually acceptable for STEM imaging and analysis. The principal methods for producing suitable thin specimens are electropolishing (metals only), ion-beam thinning (all hard materials), and ultramicrotomy (polymers, some metals, and small hard particles).[15] In the first two cases the specimen is thinned to perforation, and analysis takes place around the edge of the hole. Ultramicrotomed specimens are relatively uniform in thickness, but the fragile thin sections must be supported by a specimen grid. Hard, brittle materials also may be crushed to a powder and dispersed on a carbon film supported on a specimen grid, as in Figure 3. For microscopy of layered materials it is useful to dice the specimen to reveal a cross section of the layers before ion-beam milling. Precipitates in a metal matrix may be extracted using the extraction replica technique.

Microanalysis often places special constraints on the preparation of thin specimens beyond the general requirement to be transparent to 100-keV electrons.

CBED usually requires analysis along a particular crystallographic direction, and although specimen-tilting facilities are available in the microscope, often it is useful to orient the specimen before specimen preparation so that the thinned direction is along a crystal direction of interest. For high-quality microanalysis by EDS or EELS, the thin specimen must be free of surface films containing elements redistributed from the bulk or from contamination sources that would change the measured through-section composition. Ion-beam milling for short times usually removes such films deposited by electropolishing; however, ion-beam milling for long times may introduce elemental redistribution by differential sputtering. Specimens prepared by ultramicrotomy, extraction replication, or crushing generally do not have these compositional artifacts. Occasionally microanalysis is performed on specimens prepared by one of the latter methods while images of the same material might be obtained from electropolished or ion-beam milled specimens.

Artifacts

The major artifact typical of STEM imaging is a buildup of hydrocarbon contamination under the electron beam. This contamination appears in the bright-field image as a dark area in the shape of the scanning raster or as a dark spot if the beam has been stopped for microanalysis. Besides obscuring features in the image, contamination layers can absorb X rays from the lighter elements and increase the background in EELS analysis. Contamination can be reduced by heating the specimen in vacuum before examination, by cooling the specimen during analysis, and by improving the vacuum in the specimen chamber to better than 10^{-6} Pa (about 10^{-8} torr).

In addition to cleanliness (contamination effects), surface morphology and the alteration of composition during specimen preparation can cause serious artifacts in microanalysis. In some older instruments, the microscope itself produces undesirable high-energy X rays that excite the entire specimen, making difficult the accurate quantitation of locally changing composition. Artifacts also are observed in the EDS X-ray spectrum itself (see the article on EDS).

Conclusions

STEM can provide image resolution of thin specimens rivaling TEM, but in addition can provide simultaneous crystallographic and compositional analysis at a higher spatial resolution than any other widely-used technique. Any solid material may be examined, provided that a specimen can be prepared that is less than about 100 nm in thickness.

Future developments of this instrumentation include field emission electron sources at 200–300 kV that will allow better elemental detectability and better spatial resolution. Multiple X-ray detectors having large collection angles will also improve elemental detectability in X-ray microanalysis. The higher accelerating voltages should allow EDS X-ray and EELS compositional measurements to be made on specimens of the same thickness, instead of requiring a much thinner specimen for EELS. Electronic capture of digital diffraction pattern images should make some automation of diffraction pattern analysis possible. Under complete digital control automated experiments will be possible that cannot be accomplished manually.

Related Articles in the Encyclopedia

TEM, EDS, EELS, and SEM

References

1 *Introduction to Analytical Electron Microscopy.* (J. J. Hren, J. I. Goldstein, and D. C. Joy, eds.) Plenum, New York, 1979. Somewhat dated text that is still useful for certain chapters including those of Cowley, Steeds, and Isaacson, et al.

2 *Quantitative Microanalysis with High Spatial Resolution.* (G. W. Lorimer, M. H. Jacobs, and P. Doig, eds.) The Metals Society, London, 1981. Research papers giving results from many materials.

3 D. B. Williams. *Practical Analytical Electron Microscopy in Materials Science.* Philips Electronic Instruments, Mahwah, NJ, 1984. Concise textbook on CBED, EDS, and EELS with a pronounced "how-to" flavor.

4 *Principles of Analytical Electron Microscopy.* (D. C. Joy, A. D. Romig Jr, and J. I. Goldstein, eds.) Plenum, New York, 1986. An updated version of Reference. 1.

5 R. F. Egerton. *Electron Energy Loss Spectroscopy in the Electron Microscope.* Plenum, New York, 1986. The principle textbook on EELS.

6 *High Resolution Transmission Electron Microscopy and Associated Techniques.* (P. R. Buseck, J. M. Cowley, and L. Eyring, eds.) Oxford University Press, New York, 1988. A review covering these techniques in detail (except X-ray microanalysis) including extensive material on high-resolution TEM.

7 A. V. Crewe, J. Wall, and J. Langmore. *Science.* **168,** 1338, 1970. The classic first attempt to image single atoms with the STEM.

8 J. Mansfield. *Convergent Beam Electron Diffraction of Alloy Phases.* (by the Bristol Group, under the direction of J. Steeds, and compiled by J. Mansfield) Hilger, Bristol, 1984. This book is an atlas of CBED patterns that may be used to identify phases by comparing published patterns with experimental patterns.

9 B. F. Buxton, J. A. Eades, J. A. Steeds, and G. M. Rackham. *Phil. Trans. R. Soc. A.* **281,** 171, 1976. This paper outlined point group determination for the first time, but the major conclusions are also summarized in Williams (*op. cit.*).

10 G. Cliff and G. W. Lorimer. *J. Microscopy.* **103,** 203, 1975. This paper summarizes the Cliff-Lorimer analysis technique, but more complete reviews of the method may be found in References 1–4.

11 S. J. Pennycook. *EMSA Bulletin.* **19,** 67, 1989. A summary of compositional imaging using a high-angle annular dark-field detector in a field emission STEM instrument published by the Electron Microscopy Society of America, Box EMSA Woods Hole, MA 02543.

12 V. P. Dravid, J. A. Sutliff, A. D. Westwood, M. R. Notis, and C. E. Lyman. *Phil. Mag. A.* **61,** 417, 1990. An example of how to practically determine the space group of a phase.

13 *JCPDS-ICDD Elemental and Lattice Spacing Index* (1990). This index is available from JCPDS-International Centre for Diffraction Data, 1601 Park Lane Swarthmore, PA 19081.

14 C. E. Lyman, H. G. Stenger, and J. R. Michael. *Ultramicroscopy.* **22,** 129, 1987. This paper demonstrates high-resolution compositional imaging with the field-emission STEM.

15 *Speciman Preparation for Transmission Electron Microscopy of Materials* (J. C. Brauman, R. M. Anderson, and M. L. McDonald, eds.) MRS Symp. Proc vol. 115, Materials Research Society, Pittsburg, 1988. This conference proceedings contains many up-to-date methods as well as references to books on various aspects of specimen preparation.

3.5 EPMA

Electron Probe X-Ray Microanalysis[1]

DALE E. NEWBURY

Contents

Introduction

Electron Probe X-Ray Microanalysis (EPMA) is a spatially resolved, quantitative elemental analysis technique based on the generation of characteristic X rays by a focused beam of energetic electrons.[1–3] EPMA is used to measure the concentrations of elements (beryllium to the actinides) at levels as low as 100 parts per million (ppm) and to determine lateral distributions by mapping. The modern EPMA instrument consists of several key components:

1. An electron-optical column capable of forming a beam ranging in diameter from nm to μm and carrying a current ranging from pA to μA
2. An energy-dispersive X-ray spectrometer and at least one wavelength-dispersive X-ray spectrometer
3. An optical microscope for precise positioning of the specimen relative to the X-ray spectrometers
4. A vacuum system capable of operating at pressures ranging from 10^{-4} to 10^{-6} Pa

1. Note: This artcle is a contribution of the United States Government and is not subject to copyright.

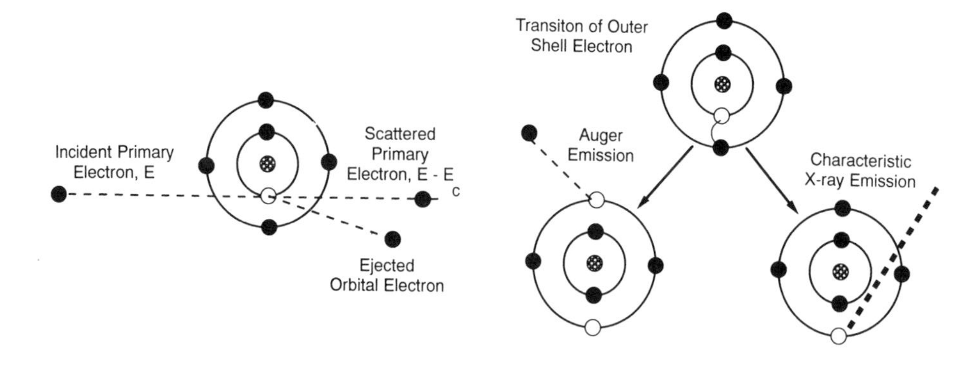

Figure 1 **Schematic of inner shell ionization and subsequent deexcitation by the Auger effect and by X-ray emission.**

5 A computer system to control the beam, spectrometers, specimen stage, and quantitative data processing.

The electron-optical performance of the EPMA system is indistinguishable from that of a conventional scanning electron microscope (SEM); thus, EPMA combines all of the imaging capabilities of a SEM with quantitative elemental analysis using both energy- and wavelength-dispersive X-ray spectrometry.[2, 3]

Physical Basis of EPMA

The physical basis of electron probe microanalysis is the generation of characteristic X rays by bombarding a solid with energetic electrons. Interactions of energetic electrons with matter include elastic scattering, which causes significant angular deviation of the electron trajectories, and inelastic scattering, which reduces the energy of the electrons and sets a limit to their range. Mechanisms of inelastic scattering include the production of secondary electrons, inner shell ionization, bremsstrahlung (continuum) X rays, long-wavelength electromagnetic radiation in the ultraviolet, visible, and infrared regions, electron–hole pairs, lattice vibrations (phonons), and electron oscillations (plasmons). The rate of energy loss depends on the electron energy and the target composition, and is typically 0.1–10 eV/nm. Inner shell ionization, illustrated in Figure 1, occurs when the beam electron ejects a bound atomic electron by transferring sufficient energy to the atom to exceed the critical excitation (or binding) energy E_c of an atomic electron. The resulting atomic vacancy is filled by a transition involving one or more outer shell electrons. The energy difference between shell levels can be manifested as an emitted X-ray photon, or the energy can be transferred to another outer shell electron, which is ejected as an Auger electron. Because the energy of the atomic shells is sharply defined, the emitted X ray or Auger electron has a precisely defined energy and is therefore characteristic of the atom species originally ionized by the beam electron.

For atoms having an atomic number greater than 10, the electron filling the inner shell vacancy may come from one of several possible subshells, each at a different energy, resulting in families of characteristic X-ray energies, e.g., the Kα, β family, the Lα, β, γ family, etc.

The beam electrons also produce X rays when they undergo deceleration in the Coulombic field of the specimen atoms, losing energy and emitting X rays known as "braking" radiation, or *bremsstrahlung.* Since this energy loss can take on any value from 0 to the entire energy carried by the high-energy electron, the bremsstrahlung X rays form a continuum background spanning all energies. Bremsstrahlung X rays are indistinguishable from characteristic X rays that may occur at the same energy. Statistical fluctuations in this bremsstrahlung background form the eventual limit that determines the ultimate sensitivity of electron probe X-ray microanalysis.

The critical parameter for X-ray generation is the overvoltage $U = E_0/E_c$, where E_0 is the incident beam energy. The intensity of characteristic X rays is given by:

$$I_{ch} = ai_B(U-1)^n \tag{1}$$

where a is a constant, i_B is the beam current, and the exponent n is approximately 1.5.

Equation (1) demonstrates that the analyst must choose a beam energy that exceeds the critical excitation energy for the species to be analyzed. In general, a value of $U > 2$ is required to achieve adequate efficiency in the production of X rays.

The volume of analysis, i.e., the diameter and depth of the analyzed region, is limited by a combination of the elastic and inelastic scattering. The maximum depth of the interaction volume is described by the Kanaya-Okayama electron range:

$$R(\mu m) = \left(\frac{0.0276A}{Z^{0.89}\rho}\right)E_0^{1.67} \tag{2}$$

where Z is the atomic number, A is the atomic weight (g/mole), ρ is the density (g/cm^3), and E_0 is the incident beam energy (keV). The maximum width of the interaction volume is similar in value to the range. When the production of X rays is considered, the range is reduced because X rays for a particular element cannot be produced below E_c for that element. Equation (2) is modified to reflect this condition:

$$R_X(\mu m) = \left(\frac{0.0276A}{Z^{0.89}\rho}\right)(E_0^{1.67} - E_c^{1.67}) \tag{3}$$

Monte Carlo electron trajectory simulations provide a pictorial view of the complex electron–specimen interaction. As shown in Figure 2a, which depicts the interac-

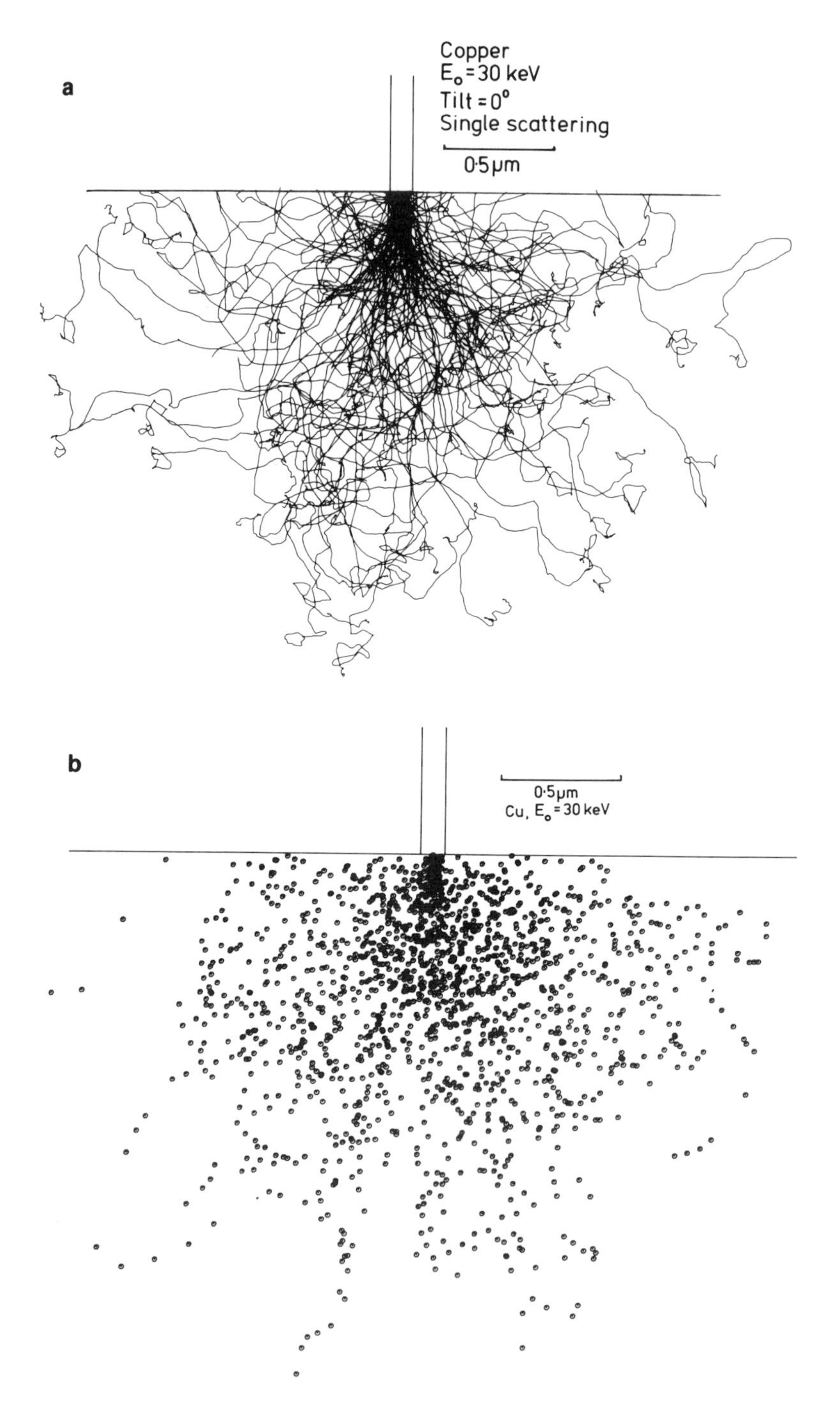

Figure 2 **(a) Monte Carlo simulation of electron trajectories in copper, beam energy 30 keV. (b) Corresponding sites of K-shell ionization.**

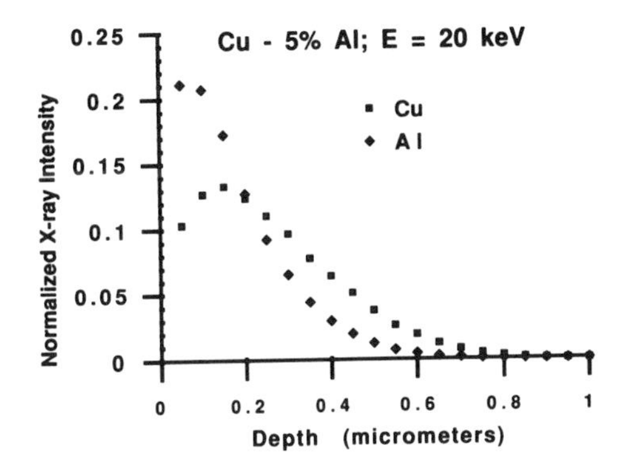

Figure 3 **Depth distribution of generation of Cu Kα X rays for an incident beam energy of 20 keV, and the effect of absorption.**

tion volume of a beam incident perpendicularly on a flat specimen of copper, the interaction volume has dimensions on the order of µm. Generally, the spread of the electrons through the specimen, not the diameter of the incident probe, is the limiting factor in determining the spatial resolution of analysis. A focused beam carrying sufficient electron current for analysis can be obtained with a diameter that is generally a factor of 5–10 smaller than the dimension of the interaction volume given by Equation (3). For example, the range for production of Cu Kα X rays at an incident beam energy of 20 keV in copper is approximately 1 µm, while a beam diameter of 100 nm (0.1 µm) would carry sufficient beam current for analysis. The production of X rays within the interaction volume is illustrated in Figure 2b. An important concept related to the interaction volume is the depth distribution of X rays, the so-called $\phi(\rho z)$ function shown in Figure 3. This function forms the basis for the development of matrix correction procedures in the quantitative analysis protocol described below.

Measurement of X Rays

The analyst has two practical means of measuring the energy distribution of X rays emitted from the specimen: energy-dispersive spectrometry and wavelength dispersive spectrometry. These two spectrometers are highly complementary; the strengths of each compensate for the weaknesses of the other, and a well-equipped electron probe instrument will have both spectrometers.

Energy Spectrometry (EDS) uses the photoelectric absorption of the X ray in a semiconductor crystal (silicon or germanium), with proportional conversion of the X-ray energy into charge through inelastic scattering of the photoelectron. The quantity of charge is measured by a sophisticated electronic circuit linked with a computer-based multichannel analyzer to collect the data. The EDS instrument is purely a line-of-sight device; any X ray emitted from the specimen into its solid angle of collection can be measured. To reduce thermal noise, the detector crystal and preamplifier are cooled to liquid nitrogen temperatures. For application to conventional electron-optical columns with a pressure on the order of 10^{-4} Pa, a thin window of beryllium, diamond, boron carbide, or a metal-coated polymer is used to protect the crystal from condensing water vapor from the microscope environment. The window material, the gold surface electrode, and an inevitable "dead layer" of partially active semiconductor just below the electrode act to absorb X rays. With optimum low-mass window materials like diamond or boron carbide, the EDS can measure X rays as low in energy as the Be K level (0.109 keV).

Wavelength Spectrometry (WDS) is based upon the phenomenon of Bragg diffraction of X rays incident on a crystal. The diffraction phenomenon is described by the expression:

$$n\lambda = 2d\sin\theta_B \tag{4}$$

where λ is the wavelength equivalent of the X-ray energy E: λ (nm) = 1.2398/E (keV), d is the crystal spacing for the diffracting planes; and θ_B is the angle at which constructive interference occurs (i.e., the Bragg angle). The WDS instrument is mechanical in nature and is a focusing device (see Figure 1). The X-ray source at the specimen, the diffracting crystal, and the gas counter to detect the X rays must follow a specific geometry (a Rowland circle) to satisfy Equation (4); accurate positioning is ensured by incorporating an optical microscope into the electron probe. To detect a wide energy range, different crystals with various d spacings are mounted on a mechanical turret, and are selectable under computer control.

EDS and WDS spectra of a multielement glass, shown in Figure 4, illustrate some major strengths and weaknesses:

1 Spectral acquisition. An EDS instrument can measure any X-ray photon that reaches the detector, and thus can continuously view the entire X-ray spectrum (0.1–20 keV), while the WDS instrument must be mechanically scanned with several crystal changes to acquire a complete spectrum over that same range. In Figure 4, for equivalent sensitivity, the WDS spectrum required approximately a factor of 10 greater acquisition time. When an unknown is examined, it is highly unlikely that a complete qualitative WDS scan will be made at every location chosen for analysis. Since EDS is always capable of measuring a complete spectrum, unexpected major and minor constituents can always be detected. For survey work of major and minor constituents in unknowns, EDS therefore has a significant advantage for rapid qualitative analysis.

2 Spectral resolution. Consider a Mn Kα photon, which has a natural peak width (the full width at half maximum, or FWHM) of approximately 1

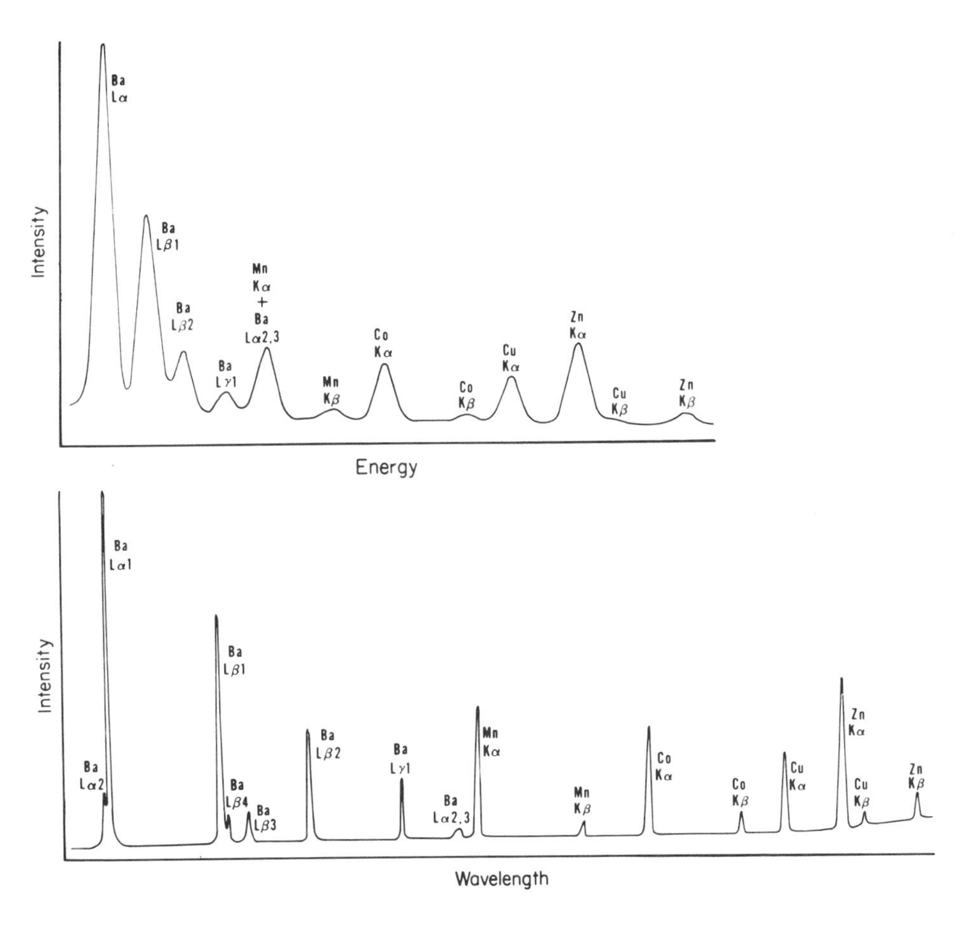

Figure 4 **Comparison of EDS and WDS spectra from a complex, multielement glass. (Courtesy of Charles Fiori, National Institute of Standards and Technology)**

eV. In WDS, the diffraction process, diffraction crystal imperfections, and the mechanical construction degrade this peak to a FWHM of approximately 10 eV. In EDS, the statistics of charge carrier creation degrade the peak to a FWHM of approximately 140 eV. The effect of spectral resolution can be seen readily in Figure 4, where a single peak in the EDS spectrum is actually composed of Mn Kα (keV) + Ba $L\alpha_{2,3}$. This double peak is easily resolved using WDS into two peaks that differ in relative intensity by a factor of 10. Although powerful methods for mathematical spectral deconvolution of peaks can be applied to EDS spectra, deconvolution is likely to produce false solutions when the peaks are separated by less than 50 eV, when the peaks are sharply different in relative intensity, and when significant statistical fluctuations exist in the spectral data, which is usually the situation in practical analysis. A classic example of an EDS interference situation encountered in practical analysis is that of S Kα, β (2.307 keV), Mo Lα, β (2.29 keV), and Pb Mα, β (2.346 keV); this interference situation is easily resolved by WDS.

3 Peak-to-background ratio and limits of detection. The peak-to-background ratio, which is a direct consequence of the spectral resolution, plays a major role in determining the limits of detection. A peak is more sharply defined with WDS than with EDS. The peak-to-background ratio is proportionally higher, since the intensity due to the characteristic radiation is spread over a smaller energy range of the X-ray continuum background. In practical analysis situations, the limit of detection for WDS is 100 ppm while that for EDS is approximately 0.1% wt. (1000 ppm) for elements having an atomic number > 11 (sodium). The detection limits are substantially poorer for light elements because of the high absorption losses in the specimen and spectrometer components. Depending on the matrix, the detection limit for light elements may be as high as 1–10% wt.

4 Dead time and limiting count rate. The pulse-processing time (dead time) is on the order of 10–50 μs for EDS and 0.5–1 μs for WDS. Because the WDS makes use of the diffraction effect to disperse the X rays before the detector, the limiting count rate applies to a single X-ray peak of interest; limiting count rates of 10^5 cps are possible and trace constituents can be measured with no dead time from major constituent peaks. The EDS instrument, which has a limiting count rate of approximately 2.5×10^4 cps, must process every X-ray photon in the range from the low-energy cutoff (approximately 100 eV) to the beam energy. The EDS dead time applies to the total spectrum rather than to a single peak of interest, making trace constituent measurement much more difficult.

5 Spectrometer positioning. The WDS instrument is a focusing optical device, with an ellipsoid of volume in focus for uniform X-ray transmission. The long axes of the ellipsoid have dimensions on the order of 100 to 1000 μm, while the short axis is approximately 10 μm. Precise positioning is vital if consistent X-ray intensity measurements are to be obtained. On a formal electron probe, this positioning is ensured by incorporating an optical microscope with its shallow depth of field and mechanically translating the specimen stage. The EDS instrument is a nonfocusing device, and it is therefore not critical to position it accurately relative to the specimen.

Qualitative Analysis

Qualitative analysis is generally assumed to be a trivial procedure consisting of matching the observed peaks to tabulated energies. Identification of peaks of major constituents is possible with a high degree of confidence, but misidentification of low-intensity peaks from minor and trace constituents is very much a possibility unless a careful strategy is followed. A good qualitative analysis strategy consists of first identifying the principal peaks of a major constituent and then locating all associated low-intensity peaks, including other X-ray family members, as well as all spectral artifacts, such as escape and sum peaks for EDS; for WDS, multiple-order lines—$n = 2, 3, 4$, etc., in Equation (4)—and satellite lines must be located. Only then should remaining peaks be identified as minor and trace constituents. Generally, the confidence with which a minor or trace constituent can be identified will be reduced because fewer peaks can be recognized above background. Absolutely no spectral smoothing should be applied in EDS spectra, since this introduces false "peaks" due to statistical fluctuations.

Quantitative Analysis

The keystone of practical quantitative electron probe microanalysis is Castaing's first approximation, which relates the concentration for a constituent in the unknown to the concentration in a standard in terms of the ratio of X-ray intensities generated in the target:[1]

$$\frac{C_i}{C_{i^*}} \sim \frac{I_i^G}{I_{i^*}^G} \tag{5}$$

where C is the mass concentration, I^G is the generated characteristic peak intensity (corrected for dead time, peak overlaps, and background), i denotes a specific constituent, and * denotes the standard, such as a pure element. What is actually measured is the ratio of emitted intensities $I_i^E / I_{i^*}^E$, the so-called *k value.* The emitted X-ray intensity ratio deviates from the generated intensity ratio in Equation (5) because of interelement or matrix effects; that is, the intensity of element i is affected by the presence of other elements j, k, l, etc., in the specimen. The origin of these matrix effects can be obtained by examining Figures 2 and 3.

Matrix Corrections

Atomic Number Effect

A significant fraction—about 15% for aluminum, 30% for copper, and 50% for gold—of the beam electron trajectories intersect the surface and escape as backscattered electrons (BSEs), as shown in Figure 2a. The majority of BSEs escape with more than half of their incident energy. Clearly, if these electrons had remained in the specimen, they would have continued to scatter inelastically to produce additional X rays; their loss reduces the total possible generated intensity. The X-ray loss due to backscattering depends strongly on the specimen's composition, and increases with average atomic number. A related effect concerns the stopping power, or rate of energy loss, of the electrons within the specimen, which is also composition dependent.

Absorption Effect

Figures 2b and 3 demonstrate that X rays are produced over a range of depth into the sample. The X rays must propagate along a finite path through the specimen to reach the detector, and are subject to photoelectric absorption and scattering, which follows an exponential relation:

$$\frac{I}{I_0} = \exp\left[\left(\frac{-\mu}{\rho}\right) \rho z \csc \psi \right] \tag{6}$$

where I is the X-ray intensity; μ/ρ is the mass absorption coefficient (cm^2/g), which depends on composition; z is the depth below the surface (cm); and ψ is the detector take-off angle, which is the angle between the detector's axis and the specimen's surface. Figure 3 shows the distribution of X-ray production with depth that escapes the specimen for Cu Kα and Al Kα in a 90% copper–10% aluminum alloy. Although the Al Kα X ray is actually produced deeper in the specimen according to Equation (3), self absorption of the X rays by the specimen limits the sampling depth to be less than that for Cu Kα X rays. The sampling depths are quite different for Cu Kα and Al Kα in this case because of the difference in the mass absorption coefficients. Note that the absorption effect becomes greater as the X rays are produced deeper in the specimen, which provides another reason to operate at the lowest practical beam energy to minimize the range of the electrons.

Fluorescence Effect

A consequence of absorption of X rays is the inner shell ionization of the absorbing atoms and the subsequent generation of characteristic X rays from the absorbing atoms, called secondary fluorescence, which raises the generated intensity over that produced by the direct action of the beam electrons. Secondary fluorescence can be induced by both characteristic and bremsstrahlung X rays. Both effects are compositionally dependent.

Quantitative Calculations

Fortunately, the physics of interaction of energetic electrons and X rays with solid matter is sufficiently well understood to permit implementation of a quantitative analysis procedure, the so-called ZAF method, which is based upon a combination of theory and empiricism, to calculate separate correction factors for each of the matrix effects. Each factor—atomic number Z, absorption A, and fluorescence F—is calculated as a ratio for the postulated specimen composition compared to the known composition of the standard. These factors multiply the measured k value to give a concentration ratio:

$$C_i = ZAF\left(\frac{I_i^E}{I_{i^*}^E}\right)C_{i^*} \tag{7}$$

Since the composition of the unknown appears in each of the correction factors, it is necessary to make an initial estimate of the composition (taken as the measured k value normalized by the sum of all k values), predict new k values from the composition and the ZAF correction factors, and iterate, testing the measured k values and the calculated k values for convergence. A closely related procedure to the ZAF method is the so-called $\phi(\rho z)$ method, which uses an analytic description of the X-ray depth distribution function determined from experimental measurements to provide a basis for calculating matrix correction factors.

A particular strength of Equation (7) is that the intensity ratio is formed between measurements of the same X-ray energy in both the unknown and standard. This procedure has significant advantages: First, there is no need to know the spectrometer's efficiency, a value that is very difficult to calibrate absolutely, since it appears as a multiplicative factor in both terms and therefore cancels. Second, an exact knowledge of the inner shell ionization cross section or fluorescence yields is not needed, since they also cancel in the ratio.

Accuracy of Electron Probe Microanalysis

Experience gained in the ZAF analysis of major and minor constituents in multielement standards analyzed against pure element standards has produced detailed error distribution histograms for quantitative EPMA. The error distribution is a normal distribution centered about 0%, with a standard deviation of approximately 2% relative. Errors as high as 10% relative are rarely encountered. There are several important caveats that must be observed to achieve errors that can be expected to lie within this distribution:

1. The characteristic peaks must be separated from the background. In EDS this is usually accomplished by mathematical filtering of the spectrum or by background modeling. In WDS, the peak is sufficiently sharp to permit background measurement by detuning the spectrometer.
2. The characteristic peaks must be deconvolved to eliminate peak interference; powerful deconvolution algorithms exist for EDS and WDS.
3. The unknown and standards must be measured under identical conditions of beam energy and spectrometer parameters. The specimen's surface must be oriented to known angles relative to the electron beam and the detector. All meas-

urements must be normalized to the same electron dose.

4 Differences in the specimen composition relative to the standard are the only source of difference in the measured X-ray intensities. There are no specimen size or shape effects influencing the electron–X-ray interactions, as in the case of particles or rough surfaces. To ensure this, the specimen must be flat and mirror-polished, with fine-scale surface topography reduced to less than 2% of the dimensions of the interaction volume. The specimen must be of sufficient thickness, typically > 50 μm, to contain the electron-excited volume as well as the full range of X-ray induced secondary fluorescence. No chemical polishing or etching can be tolerated because of the danger of modifying the near-surface composition through selective leaching.

5 The specimen must be homogeneous throughout the interaction volume sampled by the beam, since X rays of different energies originate from different depths.

6 The atomic number of the analyzed species must be 11 (sodium) or higher. For light elements from beryllium to fluorine, the low energy of the characteristic X rays leads to high absorption situations and an absorption correction factor exceeding 2. The mass absorption coefficients for the X rays of these elements are poorly known, so that large relative errors exceeding 10% are often encountered. Consequently, in fully oxidized systems, oxygen is frequently calculated indirectly by means of assumed stoichiometry for the cation species.

7 If all constituents are separately measured, or are calculated by stoichiometry, as in the case of oxygen, then the analytical total, which is the sum of all elements plus any calculated by stoichiometry, conveys useful information. If the analytical total is above 102%, a possible mistake in the measurement of the electron dose or an unexpected change in spectrometer parameters is likely. If the analytical total is significantly below 98%, an additional constituent is likely to have been missed.

When multielement standards similar in composition to the unknown are available, a purely empirical mathematical procedure, the Bence-Albee or Ziebold-Ogilvie method, can be applied. This procedure is often used in the analysis of minerals for which close standards exist.

The "standardless" approach attempts to apply first principles descriptions of X-ray production to the calculation of interelement relative sensitivities. Several of the key parameters necessary for first principles calculations are poorly known, and the accuracy of the standardless method often suffers when different X-ray families must be used in measuring several elemental constituents in a specimen.

Special Cases of Quantitative Analysis

Particles, rough surfaces, inclusions, layered film materials, and films on substrates represent special cases of considerable technological importance where the conventional quantitative analysis procedures fail. In such materials, shape and dimensions can influence the interaction of the beam electrons and the production of X rays. Several different strategies have been developed for dealing successfully with such geometric effects.[2, 3] These methods range from simple approaches, such as normalization of the analytical total to compensate for particle effects, to complex mathematical treatments, including theoretical or empirical modifications of the ZAF correction factors and rigorous Monte Carlo electron trajectory simulation of the exact geometrical boundary conditions of the specimen. Films on substrates, including multilayers, can be analyzed with corrections calculated by the $\phi(\rho z)$ method.[5] For particles and rough surfaces, errors in excess of 100% relative may be encountered with calculations uncorrected for geometrical effects; with simple normalization the maximum errors will generally be less than 25% relative. When the feature of interest is an inclusion contained in a matrix of different composition, where the lateral and depth dimensions of the inclusion are less than the X-ray range, quantitative and even meaningful qualitative analysis becomes impossible, since the interaction volume simultaneously samples both the inclusion and the surrounding matrix. In such cases, it may be necessary to mechanically or chemically separate the inclusions, and to analyze them as freestanding particles.

Compositional Mapping

Because X-ray counting rates are relatively low, it typically requires 100 seconds or more to accumulate adequate counting statistics for a quantitative analysis. As a result, the usual strategy in applying electron probe microanalysis is to make quantitative measurements at a limited collection of points. Specific analysis locations are selected with the aid of a rapid imaging technique, such as an SEM image prepared with backscattered electrons, which are sensitive to compositional variations, or with the associated optical microscope.

Often, more detailed information is needed on the distribution of a constituent. The technique of X-ray area scanning, or *dot mapping*, can provide a qualitative view of elemental distributions. As the beam is scanned in a raster pattern on the specimen, a cathode ray tube scanned in synchronism is used to display a full white dot whenever the X-ray detector (WDS or EDS) detects an X ray within a certain narrow energy range. The pattern of dots is recorded on film to produce the dot map. Dot maps are subject to the following limitations:[4]

1. The information is qualitative in nature. The area density of dots suggests local concentration differences, but the count rate at each point, which is fundamental information required for quantitation, is lost.
2. Because of count rate performance and peak-to-background, WDS is preferable to EDS for mapping, particularly for minor and trace constituents.

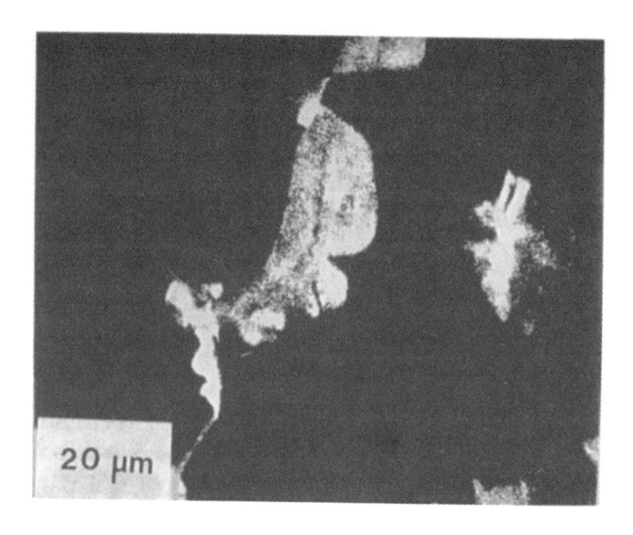

Figure 5 **X-ray area scan (dot map) showing the distribution of zinc at the grain boundaries of copper.**

3 Mapping of major constituents can be carried out in approximately 15–30 minutes of scanning per image. Minor constituents require 0.5–3 hours, and trace constituents require 3–10 hours. An example of a dot map of zinc at concentrations in copper as low as 1% is shown in Figure 5; 6 hours of scan time was needed to produce a dot map at this level.

4 While low concentrations can be mapped against a background of 0 concentration, as shown in Figure 5, dot mapping has poor contrast sensitivity. It would be difficult to map a 5% concentration modulation above a general level of 50% because there is no provision for subtracting the influence of the dominant general signal.

5 Since no background correction can be made, dot maps of minor and trace constituents are subject to possible artifacts caused by the dependence of the bremsstrahlung on composition, particularly with EDS X-ray measurement.

Quantitative compositional mapping overcomes the limitations of dot mapping by recording in computer memory the count rates for all constituents of interest at all picture elements in the scan.[4] Complete quantitative procedures are then applied to the count rate arrays: dead time correction, background correction, standardization, and matrix correction. The resulting concentration arrays are presented as images on a digital display with a gray or color scale encoding the

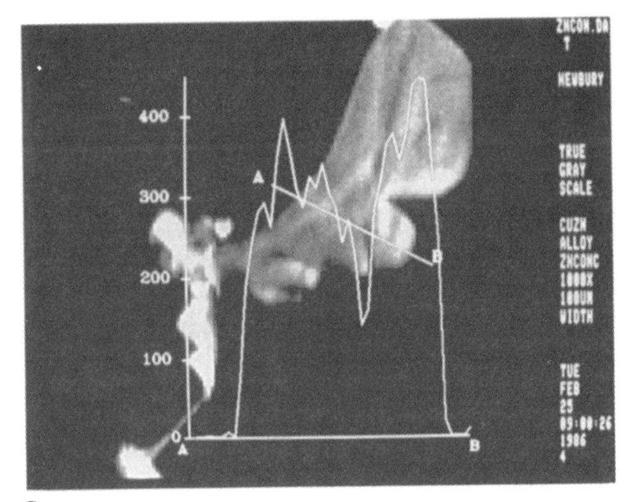

a

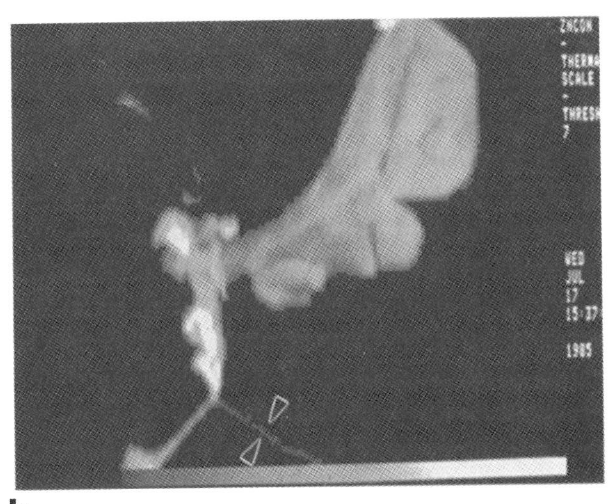

b

Figure 6 (a) Quantitative compositional map of the distribution of zinc at the grain boundaries of copper; gray scale corresponds to the concentration range 0–10% wt.; superposed compositional profile (vertical scale, 0–4.0% wt.) along the locus $\overline{AB}$. (b) Contrast enhancement applied to emphasize low-concentration structure (0.1% wt. = 1000 ppm) at arrows; image width 100 μm.

concentration information. Typical dwell times per pixel are 0.1–2 s, which restricts detection limits in imaging to approximately 500 ppm with high beam currents and WDS detection; comparable accuracy to single point analysis can be achieved in the mapping mode. An example is shown in Figure 6a of a digital compositional map of the same area as Figure 5; concentration levels as low as 0.1% wt. are visible. Digital compositional maps can be subjected to subsequent image processing to enhance features of interest. Contrast expansion of Figure 6a, shown in Figure 6b, reveals fine details in the zinc diffusion zone. This combination of images supported by complete numerical concentration values at every pixel, including the measurement statistics, is a powerful tool for solving a wide range of problems.

Conclusions

The electron probe X-ray microanalyzer provides extraordinary power for measuring the elemental composition of solid matter with μm lateral spatial resolution. The spatial resolution, limited by the spread of the beam within the specimen, permits pg samples to be measured selectively, with elemental coverage from boron to the actinides. By incorporating the imaging capability of the SEM, the electron probe X-ray microanalyzer combines morphological and compositional information.

Several future trends are evident. The great utility of true compositional maps suggests that this mode of operation will replace the now antiquated dot maps. Greater use of computer-aided imaging and analysis will increase the efficiency of the instrument in applications where large quantities of data are required. Improved long-term stability will permit the long counting times required for routine application of trace analysis. Continued development of advanced correction methods will address special difficulties raised by complex specimens such as multilayers and inhomogeneous particles. Lastly, the need for improved lateral and depth spatial resolution will lead to increased application of low beam energy analysis.

Related Articles in the Encyclopedia

EDS, XRF, and SEM

References

1 R. Castaing. *Application of Electron Probes to Local Chemical and Crystallographic Analysis.* Ph.D. thesis, University of Paris, 1951.

2 J. I. Goldstein, D. E. Newbury, P. Echlin, D. C. Joy, C. E. Fiori, and E. Lifshin. *Scanning Electron Microscopy and X-Ray Microanalysis.* Plenum, New York, 1981. References 2 and 3 are comprehensive textbooks covering all

aspects of electron probe microanalysis and the associated technique of scanning electron microscopy.

3 D. E. Newbury, D. C. Joy, P. Echlin, C. E. Fiori, and J. I. Goldstein. *Advanced Scanning Electron Microscopy and X-Ray Microanalysis.* Plenum, New York, 1986.

4 D. E. Newbury, C. E. Fiori, R. B. Marinenko, R. L. Myklebust, C. R. Swyt, and D. S. Bright. Compositional Mapping with the Electron Probe Microanalyzer. *Anal. Chem.* 1990, 62, Part I, 1159A; Part II, 1245A.

5 G. F. Bastin and H. J. M. Heijligers. Quantitative Electron Probe Microanalysis of Ultralight Elements (Boron–Oxygen). Scanning. **12,** 225, 1990.

4

STRUCTURE DETERMINATION BY DIFFRACTION AND SCATTERING

4.0 INTRODUCTION

This chapter contains articles on six techniques that provide structural information on surfaces, interfaces, and thin films. They use X rays (X-ray diffraction, XRD, and Extended X-ray Absorption Fine-Structure, EXAFS), electrons (Low-Energy Electron Diffraction, LEED, and Reflection High-Energy Electron Diffraction, RHEED), or X rays in and electrons out (Surface Extended X-ray Absorption Fine Structure, SEXAFS, and X-ray Photoelectron Diffraction, XPD). In their "usual" form, XRD and EXAFS are bulk methods, since X rays probe many microns deep, whereas the other techniques are surface sensitive. There are, however, ways to make XRD and EXAFS much more surface sensitive. For EXAFS this converts the technique into SEXAFS, which can have submonolayer sensitivity.

The techniques can be broadly classified into two groups: those which directly identify the atomic species present and then provide structural information about the identified species from diffraction or scattering effects (EXAFS, SEXAFS, and XPD); and those which are purely diffraction-based and do not directly identify the atoms involved, but give long-range order information on atomic positions from diffraction patterns (XRD, LEED, and RHEED). The latter group is only concerned with crystalline material and, for individual crystalline phases present, average unit cell dimensions, symmetries, and orientations are obtained directly from the diffraction patterns. Deviations from the average, i.e., defects of some sort, show up as a broadening of diffraction peaks, or, if they are periodic, as splittings of peaks.

XRD is the most widely used technique for general crystalline material characterization. Owing to the huge data bank available covering practically every phase of every known material (powder diffraction patterns), it is routinely possible to identify phases in polycrystalline bulk material and to determine their relative amounts from diffraction peak intensities. Phase identification for polycrystalline thin films, using standard equipment and diffraction geometries, is also possible down to thicknesses of 100 Å. For completely random polycrystalline thin films relative amounts are also easily determined. Once preferred orientations occur (texturing) this gets more difficult, requiring the collection of much more data or the introduction of more sophisticated equipment with different diffraction geometries so that the orientations can be "seen" effectively. These diffraction geometries include Grazing Incidence XRD (GIXRD), in which case the X-ray probing depth is greatly reduced. This has the effect of greatly improving surface sensitivity and allowing a depth profiling mode (50 Å to microns) by varying the incidence angle. When coupled to a synchrotron radiation X-ray source (to produce an intense, parallel X-ray beam) monolayer sensitivity can be achieved by GIXRD. GIXRD can therefore be used for the extreme situation of surface structure, or epitaxial relationships at the interfaces of films, where atomic positions can be determined to an accuracy of 0.001 Å in favorable cases.

Besides phase identification XRD is also widely used for strain and particle size determination in thin films. Both produce peak broadenings, but they are distinguishable. Compared to TEM, XRD has poor area resolution capability, although by using synchrotron radiation beam diameters of a few µm can be obtained. Defect topography in epitaxial films can be determined at this resolution.

Since LEED and RHEED use electron, instead of X ray, diffraction, the probing depths are short and surface crystallographic information is provided. Vacuum is also required. LEED uses normal incidence, with electron energies between 10 and 1000 eV, whereas RHEED (5–50 keV) uses grazing incidence and detection. The grazing geometry restricts RHEED information to the top few atomic layers, even though it is intrinsically less surface sensitive than the lower energy LEED. In both cases diffraction comes from two-dimensional rows of atoms, compared to the three-dimensional planes of atoms in XRD. In LEED the diffraction pattern (electron intensity "spots" projected back onto a phosphor screen; cf., an X-ray back-reflection Laue diffraction photograph) directly reveals the size and shape of the unit cell of the outermost, ordered atomic layer of single crystal material. To establish the locations of the atoms within the unit cell and the separation of the surface plane from the bulk requires detailed measurements of the diffraction peak intensities versus the incident electron energy (wavelength), plus a comparison to a complex theoretical treatment of the electron scattering process in the material. Because of this only relatively few laboratories use LEED this way to determine the structures of clean, terminated bulk surfaces (which are often different from the bulk) and ordered adsorbed overlayers. The major use is to qualitatively check that single crystal surfaces are "well-ordered" and to determine the unit cell symmetries and dimensions of adsorbed overlayers. In analogy to the three-dimensional situation with XRD, disorder in the surface plane broadens LEED diffraction spots, and, if periodic, splits them. Information on finite island sizes (cf., particle size by XRD), strain, step densities, etc., can be obtained from detailed studies of the peak line shapes. Owing to its grazing angle geometry it is more complex, in general, to extract unit cell parameters from RHEED. If, however, the incident beam is aligned along a major surface crystallographic direction, this can be achieved fairly simply. Again, surface disorder translates to diffraction spot broadening. In practice RHEED is used primarily to monitor *in situ* the nucleation and growth of epitaxial films (e.g., MBE). The grazing geometry leaves the surface very "open" for the deposition process. In addition it makes RHEED very sensitive to surface roughness effects, such as island formation, since the reflected beam is actually transmitted through island "asperities," giving rise to new diffraction features. In standard form neither LEED nor RHEED have much spatial resolution, beam spot sizes being a fraction of a mm. Microscopic modes exist, however, using electron beam columns. They are Low-Energy Electron Microscopy, LEEM, and Scanning Reflection Electron Microscopy, SREM. Neither technique is discussed here, since they are too specialized.

In EXAFS a tunable source of X rays (supplied by a synchrotron facility) is used to scan energies from below to above the Binding Energy, BE, of electron core levels for any atom present. An increase in absorption associated with excitation of that electron occurs at the BE. Detection of such absorption "edge jumps" (usually measured using transmission through many thousands of angstroms of material) directly identifies the atoms present. Above the absorption edge weak periodic oscillations in the absorption strength occur due to interferences between the outgoing photoelectron wave from the absorbing atom and backscattering of this wave from neighboring atoms. The oscillations thus contain local radial distribution information (within ~5 Å of the absorbing atom). The frequency (or frequencies) of the oscillations relate to the distance of the surrounding atomic shell (or shells); the amplitudes relate to the number and types of surrounding atoms. If multiple sites of the specific element concerned are present a superimposition of oscillation frequencies is obtained that can be difficult to analyze, but for single site situations accuracies to ~ 0.03 Å in bond length can be achieved using model compounds of known structure. Since local order is probed, the material need not be crystalline. High-Z elements are more easily studied than low-Z elements, because the signal strengths are higher and because, in general, the regions above their higher energy core level BEs are clear of other interfering absorption features over a large energy range. In bulk materials heavy elements can be detected and studied down to around 100 ppm. Also, for X-ray energies above ~5 keV samples can be studied under ambient pressures instead of in vacuum. The technique is widely used for bulk catalysts and in biological materials.

If the emitted electron intensity resulting from the core-level excitation process is used to detect the X-ray absorption process (in reflection), the bulk EXAFS method is turned into the surface-sensitive SEXAFS technique, where the depths probed are only a few atomic layers. Surface structures (absorption sites, bond lengths) for atoms absorbed on single crystal surfaces have been determined this way. The advantage compare to LEED is that the information is element specific, and long-range order of the adsorbate is not required. Accurate information is only obtainable, however, when the atom occupies a single unique site. Often C, O, N, S, etc., are studied. These light elements have core levels in the low-energy "soft X-ray" region so that vacuum techniques are required. For monolayer surface studies this is always true, anyway. It is also possible to use X-ray fluorescence as a detection scheme, instead of electrons. In this case the probing depth is extended to many hundreds of angstroms for the light elements, and deeper for the heavy elements.

For molecules adsorbed on surfaces strong features often appear in the spectrum at or very near the absorption edge-jumps (termed Near Edge X-ray Absorption Fine Structure, NEXAFS.) These are excitations from the core level to specific unoccupied molecular orbitals characteristic of the internal bonding of the molecule. From the response of the strength of these features to changes in the polarization of the X-ray beam, the orientation of the molecule with respect to the surface can be determined. Shifts in position of the features can also be related to internal bond length changes to an accuracy of about 0.05 Å. NEXAFS features are also sensitive to such things as oxidation states and chemical coordination. They can be used, for example, to identify different chemical groupings at polymer surfaces, or to identify reaction products at metal–semiconductor interfaces.

XPD is an extension of XPS (Chapter 5). In XPS X rays of a fixed wavelength, hv, eject core-level electrons from atoms in the sample. The kinetic energy, KE of these photoelectrons is measured, thereby allowing a determination of the core-level BEs (hv–KE), which provides an atomic identification and chemical state information in some cases. In XPD the angular distributions of these atom specific photoelectrons are measured for materials possessing long range order (usually single crystal surfaces, or for adsorption or reaction at these surfaces). The angular distributions are generated by diffraction of the photoelectron emitted from the target atom by its neighboring atoms. At the photoelectron energies involved the electron–atom scattering events peak in the forward direction, leading to the simple result that intensity maxima occur along atomic rows. Direct crystallographic information about the near surroundings of the target atom is thus obtained. By comparing detailed calculations of the scattering processes for assumed models to the data, atom spacings can be obtained to an accuracy of ~ 0.05 Å. The probing depth is the same as for XPS; a few atomic layers. Strain, alloying, island formation, or interdiffusion during epitaxial growth is observable because the angular distributions are sensitive to atoms moving to "off-site" positions. XPD has been used as a simple diagnostic tool for distinguishing growth modes in well-ordered systems.

Auger Electron Diffraction, AED, is an exact analogy to XPD, providing basically the same information. Instead of measuring the angular distribution of the ejected photoelectrons one uses the Auger electrons (Chapter 5).

4.1 XRD

X-Ray Diffraction

MICHAEL F. TONEY

Contents

Introduction

X-ray Diffraction (XRD) is a powerful technique used to uniquely identify the crystalline phases present in materials and to measure the structural properties (strain state, grain size, epitaxy, phase composition, preferred orientation, and defect structure) of these phases. XRD is also used to determine the thickness of thin films and multilayers, and atomic arrangements in amorphous materials (including polymers) and at interfaces.

XRD offers unparalleled accuracy in the measurement of atomic spacings and is the technique of choice for determining strain states in thin films. XRD is noncontact and nondestructive, which makes it ideal for in situ studies. The intensities measured with XRD can provide quantitative, accurate information on the atomic arrangements at interfaces (e.g., in multilayers). Materials composed of any element can be successfully studied with XRD, but XRD is most sensitive to high-Z elements, since the diffracted intensity from these is much larger than from low-Z elements. As a consequence, the sensitivity of XRD depends on the material of interest. With lab-based equipment, surface sensitivities down to a thickness of ~50 Å are achievable, but synchrotron radiation (because of its higher intensity) allows the characterization of much thinner films, and for many materials, monatomic layers can be analyzed. While the structure as a function of depth is not normally measured in XRD, this is possible using specialized methods.

Alternatives to XRD include transmission electron microscopy (TEM) and diffraction, Low-Energy and Reflection High-Energy Electron Diffraction (LEED and RHEED), extended X-ray Absorption Fine Structure (EXAFS), and neutron diffraction. LEED and RHEED are limited to surfaces and do not probe the bulk of thin films. The elemental sensitivity in neutron diffraction is quite different from XRD, but neutron sources are much weaker than X-ray sources. Neutrons are, however, sensitive to magnetic moments. If adequately large specimens are available, neutron diffraction is a good alternative for low-Z materials and for materials where the magnetic structure is of interest.

While XRD is nondestructive and can be used in most environments, TEM and electron diffraction are destructive techniques (due to specimen preparation methods) and require high vacuum. One of the disadvantages of XRD, compared to electron diffraction, is the low intensity of diffracted X rays, particularly for low-Z materials. Typical intensities for electron diffraction are ~10^8 times larger than for XRD. Because of small diffracted intensities, thin-film XRD generally requires large specimens (~0.5 cm) and the information acquired is an average over a large area. Usually, XRD does not provide spatial resolution, but for special applications, resolution of greater than ~10 μm can be obtained with a microfocus source and a suitably thick film ~1μm). The use of intense synchrotron X-ray radiation mitigates these two disadvantages somewhat; however, the XRD analysis of thin films with synchrotron radiation is not routine (because of the limited accessibility of synchrotron sources).

Thin-film XRD is important in many technological applications, because of its abilities to accurately determine strains and to uniquely identify the presence and composition of phases. In semiconductor and optical materials applications, XRD is used to measure the strain state, orientation, and defects in epitaxial thin films, which affect the film's electronic and optical properties. For magnetic thin films, it is used to identify phases and to determine preferred orientations, since these can determine magnetic properties. In metallurgical applications, it is used to determine strains in surface layers and thin films, which influence their mechanical properties. For packaging materials, XRD can be used to investigate diffusion and phase formation at interfaces

Basic Principles

This section briefly discusses the fundamental principles of XRD; the reader is referred to the works by Warren, Cullity, and Schwartz and Cohen for more detail.[1–3] Figure 1 shows the basic features of an XRD experiment, where the diffraction angle 2θ is the angle between the incident and diffracted X rays. In a typical

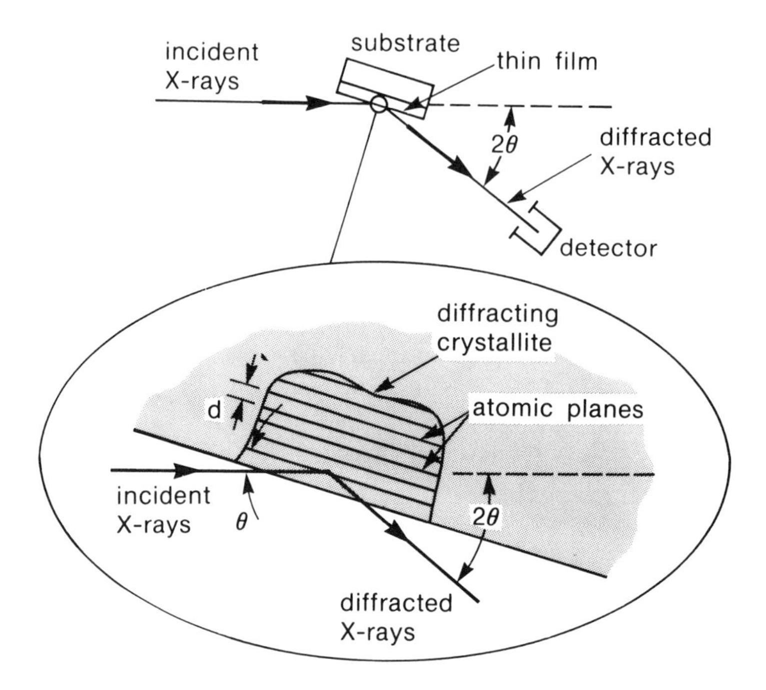

Figure 1 **Basic features of a typical XRD experiment.**

experiment, the diffracted intensity is measured as a function of 2θ and the orientation of the specimen, which yields the diffraction pattern. The X-ray wavelength λ is typically 0.7–2 Å, which corresponds to X-ray energies ($E = 12.4\ keV/\lambda$) of 6 – 17 keV.

The Directions of Diffracted X Rays

Before considering the conditions for XRD, we will briefly review some important properties of crystalline materials. Crystals consist of planes of atoms that are spaced a distance d apart (Figures 1 and 2), but can be resolved into many atomic planes, each with a different d-spacing. To distinguish between these, we introduce a coordinate system for the crystal whose unit vectors a, b, and c are the edges of the unit cell (Figure 2b). For the familiar cubic crystal, these form an orthogonal system. Any atomic plane can now be uniquely distinguished by its Miller indices. These are the three reciprocal intercepts of the plane with the a-, b-, and c-axes and are reduced to the smallest integers having the same ratio. Thus, an (hkl) plane intercepts the crystallographic axes at a/h, b/k, and c/l; examples are shown in Figure 2. The d-spacing between (hkl) planes is denoted d_{hkl}, and for cubic crystals, it is

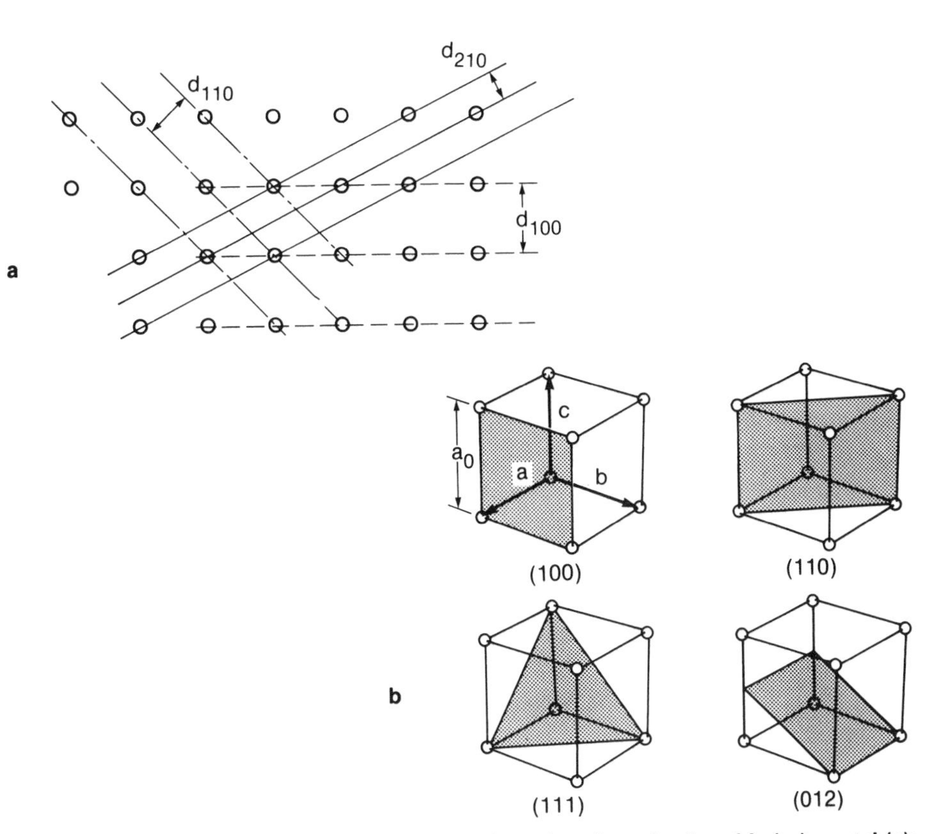

Figure 2 **Several atomic planes and their d-spacings in a simple cubic (sc) crystal (a); and Miller indices of atomic planes in an sc crystal (b). As an example consider the (012) plane. This intercepts the a-, b-, and c-axes at ∞, 1, and 1/2, respectively, and thus, h = 1/∞ = 0, k = 1/1 = 1, and l = 1/(1/2) = 2.**

$$d_{hkl} = \frac{a_0}{\sqrt{h^2 + k^2 + l^2}} \tag{1}$$

where a_0 is the lattice constant of the crystal (see Figure 2).

When there is constructive interference from X rays scattered by the atomic planes in a crystal, a diffraction peak is observed. The condition for constructive interference from planes with spacing d_{hkl} is given by Bragg's law:

$$\lambda = 2d_{hkl} \sin\theta_{hkl} \tag{2}$$

where θ_{hkl} is the angle between the atomic planes and the incident (and diffracted) X-ray beam (Figure 1). For diffraction to be observed, the detector must be positioned so the diffraction angle is $2\theta_{hkl}$, and the crystal must be oriented so that the normal to the diffracting plane is coplanar with the incident and diffracted

X rays and so that the angle between the diffracting plane and the incident X rays is equal to the Bragg angle θ_{hkl}. For a single crystal or epitaxial thin film, there is only one specimen orientation for each (hkl) plane where these diffraction conditions are satisfied.

Thin films, on the other hand, can consist of many grains or crystallites (small crystalline regions) having a distribution of orientations. If this distribution is completely random, then diffraction occurs from any crystallite that happens to have the proper orientation to satisfy the diffraction conditions. The diffracted X rays emerge as cones about the incident beam with an opening angle of $2\theta_{hkl}$, creating a "powder" diffraction pattern. Thin films are frequently in a class of materials intermediate between single crystals and powders and have fiber texture. That is, all the crystallites in the film have the same atomic planes parallel to the substrate surface, but are otherwise randomly distributed. Face-centered cubic (fcc) films often grow with (111) fiber texture: The (111) planes are parallel to the substrate plane and planes perpendicular to this—e.g., (220)—are necessarily perpendicular to the substrate but otherwise randomly distributed. The resulting diffraction pattern consists of rings about the film normal or (111) axis. Thin films may not have complete fiber texture, but may possess a preferred orientation, where most, but not all, of the crystallites have the same atomic planes parallel to the substrate. For example, while most of the fcc film might have (111) texture, a few of the crystallites might have their (200) planes parallel to the substrate.

As we have seen, the orientation of crystallites in a thin film can vary from epitaxial (or single crystalline), to complete fiber texture, to preferred orientation (incomplete fiber texture), to randomly distributed (or powder). The degree of orientation not only influences the thin-film properties but also has important consequences on the method of measurement and on the difficulty of identifying the phases present in films having multiple phases.

Intensities of Diffracted X Rays

Before considering diffracted intensities, we first must consider X-ray absorption, since this affects intensities. All materials absorb X rays. Thus, an X-ray beam is attenuated as it traverses matter. The transmitted intensity decays exponentially with the distance traveled through the specimen and the linear absorption coefficient μ describes this decrease. The absorption length (1/e decay length) is $1/\mu$, and at $\lambda = 1.54$ Å, typical values are 1mm, 66 μm, and 4μm for carbon, silicon, and iron, respectively. Except near an absorption edge,[1–3] μ increases with increasing atomic number and increasing wavelength.

Neglecting unimportant geometric factors,[1–3] the integrated X-ray intensity diffracted from a thin film is

$$I_{hkl} \propto \left|F_{hkl}\right|^{-2M} V \quad (3)$$

Here F_{hkl} is the structure factor for the (hkl) diffraction peak and is related to the atomic arrangements in the material. Specifically, F_{hkl} is the Fourier transform of the positions of the atoms in one unit cell. Each atom is weighted by its form factor, which is equal to its atomic number Z for small 2θ, but which decreases as 2θ increases. Thus, XRD is more sensitive to high-Z materials, and for low-Z materials, neutron or electron diffraction may be more suitable. The factor e^{-2M} (called the Debye-Waller factor) accounts for the reduction in intensity due to the disorder in the crystal, and the diffracting volume V depends on μ and on the film thickness. For epitaxial thin films and films with preferred orientations, the integrated intensity depends on the orientation of the specimen.

Dynamical X-Ray Diffraction

In the concepts developed above, we have used the kinematic approximation, which is valid for weak diffraction intensities arising from "imperfect" crystals. For perfect crystals (available thanks to the semiconductor industry), the diffraction intensities are large, and this approximation becomes inadequate. Thus, the dynamical theory must be used. In perfect crystals the incident X rays undergo multiple reflections from atomic planes and the dynamical theory accounts for the interference between these reflections. The attenuation in the crystal is no longer given by absorption (e.g., μ) but is determined by the way in which the multiple reflections interfere. When the diffraction conditions are satisfied, the diffracted intensity from perfect crystals is essentially the same as the incident intensity. The diffraction peak widths depend on $2\theta_{hkl}$ and F_{hkl} and are extremely small (less than ~0.001–0.005°).

Experimental Methods for Thin Film Characterization by XRD

Because the diffracting power of thin films is small, the instrumentation and techniques for thin-film XRD are designed to maximize diffracted intensities and to minimize background. There are basically two classes of measurement techniques. The first, and oldest, uses photographic film; these methods provide fast, preliminary information and yield two-dimensional data. However, progress in computers and high-power X-ray generators has lead to the widespread use of diffractometers, where the diffracted X rays are detected with photon counters. Compared to photographic methods, counters provide more accurate, quantitative data and have superior signal-to-noise ratios. Furthermore, diffractometers are easily automated and provide better angular resolution. Recently, there has been increasing use of position-sensitive detectors,[2, 3] which use parallel detection to scan a range in 2θ.

X-Ray Diffractometer Methods

The Bragg-Brentano geometry [2–4] is used widely for preferentially and randomly oriented polycrystalline films. In this geometry (Figure 3a), slits collimate the inci-

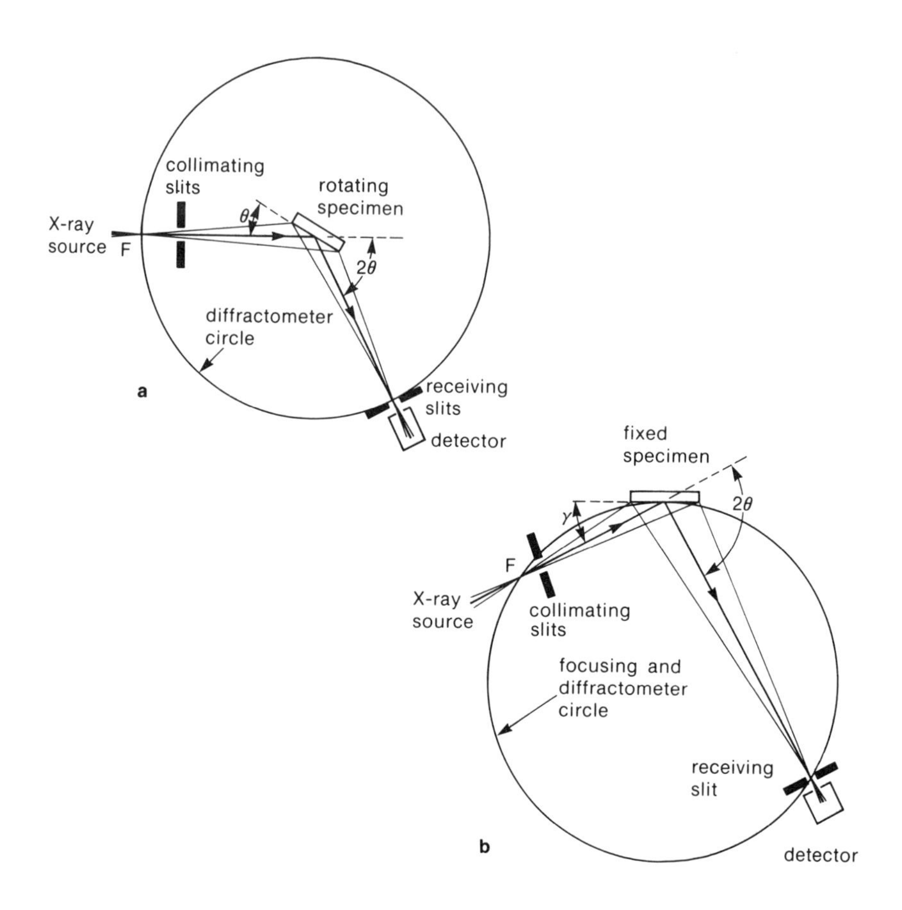

Figure 3 **Bragg-Brentano diffractometer (a); and Seemann-Bohlin diffractometer (b). The point F is either the focal point on an X-ray tube or the focal point of a focusing monochromator.**

dent X rays, which impinge on the specimen at an angle θ. After passing through receiving slits, the diffracted X rays are detected. The specimen is rotated at one-half the angular velocity of the detector. Since the incident and diffracted X rays make the same angle to the specimen surface, structural information is obtained only about (hkl) planes parallel to this surface. When the receiving slits, the specimen, and the focal point F lie on a circle, the diffracted X rays are approximately focused on the receiving slits (parafocusing), which considerably improves the sensitivity.

For the Seemann-Bohlin geometry (Figure 3b) the incident X rays impinge on a fixed specimen at a small angle $\gamma \sim 5\text{–}10°$ and the diffracted X rays are recorded by a detector that moves along the focusing circle.[2–4] This method provides good sensitivity for thin films, due to parafocusing and the large diffracting volume, which results from γ being small and the X-ray path length in the film being large (proportional to $1/\sin\gamma$). Because γ is fixed, the angle between the incident X rays and the diffracting planes changes as the detector moves through 2θ. Because only planes with the correct orientation diffract X rays, this method is most useful for polycrystalline films having random or nearly random crystallite orientations.

A Double-Crystal Diffractometer (DCD) is useful for characterizing nearly perfect, epitaxial thin films.[4–6] This geometry is similar to the Bragg-Brentano case, but the incident beam is first diffracted from a perfect single crystal (placed near F in Figure 3a). It is thus monochromatic and well collimated. This insures that the measured diffraction peak width of the specimen is narrow, thereby permitting high-resolution measurements. Typically, the detector is fixed near $2\theta_0$, the diffraction angle for the (hkl) planes of interest, and the receiving slits are open to accept a large range in 2θ. The intensity is recorded in a "rocking curve," where the specimen rotates about θ_0. Asymmetric reflections, where the (hkl) planes are not parallel to the substrate can also be measured. Since the diffraction peak widths in DCD measurements are narrow, this method enables accurate determination of very small deviations in d-spacings due to strain.

For ultrathin epitaxial films (less than ~100 Å), GrazingIncidence X-ray Diffraction (GIXD) is the preferred method[6] and has been used to characterize monolayer films. Here the incidence angle is small (~0.5°) and the X rays penetrate only ~100–200 Å into the specimen (see below). The exit angle of the diffracted X rays is also small and structural information is obtained about (hkl) planes perpendicular to the specimen surface. Thus, GIXD complements those methods where structural information is obtained about planes parallel to the surface (e.g., Bragg-Brentano and DCD).

Photographic Methods

Photographic methods[2–4] of characterizing polycrystalline thin films are used to acquire preliminary data and to determine orientational relationships. Guinier and Read cameras are common, although other methods are also used. In a Guinier camera the geometry is the same as for a Seemann-Bohlin diffractometer with a focusing monochromator, except that the "detector" is now a cylinder of film. The geometry of a Read camera is similar to the Bragg-Brentano diffractometer, but the incidence angle is fixed (~5–10°) and the cylindrical film spans a wide range in 2θ.

X-ray topography is a photographic method used to image defects in nearly perfect single crystals.[3, 7] The topograph is a map of the diffracted intensity across the specimen. There are several topographic techniques useful for thin films; [3, 7] we only note a few of their capabilities here. The Berg-Barrett and section methods are simple and give good surface sensitivity. The Lang method (scanning reflection) is more complicated, but large areas can be imaged with good surface sensitivity and spatial resolution. These methods all use a single perfect crystal—the specimen. In the double-crystal method, a reference crystal is used also to produce a monochromatic, collimated incident beam. Although more complicated, this method provides maximum sensitivity to defects and can provide good surface sensitivity and spatial resolution.

Examples of XRD Characterization of Thin Films

Phase Identification

One of the most important uses of thin-film XRD is phase identification. Although other techniques (e.g., RBS, XPS, and XRF) yield film stoichiometries, XRD provides positive phase identification. This identification is done by comparing the measured d-spacings in the diffraction pattern and, to a lesser extent, their integrated intensities with known standards in the JCPDS Powder Diffraction File (Joint Committee on Powder Diffraction Standards, Swathmore, Pennsylvania, 1986). However, thin films often have a preferred orientation, and this can cause the measured intensities to disagree with the JCPDS file, which are for random orientations. For films containing several phases, the proportion of each phase can be determined from the integrated intensities in the diffraction pattern. If the phases in the film have random orientation or almost complete fiber texture, this determination is simple.[2–4] However, if there is some preferred orientation (incomplete fiber texture), the determination of phase proportions may require integrated intensities at many specimen orientations, which is time consuming. Furthermore, for multiphase specimens, preferred orientation can make positive phase identification difficult, since the integrated intensities may not be useful for phase identification. (For example, peaks that are strong in powder patterns may be weak or completely absent in a specimen with preferred orientation). This difficulty can be particularly acute if data are available only for one specimen orientation (i.e., the Bragg-Brentano geometry) or if the phases produce many diffraction peaks.

Other excellent methods of phase identification include TEM and electron diffraction. These may be more useful for low-Z materials, ultrathin films, and for characterizing small areas, including individual grains. For multiphase films with incomplete texture, these methods and XRD are complementary, since in commonly used geometries, they probe atomic planes perpendicular and parallel to the thin film surface, respectively.

Figure 4 shows an example where XRD is used to unambiguously identify the phases in three high-T_C superconducting thin films.[8] Since the films have nearly complete fiber texture (see below), the identification was simple and was done by comparison to the diffraction patterns from bulk materials. Furthermore, from comparison to standards, the presence of a small amount of CuO is apparent in one film (Figure 4a). We also conclude that the film in Figure 4b consists of approximately equal mixtures of $Tl_2CaBa_2Cu_2O_x$ and $Tl_2Ca_2Ba_2Cu_3O_y$, since it can be reproduced by an approximately equal combination of the patterns in Figures 4a and 4c. Again, because of the strong fiber texture, this determination is straightfor-

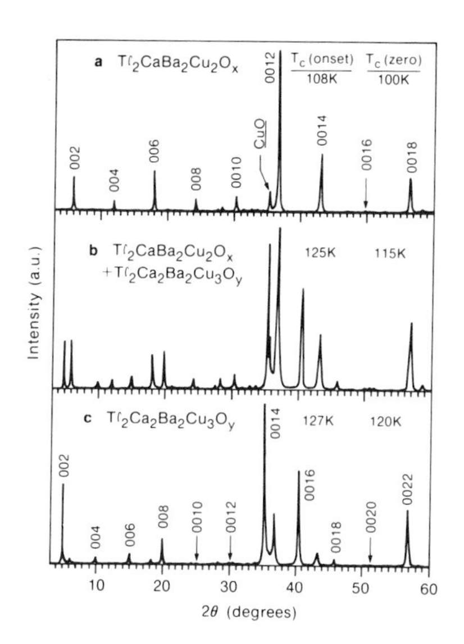

Figure 4 **Diffraction patterns (Bragg-Brentano geometry) of three superconducting thin films (~2-μm thick) annealed for different times.[8] The temperatures for 0 resistance and for the onset of superconductivity are noted.**

ward. Figure 5 shows the XRD pattern from a bilayer film of $Co_{70}Pt_{12}Cr_{18}/Cr$ used for magnetic recording.[9] Here the phase of the CoPtCr magnetic media was shown to be hexagonal close packed (hcp) by comparing the measured peaks (including some not shown) with those expected for an hcp solid solution $Co_{70}Pt_{12}Cr_{18}$.

Determination of Strain and Crystallite Size

Diffraction peak positions, and therefore, atomic spacings are accurately measured with XRD, which makes it the best method for characterizing homogeneous and inhomogeneous strains.[2–4, 6, 10] Homogeneous or uniform elastic strain shifts the diffraction peak positions, and if $d_{0,\,hkl}$ is the unstrained d-spacing, $(d_{hkl}-d_{0,\,hkl})\,/\,d_{0,\,hkl}$ is the component of elastic strain in the (hkl) direction. Figure 6 shows[5] a rocking curve of 2500-Å $Al_{0.88}Ga_{0.12}As$ film on GaAs and illustrates the superb resolution possible with XRD. From the shift in peak positions, one can calculate the difference in d-spacings between the thin film and substrate (in the (100) direction). Although this is only 0.231%, the diffraction peaks from the film and substrate are easily distinguished. Furthermore, the variation in the d-spacing of the film near the film–substrate interface is determined from modeling the data.[5]

Inhomogeneous strains vary from crystallite to crystallite or within a single crystallite and this causes a broadening of the diffraction peaks that increases with sin θ. Peak broadening is also caused by the finite size of crystallites, but here the broadening is independent of $\sin\theta$. When both crystallite size and inhomogeneous strain contribute to the peak width, these can be separately determined by careful analysis of peak shapes for several diffraction orders (e.g., (111), and (222)).[1–4] Furthermore, the diffraction peak shape can provide information on other types of imperfections (i.e., the presence, extent, and type of stacking faults).[1, 3] If there is no inhomogeneous strain, the crystallite size L is estimated from the peak width $\Delta 2\theta$ with the Scherrer formula:

$$L \sim \frac{\lambda}{(\Delta 2\theta)\cos\theta} \tag{4}$$

Using this and the data in Figure 5, one estimates L ~180 Å for the CoPtCr film.[9] Grain or crystallite size are also determined with TEM through direct imaging. Since this method is a local probe, it can provide more detailed information on imperfections than XRD. Although strain gauges can measure homogeneous strain, there is no good alternative to XRD for strain measurements.

Determination of Preferred Orientation

For polycrystalline films, the amount of preferred orientation can be estimated by comparing the integrated intensities (after correction for geometric factors) to the JCPDS file or the expression for integrated intensity (Equation (3)). If the film has (hkl) fiber texture, then in the Bragg-Brentano geometry, the (hkl) diffraction peak will have a larger relative intensity than expected. Figure 4 shows an example of nearly complete (001) fiber texture in high-T_C thin films, since only the (00l) peaks are observed.[8] For the CoPtCr media (Figure 5), the (002) peak is observed but is weak compared to the (100) and (101) peaks. Thus, the preferred orientation for the (001) axis is to lie in the plane.[9] Although this film possess incomplete fiber texture, phase identification is straightforward, since the reasonable phase possibilities (hcp, fcc, and bcc) are easily distinguished. To obtain a more quantitative determination of preferred orientation, the intensity of an (hkl) peak is measured at different specimen orientations, for example with a pole-figure goniometer.[2] Photographic methods are particularly useful for this, since they provide two-dimensional information, but are less quantitative. For epitaxial films, GIXD and DCD are used to determine thin film orientation. Preferred orientation is also measured with TEM, although less quantitatively.

Film Thickness Determination

The film thickness of epitaxial and highly textured thin films can be measured with XRD.[4, 5] Close to the usual or primary diffraction peaks there are secondary or subsidiary maxima in the diffracted intensity (see Figure 6), which are due to the finite film thickness.[1, 3] The film thickness is inversely proportional to the spacing between these maxima and is easily calculated. X-ray reflectivity is another accurate method for measuring a film's thickness.

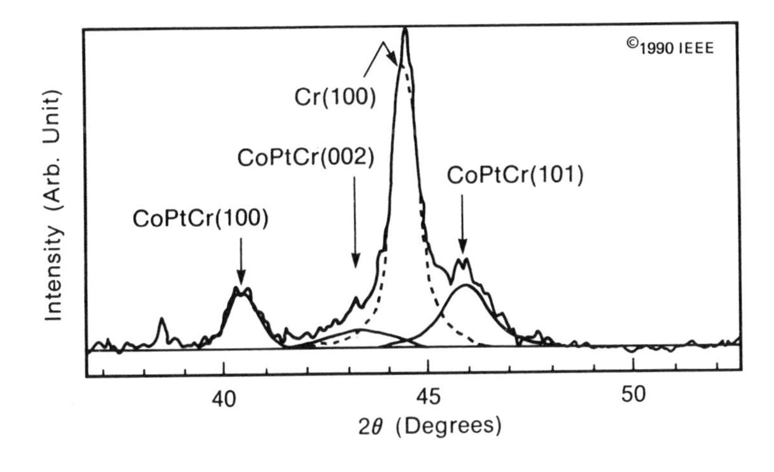

Figure 5 **Bragg-Brentano diffraction pattern for magnetic media used in a demonstration of 1-Gb/in² magnetic recording.[9] The lines show a deconvolution of the data into individual diffraction peaks, which are identified.**

Depth-Dependent Information

In most thin-film XRD analyses, depth-dependent structural information is not obtained, but recently such measurements have been performed using a grazing incidence geometry.[11] Since the refractive index for X rays[1, 3] is less than 1, X rays experience total external reflection at incidence angles less than the critical angle for total reflection (α_C). By varying the incidence angle near α_C, the penetration depth of the incident X rays is varied from ~50 Å up to several µm. Since the diffracted X rays originate from different depths, the depth-dependent structure of the specimen may be obtained from diffraction patterns taken at different incidence angles. Depth-dependent structure can also be obtained from TEM, although less quantitatively.

Interdiffusion of bilayered thin films also can be measured with XRD.[12, 13] The diffraction pattern initially consists of two peaks from the pure layers and after annealing, the diffracted intensity between these peaks grows because of interdiffusion of the layers. An analysis of this intensity yields the concentration profile, which enables a calculation of diffusion coefficients, and diffusion coefficients $\sim 10^{12}$–10^{-15} cm²/s are readily measured.[12] With the use of multilayered specimens, extremely small diffusion coefficients ($\sim 10^{-23}$ cm²/s) can be measured with XRD.[12, 13] Alternative methods of measuring concentration profiles and diffusion coefficients include depth profiling (which suffers from artifacts), RBS (which can not resolve adjacent elements in the periodic table), and radiotracer methods (which are difficult). For XRD (except for multilayered specimens), there must be a unique relationship between composition and the d-spacings in the initial films and any solid solutions or compounds that form; this permits calculation of the compo-

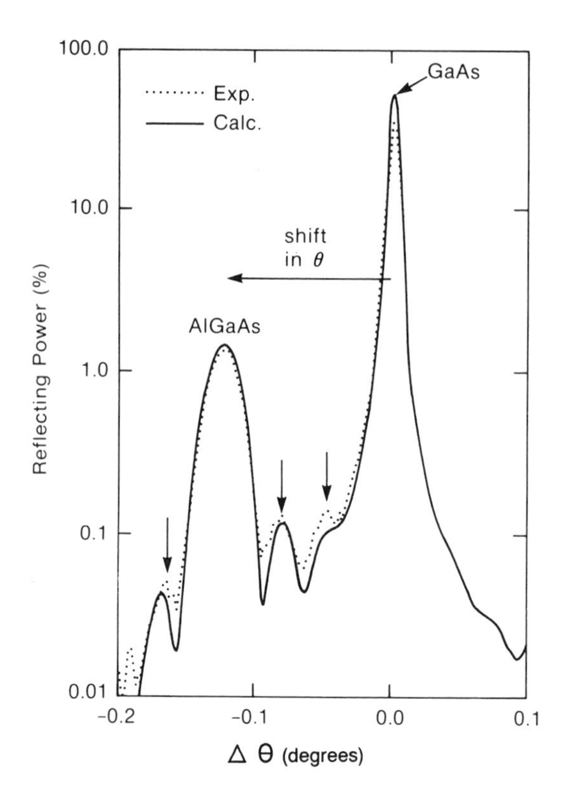

Figure 6 **DCD rocking curves—measured (dashed) and calculated (solid)—of the (400) diffraction peak from $Al_{0.88}Ga_{0.12}As$ on GaAs(100).[5] The arrows mark the subsidiary maxima.**

sition from the diffraction peak positions. On the other hand, the nondestructive nature of XRD makes it a very powerful technique for the measurement of concentration profiles.

X-Ray Topography

X-ray topography[3, 7] is used to image defects in nearly perfect epitaxial films and the surface layers of substrates. Typical defects include dislocations, precipitates, fault planes, and local blisters and cracks, and these may be present in the thin film or substrate, or may have been induced into the substrate by the presence of the film. An X-ray topograph is a image of the diffracted intensity across the specimen. Contrast is produced by local variation in the intensity because of changes in the diffraction conditions and is most often due to the strain fields associated with defects. A topograph must be interpreted with the dynamical theory. Although the lateral resolution of X-ray topography (~1 μm) is about 1000 times poorer than TEM, much smaller strains (~10^{-6}) and larger areas (~1–10 cm) can be imaged. Using conventional methods, layers as thin as 1 μm are imaged, but by using grazing incidence angles, the minimum thickness that can be imaged is about 0.1 μm. If the diffraction peaks from a thin film and substrate occur at sufficiently different 2θ, an image of the peak from the thin-film maps only the defects in the film.

Characterization of Multilayered Films

Recently, there has been remarkable progress in the controlled synthesis of multilayered materials,[6, 13] which have a repeating modulation in chemical composition. The atomic structure of multilayers, which influences their properties, is readily studied with XRD. Most XRD investigations have concentrated on structure along the growth direction and this involves measuring atomic planes parallel to the surface, often with Bragg-Brentano or DCD geometries. The wavelength of the modulation in composition (modulation wavelength) and the average strain are determined from the diffraction peak positions; phase identification and measurements of the crystallite size and the preferred orientation are performed as described earlier. A full characterization and understanding of these materials requires accurate knowledge of the interfacial roughness and the composition and position of the atomic planes in a modulation wavelength. This is difficult, since it requires careful analysis of the peak shapes and a complete set of integrated intensities. A determination of the structure in the surface plane can also be important and is performed with GIXD; however, this often is not done. Alternative characterization techniques include neutron diffraction and reflectivity, TEM, EXAFS, LEED, and RHEED, although none of these provide the quantitative detail available with XRD. RHEED, however, is used to monitor multilayer growth in situ.

Characterization of Amorphous Materials

The diffraction pattern from amorphous materials (including many polymers) is devoid of the sharp peaks characteristic of crystals and consists of broad features or halos.[1, 3] Quantitative analysis of XRD data from amorphous materials is complicated but provides important information on the local atomic structure (short-range order), including the bond lengths, the number of neighbors, and the extent of atomic correlations. Since the diffraction from amorphous materials is weak, thick specimens or synchrotron radiation is necessary, particularly for low-Z materials. Many polymers are amorphous or semicrystalline, and for polymeric materials, XRD is used to probe the structure, morphology, and degree of crystallinity.[14] TEM is a widely used alternative for amorphous materials, but is less quantitative and can damage polymers. EXAFS is also widely used and complements XRD but cannot be used for materials composed of only low-Z elements.

Conclusions

XRD is an excellenr, nondestructive method for identifying phases and characterizing the structural properties of thin films and multilayers. It is inexpensive and easy to implement. The future will see more use of GIXD and depth dependent measurements, since these provide important information and can be carried out on lab-based equipment (rather than requiring synchrotron radiation). Position sensitive detectors will continue to replace counters and photographic film.

Multilayered materials will become more important in the future, and therefore, their structural characterization using XRD will grow in importance. The use of synchrotron radiation as an analytical tool for thin film characterization will also increase. The unique characteristics of this radiation enable formerly difficult lab-based experiments to be done simply (e.g., GIXD and depth-dependent structural measurements) and permit experiments that are otherwise impossible. These include thin-film energy dispersive XRD, where the incident beam is polychromatic and the diffraction from many atomic planes is obtained simultaneously, and anomalous diffraction, where the abrupt change in the diffracting strength of an element near a particular X-ray energy (an absorption edge) is used to differentiate elements.[2, 3]

Related Articles in the Encyclopedia

TEM, EXAFS, RHEED, LEED, Neutron Diffraction

References

1 B. E. Warren. X-Ray Diffraction. Addison-Wesley, Reading, 1969. A classic text that is complete and thorough, although somewhat dated.

2 B. D. Cullity. Elements of X-Ray Diffraction. Addison-Wesley, Reading, 1978. Another classic text, which is simpler than Warren and emphasizes metallurgy and materials science. A good introduction.

3 L. H. Schwartz and J. B. Cohen. Diffraction from Materials. Springer-Verlag, Berlin, 1987. A recent text that includes X-ray, neutron, and electron diffraction, but emphasizes XRD in materials science. A good introduction and highly recommended.

4 A. Segmuller and M. Murakami. Characterization of Thin Films by X-Ray Diffraction. In: Thin Films from Free Atoms and Particles. (K.J. Klabunde, ed.) Academic Press, Orlando, 1985, p.325 A recent brief review article with many references.

5 V. S. Speriosu, M. A. Nicolet, J. L. Tandon, and Y. C. M. Yeh. Interfacial Strain in AlGaAs Layers on GaAs. J. Appl. Phys. 57, 1377, 1985.

6 A. Segmuller, I. C. Noyan, V. S. Speriosu. X-Ray Diffraction Studies of Thin Films and Multilayer Structures. Prog. Cryst. Growth and Charact. 18, 21, 1989.

7 D. K. Bowen. X-Ray Topography of Surface Layers and Thin Films. In: Advances in X-ray Analysis. (C.S. Barrett, J. V. Gilfrich, R. Jenkins, and P. K. Predecki, eds.) Plenum, New York, 1990, vol. 33, p.13.

8 W. Y. Lee, V. Y. Lee, J. Salem, T. C. Huang, R. Savoy, D. C. Bullock and S. S. P. Parkin. Superconducting TlCaBaCuO Thin Films with Zero Resistance at Temperatures of up to 120K. Appl. Phys. Lett. 53, 329, 1988.

9 T. Y. Yogi, C. Tsang, T. A. Nguyen, K. Ju, G. L. Gorman, and G. Castillo. Longitudinal Magnetic Media for 1Gb / Sq. In. Areal Density. IEEE Trans. Magn. MAG-26, 2271, 1990.

10 A. Segmuller and M. Murakami. X-Ray Diffraction Analysis of Strains and Stresses in Thin Films. In: Analytical Techniques for Thin Films. (K.N. Tu and R. Rosenberg, eds.) Academic, San Diego, 1988, p.143.

11 M. F. Toney and S. Brennan. Structural Depth Profiling of Iron Oxide Thin Films using Grazing Incidence Asymmetric Bragg X-ray Diffraction. J. Appl. Phys. 65, 4763, 1989.

12 M. Murakami, A. Segmuller, K. N. Tu. X-Ray Diffraction Analysis of Diffusion in Thin Films. In: Analytical Techniques for Thin Films. (K. N. Tu and R. Rosenberg, eds.) Academic Press, San Diego, 1988, p.201.

13 P. Dhez and C. Weisbuch. Physics, Fabrication, and Applications of Multilayered Structures. Plenum, New York, 1988.

14 M. Kakudo and N. Kasai. X-ray Diffraction by Polymers. Elsevier, Tokyo, 1972.

4.2 EXAFS

Extended X-Ray Absorption Fine Structure

MARK R. ANTONIO

Contents

Introduction

The discovery of the phenomenon that is now known as extended X-ray absorption fine structure (EXAFS) was made in the 1920s, however, it wasn't until the 1970s that two developments set the foundation for the theory and practice of EXAFS measurements.[1] The first was the demonstration of mathematical algorithms for the analysis of EXAFS data. The second was the advent of intense synchrotron radiation of X-ray wavelengths that immensely facilitated the acquisition of these data. During the past two decades, the use of EXAFS has become firmly established as a practical and powerful analytical capability for structure determination.[2–8]

EXAFS is a nondestructive, element-specific spectroscopic technique with application to all elements from lithium to uranium. It is employed as a direct probe of the atomic environment of an X-ray absorbing element and provides chemical bonding information. Although EXAFS is primarily used to determine the local structure of bulk solids (e.g., crystalline and amorphous materials), solid surfaces, and interfaces, its use is not limited to the solid state. As a structural tool, EXAFS complements the familiar X-ray diffraction technique, which is applicable only to crystalline solids. EXAFS provides an atomic-scale perspective about the X-ray absorbing element in terms of the numbers, types, and interatomic distances of neighboring atoms.

EXAFS is part of the field of X-ray absorption spectroscopy (XAS), in which a number of acronyms abound. An X-ray absorption spectrum contains EXAFS data as well as the X-ray absorption near-edge structure, XANES (alternatively called the near-edge X-ray absorption fine structure, NEXAFS). The combination of XANES (NEXAFS) and EXAFS is commonly referred to as X-ray absorption fine structure, or XAFS. In applications of EXAFS to surface science, the acronym SEXAFS, for surface-EXAFS, is used. The principles and analysis of EXAFS and SEXAFS are the same. See the article following this one for a discussion of SEXAFS and NEXAFS.

Experimental Aspects

An EXAFS experiment involves the measurement of the X-ray photoabsorption of a selected element as a function of energy above its core-shell electron binding energy. The most direct measurement of EXAFS is the transmission method, wherein the sample is placed in the X-ray beam and the incident and transmitted X-ray intensities, I_0 and I_t, respectively, are recorded (see Figure 1). The measurement of I_0 and I_t is accomplished with two ion chamber proportional counters that are gas filled (typically with nitrogen and argon) to provide about 10–20% absorption of I_0 and 80–90% absorption of I_t. As shown in Figure 1, it is useful to have a third ion chamber for simultaneous measurements of a reference material (e.g., a thin metal foil) to maintain accurate energy calibration throughout the course of experiment. For successful transmission measurements, the ideal sample thickness x is one absorption length, i.e., $x = 1/[(\mu/\rho)\rho]$; here μ/ρ is the total mass absorption coefficient and ρ is the density. Transmission EXAFS data for samples with larger absorption lengths can be seriously distorted and are not suitable for analysis.[9] Transmission EXAFS data are displayed in the form $\ln(I_0/I_t)$ versus incident X-ray energy, as shown in Figure 2.

A wide selection of metal reference foils and powder films of ideal thickness for tranmission EXAFS is available from The EXAFS Materials Company, Danville, CA, USA. The transmission method is well-suited for *in situ* measurements of materials under industrially relevant conditions of extreme temperature and controlled atmosphere. Specially designed reactors for catalysis experiments and easy-

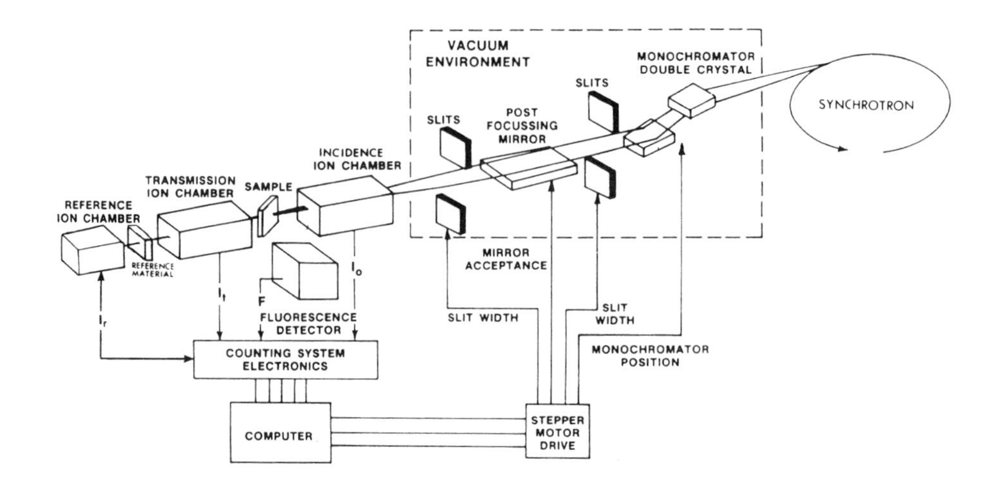

Figure 1 **Schematic view of a typical EXAFS experiment at a synchrotron radiation facility. Note that it is possible to record transmission and fluorescence EXAFS simultaneously with reference EXAFS.**

to-use detectors are commercially available from The EXAFS Company, Seattle, WA, USA.

In addition to transmission, EXAFS data can be recorded through the detection of

1 X-ray fluorescence
2 Electron yield
3 Ion yield
4 Optical luminescence
5 Photoconductivity
6 Photoacoustic signals.

The last three detection schemes apply only under very special circumstances.[3–6] Transmission EXAFS is strictly a probe of bulk structure, i.e., more than about a thousand monolayers. The electron- and ion-yield detection methods, which are used in reflection rather than transmission schemes, provide surface sensitivity, ~1–1,000 Å, and are inherently insensitive to bulk structure. X-ray fluorescence EXAFS has the widest range of sensitivity—from monolayer to bulk levels. The combination of electron or ion yield and transmission EXAFS measurements can provide structural information about the X-ray absorbing element at the surface and in the bulk, respectively, of a sample.

Without exception, the highest quality EXAFS data are acquired at synchrotron radiation facilities. There are 20 operational facilities throughout the world.[10] Each has unique instrumentation: The interested user is encouraged to contact the facil-

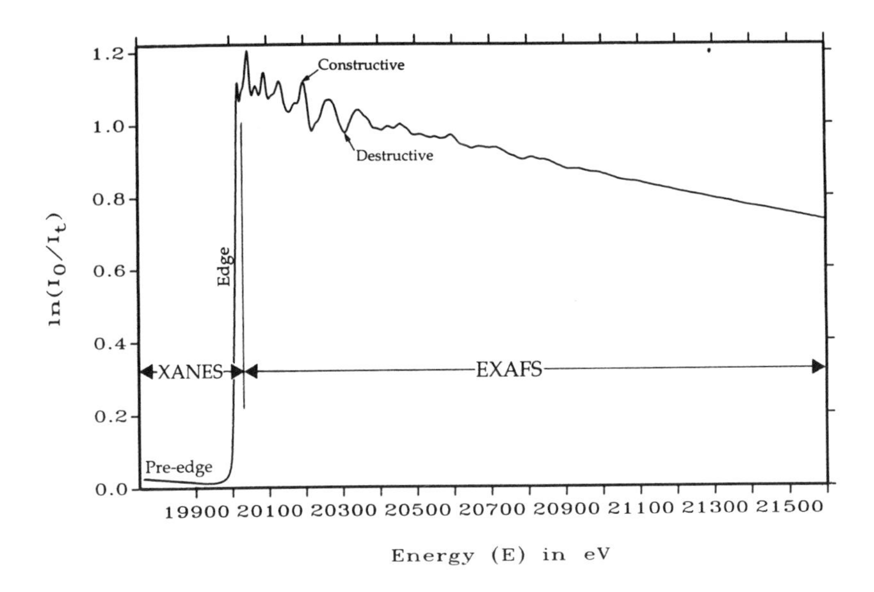

Figure 2 **Molybdenum K-edge X-ray absorption spectrum, $\ln(I_0/I_t)$ versus X-ray energy (eV), for molybdenum metal foil (25-μm thick), obtained by transmission at 77 K with synchrotron radiation. The energy-dependent constructive and destructive interference of outgoing and backscattered photoelectrons at molybdenum produces the EXAFS peaks and valleys, respectively. The pre-edge and edge structures marked here are known together as X-ray absorption near edge structure, XANES and EXAFS are provided in a new compilation of literature entitled *X-ray Absorption Fine Structure* (S.S. Hasain, ed.) Ellis Horwood, New York, 1991.**

ity for detailed information, such as is available in Gmur.[11] In general, "hard" X-ray beam lines (approximately ≥ 2,000 eV) employ flat-crystal monochromators to scan the X-ray energy over the region of interest, whereas "soft" X-ray beam lines (approximately ≤ 1,000 eV) employ grating-type monochromators for the same purpose. The monochromatization of X rays with energies between approximately 1,000 and 2,000 eV is a difficult problem—neither crystal nor grating monochromators work particularly well.

Basic Principles

Both inner-shell (K and L) and outer-shell (M, N, etc.) electrons can be excited by the absorption of X rays and by the inelastic scattering of electrons. In either instance, at an electron binding energy characteristic of an element in a sample,

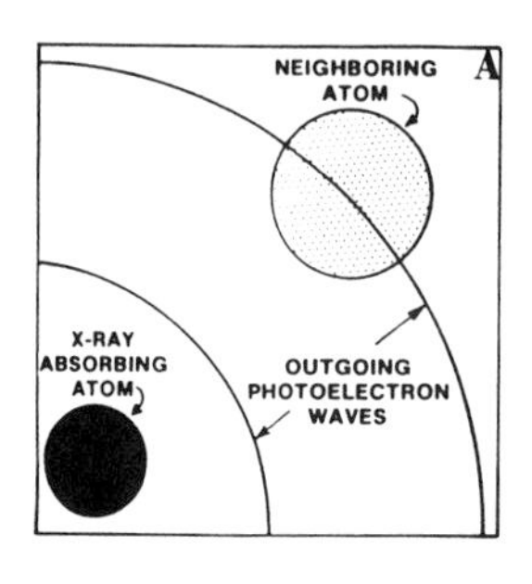

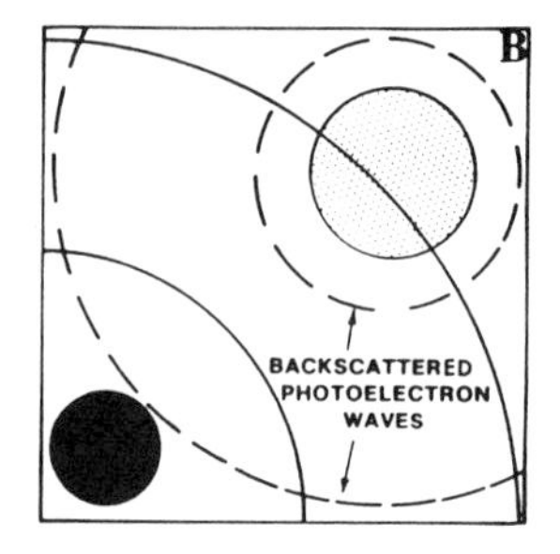

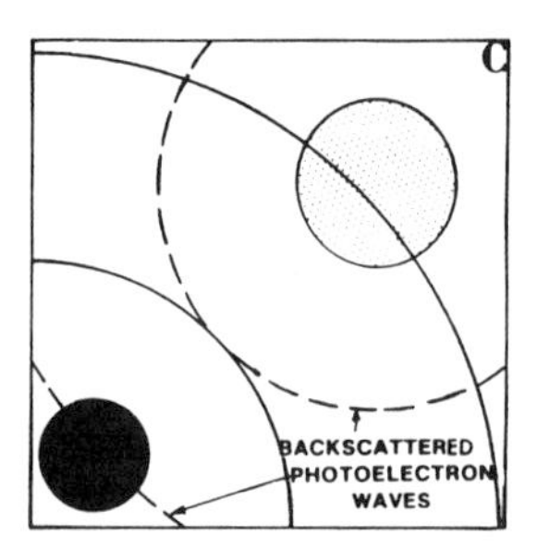

Figure 3 **Schematic illustration of the EXAFS phenomenon: (A) outgoing photoelectron (solid curve) from X-ray absorbing atom; (B) destructive interference at the absorbing atom between outgoing (solid curve) and backscattered (dashed curve) photoelectron from neighboring atom; (C) constructive interference at the absorbing atom between outgoing (solid curve) and backscattered (dashed curve) photoelectron from neighboring atom. Adapted from T. M. Hayes and J. B. Boyce. *Solid State Phys.* 37, 173, 1982.**

absorption occurs and a steeply rising absorption edge is observed. For example, molybdenum exhibits an X-ray absorption edge at 20,000 eV, which is the 1s electron binding energy (K edge) , see Figure 2. The pre-edge and edge features are collectively referred to as XANES or NEXAFS, depending upon the application. These data are valuable for probing the site symmetry and valence of the X-ray absorbing element, but will not be discussed further here.

For X-ray energies greater than the binding energy, the absorption process leads to the excitation of the core electron to the ionization continuum. The resulting photoelectron wave propagates from the X-ray absorbing atom and is scattered by the neighboring atoms, as illustrated in Figure 3. The EXAFS spectrum results from the constructive and destructive interference between the outgoing and incoming photoelectron waves at the absorbing atom. The interference gives rise to the modulatory structure (i.e., peaks and valleys) of the X-ray absorption versus incident X-ray energy, as in Figure 2. This process also makes EXAFS unique—the absorbing atom acts as both the source and detector of the interference that is the EXAFS phenomenon.

EXAFS is a probe of the structural distribution, e.g., interatomic distances, numbers of neighboring atoms (the so-called coordination number), and degree of disorder—and identity of atoms in the immediate vicinity (~5 Å) of the X-ray absorbing atom. A simplified schematic representation of several descriptive features of EXAFS is presented in Figure 4. The frequency of EXAFS oscillations is related to the distance between the X-ray absorbing atom (filled circles) and the backscattering atoms (open circles). For large interatomic distances ($R_1 > R_2$), the EXAFS has shorter periods (higher frequencies) than for small distances; see curves

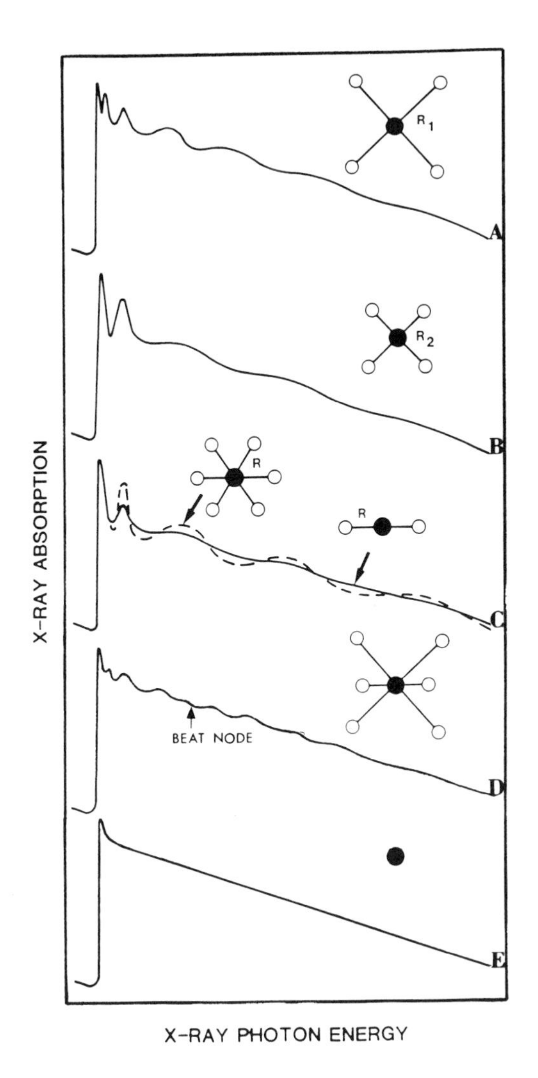

Figure 4 **Descriptive aspects of EXAFS: Curves A–E are discussed in the text. Adapted from J. Stohr. In: *Emission and Scattering Techniques: Studies of Inorganic Molecules, Solids, and Surfaces.* (P. Day, ed.) Kluwer, Norwell, MA, 1981.**

A and B, respectively, in Figure 4. The periodicity is also related to the identity of the absorbing and backscattering elements. Each has unique phase shifts.[12]

EXAFS has an energy-dependent amplitude that is just a few % of the total X-ray absorption. This amplitude is related to the number, type, and arrangement of backscattering atoms around the absorbing atom. As illustrated in Figure 4 (curve C), the EXAFS amplitude for backscattering by six neighboring atoms at a distance R is greater than that for backscattering by two of the same atoms at the same distance. The amplitude also provides information about the identity of the backscattering element—each has a unique scattering function[12]—and the number of different atomic spheres about the X-ray absorbing element. As shown in Figure 4, the EXAFS for an atom with one sphere of neighbors at a single distance exhibits a smooth sinusoidal decay (see curves A–C), whereas that for an atom with two (or more) spheres of neighbors at different distances exhibits beat nodes due to superposed EXAFS signals of different frequencies (curve D).

The EXAFS amplitude is also related to the Debye-Waller factor, which is a measure of the degree of disorder of the backscattering atoms caused by dynamic (i.e., thermal–vibrational properties) and static (i.e., inequivalence of bond lengths) effects. Separation of these two effects from the total Debye-Waller factor requires temperature-dependent EXAFS measurements. In practice, EXAFS amplitudes are larger at low temperatures than at high ones due to the reduction of atomic motion with decreasing temperature. Furthermore, the amplitude for six backscattering atoms arranged symmetrically about an absorber at some average distance is larger than that for the same number of backscattering atoms arranged randomly about an absorber at the same average distance. Static disorder about the absorbing atom causes amplitude reduction. Finally, as illustrated in Figure 4 (curve E), there is no EXAFS for an absorbing element with no near neighbors, such as for a noble gas.

Data Analysis

Because EXAFS is superposed on a smooth background absorption μ_0 it is necessary to extract the modulatory structure μ from the background, which is approximated through least-squares curve fitting of the primary experimental data with polynomial functions (i.e., $\ln(I_0/I_t)$ versus E in Figure 2).[7, 12] The EXAFS spectrum χ is obtained as $\chi = [\mu - \mu_0]/\mu_0$. Here χ, μ, and μ_0 are functions of the photoelectron wave vector $\mathbf{k}$ ($Å^{-1}$), where $k = [0.263\ (E - E_0)]^{1/2}$; E_0 is the experimental energy threshold chosen to define the energy origin of the EXAFS spectrum in k-space. That is, $k = 0$ when the incident X-ray energy E equals E_0, and the photoelectron has no kinetic energy.

EXAFS data are multiplied by k^n ($n = 1$, 2, or 3) to compensate for amplitude attenuation as a function of k, and are normalized to the magnitude of the edge jump. Normalized, background-subtracted EXAFS data, $k^n\chi(k)$ versus k (such as illustrated in Figure 5), are typically Fourier transformed without phase shift correction. Fourier transforms are an important aspect of data analysis because they relate the EXAFS function $k^n\chi(k)$ of the photoelectron wavevector $\mathbf{k}$ ($Å^{-1}$) to its complementary function $\Phi_n(r')$ of distance r'(Å). Hence, the Fourier transform provides a simple physical picture, a pseudoradial distribution function, of the environment about the X-ray absorbing element. The contributions of different coordination spheres of neighbors around the absorber appear as peaks in the Fourier transform. The Fourier transform peaks are always shifted from the true distances r to shorter ones r' due to the effect of a phase shift, which amounts to ~0.2–0.5 Å, depending upon the absorbing and backscattering atom phase functions.

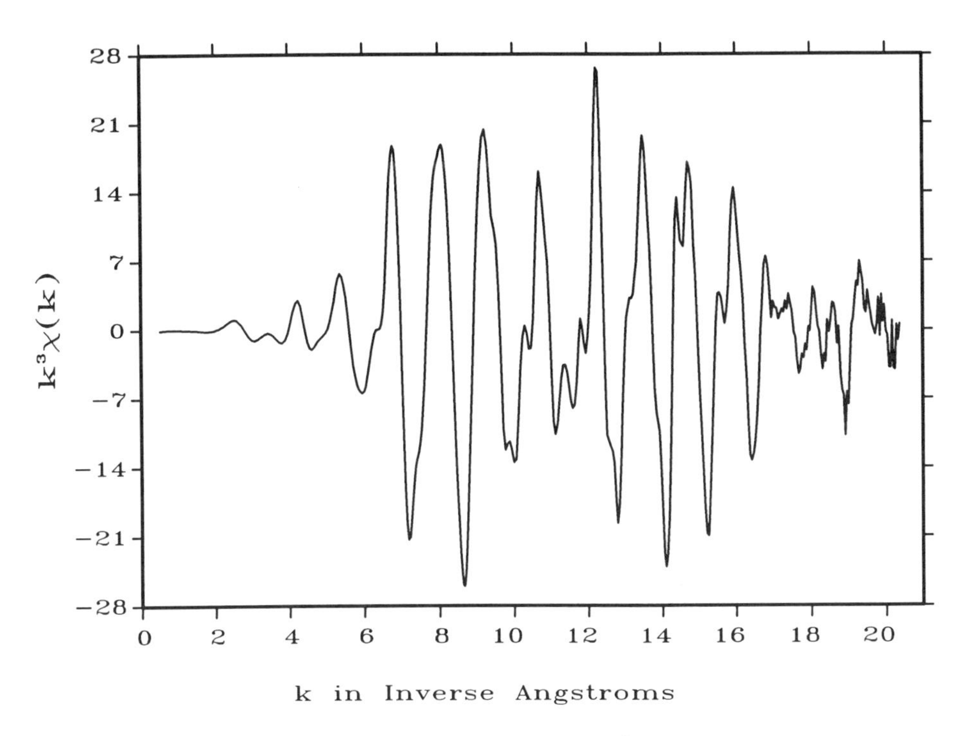

Figure 5 **Background-subtracted, normalized, and k^3-weighted Mo K-edge EXAFS, $k^3\chi(k)$ versus k (Å^{-1}), for molybdenum metal foil obtained from the primary experimental data of Figure 2 with E_0 = 20,025 eV.**

The Fourier transform of the EXAFS of Figure 5 is shown in Figure 6 as the solid curve: It has two large peaks at 2.38 and 2.78 Å as well as two small ones at 4.04 and 4.77 Å. In this example, each peak is due to Mo–Mo backscattering. The peak positions are in excellent correspondence with the crystallographically determined radial distribution for molybdenum metal foil (bcc)—with Mo–Mo interatomic distances of 2.725, 3.147, 4.450, and 5.218 Å, respectively. The Fourier transform peaks are phase shifted by ~0.39 Å from the true distances.

To extract structural parameters (e.g. interatomic distances, Debye-Waller factors, and the number of neighboring atoms) with greater accuracy than is possible from the Fourier transform data alone, nonlinear least-squares minimization techniques are applied to fit the EXAFS or Fourier transform data with a semiempirical, phenomenological model of short-range, single scattering.[7, 12] Fourier-filtered EXAFS data are well suited for the iterative refinement procedure. High-frequency noise and residual background apparent in the experimental data are effectively removed by Fourier filtering methods. These involve the isolation of the peaks of interest from the total Fourier transform with a filter function, as illustrated by the dashed curve in Figure 6. The product of the smooth filter with the real and imagi-

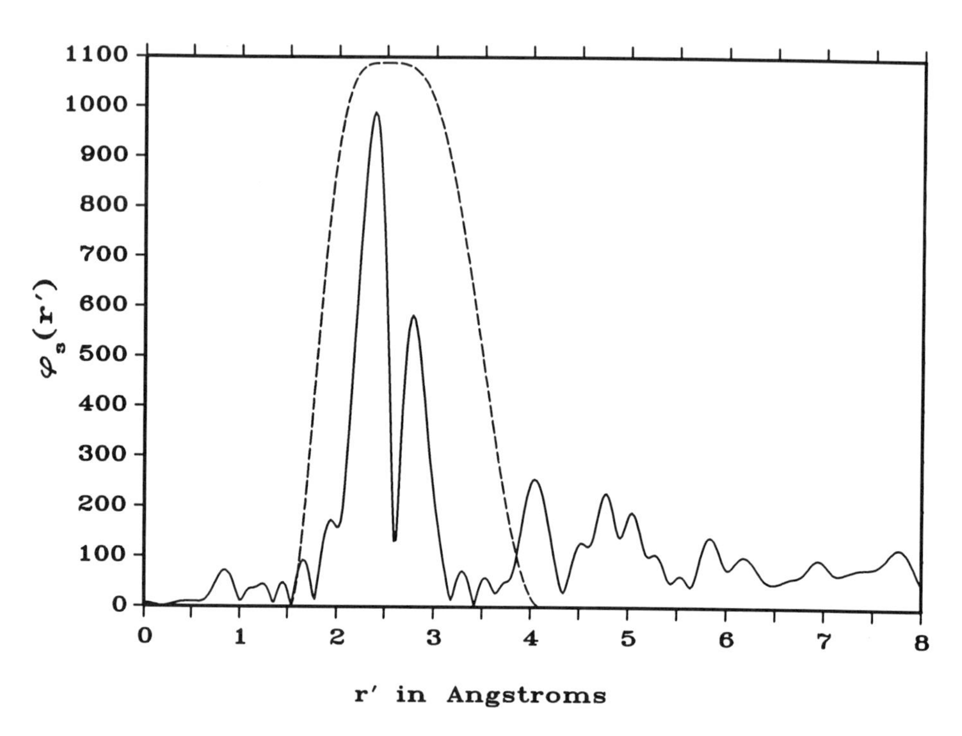

Figure 6 Fourier transform (solid curve), $\Phi_3(r')$ versus r' (Å, without phase-shift correction), of the Mo K-edge EXAFS of Figure 5 for molybdenum metal foil. The Fourier filtering window (dashed curve) is applied over the region ~1.5–4.0 Å to isolate the two nearest Mo–Mo peaks.

nary parts of the Fourier transform on the selected distance range is then Fourier inverse-transformed back to wavevector space to provide Fourier-filtered EXAFS, as illustrated by the solid curve of Figure 7. For curve fitting, phase shifts and back-scattering amplitudes are fixed during the least-squares cycles. These can be obtained readily from theoretical or, alternatively, empirical tabulations.[12] The best fit (dashed curve) to the Fourier-filtered EXAFS data (solid curve) of the first two coordination spheres of molybdenum metal is shown in Figure 7.

Capabilities and Limitations

The classical approach for determining the structures of crystalline materials is through diffraction methods, i.e., X-ray, neutron-beam, and electron-beam techniques. Diffraction data can be analyzed to yield the spatial arrangement of all the atoms in the crystal lattice. EXAFS provides a different approach to the analysis of atomic structure, based not on the diffraction of X rays by an array of atoms but rather upon the absorption of X rays by individual atoms in such an array. Herein lie the capabilities and limitations of EXAFS.

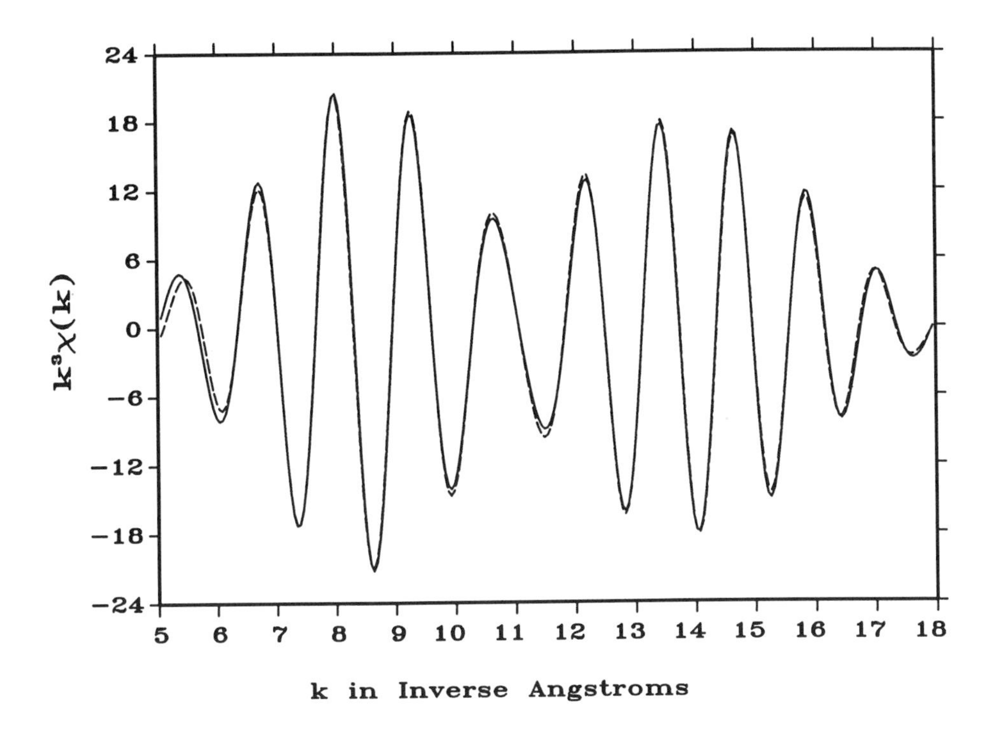

Figure 7 Fourier-filtered Mo K-edge EXAFS, $k^3\chi(k)$ versus k (Å^{-1}) (solid curve), for molybdenum metal foil obtained from the filtering region of Figure 6. This data is provided for comparison with the primary experimental EXAFS of Figure 5. The two-term Mo–Mo best fit to the filtered data with theoretical EXAFS amplitude and phase functions is shown as the dashed curve.

Because diffraction methods lack the element specificity of EXAFS and because EXAFS lacks the power of molecular-crystal structure solution of diffraction, these two techniques provide complementary information. On the one hand, diffraction is sensitive to the stereochemical short- and long-range order of atoms in specific sites averaged over the different atoms occupying those sites. On the other hand, EXAFS is sensitive to the radial short-range order of atoms about a specific element averaged over its different sites. Under favorable circumstances, stereochemical details (i.e., bond angles) may be determined from the analysis of EXAFS for both oriented and unoriented samples.[12] Furthermore, EXAFS is applicable to solutions and gases, whereas diffraction is not. One drawback of EXAFS concerns the investigation of samples wherein the absorbing element is in multiple sites or multiple phases. In either case, the results obtained are for an average environment about all of the X-ray absorbing atoms due to the element-specific site averaging of structural information. Although not common, site-selective EXAFS is possible.[3]

Unlike traditional surface science techniques (e.g., XPS, AES, and SIMS), EXAFS experiments do not routinely require ultrahigh vacuum equipment or electron- and ion-beam sources. Ultrahigh vacuum treatments and particle bombardment may alter the properties of the material under investigation. This is particularly important for accurate valence state determinations of transition metal elements that are susceptible to electron- and ion-beam reactions. Nevertheless, it is always more convenient to conduct experiments in one's own laboratory than at a synchrotron radiation facility, which is therefore a significant drawback to the EXAFS technique. These facilities seldom provide timely access to beam lines for experimentation of a proprietary nature, and the logistical problems can be overwhelming.

Although not difficult, the acquisition of EXAFS is subject to many sources of error, including those caused by poorly or improperly prepared specimens, detector nonlinearities, monochromator artifacts, energy calibration changes, inadequate signal-to-noise levels, X-ray beam induced damage, etc.[9] Furthermore, the analysis of EXAFS can be a notoriously subjective process: an accurate structure solution requires the generous use of model compounds with known structures.[7, 12]

Applications

EXAFS has been used to elucidate the structure of adsorbed atoms and small molecules on surfaces; electrode–electrolyte interfaces; electrochemically produced solution species; metals, semiconductors, and insulators; high-temperature superconductors; amorphous materials and liquid systems; catalysts; and metalloenzymes. Aspects of the applications of EXAFS to these (and other) systems are neatly summarized in References 1–9, and will not be repeated here. It is important to emphasize that EXAFS experiments are indispensable for *in situ* studies of materials, particulary catalysts[3–9] and electrochemical systems.[13] Other techniques that have been successfully employed for *in situ* electrochemical studies include ellipsometry, X-ray diffraction, X-ray standing wave detection, Mossbauer-effect spectroscopy, Fourier-transform infrared spectroscopy, UV-visible reflectance spectroscopy, Raman scattering, and radiotracer methods. Although the established electrochemical technique of cyclic voltammetry is a true *in situ* probe, it provides little direct information about atomic structure and chemical bonding. EXAFS spectroelectrochemistry is capable of providing such information.[13] In this regard, thin oxide films produced by passivation and corrosion phenomena have been the focus of numerous EXAFS investigations.

It is known that thin (~20 Å) passive films form on iron, nickel, chromium, and other metals. In aggressive environments, these films provide excellent corrosion protection to the underlying metal. The structure and composition of passive films on iron have been investigated through iron K-edge EXAFS obtained under a variety of conditions,[8, 14] yet there is still some controversy about the exact nature of passive films on iron. The consensus is that the passive film on iron is a highly disordered form of γ-FeOOH. Unfortunately, the majority of EXAFS studies of passive films have been on chemically passivated metals: Electrochemically passivated metals are of greater technological significance. In addition, the structures of passive films after attack by chloride ions and the resulting corrosion formations have yet to be thoroughly investigated with EXAFS.

Conclusions

Since the early 1970s, the unique properties of synchrotron radiation have been exploited for EXAFS experiments that would be impossible to perform with conventional sources of X-radiation. This is not surprising given that high-energy electron synchrotrons provide 10,000 times more intense continuum X-ray radiation than do X-ray tubes. Synchrotron radiation has other remarkable properties, including a broad spectral range, from the infrared through the visible, vacuum ultraviolet, and deep into the X-ray region; high polarization; natural collimation; pulsed time structure; and a small source size. As such, synchrotron radiation facilities provide the most useful sources of X-radiation available for EXAFS.

The future of EXAFS is closely tied with the operation of existing synchrotron radiation laboratories and with the development of new ones. Several facilities are now under construction throughout the world, including two in the USA (APS, Argonne, IL, and ALS, Berkeley, CA) and one in Europe (ESRF, Grenoble, France). These facilities are wholly optimized to provide the most brilliant X-ray beams possible—10,000 times more brilliant than those available at current facilities! The availability of such intense synchrotron radiation over a wide range of wavelengths will open new vistas in EXAFS and materials characterization. Major advances are anticipated to result from the accessibility to new frontiers in time, energy, and space. The tremendous brilliance will facilitate time-resolved EXAFS of processes and reactions in the microsecond time domain; high-energy resolution measurements throughout the electromagnetic spectrum; and microanalysis of materials in the submicron spatial domain, which is hundreds of times smaller than can be studied today. Finally, the new capabilities will provide unprecedented sensitivity for trace analysis of dopants and impurities.

Related Articles in the Encyclopedia

NEXAFS, EELS, LEED, Neutron Diffraction, AES, and XPS

References

1 *EXAFS Spectroscopy: Techniques and Applications.* (B. K. Teo and D. C. Joy, eds.) Plenum, New York, 1981. Contains historical items and treatments of EXELFS, the electron-scattering counterpart of EXAFS.

2 P. A. Lee, P. H. Citrin, P. Eisenberger, and B. M. Kincaid. Extended X-ray Absorption Fine Structure—Its Strengths and Limitations as a Structural Tool. *Rev. Mod. Phys.* **53,** 769, 1981.

3 *XAFS V, Proceedings of the Fifth International Conference on X-ray Absorption Fine Structure.* (J. M. de Leon, E. A. Stern, D. E. Sayers, Y. Ma, and J. J. Rehr, eds.) North-Holland, Amsterdam, 1989. Also in *Physica B.* **158,** 1989. "Report of the International Workshop on Standards and Criteria in X-ray Absorption Spectroscopy" (pp. 701–722) is essential reading.

4 *EXAFS and Near Edge Structure IV. Proceedings of the International Conference.* (P. Lagarde, D. Raoux, and J. Petiau, eds.) *J. De Physique,* **47**, Colloque C8, Suppl. 12, 1986, Volumes 1 and 2.

5 *EXAFS and Near Edge Structure III. Proceedings of an International Conference.* (K. O. Hodgson, B. Hedman, and J. E. Penner-Hahn, eds.) Springer, Berlin, 1984.

6 *EXAFS and Near Edge Structure. Proceedings of the International Conference.* (A. Bianconi, L. Incoccia, and S. Stipcich, eds.) Springer, Berlin, 1983.

7 *X-Ray Absorption. Principles, Applications, Techniques of EXAFS, SEXAFS and XANES.* (D. C. Koningsberger and R. Prins, eds.) Wiley, New York, 1988.

8 Structure of Surfaces and Interfaces as Studied Using Synchrotron Radiation. *Faraday Discussions Chem. Soc.* 89, 1990. A lively and recent account of studies in EXAFS, NEXAFS, SEXAFS, etc.

9 *Applications of Synchrotron Radiation.* (H. Winick, D. Xian, M. H. Ye, and T. Huang, eds.) Gordon and Breach, New York, 1988, Volume 4. F. W. Lytle provides (pp. 135–223) an excellent tutorial survey of experimental X-ray absorption spectroscopy.

10 H. Winick and G. P. Williams. Overview of Synchrotron Radiation Sources World-wide. *Synchrotron Radiation News.* **4,** 23, 1991.

11 *National Synchrotron Light Source User's Manual: Guide to the VUV and X-Ray Beam Lines.* (N. F. Gmur ed.) BNL informal report no. 45764, 1991.

12 B. K. Teo. *EXAFS: Basic Principles and Data Analysis.* Springer, Berlin, 1986.

13 L. R. Sharpe, W. R. Heineman, and R. C. Elder. EXAFS Spectroelectrochemistry. *Chem. Rev.* **90,** 705, 1990.

14 *Passivity of Metals and Semiconductors.* (M. Froment, ed.) Elsevier, Amsterdam, 1983.

4.3 SEXAFS / NEXAFS

Surface Extended X-Ray Absorption Fine Structure and Near Edge X-Ray Absorption Fine Structure

DAVID NORMAN

Contents

Introduction

SEXAFS is a research technique providing the most precise values obtainable for adsorbate–substrate bond lengths, plus some information on the number of nearest neighbors (coordination numbers). Other methods for determining the quantitative geometric structure of atoms at surfaces, described elsewhere in this volume (e.g., LEED, RHEED, MEIS, and RBS), work only for single-crystal substrates having atoms or molecules adsorbed in a regular pattern with long-range order within the adsorbate plane. SEXAFS does not suffer from these limitations. It is sensitive only to local order, sampling a short range within a few Å around the absorbing atom.

SEXAFS can be measured from adsorbate concentrations as low as ~0.05 monolayers in favorable circumstances, although the detection limits for routine use are several times higher. By using appropriate standards, bond lengths can be determined as precisely as ±0.01 Å in some cases. Systematic errors often make the accuracy much poorer than the precision, with more realistic estimates of ±0.03 Å or worse.

NEXAFS has become a powerful technique for probing the structure of molecules on surfaces. Observation of intense resonances near the X-ray absorption edge can indicate the type of bonding. On a flat surface, the way in which the resonances vary with angle of the specimen can be analyzed simply to give the molecular orientation, which is precise to within a few degrees. The energies of resonances allow one to estimate the intramolecular bond length, often to within ±0.05 Å. Useful NEXAFS can be seen for concentrations as low as ~0.01 monolayer in favorable cases.

The techniques can be applied to almost any adsorbate on almost any type of solid sample—metal, semiconductor or insulator. Light adsorbates—say, from C through Al—are more difficult to study than heavier ones because their absorption edges occur at low photon energies that are technically more difficult to produce. The technique samples all absorbing atoms of the same type, and averages over them, so that good structural information is obtained only when the adsorbates uniquely occupy equivalent sites. Thus it is not easy to examine clean surfaces, where the EXAFS signal from surface atoms is overwhelmed by that from the bulk. The best way to study such samples is with X rays incident on the sample at a grazing angle so that they interact only in a region close to the surface: by varying the angle, the probing depth can be changed somewhat. The reviews of SEXAFS and NEXAFS[1–5] should be consulted for more details.

Basic Principles of X-ray Absorption

The physical processes of X-ray absorption are depicted schematically in Figure 1. The energies of discrete core levels are uniquely determined by the atom type (as in XPS or AES), so tuning the photon energy to a particular core level gives an *atom-specific* probe. When the photon energy equals the binding energy of the electron in a core level, a strong increase in absorption is seen, which is known as the absorption edge. The absorbed photon gives its energy to a photoelectron that propagates as a wave. In a molecule or solid, part of this photoelectron wave may be backscattered from neighboring atoms, the backscattered wave interfering constructively or destructively with the outgoing wave. Thus one gets a spectrum of absorption as a function of photon energy that contains wiggles (EXAFS) superimposed on a smooth background. The amplitude of the EXAFS wiggles depends on the number of neighbors, the strength of their scattering and the static and dynamic disorder in their position. The frequency of the EXAFS wiggles depends on the wavevector k of the photoelectrons (related to their kinetic energy) and the distance to neighboring atoms. The frequency is inversely related to the nearest neighbor separation, with a short distance giving widely spaced wiggles and vice versa.

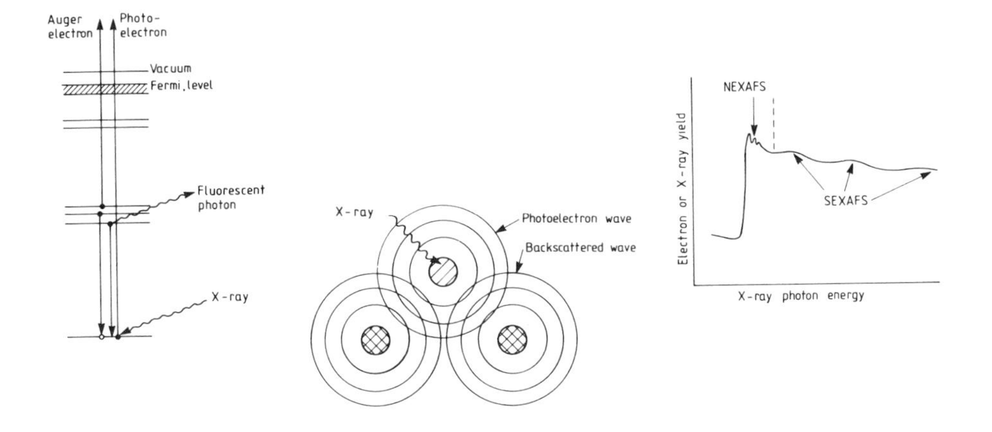

Figure 1 **Basic physical principles of X-ray absorption. As in XPS/ESCA, absorption of a photon leads to emission of a photolectron. This photoelectron, propagating as a wave, may be scattered from neighboring atoms. The backscattered wave interferes constructively or destructively with the outgoing wave, depending on its wavelength and the distance to neighboring atoms, giving wiggles in the measured absorption spectrum.**

EXAFS can be used to study surfaces or bulk samples. Ways of making the technique surface-sensitive are spelled out below. EXAFS gives a spherical average of information in a shell around an absorbing atom. For an anisotropic sample with a polarized photon beam, one gets a searchlight effect, where neighbors in directions along that of the polarization vector **E** (perpendicular to the direction of the X rays) are selectively picked out. For studies on flat surfaces the angular variation of the EXAFS intensity is one of the best methods of identifying an adsorption site. The form of the backscattering amplitude depends on atomic number, differing between atoms in different rows of the periodic table,[5] and this helps one to determine which atoms in a compound are nearest neighbors.

Phase Shifts

When an electron scatters from an atom, its phase is changed so that the reflected wave is not in phase with the incoming wave. This changes the interference pattern and hence the apparent distance between the two atoms. Knowledge of this phase shift is the key to getting precise bond lengths from SEXAFS. Phase shifts depend mainly on which atoms are involved, not on their detailed chemical environment, and should therefore be transferable from a known system to unknown systems. The phase shifts may be obtained from theoretical calculations, and there are published tabulations, but practically it is desirable to check the phase shifts using model compounds: the idea is to take a sample of known composition and crystallography, measure its EXAFS spectrum and analyze it to determine a phase shift ϕ. The model compound should ideally contain the same absorber and backscatterer atoms as the unknown, and in the same chemical state. If this is not possible, the next best option is to use a model whose absorber and a backscatterer are neighboring elements in the periodic table to those in the unknown sample, although for highest precision the backscatterer should be the same as in the unknown.

One must be sure of the purity of the model compound. It may have deteriorated (for example, by reaction or water absorption), its surface may not have the same composition as the bulk, or it may not be of the correct crystallographic phase. It is tempting to use single crystals to be sure of the geometric structure, but noncubic crystals give angle-dependent spectra. The crystallography of any compound should be checked with XRD.

Experimental Details

There are several ways to make a SEXAF/NEXAFS measurement surface sensitive.

1. By using dispersed samples, the surface-to-bulk ratio is increased, and standard methods of studying "bulk" samples will work (see the article on EXAFS).
2. By making the X rays incident on the sample at shallow angles (usually a fraction of a degree), they see only the near-surface region, some 20–50 Å deep. The angle of incidence can be varied, allowing crude depth profiling, but the penetration is crucially dependent on the flatness of the reflecting surface, and large homogeneous samples are needed. This is potentially a useful technique for studying buried interfaces, where the signal will come predominantly from the interface if the substrate is more dense than the overlayer. This method has been little tried in the soft X-ray region but should work well, since the critical angle is larger than for hard X rays.
3. Since X-ray absorption is an atom-specific process, any atoms known to be, or deliberately placed, on a solid consisting of different atoms can be studied with high sensitivity.
4. The absorption may be monitored via a secondary decay process that is surface-sensitive, such as the emission of Auger electrons, which have a well-defined energy and a short mean free path.

X-Ray Sources

The only X-ray source with sufficient intensity for surface measurements is synchrotron radiation. Synchrotron radiation is white light, including all wavelengths from the infrared to X rays. A spectroscopy experiment needs a particular wavelength (photon energy) to be selected with a monochromator and scanned through the spectrum. For EXAFS, a range is needed of at least 300 eV above the absorption edge that does not contain any other edges, such as those from coadsorbates, the substrate, or from higher order light (unwanted X rays from the monochromator with two, three, or more times the desired energy). NEXAFS needs a clear range of perhaps 25–30 eV above the edge. Perusal of a table of energy levels is essential.[6] Photon energies from about 4 keV to 15 keV are easiest to use, where X-ray windows allow sample chambers to be separated from the monochromator. Energies below about 1800 eV are technically the most difficult, requiring ultrahigh vacuum monochromators directly connected to the sample chamber. K edges are easiest to interpret, but $L_{2,3}$ edges can be used: line widths are much broader at L_1 edges, and states such as M_5 may have an absorption edge too wide to be usable for EXAFS.

Detection Methods

The experiment consists of measuring the intensity of photons incident on the sample, and the proportion of them that is absorbed. Most SEXAFS experiments detect the X-ray absorption coefficient indirectly by measuring the fluorescence or Auger emission that follows photon absorption. (See the articles on AES and XRF.) The various electron or photon detection schemes should be tested to see which one gives the best data in each case. Measuring all electrons, the total electron yield (TEY), or those in a selected bandpass, the partial electron yield (PEY), will give higher signals but poorer sensitivity than the Auger electron yield (AEY). Fluorescence yields (FYs) are low for light elements, so their measurement usually gives weak signals, but the background signal is usually low, in which case FY will give high sensitivity. FY is the technique of choice for insulating samples that may charge up and confuse electron detection. FY also allows for experiments in which the sample is in an environment other than the high vacuum needed for electrons. With suitable windows, surface reactions may be followed *in situ*, for instance in a high-pressure chamber or an electrochemical cell, although this type of work is yet in its infancy.

Electron Excitation

The advantages of SEXAFS/ NEXAFS can be negated by the inconvenience of having to travel to synchrotron radiation centers to perform the experiments. This has led to attempts to exploit EXAFS-like phenomena in laboratory-based techniques, especially using electron beams. Despite doubts over the theory there appears to be good experimental evidence[7] that electron energy loss fine structure (EELFS) yields structural information in an identical manner to EXAFS. However, few EELFS experiments have been performed, and the technique appears to be more taxing than SEXAFS.

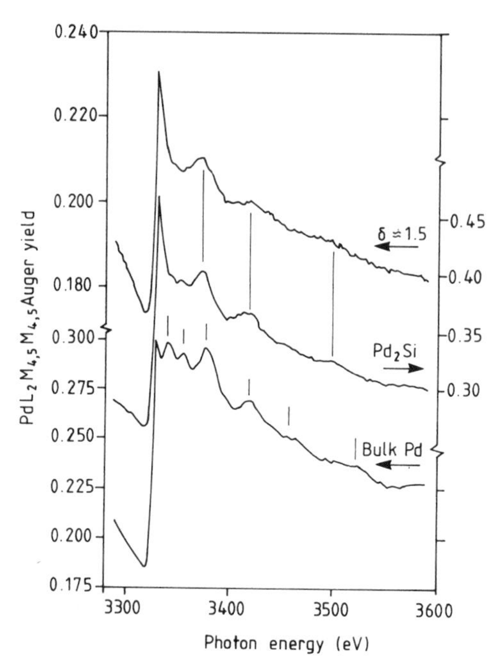

Figure 2 **Surface EXAFS spectra above the Pd L_2-edge for a 1.5 monolayer evaporated film of Pd on Si(111) and for bulk palladium silicide, Pd_2Si and metallic Pd.**

SEXAFS Data Analysis and Examples

Often a comparison of raw data directly yields useful structural information. An example is given in Figure 2, which shows SEXAFS spectra[8] above the palladium L_2 edge for 1.5 monolayers of Pd evaporated onto a Si (111) surface, along with pure Pd and the bulk compound Pd_2Si. It is clear just from looking at the spectra and without detailed analysis that the thin layer of Pd reacts to give a surface compound similar to the palladium silicide and completely different from the metallic Pd. By contrast, a thin layer of silver, studied in the same experiment, remains as a metallic Ag overlayer, as judged from its SEXAFS wiggles.

Fourier Transformation

One of the major advantages of SEXAFS over other surface structural techniques is that, provided that single scattering applies (see below), one can go directly from the experimental spectrum, via Fourier transformation, to a value for bond length. The Fourier transform gives a real space distribution with peaks in $|F(R)|$ at distances $R - \phi$. Addition of the phase shift, ϕ, then gives the true interatomic distance. Figure 3 shows how this method[9] is applied to obtain the O–Ni distance in the half-monolayer structure of oxygen absorbed on Ni (100). The data, after back-

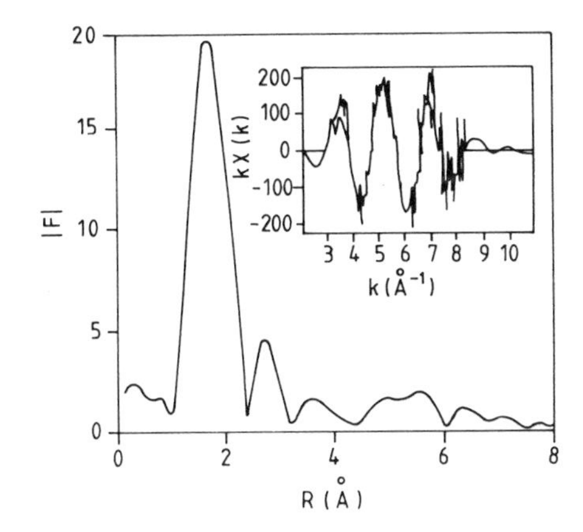

Figure 3 **The modulus of the Fourier transform of the SEXAFS spectrum for the half-monolayer coverage on Ni(100) The SEXAFS spectrum itself is shown in the inset with the background removed.**

ground subtraction, yield a Fourier transform dominated by a single peak at $R \approx 1.73$ Å. Correcting for the phase shift derived from bulk NiO, a nearest neighbor distance of $R_{O-Ni} \approx 1.98$ Å is obtained.

Fourier transforms cannot be used if shells are too close together, the minimum separation ΔR being set by the energy range above the absorption edge over which data are taken, typically $\approx$0.2 Å for SEXAFS. A useful application of the Fourier technique is to filter high-frequency noise from a spectrum. This is done by putting a window around a peak in *R*-space and transforming back into *k*-space: each shell may be filtered and analyzed separately, although answers should always be checked against the original unfiltered spectrum.

Curve Fitting

The beauty of using photons is that their absorption is easily understood and exactly calculable, so that structural analysis can be based on comparisons of experimental data and calculated spectra. Statistical confidence limits can easily be computed, although the systematic errors will often be much greater than the random errors. An example of data analysis by curve fitting is depicted in Figure 4 for the system of ⅔ monolayer of Cl on Ag (111).[10] The nearest neighbor Cl–Ag (2.70 Å) and Cl–Cl (2.89 Å) shells are so close in distance that they cannot be separated in a Fourier transform approach, but they are easily detected here by the fact that their atomic backscattering factors vary differently with energy, thus influencing the overall shape of the spectrum.

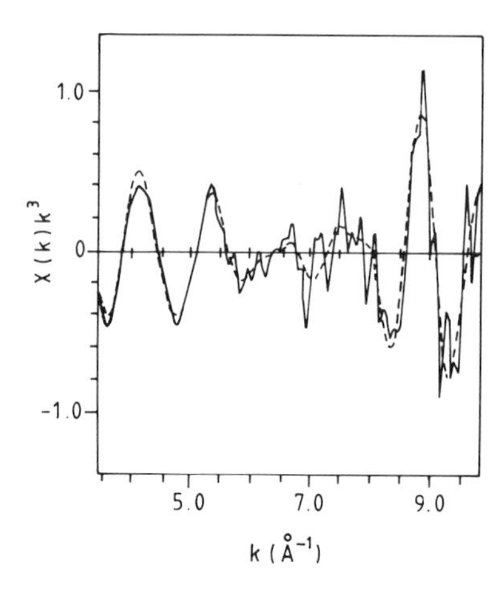

Figure 4 **The EXAFS function $\chi(k)$, weighted by k^3, experimental data for ⅔ monolayer of Cl on Ag(111) (solid line), with the best theoretical fit (dashed line) from the least-squares curve fitting method with neighbors as distances of 2.70 Å (Ag), 2.89 Å (CL), 3.95 Å (Ag) and 5.00 Å (Cl).**

Complications

The simple theory assumes single scattering only, in which electrons go out only from the absorber atom to a backscatterer and back, rather than undertaking a journey involving two or more scattering atoms. Such multiple scattering may sometimes be important in EXAFS, especially when atoms are close to collinear, giving wrong distances and coordination numbers. With modern, exact theories of EXAFS one can deal with multiple scattering, but it is complicated and time-consuming, and a unique analysis may be impossible. However, nearest neighbor information can never be affected by multiple scattering, since there is no possible electron path shorter than the direct single scattering route.

EXAFS analysis usually assumes a shell of neighbors at a certain distance, with a Gaussian (normal) distribution around that distance to cope with the effects of disorder, both static (positional) and dynamic (vibrational). Static disorder arises where, even at zero temperature, a range of sites is occupied, as found particularly with amorphous or glassy samples. EXAFS samples directly the distance between nearby atoms and thus measures correlated motion, giving a disorder (Debye–Waller) factor smaller than that derived from long-range diffraction techniques like XRD or LEED. Vibrational amplitudes at a surface usually differ from those in the bulk, and SEXAFS spectra measured at different angles have been used to reveal surface dynamics, resolving vibrations parallel and perpendicular to a single-crystal surface.

The assumption of harmonic vibrations and a Gaussian distribution of neighbors is not always valid. Anharmonic vibrations can lead to an incorrect determination of distance, with an apparent mean distance that is shorter than the real value. Measurements should preferably be carried out at low temperatures, and ideally at a range of temperatures, to check for anharmonicity. Model compounds should be measured at the same temperature as the unknown system. It is possible to obtain the real, non-Gaussian, distribution of neighbors from EXAFS, but a model for the distribution is needed[11] and inevitably more parameters are introduced.

Some of these complications can lead to an incorrect structural analysis. For instance, it can be difficult to tell whether one's sample has many nearest neighbors with large disorder or fewer neighbors more tightly defined. Analysis routines are available at almost all synchrotron radiation centers: curve fitting may be the best method because most of the factors affecting the spectrum vary with energy in a different way and this k-dependence allows them to be separated out. A curved-wave computational scheme can be especially useful for analyzing data closer to the absorption edge.

NEXAFS Data Analysis and Examples

Chemical Shifts and Pre-Edge Features

The absorption edge occurs when the photon energy is equal to the binding energy of an electron core level. Shifts in the position of the edge are caused by small differences in the chemical environment, as in ESCA (XPS). If one needs to know the exact energy of the edge, perhaps for comparison with other published data, then a model compound with a calibrated energy should be measured under the same conditions as the unknown. Features may be seen before the absorption edge, most obviously in transition metals and their compounds. These small peaks are characteristic of local coordination (octahedral, tetrahedral or whatever); their intensity increases with oxidation state.

Atomic Adsorbates

The NEXAFS region near an absorption edge is usually discarded in an EXAFS analysis because the strong scattering and longer mean free path of the excited photoelectron give rise to sizable multiple-scattering corrections. For several atomic adsorbates NEXAFS has been modeled by complicated calculations, which show that scattering involving around 30 atoms, to a distance >5.0 Å from the absorbing atom, contributes to the spectrum. This makes interpretation difficult and not useful for practical purposes, except possibly for fingerprinting different adsorption states.

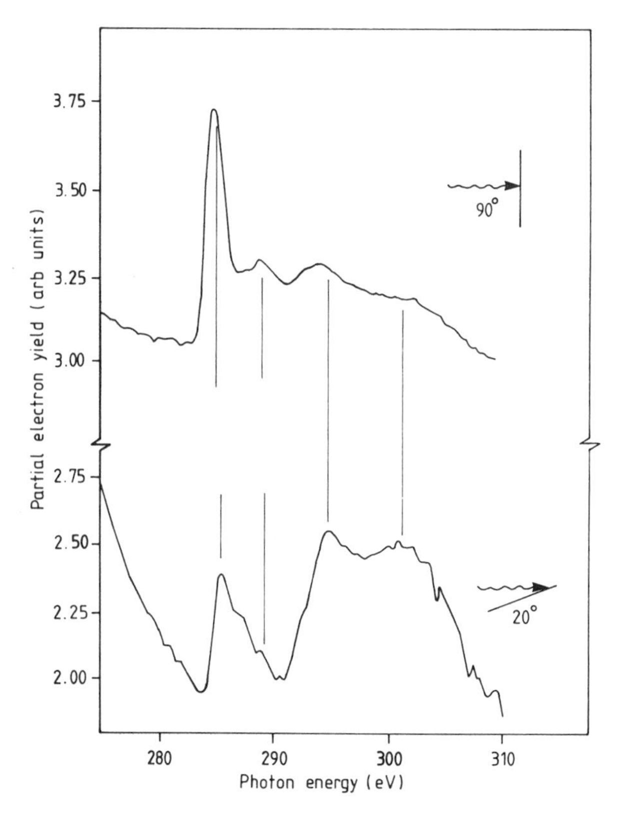

Figure 5 **NEXAFS spectra above the C K-edge for a saturation coverage of pyridine C_5H_5N on Pt(111), measured at two different polarisation angles with the X-ray beam at normal incidence and at 20° to the sample surface.**

Molecular Adsorbates—Orientation

For molecules, NEXAFS often contains intense resonances that dwarf the effects of atomic scattering in the spectrum. These resonances arise from states that are localized in space within the molecule, rather than being spread out and shared between various atoms: they are thus mainly characteristic of the molecule itself and only weakly affected by differences in the way the molecule may be bonded to the surface. Despite the technical difficulties, most NEXAFS work has been done at the carbon K-edge. An example is depicted in Figure 5, which shows NEXAFS for a saturation coverage of pyridine C_5H_5N on Pt (111), measured at different angles to the photon beam.[12] Peak A is identified as a π resonance, arising from transitions from the C 1s state to the unfilled π^* molecular orbital. Peak B comes from CO impurity. Peaks C and D are transitions to σ shape resonances that lie in the plane of the molecule. The variation of intensity of the π and σ resonances with polarization angle gives the molecular orientation, each peak being maximized when the polarization vector **E** lies along the direction of the orbital. The π intensity is greatest when **E** is parallel to the surface ($\theta = 90°$), so the π orbitals must lie parallel to the surface. Therefore the pyridine molecule must stand upright on the Pt (111) surface. NEXAFS alone tells us only the orientation with respect to the top plane of the substrate, not the detailed bonding to the individual atoms, nor which end of the molecule is next to the surface: this detailed geometry must be determined from other techniques.

There may be deviations from the perfect angular dependence due to partial polarization of the X rays or to a tilted molecule. This can be investigated by analysis of the intensities of the resonances as a function of angle. Measuring the intensity of NEXAFS peaks is not always straightforward, and one has to be careful in removing experimental artifacts from the spectrum and in subtracting the atomic absorption background, for which various models now exist. Detailed analysis is not always needed, for instance the mere observation of a π resonance can be chemically useful in distinguishing between π and di-σ bonding of ethylene on surfaces. NEXAFS can be applied to large molecules, such as polymers and Langmuir-Blodgett films. The spectra of polymers, such as those[13] depicted in Figure 6, contain a wealth of detail and it is beyond the current state of knowledge to assign all the peaks. However, the sharper, lower lying ones are attributed to π^* molecular orbital states. Changes in these features were observed after deposition of submonolayer amounts of chromium, from which it is deduced that the carbonyl groups on the polymers are the sites for initial interaction with the metal overlayer. It has been suggested[4] that most examples of molecular adsorbate NEXAFS may be analyzed with quite simple models that decompose complex molecules into building blocks of diatoms or rings.

Intramolecular Bond Length

The energies of shape resonances often seem inversely related to the intramolecular bond length, with a long bond giving a σ resonance close to threshold and a shorter bond showing a peak at higher energy.[4] This effect has been demonstrated for many small molecules, although some do not fit the general trend. A mathematical relationship has been derived to allow estimates of bond length, but with the current state of knowledge it seems safest to restrict its use to diatomic molecules or ligands. With this procedure, intramolecular chemisorption bond lengths can be determined to an accuracy of ± 0.05 Å.

Conclusions

X-ray absorption spectroscopy is an important part of the armory of techniques for examining pure and applied problems in surface physics and chemistry. The basic physical principles are well understood, and the experimental methods and data analysis have advanced to sophisticated levels, allowing difficult problems to be solved. For some scientists the inconvenience of having to visit synchrotron radia-

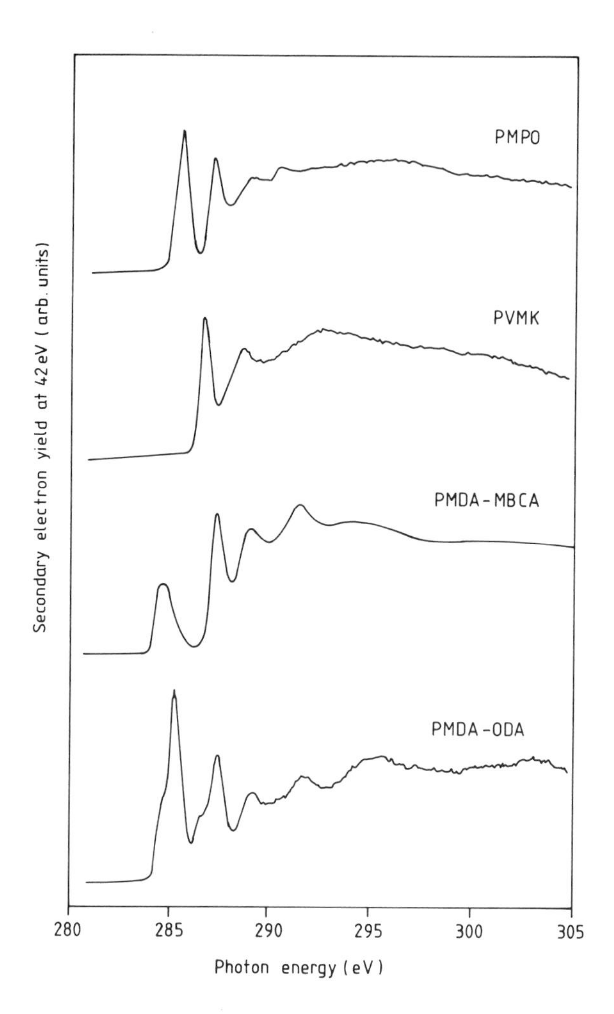

Figure 6 **NEXAFS spectra above the C K-edge for the polymers PMPO poly (dimethyl phenylene oxide), PVMK poly (vinyl methyl ketone), PMDA-MBCA PI poly (pyromellitimido 4, 4-methylene bis-cyclohexyl amine) and PMDA-ODA PI poly (pyromellitimido oxydianiline).**

tion centers is outweighed by the unique surface structural information obtainable from SEXAFS/NEXAFS. Nevertheless, although they are powerful techniques in the hands of specialists, it is difficult to foresee their routine use as analytical tools.

A database of model compound spectra and a better understanding of complex molecules would help the inexperienced practitioner. More dilute species could be studied by brighter synchrotron radiation sources. An obvious experimental improvement would be to use a polychromatic energy-dispersive arrangement for speedier data collection. In such a scheme the X rays are dispersed across a sample so that photons having a range of energies strike the specimen, and a detection method has to be used that preserves the spatial distribution of the emitted electrons. Currently available photon fluxes are such that collection times less than or about one second should then be obtainable for a NEXAFS spectrum.

Related Articles in the Encyclopedia

AES, EXAFS, LEED/RHEED, XPS, and XRF

References

1. D. Norman. *J. Phys. C: Solid State Phys.* **19,** 3273, 1986. Reprinted with an appendix bringing it up to date in 1990 as pp. 197–242 in *Current Topics in Condensed Matter Spectroscopy.* Adam Hilger, 1990. An extensive review of SEXAFS and NEXAFS, concentrating on physical principles.
2. P. H. Citrin. *J. Phys. Coll.* **C8,** 437, 1986. Reviews all SEXAFS work up to 1986, with personal comments by the author.
3. *X-Ray Absorption: Principles, Applications, Techniques of EXAFS, SEXAFS and XANES.* (D.C. Koningsberger and R. Prins, Eds.) Wiley, New York, 1988. The best book on the subject. Especially relevant is the chapter by J. Stöhr, which is a comprehensive and readable review of SEXAFS.
4. J. Stöhr. *NEXAFS Spectroscopy.* Springer-Verlag, New York, 1992. A book reviewing everything about NEXAFS.
5. D. Norman. In *Physical Methods of Chemistry.* Wiley–Interscience, New York, in press. Practical guide with emphasis on chemical applications.
6. *The X-Ray Data Booklet.* Lawrence Berkeley Laboratory, Berkeley, is an excellent source of information.
7. M. de Crescenzi. *Surf. Sci.* **162,** 838, 1985.
8. J. Stöhr and R. Jaeger. *J. Vac. Sci. Technol.* **21,** 619, 1982.
9. L. Wenzel, D. Arvanitis, W. Daum, H. H. Rotermund, J. Stöhr, K. Baberschke, and H. Ibach. *Phys. Rev. B.* **36,** 7689, 1987.
10. G. M. Lamble, R. S. Brooks, S. Ferrer, D. A. King, and D. Norman. *Phys. Rev. B.* **34,** 2975, 1986.
11. T. M. Hayes and J. B. Boyce. *Solid State Phys.* **37,** 173, 1982. Good background reading on EXAFS.
12. A. L. Johnson, E. L. Muetterties, and J. Stöhr. *J. Chem. Soc.* **105,** 7183, 1983.
13. J. L. Jordan-Sweet, C. A. Kovac, M. J. Goldberg, and J. F. Morar. *J. Chem. Phys.* **89,** 2482, 1988.

4.4 XPD and AED

X-Ray Photoelectron and Auger Electron Diffraction

BRENT D. HERMSMEIER

Contents

Introduction

X-ray Photoelectron Diffraction (XPD) and Auger Electron Diffraction (AED) are well-established techniques for obtaining structural information on chemically specific species in the surface regions of solids. Historically, the first XPD effects were observed in single crystals some 20 years ago by Siegbahn et al.[1] and Fadley and Bergstrom.[2] A short time later, others[3] began to quantify the technique as a surface and surface–adsorbate structural tool and, more recently, Egelhoff[4] applied AED to metal-on-metal systems to determine growth modes and shallow interface structures, which has strongly influenced the current expansion to materials research. In general, these pioneering studies have introduced XPD and AED to areas like adsorbate site symmetry, overlayer growth modes, surface structural quality, and element depth distributions,[5, 6] any one of which may be a key to understanding the chemistry or physics behind a measured response. Studies as widely separated as the initial stages of metallic corrosion, interface behavior in epitaxial thin films, and semiconductor surface segregation have already profited from XPD and AED experiments. A broad range of research communities—catalysis, semiconductor, corrosion, material science, magnetic thin film, and packaging, stand to benefit from these diffraction studies, since surfaces and interfaces govern many important interactions of interest. XPD and AED are used primarily as research tools, but, given the hundreds of XPS and AES systems already in use by the aforementioned research communities, their move to more applied areas is certain.

Adaptation of existing XPS and AES instruments into XPD and AED instruments is straightforward for spectrometers that are equipped with an angle resolved analyzer. Traditionally, XPD and AED instruments were developed by individuals to address their specific questions and to test the limits of the technique itself. Today, surface science instrument companies are beginning to market XPD and AED capabilities as part of their multi-technique spectroscopic systems. This approach has great potential for solving both a broad range of problems as is typically found in industrial laboratories and in studies that intensely focus on atomic details, as ins often found in university laboratories. Key to obtaining quality results, whichever the mode of operation, are in the speed of data acquisition, the angular and energy resolution, the accessible angular range, and the capability to meaningfully manipulate the data.

The reader is urged to review the XPS and AES articles in this Encyclopedia to obtain an adequate introduction to these techniques, since XPD and AED are actually their by-products. In principle two additions to XPS and AES are needed to perform diffraction studies, an automated two-axis sample goniometer and an angle-resolved analyzer. Ultrahigh-vacuum conditions are necessary to maintain surface cleanliness. Standard surface cleaning capabilities such as specimen heating and Ar^+ sputtering, usually followed by sample annealing, are often needed. Sample size is rarely an issue, especially in AED where the analysis area may be as small as 300 Å, using electron field emitter sources.

Excellent reviews of XPD and AED have been published by Fadley[5, 6] and are strongly recommended for readers needing information beyond that delineated here.

Basic Principles

The diffraction mechanisms in XPD and AED are virtually identical; this section will focus on only one of these techniques, with the understanding that any conclusions drawn apply equally to both methods, except where stated otherwise. XPD will be the technique discussed, given some of the advantages it has over AED, such as reduced sample degradation for ionic and organic materials, quantification of chemical states and, for conditions usually encountered at synchrotron radiation facilities, its dependence on the polarization of the X rays. For more details on the excitation process the reader is urged to review the relevant articles in the Encyclopedia and appropriate references in Fadley.[5, 6]

Scattering Concept

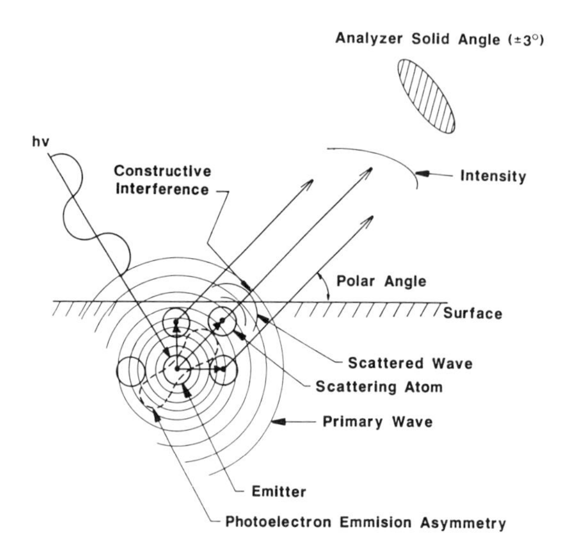

Figure 1 **Simplistic schematic illustration of the scattering mechanism upon which X-ray photoelectron diffraction (XPD) is based. An intensity increase is expected in the forward scattering direction, where the scattered and primary waves constructively interfere.**

XPD is a photoelectron scattering process that begins with the emission of a spherical electron wave created by the absorption of a photon at a given site. This site selectivity allows XPD to focus on specific elements or even on different chemical states of the same element when acquiring diffraction data. The excitation process obeys dipole selection rules, which under special conditions may be used to enhance regions or directions of interest by taking advantage of photoelectron emission asymmetries in the emission process; for example, enhanced surface or bulk sensitivity can be obtained by aligning the light's electric field vector to be parallel or perpendicular to the sample's surface plane, respectively. This flexibility, unfortunately, is not available in most spectrometers because the angle between the excitation source and the analyzer is fixed. The spherical photoelectron wave propagates from the emitter, scatters off neighboring atoms, and decays in amplitude as $1/r$. The scattering events modify the photoelectron intensity reaching the detector relative to that of the unscattered portion, or primary wave. A physical picture of this is given in Figure 1, where a spherical wave propagating outward from the emitter passes through a scattering potential to produce a spherically scattered wave that is nearly in phase with the primary wave in the forward direction. Since the primary and scattered waves have only a slight phase shift in this direction the two waves can be thought of as constructively interfering. However, since both waves are spherical and have different origins, they will tend to interfere destructively in off-forward directions. This interference process results in increased intensity for emission geometries in which rows of closely spaced atoms become aligned with the entrance axis to the analyzer.

The diffraction pattern itself is simply the mapping of the intensity variations as a function of polar or azimuthal angles. Angular scans can be obtained by rotating the sample while leaving the analyzer fixed, or by rotating the analyzer and leaving the sample fixed. In either case, an intensity profile is obtained for a given core-level transition. The recipe is therefore quite simple. First, to determine whether the sample is structurally ordered, one simply looks for any intensity variation as a function of angle; if found it can be concluded that there exists some kind of order over the probing length of the photoelectron. Second, to identify a low-index crystallographic direction, one looks for maxima along various logical or predetermined directions, such as those associated with cleavage planes or previously identified by X-ray diffraction. Usually this confirmation can be done by directly monitoring the intensity with a ratemeter if the specimen is reasonably well ordered. Third, to determine the symmetry of the structure, often the main goal in XPD, one collects several polar and azimuthal scans to correlate the appearance of diffraction peaks at measured angles to suspected Bravais lattice structures having near-neighbor atoms at similar angles. Fourth, one monitors changes in the diffraction features as a function, for example, of sample temperature or of overlayer thickness. This is particularly informative when comparing absorbate or overlayer symmetries with that of the substrate in a fingerprint analysis mode. Although the intensities are dominated by forward scattering processes, a detailed understanding must consider contributions to the detected intensity by all of the scattering atoms within several lattice constants of the emitter. To simulate such a scattering process, a kinematical or single-scattering approach is sufficient if the electron's kinetic energy is greater than 150 eV, and if it is not necessary to fully understand the fine structure in the diffraction pattern.[5, 6] Complicated multiple-scattering calculations can also aid in the quantification of XPD data by more accurately addressing intensity anisotropies and improving identification of the fine structure. But more often than not, multiple-scattering effects contribute little to the basic understanding of the structure, and thus will not be discussed further. Single-scattering results will be displayed along with experimental data, and compared to geometric arguments for resemblance.

Unlike more common electron diffraction methods, such as LEED and RHEED, XPD is dominated by near-neighbor interactions averaged over a very short time scale. The $1/r$ decay of a spherical wave, coupled with a short mean free path for electrons in solids (due to inelastic scattering energy losses), uniquely allows XPD to probe the local, or short-range, order about an emitter. This is simplified further by the incredibly short times involved in the scattering process. Only 10^{-17} seconds are required for a photoelectron to experience a scattering event while typical crystal fluctuations, which can average out short-range order effects, occur on a much longer time scale. In essence, the scattering events can be thought of as a snapshot of the local crystal order.

A brief comment needs to made to clear up some misleading information that has entered the literature concerning the scattering mechanism upon which XPD and AED are based. Frank et al.[7] have proposed what they believe to be a new technique capable of solving surface structure problems. The technique, which they have termed angular distribution Auger microscopy (ADAM), is based on an argument claiming that emitted Auger electrons are solid particles that are blocked by neighboring atoms in the solid. This interpretation is in direct contradiction with the scattering picture presented in the basic principles section which is based on the quantum mechanical wave nature of electrons, a picture for which there is overwhelming evidence from theoretical and experimental comparisons. In light of this, no further consideration will be given to ADAM in this article.

Experimental Details

The specimen to be studied usually will be an ordered solid, such as a single crystal or a textured sample, that is rigidly mounted face-up on the goniometer. Prior knowledge of the crystal orientation is greatly beneficial. In most instances the samples or substrates are aligned with a low-index direction normal to the surface by means of standard X-ray diffraction methods, e.g., by back-reflection Laue or θ–2θ scans. The surface plane of the sample should lie in, or as close as possible to, the polar rotation axis; the maximum offset should be 2.5 mm. An often more critical alignment is that needed to get the surface normal and the azimuthal rotation axis to coincide. This will minimize crystal "wobble," thus minimizing ambiguous diffraction effects (usually apparent as a sloping background) that are accentuated at grazing emission angles, where signal intensities are dominated by an instrumental function highly responsive to slight changes in polar angle. Proper azimuthal alignment is obtained by centering a surface-reflected laser beam to within $\pm$ 0.25° during azimuthal rotation. Polar rotation should have a range of at least 120° to include both grazing and surface-normal emissions directions and the azimuthal range should have a minimum of 200°, preferably 380°, to allow for the full rotational symmetry.

Since diffraction data is angle dependent, an angle-resolved analyzer is necessary to discriminate electron trajectories, allowing only those electrons with similar emission directions to reach the detector. The practical upper limit for the acceptance solid-angle is approximately $\pm$ 12°, with $\pm$ 3° being a more common value. (There have been some high angular-resolution studies done at $\pm$ 1.5° showing diffraction fine structure that led to a more quantitative description of the observed structure, however this will not be necessary for routine structure determinations.) Increasing the angular resolution is usually a straightforward task that involves the physical placement of an aperture or an array of cylinders in front of the electron-analyzing optics. Or, if an electron lens is included, voltage adjustments to the lens elements may act to effectively reduce the acceptance solid angle. This, as an example, has been done by the author on a VG MICROLAB II spectrometer, which has a two-element lens. Here, when the front elements are powered and the back elements are grounded, a ± 12° solid angle results, while a ± 4° solid angle may be achieved by powering the back elements and grounding the front. In either mode, the implementation of a reduced solid angle and sample rotation is a relatively easy and inexpensive procedure, and may be accomplished with many existing electron spectrometers.

Illustrative Examples

Surface adsorbates

From environmental to packaging to catalysis issues, the need to understand how molecules interact chemically and bond to a surface is paramount. XPD is an extremely good candidate for investigating adsorbate–substrate interactions because chemical shifts in the core-level transitions can lead to the identification of a specific species, and the scattering of core-level photoelectrons can lead to the determination of the structure in which they exist. Consider the initial interaction of gaseous CO on room temperature Fe (001).[8] At this temperature Fe (001) has a bcc lattice structure with a fourfold symmetric surface. At a coverage of less than a monolayer, it was known that the CO adsorbs to the Fe, residing in fourfold hollow sites with only the C making direct contact with the Fe. The orientation of the C-O bond remained a question. It was proposed that this early stage of CO coverage on Fe (001) produced an intermediary state for the dissociation of the C and O, since that the CO bond was tilted with respect to the surface normal, unlike the upright orientation that CO was found to possess on Ni.[5] Although near-edge X-ray adsorption fine structure (NEXAFS) results measured a tilt in the CO bond, the results were not very quantitative regarding the exact angle of the tilt. XPD, on the other hand, gave the CO bond angle as $35 \pm 2°$ relative to the surface, as determined from the large forward scattering peak depicted in by the solid line Figure 2a along the (100) azimuth. Here the ordinate is plotted as the C 1s intensity divided by the O 1s intensity. Plotting this ratio effectively removes instrumental contributions to the diffraction pattern, the oxygen atoms have no atoms above them from which their photoelectrons can scatter and thus should be featureless. The azimuthal scan shown in Figure 2b was taken at a polar angle of 35° to enhance the C 1s diffraction signal. From the fourfold symmetry and knowledge of the crystallographic orientation of the Fe, it is clear that the tilt direction lies in the <100> planes, as depicted in Figure 2c; the absence of a diffraction peak in the $[1\bar{1}0]$ polar scan shown by the dashed line in Figure 2a helps to confirm this.

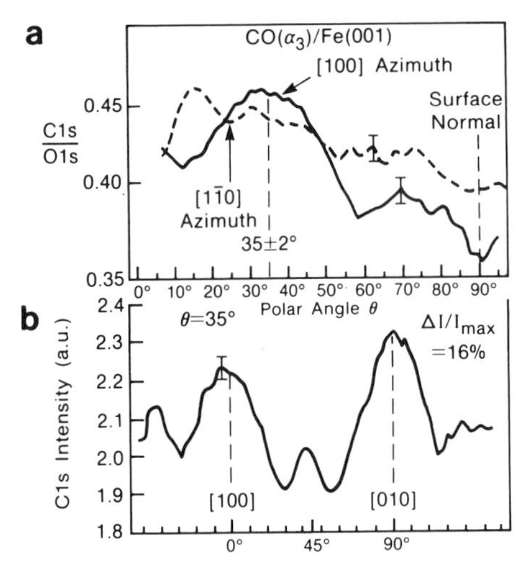

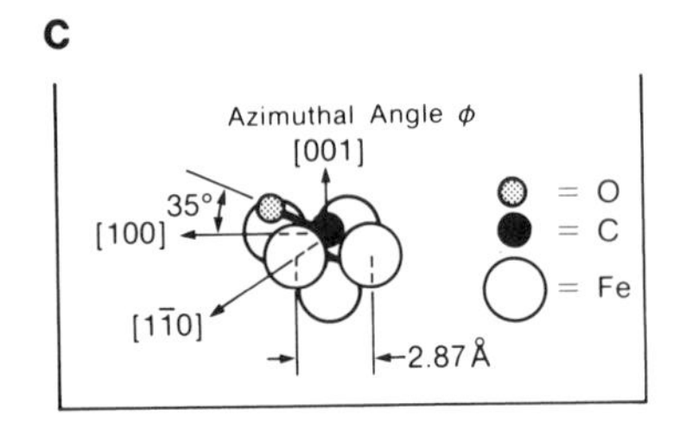

Figure 2 **Experimental data from an early stage of CO adsorbed on Fe (001) known as the α_3 state: polar scans (a) of the C 1s–O 1s intensity ratio taken in two Fe (001) azimuthal planes, the (100) and the (1$\bar{1}$0) (the C 1s and O 1s electron kinetic energies are 1202 eV and 955 eV, respectively); C 1s azimuthal scan (b) taken at the polar angle of maximum intensity in (a); and geometry (c) deduced from the data.**

The intensity anisotropies in Figure 2b associated with the <100> diffraction peaks are similar in magnitude. This was found to be true also for the other two quadrants not shown here and suggests that there is no preferred tilt of the CO molecule into any one of the four quadrants. This also demonstrates the high level of sensitivity one can expect for an XPD pattern, considering that less than a quarter monolayer of (low-Z) CO molecules, i.e., only those with C–O bonds pointing in the direction presented in Figure 2b, contribute to a 16% anisotropy. To put this into perspective, a "perfect" single crystal will yield an anisotropy of about 50%. Since the diffraction features are dominated by near-neighbor scattering events, a 16% intensity change is not too surprising and further suggests that even though the CO molecules are tilted along four different directions they are highly ordered along each of them. This presents some practical experimental assurance of the sensitivity that can be expected from XPD or AED.

Overlayers

Perhaps the best examples to illustrate the analysis strength of XPD and AED are the epitaxial growth modes of deposited overlayers. Here, the structure and chemistry of an overlayer, or the new interface, will influence the properties of the film. To control such effects, an understanding of the basic structure and chemistry is essential. Epitaxial Cu on Ni (001) is an excellent example for demonstrating the

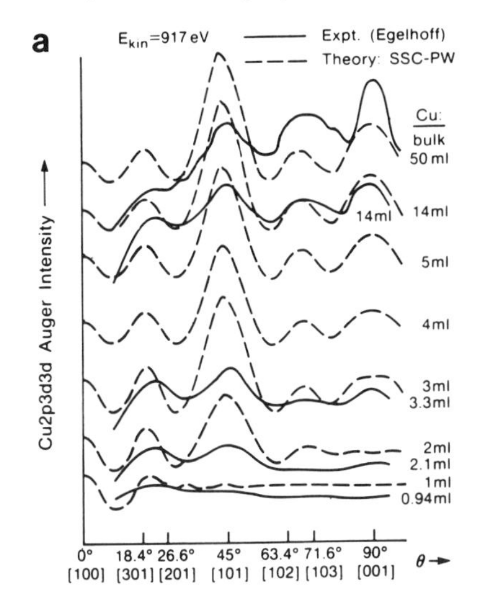

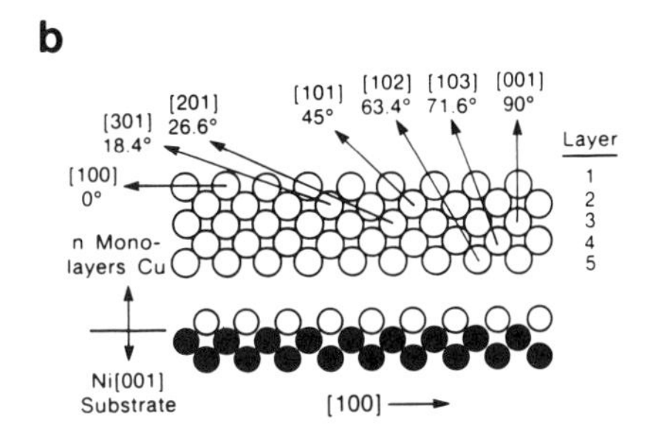

Figure 3 **Experimental and calculated results (a) for epitaxial Cu on Ni (001). The solid lines represent experimental data at the Cu coverage indicated and the dashed lines represent single-scattering cluster calculations assuming a plane wave final state for the Cu *LMM* Auger electron; A schematic representation (b) of the Ni (010) plane with 1–5 monolayers of Cu on top. The arrows indicate directions in which forward scattering events should produce diffraction peaks in (a).**

types of structural information that can be obtained for a metal-on-metal system. Figure 3a shows several experimental polar diffraction patterns obtained using the Cu *LMM* Auger transition, which are compared to single-scattering cluster calculations that use plane waves to represent the emitted electrons.[9] XPD patterns using the Cu $^2P_{3/2}$ core level would look virtually identical. Figure 3b gives a diagrammatic sketch to help interpret the origins of the diffraction features. The diagram should be viewed, layer by layer, from the top down; the angles associated with each arrow, when in full view, indicate the directions in which a diffraction peak should appear. For example, after the second monolayer is formed three arrows are in full view, at angles of 0°, 18°, and 45°. The first arrow represents an emission direction in which it is physically impossible to collect data. The other two arrows appear in the experimental data of Figure 3a only after a second monolayer is deposited. The appearance of these diffraction peaks with the deposition of the second monolayer is consistent with intensity maxima occurring along directions having neighboring atoms. This is then confirmed by the sudden appearance of the diffraction peak at 90° after the deposition of the third layer.

Although the interpretation presented is conceptually inviting, caution must be taken not to extend it beyond what is reliable. For this reason, results from single-scattering cluster calculations are included in Figure 3. By calculating the intensity from a fabricated atomic cluster, in which atoms can be removed from the cluster at will, a very accurate understanding of the relative contributions for various atoms can be made. Indeed, the peaks at 45° and 90° result almost entirely from scattering events with the near-neighbor atoms along these directions, whereas the peak at 70° gets its intensity not only from the atoms along this direction but, to some degree, also from the near-neighbor atoms at 90° and 45°. One should be careful when attributing diffraction intensities to atoms in the solid that are more than a few neighbors away.

Polar XPD data taken in a low-index azimuthal plane creates the opportunity to directly obtain the *c:a* axis ratio for measuring induced lattice strain. This has been done by Chambers et al.[10] for the Cu–Ni system mentioned above and, more recently, by the author et al.[10] for thin Co layers in a Co–Pt superlattice. The induced strain in the overlayer alters the angles, excluding the 90° peak, at which the diffraction signals appear. The change in angle leads to a direct determination of the type and degree of strain present in the sample.

Special Topics

As is common to most techniques, XPD and AED have spawned related techniques that are too specialized to present in detail here, but the reader should be at least aware of some of the more interesting ones, which show future possibilities.

Holography

It is the goal of XPD and AED to give chemically specific structural information in ordered samples. The presentation of such data has recently been simplified from angular plots to real space images. The mechanics of the data collection procedure remains the same as in normal XPD or AED, with the additional need to acquire data over the full hemisphere of angular space above the surface, or a significant fraction thereof. The difference emerges in the final treatment, which involves a Helmholtz-Kirchoff Fourier transformation. The result is an image that apparently allows one to display the surface structure from numerous perspectives and to directly measure the bond distances in the solid.[11] It is important to point out, however, that using the preceding scattering theories one can get exactly the same information, minus the friendly display.

Spin-Polarized Photoelectron Diffraction

This article has been devoted to understanding the structural order at surfaces and interfaces through XPD and AED. However, these techniques are not limited to structural studies. An example can be found in studies of surface magnetic ordering. Here XPD is thought to offer a unique approach by taking advantage of the physics involved in the emission of outer core-level photoelectrons from magnetic atoms and in the electron-atom scattering process. Simply put, such photoelectrons will have multiple final states separated in energy (several eV) and spin. An energy-dispersive analyzer can thus be used to monitor spin-polarized signals. The scattering process consisting of two contributions, nonmagnetic (Coulombic) and magnetic (exchange), will affect a spin-parallel photoelectron differently from an antiparallel one. It will thus give rise to spin-polarized XPD or spin-polarized photoelectron diffraction (SPPD).[12] The simplest approach for understanding SPPD is to monitor spin-split photoelectron peaks as a function of crystal temperature, as opposed to angle (as done in XPD). A relative change in intensity, beyond that expected by lattice responses to the temperature (such as Debye-Waller effects) can be attributed to a change in the magnetic order of the solid. As has been observed, a sudden change occurs in the relative photoelectron intensities at temperatures well above the bulk ordering temperature that is speculated to be a surface short-range ordering temperature. As a final point, since the spin-polarized beam of photoelectrons is referenced internally to the sample and not the spectrometer, as in numerous other spin-polarized electron experiments, it appears to be able to measure antiferromagnetic as well as ferromagnetic materials.

Surface Melting

The ability of XPD and AED to measure the short-range order of materials on a very short time scale opens the door for surface order–disorder transition studies, such as the surface solid-to-"liquid" transition temperature, as has already been done for Pb and Ge. In the case of bulk Ge,[13] a melting temperature of 1210 K was found. While monitoring core-level XPD photoelectron azimuthal scans as a function of increasing temperature, the surface was found to show an order–disorder temperature 160° below that of the bulk.

Valence Bands

The ambiguous nature of valence band peaks are well known, especially when further complicated by having a multicomponent system, such as compounds and alloys. XPD has been used to offer unique insight in addressing this problem. The process is straightforward. By directly comparing the degree of similarity between core-level diffraction patterns and diffraction patterns of deconvolved valence band peaks, one can immediately determine which peaks are due to which elements.[14] From here, the total density of states can be separated then into its partial densities of states. This has important implications, as the use of valence-band regions for identifying complicated compounds increases.

Conclusions

There is a clear need in materials science to achieve a more quantitative understanding of chemical and structural properties of surfaces and interfaces. XPD and AED offer both with a relatively small investment in time and equipment, if one is using XPS or AES already. With a foundation in spectroscopy, XPD and AED are well suited for elemental analysis; XPD is capable even of giving quantitative chemical state information in many cases. Once the element or elements are identified, information on structure can be gathered. The approach emphasized in this article is the simplest and most straightforward, and consists of correlating the main experimental diffraction peaks obtained from several diffraction scans with near-neighbor atomic positions, leading to a determination of the lattice type. When an overlayer is deposited onto a substrate, there is the unique ability to obtain diffraction patterns from both simultaneously—measuring the quality of the lattice matching as well as the structure. As a further step, exact atomic positions can be obtained by comparing the experimental diffraction data with scattering theory. The information gained for a given amount of effort is substantial.

Hundreds of electron spectrometers are already being used to pursue surface chemical information, many on crystalline and textured materials, and it seems only reasonable that a number of these could benefit further from the addition of structural analysis capabilities. Should this addition be realized, one would expect that XPD and AED will move more from the basic research environments into applied areas. An example of this can be seen with the future growth of thin-film technology where XPD and AED may serve a valuable function in quality control issues by monitoring the degree of surface texturing or interface diffusion. Whatever the future may be, surface structure and chemistry will inevitably play a key role in materials research. XPD and AED are good candidates for doing both.

Related Articles in the Encyclopedia

XPS, AES, EXAFS, SEXAFS, LEED, RHEED.

References

1. K. Siegbahn, U. Gelius, H. Siegbahn, and E. Olsen. *Phys. Lett.* **32A,** 221, 1970.
2. C. S. Fadley and S. A. L. Bergstrom. *Phys. Lett.* **35A,** 375, 1971.
3. S. Kono, C. S. Fadley, N. F. T. Hall and Z. Hussein. *Phys. Rev. Lett.* **41,** 117,1978; S. D. Kevan, D. H. Rosenblatt, D. Denley, B.-C. Lu, and D. A. Shirley. *Phys. Rev. Lett.* 41,1565 (1978); D. P. Woodruff, D. Norman, B. W. Holland, N. V. Smith, H. H. Farrell and M. M. Traum. *Phys. Rev. Lett.* **41,** 1130, 1978.

4 W. F. Egelhoff. *Phys. Rev.* **B30,** 1052, 1984; R. A. Armstrong and W. E. Egelhoff. *Surf. Sci.* **154,** L225, 1985.

5 C. S. Fadley. *Prog. Surf. Sci.* **16,** 275, 1984; C. S. Fadley. *Physica Scripta.* **T17,** 39, 1987. In this and the following reference, the author gives a detailed overview of XPD and AED, including a quantitative description of the scattering theory behind the diffraction effect, instrumental considerations, and comparisons with related techniques like extended X-ray absorption fine structure (EXAFS) and low-energy electron diffraction (LEED). Reference 6, especially, has a comprehensive profile on recent XPD and AED experiments that have strongly contributed to the field.

6 C. S. Fadley. in *Synchrotron Radiation Research: Advances in Surface Science* (R. Z. Bachrach, ed.) Plenum Press, New York, 1989.

7 D. G. Frank, N. Batina, T. Golden, F. Lu, and A. T. Hubbard. *Science.* **247,** 182, 1990.

8 R. S. Saiki, G. S. Herman, M. Yamada, J. Osterwalder, and C. S. Fadley. *Phys. Rev. Lett.* **63,** 283, 1989.

9 W. F. Egelhoff. *Phys. Rev.* **B30,** 1052, 1984; E. L. Bullock and C. S. Fadley. *Phys. Rev.* **B31,** 1212 (1985).

10 S. C. Chambers, H. W. Chen, I. M. Vitomirov, S. B. Anderson, and J. H. Weaver. *Phys. Rev.* **B33,** 8810, 1986; B. Hermsmeier, R. F. C. Farrow, C. H. Lee, E. E. Marinero, C. J. Lin, R. F. Marks, and C. J. Chien. *J. Magn. Mat.* 1990, in press.

11 G. R. Harp, D. K. Saldin, and B. P. Tonner. *Phys. Rev.* **B42,** 9199, 1990.

12 B. Hermsmeier, J. Osterwalder, D. J. Friedman, B. Sinkovic, T. Tran, and C. S. Fadley. *Phys. Rev.* **B42,** 11895, 1990.

13 U. Breuer, O. Knauff, and H. P. Bonzel. *J. Vac. Sci. Technol.* **A8,** 2489, 1989; D. J. Friedman, T. Tran, and C. S. Fadley. *The Structure of Surfaces* (S. Y. Tong, M. A. Van Hove, X. Xide, and K. Takayanagi, eds.) Springer-Verlag, Berlin, in press.

14 A. Stuck, J. Osterwalder, T. Greber, S. Hüfner, and L. Schlapbach. *Phys. Rev. Lett.* **65,** 3029, 1990.

4.5 LEED

Low-Energy Electron Diffraction

MAX G. LAGALLY

Contents

Introduction

Low-energy electron diffraction (LEED) is a technique for investigating the crystallography of surfaces and overlayers or films adsorbed on surfaces. LEED is generally performed with electron energies of 10–1000 eV. The limited penetration of electrons in this energy range provides the sensitivity to the surface. Diffraction of electrons occurs because of the periodic arrangement of atoms in the surface. This periodic arrangement can be conceptualized as parallel rows of atoms analogous to grating lines in a diffraction grating. Thus, diffraction in LEED occurs from rows of atoms, just as in its three-dimensional (3D) counterpart, X-ray diffraction, which can be considered as occurring from planes. The diffracted beams emanate from the crystal surface in directions satisfying interference conditions from these rows of atoms. The diffraction pattern and the intensity distribution in the diffracted beams can provide information on the positions of atoms in the surface and on the existence of various kinds of crystallographic disorder in the periodic arrangement of surface atoms.[1–7] In its most elementary form, LEED can be used to test for the existence of overlayer phases having a two-dimensional (2D) crystal structure different from the surface on which they are adsorbed and to test whether a surface phase is ordered or disordered.

Although LEED is the best-known and most widely used surface crystallographic technique, other diffraction techniques can provide information on the surface structure. Because of its convenient configuration, reflection high-energy electron diffraction (RHEED) is used widely in the epitaxial growth of films. X-ray diffraction can also be used for surface crystallography under appropriate conditions. Thermal-energy atom diffraction (TEAS) is a powerful technique for looking at disorder on surfaces because of the high sensitivity of low-energy atoms to isolated surface atoms. Surface extended absorption fine structure (SEXAFS) and photoelectron (XPD) diffraction are specialized techniques that also can give structural information on surfaces. RHEED, XRD, XPD, and SEXAFS are covered in separate articles.

Basic Principles

Surface Crystallography Vocabulary

If an imaginary plane is drawn somewhere through a perfectly periodic 3D crystal, and the two halves of the crystal are separated along this plane, ideal surfaces are formed. If the imaginary plane corresponds to an (*hkl*) plane in the bulk crystal, the surface is defined as an (*hkl*) surface, using the usual Miller indices. The periodic arrangement of atoms in the surface can be viewed as a 2D lattice; that is, every point in this arrangement can be reached by a translation vector. The smallest translation vectors define the unit mesh, the 2D analog of the unit cell. Primitive unit meshes contain one lattice point per mesh; nonprimitive meshes, more than one lattice point per mesh. Figure 1a shows the five 2D Bravais nets. The unit mesh vectors are conventionally defined, as shown, with the angle between the unit vectors $\geq 90°$, **a** denoting the shorter unit vector, and **b** (aligned horizontally) the longer one. A 2D lattice may also have a basis. The lattice is defined by those points that can be reached by a translation vector. The basis is the conformation of atoms around each of these points.

The arrangement of lattice points in a 2D lattice can be visualized as sets of parallel rows. The orientation of these rows can be defined by 2D Miller indices (*hk* see Figure 1b). Inter-row distances can be expressed in terms of 2D Miller indices, analogous to the notation for 3D crystals.

In the discussion so far, an ideal termination of the bulk crystal has been assumed at the surface; that is, the positions of atoms in the surface have been assumed to be the same as what they would have been in the bulk before the surface was created. This may not be true. Reconstruction, a rearrangement of atoms in the surface and near-surface layers, occurs frequently. It is caused by an attempt of the surface to lower its free energy by eliminating broken bonds. The atomic layers par-

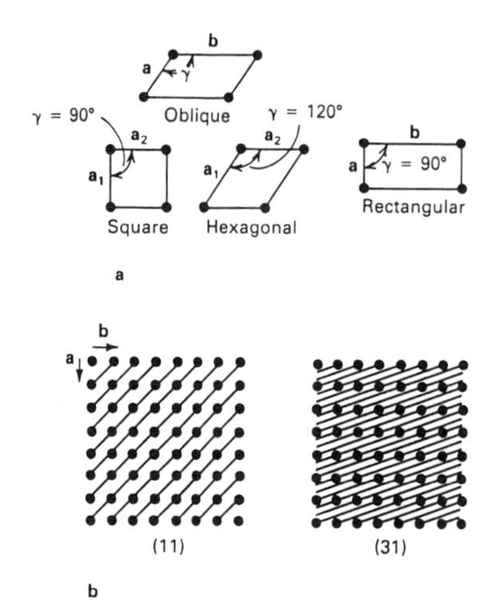

Figure 1 **Unit meshes and 2D Miller indices. (a) Examples of 2D Bravais nets; (b) examples of families of rows with Miller indices referenced to the unit mesh vectors. (11) and (31) families of rows are shown.**

ticipating in this reconstruction can be considered as a different phase having a different periodic arrangement of atoms. Even when no reconstruction occurs, adsorption of a foreign species onto the substrate surface from the ambient gas, from a deliberately created beam of atoms, or as a consequence of segregation of an impurity out of the bulk creates a surface phase termed an overlayer. An overlayer can be a fraction of a single atomic layer or several atom layers forming a thin film. The overlayer phase has its own crystal structure, which may or may not be related in a simple manner to that of the substrate surface. Examples are shown in Figure 2. An overlayer may be commensurate or incommensurate with the substrate. Commensurate overlayers have unit meshes that are related by simple rational numbers to those of the substrates on which they are adsorbed. Incommensurate layers do not. The unit mesh of overlayers is defined as a multiple of that unit mesh of the substrate surface that would be produced by the ideal termination of the bulk lattice. Thus, (100) and (111) surfaces of an fcc crystal have square- and parallelogram-shaped primitive unit meshes, respectively (Figure 1a), both of which would be considered (1×1) unit meshes. Commensurate overlayer unit meshes are then $p(m \times n)$ or $c(m \times n)$ (Figures 2a and 2b, respectively), where p and c refer to primitive and centered overlayer unit meshes relative to the primitive substrate unit mesh, and m and n are constants. An example of a complete unit mesh description of an overlayer on a particular substrate is W(110) $p(2 \times 1)$ O, which describes an oxygen overlayer adsorbed on the (110) surface of tungsten having a unit mesh twice that of the primitive W(110) unit mesh in the **a** direction and the same as the W(110) unit mesh in the **b** direction.

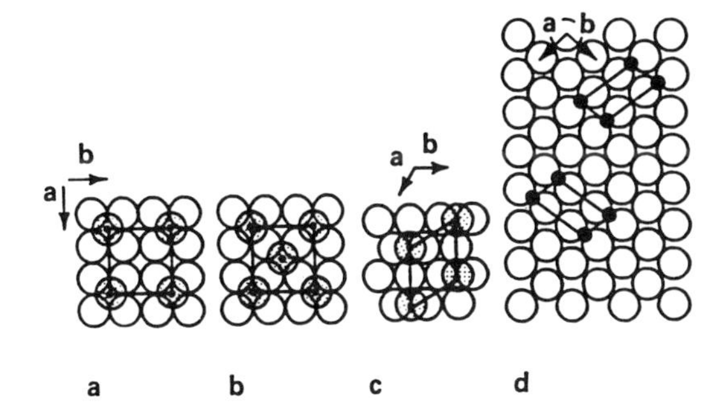

Figure 2 Examples of overlayer structures with appropriate notation. (a) fcc (100) p(2 × 2); (b) fcc (100) c(2 × 2); (c) fcc (111) p($\sqrt{3}$ × $\sqrt{3}$)R30°; (d) bcc (110) p(2 × 1) and bcc (110) p(1 × 2). The two orientations for the unit mesh in (d) are both possible because of the symmetry of the substrate; they have the same free energy and are called degenerate.

The overlayer unit mesh is sometimes rotated relative to that of the substrate (Figure 2c). An example of the notation for such a structure is Ni (111) ($\sqrt{3} \times \sqrt{3}$) R30° O, which describes an oxygen overlayer on Ni (111) whose unit mesh is $\sqrt{3}$ times as large as the substrate mesh in both unit mesh directions and is rotated by 30° relative to the substrate mesh. The symmetry of the substrate is often low, permitting the formation of energetically equivalent structures rotated relative to each other or translated by some specific amount, as shown in Figure 2d. Domains having symmetry related structure are called rotational or translational *antiphase domains* and are degenerate; that is, they have the same free energy, and there is no preference for the formation of one over the other. Overlayers with ($m \times n$) meshes generally also form ($n \times m$) meshes; for example, p(2 × 1) → p(1 × 2) and c(2 × 4) → c(4 × 2) (Figure 2d). Most low-symmetry surfaces allow degenerate overlayer or reconstruction domains.

Diffraction From Surfaces

Diffraction from surfaces can be viewed most simply as the scattering of waves from families of lattice rows that connect scattering centers lying in a single plane (Figure 3). If a wave with wavelength λ is permitted to fall at an angle of incidence θ_0 onto a family of rows separated from each other by a distance d_{hk}, the 2D Laue condition can be calculated for constructive interference between incoming and outgoing waves by considering the difference in paths traveled by the rays striking two adjacent rows of atoms:

$$m\lambda = d_{hk}(\sin\theta_{mhmk} - \sin\theta_0) \qquad (1)$$

where the difference in paths traveled by two adjacent rays is m wavelengths. The wavelength λ of electrons (in Å) is related to their energy E (given in eV) by

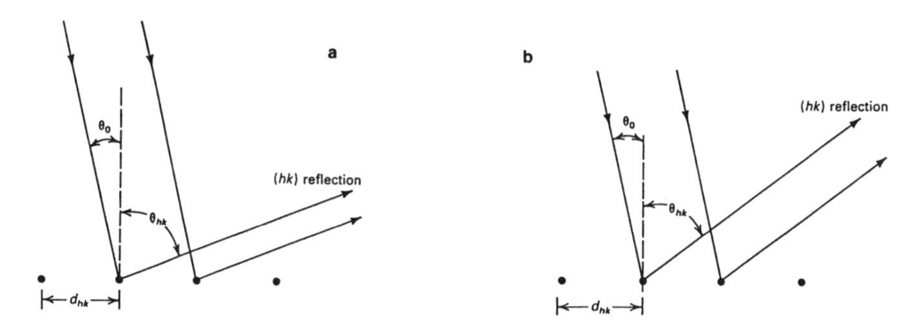

Figure 3 **Diffraction from a family of rows spaced d_{hk} apart and its dependence on wavelength. Each solid circle represents a row of atoms in the plane of the paper. Rays with wavelength λ fall on this family of rows at an angle θ_0. Interference maxima occur at angles θ_0 that satisfy Equation (2). Only the first-order reflection is shown. (a) Longer wavelength; (b) shorter wavelength.**

$$\lambda = \sqrt{\frac{150}{E}} \tag{2}$$

For a fixed incident-beam energy, that is, a fixed wavelength, and a fixed angle of incidence, each family of (*hk*) rows diffracts radiation at the appropriate exit angle, θ_{hk}. If a fluorescent screen or other detector is positioned to intercept these scattered beams, a diffraction pattern having the symmetry of the surface or overlayer unit mesh will be observed (Figure 4). At the center of the pattern will be the (00) beam, which has a path difference of zero and is therefore not sensitive to d_{hk}, that is, to the relative lateral positions of the surface atoms. Around it will be the first-order diffracted beams ($m = 1$) from each family of possible rows that can be drawn through the surface atoms, for example, (10), (01), (11), (21), and so on. In addition, there will be higher-order reflections ($m \geq 1$) at larger angles. For example, the (20) and (30) reflections will fall on the extension of a line connecting the (00) and (10) beams; the (00), (11), (22), (33), etc., beams will be collinear, and so forth. Diffracted beams are indexed with the (*hk*) notation of the families of rows from which they are diffracted. Thus, the (10) beam is scattered from (10) rows in the direction perpendicular to these rows. The distance of the (*hk*) reflection from the mirror reflection, the (00) beam, is inversely related to the (*hk*) inter-row distance d_{hk}; see Equation (1).

If the energy of the incident beam or the angle of incidence is changed, the diffraction angles θ_{hk} adjust to continue to satisfy the Laue conditions. For example, if the energy is increased, the entire pattern will appear to shrink around the (00) beam because θ_{hk} becomes smaller.

Any periodicity in the surface or overlayer different from that of the clean unreconstructed surface produces additional diffracted beams. A unit mesh larger than

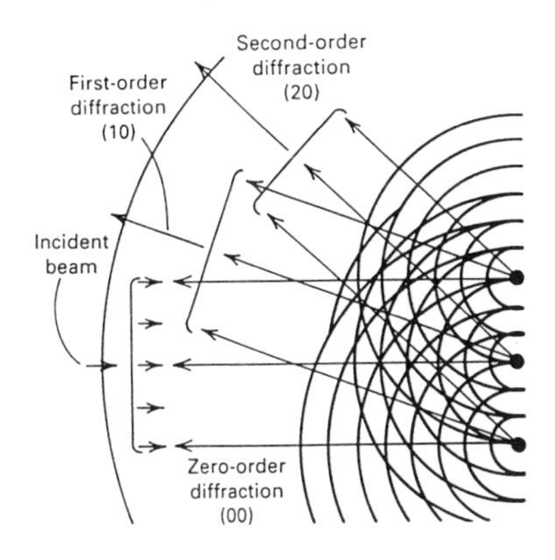

Figure 4 **Interference pattern created when regularly spaced atoms scatter an incident plane wave. A spherical wave emanates from each atom; diffracted beams form at the directions of constructive interference between these waves. The mirror reflection—the (00) beam—and the first- and second-order diffracted beams are shown.**

that of the substrate surface yields diffracted beams that are nearer each other; i.e., if d_{hk} is larger, θ_{hk} is smaller; see Equation (1).

The diffraction conditions can be depicted most easily using a reciprocal-lattice and Ewald construction (Figure 5).[4] The Ewald sphere provides a schematic description of the conservation of energy; that is, since diffraction involves elastic scattering, the incident and exiting beams have the same energy. Its radius is inversely proportional to the wavelength λ. The reciprocal lattice for a single layer of atoms consisting of families of rows is a set of rods (*hk*) normal to the crystal surface having spacing $2\pi/d_{hk}$. The intersection of the Ewald sphere and the reciprocal-lattice rods is a graphical solution of the Laue equation, Equation (1), and therefore yields the diffraction pattern. As the energy of the incident beam or the angle of incidence is varied, the radius of the Ewald sphere or its orientation relative to the rods changes, consequently also changing the points of intersection with the rods. The directions of the outgoing vectors define the directions of the diffracted beams.

The discussion of diffraction so far has made no reference to the size of the 2D "grating." It has been assumed that the grating is infinite. In analogy with optical or X-ray diffraction, finite sizes of the ordered regions on the surface (finite-sized gratings) broaden the diffracted beams. From an analysis of the diffracted-beam shapes, the types of structural disorder in the surface region can be identified and quantified.[7–9]

Surface sensitivity in LEED is provided by the limited mean free path for inelastic scattering of slow electrons. This mean free path is the distance traveled by an electron in the solid before it collides inelastically, loses energy, and thus becomes

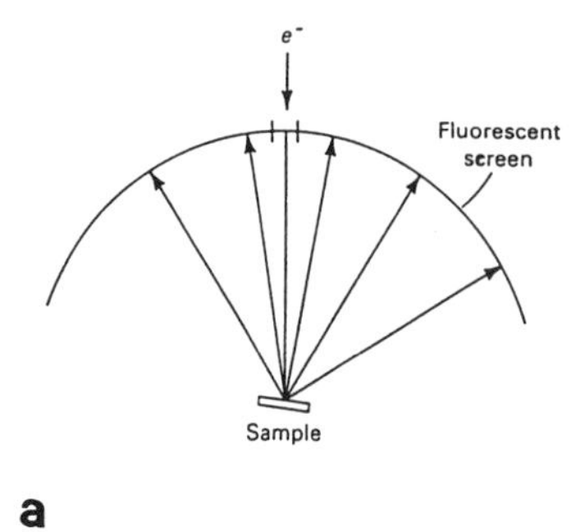

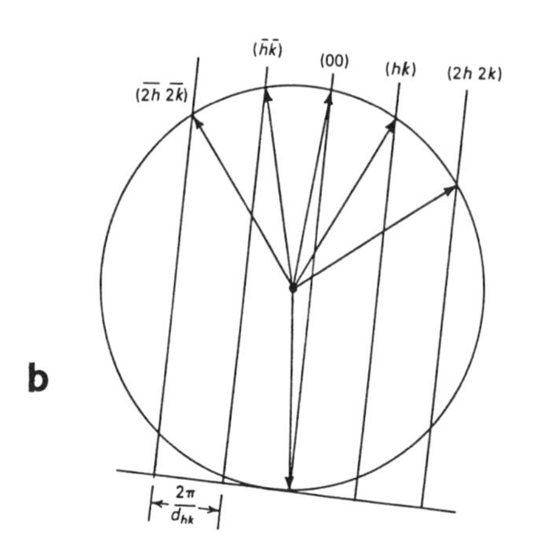

Figure 5 **Reciprocal-lattice and Ewald construction corresponding to LEED and comparison to real-space picture. (a) Real-space schematic diagram of diffraction from a surface. The electron beam is incident on the sample along the direction given by e^-. The five diffracted beams represent the $(\overline{2h}\ \overline{2k})$, $(\overline{hk})$, (00), (hk), and $(2h\ 2k)$ beams from a family of rows (hk) with spacing d_{hk}. (b) The corresponding cut through the reciprocal-lattice and the Ewald construction. The reciprocal-lattice rods are normal to the crystal surface, given by the nearly horizontal line. They represent the same set of beams as in (a). The Ewald sphere always intersects the origin of the reciprocal lattice at the point of incidence of the incoming ray. Directions of the diffracted beams depend on the radius ($2\pi/\lambda$) of the Ewald sphere. Similarly, as the orientation of the crystal surface is changed in relation to the incident beam, the reciprocal lattice rotates about its origin.**

lost for diffraction. The mean free path in the energy range of LEED is from 4 to 20 Å. Because a layer spacing is of the order of 3 Å, these slow electrons probe only a few atomic layers. The inelastic scattering of slow electrons is discussed further in other articles, e.g., AES and XPS.

Diffraction Measurements

Figure 6 shows a schematic diagram of a typical LEED system.[10] The electron gun can produce a monoenergetic beam of approximately 10–1000 eV, with beam sizes typically a fraction of a mm, but ranging in newer instruments down to ~10 µm. The goniometer generally allows two sample motions— rocking about an axis in the plane of the crystal surface and rotation about an axis normal to the crystal surface—as well as heating and sometimes cooling of the sample and temperature measurement capability. The detector consists usually of concentric grids and a fluorescent screen. The grids filter inelastically scattered electrons that also emanate

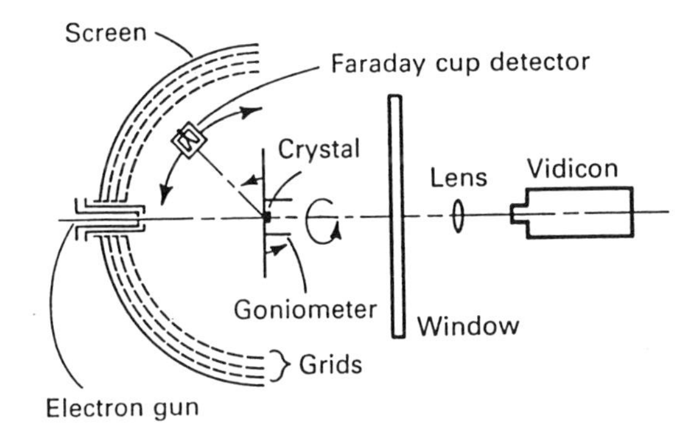

Figure 6 **LEED diffractometer. The vidicon camera can be interfaced with a computer to record the diffraction pattern displayed on the fluorescent screen. The Faraday collector may contain a channeltron electron multiplier for added sensitivity.**

from the crystal surface. For accurate measurement of beam currents and beam profiles a Faraday cup detector can be used. The Faraday cup is biased to exclude inelastically scattered electrons. Modern detectors incorporate a channeltron electron multiplier or a resistive anode encoder in the detector, depending on the need for sensitivity or resolution.[10]

The simplest diffraction measurement is the determination of the surface or overlayer unit mesh size and shape. This can be performed by inspection of the diffraction pattern at any energy of the incident beam (see Figure 4). The determination is simplest if the electron beam is incident normal to the surface, because the symmetry of the pattern is then preserved. The diffraction pattern determines only the size and shape of the unit mesh. The positions of atoms in the surface cannot be determined from visual inspection of the diffraction pattern, but must be obtained from an analysis of the intensities of the diffracted beams. Generally, the intensity in a diffracted beam is measured as a function of the incident-beam energy at several diffraction geometries. These "intensity-versus-energy" curves are then compared to model calculations.[6]

Diffraction is useful whenever there is a distinct phase relationship between scattering units. The greater the order, the better defined are the diffraction features. For example, the reciprocal lattice of a 3D crystal is a set of points, because three Laue conditions have to be exactly satisfied. The diffraction pattern is a set of sharp spots. If disorder is introduced into the structure, the spots broaden and weaken. Two-dimensional structures give diffraction "rods," because only two Laue conditions have to be satisfied. The diffraction pattern is again a set of sharp spots, because the Ewald sphere cuts these rods at precise places. Disorder in the plane broadens the rods and, hence, the diffraction spots in x and y. The existence of streaks, broad spots, and additional diffuse intensity in the pattern is a common occurrence indicating that the overlayer or surface structure is not perfect. The angular profile of diffracted beams provides information on disorder and finite-size effects in surfaces analogous to particle size broadening in X-ray diffraction of bulk materials. The measurement of LEED angular profiles as a function of diffraction parameters, such as energy, angles, and beam order, makes it possible to distinguish types of surface disorder, including finite island sizes, strain, crystal mosaic, monatomic and multiatomic steps at the surface, and regular or irregular domain or antiphase boundaries.[7–9] To measure a profile, a detector with a narrow slit or a point aperture must be used to minimize the influence of instrumental broadening.[10]

Examples of Uses

Surface-sensitive diffraction require a high vacuum because surface contamination rates from ambient gases are rapid. Contaminated surfaces are generally disordered and generally do not provide diffraction patterns. Because the beam penetration is so small, even a few layers of adsorbed gas or oxide can mask the underlying pattern. The ideal surface preparation method is cleavage in vacuum. Cleavage exposes an internal interface that has not been subjected to ambient atmospheric contamination. However, cleavage is limited to a few crystals and generally to only one surface orientation of these crystals. Samples not cleaved in vacuum must be cut from a single crystal, polished, and oriented carefully to the desired surface using a Laue diffraction camera. Standard polishing procedures are used.

LEED can be used to determine the atomic structure of surfaces, surface structural disorder, and to some extent, surface morphology, as well as changes in structure with time, temperature, and externally controlled conditions like deposition or chemical reaction. Some examples are briefly discussed here:

1. Surface order and cleanliness. The most common use of LEED is in association with other surface analysis or surface research methods, to check surface structural order and cleanliness. A visual display of the diffraction pattern is used. An expected pattern, with sharp diffracted beams, is an indication that the surface is clean and well ordered.

2. Surface atomic structure. The integrated intensity of several diffracted beams is measured as a function of electron beam energy for different angles of incidence. The measurements are fitted with a model calculation that includes multiple scattering. The atomic coordinates of the surface atoms are extracted.[6] (See also the article on EXAFS.)

3. Step density. For surfaces containing steps, both the step height and step density can be determined by measuring the angular profile of diffracted beams (the shape of the beam in angle) at specific conditions of energy and angle. This measurement can be static or dynamic, the latter if the surface is evolving with

time, as a function of temperature or adatom coverage. The analysis is straightforward in terms of kinematic theory.[7–9] Similar measurements can identify surface strain, the existence of structural domains or islands, and crystal mosaic.[7] (For crystal mosaic see also the article on XRD.)

4. Phase transitions in overlayers or surfaces. The structure of surface layers may undergo a transition with temperature or coverage. Observation of changes in the diffraction pattern gives a qualitative analysis of a phase transition. Measurement of the intensity and the shape of the profile gives a quantitative analysis of phase boundaries and the influence of finite sizes on the transition.[7]
5. Dynamics of phase transitions, ordering, disordering, and growth. Measurement of the time evolution of the diffracted intensity and the shape of diffracted beams gives information on the kinetic mechanisms of ordering and disordering, growth laws, and finite-size effects.[9] Transitions from layered to cluster growth, as well as transitions from amorphous to crystalline phases, can be investigated. Activation energies for diffusion and domain boundary motion can, in principle, be extracted.
6. Thermal properties of overlayer atoms. Measurement of the intensity of any diffracted beam with temperature and its angular profile can be interpreted in terms of a surface-atom Debye–Waller factor and phonon scattering.[4] Mean-square vibrational amplitudes of surface atoms can be extracted. The measurement must be made away from the parameter space at which phase transitions occur.
7. Chemical reactions of surfaces. Diffraction can be used qualitatively to identify different surface phases resulting from adsorption and chemical reaction at surfaces.[3] Reaction rates can be investigated by following the evolution of diffracted beam intensities.
8. Grain size in textured films. For films having a preferred growth direction—e.g., (111)—LEED can be used to determine the preferred direction and the grain size parallel to the surface. The preferred direction is obtained from the symmetry of the diffraction patterns, while the grain size is obtained from the shape in angle of diffracted beams.

Limitations of LEED

Surface-sensitive diffraction is, for the most part, restricted to analysis of surfaces of single crystals and overlayers and films on such surfaces. If a polycrystalline sample is illuminated using a beam of low-energy electrons, each crystallite surface exposed will create its own diffraction pattern, all of which will be superimposed on the fluorescent screen detector. If more than a few orientations are illuminated by the beam, the pattern becomes too complicated to analyze. However, if the size of the

LEED beam is reduced to less than the sample grain size, individual grains in a polycrystal can be investigated. It is difficult to make low-energy electron beams this small, but in some materials large grain sizes can be achieved in polycrystals.

Because electrons are charged, only materials having reasonable conductivity can be investigated. Insulators and ceramics pose difficulties because charge accumulates on the sample and eventually prevents the incident beam from striking the surface. Most insulators can be investigated with special techniques, such as providing surface conduction paths near the analysis area or using very thin samples mounted on a conducting plate.

The use of LEED for quantitative determinations of atomic positions in surfaces is complicated by the multiple scattering of the electrons, a process that is highly sensitive to scattering geometry. An accurate goniometer and detection scheme are required to measure intensities for which the scattering geometry is precisely known. A detector with high sensitivity is required to measure the frequently small intensities in this type of measurement, but high resolution is not necessary. Massive calculations that include multiple scattering are necessary to fit data to extract atomic positions.

Measurements of surface disorder require a high resolving power (the ability to distinguish two close-lying points in the diffraction pattern). Quantitative measurements of surface disorder are limited in the following manner: the worse the resolving power, the smaller the maximum scale of surface disorder that can be detected. For example, if the maximum resolvable distance of the diffractometer is 100 Å, then a surface that has steps spaced more than 100 Å apart will look perfect to the instrument. The theoretical analysis of disorder is much simpler than that for atomic positions.

Conclusions

LEED is the most powerful, most widely used, and most developed technique for the investigation of periodic surface structures. It is a standard tool in the surface analysis of single-crystal surfaces. It is used very commonly as a method to check surface order. The evolution of the technique is toward greater use to investigate surface disorder. Progress in atomic-structure determination is focused on improving calculations for complex molecular surface structures.

Preparation of this article was supported by the National Science Foundation, Solid State Chemistry Program.

Related Articles in the Encyclopedia

RHEED, XRD, XPD, and EXAFS

References

1 J. J. Lander. *Prog. Solid State Chem.* **2,** 26, 1965.

2 P. J. Estrup. In *Modern Diffraction and Imaging Techniques in Materials Science.* (S. Amelinckx, R. Gevers, G. Remaut, and J. Van Landuyt, Eds.) North-Holland, 1970; and P. J. Estrup and E. G. McRae. *Surf. Sci.* **25,** 1, 1971.

3 G. A. Somorjai and H. H. Farrell. *Adv. Chem. Phys.* **20,** 215, 1972. Contains listings of overlayer phases that have been observed to that date.

4 M. B. Webb and M. G. Lagally. *Solid State Phys.* **28,** 301, 1973. A detailed discussion at a more advanced level.

5 G. Ertl and J. Küppers. *Low-Energy Electrons and Surface Chemistry.* Verlag Chemie, Weinheim, 1974, Chps. 9 and 10. An introductory treatment of diffraction.

6 J. B. Pendry. *Low-Energy Electron Diffraction.* Academic Press, New York, 1974. Theoretical treatment, principally on surface atomic structure determination.

7 M. G. Lagally. In *Methods of Experimental Physics—Surfaces.* (R. L. Park and M. G. Lagally, Eds.) Academic Press, New York, 1985, Vol. 22, Chp. 5.

8 M. Henzler. *Top. Curr. Phys.* **4,** 117, 1977. Basic information on use of LEED to analyze steps and step disorder on surfaces.

9 M. G. Lagally. In *Reflection High-Energy Electron Diffraction and Reflection Electron Imaging of Surfaces.* (P. K. Larsen and P. J. Dobson, Eds.) Plenum, New York, 1989.

10 M. G. Lagally and J. A. Martin. *Rev Sci. Instr.* **54,** 1273,1983. A review of LEED instrumentation.

4.6 RHEED

Reflection High-Energy Electron Diffraction

DONALD E. SAVAGE

Contents

Introduction

Reflection High-Energy Electron Diffraction (RHEED) is a technique for probing the surface structures of solids in ultrahigh vacuum (UHV). Since it is a diffraction-based technique, it is sensitive to order in solids and is ideally suited for the study of clean, well-ordered single-crystal surfaces. In special cases it can be used to study clean polycrystalline samples as well. It gives essentially no information on the structure of amorphous surfaces, which makes it unsuitable for use on sputter-cleaned samples or in conjunction with sputter depth profiling, unless the sample can be recrystallized by annealing.

The area of a surface sampled in RHEED is determined by the primary beam size at the specimen. The typical electron gun used for RHEED focuses the electrons to a spot on the order of 0.2 mm in diameter. The diffraction pattern should be interpreted as arising from a spatial average over an area whose width is the beam diameter and whose length is the beam diameter divided by the sine of the incidence angle. Of course, high-energy electrons can be focused to a much smaller spot, on the order of several Å. The detection of a RHEED feature as an electron beam is rastered along the surface is the basis for Scanning Reflection Electron Microcopy (SREM). The surface sensitivity of RHEED comes from the strong interaction between electrons and matter. Electrons with RHEED energies and a grazing angle of incidence will scatter elastically from only the first few atomic layers. One disadvantage of using electrons is that the sample must be sufficiently conducting so as not to build up charge and deflect the primary beam. Also, electron-sensitive materials can be damaged during a measurement.

When used to examine a crystal surface, RHEED gives information on the surface crystal structure, the surface orientation, and the degree of surface roughness. Evidence for surface reconstruction (the rearrangement of surface atoms to minimize the surface energy) is obtained directly. In addition, when RHEED is used to study film growth on crystalline surfaces, it gives information on the deposited material's growth mode (i.e., whether it grows layer-by-layer or as three-dimensional (3D) crystallites), the crystal structure, the film's orientation with respect to the substrate, and the growth rate for films that grow layer-by-layer. RHEED is particularly useful in studying structure changes dynamically, i.e., as function of temperature or time. Combined with its open geometry (the area normal to the surface will have a direct line of sight to deposition sources) this feature makes it one of the most commonly used techniques for monitoring structural changes during molecular beam epitaxy (MBE).

Other diffraction techniques whose principles are similar to RHEED include Low-Energy Electron Diffraction (LEED), Thermal-Energy Atom Scattering (TEAS), and X-ray Diffraction. Of these, LEED is the most similar to RHEED, differing mainly by its normal-incidence scattering geometry. Atom scattering differs from RHEED in that it is more surface sensitive than electron diffraction because atoms scatter off the outer electronic shells of atoms in the surface. This makes TEAS very sensitive to defects like atomic steps, but corrugations due to the regular positions of atoms in the surface are difficult to observe. X-ray diffraction is the classic technique for measuring bulk crystal structure. With the use of high-brightness X-ray sources, the surface structure also can be determined by grazing incidence methods.

Basic Principles

The underlying principle of RHEED is that particles of matter have a wave character. This idea was postulated by de Broglie in (1924). He argued that since photons behave as particles, then particles should exhibit wavelike behavior as well. He predicted that a particle's wavelength is Planck's constant h divided by its momentum. The postulate was confirmed by Davisson and Germer's experiments in 1928, which demonstrated the diffraction of low-energy electrons from Ni.[2]

For nonrelativistic electrons, the wavelength (in Å) can be written

$$\lambda = \sqrt{\frac{150.4}{E}} \qquad (1)$$

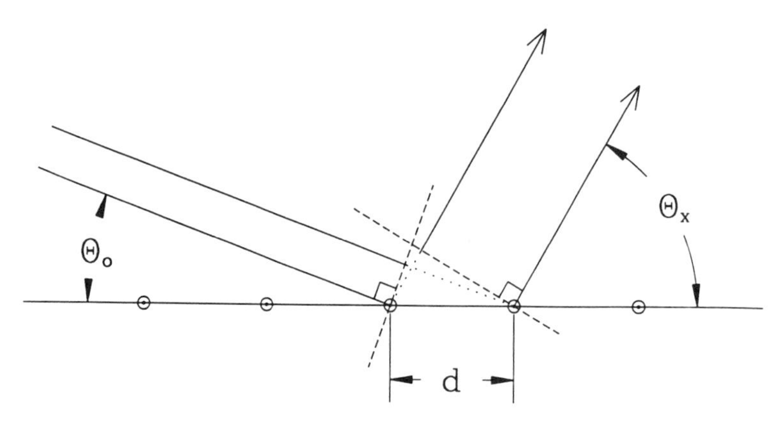

Figure 1 **Plane wave scattering from two consecutive lines of a one-dimensional diffraction grating. The wave scatters in-phase when the path difference (the difference in length of the two dotted sections) equals an integral number of wavelengths.**

where *E* is the energy of the electron (in electron volts). For a primary beam energy of 5–50 keV this relationship gives λ ranging from 0.17 to 0.055 Å.

Most of the features of a RHEED pattern can be understood qualitatively with the use of kinematic scattering theory, i.e., by considering the single scattering of plane waves off a collection of objects (in this case atoms). In the kinematic limit, the scattering of electrons from a single-crystal surface can be treated in the same way as the scattering of monochromatic photons from a two-dimensional (2D) diffraction grating. Given the wavelength and angle of incidence of the source radiation, the angles at which diffracted beams are scattered will depend on the grating line spacing (i.e., the atomic row spacing). Consider a plane wave of wavelength λ incident on a one-dimensional (1D) grating with line spacing *d*, as shown in Figure 1. The diffraction maxima occur when successive rays scatter in-phase, i.e., when their path difference is an integral number of wavelengths. The grating equation

$$n\lambda = d(\cos\theta_x - \cos\theta_0) \tag{2}$$

defines conditions where the maxima occur. Using Equation (2) one can show that to see higher order diffraction maxima when scattering from atomic rows, their relatively close spacing requires that the source wavelength be short.

The analogy of a crystal surface as a diffraction grating also suggests how surface defects can be probed. Recall that for a diffraction grating the width of a diffracted peak will decrease as the number of lines in the grating is increased.[3] This observation can be used in interpreting the shape of RHEED spots. Defects on a crystal surface can limit the number of atomic rows that scatter coherently, thereby broadening RHEED features.

In a diffraction experiment one observes the location and shapes of the diffracted beams (the diffraction pattern), which can be related to the real-space structure using kinematic diffraction theory.[4] Here, the theory is summarized as a set of rules relating the symmetry and the separation of diffracted beams to the symmetry and separation of the scatterers.

The location of a diffracted beam can be defined by specifying the magnitudes and the directions of the incoming and outgoing waves. This can be written in a shorthand notation using the momentum transfer vector **S**, where $\mathbf{S} = \mathbf{K}_{out} - \mathbf{K}_{in}$. The vector **K** is the wave propagation vector and has units of inverse length. Its magnitude is $2\pi/\lambda$, and it points along the direction of wave propagation. The vector **S** can be thought of as the change in momentum of a wave upon scattering. The periodicity of the scatterers will constrain the values of **S** where in-phase scattering occurs.

The dimensionality of the diffraction problem will have a strong effect on how the diffraction pattern appears. For example in a 1D problem, e.g., diffraction from a single line of atoms spaced *d* apart, only the component of **S** in the direction along the line is constrained. For a 2D problem, e.g., the one encountered in RHEED, two components of **S** in the plane of the surface are constrained. For a 3D problem, e.g., X-ray scattering from a bulk crystal, three components of **S** are constrained.

For a given structure, the values of **S** at which in-phase scattering occurs can be plotted; these values make up the reciprocal lattice. The separation of the diffraction maxima is inversely proportional to the separation of the scatterers. In one dimension, the reciprocal lattice is a series of planes, perpendicular to the line of scatterers, spaced $2\pi/d$ apart. In two dimensions, the lattice is a 2D array of infinite rods perpendicular to the 2D plane. The rod spacings are equal to $2\pi/$(atomic row spacings). In three dimensions, the lattice is a 3D lattice of points whose separation is inversely related to the separation of crystal planes.

Kinematic Diffraction from a 2D Plane

Electrons having energies and incident angles typical of RHEED can be treated as nearly nonpenetrating. As a result, atoms in the outermost plane are responsible for most of the scattering, and the resulting reciprocal lattice will be an array of rods perpendicular to the surface plane.

The symmetry and spacing of the 2D reciprocal-lattice mesh (the view looking down upon the array of reciprocal-lattice rods) will depend on the symmetry and spacing of atomic rows in the surface. Consider a crystal plane with lattice points located on the parallelogram shown in Figure 2a. The corresponding reciprocal mesh is shown in Figure 2b. One can construct this reciprocal-mesh by defining two nonorthogonal mesh vectors $\mathbf{A}^*$ and $\mathbf{B}^*$, whose lengths are equal to 2π divided by the separation between two adjacent atomic rows, and whose directions lie in the plane of the surface and are perpendicular to those rows. The (h, k) reciprocal-lattice rod is located by the vector $\mathbf{G}_{hk} = h\mathbf{A}^* + k\mathbf{B}^*$, where h and k are integers. Note

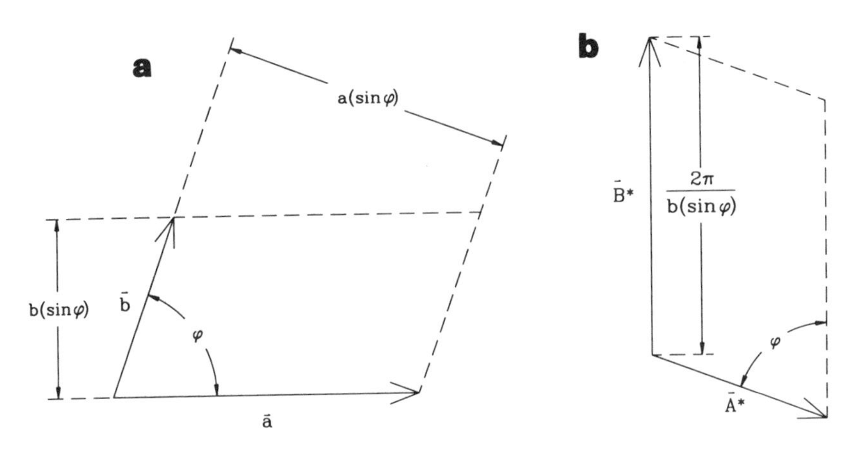

Figure 2 **View looking down on the real-space mesh (a) and the corresponding view of the reciprocal-space mesh (b) for a crystal plane with a nonrectangular lattice. The reciprocal-space mesh resembles the real-space mesh, but rotated 90°. Note that the magnitude of the reciprocal lattice vectors is inversely related to the spacing of atomic rows.**

that the symmetry of the surface reciprocal-mesh is the same as the symmetry of the surface lattice rotated by 90°. This result will always hold.

A diffraction pattern will be influenced strongly by the direction of the primary beam relative to the surface. The grazing angles used in RHEED can make the interpretation of the pattern difficult. For a specific choice of $\mathbf{K}_{in}$, i.e., primary beam energy and direction, the direction in which beams will be diffracted can be determined using the Ewald construction. This is a graphical construction that uses conservation of energy (for elastic scattering $|\mathbf{K}|$ is conserved) and momentum ($\mathbf{K}_{out}$ must lie on the reciprocal lattice). It is done by drawing $\mathbf{K}_{in}$ terminating at the origin of reciprocal-space. A sphere of radius $|\mathbf{K}|$ centered at the origin of $\mathbf{K}_{in}$ gives the locus of all possible scattered waves that conserve energy. The intersection of the sphere with the reciprocal-lattice locates the diffracted beams.

Consider as an example the RHEED pattern from GaAs (110). Gallium Arsenide can be cleaved along (110) planes; the resulting surface is nearly perfect. In Figure 3a the surface real-space lattice and the reciprocal-space mesh of GaAs (110) are shown. For GaAs (110), the surface real-space net is rectangular and has a 2-atom basis; only the lattice points (not the atom locations) are depicted. Figure 3b shows a portion of the Ewald construction for a primary beam incident along an azimuth parallel to the real-space lattice rows (spaced $|b|$ apart), i.e., in the **a** direction, also called an [001] direction, and 20° from grazing. Note that the intersections of rows of reciprocal lattice rods with the Ewald sphere lie on circles similar to circles of constant latitude on a globe. If the diffracted beams are pro-

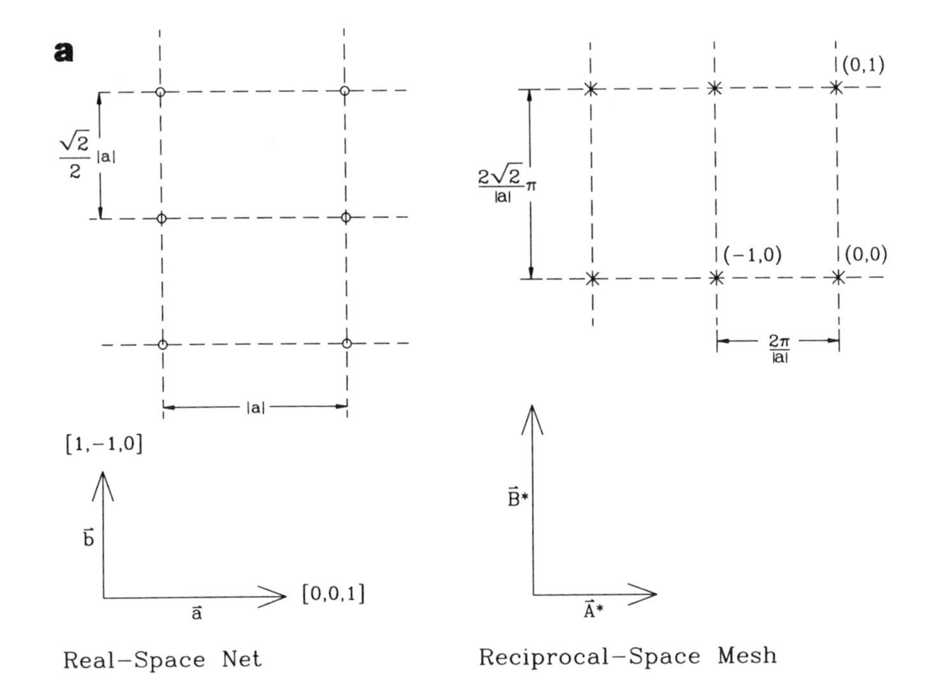

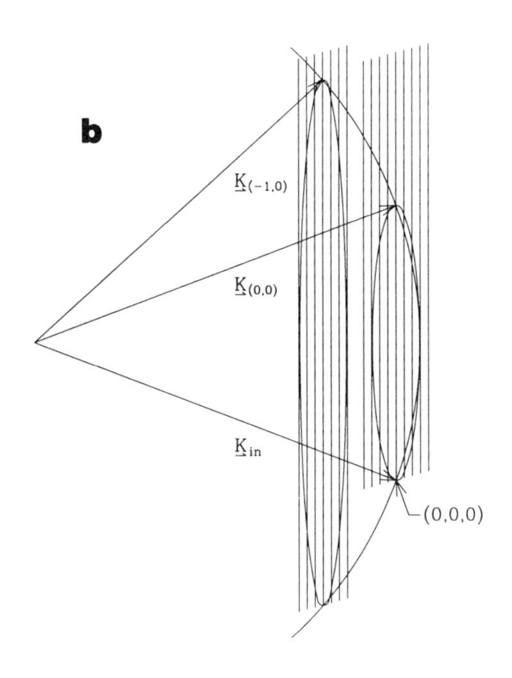

Figure 3 **(a) Real-space lattice and reciprocal-space mesh for the GaAs (110) plane. (b) Ewald construction for this surface with a primary beam incident along the a direction (the [001] azimuth) and elevated 20° from grazing.**

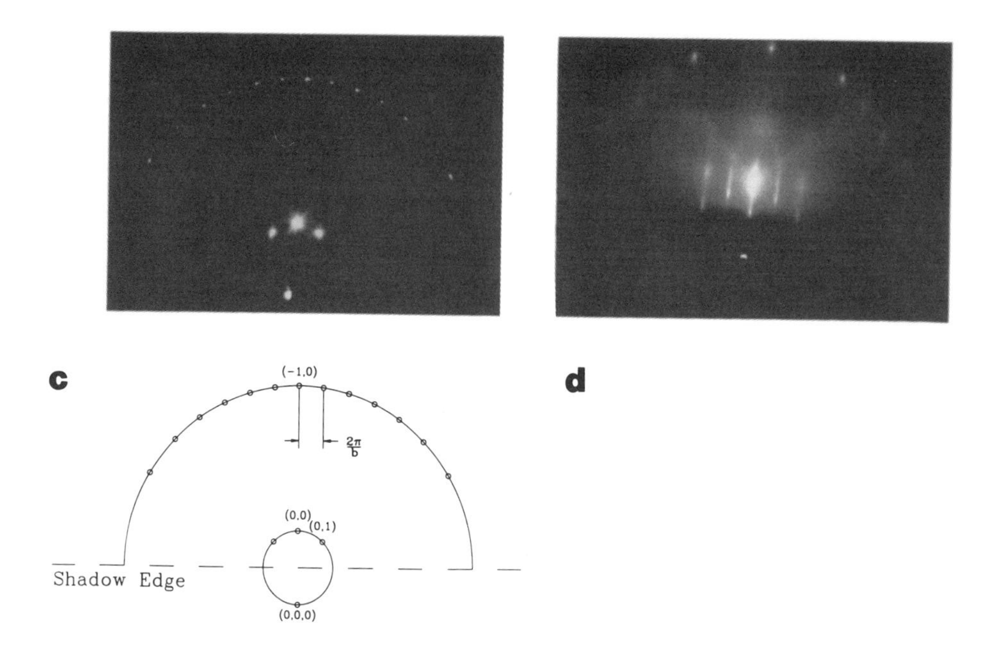

Figure 3 (c) Photograph of a RHEED pattern from cleaved GaAs(110) obtained using a 10 KeV beam incident along the [001] azimuth at an angle of 2.4°—the spots lie along arcs (the inner is the 0th Laue circle, and the outer is the first Laue circle). The pattern is indexed in the drawing below the photo. (d) Photograph of the RHEED pattern after the sample has been sputter etched with 2-KeV Argon ions and subsequently annealed at 500° C for 10 minutes. The pattern is streaky, indicating the presence of atomic steps on the surface.

jected forward onto a phosphor screen, they will appear as spots lying on semicircles in the same way that lines of constant latitude will inscribe circles if projected onto a screen located at the north pole of a globe. These rings are called Laue circles and are labeled 0, 1, etc., starting with the innermost. For this example, the 0th Laue zone contains the set of (0, *k*) reciprocal lattice rods. Note that if the crystal is rotated, thereby changing the orientation of primary beam azimuth so that it is no longer parallel to a set of rows, the projected diffracted beams will no longer lie on semicircles, and the diffraction pattern will appear skewed.

Figure 3c shows a photograph of a RHEED pattern from a cleaved GaAs (110) surface obtained with a 10-keV primary beam directed along an [001] azimuth. The diffracted maxima appear as spots lying along circles, as predicted, and are indexed in the view to the right. The diffracted spots of the 0th and 1st Laue circles are shown. One can tell from this view that the symmetry of the GaAs (110) surface net is orthogonal because the spots in the 1st Laue circle lie directly above those in the 0th. If the surface net had been a parallelogram, as in Figure 2b, the spots in the 1st Laue circle would be displaced. The spacing of the (0, *k*) rows is obtained from the spacing of the spots on the Laue circle, and equals $2\pi/b$. The spacing of the (h, 0) rows (the lattice rows perpendicular to the [001] direction) can be obtained from the spacing between Laue circles, and equals $2\pi/a$. In practice, it is more common to rotate the sample until the primary beam is parallel to the (h, 0) atomic rows (for an orthogonal net, rotation by 90°) and to obtain the row spacing from the lateral spacing of the spots in that pattern. Some other the features to note in the pattern are the so-called transmitted beam, the shadow edge, and the specular reflection. The transmitted beam is the bright spot at the bottom of the photo. It arises when part of the incident beam misses the sample and strikes the phosphor screen directly. This is done intentionally by moving the sample partly out of the beam. The spot, also called the (000) beam, is a useful reference point because it locates the origin of reciprocal space. The spot directly above the transmitted spot is the specular reflection. Halfway between them is the shadow edge, below which no scattered electrons can reach the phosphor screen.

Figure 3d is a photograph of a cleaved GaAs (110) sample taken under conditions similar to Figure 3c but after the sample had been sputter etched with 2-keV Ar ions and subsequently annealed at 500° C for 10 minutes. Because the treatment resulted in a less intense diffraction pattern, the exposure time for this photograph was increased. This RHEED pattern shows streaks instead of spots, obscuring the location of the specular reflection. The separation of the streaks equals $2\pi/b$, as in the freshly cleaved surface, but the angular separation from circle to circle is no longer clear. The streaks arise because atomic steps are created during the sample treatment that limit the long-range order on the surface and broaden the reciprocal-lattice rods. Their intersections with the Ewald sphere produce the elongated streaks. Defects in the surface also give rise to a diffuse background intensity, making the shadow edge more visible. The effect of disorder on the diffraction pattern will be discussed more in the section on nonideal surfaces.

Applications

Surface Reconstruction

The size and symmetry of the surface lattice, which can differ from the bulk termination, is determined directly from the symmetry and spacing of beams in the diffraction pattern.[6] Because surface atoms have a smaller coordination number (i.e., fewer nearest neighbors) than bulk atoms they may move or rebond with their neighbors to lower the surface energy. If this occurs, the surface real-space lattice will become larger; i.e., one must move a longer distance on the surface before repeating the structure. As a result additional reflections, called superlattice reflections, appear at fractional spacings between integral-order reflections. One example of this is the (7 × 7) reconstruction of Si (111). The real-space lattice for this reconstructed surface is 7 times larger than the bulk termination. If the sample is well prepared, 6 fractional-order spots are observed between the integral-order spots on a Laue circle and, more importantly, 6 fractional-order Laue circles are observed.[7] In MBE, one of the uses of RHEED is to monitor the surface reconstruction during deposition to obtain optimal growth conditions. An example is in the growth of GaAs on GaAs (100), where different reconstructions are observed depending on the surface stoichiometry. Deposition parameters can be varied to obtain the desired reconstruction.

Nonideal Surfaces

All real surfaces will contain defects of some kind. A crystalline surface must at the very least contain vacancies. In addition, atomic steps, facets, strain, and crystalline subgrain boundaries all can be present, and each will limit the long-range order on the surface. In practice, it is quite difficult to prepare an atomically flat surface.

Because defects limit the order on a surface, they will alter the diffraction pattern, primarily by broadening diffracted beams.[8] Methods have been developed, mostly in the LEED literature, to analyze the shape of diffracted beams to gain information on step distributions on surfaces. These methods apply equally well to RHEED.

A statement that one will find in the MBE literature is that a smooth surface will give a "streaky" RHEED pattern while a rough surface will give a "spotty" one. This is in apparent contradiction to our picture of an ideal RHEED pattern. These statements can be reconciled if the influence of defects is included. A spotty RHEED pattern that appears different from spots along a Laue circle will result from transmission of the primary beam through asperities rising above the plane of a rough surface. This type of pattern is described below. A flat surface, one without asperities but having a high density of defects that limit order within a plane, will give rise to a streaky RHEED pattern. If a smooth and flat surface is desired, a streaky pattern is preferable to a transmission pattern, but a surface giving a streaky pattern is not ideally flat or smooth.

Limiting long-range order within the plane of the surface will broaden the reciprocal lattice rods uniformly. For a plane with dimensions $L \times L$, the width of a rod will be approximately equal to $2\pi/L$. The streaking is caused by the grazing geometry of RHEED. The Ewald sphere will cut a rod at an angle such that the length is approximately $1/(\sin\theta_x)$ times the width, where θ_x is the angle the exit beam makes with the surface. For $\theta_x \sim 3°$, this is a factor of 20.

What gives rise to streaks in a RHEED pattern from a real surface? For integral-order beams, the explanation is atomic steps. Atomic steps will be present on nearly all crystalline surfaces. At the very least a step density sufficient to account for any misorientation of the sample from perfectly flat must be included. Diffraction is sensitive to atomic steps.[9] They will show up in the RHEED pattern as streaking or as splitting of the diffracted beam at certain diffraction conditions that depend on the path difference of a wave scattered from atomic planes displaced by an atomic step height. If the path difference is an odd multiple of $\lambda/2$, the waves scattered from each plane will destructively interfere. For such a diffraction condition (called an out-of-phase condition) the surface will appear to be made up of finite sized patches and the reciprocal lattice rod will be broadened. In contrast, if the path length differs by an integral multiple of λ, the surface will appear perfect to that wave and the no broadening is observed. This is called an in-phase condition. The width of a reciprocal-lattice rod at the out-of-phase condition is proportional to the step density on the surface. Another type of disorder that will broaden superlattice rods only, is the presence of antiphase domains. These domains occur because patches of reconstructed surface can nucleate in positions shifted from one another. Electrons will scatter incoherently from them, each domain acting as if it were a 2D crystal of finite size. This broadening will be independent of the position along the superlattice rod; i.e., there are no in-phase or out-of-phase positions for this kind of disorder.

Transmission features will be present if the surface is sufficiently rough. These arise because electrons with RHEED energies can penetrate on the order of 100 Å through a solid before inelastically scattering. In RHEED, surface sensitivity is obtained in part from the grazing geometry; i.e., electrons must travel a long distance before seeing planes deep in the bulk. For a rough surface, asperities rising above the surface will be struck at a less glancing angle and if the asperity is thin enough in the primary-beam direction, transmission will occur. Transmitted electrons will see the additional periodicity of the atomic planes below the surface. As a result, a constraint on **S** in the direction perpendicular to the surface is added and the reciprocal lattice will be an array of points instead of rods. Because transmission can occur only through thin objects, the 3D reciprocal-lattice points from the asperities will be elongated in the direction of the primary beam. The Ewald sphere will intersect a number of the reciprocal-lattice points in the plane containing the 0th Laue circle (called the 0th Laue zone), giving rise to a pattern that will appear as a regular array of spots. It appears as a projection of the reciprocal lattice plane that is nearly perpendicular to the incident beam. The spacing of the spots is inversely related to the spacing of planes in the crystal.

In addition to differences in their appearances, a practical way to differentiate between transmission and reflection features is by observing the diffraction pattern while changing the incident azimuth. The intensity of transmission features will change as the azimuth is changed, but their location will not. In contrast, reflection features (other than the (0, 0) beam) will move continuously either closer to or farther away from the shadow edge as the azimuth is changed, a result of the intersection of the Ewald sphere with a rod moving up or down as the azimuth is changed.

Film Growth

One of the main uses of RHEED is to monitor crystal structure during film growth in ultrahigh vacuum. Its ability to distinguish between 2D and 3D structure gives direct evidence on the growth mode of a film. The onset of clustering is associated with the appearance of transmission features. In 2D film growth, changes in the surface lattice periodicity (reconstruction) show up in the appearance or disappearance of superlattice features,[10] while roughness shows up as a change in the shape of the RHEED features. In addition, changes in the separation between RHEED streaks, which relate directly to changes in the surface lattice constant, can be used to follow strain relaxation.

In films that grow 2D for many layers, "intensity oscillations" have been observed for certain growth conditions using RHEED[11] and LEED. Observation is made by monitoring the intensity of a diffracted beam as a function of time during growth. The period of an oscillation corresponds to the time it takes to deposit a monolayer. In practice, oscillations are frequently used to calibrate deposition rates.

Oscillations arise from periodic changes in the surface structure as a deposited layer nucleates, grows, and then fills in the previously deposited layer. They can be understood qualitatively within the framework developed in the discussion of atomic steps. As a material is deposited, 2D islands nucleate, increasing the step density. This causes diffraction features to become smeared out, reducing their intensity at the location of the detector. As the surface becomes covered, the intensity recovers. This process is repeated, with continued deposition causing the intensity to oscillate. Further evidence for this interpretation has been obtained from the analysis of RHEED diffracted-beam profiles for the growth of GaAs on GaAs (100), where the shape of diffracted beams has been modeled to extract step densities.[12]

While kinematic diffraction theory describes intensity oscillations adequately in some cases, there are problems with it when it is used to analyze RHEED measurements. The period of the oscillations is correctly predicted, but not necessarily the phase. In spite of these complications, intensity oscillations are evidence for periodic changes in the surface structure.

For deposited materials that are not perfectly lattice-matched to a substrate, strain will build up in the deposited layer until eventually it becomes energetically favorable for 3D clusters of the deposited material's relaxed crystal structure to form. Some materials will not wet a substrate at all, and grow as 3D clusters at submonolayer coverages. RHEED can give information on cluster orientations, shapes, and sizes. Figure 4a is a photograph of a RHEED pattern obtained for 4 monolayers of In deposited on cleaved GaAs (110). The regular array of spots is clearly a transmission pattern and indicates that the deposited In has formed 3D clusters. The pattern can be indexed with the symmetry and spacing of an Indium [110] reciprocal lattice plane, as shown in Figure 4b. Since all spots are accounted for, clusters present are oriented in the same direction with respect to the substrate. Different orientations would show up as additional arrays of spots, and in the limit of a random orientation the pattern would appear as continuous rings centered about (0,0,0). Information on cluster shape can be extracted if the clusters have a

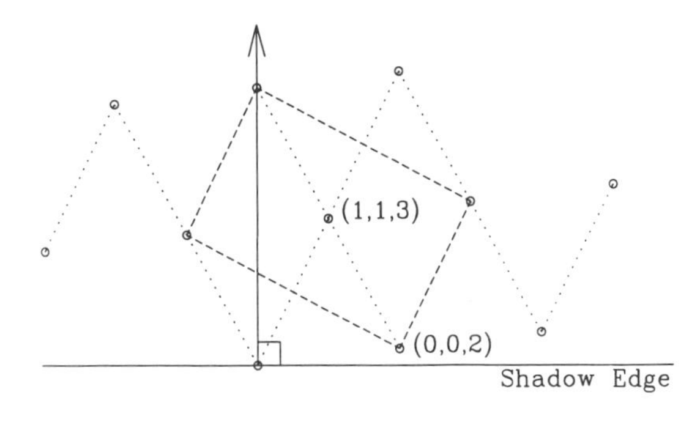

Figure 4 **(a) Photograph of a RHEED pattern for 4 monolayers of In deposited on cleaved GaAs (110). The primary beam is incident along the GaAs [110] azimuth. The regular array of spots are a result of transmission diffraction through indium 3D clusters. The streaks of intensity connecting the spots are from the reflection pattern off facets of the In clusters. (b) Spots in the photo can be indexed as the In [$\bar{1}$10] reciprocal-lattice plane oriented so that the In (113) real-space plane is in contact with the substrate.**

preferred orientation and are bounded by crystal facets. In Figure 5a the streaks connecting the spots are reflection features from In facet planes. (They are from (111)- and (001)-type facets.) Cluster size information is contained in the shape of the transmitted features and attempts have been made to extract quantitative values.[13]

Conclusions

RHEED is a powerful tool for studying the surface structure of crystalline samples in vacuum. Information on the surface symmetry, atomic-row spacing, and evidence of surface roughness are contained in the RHEED pattern. The appearance of the RHEED pattern can be understood qualitatively using simple kinematic scattering theory. When used in concert with MBE, a great deal of information on film growth can be obtained.

The time evolution of the RHEED pattern during film growth or during post-growth annealing is a subject of current interest. RHEED intensity oscillations have been observed to damp out over time during deposition. If growth is interrupted the RHEED pattern recovers. Initial attempts have been made to extract surface diffusion coefficients and activation energies by measuring the rate of such processes at different temperatures.

RHEED intensities cannot be explained using the kinematic theory. Dynamical scattering models of RHEED intensities are being developed.[14] With them one will be able to obtain positions of the surface atoms within the surface unit cell. At this writing, such modeling has been done primarily for LEED.

RHEED differs from LEED because of its grazing geometry and higher electron energies. It is the grazing geometry that allows RHEED to be used in concert with film growth. In addition, both reflected and forward scattered (transmitted) electrons can be observed with RHEED. The latter are not detected with LEED, making it less suited for the study of 3D roughness, i.e., clusters or very rough surfaces.

Related Articles in the Encyclopedia

LEED and XRD

References

1 L. de Broglie. Dissertation, Paris, 1924.

2 C. J. Davisson and L. H. Germer. *Phys. Rev.* **30,** 705, 1927.

3 For a basic treatment of interference and diffraction from a 1D grating, see E. Hecht. *Optics.* Addison-Wesley, Reading, 1987, Chapter 10.

4 For a review on the formulation of kinematic diffraction theory with emphasis on the scattering of low-energy electrons, see M. G. Lagally and M. B. Webb. In: *Solid State Physics.* (H. Ehrenreich, F. Seitz, and D. Turnbull, eds.) Academic, New York, 1973, Volume 28.

5 For more detail on the geometrical relationship between the RHEED pattern and the surface crystal structure, see J. E. Mahan, K. M. Geib, G. Y. Robinson, and R. G. Long. *J. Vac. Sci. Technol.* **A8,** 3692, 1990.

6 For a review of how various surface reconstructions appear in a diffraction pattern, see M. A. Van Hove, W. H. Weinberg, and C. -M. Chan. *Low-Energy Electron Diffraction.* Springer, Berlin, 1986, Chapter 3.

7 S. Hasegawa, H. Daimon, and S. Ino. *Surf. Sci.* **187,** 138, 1987.

8 For a review of how defects manifest themselves in a LEED experiment, see M. Henzler. In: *Electron Spectroscopy for Surface Analysis.* (H. I. Ibach, ed.) Springer, Berlin, 1977.

9 For more information on kinematic treatment of diffraction from stepped surfaces, see M. G. Lagally, D. E. Savage, and M. C. Tringides. In: *Reflection High-Energy Electron Diffraction and Reflection Electron Imaging of Surfaces.* NATO ASI Series B, Plenum, New York, 1988, Volume 188.

10 Surface phase transformations and surface chemical reactions are followed by studying the time evolution of superlattice beams originating from monolayer or submonolayer films. See, for example, Chapters 8–10 in *Low-Energy* Van Hove et al. (*op cit.*).

11 For an overview on RHEED intensity oscillations, see B. A. Joyce, J. H. Neave, J. Zhang, and P. J. Dobson. In: *Reflection* NATO (*op cit.*).

12 P. I. Cohen, P. R. Pukite, J. M. Van Hove, and C. S. Lent. *J. Vac. Sci. Tech nol.* **A4,** 1251, 1986.

13 D. E. Savage and M. G. Lagally. *J. Vac. Sci. Technol.* **B4,** 943, 1986.

14 J. L. Beeby. *Reflection* NATO (*op cit.*).

5

ELECTRON EMISSION SPECTROSCOPIES

5.0 INTRODUCTION

In this chapter we have collected together those techniques in which one measures the energy distribution of electrons ejected from a material. In all four techniques covered electronic energy level excitations are involved, providing atomic or chemical state identification, or both. All are also true surface techniques, since the energies of the electrons concerned fall in the range where they travel can only very short distances without being inelastically scattered. These techniques are all sensitive to less than monolayer amounts of material and none have probing depths greater than about 50 Å without using sputter profiling.

The first two techniques discussed, X-Ray Photoelectron Spectroscopy, XPS, (also known as Electron Spectroscopy for Chemical Analysis, ESCA) and Ultraviolet Photoelectron Spectroscopy, UPS, are very closely related. XPS involves soft X rays (usually 1486 eV, from an Al anode) ejecting photoelectrons from the sample. Electrons originating from the core levels identify the elements present from their Binding Energies, BE. Small "chemical shifts" in the BEs provide additional chemical state information. The relative concentrations of the different elements present can be determined from relative peak intensities. XPS identifies all elements except hydrogen and helium from a depth ranging from around 2 monolayers to 25 monolayers. Typical values for XPS peaks in the 500–1400 eV kinetic energy range are 5 to 10 monolayers.

The strengths of XPS are its good quantification, its excellent chemical state determination capabilities, its applicability to a wide variety of materials from biological materials to metals, and its generally nondestructive nature. XPS's weaknesses are its lack of good spatial resolution (70 μm), only moderate absolute sensitivity (typically 0.1 at. %), and its inability to detect hydrogen. Commercial XPS instruments are usually fully UHV compatible and equipped with accessories, including a sputter profile gun. Costs vary from $250,000 to $600,000, or higher if other major techniques are included.

UPS differs from XPS only in that it uses lower energy radiation to eject photoelectrons, typically the 21.2-eV and 40.8-eV radiation from a He discharge lamp, or up to 200 eV at synchrotron facilities. The usual way to perform UPS is to add a He lamp to an existing XPS system, at about an incremental cost of $30,000. Most activity using UPS is in the detailed study of valence levels for electronic structure information. For materials analysis it is primarily useful as an adjunct to XPS to look at the low-lying core levels that can be accessed by the lower energy UPS radiation sources. There are several advantages in doing this: a greater surface sensitivity because the electron kinetic energies are lower, better energy resolution because the source has a narrower line width, and the possibility of improved lateral resolution using synchrotron sources.

Auger Electron Spectroscopy, AES, is also closely related to XPS. The hole left in a core level after the XPS process, is filled by an electron dropping from a less tightly bound level. The energy released can be used to eject another electron, the Auger electron, whose energy depends only on the energy levels involved and not on whatever made the initial core hole. This allows electrons, rather than X rays, to be used to create the initial core hole, unlike XPS. Since all the energy levels involved are either core or valence levels, however, the type of information supplied, like XPS, is elemental identification from peak positions and chemical state information from chemical shifts and line shapes. The depths probed are also similar to XPS. Dedicated AES systems for materials analysis, which are of similar cost to XPS instruments, have electron optics columns producing finely focused, scannable electron beams of up to 30 kV energy and beam spot sizes as small as 200 Å, a great advantage over XPS. AES could have been discussed in Chapter 3 along with STEM, EMPA, etc. When the incident beam is scanned over the sample (Scanning Auger Microprobe, SAM) mapping at high spatial resolution is obtained. For various reasons the area analyzed is always larger than the spot size, the practical limit to SAM being in the 300–1000 Å range. Another advantage of AES over XPS is speed, since higher electron beam currents can be used. There are major disadvantages to using electrons, however. Beam damage is often severe, particularly for organics, where desorption or decomposition often occurs under the beam. Sample charging for insulators is also a problem. Overall, the two techniques are about equally widespread and are the dominant methods for nontrace level analysis at surfaces. AES is the choice for inorganic systems where high spatial resolution is needed (e.g., semiconductor devices) and XPS should be one's choice otherwise. Combined systems are quite common.

Reflected Electron Energy-Loss Spectroscopy, REELS, is a specialized adjunct to AES, just as UPS is to XPS. A small fraction of the primary incident beam in AES is reflected from the sample surface after suffering discrete energy losses by exciting core or valence electrons in the sample. This fraction comprises the electron energy-loss electrons, and the values of the losses provide elemental and chemical state information (the Core Electron Energy-Loss Spectra, CEELS) and valence band information (the Valence Electron Energy-Loss Spectra, VEELS). The process is identical to the transmission EELS discussed in Chapter 3, except that here it is used in reflection, (hence REELS, reflection EELS), and it is most useful at very low beam energy (e.g., 100 eV) where the probing depth is at a very short minimum (as in UPS). Using the rather high-intensity VEELS signals, a spatial resolution of a few microns can be obtained in mapping mode at 100-eV beam energy. This can be improved to 100 nm at 2-keV beam energy, but the probing depth is now the same as for XPS and AES. Like UPS, VEELS suffers in that there is no direct elemental analysis using valence region transitions, and that peaks are often overlapped. The technique is free on any AES instrument and has been used to map metal hydride phases in metals and oxides at grain boundaries at the 100-nm spatial resolution level.

5.1 XPS

X-Ray Photoelectron Spectroscopy

C. R. BRUNDLE

Contents

Introduction

The photoelectric process, discovered in the early 1900s, was developed for analytical use in the 1960s, largely due to the pioneering work of Kai Siegbahn's group.[1] Important steps were the development of better electron spectrometers, the realization that electron binding energies were sensitive to the chemical state of the atom, and that the technique was surface sensitive. This surface sensitivity, combined with quantitative and chemical state analysis capabilities have made XPS the most broadly applicable general surface analysis technique today. It can detect all elements except hydrogen and helium with a sensitivity variation across the periodic table of only about 30. Samples can be gaseous, liquid, or solid, but the vast majority of electron spectrometers are designed to deal with solids. The depth of the solid material sampled varies from the top 2 atomic layers to 15–20 layers. The area examined can be as large as 1 cm × 1 cm or as small as 70 μm × 70 μm (10-μm diameter spots may be achieved with very specialized equipment). It is applicable to biological, organic, and polymeric materials through metals, ceramics, and semiconductors. Smooth, flat samples are preferable but engineering samples and even powders can be handled. It is a nondestructive technique. Though there are some cases where the X-ray beam damage is significant (especially for organic materials), XPS is the least destructive of all the electron or ion spectroscopy techniques. It has relatively poor spatial resolution, compared to electron-impact and ion-impact techniques. It is also not suitable for trace analysis, the absolute sensitivity being between 0.01–0.3% at., depending on the element. XPS can be a slow technique if the extent of chemical detail to be extracted is large. Analysis times may vary from a few minutes to many hours.

There are thousands of commercial spectrometers in use today in materials analysis, chemistry, and physics laboratories. The largest concentrations are in the US and Japan. They are used in universities, the semiconductor and computer industries, and the oil, chemical, metallurgical, and pharmaceutical industries.

Instruments combining XPS with one or more additional surface techniques are not uncommon. Such combinations use up relatively little extra space but cost more.

Basic Principles

Background

A photon of sufficiently short wavelength (i.e., high energy) can ionize an atom, producing an ejected free electron. The kinetic energy *KE* of the electron (the photoelectron) depends on the energy of the photon $h\nu$ expressed by the Einstein photoelectric law:

$$KE = h\nu - BE \tag{1}$$

where *BE* is the binding energy of the particular electron to the atom concerned. All of photoelectron spectroscopy is based on Equation (1). Since $h\nu$ is known, a measurement of *KE* determines *BE*. The usefulness of determining *BE* for materials analysis is obvious when we remember the way in which the electron shells of an atom are built up. The number of electrons in a neutral atom equals the number of protons in the nucleus. The electrons, arranged in orbitals around the nucleus, are bound to the nucleus by electrostatic attraction. Only two electrons, of opposite spin, may occupy each orbital. The energy levels (or eigenvalues ε) of each orbital are discrete and are different for the same orbital in different atoms because the electrostatic attraction to the different nuclei (i.e., to a different number of protons) is different. To a first approximation, the *BE* of an electron, as determined by the amount of energy required to remove it from the atom, is equal to the ε value (this would be exactly true if, when removing an electron, all the other electrons did not

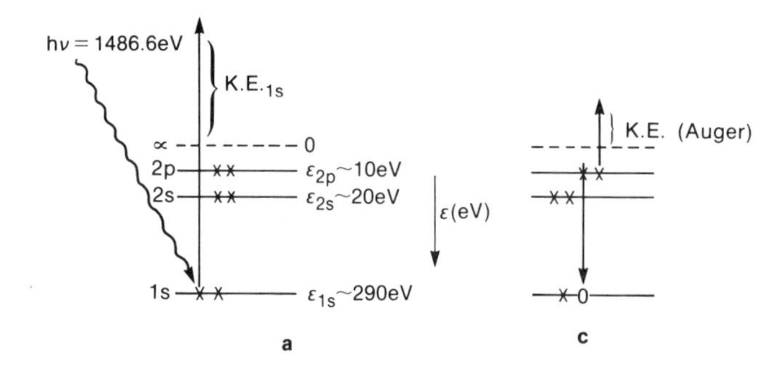

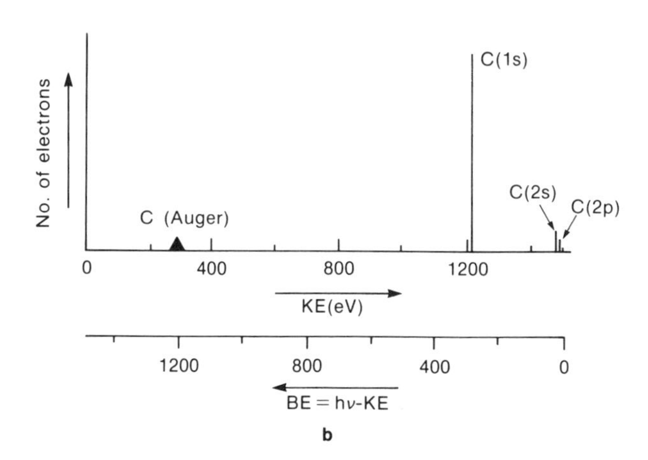

Figure 1 **(a) Schematic representation of the electronic energy levels of a C atom and the photoionization of a C 1s electron. (b) Schematic of the *KE* energy distribution of photoelectrons ejected from an ensemble of C atoms subjected to 1486.6-eV X rays.(c) Auger emission relaxation process for the C 1s hole-state produced in (a).**

respond in any way). So, by experimentally determining a *BE*, one is approximately determining an ε value, which is specific to the atom concerned, thereby identifying that atom.

Photoelectron Process and Spectrum

Consider what happens if, for example, an ensemble of carbon atoms is subjected to X rays of 1486.6 eV energy (the usual X-ray source in commercial XPS instruments). A carbon atom has 6 electrons, two each in the 1s, 2s, and 2p orbitals, usually written as C $1s^2\,2s^2\,2p^2$. The energy level diagram of Figure 1a represents this electronic structure. The photoelectron process for removing an electron from the

1s level, the most strongly bound level, is schematically shown. Alternatively, for any individual C atom, a 2s or a 2p electron might be removed. In an ensemble of C atoms, all three processes will occur, and three groups of photoelectrons with three different *KE*s will therefore be produced, as shown in Figure 1b where the *KE* distribution (the number of ejected photoelectrons versus the kinetic energy)—the photoelectron spectrum—is plotted. Using Equation (1), a *BE* scale can be substituted for the *KE* scale, and a direct experimental determination of the electronic energy levels in the carbon atom has been obtained. Notice that the peak intensities in Figure 1b are not identical because the probability for photoejection from each orbital (called the photoionization cross section, σ) is different. The probability also varies for a given orbital (e.g., a 1s orbital) in different atoms and depends on the X-ray energy used. For carbon atoms, using a 1486.6-eV X ray, the cross section for the 1s level, $\sigma_{C\ 1s}$ is greater than $\sigma_{C\ 2s}$ or $\sigma_{C\ 2p}$, and therefore the C 1s XPS peak is largest, as in Figure 1b.

Thus, the number of peaks in the spectrum corresponds to the number of occupied energy levels in the atoms whose *BE*s are lower than the X-ray energy $h\nu$; the position of the peaks directly measures the *BE*s of the electrons in the orbitals and identifies the atom concerned; the intensities of the peaks depend on the number of atoms present and on the σ values for the orbital concerned. All these statements depend on the idea that electrons behave independently of each other. This is only an approximation. When the approximation breaks down, additional features can be created in the spectrum, owing to the involvement of some of the passive electrons (those not being photoejected).

Analysis Capabilities

Elemental Analysis

The electron energy levels of an atom can be divided into two types: core levels, which are tightly bound to the nucleus, and valence levels, which are only weakly bound. For the carbon atom of Figure 1, the C 1s level is a core level and the C 2s and 2p levels are valence levels. The valence levels of an atom are the ones that interact with the valence levels of other atoms to form chemical bonds in molecules and compounds. Their character and energy is changed markedly by this process, becoming characteristic of the new species formed. The study of these valence levels is the basis of ultraviolet photoelectron spectroscopy (UPS) discussed in another article in this encyclopedia. The core-level electrons of an atom have energies that are nearly independent of the chemical species in which the atom is bound, since they are not involved in the bonding process. Thus, in nickel carbide, the C 1s *BE* is within a few eV of its value for elemental carbon, and the Ni 2p *BE* is within a few eV of its value for Ni metal. The identification of core-level *BE*s thus provides unique signatures of the elements. All elements in the periodic table can be identified in this manner, except for H and He, which have no core levels. Approximate

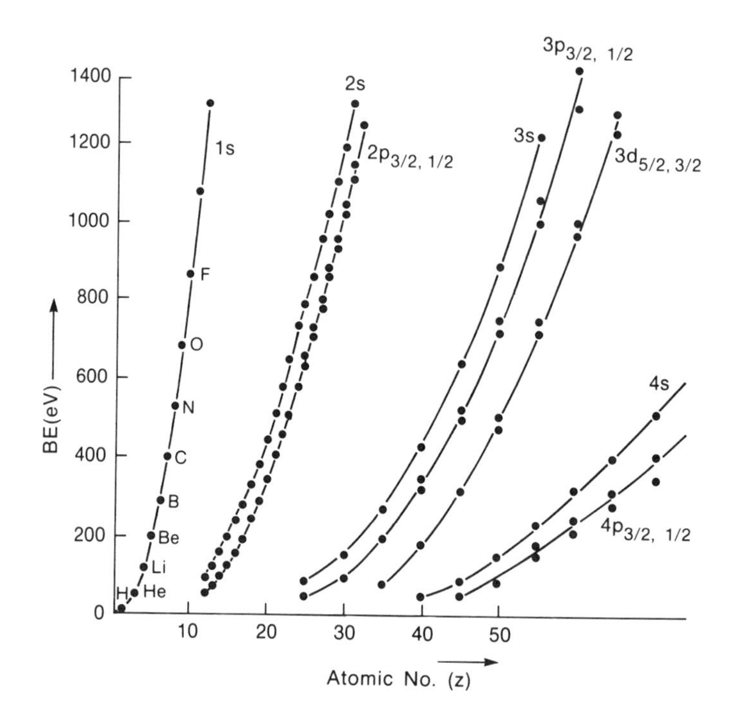

Figure 2 **Approximate *BE*s of the different electron shells as a function of atomic number *Z* of the atom concerned, up to the 1486.6-eV limit accessible by Al Kα radiation.** [2]

*BE*s of the electrons in all the elements in the period table up to $Z = 70$ are plotted in Figure 2, as a function of their atomic number *Z*, up to the usual 1486.6-eV accessibility limit.[2] Chance overlaps of *BE* values from core levels of different elements can usually be resolved by looking for other core levels of the element in doubt.

Quantitative analysis, yielding relative atomic concentrations, requires the measurement of relative peak intensities, combined with a knowledge of σ, plus any experimental artifacts that affect intensities. Cross section values are known from well-established calculations,[3] or from experimental measurements of relative peak areas on materials of known composition (standards).[4] A more practical problem is in correctly determining the experimental peak areas owing to variations in peak widths and line shapes, the presence of subsidiary features (often caused by the breakdown of the independent electron model), and the difficulty of correctly subtracting a large background in the case of solids. There are also instrumental effects to account for because electrons of different *KE* are not transmitted with equal efficiency through the electron energy analyzer. This is best dealt with by calibrating the instrument using local standards, i.e., measuring relative peak areas for stan-

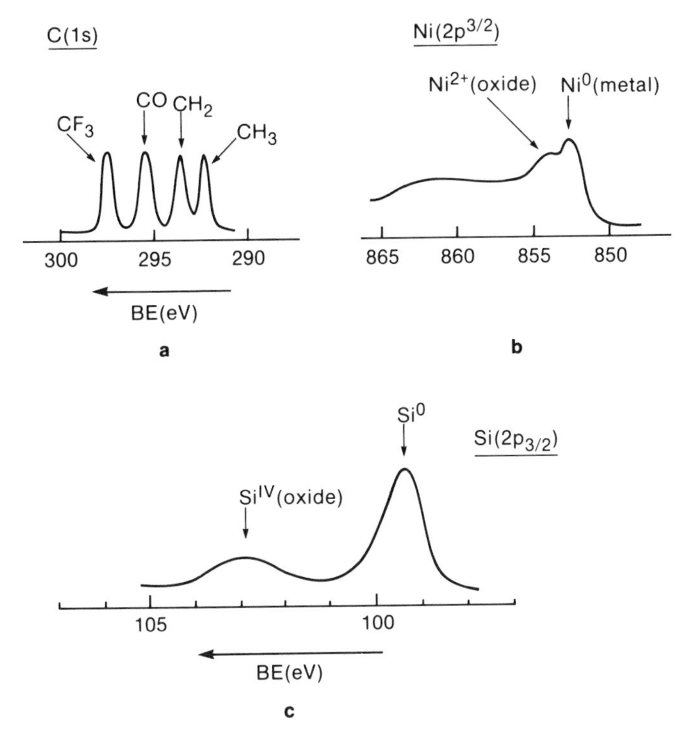

Figure 3 **(a) C 1s XPS spectrum from gaseous $CF_3COCH_2CH_3$.[1] (b) Ni $2p_{3/2}$ XPS spectrum from a mixed Ni metal/Ni metal oxide system. (c) Si $2p_{3/2}$ XPS spectrum from a mixed Si/SiO_2 system.**

dards of known composition in the same instrument to be used for the samples of unknown composition. Taking all the above into account, the uncertainty in quantification in XPS can vary from a few percent in favorable cases to as high as 30% for others. Practitioners generally know which core levels and which types of materials are the most reliable, and in general, relative differences in composition of closely related samples can be determined with much greater accuracy than absolute compositions.

Chemical State Analysis

Though a core level *BE* is approximately constant for an atom in different chemical environments, it is not exactly constant. Figure 3a shows the C 1s part of the XPS spectrum of the molecule $CF_3COCH_2CH_3$. Four separated peaks corresponding to the four inequivalent carbon atoms are present.[1] The chemical shift range ΔBE covering the four peaks is about 8 eV compared to the *BE* of ~290 eV, or ~3%. The carbon atom with the highest positive charge on it, the carbon of the CF_3 group, has the highest *BE*. This trend of high positive charge and high *BE* is in accordance

Element	Oxidation state	Chemical shift from zero-valent state
Ni	Ni^{2+}	~2.2 eV
Fe	Fe^{2+}	~3.0 eV
	Fe^{3+}	~4.1 eV
Ti	Ti^{4+}	~6.0 eV
Si	Si^{4+}	~4.0 eV
Al	Al^{3+}	~2.0 eV
Cu	Cu^{+}	~0.0 eV
	Cu^{2+}	~1.5 eV
Zn	Zn^{2+}	~0 eV
W	W^{4+}	2 eV
	W^{6+}	4 eV

Table 1 **Typical chemical shift values for XPS core levels.**

with the simplest classical electrostatic representation of the atom as a sphere of radius *r* with a valence charge *q* on its surface. The potential inside the sphere q/r is felt by the 1s electrons. If q increases, the *BE* of the 1s level increases, and vice versa. This picture is a gross oversimplification because electrons are not so well separated in space, but the general idea that the *BE* increases with increasing charge on the atom holds in the majority of cases. Table 1 lists the approximate chemical shifts found for the different oxidation states of various metals and semiconductors. The typical range is 1 to several eV, though in some important cases (e.g., Cu and Zn) it is very small. Typical spectra illustrating these chemical shifts for a mixed Ni metal/nickel oxide system and a mixed silicon/silicon dioxide system are shown in Figures 3b and 3c.

The spectra of Figure 3 illustrate two further points. All the C 1s peaks in Figure 3a are of equal intensity because there are an equal number of each type of C atom present. So, when comparing relative intensities of the same atomic core level to get composition data, we do not need to consider the photoionization cross section. Therefore, Figure 3c immediately reveals that there is four times as much elemental Si present as SiO_2 in the Si 2p spectrum. The second point is that the chemical shift range is poor compared to the widths of the peaks, especially for the solids in Figures 3b and 3c. Thus, not all chemically inequivalent atoms can be distinguished this way. For example, Cu^0 (metal) is not distinguishable from Cu^+ in Cu_2O, and Zn^0 is not distinguishable from Zn^{2+} (e.g., in ZnO).

More Complex Effects

In reality, while the photoelectron is leaving the atom, the other electrons respond to the hole being created. The responses, known as *final state effects*, often lead to additional features in the XPS spectrum, some of which are useful analytically.

An effect that always occurs is a lowering of the total energy of the ion due to the relaxation of the remaining electrons towards the hole. This allows the outgoing photoelectron to carry away greater *KE*, i.e., the *BE* determined is always lower than ε. This needs to be considered when comparing theoretical ε values to experimental *BE*s, i.e., for detailed interpretation of electronic structure effects, but is not generally used analytically.

Spin–orbit splitting results from a coupling of the spin of the unpaired electron left behind in the orbital from which its partner has been photoejected with the angular momentum of that orbital, giving two possible different energy final states (spin up or spin down). It occurs for all levels except s levels, which have no orbital angular momentum (being spherical), turning single peaks into doublet peaks. The splitting increases with *Z*, as can be seen from Figure 2 in, for example, the $2p_{3/2}$ and $2p_{1/2}$ spin–orbit split components of the 2p level. The only analytical usefulness is that the splitting increases the number of XPS peaks per atom in a completely known way, which can help when overlaps occur.

Some elements, particularly the transition metals, have unpaired electron spins in their valence levels. The degree of unpairing is strongly affected by the bonding process to other atoms. An unpaired core-electron remaining after the photoemission process will couple to any unpaired spin in the valence level, again leading to more than one final state and peak splitting, called multiplet splitting (weaker than the equivalent spin-orbital splitting). Since the degree of unpaired electron spin in the valence levels is strongly affected by chemical bonding, so is the size of the multiplet splitting. For example, the Cr (3s) level of the Cr^{III} ion of Cr_2O_3 is split by 4.2 eV, whereas in the more covalent compound Cr_2S_3 the splitting is 3.2 eV, allowing distinction of Cr^{III} in the two compounds.[5]

While a core-electron is being ejected, there is some probability that a valence electron will be simultaneously excited to an empty orbital level during the relaxation process, Figure 4b. If this shake-up process occurs, the photoelectron must be ejected with less energy, shifting the XPS peak to apparently higher *BE* than for a case where shake-up doesn't occur, as shown in Figure 4c. These "shake-up satellites" in the spectrum are usually weak because the probability of their occurrence is low, but in some cases they can become as strong as the "main" peak. Shake-up structure can provide chemical state identification because the valence levels are involved. A typical example is given in Figure 4d. The ion Cu^{2+} (in CuO) is distin-

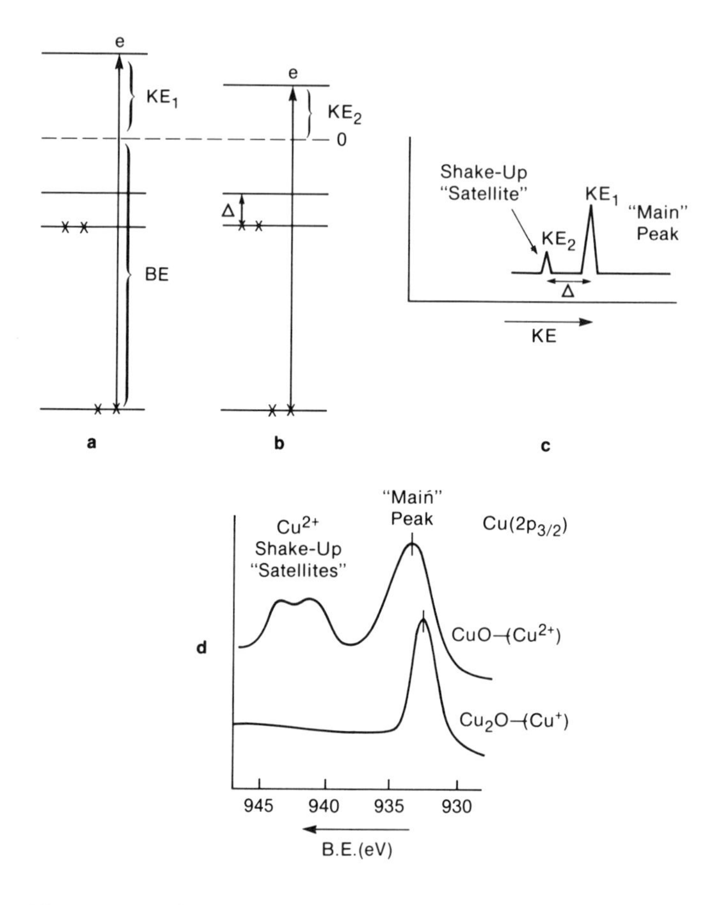

Figure 4 **Schematic electron energy level diagram: (a) of a core-level photoelectron ejection process (one electron process); (b) core-level photoelectron ejection process with shake-up (two- electron process); (c) schematic XPS spectrum from (a) plus (b); (d) Cu $2p_{3/2}$ XPS spectrum for Cu^+ in Cu_2O and Cu^{2+} in CuO. The latter shows strong shake-up features.**

guishable from Cu^+ (in Cu_2O) by the presence of the very characteristic strong Cu 2p shake-up structure for Cu^{2+}. The chemical shift between Cu^{2+} and Cu^+ could also be used for identification, provided accurate *BE*s are measured. It is sometimes an advantage not to have to rely on accurate *BE*s, for instance, when comparing data of different laboratories or if there is a problem establishing an accurate value because of sample charging. In such cases the "fingerprinting" pattern identification of a main peak plus its satellites, as in Figure 4d, is particularly useful.

After the photoemission process is over, the core-hole left behind can eventually be filled by an electron dropping into it from another orbital, as shown in Figure 1c for the example of carbon. The energy released, in this example $\varepsilon_{1s} - \varepsilon_{2p}$, may be sufficient to eject another electron. The example of a 2p electron being ejected is shown. This is called Auger electron emission and the approximate *KE* of the ejected Auger electron will be

$$KE(\text{Auger}) = (\varepsilon_{1s} - \varepsilon_{2p}) - \varepsilon_{2p} \quad (2)$$

The value is characteristic of the atomic energy levels involved and, therefore, also provides a direct element identification (see the article on AES). The *KE* (Auger) is independent of the X-ray energy $h\nu$ and therefore it is not necessary to use monochromatic X rays to perform Auger spectroscopy. Therefore, the usual way Auger spectroscopy is performed is to use high- energy electron beams to make the coreholes, as discussed in the AES article. We mention the process here, however, because when doing XPS the allowable Auger process peaks are superimposed on the spectrum, and they can be used as an additional means of element analysis. Also, in many cases, chemical shifts of Auger peaks, which have a similar origin to XPS core-level shifts, are larger, allowing chemical state identification in cases where it is not possible directly from the XPS core levels. For example, Zn^{2+} can be distinguished from Zn^0 by a 3-eV shift in Auger peak *KE*, whereas it was mentioned earlier that the two species were not distinguishable using XPS core levels.

Surface Sensitivity

Electrons in XPS can travel only short distances through solids before losing energy in collisions with atoms. This inelastic scattering process, shown schematically in Figure 5a, is the reason for the surface sensitivity of XPS. Photoelectrons ejected from atoms "very near" the surface escape unscattered and appear in the XPS peaks. Electrons originating from deeper have correspondingly reduced chances of escaping unscattered and mostly end up in the background at lower *KE* after the XPS peak, as in Figure 5b. Thus, the peaks come mostly from atoms near the surface, the background mostly from the bulk.

If I_0 is the flux of electrons originating at depth *d*, the flux emerging without being scattered, I_d, exponentially decreases with depth according to

$$I_d = I_0^{\frac{-d}{\lambda_e \sin\theta}} \quad (3)$$

where θ is the angle of electron emission and $d/\sin\theta$ is the distance travelled through the solid at that angle. The quantity λ_e is called the *inelastic mean free path length.* The value of λ_e, which determines quantitatively exactly how surface sensitive the measurement is, depends on the *KE* of the electron and the material through which it travels. Empirical relationships between λ_e and *KE* are plotted in Figure 6 for elements and for compounds.[6] They are meant as rough guides because values can vary considerably (by a factor of almost 4), depending on what element

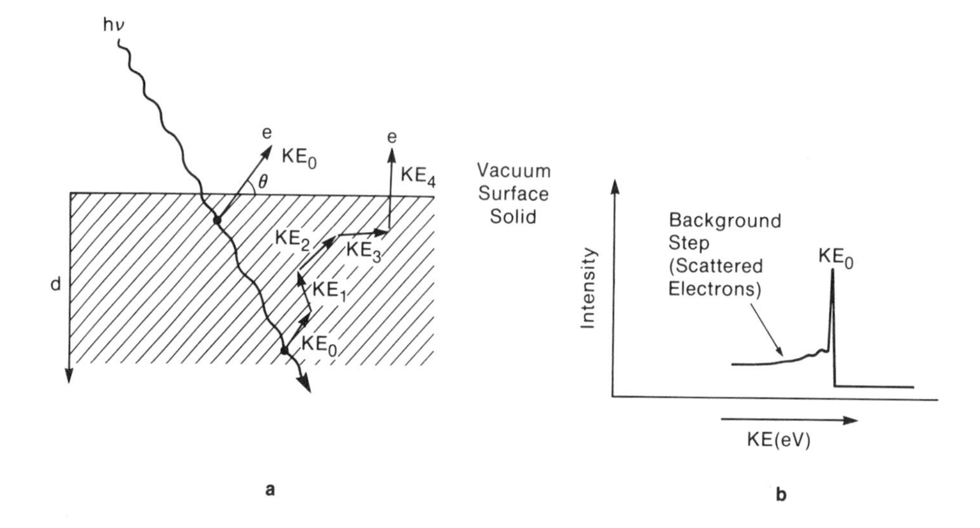

Figure 5 **(a) Schematic of inelastic electron scattering occurring as a photoelectron, initial energy KE_0, tries to escape the solid, starting at different depths. $KE_4 < KE_3 < KE_2 < KE_1 < KE_0$. (b) *KE* energy distribution (i.e., electron spectrum) obtained due to the inelastic scattering in (a). Note that the peak, at E_0, must come mainly from the surface region, and the background step, consisting of the lower energy scattered electrons, from the bulk.**

or compound is involved. Substituting λ_e values from the curves into Equation (3) tells us that for normal emission ($\theta = 90°$) using a 200-eV *KE* XPS peak, 90% of the signal originates from the top ~25 Å, for elements. For a 1400-eV peak the depth is ~60 Å. The numbers are about twice as big for compounds. Thus, the depth probed by XPS varies strongly depending on the XPS peaks used and the material involved. The depth probed can also be made smaller for any given XPS peak and material by detecting at grazing emission angle θ. For smooth surfaces, values down to 10° are practical, for which the depth probed is reduced by a factor of 1/sin 10, or ~6, compared to 90°, from Equation (3). Varying *KE* or θ are important practical ways of distinguishing what is in the outermost atomic layers from what is underneath.

Instrumentation

An XPS spectrometer schematic is shown in Figure 7. The X-ray source is usually an Al- or Mg-coated anode struck by electrons from a high voltage (10–15 kV) Alkα or Mgkα radiation lines produced at energies of 1486.6 eV and 1256.6 eV, with line widths of about 1 eV. The X rays flood a large area (~1 cm^2). The beam's spot size can be improved to about 100-μm diameter by focusing the electron beam

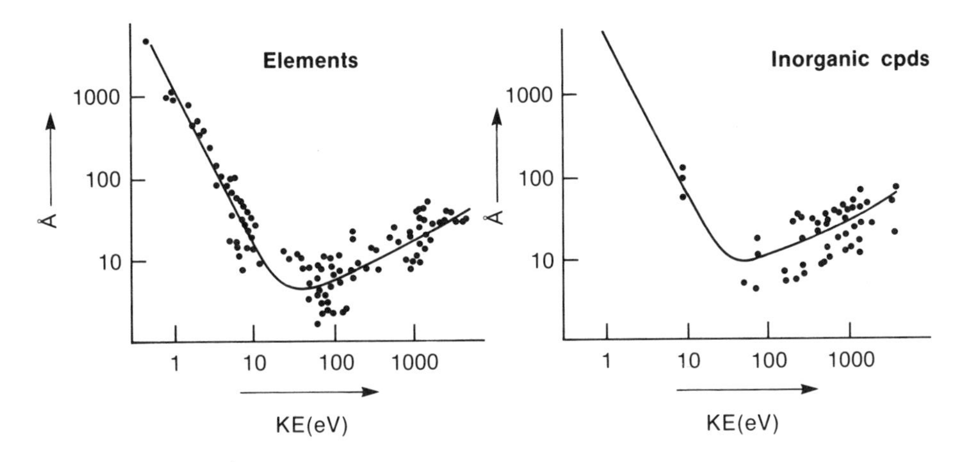

Figure 6 **Mean free path lengths λ_e as a function of *KE*, determined for (a) metals and (b) inorganic compounds.[6]**

onto the anode and passing the X rays through an X-ray monochromator. The latter also improves line widths to between 0.5 and 0.25 eV, leading to higher resolution spectra (thus improving the chemical state identification process) and removing an unwanted X-ray background at lower energies.

Practical limits to the shape and size of samples are set by commercial equipment design. Some will take only small samples (e.g., 1 cm × 1 cm) while others can handle whole 8-in computer disks. Flat samples improve signal strength and allow quantitative θ variation, but rough samples and powders are also routinely handled. Insulating samples may charge under the X-ray beam, resulting in inaccurate *BE* determinations or spectra distorted beyond use. The problem can usually be mitigated by use of a low-energy electron flood gun to neutralize the charge, provided this does not damage the sample.

The electron lenses slow the electrons before entering the analyzer, improving energy resolution. They are also used to define an analyzed area on the sample from which electrons are received into the analyzer and, in one commercial design, to image the sample through the analyzer with 10-μm resolution. Older instruments may have slits instead of lenses. The most popular analyzer is the hemispherical sector, which consists of two concentric hemispheres with a voltage applied between them. This type of analyzer is naturally suited to varying θ by rotating the sample, Figure 7. The XPS spectrum is produced by varying the voltages on the lenses and the analyzer so that the trajectories of electrons ejected from the sample at different energies are brought, in turn, to a focus at the analyzer exit slit. A channeltron type electron multiplier behind the exit slit of the analyzer amplifiers individual electrons by 10^5–10^6, and each such pulse is fed to external conventional pulse counting electronics and on into a computer. The computer also controls the lens and

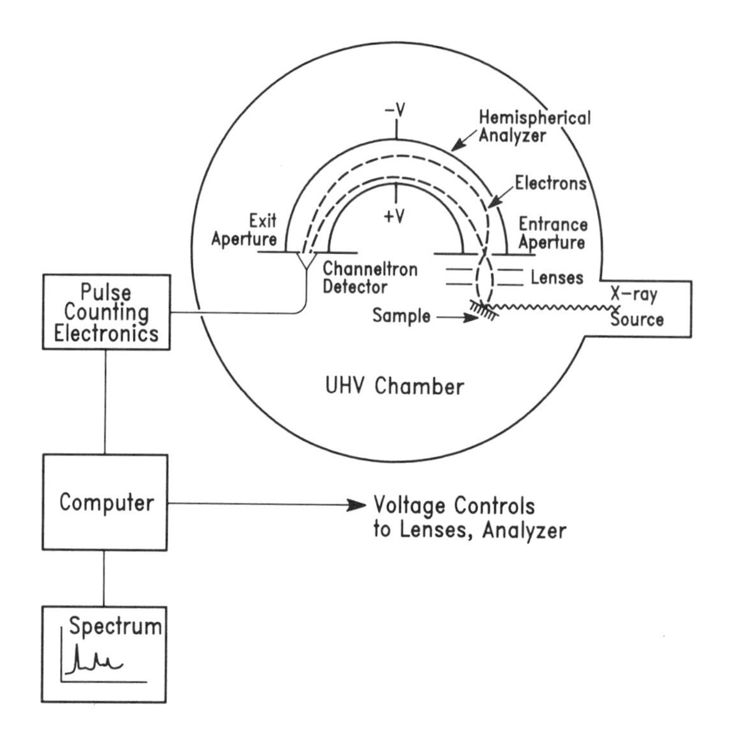

Figure 7 **Schematic of a typical electron spectrometer showing all the necessary components. A hemispherical electrostatic electron energy analyser is depicted.**

analyzer voltages. A plot of electron pulses counted against analyzer–lens voltage gives the photoelectron spectrum. More sophisticated detection schemes replace the exit slit–multiplier arrangement with a multichannel array detector. This is the modern equivalent of a photographic plate, allowing simultaneous detection of a range of *KE*s, thereby speeding up the detection procedure.

Commercial spectrometers are usually bakeable, can reach ultrahigh-vacuum pressures of better than 10^{-9} Torr, and have fast-entry load-lock systems for inserting samples. The reason for the ultrahigh-vacuum design, which increases cost considerably, is that reactive surfaces, e.g., clean metals, contaminate rapidly in poor vacuum (1 atomic layer in 1 s at 10^{-6} Torr). If the purpose of the spectrometer is to always look at as-inserted samples, which are already contaminated, or to examine rather unreactive surfaces (e.g., polymers) vacuum conditions can be relaxed considerably.

Applications

XPS is routinely used in industry and research whenever elemental or chemical state analysis is needed at surfaces and interfaces and the spatial resolution requirements are not demanding (greater than 150 μm). If the analysis is related specifically to the top 10 or so atomic layers of air-exposed sample, the sample is simply inserted and data taken. Examples where this might be appropriate include: examination for and identification of surface contaminants; evaluation of materials processing steps, such as cleaning procedures, plasma etching, thermal oxidation, silicide thin-film formation; evaluation of thin-film coatings or lubricants (thickness–quantity, chemical composition); failure analysis for adhesion between components, air oxidation, corrosion, or other environmental degradation problems, tribological (wear) activity; effectiveness of surface treatments of polymers and plastics; surface composition differences for alloys; examination of catalyst surfaces before and after use, after "activation" procedures, and unexplained failures.

Figure 3c was used to illustrate that Si^{IV} could be distinguished from Si^{0} by the Si 2p chemical shift. The spectrum is actually appropriate for an oxidized Si wafer having an ~10-Å SiO_2 overlayer. That the SiO_2 is an overlayer can easily be proved by decreasing θ to increase the surface sensitivity; the Si^{0} signal will decrease relative to the Si^{IV} signal. The 10-Å thickness can be determined from the Si^{IV}/Si^{0} ratio and Equation (3), using the appropriate λ_e value. That the overlayer is SiO_2 and not some other Si^{IV} compound is easily verified by observing the correct position (*BE*) and intensity of the O 1s peak plus the absence of other element peaks. If the sample has been exposed to moisture, including laboratory air, the outermost atomic layer will actually be hydroxide, not oxide. This is easily recognized since there is a chemical shift between OH and O in the O 1s peak position.

Figure 8 shows a typical example where surface modification to a polymer can be followed.[7] High-density polyethylene $(CH_2CH_2)_n$ was surface-fluorinated in a dilute fluorine–nitrogen mixture. Spectrum A was obtained after only 0.5 s treatment. A F 1s signal corresponding to about a monolayer has appeared, and CF formation is obvious from the chemically shifted shoulder on the C 1s peak at the standard CF position. After 30 s reaction, the F 1s / C 1s ratio indicates (spectrum B) that the reaction has proceeded to about 30 Å depth, and that CF_2 formation has occurred, judging by the appearance of the C 1s peak at 291 eV. Angular studies and more detailed line shape and relative intensity analysis, compared to standards, showed that for the 0.5-s case, the top monolayer is mainly polyvinyl fluoride $(CFHCH_2)_n$, whereas after 30 s polytrifluoroethylene $(CF_2CFH)_n$ dominates in the top two layers. While this is a rather aggressive example of surface treatment of polymers, similar types of modifications frequently are studied using XPS. An equivalent example in the semiconductor area would be the etching processes of Si/SiO_2 in CF_4/O_2 mixtures, where varying the CF_4/O_2 ratio changes the relative etching rates of Si and SiO_2, and also produces different and varying amounts of residues at the wafer's surface.

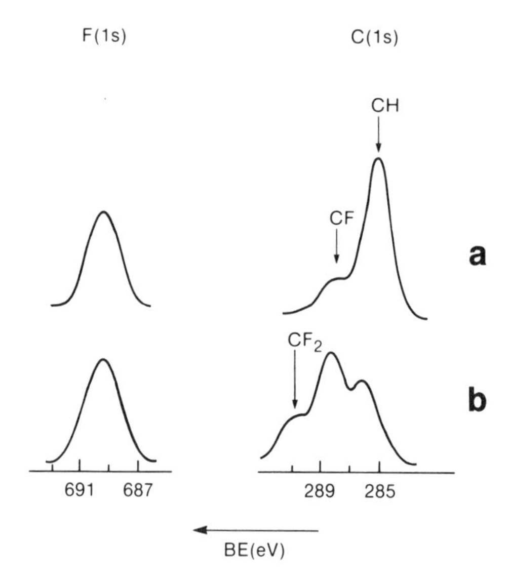

Figure 8 **XPS spectrum in the C 1s and F 1s regions of polyethylene $(CH_2)_n$, treated with a dilute F_2/N_2 gaseous mixture for (a) 0.5 sec, and (b) 30 sec.[7]**

In many applications the problem or property concerned is not related just to the top 10 or so atomic layers. Information from deeper regions is required for a number of reasons: A thick contaminant layer, caused by air exposure, may have covered up the surface of interest; the material may be a layered structure in which the buried interfaces are important; the composition modulation with depth may be important, etc. In such cases, the 2–15 atomic layer depth resolution attainable in XPS by varying θ is insufficient, and some physical means of stripping the surface while taking data, or prior to taking data, is required. This problem is common to all very surface sensitive spectroscopies. The most widely used method is argon ion sputtering, done inside the spectrometer while taking data. It can be used to depths of μm, but is most effective and generally used over much shorter distances (hundreds and thousands of Å) because it can be a slow process and because sputtering introduces artifacts that get worse as the sputtered depth increases.[8] These include interfacial mixing caused by the movement of atoms under the Ar^+ beam, elemental composition alteration caused by preferential sputtering of one element versus another, and chemical changes caused by bonds being broken by the sputtering process.

If the interface or depth of interest is beyond the capability of sputtering, one can try polishing down, sectioning, or chemical etching the sample before insertion.

The effectiveness of this approach varies enormously, depending on the material, as does the extent of the damaged region left at the surface after this preparation treatment.

In some cases, the problem or property of interest can be addressed only by performing experiments inside the spectrometer. For instance, metallic or alloy embrittlement can be studied by fracturing samples in ultrahigh vacuum so that the fractured sample surface, which may reveal why the fracture occurred in that region, can be examined without air exposure. Another example is the simulation of processing steps where exposure to air does not occur, such as many vacuum deposition steps in the semiconductor and thin-film industries. Studying the progressive effects of oxidation on metals or alloys inside the spectrometer is a fairly well-established procedure and even electrochemical cells are now coupled to XPS systems to examine electrode surfaces without air exposure. Sometimes materials being processed can be capped by deposition of inert material in the processing equipment (e.g., Ag, Au, or in GaAs work, arsenic oxide), which is then removed again by sputtering or heating after transfer to the XPS spectrometer. Finally, attempts are sometimes made to use "vacuum transfer suitcases" to avoid air exposure during transfer.

Comparison with other Techniques

XPS, AES, and SIMS are the three dominant surface analysis techniques. XPS and AES are quite similar in depth probed, elemental analysis capabilities, and absolute sensitivity. The main XPS advantages are its more developed chemical state analysis capability, somewhat more accurate elemental analysis, and far fewer problems with induced sample damage and charging effects for insulators. AES has the advantage of much higher spatial resolutions (hundreds of Å compared to tens of µm), and speed. Neither is good at trace analysis, which is one of the strengths of SIMS (and related techniques). SIMS also detects H, which neither AES nor XPS do, and probes even less deeply at the surface, but is an intrinsically destructive technique. Spatial resolution is intermediate between AES and XPS. ISS is the fourth spectroscopy generally considered in the "true surface analysis" category. It is much less used, partly owing to lack of commercial instrumentation, but mainly because it is limited to elemental analysis with rather poor spectral distinction between some elements. It is, however, the most surface sensitive elemental analysis technique, seeing only the top atomic layer. With the exception of EELS and HREELS, all other spectroscopies used for surface analysis are much less surface sensitive than the above four. HREELS is a vibrational technique supplying chemical functional group information, not elemental analysis, and EELS is a rarely used and specialized technique, which, however, can detect hydrogen.

Conclusions

XPS has developed into the most generally used of the truly surface sensitive techniques, being applied now routinely for elemental and chemical state analysis over a range of materials in a wide variety of technological and chemical industries. Its main current limitations are the lack of high spatial resolutions and relatively poor absolute sensitivity (i.e., it is not a trace element analysis technique). Recently introduced advances in commercial equipment have improved speed and sensitivity by using rotating anode X-ray sources (more photons) and parallel detection schemes. Spot sizes have been reduced from about 150 μm, where they have languished for several years, to 75 μm. Spot sizes of 10 μm have been achieved, and recently anounced commercial instruments offer these capabilities. When used in conjunction with focused synchrotron radiation in various "photoelectron microscope" modes higher resolution is obtainable. Routinely available 1 μm XPS resolution in laboratory-based equipment would be a major breakthrough, and should be expected within the next three years.

Special, fully automated one-task XPS instruments are beginning to appear and will find their way into both quality control laboratories and process control on production lines before long.

More detailed discussions of XPS can be found in references 4–12, which encompass some of the major reference texts in this area.

Related Articles in the Encyclopedia

UPS, AES, SIMS, and ISS

References

1 K. Siegbahn et al. *ESCA: Atomic, Molecular, and Solid State Structure Studied by Means of Electron Spectroscopy.* Nova Acta Regime Soc. Sci., Upsaliensis, 1967, Series IV, Volume 20; and K. Siegbahn et al. *ESCA Applied to Free Molecules.* North Holland, Amsterdam, 1969. These two volumes, which cover the pioneering work of K.Siegbahn and coworkers in developing and applying XPS, are primarily concerned with chemical structure identification of molecular materials and do not specifically address surface analysis.

2 Charts such as this, but in more detail, are provided by all the XPS instrument manufacturers. They are based on extensive collections of data, much of which comes from Reference 1.

3 J. H. Scofield. *J. Electron Spect.* **8,** 129, 1976. This is the standard quoted reference for photoionization cross sections at 1487 eV. It is actually one of the most heavily cited references in physical science. The calculations are published in tabular form for all electron level of all elements.

See, for example, S. Evans et al. *J. Electr. Spectr.* **14,** 341, 1978. Relative experimental ratios of cross sections for the most intense peaks of most elements are given.

5 J. C. Carver, G. K. Schweitzer, and T. A. Carlson. *J. Chem. Phys.* **57,** 973, 1972. This paper deals with multiplet splitting effects, and their use in distinguishing different element states, in transition metal complexes.

6 M. P. Seah and W. A. Dench. *Surf. Interface Anal.* **1,** 1, 1979. Of the many compilations of measured mean free path length versus *KE,* this is the most thorough, readable, and useful.

7 D. T. Clark, W. J. Feast, W. K. R. Musgrave, and I. Ritchie. *J. Polym. Sci. Polym. Chem.* **13,** 857, 1975. One of many papers from Clark's group of this era which deal with all aspects of XPS of polymers.

8 See the article on surface roughness in Chapter 12.

9 The book series *Electron Spectroscopy: Theory, Techniques, and Applications,* edited by C. R. Brundle and A. D. Baker, published by Academic Press has a number of chapters in its 5 volumes which are useful for those wanting to learn about the analytical use of XPS: In Volume 1, *An Introduction to Electron Spectroscopy* (Baker and Brundle); in Volume 2, *Basic Concepts of XPS* (Fadley); in Volume 3, *Analytical Applications of XPS* (Briggs); and in Volume 4, *XPS for the Investigation of Polymeric Materials* (Dilks).

10 T. A. Carlson, *Photoelectron and Auger Spectroscopy,* Plenum, 1975. A complete and largely readable treatment of both subjects.

11 *Practical Surface Analysis,* edited by D. Briggs and M. P. Seah, published by J. Wiley; *Handbook of XPS and UPS,* edited by D. Briggs. Both contain extensive discussion on use of XPS for surface and material analysis.

12 *Handbook of XPS,* C. D. Wagner, published by PHI (Perkin Elmer). This is a book of XPS data, invaluable as a standard reference source.

5.2 UPS

Ultraviolet Photoelectron Spectroscopy

C. R. BRUNDLE

Contents

Introduction

The photoelectric process, which was discovered in the early 1900s, was developed as a means of studying the electronic structure of molecules in the gas phase in the early 1960s, largely owing to the pioneering work of D. W. Turner's group.[1] A major step was the introduction of the He resonance discharge lamp as a laboratory photon source, which provides monochromatic 21.2-eV light. In conjunction with the introduction of high resolution electron energy analyzers, this enables the binding energies (BE) of all the electron energy levels below 21.2 eV to be accurately determined with sufficient spectral resolution to resolve even vibrational excitations. Coupled with theoretical calculations, these measurements provide information on the bonding characteristics of the valence-level electrons that hold molecules together. The area has become known as ultraviolet photoelectron spectroscopy (UPS) because the photon energies used (21.2 eV and lower) are in the vacuum ultraviolet (UV) part of the light spectrum. It is also known as molecular photoelectron spectroscopy, because of its ability to provide molecular bonding information.

In parallel with these developments for studying molecules, the same technique was being developed independently to study solids; particularly metals and semiconductors.[2] This branch of the technique is usually known as UV photoemission. Here the electronic structure of the solid (the band structure for metals and semiconductors) was the interest. Since the technique is sensitive to only the top few atomic layers, the electronic structure of the surface, which in general can be different from that of the bulk, is actually obtained. The two branches of UPS, gas-phase and solid-surface studies, come together when adsorption and reaction of molecules at surfaces is studied.[3, 4]

Though commercial UPS instruments were sold in the 1970s, for gas-phase work, none are sold today. Since the only additional item required to perform UPS on an XPS instrument is a He source, this is usually how UPS is performed in the laboratory. An alternative, more specialized approach, is to couple an electron spectrometer to the beam-line monochromator of a synchrotron facility. This provides a tunable source of light, usually between around 10 eV and 200 eV, though many beam lines can obtain much higher energies. This approach can provide a number of advantages, including variable surface sensitivity and access to core levels up to the photon energy used, at much higher resolution than obtainable by laboratory XPS instruments. Even using a laboratory UPS source, such as a He resonance lamp, some low-lying core levels are accessible. When using either synchrotron or laboratory sources to access core levels, all the materials surface analysis capabilities of XPS described in the preceding article become available.

Basic Principles

The photoionization process and the way it is used to measure BEs of electrons to atoms is described in the article on XPS and will not be repeated here. Instead, we will concentrate on the differences between the characteristics of core-level BEs, described in the XPS article, and those of valence-level BEs. In Figure 1a the electron energy-level diagram for a CO molecule is shown, schematically illustrating how the atomic levels of the C and O atom interact to form the CO molecule. The important point to note is that whereas the BEs of the C 1s and O 1s core levels remain characteristic of the atoms when the CO molecule is formed (the basis of the use of XPS as an elemental analysis tool), the C 2p and O 2p valence levels are no longer characteristic of the individual atoms, but have combined to form a new set of molecular orbitals entirely characteristic of the CO molecule. Therefore, the UPS valence-band spectrum of the CO molecule, Figure 1b, is also entirely characteristic of the molecule, the individual presence of a C atom and an O atom no longer being recognizable. For a solid, such as metallic Ni, the valence-level electrons are smeared out into a band, as can be seen in the UPS spectrum of Ni (Figure 2a). For molecules adsorbed on surfaces there is also a smearing out of structure. For example, Figure 2b shows a monolayer of CO adsorbed on an Ni surface.

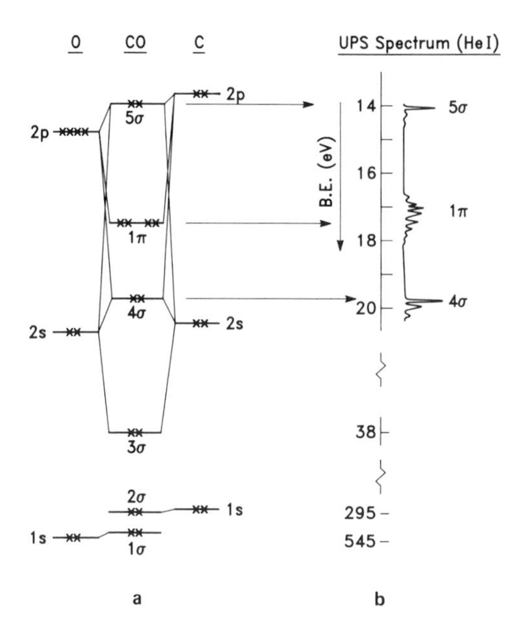

Figure 1 **(a) Electron energy diagram for the CO molecule, illustrating how the molecular orbitals are constructed from the atomic levels. (b) He I UPS spectrum of CO.[1]**

Analytical Capabilities

As stated earlier, the major use of UPS is not for materials analysis purposes but for electronic structure studies. There are analysis capabilities, however. We will consider these in two parts: those involving the electron valence energy levels and those involving low-lying core levels accessible to UPS photon energies (including synchrotron sources). Then we will answer the question "why use UPS if XPS is available?"

Valence Levels

The spectrum of Figure 1b is a fingerprint of the presence of a CO molecule, since it is different in detail from that of any other molecule. UPS can therefore be used to identify molecules, either in the gas phase or present at surfaces, provided a data bank of molecular spectra is available, and provided that the spectral features are sufficiently well resolved to distinguish between molecules. By now the gas phase spectra of most molecules have been recorded and can be found in the literature.[1, 5] Since one is using a pattern of peaks spread over only a few eV for identification purposes, mixtures of molecules present will produce overlapping patterns. How well mixtures can be analyzed depends, obviously, on how well overlapping peaks can be resolved. For molecules with well-resolved fine structure (vibrational) in the spectra (see Figure 1b), this can be done much more successfully than for the broad,

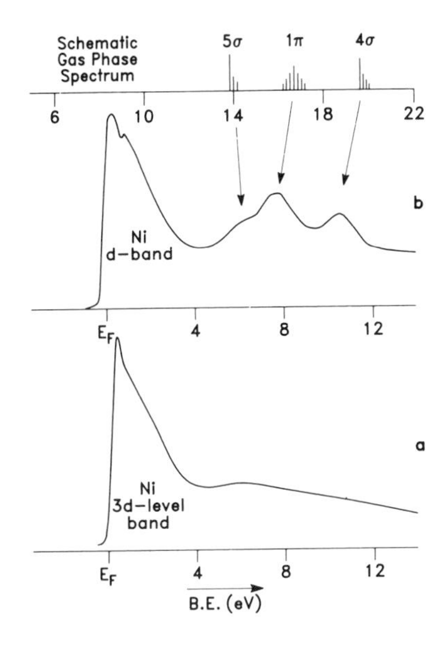

Figure 2 **(a) He II UPS spectrum of a Ni surface.[4] (b) He II UPS spectrum of a CO monolayer adsorbed on a Ni surface.[4] Note the broadening and relative binding-energy changes of the CO levels compared to the gas phase spectrum. Gas-phase binding energies were measured with respect to the vacuum level; solid state binding energies relative to the Fermi level E_F.**

unresolved bands found for solid surfaces (see Figure 2b). For solids that have electronic structure characteristics in between those of molecules and metals, such as polymers, ionic compounds, or molecules adsorbed on surfaces (Figure 2b), enough of the individual molecular-like structure of the spectra often remains for the valence levels to be used for fingerprinting purposes. Reactions between molecules and surfaces often can be fingerprinted also. For example, in Figure 3 the UPS differences between molecular H_2O on a metal, and its only possible dissociation fragments, OH and atomic O, are schematically illustrated.

The examples of valence-level spectra given so far, for solid surfaces, i.e., those in Figures 2a, 2b, and 3, are all *angle-integrated* spectra; that is, electrons emitted over a wide solid angle of emission are collected and displayed. In fact, the energy distribution of photoemitted electrons from solids varies somewhat depending on the direction of emission and if data is taken in an angular-resolved mode, that is, for specific directions for the photon beam and the photoemitted electrons, detailed information about the three-dimensional (3D) band structure of the solid, or the two-dimensional (2D) band structure of an adsorbate overlayer may be obtained, together with information on the geometric orientation of such adsorbate mole-

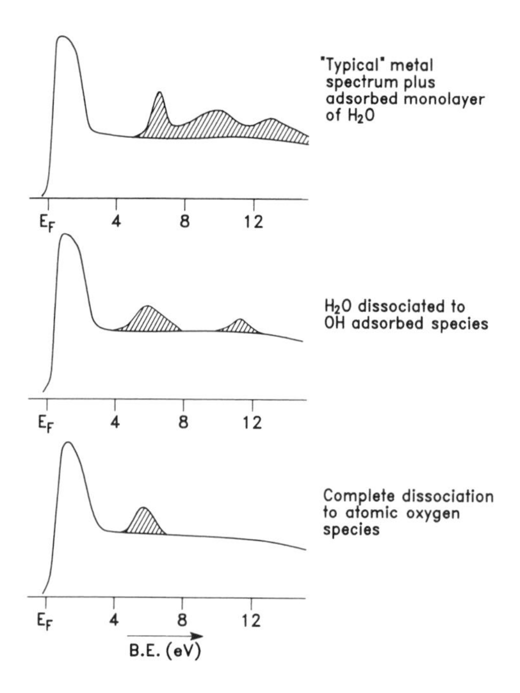

Figure 3 **Schematic spectra of H_2O, OH, and atomic O adsorbed on a metal surface illustrate how molecules can be distinguished from their reactor products by fingerprinting.**

cules. To properly exploit the technique requires also variation of the photon energy, $h\nu$ (therefore requiring synchrotron radiation) and the polarization of the radiation (s and p, naturally available from the synchrotron source). Basically, recording the UPS spectrum while varying all these parameters (angle, photon energy, and polarization) picks out specific parts of the density of states. A fuller description of this type of work[6] is beyond the scope of this article and is not particularly relevant to materials analysis, except for the fact that molecular orientation at surfaces can be determined. This property is, however, restricted to situations with long-range order, i.e., 2D arrays of molecules on single-crystal surfaces.

Low-Lying Accessible Core Levels

Table 1 lists core levels and their BEs for elements commonly used in technology, which are sufficiently sharp and intense, and which are accessible to laboratory He I or He II sources (21.2-eV or 40.8-eV photon energy) or to synchrotron sources (up to 200 eV or higher). The analytical approaches are the same as described in the XPS article. For example, in that article examples were given of Si 2p spectra obtained using a laboratory Al Kα X-ray source at 1486-eV photon energy. The

Element	Core level	Approximate binding energy (eV)	Usable r radiation
Al	2p	72	S
Si	2p	100	S
S	2p	164	S
Zn	3d	9	He I, He II, S
Ga	3d	18	He I, He II, S
Ge	3d	29	He II, S
As	3d	41	S
Ta	4f	25	He II, S
W	4f	34	S
Ir	4f	60	S
Pt	4f	70	S
Au	4f	84	S
Hg	4f	99	S
	5d	7	He I, He II, S
Pb	4f	138	S

Table 1 **Narrow, intense core levels of some elements commonly used in technological materials that are accessible to He I / He II radiation, or synchrotron radiation below 200 eV.**

Si 2p line, at about 100 eV BE, is also easily accessible at most synchrotron sources but cannot, of course, be observed using He I and He II radiation. On the other hand, the Zn 3d and Hg 4f lines can be observed quite readily by He I radiation (see Table 1) and the elements identified in this way. Quantitative analysis using relative peak intensities is performed exactly as in XPS, but the photoionization cross sections σ are very different at UPS photon energies, compared to Al Kα energies, and tabulated or calculated values are not so readily available. Quantitation, therefore, usually has to be done using local standards.

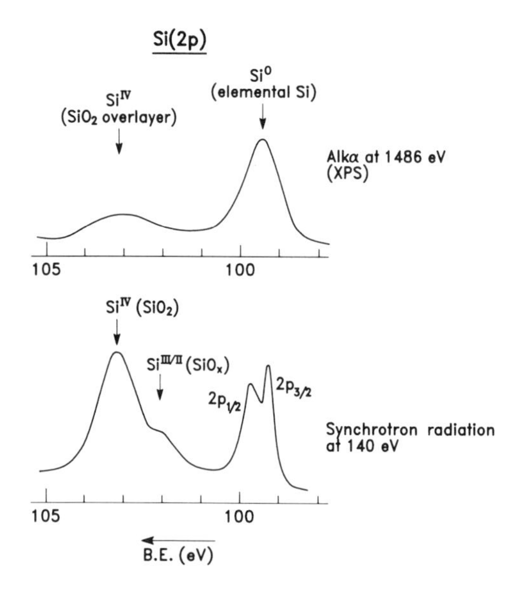

Figure 4 **Schematic comparison of the Si 2p spectra of an Si/SiO_2 interface taken using Al K radiation at 1486 eV and synchrotron radiation at 40 eV photon energy. Note the greater surface sensitivity and higher resolution in the synchrotron case.**

Why Use UPS for Analysis?

Since all the valence levels and core levels that are accessible to UPS photon sources are also accessible to XPS, what are the reasons for ever wanting to use laboratory He sources or synchrotron radiation? There are at least four significant differences that can be important analytically in special circumstances. First, the surface sensitivity is usually greater in UPS because for a given energy level being examined, the lower photon energy sources in UPS yield ejected photoelectrons having lower kinetic energies. For example, the Si 2p signal of Figure 3 in the XPS article consists of electrons having a kinetic energy 1486–100 eV = 1386 eV. If the Si 2p spectrum were recorded using 140 eV synchrotron photons, the kinetic energy would be 140–100 eV = 40 eV. Looking at the inelastic mean-free path length diagram of Figure 6 in the XPS article, one can see that 40-eV photoelectrons have about one-third the inelastic scattering length of 1400-eV electrons. Therefore the synchrotron recorded signal would be roughly three times as surface sensitive, as illustrated in Figure 4 where the XPS SiO_2 / Si spectrum is schematically compared for 1486 eV and 140 eV photon sources. The SiO_2 part of the Si 2p signal is much stronger in the synchrotron spectrum and therefore much thinner layers will be more easily detectable.

Secondly, spectral resolution can be significantly higher for UPS or synchrotron data, compared to XPS. This is simply a consequence of UPS (synchrotron) sources having narrower line widths than laboratory X-ray sources. Thus, whereas the XPS recorded Si 2p signal of Figure 4 has a width of about 1 eV, the individual $2p_{3/2}$ and $2p_{1/2}$ components of the synchrotron recorded signal are only about 0.25 eV wide. Whether this resolution improvement can be achieved in any individual case depends on the natural line width of the particular core level concerned. Si 2p, W 4f, Al 2p, Pt 4f, and Au 4f are all examples of narrow core lines, where a large resolution improvement would occur using synchrotron sources, allowing small chemical shifts corresponding to chemically distinct species to be more easily seen. For valence levels, higher resolution is also an obvious advantage since, as described earlier, one is usually looking at several lines or bands, which may overlap significantly. Two additional practical points about resolution also should be noted. The spectral resolution of the gratings used to monochromatize synchrotron radiation gets worse as the photon energy gets higher, so the resolution advantage of synchrotron radiation decreases as one goes to high BE core levels. Second, monochromators can be used with laboratory X-ray sources, improving XPS resolution significantly, but not to the degree achievable in UPS or synchrotron work.

The third significant difference between UPS and XPS, from an analytical capability point of view, concerns signal strength. To zeroth order, σ values are a maximum for photon energies just above photoionization threshold, and then decrease strongly as the photon energy is increased, so valence levels in particular have much greater σ values using UPS or synchrotron sources, compared to XPS. When coupled with the high photon fluxes available from such sources, this results in greater absolute sensitivity for UPS or synchrotron spectra.

Taking these differences together, one can see that all three work in favor of UPS or synchrotron compared to XPS when trying to observe very thin layers of chemically distinct material at the surface of a bulk material: improved surface sensitivity; improved resolution allowing small surface chemically shifted components in a spectrum to be distinguished from the underlying bulk signal; and improved absolute sensitivity. As a practical matter, one has to ask whether the core levels one wants to use are even accessible to UPS or synchrotron and whether the need to go to a national facility on a very access-limited basis can compare to day-in, day-out laboratory operations. For UPS using He I and He II radiation sources the addition of these to existing XPS system is not excessively costly and is then always there to provide additional capability useful for specific materials and problems.

The final difference between UPS or synchrotron capabilities and XPS, from an analytical point of view, is in lateral resolution. Modern laboratory XPS small-spot instruments can look at areas down to 30–150 μm, depending on the particular instrument, with one very specialized instrument offering imaging capabilities at 10-μm resolution, but with degraded spectroscopy capabilities.[8] For UPS and synchrotron radiation, much higher spatial resolution can be achieved, partly because the lower kinetic energy of the photoelectron lends itself better to imaging schemes and partly because of efforts to focus synchrotron radiation to small spot sizes. The potential for a true photoelectron microscope with sub 1000-Å resolution therefore exists, but it has not been realized in any practical sense yet.

Conclusions

UPS, if defined as the use of He I, He II, or other laboratory low-energy radiation sources (<50 eV), has rather limited materials surface analysis capabilities. Valence and core electron energy levels below the energy of the radiation source used can be accessed and the main materials analysis role is in providing higher resolution and high surface sensitivity data as a supplement to XPS data, usually for the purpose of learning more about the chemical bonding state at a surface. Angle-resolved UPS can supply molecular orientation geometric information for ordered structures on single crystal surfaces, but its main use is to provide detailed band structure information.

Synchrotron radiation can be used to provide the same information, but also has the great advantage of a wider, tunable, photon energy range. This allows one to access some core levels at higher resolution and surface sensitivity than can be done by XPS. The variable energy source also allows one to vary the surface sensitivity by varying the kinetic energy of the ejected photoelectrons, thereby creating a depth profiling capability. Most synchrotron photoemission work to date has involved fundamental studies of solid state physics and chemistry, rather than materials analysis, albeit on such technologically important materials as Si, GaAs, and CdTeHg. Some quite applied work has been done related to the processing of these materials, such as studying the effects of cleaning procedures on residual surface contaminants, and studying reactive ion-etching mechanisms.[9] The major drawback of synchrotron radiation is that it is largely unavailable to the analytical community and is an unreliable photon source for those who do have access. As the number of synchrotron facilities increase and as they become more the domain of people wanting to use them as dedicated light sources, rather than in high-energy physics collision experiments, the situation for materials analysis will improve and the advantages over laboratory-based XPS will be more exploitable. Synchrotron radiation will never replace laboratory-based XPS, however, and it should be regarded as complementary, with advantages to be exploited when really needed. High spatial resolution photoelectron microscopy is likely to become one such area.

Related Articles in the Encyclopedia

XPS and SEXAFS

References

1 D. W. Turner, C. Baker, A. D. Baker, and C. R. Brundle. *Molecular Photoelectron Spectroscopy.* Wiley, London, 1970. This volume presents a brief

introduction to the principles of UPS and a large collection of spectra on small molecules, together with their interpretation in terms of the electronic structure and bonding of the molecules.

2 W. E. Spicer. In *Survey of Phenomena in Ionized Gases.* International Atomic Energy Agency, Vienna, 1968, p. 271. A review of the early photoemission work on solids by the pioneering group in this area.

3 D. Menzel. *J. Vac. Sci. Tech.* **12,** 313, 1975. A review of the applications of UPS to the adsorption of molecules at metal surfaces.

4 C. R. Brundle. In *Molecular Spectroscopy.* (A. R. West, Ed.) Heyden, London, 1976. This review discusses both the use of XPS and UPS in studying adsorption and reactions at surfaces.

5 K. Kimura, S. Katsumata, Y. Achita, Y. Yamazaki, and S. Iwata. *Handbook of He I Photoelectron Spectra of Fundamental Organic Molecules.* Halsted Press, New York, 1981. This volume collects together spectra and interpretation for 200 organic molecules.

6 *Photoemission in Solids.* (L. Ley and M. Cardona, Eds.) Springer-Verlag, New York, 1978 and 1979, Vols. 1 and 2.

7 N. V. Smith and F. J. Himpsel. In *Handbook on Synchrotron Radiation.* (E. E. Koch, Ed.) North Holland, New York, 1983, Vol. 1b.

8 P. Coxon, J. Krizek, M. Humpherson, and I. R. M. Wardell. *J. Elec. Spec.* 821, 1990.

9 J. A. Yarmoff and F. R. McFeely, *Surface Science* **184**, 389, 1987.

5.3 AES

Auger Electron Spectroscopy

Y.E. STRAUSSER

Contents

Introduction

Auger electron spectroscopy (AES) is a technique used to identify the elemental composition, and in many cases, the chemical bonding of the atoms in the surface region of solid samples. It can be combined with ion-beam sputtering to remove material from the surface and to continue to monitor the composition and chemistry of the remaining surface as this surface moves into the sample. It uses an electron beam as a probe of the sample surface and its output is the energy distribution of the secondary electrons released by the probe beam from the sample, although only the Auger electron component of the secondaries is used in the analysis.

Auger electron spectroscopy is the most frequently used surface, thin-film, or interface compositional analysis technique. This is because of its very versatile combination of attributes. It has surface specificity—a sampling depth that varies between 5 and 100 Å depending upon the energy of the Auger electrons measured and the signal-to-noise ratio in the spectrum. It has good lateral spatial resolution, which can be as low as 300 Å, depending on the electron gun used and the sample material. It has very good depth resolution, as low as 20 Å depending on the characteristics of the ion beam used for sputtering. It has a good absolute detectability, as low as 100 ppm for most elements under good conditions. It can produce a three-dimensional map of the composition and chemistry of a volume of a sample that is tens of μm thick and hundreds of μm on a side.

On the other hand, AES cannot detect H or He. It does not do nondestructive depth profiling. It uses an electron beam as a probe, which can be destructive to some samples. It requires the sample to be put into and to be compatible with high vacuum. Some nonconducting samples charge under electron beam probing and cannot be analyzed. The sputtering process can alter the surface composition and thereby give misleading results. It does turn out to be the technique of choice, in its area, much of the time. The purpose of this article is to make clear what it can and cannot do and how to get the most information from it.

The Auger process, which produces an energetic electron in a radiationless atomic transition, was first described by Pierre Auger in 1923.[1] The detection of Auger electrons in the secondary electron energy spectra produced by electron bombardment of solid samples was reported by J. J. Lander in 1953.[2] Its use in an analytical technique to characterize solid surfaces was made practical by Larry Harris' analog detection circuitry in 1967.[3] From that time the technology developed very rapidly, and the technique gained momentum through the 1970s and 1980s.

As the technique developed so did the instrumentation. The hardware development has taken advantage of improvements in ultrahigh vacuum technology and computerization. Systems are available having 300-Å diameter field emission electron beams; user-friendly, rapidly attained ultrahigh vacuum; and complete computer control of the system. At the other end of the price range are components that can be "plugged in" to various deposition and processing systems to provide *in-situ* surface characterization.

AES, X-Ray Photoelectron Spectroscopy (XPS), Secondary Ion Mass Spectroscopy (SIMS), and Rutherford Backscattering Spectroscopy (RBS) have become the standard set of surface, thin-film, and interface analysis tools. Each has its own strengths, and mostly they are complementary. XPS uses X rays as a probe, which are usually less damaging to the surface than the electron beam of Auger but which can't be focused to give high lateral spatial resolution. XPS is also more often selected to determine chemical information. SIMS can detect H and He and has a much higher absolute sensitivity in many cases, but seldom gives any chemical information and, by its nature, has to remove material to do its analysis. RBS readily produces good quantitative results and does nondestructive depth profiling, but it lacks the absolute sensitivity of Auger to many of the important elements and its depth resolution is not as good as Auger can produce, in many cases.

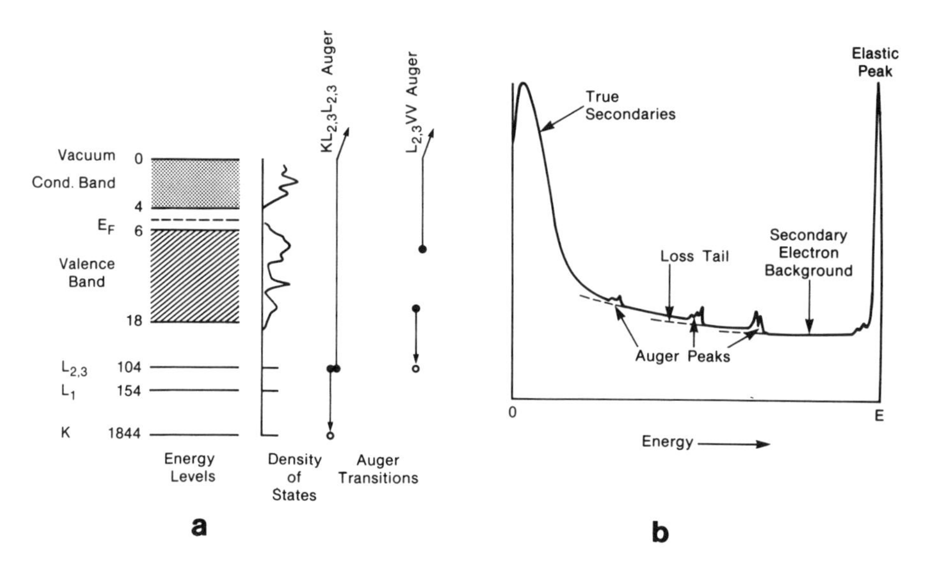

Figure 1 **(a) Energy level diagram of solid Si, including the density of states of the valence and conduction bands, a schematic representation of the Si $KL_{2,3}L_{2,3}$ Auger transitions, and a subsequent LVV Auger transition. (b) The complete secondary electron energy distribution produced by the interaction of a primary electron beam of energy *E* with a solid surface. The true secondary peak, the elastic peak, and some Auger peaks are shown. Also shown are the secondary background and the loss tail contributions to the background from each of the Auger peaks.**

Basic Principles of Auger

The basic Auger process involves the production of an atomic inner shell vacancy, usually by electron bombardment, and the decay of the atom from this excited state by an electronic rearrangement and emission of an energetic electron rather than by emission of electromagnetic radiation. For example, as illustrated in Figure 1a, if a Si surface is bombarded by 5-keV electrons, some of the Si atoms will lose electrons from their K shell, whose binding energy is ~1.8 keV. The K shell vacancy will typically be filled by the decay of an electron from one of the L subshells, let's say the $L_{2,3}$ shell, which has a binding energy of 104 eV. This leaves an energy excess of 1.7 keV. This is sometimes relieved by the emission of a 1.7-keV X ray, which is the basis for the EDS and WDS techniques used in the SEM. Most of the time, however, it is relieved by the ejection of another $L_{2,3}$ shell electron that overcomes its 0.1-keV binding energy and carries off the remaining 1.6 keV of energy. This characteristic energy is the basis for the identification of this electron as having come from a Si atom in the sample. This electron is called a Si $KL_{2,3}L_{2,3}$ Auger electron and the process is called a KLL Auger transition. This process leaves the atom with 2 vacancies in the $L_{2,3}$ shell that may further decay by Auger processes involving electrons from the Si M shell, which is also the valence band, and thus these Auger transitions are called LVV transitions. The two valence-band electrons involved in an LVV transition may come from any two energy states in the band, although they will most probably come from near peaks in the valence-band density of states, and thus the shape of the LVV "peak" is derived from a self convolution of the valence-band density of states, and the width of the LVV peak is twice the width of the valence band.

The complete description of the number of Auger electrons that are detected in the energy distribution of electrons coming from a surface under bombardment by a primary electron beam contains many factors. They can be separated into contributions from four basic processes, the creation of inner shell vacancies in atoms of the sample, the emission of electrons as a result of Auger processes resulting from these inner shell vacancies, the transport of those electrons out of the sample, and the detection and measurement of the energy distribution of the electrons coming from the sample.

In fact, Auger electrons are generated in transitions back to the ground state of atoms with inner shell vacancies, no matter what process produced the inner shell vacancy. Auger peaks are therefore observed in electron energy spectra generated by electron excitation, X-ray excitation, and ion excitation, as well as in certain nuclear reactions. The technique usually referred to as Auger electron spectroscopy uses excitation by an electron beam. The spectra produced by X-ray excitation in XPS routinely also include Auger peaks mixed in with the photoelectron peaks. Ion beam-induced Auger peaks occur, at times, during the depth profiling mode of analysis in AES.

Production of Inner Shell Vacancies

The probability (cross section) that a high-energy incident electron will produce a particular inner shell vacancy in a certain element is a function of the ratio of the primary electron energy to the binding energy of the electrons in that shell. In general the cross section rises steeply from 0 at a ratio of 1 to a maximum at a ratio in the range from 3 to 6 and then decreases gradually as the ratio increases further. As an example, the Si K shell binding energy is 1844 eV. To get the maximum yield of Si K shell vacancies, and therefore Si KLL Auger electrons, a primary electron-beam energy of 5.5–11.0 keV should be used. On the other hand if better surface sensitivity is needed (see below) the low-energy Si LVV transition is preferred. The Si L shell binding energies are 154 and 104 eV, so the primary beam energy would be optimized at 0.3–0.9 keV for these transitions.

Auger Electron Emission

Once an inner shell vacancy is created in an atom the atom may then return toward its ground state via emission of a characteristic X ray or through a radiationless Auger transition. The probability of X-ray emission is called the fluorescence yield.

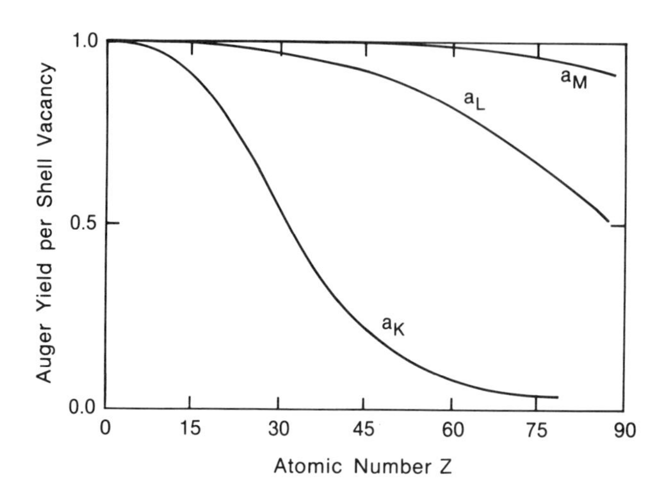

Figure 2 **Percentage of inner shell vacancies resulting in Auger electron emission for holes in the K, L, and M shells.**

The Auger yield is 1 minus the fluorescence yield, since these are the only two options. Figure 2 shows the Auger yield as a function of atomic number for initial vacancies in the K, L, and M shells. It is clear that Auger emission is the preferred decay mechanism for K shell vacancies in the low atomic number elements, and for L and M shell initial vacancies for all elements. By properly selecting the Auger transition to monitor, all elements (except H and He) can be detected using Auger transitions that have a 90% or higher Auger yield per initial vacancy.

Electron Transport to the Surface

As the various electrons, including Auger electrons, resulting from primary electron bombardment diffuse through the sample and to the surface many scattering events occur. The inelastic collisions have the effect of smoothing the energy distribution of these electrons and result in a power law energy distribution[4] at energies between the elastic peak and the "true secondary" peak, which occur at the high-energy and low-energy end of the distribution, respectively. This produces a background, as shown in Figure 1b, on which the Auger peaks are superimposed, that can be modeled and removed (see below). Inelastic collisions also have the effect of removing some of the Auger electrons from their characteristic energy position in an Auger peak and transferring them to lower energies as part of the "loss tail," which starts at the low-energy side of the Auger peak and extends all the way to zero energy.

The inelastic collision process is characterized by an inelastic mean free path, which is the distance traveled after which only $1/e$ of the Auger electrons maintain their initial energy. This is very important because only the electrons that escape the sample with their characteristic Auger energy are useful in identifying the atoms in the sample. This process gives the technique its surface specificity. This inelastic mean free path is a function, primarily, of the energy of the electron and, secondarily, of the material through which the electron is traveling. Figure 6 in the XPS article shows many measurements of the inelastic mean free path in various materials and over a wide range of energies, and an estimate of a universal (valid for all materials) inelastic mean free path curve versus energy.

The minimum in the mean free path curve, at around 80 eV, is the energy at which electrons travel the shortest distance before suffering an energy-altering scattering event. Thus Auger electrons that happen to have their energy in this vicinity will be those that will have the thinnest sampling depth at the surface. For example, while Si LVV Auger electrons from oxidized Si (at approximately 78-eV) are generated at depths ranging from the top monolayer to nearly a μm from a primary electron beam with a typical 5-keV energy, 63% of the electrons that escape without losing any energy come from the top 5 Å of the sample. Furthermore, 87% are contributed by the top 10 Å of the sample and 95% have been produced in the top 15 Å of material. The depth from which there is no longer any signal contribution is ultimately determined by the signal-to-noise ratio in the measured spectrum. If a 5% signal variation is accurately measurable then atoms 3 mean free paths down contribute to the measurement. If 2% of the signal is well above the noise level then atoms at a depth of 4 mean free paths contribute to the measurement.

Secondary Electron Collection

As the electrons leave the surface they move in a cosine-shaped intensity distribution away from the analysis point and travel in straight lines until they enter the energy analyzer. The entrance slit of the energy analyzer determines the percentage that are collected, but it is typically just under 20% for the most commonly used energy analyzer, the cylindrical mirror analyzer (CMA). Once in the energy analyzer more electrons are lost by scattering at grids and the CMA transmission is typically 60%.

Information in Auger Spectra

Using the best procedures during data acquisition produces spectra with the maximum available information content. Once spectra are recorded that contain the information that is sought using the best procedures for extracting the information from the data is important to maximize the value of the analysis. This section will consider the procedures for data acquisition and the extraction of various types of information available from the data.

Data Acquisition

For primarily historical reasons people have come to consider Auger spectra as having the form, $dN(E)/dE$ versus E, where $N(E)$ is the energy distribution of the sec-

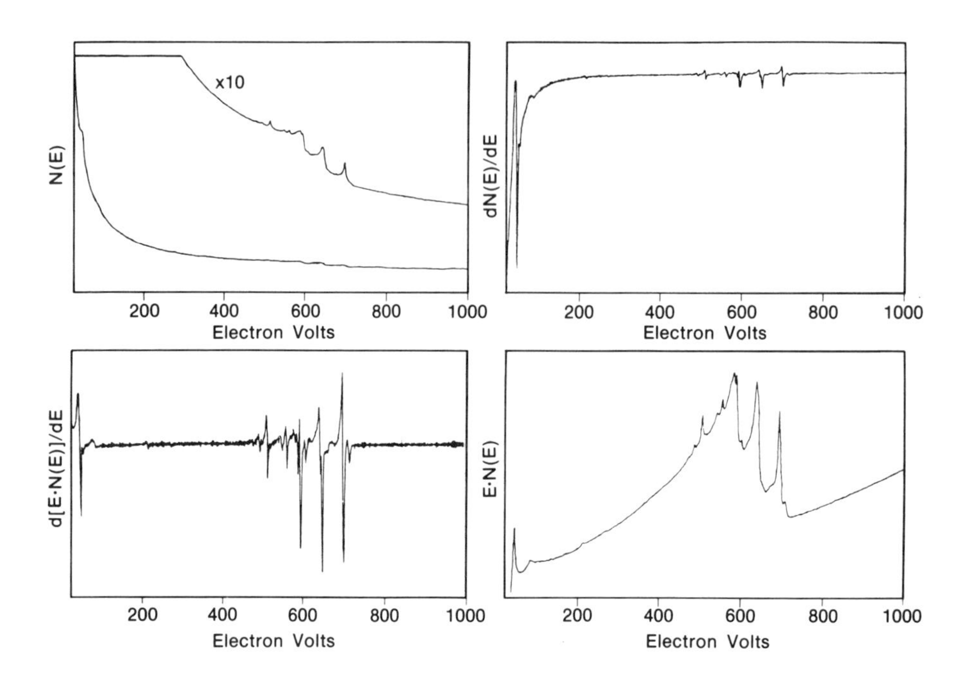

Figure 3 **The *N(E)*, *dN(E)/dE*, *dEN(E/dE*, and *EN(E)* forms of secondary electron energy spectra from a slightly contaminated Fe surface.**

ondary electrons being detected and E is their energy. This came about because of the properties of various energy analyzers used and because of peculiarities of the analog electronics used to run them. Spectra in this form were acquired by adding an AC component to the energy-selecting voltage of the energy analyzers (a modulation) and detecting the signal with a lock-in amplifier.[5] This led to the signal being acquired in the differential mode, $dN(E)/dE$ versus E, instead of $N(E)$ versus E. These forms of acquired spectra are shown in Figure 3. With the advent of the CMA and computer-controlled digital signal acquisition, which can be coupled with either pulse counting or voltage-to-frequency conversion for decoupling the signal from the high positive collection voltage, it has finally become practical to discard the modulation and the lock-in amplifier in signal acquisition, as is done in Figure 3(bottom right panel). Acquiring data directly in $N(E)$ (or $E \times N(E)$) form, followed by subsequent mathematical processing, provides six valuable advantages:

1 There is an improved signal-to-noise ratio in the raw data. This can be seen in the $E \times N(E)$ form of data in Figure 3.

2 The energy analyzer is always operated at its best energy resolution.

3 The measured Auger signal is proportional to the number of atoms sampled. In the derivative mode of data acquisition this is frequently not the case, for example, if an inappropriate modulation voltage is used or if the line shape has

changed due to a change in chemical environment.

4 The physical information in the line shape is immediately available for observation.

5 Peak overlaps can be eliminated simply by peak fitting and subtraction.

6 Loss tail analysis can be applied to the data. (This procedure is discussed below.)

Thus it is best to acquire and store the data in the simplest and least-processed form possible.

Extracting Information From the Data

There are at least four kinds of information available from an Auger spectrum. The simplest and by far most frequently used is qualitative information, indicating which elements are present within the sampling volume of the measurement. Next there is quantitative information, which requires a little more care during acquisition to make it extractable, and a little more effort to extract it, but which tells how much of each of the elements is present. Third, there is chemical information which shows the chemical state in which these elements are present. Last, but by far the least used, there is information on the electronic structure of the material, such as the valance-band density of states that is folded into the line shape of transitions involving valance-band electrons. There are considerations to keep in mind in extracting each of these kinds of information.

Qualitative Information

Qualitative information can be extracted from Auger spectra quite simply, by a trained eye or by reference to one of the available Auger charts, tables of energies, or handbooks of spectra. The most basic identification is done from the energies of the major peaks in the spectrum. The next level of filtration is done from the peak intensity ratios in the patterns of peaks in the spectra of the elements present. One of the charts of Auger peak energies available is shown in Figure 4. The useful Auger spectra of the elements fall into groups according to the transition type, KLL, LMM, MNN, etc. If you look across the chart, following a given energy, it is clear that there are many possibilities for intermixing of patterns from different elements, but there are few direct peak overlaps. Generally, if there are peaks from two elements that interfere, there are other peaks from both those elements that do not overlap. One of the most difficult exceptions to this rule is in the case of B and Cl: B has only one peak, a KLL peak at 180 eV. Cl has an LMM peak at 180 eV and its KLL peaks are at 2200–2400 eV, high enough that they are seldom recorded. If there is a real uncertainty as to which of these elements is present, it is necessary to look for the latter peaks.

Peak overlaps that totally obscure one of the elements in the spectrum have been shown to be separable.[6] A Co–Ni alloy film under a Cu film is a combination that produces a spectrum where the Ni peaks are all overlapped by Cu or Co peaks, or

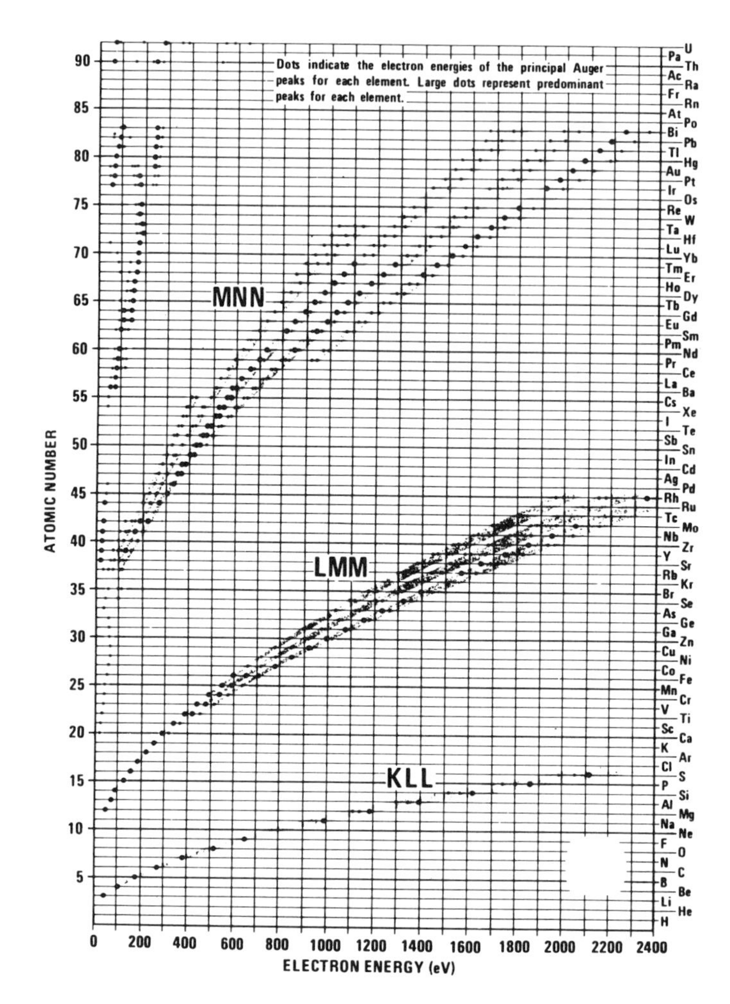

Figure 4 **One of the numerous available charts of Auger electron energies of the elements.**

both. The intensities in the Cu and Co patterns show that another element is present. With the use of background subtraction, standard spectra, and peak fitting and subtraction, the Ni spectrum was uncovered and identified, and even quantitative information, with identified accuracy limitations, was obtained.

When listing the elements present from qualitative analysis, the issues of sensitivity and signal-to-noise level arise. The minimum amount of an element that must be present to be detected in an Auger spectrum is a function of a number of variables. Some of these are determined by the element, such as its ionization cross section at the primary energy being used, the Auger yield from its most prolific inner shell vacancy, the energy of its Auger electron (since this determines the electrons' mean free path for escape from the solid), etc. Other variables are under the control of the measurement parameters, such as the primary beam energy and current, the energy resolution of the energy analyzer, the angle of incidence of the primary beam onto the sample and the acceptance angle of the energy analyzer. These variables can, to a certain degree, be controlled to yield the maximum signal-to-noise ratio for the element of interest. When these parameters are optimized the detection limit for most elements is on the order of a few times $10^{18}/cm^3$ homogeneously distributed, or about 1 atom in 10,000.

Quantitative Information

The number of Auger electrons from a particular element emitted from a volume of material under electron bombardment is proportional to the number of atoms of that element in the volume. However it is seldom possible to make a basic, first principles calculation of the concentration of a particular species from an Auger spectrum. Instead, sensitivity factors are used to account for the unknown parameters in the measurement and applied to the signals of all of the species present which are then summed and each divided by the total to calculate the relative atomic percentages present.

Of the total number of Auger electrons emitted only a fraction escapes the sample without energy loss. The rest become part of the loss tail on the low-energy side of the Auger peak extending to zero energy and contribute to the background under all of the lower energy Auger peaks in the spectrum. This process must be taken into account when using a sensitivity factor for a particular Auger system. Sensitivity factors are usually taken from pure elemental samples or pure compound samples. This means that the element is homogeneously distributed in the standard. If this is not true in the unknown sample, the percentage of Auger electrons that escape the sample without energy loss changes. If the element is concentrated at the surface, fewer Auger electrons will suffer energy loss; if it is concentrated in a layer beneath another film, more Auger electrons will suffer energy loss before they escape the sample. This can be seen in Figure 5, which shows oxygen in a homogeneous SiO_2 film, in a surface oxide on Si, and from an SiO_2 film under a layer of Si. An oxygen sensitivity factor determined from a homogeneous sample would not properly represent the oxygen concentration in the lower two spectra of Figure 5.

Sensitivity factors should be measured on the same energy analyzer, at the same energy resolution, at the same primary electron beam energy, and at the same sample orientation to the electron beam and energy analyzer, as the spectra to which they are applied. Only when these precautions are taken can any sort of quantitative accuracy be expected. Even with these precautions the oxygen example discussed above and shown in Figure 5 would present a problem. The most direct way to prevent this problem is by the process referred to above as "loss tail analysis." This involves comparing the ratios of the peak heights to the loss tail heights, on background subtracted spectra, from the spectrum of the unknown sample and the

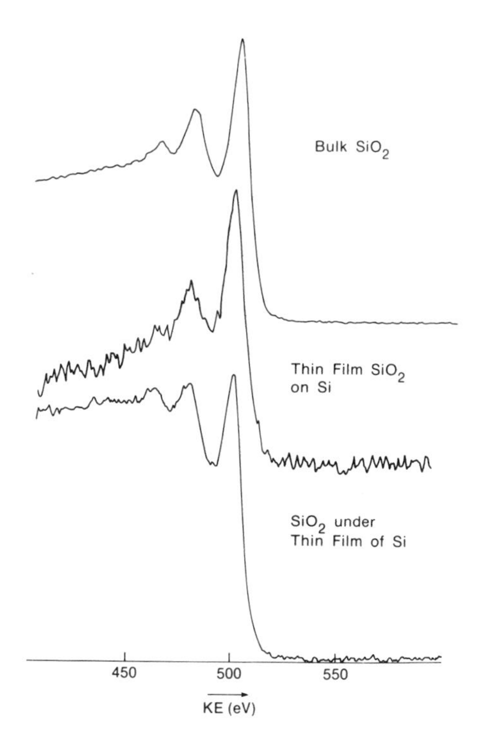

Figure 5 **Oxygen spectra from bulk SiO_2, a thin film of SiO_2 on Si, and SiO_2 under a thin film of Si. These spectra have had their background removed, and so the loss tail can be seen as the height of the spectra at energies below the peaks.**

spectrum from which the sensitivity factor was determined. When these ratios are equal the same degree of depth homogeneity of the element in question is assured.

Chemical Information

There is a great deal of chemical information in the line shapes and chemical shifts of peaks in Auger spectra. XPS is generally considered to be a more appropriate tool to determine chemistry in a sample. It is true that the photoelectron lines used in XPS are typically narrower and that therefore smaller chemically induced energy shifts can be detected. Moreover, the energy analyzers used in XPS often have better energy resolution. However, it is also true that the chemically induced energy shifts in Auger peaks are usually larger than the corresponding shifts in photoelectron peaks.[7]

Chemical information is present in Auger spectra in two forms; a shift in the energy of the peak maximum and sometimes as a change in the line shape of the Auger peak. Line shape changes are greatest in transitions involving valance-band electrons, such as the LVV transition in Si. Since this line shape is just a weighted

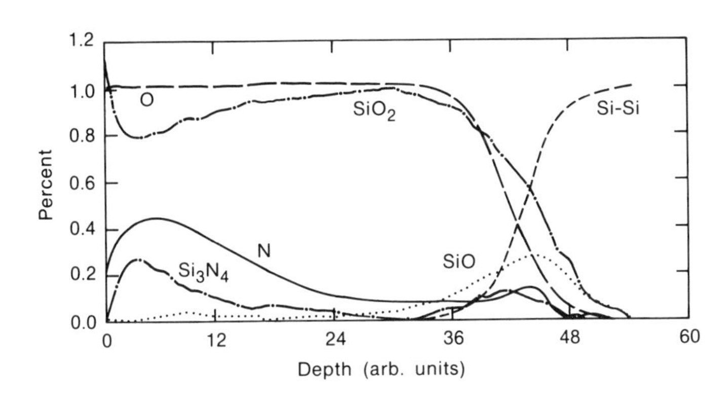

Figure 6 **Depth profile of an SiO_2 film that had been nitrided by exposure to ammonia. The N and O profiles are shown, along with curves of the percentages of the Si present that is bonded as SiO_2, Si_3N_4, "SiO," and in Si–Si bonds.**

self convolution of the valance-band density of states, the line shape varies considerably among spectra originating from Si atoms bound to other Si atoms, Si atoms bound to oxygen atoms, Si atoms bound to nitrogen atoms, etc. The Si KLL spectra are also sensitive to this chemistry, but they manifest it primarily in energy shifts; 7 eV between Si–Si bonding and Si–O bonding, and smaller shifts from the Si–Si peak for other bonding. The KLL spectra also differ in loss tail heights and in some of the plasma loss peaks that are sometimes present.

As an example of the use of AES to obtain chemical, as well as elemental, information, the depth profiling of a nitrided silicon dioxide layer on a silicon substrate is shown in Figure 6. Using the linearized secondary electron cascade background subtraction technique[4] and peak fitting of chemical line shape standards, the chemistry in the depth profile of the nitrided silicon dioxide layer was determined and is shown in Figure 6. This profile includes information on the percentage of the Si atoms that are bound in each of the chemistries present as a function of the depth in the film.

Methods for Surface and Thin-Film Characterization

AES analysis is done in one of four modes of analysis. The simplest, most direct, and most often used mode of operation of an Auger spectrometer is the point analysis mode, in which the primary electron beam is positioned on the area of interest on the sample and an Auger survey spectrum is taken. The next most often used mode of analysis is the depth profiling mode. The additional feature in this mode is that an ion beam is directed onto the same area that is being Auger analyzed. The ion beam sputters material off the surface so that the analysis measures the variation, in depth, of the composition of the new surfaces, which are being continuously uncovered. In this mode the Auger data may be acquired in the same, wide energy sweep survey spectrum as in a point analysis, or it may be taken in any number of narrow energy scan windows whose energies are selected to monitor Auger peaks of particular elements of interest.

The results shown in Figure 6 above are an example of this mode of analysis, but include additional information on the chemical states of the Si. The third most frequently used mode of analysis is the Auger mapping mode, in which an Auger peak of a particular element is monitored while the primary electron beam is raster scanned over an area. This mode determines the spatial distribution, across the surface, of the element of interest, rather than in depth, as depth profiling does. Of course, the second and third modes can be combined to produce a three-dimensional spatial distribution of the element. The fourth operational mode is just a subset of the third mode; a line scan of the primary beam is done across a region of interest, instead of rastering over an area.

Artifacts That Require Caution

Many artifacts may be present in Auger spectra. Some are caused by the primary electrons interacting with the sample in ways other than creating inner shell vacancies. This can result in removal of species from the surface, through processes like electron-stimulated desorption, or in peaks in the energy distributions that look like Auger peaks but that are photoelectron peaks, ionization loss peaks, or peaks from other processes. Some artifacts are caused by the secondary electrons interacting with the sample on their way out. This can produce peaks due to plasmon excitation processes or can change the detected peak intensities via diffraction processes. During depth profiling, some peaks are caused by the ion-beam interacting with the sample in ways other than simply uniformly removing material. Crystallographic and shadowing effects can produce roughness that increases as depth profiling proceeds and increasingly degrades depth resolution in the profile. Certain ion-beam conditions in combination with certain materials produce ion-induced Auger peaks that can interfere with quantitative accuracy. Variation of sputtering yields among the elements on a surface can artificially change the surface composition.

Conclusions

AES is an important, widely used technique for surface, interface, and thin-film analysis of materials not strongly affected by electron beams. It continues to be improved through advancements in both systems and technique. Higher spatial resolution hardware continues to be developed, along with more rapid data acquisition and processing. Quantitative accuracy is benefiting from improved understanding. Areas of application are expanding through the study of new materials problems and also by new applications in low-energy electron microscopy and in measurement of surface atom geometries by observing shadowing and diffraction effects on angular distributions of Auger electrons leaving surfaces (AED).

Related Articles in the Encyclopedia

XPS, XPD/AED, SEM, EDS, EPMA, SIMS, and RBS

References

1 P. Auger. *Compt. Rend.* **177,** 169, 1923; *ibid.* **180,** 65, 1925.

2 J. J. Lander. *Phys. Rev.* **91,** 1382, 1953.

3 L. A. Harris. *J. Appl. Phys.* **39,** 3; *ibid.* 1419, 1968.

4 E. N. Sickafus. *Phys. Rev. B.* **16,** 1436, 1977; ibid. **16,** 1448, 1977; and E. N. Sickafus and C. Kukla. *Phys. Rev. B* **19,** 4056, 1979.

5 M. P. Seah. In *Methods of Surface Analysis.* (J. M. Walls, Ed.) Cambridge University Press, 1989.

6 Y. E. Strausser, D. Franklin and P. Courtney. *Thin Solid Films.* **84,** 145, 1981.

7 C. D. Wagner. *J. Elect. Spect. Related Phen.* **10,** 305, 1977.

5.4 REELS

Reflected Electron Energy-loss Spectroscopy

ALBERT J. BEVOLO

Contents

Introduction

Reflected Electron Energy-Loss Spectroscopy (REELS) has elemental sensitivities on the order of a few tenths of a percent, phase discrimination at the few-percent level, operator controllable depth resolution from several nm to 0.07 nm, and a lateral resolution as low as 100 nm.

REELS can detect any element from hydrogen to uranium and can discriminate between various phases,[1] such as SnO and SnO_2, or diamond and graphite. By varying the primary electron beam energy E_0, the probing depth can be varied from a minimum of about 0.07 nm to a maximum of 10 nm, where these limits are somewhat sample dependent. The best probing depth is at least twice as good as any other surface technique except ISS, to which it compares favorably with the added advantage of a spatial resolution of a few microns. The lateral resolution is limited only by technological factors that involve producing small electron beam spot sizes at energies below 3 keV, rather than fundamental beam–solid interactions like rediffused primary electrons that limit the lateral resolution of SAM, EDS, and SEM techniques.[2]

The principal applications of REELS are thin-film growth studies and gas–surface reactions in the few-monolayer regime when chemical state information is required. In its high spatial resolution mode it has been used to detect submicron metal hydride phases and to characterize surface segregation and diffusion as a function of grain boundary orientation. REELS is not nearly as commonly used as AES or XPS.

Basic Principles

It is a fundamental principle of quantum mechanics that electrons bound in an atom can have only discrete energy values. Thus, when an electron strikes an atom its electrons can absorb energy from the incident electron in specific, discrete amounts. As a result the scattered incident electron can lose energy only in specific amounts. In EELS an incident electron beam of energy E_0 bombards an atom or collection of atoms. After the interaction the energy loss E of the scattered electron beam is measured. Since the electronic energy states of different elements, and of a single element in different chemical environments, are unique, the emitted beam will contain information about the composition and chemistry of the specimen.

EELS is an electron-in–electron-out technique that has two forms: The emitted electrons can be analyzed after transmission through very thin (≤ 100 nm) specimens or they can be analyzed after reflection from thick specimens. For samples thinned to 100 nm the transmission mode of EELS yields a lateral resolution of a few nm, but for specimens used in the reflected mode the best lateral resolution (as of this writing) is 100 nm. Transmission electron energy-loss spectra are obtained on STEM or TEM instruments and are covered in Chapter 3. Within the reflected mode there are two major versions distinguished by their energy resolution. The high-energy resolution EELS (HREELS) has a resolution in the meV range, suitable for molecular vibrational excitations and is covered in Chapter 8. The low-energy resolution reflected EELS (REELS) has a typical resolution of 1 eV, sufficient to resolve electronic excitations like plasmons, interband transitions, or core-level excitations. REELS currently has a lateral resolution of 100 nm, while HREELS has a resolution in the mm range. HREELS and REELS, because of their high surface sensitivity, require ultrahigh vacuum, while transmission EELS requires only high vacuum. Only REELS and transmission EELS exhibit extended fine structure suitable for atom position determinations. This article considers only REELS.

Consider Figure 1a, which shows the electronic energy states of a solid having broadened valence and conduction bands as well as sharp core-level states X, Y, and Z. An incoming electron with energy E_0 may excite an electron from any occupied state to any unoccupied state, where the Fermi energy E_F separates the two

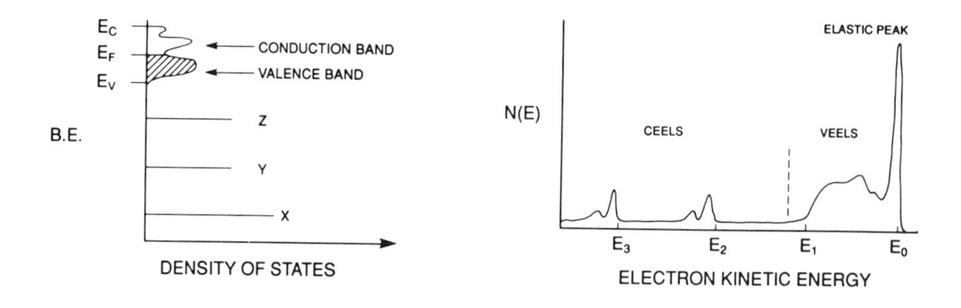

Figure 1 **Representation of a typical density of electron states for a metal having *X, Y,* and *Z* core levels (top); and REELS spectrum expected from metal shown in top panel (bottom).**

types of electronic states: If $E \leq E_F$ then that state is occupied (e.g., core levels or the valence band); if $E \geq E_F$ then that state is unoccupied (e.g., conduction-band states). The energy range over which a solid can absorb energy is the convolution of the energy spread of the initial, occupied state with that of the final, unoccupied state. For both interband transitions, defined as valence-to-conduction band excitations and for core-level transitions, defined as core-level-to-conduction band transitions, the final state is the relatively broad conduction band. Since core-level states are narrow, the line shape of the energy-loss spectra after a core-level excitation reflects the conduction-band density of states. Each element in the solid, chosen by virtue of the core level involved, can be probed for chemical state information much like AES, except AES probes the occupied valence-band density of states while core-level REELS (CEELS) probes the unoccupied conduction-band density of states. Peaks can occur in CEELS over the whole range of energy below E_0. On the other hand, for an interband transition the maximum electron energy loss is given by the energy difference between the bottom of the valence band and the top of the conduction band, which for most materials is 10–40 eV. For metals the minimum energy loss can be as low as zero while for semiconductors and insulators the minimum energy loss is the band gap energy. Since both the initial and final states of an interband transition are involved in chemical bonding, it is expected that the interband REELS spectra will be very sensitive to chemical changes. However, because all states in the conduction and valence bands are strongly mixed, interband transitions cannot be identified easily with a particular element in the solid, as can be done for CEELS. This global character of interband transitions is the same as for valence band XPS, UPS, or optical absorption spectra.

Valence electrons also can be excited by interacting with the electron beam to produce a collective, longitudinal charge density oscillation called a plasmon. Plasmons can exist only in solids and liquids, and not in gases because they require electronic states with a strong overlap between atoms. Even insulators can exhibit plasmons, because plasmons do not require electrons at E_F. Plasmon energies range from a few eV to about 35 eV, with most in the range 10–25 eV. In a free-electron metal, the plasmon energy is proportional to the square root of the electron density and so is relatively insensitive to chemical changes. Three-dimensional oscillations within a solid are called bulk plasmons, while two-dimensional oscillations on the surface are called surface plasmons.

Suppose a solid with an energy level scheme as in Figure 1 is bombarded by an electron beam of energy E_0 where $|E_y| \leq E_0 \leq |E_x|$ and E_x (E_y) is the binding energy of the core level $X(Y)$. Most of the incident electrons are reflected from the sample surface without energy loss and produce a large peak at E_0 called the *elastic* peak. The incident electrons that scatter from the various occupied states form the REELS spectra shown in Figure 1b. Peaks at energy E_0–E_y and E_0–E_x are due to CEELS excitation, their line shapes reflect the conduction-band density of states. Since the transitions occur in the presence of the empty core level, the line shape in reality reflects the conduction-band density of states in the presence of the core hole. Such a density of states may not be the same as the ground-state density of states that controls the chemical properties of the material, but changes in chemical environment will still result in changes in the excited states. Since the interband and plasmon region involves valence electrons, it is called the valence EELS (VEELS), which with CEELS constitutes a REELS spectrum.

Because both plasmons and interband transitions involve valence electrons, sum rules couple their relative intensities and energies in complex ways. If there are sharp, intense peaks in the valence and conduction-band density of states, then the energy of most interband peaks are well defined and very intense. Such is the case for the 3d-, 4d- and 5d-transition metals and the rare earths, with their highly localized d- and f-bands in both valence and conduction bands. Because interband transitions act to dampen the plasmon oscillation they can change the intensity and energy of a plasmon peak if the chemical environment has changed, even if the electron density does not change. Such effects are much less evident for the free electron-like metals, such as Al, Sn and Mg, where VEELS spectra are dominated by the plasmon peaks. An excellent discussion of the effect was given some time ago by C. Powell[3] and should be consulted carefully before interpreting plasmon energy shifts purely on the basis of electron density changes.

Common Modes of Analysis and Examples

Perhaps the most common use for REELS is to monitor gas–solid reactions that produce surface films at a total coverage of less than a few monolayers. When E_0 is a few hundred eV, the surface sensitivity of REELS is such that over 90% of the signal originates in the topmost monolayer of the sample. A particularly powerful application in this case involves the determination of whether a single phase of variable composition occurs on the top layer or whether islands occur; that is, whether two surface phases are present simultaneously. In the former case the plasmon peak of the substrate will remain as one peak but shift in energy with coverage, while in the latter case a new peak from the island phase will appear in addition to the substrate peak. The substrate peak will decrease in intensity with increased coverage by the island phase but it will not shift in energy, so that the growth of the island phase can be monitored even if the islands have lateral dimensions much smaller than the incident beam spot size.

The degree of surface cleanliness or even ordering can be determined by REELS, especially from the intense VEELS signals. The relative intensity of the surface and bulk plasmon peaks is often more sensitive to surface contamination than AES, especially for elements like Al, which have intense plasmon peaks. Semiconductor surfaces often have surface states due to dangling bonds that are unique to each crystal orientation, which have been used in the case of Si and GaAs to follow *in situ* the formation of metal contacts and to resolve such issues as Fermi-level pinning and its role in Schottky barrier heights.

Fine structure extending several hundred eV in kinetic energy below a CEELS peak, analogous to EXAFS, have been observed in REELS. Bond lengths of adsorbed species can be determined from Surface Electron Energy-Loss Fine Structure (SEELFS)[4] using a modified EXAFS formalism.

Analysis of CEELS[5] line shapes often show chemical shifts that have been used to study FeB alloys after recrystallization, C–H bonding in diamondlike films and multiple oxidation states.

With the advent of SAM instruments it soon was shown that they could be operated as REELS-mapping microprobes using a technique called Reflected Electron Energy-Loss Microscopy (REELM). The strong VEELS signals can compensate for the reduced currents required to maintain E_0 below the pass energy of a CMA, e.g., 3 keV. As a result, maps of very high contrast can be produced in just a few minutes, or maps with a lateral resolution of 100 nm can be produced by further reducing the electron current. If E_0 is set to a few hundred eV, to optimize the surface sensitivity, modern SAM instruments can produce spot sizes of a few microns sufficient to generate good REELM images.

Figure 2 shows SEM and REELM micrographs of a sample containing ScH_2 and Sc(H), the solid solution of hydrogen in scandium. Since SEM only reveals topographic information and not composition, it is not possible to distinguish between these two phases. AES cannot distinguish ScH_2 from Sc(H). Only VEELS spectra for ScH_2 and Sc(H) are sufficiently different to map the true location of ScH_2 (dark areas of the REELM image) and Sc(H) (bright areas of the REELM image). REELM is the technique of choice for the detection of metal hydrides in bulk specimens at a lateral resolution of 100 nm.

Other applications of REELM include monitoring variations like oxidation, segregation, and hydration in the surface chemistry of polycrystalline materials.[6–9] Differences of 1/10 of a monolayer in oxygen coverage due to variations in grain

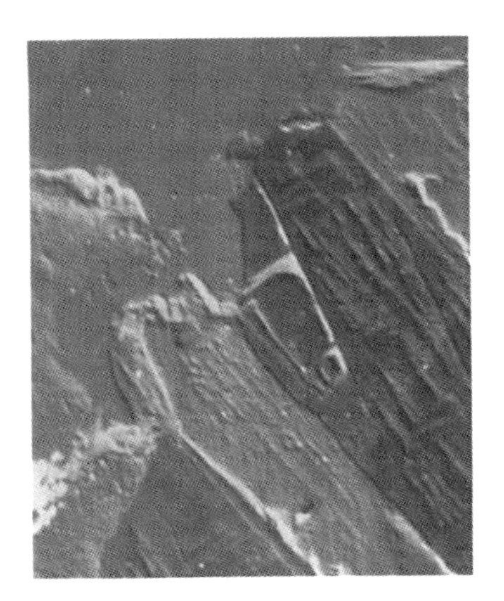

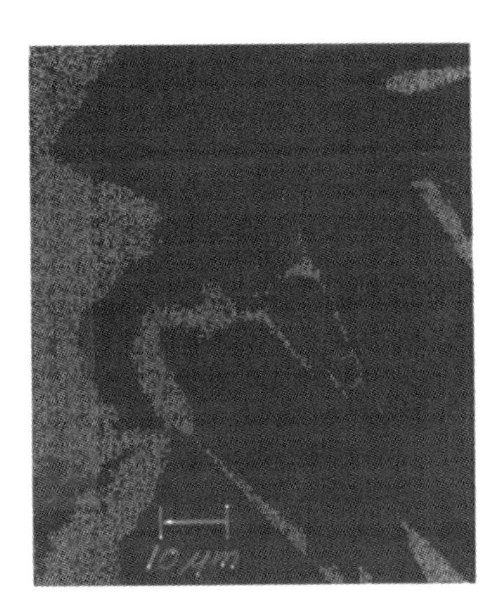

Figure 2 SEM (left) and REELM (right) micrographs of a hydrogenated scandium sample. Only the REELM image correctly identifies the scandium solid solution phase (bright) in the presence of the scandium dihydride phase (black).

boundary orientation can be displayed in high-contrast REELM images with a lateral resolution of about 1 μm.

Sample Requirements

Samples used in REELS must be ultrahigh-vacuum compatible solids or liquids, but they may be metals, semiconductors or insulators. Because REELS detects a reflected primary electron, rather than a secondary electron like an Auger electron, surface charging will not affect the electron's detected kinetic energy. As a result, insulating surfaces, even if charged, will generate good REELS signals. To avoid severe charging from the much larger number of secondary electrons it is sufficient to make the flat areas of an insulator about the same size as the incident beam spot size. By adjusting the primary beam energy and angle of incidence, zero absorbed current can be obtained.

Artifacts

VEELS spectra are limited in practice to the relatively narrow energy range of about 30 eV over which plasmons or interband transitions can occur. In contrast to AES, XPS, or even CEELS, where excitations can occur over hundreds of eV, the probability of spectral overlap is much higher for VEELS. It is fortunate that most

REELS spectrometers are in fact Auger spectrometers, so that elemental identifications can be made easily.

REELS data are commonly displayed as *N*(*E*), *dN*/ *dE* or second derivatives of *N*(*E*). The *N*(*E*) mode has the advantage that the background is not lost, as it is for either of the derivative modes, but the relatively weak CEELS signals are usually dwarfed by the background and so require some level of differentiation to enhance the weak, but sharp, CEELS features. However, the signal-to-noise ratio is degraded by successive differentiations so that the ultimate detectability is worsened. REELS spectra acquired by lock-in detectors can naturally produce either the first- or second-derivative spectra, while those with *N*(*E*) outputs usually have provisions to mathematically produce the derivative format. For the strong VEELS signals, the second derivative has the advantage that the peaks occur at the same energy as they do in the *N*(*E*) spectra, while those from the first derivative do not. However, a closely spaced, intense pair of *N*(*E*) excitations will appear as three peaks in the second-derivative mode. It is the author's judgment that the best overall display mode is the first derivative.

Not only do the new and old surfaces produce surface plasmons in the island-growth mode, but the interface between the growing film and the substrate is also capable of producing an interphase plasmon excitation. Typically an interphase plasmon will appear at an energy intermediate between the surface plasmons of the two phases. Its intensity will grow as the island phase grows laterally but will eventually disappear as the interface retreats below the thickening island layer.

Sometimes it is possible to distinguish surface and bulk plasmons by lowering E_0 so that the bulk plasmon will decrease in intensity more rapidly than the surface plasmon. However both surface states and interband transitions can show the same behavior.

Instrumentation

An Auger spectrometer or scanning Auger microprobe can be operated as a REELS spectrometer or Reflected EELS Microprobe (REELM) instrument at no additional cost in hardware or software. In contrast to AES, REELS requires that the incident electron beam energy E_0 be less than the pass energy of the analyzer, usually less than 3 keV. Also, to achieve a reasonable energy resolution, REELS must have E_0 less than about 500–1000 eV for the Cylindrical Mirror Analyzers (CMA) typical of most AES instruments. Incident electron beams with E_0 in this range have considerably larger spot sizes or lower currents than those of the 5–20 keV beams used in AES. Electronic processes such as core-level excitations, plasmons, and interband transitions have energy widths of the order of 1 eV. Because deviations from E_0 produce chromatic aberrations in the focusing of fine-spot electron

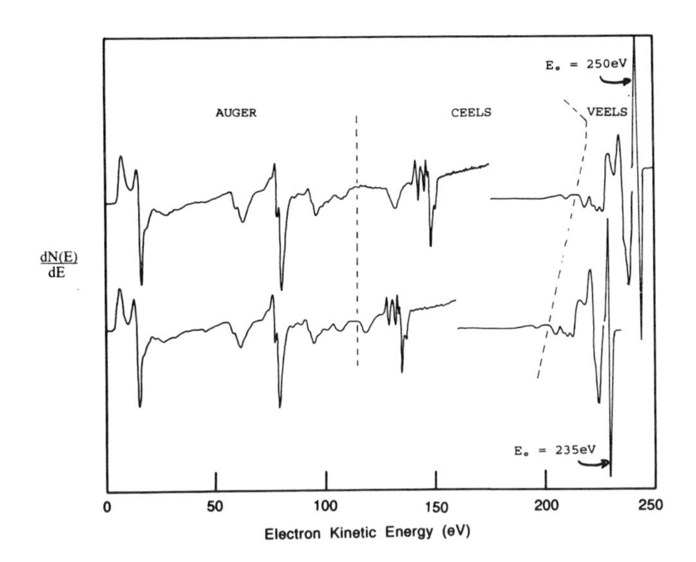

Figure 3 **First-derivative electron emission spectra from pure lanthanum taken with primary electron beams having energies of 250 and 235 eV showing the unshifted Auger peaks and the shifted REELS peaks.**

beams any beam with a spot size of 100 μm or less is sufficiently monoenergetic for REELS.[10]

Comparison With Other Techniques

In addition to reflected primary electrons there are three other types of emitted electrons: true secondaries, Auger electrons, and back scattered electrons. True secondaries are valence electrons (see Figure 1) emitted into a very intense narrow peak a few eV in kinetic energy, independent of E_0 or material composition. They are used to form SEM images that reveal the topography of the surface. Auger electrons are also fixed in energy independent of E_0, but occur over a wide energy range that can overlap the CEELS spectrum. An Auger peak and a CEELS peak can be distinguished by changing E_0 slightly, say, by 10 eV. Any peak that moves by the same 10 eV must be a CEELS peak and any peak that does not is an Auger peak. This effect is illustrated in Figure 3. Finally, backscattered electrons are all those electrons that are emitted following multiple inelastic collisions, and they form a relatively smooth background that depends on the angle of incidence of the primary beam and the average atomic number Z of the sample, but less so on E_0 .

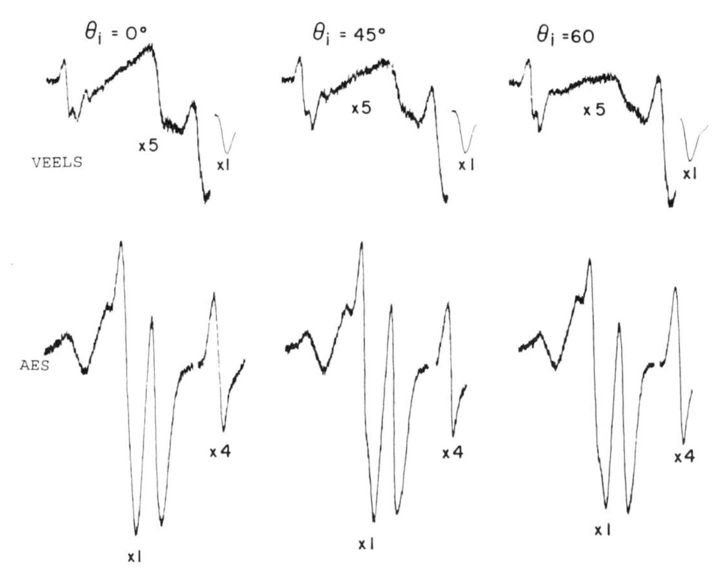

Figure 4 **VEELS and Auger spectra for tilt angles of 0°, 45°, and 60° taken from a tin sample covered by a 0.5-nm oxide layer. The doublet AES peaks are the Sn (410) peaks while the singlet AES peak is the O (510) taken with the same gain. VEELS peaks are oxide related, while the Sn (410) peak is due primarily to the metallic tin beneath the oxide, illustrating the superior depth resolution of VEELS.**

In VEELS, because E_0 and $E_0 - E$ are nearly the same, both can be tuned to the minimum in the inelastic mean free path near 200 eV, and it is then possible to obtain probing depths such that 90% of the signal comes from the top monolayer at high angles of incidence. In AES, E_0 is typically much higher, so the penetration depth of the incident beam is very large compared to the Auger escape depth. As a result tilting the specimen has little effect and at most 50% of the Auger signal comes from the top monolayer.

An example of the superior surface sensitivity of REELS compared to AES is shown in Figure 4, where $E_0 = 75$ eV for the VEELS spectra, and $E_0 = 3$ keV for the AES spectra. Both sets of first-derivative data were taken as a function of Θ_i from a sample of pure tin that had been oxidized to a thickness of 0.50 nm. The two AES spectra at each tilt angle represent the Sn (410) and O (510) AES spectra. All of the VEELS spectra (even at 0 tilt angle) are dominated by oxide-derived features, while the Sn (410) Auger peak, even at $\Theta_i = 60°$, is dominated by the metallic substrate. This work is an example of one of the most common uses for REELS, namely, investigations of the very earliest stages of gas–solid interactions. One final note, VEELS was able to distinguish SnO from SnO_2 because of their different plasmon energies but AES could not because there was no difference in the net core-level energy shift.[1]

Except for hydrogen, both SAM and EDS can be used to map elemental distributions in addition to REELM. However the role of rediffused primary electrons, present in bulk specimens thicker than 100 nm, must be understood. For both SAM and EDS, $E_0 \geq 10$ keV, so that for a specimen with an atomic number above twenty the incident electrons will diffuse inside the solid over distances of about 1 μm, even if the incident beam is smaller than 1 nm. These electrons can generate X rays within the specimen and Auger electrons at the surface of the specimen over distances of 1 μm. As a result some fraction (about 20 – 50% for AES and about 90% for EDS) of the signal comes from portions of the specimen not directly irradiated by the incident beam. By contrast, VEELS spectra are taken with $E_0 \leq 2$ keV and have peaks within 50 eV of E_0. Even though the number of back scattered electrons is still high at 2 keV (compared to 10 keV) their lateral distribution is much smaller. More importantly, any back scattered electron with an energy less than E_0- 50 eV (which is nearly all of them) cannot produce a VEELS peak within 50 eV of E_0. As a result, REELM has a lateral resolution independent of the back scattered electrons, while SAM and EDS have lateral resolutions limited by fundamental beam–solid interactions. As a consequence only the lateral resolution of REELM will benefit from any future reduction in electron beam spot sizes.

Conclusions

REELS will continue to be an important surface analytical tool having special features, such as very high surface sensitivity over lateral distances of the order of a few μm and a lateral resolution that is uniquely immune from back scattered electron effects that degrade the lateral resolution of SAM, SEM and EDS. Its universal availability on all types of electron-excited Auger spectrometers is appealing. However in its high-intensity VEELS-form spectral overlap problems prevent widespread application of REELS.

Future trends will include studies of grain-dependent surface adsorption phenomena, such as gas–solid reactions and surface segregation. More frequent use of the element-specific CEELS version of REELM to complement SAM in probing the conduction-band density of states should occur. As commercially available SAM instruments improve their spot sizes, especially at low E_0 with field emission sources, REELM will be possible at lateral resolutions approaching 10 nm without back scattered electron problems.

Related Articles in the Encyclopedia

AES, EDS, HREELS, SEM, TEM, STEM, XPS, EPMA

References

1 A. J. Bevolo, J. D. Verhoeven, and M. Noack. *Surf. Sci.* **134,** 499, 1983. Comparison of VEELS and AES analysis of the early stages of the oxidation of solid and liquid tin. Illustrates one of the main uses for REELS.

2 A. J. Bevolo. *Scanning Electron Microscopy.* 1985, vol. 4, p. 1449. (Scanning Electron Microscopy, Inc. Elk Grove Village, IL) Thorough exposition of the principles and applications of reflected electron energy-loss microscopy (REELM) as well as a comparison to other techniques, such as SAM, EDS and SEM.

3 C. J. Powell. *Opt. Soc. Amer.* **59,** 738, 1969. Excellent presentation of the interaction between interband and plasmon peaks that is often overlooked in REELS spectral analysis.

4 M. De Crescenzi. *Phys. Rev. Letts.* **30,** 1949, 1987. Use of surface electron energy-loss fine structure (SEELFS) to determine oxygen–nickel bond length changes for oxygen absorbed on Ni (100) on a function of coverage from 0 to 1.0 monolayer.

5 P. N. Ross Jr. and K. A. Gaugler. *Surf. Sci.* **122,** L579, 1982. Excellent example of chemical state information in CEELS.

6 J. Ghijsen. *Surf. Sci.* **126,** 177, 1983. REELS spectra of pure Mg versus primary beam energy showing relative intensities of the bulk and surface plasmons.

7 H. Ibach and J.E. Rowe. *Phys. Rev.* **89,** 1951, 1974. Classic example of O_2 adsorption Si (111) and Si (100) showing use of VEELS for structure and chemistry analysis.

8 A. J. Bevolo, M. L. Albers, H. R. Shanks, and J. Shinar. *J. Appl. Phys.* **62,** 1240, 1987. VEELS in fixed-spot mode to depth profile hydrogen in amorphous silicon films to determine hydrogen mobility at elevated temperatures.

9 P. Braun. *Surf. Sci.* **126,** 714, 1983. VEELS study of bulk and surface plasmon energies across Al–Mg alloy phase diagram.

10 B. Schroder. *Ninth Conference on Electron Microscopy.* 1978, vol. 1, p. 534. (Microscopial Society of Canada, Toronto, Canada) Unique combination of E_0 = 30 keV used in HREELS study of *a*-Si (H) films with meV energy resolution.

6

X-RAY EMISSION TECHNIQUES

6.0 INTRODUCTION

Three techniques involving the use of X-ray emission to obtain quantitative elemental analysis of materials are described in this chapter. They are X-Ray Fluorescence, XRF, Total Reflection X-Ray Fluorescence, TXRF, and Particle-Induced X-Ray Emission, PIXE. XRF and TXRF use laboratory X-ray tubes to excite the emission. PIXE uses high-energy ions from a particle accelerator.

The X-ray emission process following the excitation is the same in all three cases, as it is also for the electron-induced X-ray emission methods (EDS and EMPA) described in Chapter 3. The electron core hole produced by the excitation is filled by an electron falling from a shallower level, the excess energy produced being released as an emitted X ray with a wavelength characteristic of the atomic energy levels involved. Thus elemental identification is provided and quantification can be obtained from intensities. The practical differences between the techniques come from the consequences of using the different excitation sources.

In XRF all elements having $Z > 3$ can be detected using WDS (see Chapter 3), though a range of excitation tubes and analyzing crystal monochromators is needed. If EDS (see Chapter 3) is used spectral resolution is much poorer, resulting in overlapping peaks and a reduced ability to distinguish some elements at low concentrations. Also only $Z > 10$ can usually be detected. The sample is examined in air, usually with an unfocused X-ray beam without lateral resolution, though special microbeam systems down to 10 μm do exist. For normal usage X-ray penetration will be many microns, resulting in essentially bulk analysis. With special equipment angles of incidence close to grazing can be used, reducing the probing depth considerably (down to 1000 Å). Large-area flat samples are then needed.

Atomic detection limits are around 0.1% for $Z > 15$ with a few percent accuracy. Detection limits for light elements are much poorer since they go roughly as Z^2. The quantification algorithms account for several correct ion factors. XRF has been widely used for solids, powders, and liquids, and more recently for simultaneous determination of both composition and thickness of single and multilayer thin films in the range of a few hundred angstroms to several microns. Such determinations are computer iterative fits to assumed models, however, and so are not necessarily unique.

In TXRF the angle of incidence of excitation is reduced to a few mrad. which is below the total reflection angle. Only a few tens of angstroms contribute to the signal under these conditions. The X-ray optics conditions are stringent for this approach, however, and very flat, large-area samples are needed. The method is therefore readily tailored to Si and GaAs wafers, where it is becoming widely used to monitor homogeneous surface impurities at concentrations down to 10^{10}–10^{11} atoms/cm^2 for heavy metals and 10^{12}–10^{13} atoms/cm^2 for elements such as Si, S, and Cl on GaAs. The method is also useful for thin-film interfaces and multilayers. A variation of the technique is vapor phase decomposition, VPD, where the surface of a Si wafer is dissolved in HF. The resulting solution of impurities is evaporated in the center of the wafer which is then analyzed by TXRF. The sensitivity can be improved two to three orders of magnitude this way. The cost of TXRF can be up to $600,000 compared to the more modest $50,000 to $300,000 for XRF.

PIXE is an adjunct to RBS (see Chapter 9), using the same particle beam (H or He ions usually) and the same analysis chamber with an added EDS detector. Probing depths for a 2–4-MeV beam (the usual energies) are microns, similar to XRF, but control is possible by going to higher or lower energies. Often there is no lateral resolution but microbeam systems down to a few microns spot size do exist. PIXE tends to be used more for surface layer and thin-film analysis than for bulk samples. Its major use has been in the biomedical area, where it has some advantages over XRF in the microbeam mode. Insulators can be a problem because of build up of high charges, leading to discharges. PIXE's strength should be its complementary nature to RBS, since it provides unambiguous identification for some elements that are hard to separate by RBS; it is more sensitive to low-Z elements in high-Z materials; and it is better at trace analysis in general. RBS, on the other hand, provides a quantitative depth profile of the major constituents, provided they are resolvable. Altogether, however, PIXE usage is only about 1% that of XRF.

One should compare capabilities to the electron beam X-ray emission methods of Chapter 3. The major difference is the higher lateral resolution with electron beams and the associated mapping capabilities. Another difference is the shorter probing depth possible with electrons, except when compared to the specialized TXRF method. Comparing electron-beam EDS to X-ray/particle EDS or electron-beam WDS to X-ray/particle WDS, the electron beams have poorer detection limits because of the greater X-ray background associated with electron excitation. The electron-beam methods are always done in vacuum, of course, whereas XRF and TXRF are done in air. XRF in the WDX mode dominates in terms of the number of systems in use in the world (about 10,000).

In principle all the X-ray emission methods can give chemical state information from small shifts and line shape changes (cf., XPS and AES in Chapter 5). Though done for molecular studies to derive electronic structure information, this type of work is rarely done for materials analysis. The reasons are: the instrumental resolution of commercial systems is not adequate; and the emission lines routinely used for elemental analysis are often not those most useful for chemical shift meas-urements. The latter generally involve shallower levels (narrower natural line widths), meaning longer wavelength (softer) X-ray emission.

6.1 XRF

X-Ray Fluorescence

TING C. HUANG

Contents

Introduction

X-Ray Fluorescence (XRF) is a nondestructive method used for elemental analysis of materials. An X-ray source is used to irradiate the specimen and to cause the elements in the specimen to emit (or fluoresce) their characteristic X rays. A detector system is used to measure the positions of the fluorescent X-ray peaks for qualitative identification of the elements present, and to measure the intensities of the peaks for quantitative determination of the composition. All elements but low-Z elements—H, He, and Li—can be routinely analyzed by XRF.

Since the 1950s XRF has been used extensively for the analysis of solids, powders, and liquids. The technique was extended to analyze thin-film materials in the 1970s. XRF can be used routinely for the simultaneous determination of elemental composition and thickness of thin films. The technique is nondestructive, rapid, precise, and potentially very accurate. The results are in good agreement with other elemental analysis techniques including wet chemical, electron-beam excitation techniques, etc.

Basic Principles

The fundamental principles of XRF can be found in the literature.[1–3] Briefly, X rays are electromagnetic radiation of very high energy (or short wavelength). The unit of measurement for X rays is the angstrom (Å), which is equal to 10^{-8} cm. When an X-ray photon strikes an atom and knocks out an inner shell electron, if the incident photon has energy greater than the binding energy of the inner shell electron, a readjustment occurs in the atom by filling the inner shell vacancy with one of the outer electrons and simultaneously emitting an X-ray photon. The emitted photon (or fluorescent radiation) has the characteristic energy of the difference between the binding energies of the inner and the outer shells. The penetration depth of a high-energy photon into a material is normally in the μm range. (Another method commonly used to produce X rays is electron-beam excitation; the penetration depth of an electron beam is about an order of magnitude smaller than that of X rays. See the articles on EDS and EPMA.)

Measurements of the characteristic X-ray line spectra of a number of elements were first reported by H. G. J. Moseley in 1913. He found that the square root of the frequency of the various X-ray lines exhibited a linear relationship with the atomic number of the element emitting the lines. This fundamental "Moseley law" shows that each element has a characteristic X-ray spectrum and that the wavelengths vary in a regular fashion form one element to another. The wavelengths decrease as the atomic numbers of the elements increase. In addition to the spectra of pure elements, Moseley obtained the spectrum of brass, which showed strong Cu and weak Zn X-ray lines; this was the first XRF analysis. The use of XRF for routine spectrochemical analysis of materials was not carried out, however, until the introduction of modern X-ray equipment in the late 1940s.

Instrumentation

The instrumentation required to carry out XRF measurements normally comprises three major portions: the primary X-ray source, the crystal spectrometer, and the detection system. A schematic X-ray experiment is shown in Figure 1. Fluorescent X rays emitted from the specimen are caused by high-energy (or short-wavelength) incident X rays generated by the X-ray tube. The fluorescent X rays from the specimen travel in a certain direction, pass through the primary collimator. The analyzing crystal, oriented to reflect from a set of crystal planes of known *d*-spacing, reflects one X-ray wavelength (λ) at a given angle (θ) in accordance with Bragg's law: $n\lambda = 2d\sin\theta$, where n is a small positive integer giving the order of reflection. By rotating the analyzing crystal at one-half the angular speed of the detector, the various wavelengths from the fluorescent X rays are reflected one by one as the analyzing crystal makes the proper angle θ for each wavelength. The intensity of at each wavelength is then recorded by the detector. This procedure is known also as the

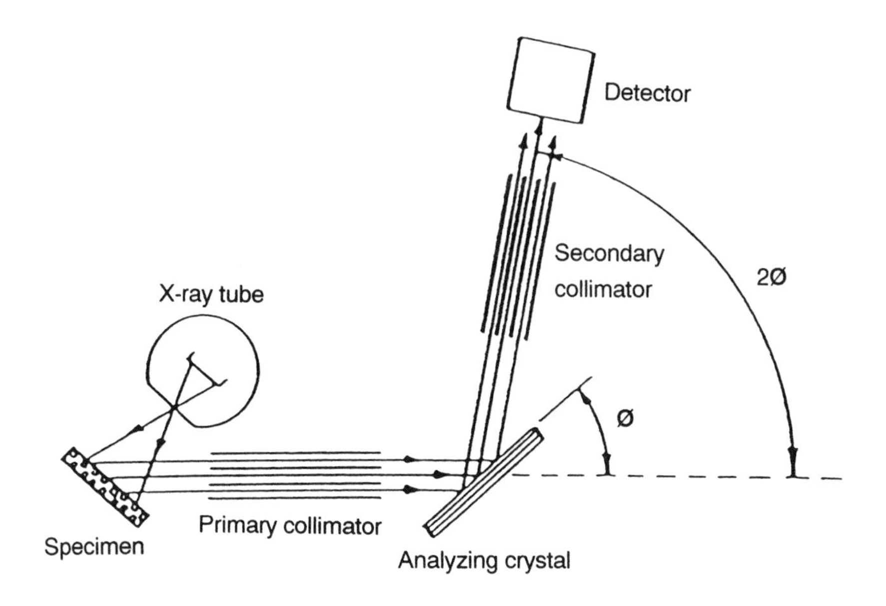

Figure 1 Schematic of XRF experiment.

wavelength-dispersive method. (The wavelength-dispersive method is used extensively in EPMA, see the EPMA article in this volume.)

X-RaySources

A sealed X-ray tube having a W, Cu, Rh, Mo, Ag, or Cr target is commonly used as the primary X-ray source to directly excite the specimen. A secondary target material located outside the X-ray tube is used sometimes to excite fluorescence. This has the advantages of selecting the most efficient energy close to the absorption edge of the element to be analyzed and of reducing (or not exciting) interfering elements. (The intensity is much reduced, however.) X-ray sources, including synchrotron radiation and radioactive isotopes like ^{55}Fe (which emits Mn *K* X rays) and AM-241 (Np *L* X rays) are used in place of an X-ray tube in some applications.

Analyzing Crystals

Crystals commonly used in XRF are: LiF (200) and (220), which have 2*d*-spacings of 4.028 and 2.848 Å, respectively; pyrolytic graphite (002), spacing 6.715 Å; PET(002), spacing 8.742 Å; TAP(001), spacing 25.7 Å; and synthetic multilayers of W/Si, W/C, V/C, Ni/C, and Mo/B_4C, spacing 55–160 Å. The lowest-*Z* element that can be detected and reflected efficiently depends on the *d*-spacing of the analyzing crystal selected. The crystals are usually mosaic, and each reflection is spread over a small angular range. It is thus important that the crystal used be of good quality to obtain intensive and sharp XRF peaks. The angular spread of the

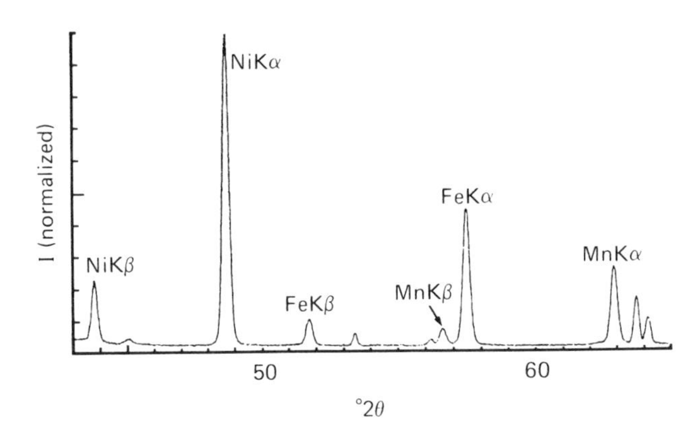

Figure 2 **XRF spectrum of MnFe/NiFe thin film.**

peaks, or the dispersion, $d\theta/d\lambda = n/(2d\cos\theta)$, increases with decreasing d. The dispersion thus can be increased by selecting a crystal with a smaller d.

X-Ray Detection Systems

The detectors generally used are scintillation counters having thin Be windows and NaI–Tl crystals for short wavelengths (above 3 Å or 4 keV), and gas-flow proportional counters having very low absorbing windows and Ar/CH_4 gas for long wavelengths (below 2 Å or 6 keV). A single-channel pulse amplitude analyzer is used to accept fluorescent X rays within a selected wavelength range to improve peak-to-background ratios and to eliminate unwanted high-order reflections.

The counting times required for measurement range between a few seconds and several minutes per element, depending on specimen characteristics and the desired precision.

A typical XRF spectrum of a FeMn/NiFe thin film is plotted in Figure 2. The $K\alpha$ and $K\beta$ XRF fluorescent peaks from the film are identified, and the remaining peaks are those from the spectrum of the X-ray tube. The experimental conditions included a Mo target X-ray tube operated at 45 kV, a LiF (200) analyzing crystal, and a scintillation counter with a single-channel pulse amplitude analyzer. The energy resolution of the Mn Kα peak at 5.89 keV was 24 eV, compared to 145 eV for a Si (Li) solid-state energy-dispersive system (see EDS article). The high spectral resolution of the wavelength-dispersive method made possible the measurements of Ni, Fe, and Mn free of interference from adjacent peaks.

Analytical Capabilities

Elemental Depth Profiling

The X-ray penetration depth in a material depends on the angle of incidence. It increases from a few tens of Å near the total reflection region to several μm at large incidence angles (a few tens of degrees). The XRF beam, which originates from variable depths, can be used for elemental depth analysis. For example, the grazing incidence XRF method has been used for studies of concentration profiles of a dissolved polymer near the air/liquid interface,[4] Langmuir-Blodgett multilayers,[5] and multiple-layer films on substrates.[6] This type of analysis requires a parallel-incidence beam geometry, which currently is not possible with a conventional spectrometer.

Chemical State Analysis

The XRF wavelengths and relative intensities of a given element are constant to first approximation. Small changes may occur when the distribution of the outer (or valence) electron changes. A major area of research in XRF involves the use of "soft" X-ray emission (or long-wavelength XRF) spectra for chemical state analysis. Soft X-ray peaks often exhibit fine structure, which is a direct indication of the electronic structure (or chemical bonding) around the emitting atom. Thus the shift in peak position, change in intensity distribution, or appearance of additional peaks can be correlated with a variety of chemical factors, including the oxidation state, coordination number, nature of covalently bound ligands, etc. The equipment required for soft X-ray analysis is almost identical to that required for conventional XRF, with one major exception. Since it is a study of transitions involving the outer orbits and therefore long wavelengths, soft X-ray analysis employs a long-wavelength X-ray source such as Al (8.34 Å for Al $K\alpha$) or Cu (13.36 Å for Cu $L\alpha$). Special analyzing crystals or gratings for measuring wavelengths in the range 10–100 Å also are needed.[7]

Quantitative Analysis

In addition to qualitative identification of the elements present, XRF can be used to determine quantitative elemental compositions and layer thicknesses of thin films. In quantitative analysis the observed intensities must be corrected for various factors, including the spectral intensity distribution of the incident X rays, fluorescent yields, matrix enhancements and absorptions, etc. Two general methods used for making these corrections are the empirical parameters method and the fundamental parameters methods.

The empirical parameters method uses simple mathematical approximation equations, whose coefficients (empirical parameters) are predetermined from the experimental intensities and known compositions and thicknesses of thin-film standards. A large number of standards are needed for the predetermination of the empirical parameters before actual analysis of an unknown is possible. Because of the difficulty in obtaining properly calibrated thin-film standards with either the same composition or thickness as the unknown, the use of the empirical parameters method for the routine XRF analysis of thin films is very limited.

The fundamental parameters method uses XRF equations derived directly from first principles. Primary and secondary excitations are taken into account. Primary excitations are caused directly by the incident X rays from the X-ray source, while the secondary excitations are caused by other elements in the same film, whose primary fluorescent X-ray radiation has sufficient energy to excite the characteristic radiation of the analyzed element. Higher order excitations are generally considered insignificant because of their much lower intensities. XRF equations relate intensity, composition, and thickness through physical constants (fundamental parameters) like fluorescent yields, atomic transition probabilities, absorption coefficients, etc. For example, the XRF equations for single-layer films were reported by Laguitton and Parrish,[8] and for multiple-layer films by Mantler.[9] The equations for thin films are very complex, and the values of composition and thickness cannot be determined directly from the observed intensities. They are obtained by computer iteration using either linear or hyperbolic approximation algorithm. The fundamental parameters technique is suitable for the analysis of thin films because it requires a minimum number of pure or mixed element and bulk or thin-film standards.

Applications

The principle application of XRF thin-film analysis is in the simultaneous determination of composition and thickness. The technique has been used for the routine analysis of single-layer films[8] since 1977 and multiple-layer films[10] since 1986. Two main sources of publications in the fields are the annual volumes of *Advances in X-Ray Analysis* by Plenum Press, New York, and the *Journal of X-Ray Spectrometry* by Heyden and Sons, London. Typical examples on the analysis of single-layer films and multiple-layer films are used to illustrate the capabilities of the technique.

Single-Layer Films

Evaporated FeNi films with a large range of compositions were selected because of the strong absorption of Ni and enhancement of Fe *K* X rays in the films. XRF compositions of 7 FeNi films deposited on quartz substrates are listed in Table 1 and are compared to those obtained by the Atomic Absorption Spectroscopy (AAS) and the Electron Probe Microanalysis (EPMA). Since the strong X-ray absorption and enhancement effects are severe for both XRF and EPMA but not present in AAS, a comparison between the XRF results and the two non-XRF techniques provide a useful evaluation of XRF.[11, 12] As shown in Table 1, there is good agreement between results of XRF and AAS or EPMA, and the average deviation is 0.9% between XRF and AAS and is 1.1% between XRF and EPMA. It is worth noting that the compositions of more than half of the 7 FeNi films obtained by XRF, AAS, and EPMA are significantly different from the intended compositions (see values inside the parentheses listed in column 1 of Table 1). The discrepancy shows the

Film	XRF	AAS	EPMA
Fe (5)–Ni (95)	4.2	5.0	2.5
Fe (10)–Ni (90)	9.2	9.0	6.2
Fe (20)–Ni (80)	19.4	19.2	19.4
Fe (34)–Ni (66)	47.3	48.4	44.5
Fe (50)–Ni (50)	59.1	61.7	59.1
Fe (66)–Ni (34)	78.9	79.8	78.4
Fe (80)–Ni (20)	89.2	89.6	89.2

Table 1 **Fe concentrations (% wt.) for FeNi films.**

risk of using intended composition and the important of determining composition experimentally by XRF or other reliable techniques.

The volume density ρ and thickness t of a film appear together as a single parameter ρt in the XRF equations, the value of ρt, the areal density (not the thickness) is determined directly by iteration. From the areal density, the film thickness can be calculated when the volume density is known experimentally or theoretically. Using the volume densities calculated from the film composition and the published volume densities of pure elements, the thicknesses of 12 $Fe_{20}Ni_{80}$ films were calculated from the XRF areal densities and are compared to those obtained by a nonXRF technique (i.e., AAS or a deposition monitor). As shown in Table 2, good agreement between XRF and non-XRF thicknesses are obtained with average and maximum deviations of 2.95% and 6.7%, respectively (see the last column of Table 2). The volume density can also be calculated from the XRF areal density when the thickness of a film is known. For example, the volume densities of 8 $Fe_{19}Ni_{81}$ permalloy films with known thicknesses of 50–10,000 Å were calculated from the XRF areal densities. The calculation shows that the volume density of the permalloy is not constant and changes systematically with the film's thickness. It is equal to the bulk value of 8.75 g/cm^3 for films of 1000 Å or greater thickness, decreases to 94% of the bulk value for the 500-Å film, and to 81% for the 50-Å film.[12]

Multiple-layer Films

XRF analysis of multiple-layer films is very complex because of the presence of XRF absorption and enhancement effects, not only between elements in the same layer but also between all layers in the film. Equations for the calculation of XRF intensities for multiple-layer films are available from the literature.[9, 13] Proper correc-

Film	XRF	Non-XRF[a]	Δ/XRF (%)[b]
1	825	848	2.8
2	858	848	1.2
3	1117	1142	2.2
4	3180	2967	6.7
5	3215	3011	6.3
6	3558	3473	2.4
7	3579	3524	1.5
8	4090	4070	0.5
9	5533	5452	1.5
10	5550	5237	5.6
11	5601	5655	1.0
12	6283	6053	3.7

a. Either AAS or monitor.
b. Δ = |XRF – non-XRF|.

Table 2 Thicknesses (Å) for $Fe_{20}Ni_{80}$ films.

tions for intralayer and interlayer effects are essential for a successful XRF analysis of multiple-layer films. The accuracy of XRF compositions and thicknesses for multiple-layer films was found to be equal to those for single-layer films.

For example, XRF was used successfully to analyze two triple-layer films of Cr, Cu, and FeNi deposited on Si substrates.[10, 12, 14] The two films, T1 and T2, have identical individual Cr, Cu and FeNi layers but different order. In T1, the FeNi layer is on top, the Cu layer in the middle, and the Cr layer at the bottom; in T2, the positions of the Cr and FeNi layers are reversed, with Cr on top and FeNi at the bottom; meanwhile the Cu layer remains in the middle. Because of this reversal of layer order, interlayer absorption and enhancement effects are grossly different between these two films. This led to large differences in the observed intensities between these two films. The differences between T1 and T2 were –17%, +2%, +20%, and +15%, respectively, for the Cr, Cu, Fe, and Ni Kα observed intensities.[12] Using the same set of observed XRF intensities, the results obtained by two different analysis programs: LAMA-III from the US[12] and DF270 from Japan[14] are essentially the same within a relative deviation of 0.2% in composition and 1% in

Parameter	Triple layer		Single layer		
	T1 (FeNi/Cu/Cr)	T2 (Cr/Cu/FeNi)	S1 (Cr)	S2 (FeNi)	S3 (Cu)
Fe (% wt.)	10.25	10.25	—	10.50	—
Ni (% wt.)	89.75	89.75	—	89.50	—
t_{Cr} (Å)	1652	1698	1674	—	—
t_{FeNi}(Å)	2121	2048	—	2115	—
t_{Cu}(Å)	2470	2457	—	—	2416

Table 3 **XRF results for films of Cr, FeNi, and Cu.**

thickness. The results obtained by the LAMA-III program are listed in Table 3. In spite of the large differences in the observed intensities of Cr, Fe and Ni, the compositions and thickness of all three layers determined by XRF are essentially the same for T1 and T2. For comparison, an XRF analysis was also done on three single-layer Cr, FeNi, and Cu films (S1, S2, and S3) prepared under identical deposition conditions to the two triple-layer films. As shown in Table 3, good agreement was obtained between the single- and triple-layer films. This indicates that the severe interlayer enhancement and absorption effects observed in T1 and T2 were corrected properly. It is also worth noting that the deviations between the results of the triple- and single-layer films are within the accuracy reported for the single-layer films.

In multiple-layer thin films, it is possible that some of the elements may be present simultaneously in two or more layers. XRF analysis of this type of film can be complicated and cannot be made solely from their observed intensities. Additional information, such as the compositions or thickness of some of the layers is needed. The amount of additional non-XRF information required depends on the complexity of the film. For example, in the analysis of a FeMn/NiFe double-layer film, the additional information needed can be the composition or thickness of either the FeMn or NiFe layer. Using the composition or thickness of one of the film predetermined from a single-layer film deposited under identical conditions, XRF analysis of the FeMn/NiFe film was successful.[12]

Related Techniques

XRF is closely related to the EPMA, energy-dispersive X-Ray Spectroscopy (EDS), and total reflection X-Ray Fluorescence (TRXF), which are described elsewhere in this encyclopedia. Brief comparisons between XRF and each of these three techniques are given below.

EPMA

Both XRF and EPMA are used for elemental analysis of thin films. XRF uses a non-focusing X-ray source, while EPMA uses a focusing electron beam to generate fluorescent X rays. XRF gives information over a large area, up to cm in diameter, while EPMA samples small spots, μm in size. An important use of EPMA is in point-to-point analysis of elemental distribution. Microanalysis on a sub-μm scale can be done with electron microscopes. The penetration depth for an X-ray beam is normally in the 10-μm range, while it is around 1 μm for an electron beam. There is, therefore, also a difference in the depth of material analyzed by XRF and EPMA.

EDS

EDS is another widely used elemental analysis technique and employs a solid state detector with a multichannel analyzer to detect and resolve fluorescent X rays according to their energies. EDS uses either X rays or an electron beam as a source to excite fluorescence. Unlike XRF, which uses the wavelength-dispersive method to record X-ray intensities one by one, EDS collects all the fluorescent X rays from a specimen simultaneously. A limitation of EDS is its energy resolution, which is an order of magnitude poorer than that of the wavelength-dispersive method. For example, the $K\alpha$ peaks of transition elements overlap the $K\beta$ peaks of the next lighter element, which cause analytical difficulties. The poorer resolution also causes relatively lower peak-to-background ratios in EDS data.

TXRF

XRF at large incident angles, as described in this article, is normally used for elemental analysis of major concentrations of 0.1% or higher. Total Reflection X-Ray Fluorescence (TXRF) with grazing-incidence angles of a few tenths of a degree is used for trace-element analysis. Detectable limits down to 10^9 atoms/cm^2 are now attainable using a monochromatic X-ray source. Examples of the use of this technique in wafer technology are given in the article on TXRF in this volume.

Conclusions

XRF is one of the most powerful analysis technique for the elemental-composition and layer-thickness determination of thin-film materials. The technique is nondestructive, inexpensive, rapid, precise and potentially very accurate. XRF characterization of thin films is important for the research, development, and manufacture of electronic, magnetic, optical, semiconducting, superconducting, and other types of high-technology materials. Future development is expected in the area of microbeam XRF, scanning XRF microscopy, grazing-incidence XRF analysis of surfaces and buried interfaces, long-wavelength XRF and chemical state analysis, and synchrotron XRF.

Related Articles in the Encyclopedia

EPMA, EDS, and TXRF

References

1 L. S. Birks. *X-Ray Spectrochemical Analysis.* Second Edition, Wiley, New York, 1969. A brief introduction to XRF, it will be useful to those who are interested in knowing enough about the technique to be able to use it for routine analysis. A separate chapter on EPMA also is included.

2 E. P. Bertin. *Principles and Practice X-Ray Spectrometric Analysis.* Plenum, New York, 1970. A practical textbook that also serves as a laboratory handbook, although somewhat dated.

3 R. Jenkins. *An Introduction to X-Ray Spectrometry.* Heyden, London, 1974. A good introduction to XRF instrumentation, qualitative and quantitative analyses, and chemical-bonding studies.

4 J. M. Bloch, M. Sansone, F. Rondelez, D. G. Peiffer, P. Pincus, M. W. Kim, and P. M. Eisenberger. *Phys. Rev. Lett.* **54,** 1039, 1985.

5 M. J. Bedzyk, G. M. Bommarito, and J. S. Schidkraut. *Phys. Rev. Lett.* **62,** 1376, 1989.

6 D. K. G. de Boer. In *Advances in X-Ray Analysis.* (C. S. Barrett et al., Eds.) Plenum, New York, 1991, Vol. 34, p. 35.

7 B. L. Henke, J. B. Newkirk, and G. R. Mallet, Eds. *Advances in X-Ray Analysis.* Plenum, New York, 1970, Vol. 13. The proceedings of the 18th Annual Denver Conference on Applications of X-ray Analysis; the central theme of the Conference was interactions and applications of low-energy X-rays.

8 D. Laguitton and W. Parrish. *Anal. Chem.* **49,** 1152, 1977.

9 M. Mantler. *Anal. Chim. Acta.* **188,** 25, 1986.

10 T. C. Huang and W. Parrish. In *Advances in X-Ray Analysis.* (C. S. Barrett, et al., Eds.) Plenum, New York, 1986, Vol. 29, p. 395.

11 T. C. Huang and W. Parrish. In *Advances in X-Ray Analysis.* (G. J. McCarthy et al., Eds.) Plenum, New York, 1979, Vol. 22, p. 43.

12 T. C. Huang. *Thin Solid Films.* **157,** 283, 1988; and *X-Ray Spect.* **20,** 29, 1991.

13 D. K. G. de Boer. *X-ray Spect.* **19,** 145, 1990.

14 Y. Kataoka and T. Arai. In *Advances in X-Ray Analysis.* (C. S. Barrett et al., Eds.) Plenum, New York, 1990, Vol. 33, p. 220.

6.2 TXRF

Total Reflection X-Ray Fluorescence Analysis

PETER EICHINGER

Contents

Introduction

X-Ray Fluorescence analysis (XRF) is a well-established instrumental technique for quantitative analysis of the composition of solids. It is basically a bulk evaluation method, its analytical depth being determined by the penetration depth of the impinging X-ray radiation and the escape depth of the characteristic fluorescence quanta. Sensitivities in the ppma range are obtained, and the analysis of the emitted radiation is mostly performed using crystal spectrometers, i.e., by wavelength-dispersive spectroscopy. XRF is applied to a wide range of materials, among them metals, alloys, minerals, and ceramics.

In the total reflection mode, with the X rays impinging at a grazing angle onto a specular solid surface, interference between the incident and the reflected X-ray waves limits the excitation depth to a few monatomic layers in which the radiation intensity is locally concentrated. Accordingly this surface sheet, which has a depth of a few nm, is strongly excited, giving rise to an intensive emission of fluorescence quanta. The bulk of the solid is virtually decoupled by the total reflection, leading to a suppression of matrix background fluorescence radiation. The high sensitivity of TXRF for surface impurities is a result of both effects: compression of the X-ray intensity in the surface sheet, and suppression of the bulk fluorescence background.

TXRF is essentially a surface-analytical technique, used to detect trace amounts of impurities on specular surfaces. Until a few years ago, its application was limited to the analysis of liquids that have been pipetted in microliter volumes on flat quartz substrates and allowed to dry. Subsequent TXRF of the droplet residue presents an attractive, multielement analysis with sensitivities down to the pg level. The main applications of this branch of TXRF are in environmental research. In recent years the application of TXRF has been expanded to semiconductor technology, with its stringent demands for surface purity, especially with respect to heavy-metal contamination.[1] In this application, the semiconductor substrate is directly subjected to TXRF. Detection limits on the order of 10^{10} metal atoms per cm^2, corresponding to 10 ppma of a monatomic surface layer, are obtained on silicon wafers using a monochromatic X-ray source. The following sections focus on the instrumentation and application of TXRF to semiconductor substrates that are usually electrochemically polished and thus provide ideal conditions for TXRF; the wafer's relatively large diameter allows for automatic adjustment of the critical angle. Dedicated wafer surface analyzers are on the verge of becoming routine monitoring tools in the semiconductor industry.

Principles of Direct TXRF

The primary X-ray beam is directed onto the solid surface in grazing incidence. The angle of incidence is kept below the critical angle at which total reflection occurs. The critical angle is given by

$$\phi_C = 3.72 \times 10^{-11} \frac{\sqrt{n_e}}{E} \tag{1}$$

where ϕ_C is the critical angle (mrad), n_e is the electron density (cm^{-3}), and E is the quantum energy (keV). The critical angle ϕ_C is inversely proportional to the energy of the X-ray quanta, and increases with the square root of the electron density of the solid. For molybdenum K_α-radiation and a silicon surface, ϕ_C is 1.8 mrad.

The angular dependence of the fluorescence yield in the neighborhood of the critical angle should be considered in detail to establish the chemical nature of surface impurities, as well as for quantitation in terms of their concentrations (Figure 1).

Agglomerated impurities, such as particles or droplet residues, do not participate in the interference phenomenon leading to total reflection; their fluorescence intensity is independent of the angle of incidence below the critical angle, and drops by a factor of 2 if the critical angle is surpassed due to the disappearance of the reflected component in the exciting beam (*nonreflecting* impurities and *residues*).

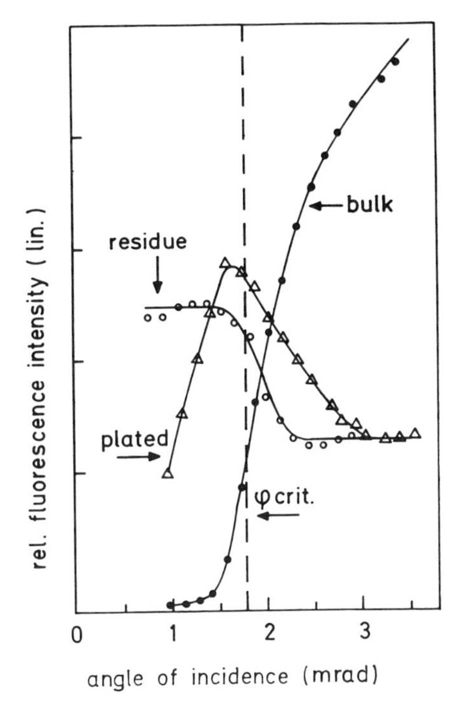

Figure 1 **Experimental curves for the angular dependence of the fluorescence intensity from plated or sputtered submonatomic Ni layers (open triangles), layers produced by the evaporation of a Ni salt solution (open circles), and the silicon substrate (filled circles).**

On the other hand, impurities that are homogeneously distributed through a submonatomic layer within the surface, such as electrochemically plated, sputtered, or evaporated atoms, are part of the reflecting surface and their fluorescent yield shows a pronounced dependence on the incidence angle. These *reflecting* or *plated* impurities exhibit basically the same angular dependence below the critical angle as the matrix fluorescence from the bulk silicon, but they peak at the critical angle.

The plated-type impurities are most commonly encountered with semiconductor substrates; they originate, for example, from wet chemical processing steps. It is apparent from Figure 1 that a precise control of the angle of incidence is an essential feature of TXRF instrumentation.

X-Ray Sources

Sealed conventional fine structure tubes with Mo, W, Cu, or Cr anodes are used as primary X-ray sources, as well as rotating anode tubes, or synchrotron radiation. The maximum energy of the X-ray quanta determines the range of elements acces-

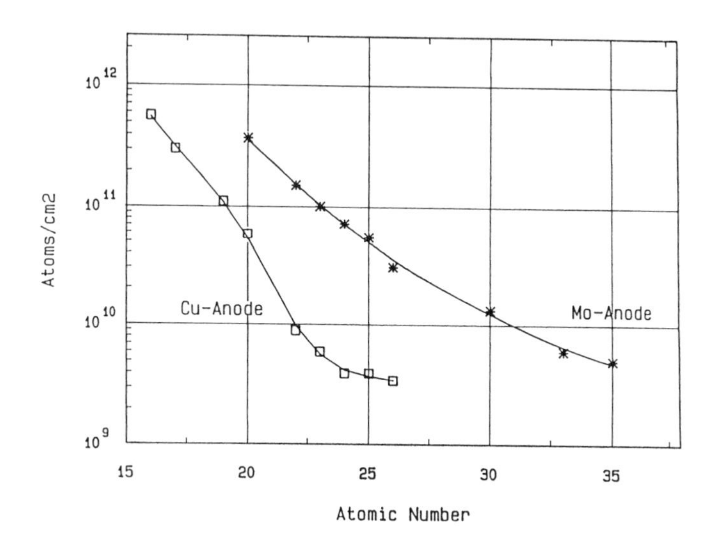

Figure 2 **Detection limits obtained with Cu and Mo anodes in conjunction with a monochromator.[2]**

sible for analysis and the detection limits of the respective elements, as shown in Figure 2 for monochromatized radiation from a Cu and Mo anode. In this example, Fe ($Z = 26$) can be detected at a level below 10^{10} atoms/cm^2 using the Cu anode, but Cu is not detectable.

In modern TXRF instrumentation, the primary radiation from the X-ray tube is filtered or monochromatized to reduce the background originating primarily from bremsstrahlung quanta with higher energy than the main characteristic line for the anode material. The higher energy radiation does not fulfill the critical angle condition for total reflection and penetrates into the substrate, thus adding scattered radiation. Energy filtering is achieved using multilayer interference or crystal diffraction.

VPD-TXRF

The term *direct TXRF* refers to surface impurity analysis with no surface preparation, as described above, achieving detection limits of 10^{10}–10^{11} cm^{-2} for heavy-metal atoms on the silicon surface. The increasing complexity of integrated circuits fabricated from silicon wafers will demand even greater surface purity in the future, with accordingly better detection limits in analytical techniques. Detection limits of less than 10^9 cm^{-2} can be achieved, for example, for Fe, using a preconcentration technique known as Vapor Phase Decomposition (VPD).

The VPD method originally was developed to determine metal trace impurities on thermally oxidized or bare silicon surfaces in combination with atomic absorp-

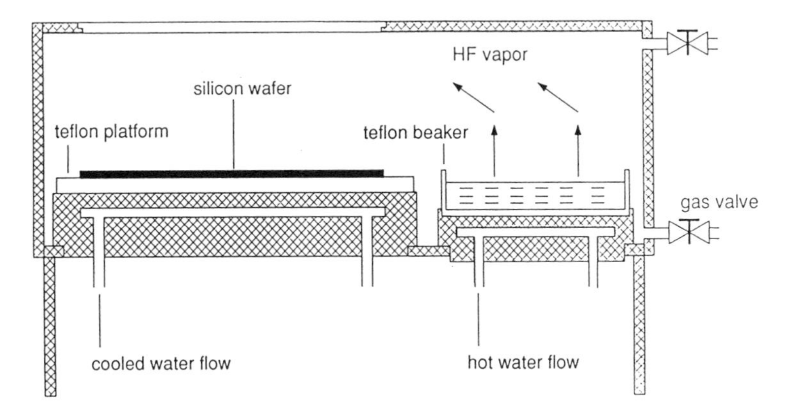

Figure 3 **Schematic arrangement for vapor phase decomposition (VPD) applied to silicon wafers.**

tion spectroscopy (VPD-AAS). The silicon wafer is exposed to the vapor of hydrofluoric acid, which dissolves the SiO_2 surface layer (native or thermal oxide) according to the reaction:

$$SiO_2 + 6HF \rightarrow H_2SiF_6 + 2H_2O \qquad (2)$$

The impurities on the surface are contained in the resulting water droplet or moisture film, and are collected *in situ* for further investigation by scanning the surface with an auxiliary water droplet (e.g., 50 μl). The VPD residue is allowed to dry in the center of the wafer and subjected to TXRF analysis. A schematic of a VPD reactor is shown in Figure 3.

With VPD preconcentration, the angular dependence of the impurity fluorescence yield follows the curve for residue impurities, as shown in Figure 1, in contrast to the plated-impurity case using direct TXRF.

The sensitivity enhancement achieved by VPD is determined by the ratio of the substrate area to the area of the detector aperture (analyzed area), provided there is full collection of the impurities. This has been demonstrated for Fe and Zn. For Cu and Au, however, only a small percentage can be collected using this technique,[3] due to electrochemical plating. An example comparing direct TXRF with VPD-TXRF on the same substrate is shown in Figure 4.

Semiconductor Applications

In silicon integrated circuit technology, TXRF analysis is applied as a diagnostic tool for heavy-metal contamination in a variety of process steps, including incoming wafer control, preoxidation cleaning, and dry processing equipment evaluation. As an example, Figure 5 shows the effect of applying a standard cleaning to silicon

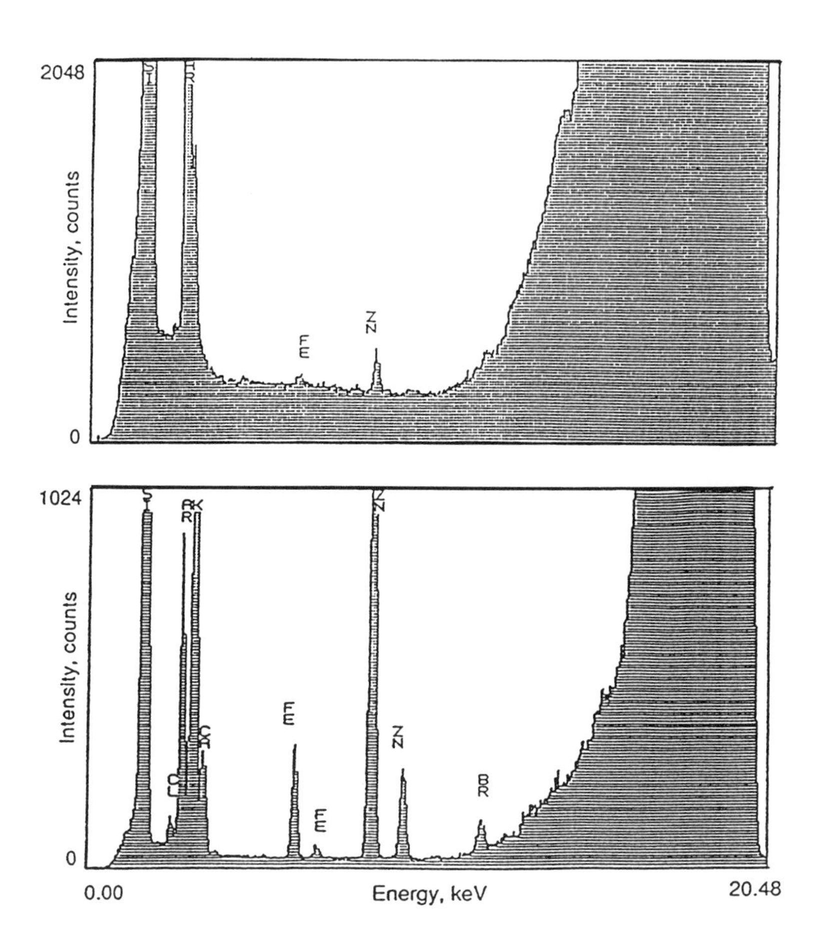

Figure 4 **Direct TXRF (upper spectrum, recording time 3000 s) and VPD-TXRF (lower spectrum, recording time 300 s) on a silicon wafer surface. The sensitivity enhancement for Zn and Fe is two orders of magnitude. The measurements were made with a nonmonochromatized instrument.**

wafers received from a commercial vendor: The contamination is similar for wafers 1, 2, and 3, with K, Ca, Fe, and Zn as the predominant metal impurities in concentrations of 10^{11}–10^{12} atoms/cm^2. Cleaning on wafers 4, 5, and 6 removes all metals to a level of less than 10^{10} cm^{-2}, except for Fe which is still detectable. The deposition of Br is due to the cleaning solution and is not considered harmful. The analysis has been performed using VPD-TXRF.

With gallium arsenide, additional elements, such as Si, S, and Cl, are of interest because of their doping character. Impurity levels on the order of 10^{12} cm^{-2} are encountered with commercial substrates, which can be readily assessed using direct TXRF.[4] VPD-TXRF is not possible in this case because of the lack of a native oxide layer on gallium arsenide.

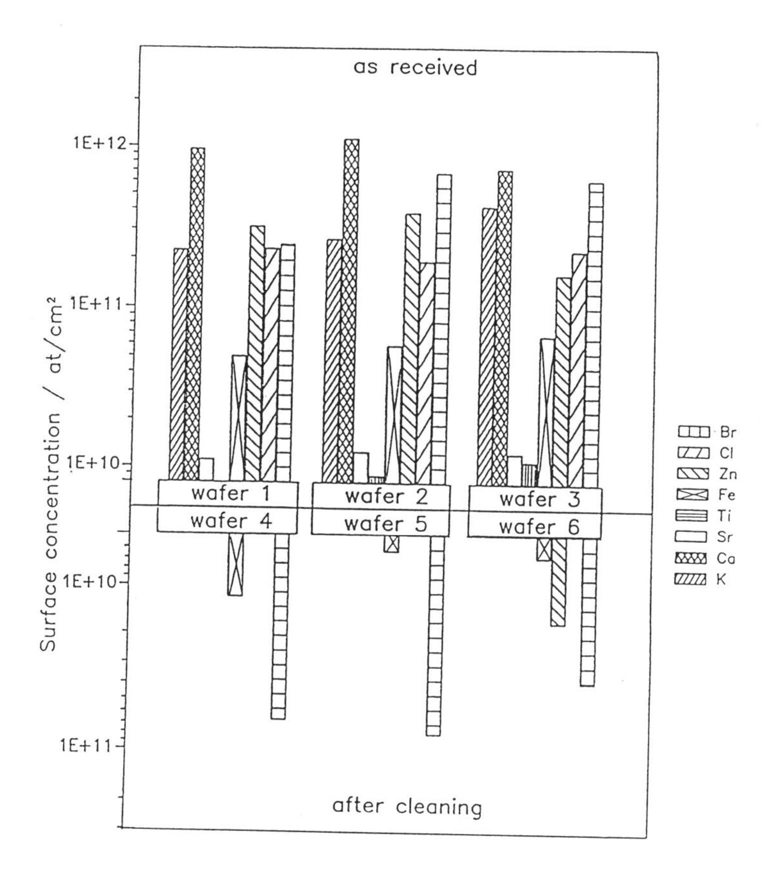

Figure 5 **Effect of cleaning on the surface purity of silicon wafers, as measured by VPD-TXRF.**

Comparative Techniques

Atomic absorption spectroscopy of VPD solutions (VPD-AAS) and instrumental neutron activation analysis (INAA) offer similar detection limits for metallic impurities with silicon substrates. The main advantage of TXRF, compared to VPD-AAS, is its multielement capability; AAS is a sequential technique that requires a specific lamp to detect each element. Furthermore, the problem of blank values is of little importance with TXRF because no handling of the analytical solution is involved. On the other hand, adequately sensitive detection of sodium is possible only by using VPD-AAS. INAA is basically a bulk analysis technique, while TXRF is sensitive only to the surface. In addition, TXRF is fast, with an typical analysis time of 1000 s; turn-around times for INAA are on the order of weeks. Gallium arsenide surfaces can be analyzed neither by AAS nor by INAA.

Conclusions

Triggered by the purity demands of silicon integrated circuit technology, TXRF has seen a rapid development in its application to solid surfaces during the last seven years, which is reflected in the availability of a variety of commercial instruments and services today. The investigation of surface cleanliness, however, does not exhaust the inherent capabilities of TXRF: From the detailed angular dependence of the fluorescence yields around the critical angle for total reflection, information may be obtained about thin films, interfaces, or multilayer structures in the future. An overview of these trends can be found in *Proceedings of the International Workshop on Total Reflection X-Ray Fluorescence.*[5] It is also possible, in principle, to obtain chemical state information, along with elemental analysis. This requires the use of high-energy resolution techniques to detect small shifts in line positions in the emitted fluorescence. In the soft X-ray region, instrumentation of this type is not commercially available.

Related Articles in the Encyclopedia

XRF and NAA

References

1 P. Eichinger, H. J. Rath, and H. Schwenke. In: *Semiconductor Fabrication: Technology and Metrology. ASTM STP 990.* (D. C. Gupta, ed.) American Society for Testing and Materials, 305, 1989.

2 U. Weisbrod, R. Gutschke, J. Knoth, and H. Schwenke. *Fresenius J. Anal. Chem.* 1991, in press.

3 C. Neumann and P. Eichinger. *Spectrochimica Acta B At. Spect.* **46,** Vol. 10, 1369, 1991.

4 R. S. Hockett, J. Metz, and J. P. Tower. *Proceedings of the Fifth Conference on Semi-Insulating III-V Materials.* Toronto, 1990.

5 *Proceedings of the International Workshop on Total Reflection X-Ray Fluorescence.* Vienna, 1990 (same as Reference 3).

6.3 PIXE

Particle-Induced X-Ray Emission

RONALD G. MUSKET

Contents

Introduction

Particle-Induced X-Ray emission (PIXE) is a quantitative, nondestructive analysis technique that relies on the spectrometry of characteristic X rays emitted when high-energy particles (~0.3–10 MeV) ionize atoms of a specimen. PIXE provides simultaneous analysis of many elements with sensitivity and detection limits that compare very favorably with other techniques. Since the first quantitative measurements of thin metal films in the late 1960s, PIXE has been applied successfully in a variety of fields, including corrosion and oxidation, semiconductors, metallurgy, thin films, geoscience, air pollution and atmospheric science, biology, medicine, art, archaeology, water analysis, and forensic science. During this 25-year period, PIXE has matured and developed into a routine analytical tool for many applications. A recent survey of over a hundred PIXE systems throughout the world revealed that the main areas of application are currently biomedicine (a major application for 40% of the systems), materials (30%) and aerosols (20%).[1] A detailed discussion of PIXE is presented in the recent book by Johansson and Campbell,[2] which was a major reference for this article.

Generally the particles used for PIXE are protons and helium ions. PIXE is one of three techniques that rely on the spectrometry of X rays emitted during irradiation of a specimen. The other techniques use irradiation by electrons (electron microprobe analysis, EMPA, and energy-dispersive X rays, EDS) and photons (X-Ray Fluorescence, XRF). In principle, each of these techniques can be used to analyze simultaneously for a large range of elements—from lithium to uranium. For simultaneous, multi-elemental determinations using a standard energy-dispersive, X-ray spectrometer, the range of elements is reduced to those with atomic number $Z > 11$. Analysis for elements with $Z > 5$ can be performed with windowless or high transmission-windowed detectors. Wavelength-dispersive detection systems can be used for high-resolution X-ray spectrometry of, at most, a few elements at a time; however, the improved resolution yields information on the chemical bonding of the element monitored. In this article only the results from the widely used lithium-drifted, silicon—Si(Li)—energy-dispersive spectrometers will be discussed. (See also the article on EDS.)

Compared to EDS, which uses 10–100 keV electrons, PIXE provides orders-of-magnitude improvement in the detection limits for trace elements. This is a consequence of the much reduced background associated with the deceleration of ions (called *bremsstrahlung*) compared to that generated by the stopping of the electrons, and of the similarity of the cross sections for ionizing atoms by ions and electrons. Detailed comparison of PIXE with XRF showed that PIXE should be preferred for the analysis of thin samples, surface layers, and samples with limited amounts of materials.[3] XRF is better for bulk analysis and thick specimens because the somewhat shallow penetration of the ions (e.g., tens of μm for protons) limits the analytical volume in PIXE.

Basic Principles

The X-ray spectrum observed in PIXE depends on the occurrence of several processes in the specimen. An ion is slowed by small inelastic scatterings with the electrons of the material, and it's energy is continuously reduced as a function of depth (see also the articles on RBS and ERS, where this part of the process is identical). The probability of ionizing an atomic shell of an element at a given depth of the material is proportional to the product of the cross section for subshell ionization by the ion at the reduced energy, the fluorescence yield, and the concentration of the element at the depth. The probability for X-ray emission from the ionized subshell is given by the fluorescence yield. The escape of X rays from the specimen and their detection by the spectrometer are controlled by the photoelectric absorption processes in the material and the energy-dependent efficiency of the spectrometer.

Interactions of Ions With Materials

After monoenergetic protons and helium ions having energies between about 0.3 and 10 MeV enter a material, they begin slowing down by inelastically scattering with electrons and elastically scattering with atomic nuclei. The statistical nature of this slowing process leads to a distribution of implanted ions about a mean depth called the projected range R_P, which has a standard deviation ΔR_P. These losses and ranges can be evaluated for various combinations of incident ion and target material using well-developed calculational procedures, such as the Monte Carlo code called TRIM (transport of ions in materials).[4]

When the velocity of the ions is much greater than that of the bound electrons, interactions with the electrons dominate and the ion path can be considered to be a straight line. At any depth associated with the straight-line part of the trajectory, the number of ions is preserved because only about one ion in a million is backscattered; however, their energies decrease slowly and spread increasingly about the average as a result of interactions with the electrons. This energy regime corresponds to that of the dominant X-ray production cross sections; thus modeling the source term for X rays is much simpler for ions than for electrons, which undergo strong deviations from their initial flight path as a result of collisions with the electrons of the target.

X-Ray Production

Although there have been various theoretical schemes for calculating cross sections for inner-shell ionizations by protons and helium ions, many PIXE workers now use the *K*- and *L*-shell cross sections calculated using the ECPSSR method. This method involves a series of modifications to the plane-wave Born approximation, which uses perturbation theory to describe the transition from an initial state consisting of a plane-wave projectile and a bound atomic electron to a final state consisting of a plane-wave particle and an ejected continuum electron. The ECPSSR method includes the deflection and velocity changes of the projectile caused by energy losses, the Coulomb field of the target nucleus, perturbation of the atomic stationary states, and relativistic effects.[5]

A tabulation[6] of the ECPSSR cross sections for proton and helium-ion ionization of *K* and *L* levels in atoms can be used for calculations related to PIXE measurements. Some representative X-ray production cross sections, which are the product of the ionization cross sections and the fluorescence yields, are displayed in Figure 1. Although these *K*-shell cross sections have been found to agree with available experimental values within 10%, which is adequate for standardless PIXE, the accuracy of the *L*-shell cross sections is limited mainly by the uncertainties in the various *L*-shell fluorescence yields. Knowledge of these yields is necessary to convert X-ray ionization cross sections to production cross sections. Of course, these same uncertainties apply to the EMPA, EDS, and XRF techniques. The *M*-shell situation is even more complicated.

The production of characteristic X rays is determined by the cross sections discussed above, but the observed X-ray spectra include both these characteristic peaks and a continuous background radiation. A detailed investigation of the origin of

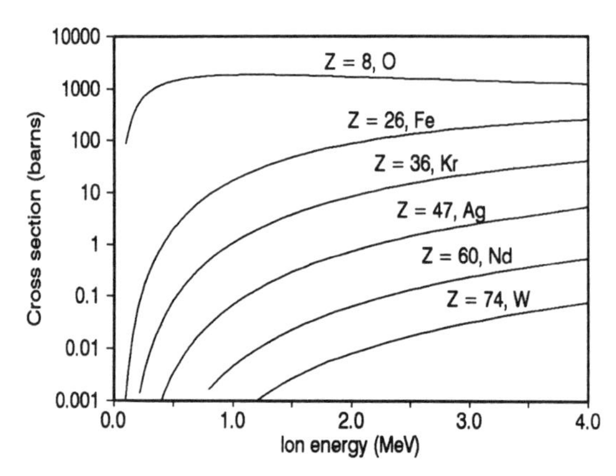

Figure 1 **Calculated *K* X-ray production cross sections for protons using the tabulated ECPSSR ionization cross sections of Cohen and Harrigan,[6] and the fluorescence yields calculated as in Johansson et al.[2] (1 barn = 10^{-24} cm^2).**

this background radiation has shown that the dominant source is the bremsstrahlung radiation emitted by the energetic electrons ejected by the ions.[7] The contributions to the background from electron and proton bremsstrahlung radiation caused by 3-MeV protons are shown in Figure 2. Deviations of the experimental results from the calculated curves for X-ray energies above 10 keV probably represent the effects of Compton scattering of γ rays from excited nuclear states, which were not accounted for in the calculations. In a classical sense, the maximum energy T_m that a 3-MeV proton can transfer to a free electron is 6.5 keV. Thus, in Figure 2 the bremsstrahlung radiation is most intense below T_m and decreases rapidly at higher energies.

From an analytical point of view, this discussion implies that changing the ion energy will not improve the characteristic-to-background (C/B) ratio for X rays having energies below about T_m because both the dominant bremsstrahlung background and the characteristic X rays result from essentially the same ionization processes. However, reducing the ion energy will shift the electron bremsstrahlung radiation to lower energies and have the effect of improving the C/B ratio (i.e., improving the detection limit) for X-ray peaks at energies above T_m. Many PIXE workers prefer 2–3 MeV protons because they provide a reasonable compromise between the characteristic X-ray production rate and the C/B ratio, while limiting the level of background from nuclear reactions. In fact, most modern ion accelerators used for materials analysis can provide protons with maximum energies of 2–4 MeV.

Detection Limits

In PIXE the X-ray spectrum represents the integral of X-ray production along the path length of the decelerating ion, as mediated by X-ray absorption in the mate-

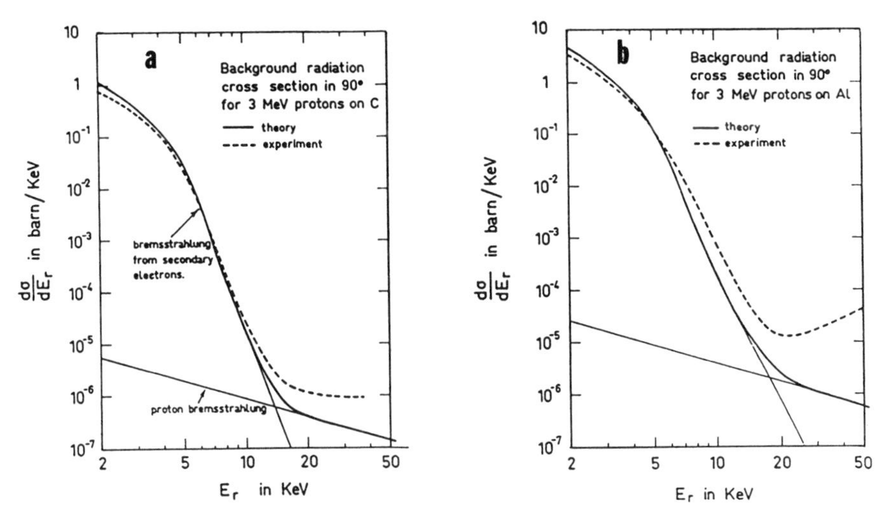

Figure 2 **Experimental and calculated background radiation production cross sections for (a) 360 μg/cm² plastic foil and (b) 200 μg/cm² Al foil.[7]**

rial. Consequently, it is convenient to consider trace analysis for three different cases: thin, free-standing specimens; surface layers (e.g., oxides or coatings) on thick specimens; and thick or bulk specimens.

Specimens are considered thin if an ion loses an insignificant amount of energy during its passage through the foil and if X-ray absorption by the specimen may be neglected. Under these circumstances, the yield of the characteristic X rays can be determined using the ionization cross sections for the energy of the incident ion, and detailed knowledge of the complete composition of the specimen is not needed to make corrections for the particle's energy loss or the absorption of X rays. As shown in Figure 3,[8] the detection limits for various elements in thin specimens depends on the host matrix. About 0.1 weight part per million (wppm) of elements with atomic numbers near 35 and near 80 can be detected in carbon. Thus, less than 10^{-12} g could be detected in or on a 100 μg/cm² carbon foil using a 1-mm² beam of 3-MeV protons.

The detection of impurities or surface layers (e.g., oxides) on thick specimens is a special situation. Although the X-ray production and absorption assumptions used for thin specimens apply, the X-ray spectra are complicated by the background and characteristic X rays generated in the thick specimen. Consequently, the absolute detection limits are not as good as those given above for thin specimens. However, the detection limits compare very favorably with other surface analysis techniques, and the results can be quantified easily. To date there has not been any systematic study of the detection limits for elements on surfaces; however, representative studies have shown that detectable surface concentrations for carbon and

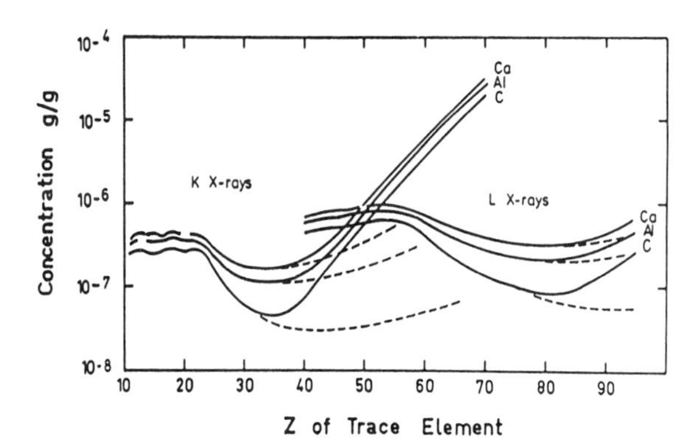

Figure 3 **Calculated detection limits for trace elements in 1 mg/cm² specimens of carbon, aluminum, and calcium (100 μC of 3-MeV protons).[8] The dashed curves represent the detection limits if the background radiation is due only to secondary electron bremsstrahlung.**

oxygen are about 100 ng C/cm² on iron using 5-MeV He[9] and 30 ng O/cm² on beryllium using 2-MeV He.[10]

In thick specimens, the particles ionize atoms along essentially their entire path in the specimen, and calculation of the characteristic X-ray production requires integration of energy-dependent cross sections over all ion energies from the incident energy to 0 and correction for the absorption of the X rays. Detection limits have been estimated for thick targets when the characteristic *K* X-ray signal occurs at an energy greater than the bremsstrahlung background (Figure 4). For thick targets,[11] limits below 100 ppm are achievable for elements with $Z > 20$ in most matrices and can be below 1 ppm for elements near $Z = 35$ in low-Z matrices; for elements with $Z < 20$, the limits are no better than 100 ppm in most matrices, but can be considerably better in low-Z matrices. For example, a detection limit of 10 ppm for oxygen in beryllium has been demonstrated.[10]

Modes of Analysis

Thin, Free-Standing Specimens

Whenever the appropriate specimens can be prepared, this mode is normally the one preferred for trace-element analysis in geoscience, air pollution and atmospheric science, biology, medicine, water analysis, and forensic science. In this case, the ions pass through the specimen with negligible energy loss and there is minimal absorption of X rays.

Surface Layers on Bulk Specimens

Included in this class of thin surface films are oxides, corrosion, contamination, and deposited layers. Although the presence of the bulk specimen results in increased

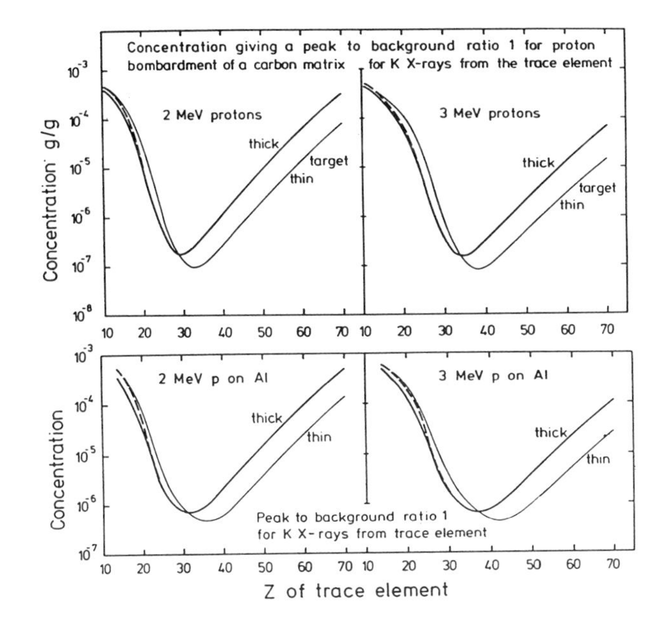

Figure 4 **Calculated minimum concentration of a trace element in thin and thick carbon and aluminum specimens for 2- and 3-MeV protons.[8]. The dashed curves show the effects of X-ray absorption on detection limits for thick targets (take-off angle of 45°).**

background radiation compared to that for thin, free standing specimens, the detection limits can be sufficiently low to permit calibration of "true" surface analysis techniques, such as Auger electron spectroscopy (AES). In this sense, PIXE fills the "quantitative gap" that exists in surface and thin-film analysis. As an example, Figure 5[12] shows the helium-induced X-ray spectrum for an anodized tantalum specimen typical of those used to define the sputtering conditions for AES depth profiling experiments. This represents the simplest surface-layer case because the element of interest in the surface layer (i.e., oxygen) is not measurable in the bulk. Thus, the O–*K* X-ray signal can be directly related to the surface concentration of the oxygen.

Bulk Material

Although XRF is generally the X-ray spectrometry method of choice for analysis of major and trace elements in bulk specimens, useful PIXE measurements can be made. A detailed review of the main considerations for thick-target PIXE[11] provides guidance for trace analysis with known and unknown matrices and bulk analysis when the constituents are unknown. Campbell and Cookson[11] also discuss the increased importance of secondary fluorescence and geometrical accuracy for bulk measurements.

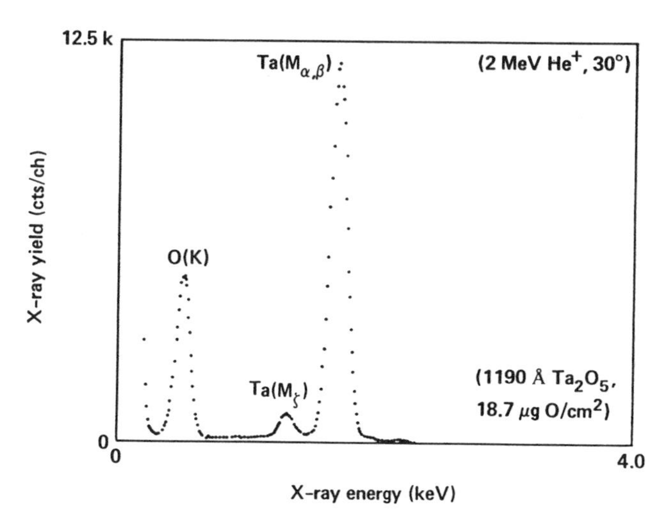

Figure 5 **Helium-ion induced X-ray spectrum from anodized tantalum (fluence = 1.5×10^{15} He$^+$/cm^2).[12]**

Depth Profiling of Surface Layers on Specimens

Rutherford Backscattering (RBS) provides quantitative, nondestructive elemental depth profiles with depth resolutions sufficient to satisfy many requirements; however, it is generally restricted to the analysis of elements heavier than those in the substrate. The major reason for considering depth profiling using PIXE is to remove this restrictive condition and provide quantitative, nondestructive depth profiles for all elements yielding detectable characteristic X rays (i.e., $Z > 5$ for Si(Li) detectors).

Because a PIXE spectrum represents the integral of all the X rays created along the particle's path, a single PIXE measurement does not provide any depth profile information. All attempts to obtain general depth profiles using PIXE have involved multiple measurements that varied either the beam energy or the angle between the beam and the target, and have compared the results to those calculated for assumed elemental distributions. Profiles measured in a few special cases suggest that the depth resolution by nondestructive PIXE is only about 100 nm and that the absolute concentration values can have errors of 10–50%.

Although depth profiling using nondestructive PIXE does not appear promising, PIXE in combination with RBS or with low-energy ion sputtering offers the possibility of quantitative depth profiles for elements with $Z > 5$. Because X rays and backscattered particles emanate from the specimen during ion irradiation, both should be detected. In fact, simultaneously performing PIXE and RBS eliminates ambiguities in the interpretation and modeling of the RBS results.[12] As an example, consider RBS results from high-Z materials with surface layers containing low-Z elements. In such cases, the RBS spectra are dominated by scattering from the high-

Z elements, with some indication of compositional changes with depth; PIXE provides unequivocal identification of the elements present and, with appropriate calibrations, the absolute areal densities of these elements.

Although not commonly available, the combination of PIXE and low-energy ion sputtering can provide quantitative, destructive depth profiles. PIXE measurements taken after each sputtering period give the quantity of each element remaining and, by difference, the number of atoms (of each element) removed during the sputtering period. Consequently, this combination can yield both the elemental concentrations and the conversion of the sputtering time or fluence to a depth scale. The achievable depth resolution would be determined ultimately by the precision of the PIXE results or nonuniform sputtering effects and could be about 1 nm at the surfaces of specimens.

Other Modes of Analysis

For the preceding modes, the discussion implicitly assumed the "normal" conditions for PIXE analysis (i.e., few mm-diameter beam, approximately constant beam current, and specimen in vacuum), and ignored the crystallographic nature of the specimen. However, some of the most interesting PIXE results have been obtained using other modes.

Reducing the ion beam to a small spot (1–10 μm) and scanning this microbeam across the specimen yields lateral concentration maps analogous to those obtainable with an electron microprobe, but with the advantages inherent in particle-induced spectra and with the potential of much higher spatial resolution, because the particle beam spot and the X-ray source volume have essentially the same lateral dimensions due to the nearly straight-line trajectory of the particles over their X-ray producing range.

External-beam PIXE refers to measurements with the specimen removed from a vacuum environment. This mode permits the analysis of large or volatile specimens and consists of allowing the particle beam to exit, through a thin window, the vacuum of the beam line and impinge on the specimen held at atmospheric pressure of air or other gases (e.g., helium).

The combination of PIXE with channeling of ions through the open directions (i.e., axial or planar channels) of monocrystalline materials can be used to determine the location of impurity atoms in either interstitial or substitutional sites or to assess the extent of lattice imperfections in the near-surface region. Channeling measurements are usually performed with RBS, but PIXE offers distinct advantages over RBS. First, because the PIXE cross sections are larger, the measurements can be performed with lower beam intensities or fluences. In addition, PIXE can be used to locate light elements in a matrix of heavy elements.

On-demand beam pulsing has been shown to be effective for eliminating pulse pileup in the X-ray detection system, minimizing the energy dissipated in delicate specimens, yet maximizing the data throughput of the overall system. In essence, the on-demand pulsing system consists of deflecting the beam off the specimen when the detector's amplifying system begins to process a pulse and returning the beam to the specimen at the end of the pulse-processing time.[13]

Quantification

Quantification of raw PIXE spectra involves identifying all the peaks, determining the net counts under each peak, and correcting for the energy-dependence of the ionization cross sections as a function of depth in the material; the absorption of the X rays in the material; the production of secondary fluorescence; and the fraction of the X rays detected. In general, high accuracy in PIXE is achieved only for thin or homogeneous, thick specimens. Although PIXE spectra are routinely analyzed using least-squares fitting codes, the accuracies are ultimately limited by the accuracies of the fundamental data bases for ionization cross sections, fluorescence yields, ion stopping cross sections, and X-ray absorption effects, and by the calibration procedures used to determine the energy dependence of the fraction of the X rays detected.

A detailed comparison of the analysis of the same set of PIXE data using five different spectral fitting programs yielded remarkably good agreement for the results obtained.[14] The five least-squares fitting codes were used on PIXE spectra from a set of thin biological, environmental, and geological specimens. The results were the same within 5%, which suggests that PIXE analysis has matured to quite an acceptable level.

Similar accuracies have been found for thick, homogeneous, complex specimens when corrections for secondary excitation are also included. With appropriate standards, total accuracies of 2% have been demonstrated. Because the determination of the lighter elements (i.e., $5 < Z < 15$) are more sensitive to the uncertainties in the data base items listed above, less accuracy should be expected for these elements.

Artifacts

The major artifacts contributing to uncertainties in PIXE results stem from effects caused by bombardment of nonideal specimens, particularly thick specimens. The ideal thick specimen would be a homogeneous, smooth electrical conductor that does not change during bombardment. Except for rather simple, well-defined layered structures (e.g., surface oxide layers), specimens having compositional variations with depth yield spectra whose analyses can have large inaccuracies.

Changes in the composition of a specimen over the analyzed depth can be caused by beam heating or by beam charging of the specimen. Beam heating can lead to selective vaporization of some elements or diffusional redistribution of the elements. If the surface charges up to some potential, then electric-field enhanced diffusion can selectively redistribute certain elements. Beam-heating effects usually can be mitigated by lowering beam intensities or by cooling the specimen. Electric-field enhanced diffusion can be controlled using the procedures described below for eliminating specimen charging by the beam.

Ion bombardment of electrically insulating specimens can lead to a surface charge giving a potential at the beam spot that can be significant (e.g., from a few kV to > 10 kV). Such potentials can cause surface discharges between the spot and the closest grounded conductor. The PIXE spectra would then contain characteristic and bremsstrahlung X rays excited by electrons participating in the discharges. Thus, the accuracy for the characteristic peaks would be reduced, and the relatively large bremsstrahlung radiation would hide the characteristic X-ray peaks of some trace elements. In addition, surface charging precludes accurate current integration from the specimen current, even when electron suppression is used. Standard procedures to eliminate or minimize these effects include neutralizing the surface with electrons from a hot filament or a thin metal or carbon foil placed in front of the specimen, but out of view of the X-ray detector; coating the specimen with a thin, conducting layer that either covers the bombarded spot or encircles the spot and leads to a conductor; and introducing a partial pressure of a gas (e.g., helium) into the chamber and letting ionization of the gas by the beam provide a conducting path to discharge the specimen.

The geometrical arrangement of the beam axis, the specimen normal, and the detection angle must be well known to obtain accurate PIXE results. A rough surface affects both the ion range and the X-ray absorption. The impact on accuracy will depend on the element of interest and the matrix. The largest effects occur for situations involving high X-ray absorption coefficients. As an example, yield variations of about 25% for sulfur in iron have been calculated for 10-μm grooves, when the groove direction is perpendicular to the detector axis.[15] Of course, minimal effects are expected when the detector views the specimen along the grooves.

Conclusions

Within the last 5–10 years PIXE, using protons and helium ions, has matured into a well-developed analysis technique with a variety of modes of operation. PIXE can provide quantitative, nondestructive, and fast analysis of essentially all elements. It is an ideal complement to other techniques (e.g., Rutherford backscattering) that are based on the spectroscopy of particles emitted during the interaction of MeV ion beams with the surface regions of materials, because

1. X rays also are emitted from the bombarded specimen.
2. Characteristic X rays provide an unambiguous identification of the elements present.
3. Quantitative analysis of low-*Z* elements on or in high-*Z* materials can be performed.

PIXE detection limits for surface layers on bulk specimens are sufficiently low to permit calibration of true surface analysis techniques (e.g., Auger electron spectroscopy).

Development of several existing trends can be expected to enhance the value of PIXE in the future. Analyses for low-Z elements (e.g., $5 < Z < 15$) should receive increased attention because of the availability of Si(Li) detectors with windows that transmit a large fraction of < 1-keV X, rays yet support atmospheric pressure. Appropriate modifications to the existing spectral fitting programs will be required to make quantification for low-Z elements a routine procedure. Since micro-PIXE equipment has just recently become available commercially, a proliferation of micro-PIXE capabilities can be anticipated. The demonstrated compatibility and usefulness of simultaneous PIXE and RBS should lead to routine applications of such measurements. With specific regard to surface analysis, arrangements combining PIXE with low-energy sputtering will allow quantitative, high-resolution depth profiling of materials. In addition, combinations of PIXE with true surface analysis techniques (e.g., AES and XPS) will improve greatly the quantitative accuracy of these techniques.

This work was performed under the auspices of the US DOE by Lawrence Livermore National Laboratory under contract no. W-7405-Eng-48.

Related Articles in the Encyclopedia

RBS, ERS, XRF, EDS, and EPMA

References

1 T. A. Cahill and J. Miranda, private communication, 1991.

2 S. A. E. Johansson and J. L. Campbell. *PIXE: A Novel Technique for Elemental Analysis.* John Wiley, Chichester, 1988. A detailed introduction to PIXE and the relevant literature.

3 J. P. Willis. *Nucl. Instr. Meth. Phys. Res.* **B35,** 378, 1988. A critical comparison of PIXE with XRF.

4 J. F. Ziegler, J. P. Biersack, and U. Littmark. *The Stopping and Range of Ions in Solids.* Pergamon Press, New York, 1985. Source book for TRIM calculations.

5 W. Brandt and G. Lapicki. *Phys. Rev.* **A20,** 465, 1979; **A23,** 1717, 1981. Original papers on ECPSSR calculations of cross sections.

6 D. D. Cohen and M. Harrigan. *Atomic Data and Nuclear Data Tables.* **33,** 255, 1985. Tabulated ECPSSR cross sections for protons and He^+.

7 F. Folkmann, C. Gaarde, T. Huus, and K. Kemp. *Nucl. Instr. Meth.* **116,** 487, 1974. First theoretical estimates of background in PIXE.

8 F. Folkmann. In: *Ion Beam Surface Layer Analysis.*(O. Meyer, G. Linker, and O. Keppeler, eds.) Plenum, New York, 1976, pgs. 695 and 747. Early experimental evaluation of X-ray background in PIXE.

9 R. G. Musket.In: *Advances in X-Ray Analysis.* (G.J. McCarthy et al., eds.) Plenum Publishing, 1979, volume 22, p.401. PIXE of C on and in Fe.

10 R. G. Musket. *Nucl. Instr.Meth. Phys. Res.* **B24/25,** 698, 1987. PIXE of O on and in Be.

11 J. L. Campbell and J. A. Cookson. *Nucl. Instr. Meth. Phys. Res.* **B3,** 185, 1984. A review of thick-target PIXE.

12 R. G. Musket. *Nucl. Instr. Meth. Phys. Res.* **218,** 420, 1983. Examples of simultaneous PIXE and RBS.

13 X. Zeng and X. Li. *Nucl. Instr. Meth. Phys. Res.* **B22,** 99, 1987. Details of fast, transistorized on-demand beam for PIXE.

14 J. L. Campbell, W. Maenhaut, E. Bombelka, E. Clayton, K. Malmqvist, J. A. Maxwell, J. Pallon, and J. Vandenhaute. *Nucl. Instr. Meth. Phys. Res.* **B14,** 204, 1986. Comparison of five PIXE spectral processing techniques.

15 J. A. Cookson and J. L. Campbell. *Nucl. Instr. Meth. Phys. Res.* **216**, 489, 1983. Calculated effects of surface roughness on thick-target PIXE.

7

VISIBLE / UV EMISSION, REFLECTION, AND ABSORPTION

7.0 INTRODUCTION

In this chapter, three techniques using visible (and UV) light to probe the near surface regions of solids are described. In two of them, Photoluminescence, PL, and Modulation Spectroscopy, electronic transitions between valence and conduction bands excited by the incident light are used. Modulation Spectroscopy is simply a specialized way of recording the *absorption spectrum* as the wavelength of the incident light is scanned. The derivative spectrum is recorded by phase-sensitive detection as the temperature, electric field, or stress of the sample is modulated, improving sensitivity to small spectral changes. PL, on the other hand, looks at the *emission spectrum* as the excited electronic states induced by the incident light decay back to lower states. The third technique, Variable Angle Spectroscopic Ellipsometry, VASE, involves measuring changes in intensity of *polarized light* reflected from interfaces as a function of incident wavelength and angle.

All three techniques probe 500 Å to 1 μm or so in depth for opaque materials, depending on the penetration depth of the incident light. For transparent materials, essentially bulk properties are measured by PL and Modulation Spectroscopy. All three techniques can be performed in ambient atmosphere, since visible light is used both as incident probe and signal.

In Modulation Spectroscopy, which is mostly used to characterize semiconductor materials, the peak positions, intensities and widths of features in the absorption spectrum are monitored. The positions, particularly the band edge (which defines the band gap), are the most useful, allowing determination of alloy concentration, strain, and damage, and identification of impurities. Absolute sensitivities are good enough to detect monolayer concentrations. Lateral resolution down to 100 μm can be achieved with suitable optics. Modulation Spectroscopy can sometimes be used as a screening method for device performance when there are features in the spectrum (positions, widths, or intensities) that correlate with performance. It can also be used as an *in situ* monitor of structural perfection while growing epitaxial material.

PL is currently more widely used than Modulation Spectroscopy in semiconductor and insulator characterization though it basically accesses some of the same information. It is also widely used outside the realm of semiconductor materials science, for example in pharmaceuticals, biochemistry, and medicine, where it is known under the general name of fluorometry. In PL of semiconductors, one monitors emission from the bottom of the conduction band to the top of the valence band. All the properties that can be determined from band-edge movement in Modulation Spectroscopy can also be determined by PL—for example, strain, damage, alloy composition, and the perfection of growing surfaces. Impurities or dopants can be detected with high sensitivity when there are transitions to impurity levels within the band gap. To sharpen transitions, and therefore improve resolution and sensitivity, PL is usually performed at cryogenic (liquid helium) temperatures. Under these conditions PL can detect species down to 10^{10}–10^{14} cm^3, depending on the particular species. Spatial resolution down to 1 μm can be obtained for fixed-wavelength laser incident PL. Instrumentation for PL and Modulation Spectroscopy is quite similar, often being constructed so that both can be performed. There is no complete "commercial system." One usually builds up a system using commercial components (light source, monochromator, detector, etc.). Costs vary from a minimum of $10,000 for primitive detection of PL to $250,000 for a system that can do everything. PL has the same physical basis as Cathodoluminescence (CL) discussed in Chapter 3. The practical differences caused by the use of an electron beam are: probing depths can be shorter (down to 100 Å); spatial resolution is better (down to 1000 Å); beam damage may occur, and a high-vacuum system is necessary.

VASE monitors the intensity of polarized light after it has been transmitted through a thin film to an interface, reflected, and transmitted back through the film. The film thickness (in the 1 nm to 1 μm range), the optical constants, and the interface roughness (if less than 100 nm) can all be extracted if enough measurements at different angles of incidence are made. However, these parameters are derived from a fit to an assumed model (thickness and optical constants), which is not a unique procedure, leaving room for gross error in the wrong hands. Planar interfaces are needed over the lateral area probed, which is usually 1 mm^2 but which can be as small as 100 μm^2. VASE is used primarily for surface coatings on semiconductors and dielectrics, optical coatings, and multilayer thin-film structures.

7.1 PL

Photoluminescence

CARL COLVARD

Contents

Introduction

Luminescence refers to the emission of light by a material through any process other than blackbody radiation. The term *Photoluminescence* (PL) narrows this down to any emission of light that results from optical stimulation. Photoluminescence is apparent in everyday life, for example, in the brightness of white paper or shirts (often treated with fluorescent whiteners to make them literally glow) or in the light from the coating on a fluorescent lamp. The detection and analysis of this emission is widely used as an analytical tool due to its sensitivity, simplicity, and low cost. Sensitivity is one of the strengths of the PL technique, allowing very small quantities (nanograms) or low concentrations (parts-per-trillion) of material to be analyzed. Precise quantitative concentration determinations are difficult unless conditions can be carefully controlled, and many applications of PL are primarily qualitative.

PL is often referred to as fluorescence spectrometry or fluorometry, especially when applied to molecular systems. Uses for PL are found in many fields, including environmental research, pharmaceutical and food analysis, forensics, pesticide studies, medicine, biochemistry, and semiconductors and materials research. PL can be used as a tool for quantification, particularly for organic materials, wherein the compound of interest can be dissolved in an appropriate solvent and examined either as a liquid in a cuvette or deposited onto a solid surface like silica gel, alumina, or filter paper. Qualitative analysis of emission spectra is used to detect the presence of trace contaminants or to monitor the progress of reactions. Molecular applications include thin-layer chromatography (TLC) spot analysis, the detection of aromatic compounds, and studies of protein structure and membranes. Polymers are studied with regard to intramolecular energy transfer processes, conformation, configuration, stabilization, and radiation damage.

Many inorganic solids lend themselves to study by PL, to probe their intrinsic properties and to look at impurities and defects. Such materials include alkali-halides, semiconductors, crystalline ceramics, and glasses. In opaque materials PL is particularly surface sensitive, being restricted by the optical penetration depth and carrier diffusion length to a region of 0.05 to several μm beneath the surface.

Emission spectra of impurity levels are used to monitor dopants in III-V, II-VI, and group IV compounds, as well as in dilute magnetic and other chalcogenide semiconductors. PL efficiency can be used to provide a measure of surface damage due to sputtering, polishing, or ion bombardment, and it is strongly affected by structural imperfections arising during the growth of films like SiC and diamond. Coupled with models of crystalline band structure, PL is a powerful tool for monitoring the dimensions and other properties of semiconductor superlattices and quantum wells (man-made layered structures with angstrom-scale dimensions). The ability to work with low light levels makes it well suited to measurements on thin epitaxial layers.

Basic Principles

In PL, a material gains energy by absorbing light at some wavelength by promoting an electron from a low to a higher energy level. This may be described as making a transition from the ground state to an excited state of an atom or molecule, or from the valence band to the conduction band of a semiconductor crystal (electron–hole pair creation). The system then undergoes a nonradiative internal relaxation involving interaction with crystalline or molecular vibrational and rotational modes, and the excited electron moves to a more stable excited level, such as the bottom of the conduction band or the lowest vibrational molecular state. (See Figure 1.)

If the cross-coupling is strong enough this may include a transition to a lower electronic level, such as an excited triplet state, a lower energy indirect conduction band, or a localized impurity level. A common occurrence in insulators and semiconductors is the formation of a bound state between an electron and a hole (called

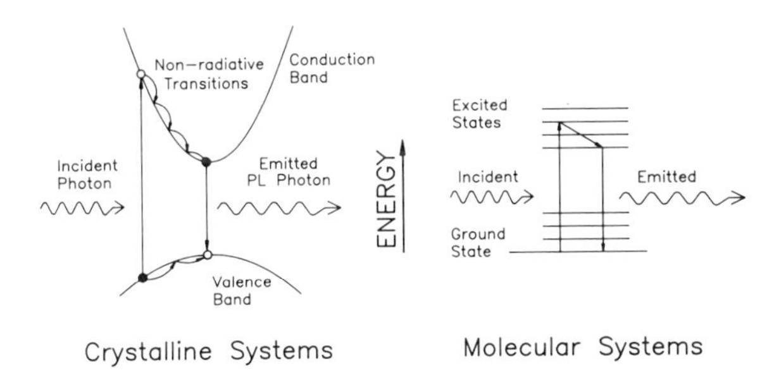

Figure 1 **Schematic of PL from the standpoint of semiconductor or crystalline systems (left) and molecular systems (right).**

an exciton) or involving a defect or impurity (electron bound to acceptor, exciton bound to vacancy, etc.).

After a system-dependent characteristic lifetime in the excited state, which may last from picoseconds to many seconds, the electronic system will return to the ground state. In luminescent materials some or all of the energy released during this final transition is in the form of light, in which case the relaxation is said to be radiative. The wavelength of this emission is longer than that of the incident light. This emitted light is detected as photoluminescence, and the spectral dependence of its intensity is analyzed to provide information about the properties of the material. The time dependence of the emission can also be measured to provide information about energy level coupling and lifetimes. In molecular systems, we use different terminology to distinguish between certain PL processes that tend to be fast (sub-microsecond), whose emission we call fluorescence, and other, slower ones (10^{-4} s to 10 s) which are said to generate phosphorescence.

The light involved in PL excitation and emission usually falls in the range 0.6–6 eV (roughly 200–2000 nm). Many electronic transitions of interest lie in this range, and efficient sources and detectors for these wavelengths are available. To probe higher energy transitions, UPS, XPS, and Auger techniques become useful. X-ray fluorescence is technically a high-energy form of PL involving X rays and core electrons instead of visible photons and valence electrons. Although lower energy intraband, vibrational, and molecular rotational processes may participate in PL, they are studied more effectively by Raman scattering and IR absorption.

Since the excited electronic distribution approaches thermal equilibrium with the lattice before recombining, only features within an energy range of ~kT of the lowest excited level (the band edge in semiconductors) are seen in a typical PL *emission* spectrum. It is possible, however, to monitor the intensity of the PL as a function of the wavelength of the *incident* light. In this way the emission is used as a probe of the absorption, showing additional energy levels above the band gap. Examples are given below.

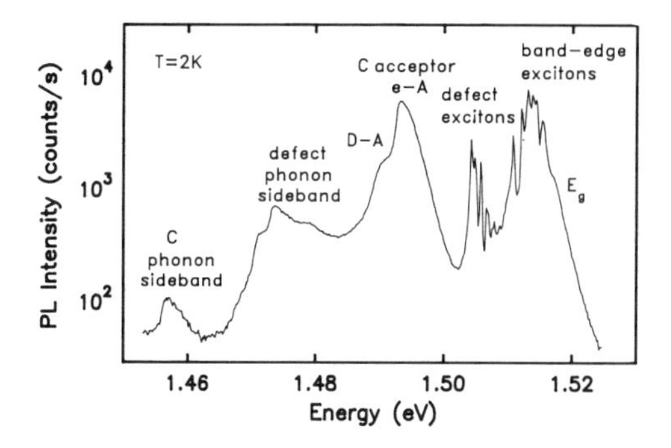

Figure 2 **PL spectra of MBE grown GaAs at 2 K near the fundamental gap, showing C-acceptor peak on a semilog scale.**

Scanning a range of wavelengths gives an emission spectrum that is characterized by the intensity, line shape, line width, number, and energy of the spectral peaks. Depending on the desired information, several spectra may be taken as a function of some external perturbation on the sample, such as temperature, pressure, or doping variation, magnetic or electric field, or polarization and direction of the incident or emitted light relative to the crystal axes.

The features of the spectrum are then converted into sample parameters using an appropriate model of the PL process. A sampling of some of the information derived from spectral features is given in Table 1.

A wide variety of different mechanisms may participate in the PL process and influence the interpretation of a spectrum. At room temperature, PL emission is thermally broadened. As the temperature is lowered, features tend to become sharper, and PL is often stronger due to fewer nonradiative channels. Low temperatures are typically used to study phosphorescence in organic materials or to identify particular impurities in semiconductors.

Figure 2 shows spectra from high-purity epitaxial GaAs ($N_A < 10^{14}$ cm^{-3}) at liquid helium temperature. The higher energy part of the spectrum is dominated by electron–hole bound pairs. Just below 1.5 eV one sees the transition from the conduction band to an acceptor impurity (*e–A*). The impurity is identified as carbon from its appearance at an energy below the band gap equal to the carbon binding energy. A related transition from the acceptor to an unidentified donor state (*D–A*) and a sideband lower in energy by one LO-phonon are also visible. Electrons bound to sites with deeper levels, such as oxygen in GaAs, tend to recombine nonradiatively and are not easily seen in PL.

PL is generally most useful in semiconductors if their band gap is direct, i.e., if the extrema of the conduction and valence bands have the same crystal momentum, and optical transitions are momentum-allowed. Especially at low temperatures,

Spectral feature	Sample parameter
Peak energy	Compound identification
	Band gap/electronic levels
	Impurity or exciton binding energy
	Quantum well width
	Impurity species
	Alloy composition
	Internal strain
	Fermi energy
Peak width	Structural and chemical "quality"
	Quantum well interface roughness
	Carrier or doping density
Slope of high-energy tail	Electron temperature
Polarization	Rotational relaxation times
	Viscosity
Peak intensity	Relative quantity
	Molecular weight
	Polymer conformation
	Radiative efficiency
	Surface damage
	Excited state lifetime
	Impurity or defect concentration

Table 1 **Examples of sample parameters extracted from PL spectral data. Many rely on a model of the electronic levels of the particular system or comparison to standards.**

localized bound states and phonon assistance allow certain PL transitions to appear even in materials with an indirect band gap, where luminescence would normally not be expected. For this reason bound exciton PL can be used to identify shallow donors and acceptors in indirect GaP, as well as direct materials such as GaAs and

InP, in the range 10^{13}–10^{14} cm^{-3}. Boron, phosphorus, and other shallow impurities can be detected in silicon in concentrations[1] approaching 10^{11} cm^{-3}. Copper contamination at Si surfaces has been detected down to 10^{10} cm^{-3} levels.[2]

Common Modes of Analysis and Examples

Applications of PL are quite varied. They include compositional analysis, trace impurity detection, spatial mapping, structural determination (crystallinity, bonding, layering), and the study of energy-transfer mechanisms. The examples given below emphasize semiconductor and insulator applications, in part because these areas have received the most attention with respect to surface-related properties (i.e., thin films, roughness, surface treatment, interfaces), as opposed to primarily bulk properties. The examples are grouped to illustrate four different modes for collecting and analyzing PL data: spectral emission analysis, excitation spectroscopy, time-resolved analysis, and spatial mapping.

Spectral Emission Analysis

The most common configuration for PL studies is to excite the luminescence with fixed-wavelength light and to measure the intensity of the PL emission at a single wavelength or over a range of wavelengths. The emission characteristics, either spectral features or intensity changes, are then analyzed to provide sample information as described above.

As an example, PL can be used to precisely measure the alloy composition x of a number of direct-gap III-V semiconductor compounds such as $Al_xGa_{1-x}As$, $In_xGa_{1-x}As$, and $GaAs_xP_{1-x}$, since the band gap is directly related to x. This is possible in extremely thin layers that would be difficult to measure by other techniques. A calibration curve of composition versus band gap is used for quantification. Cooling the sample to cryogenic temperatures can narrow the peaks and enhance the precision. A precision of 1 meV in bandgap peak position corresponds to a value of 0.001 for x in $Al_xGa_{1-x}As$, which may be useful for comparative purposes even if it exceeds the accuracy of the x-versus-bandgap calibration.

High-purity compounds may be studied at liquid He temperatures to assess the sample's quality, as in Figure 2. Trace impurities give rise to spectral peaks, which can sometimes be identified by their binding energies. The application of a magnetic field for magnetophotoluminescence can aid this identification by introducing extra field-dependent transitions that are characteristic of the specific impurity.[3] Examples of identifiable impurities in GaAs, down to around 10^{13} cm^{-3}, are C, Si, Be, Mn, and Zn. Transition-metal impurities give rise to discrete energy transitions within the band gap. Peak shifts and splitting of the acceptor-bound exciton lines can be used to measure strain. In heavily Be-doped GaAs and some quantum two-dimensional (2D) structures, the Fermi edge is apparent in the spectra, and its position can be converted into carrier concentration.

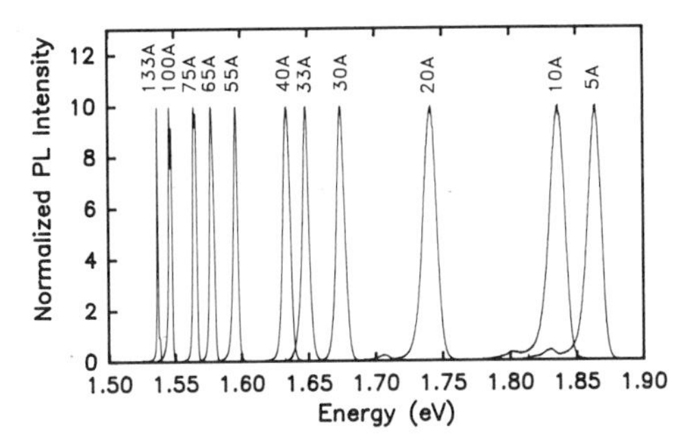

Figure 3 **Composite plot of 2 K excitonic spectra from 11 GaAs/$Al_{0.3}Ga_{0.7}As$ quantum wells with different thicknesses. The well width of each is given next to its emission peak.**

A common use of PL peak energies is to monitor the width of quantum well structures. Figure 3 shows a composite plot of GaAs quantum wells surrounded by $Al_{0.3}Ga_{0.7}As$ barriers, with well widths varying from 13 nm to 0.5 nm, the last being only two atomic layers thick. Each of these extremely thin layers gives rise to a narrow PL peak at an energy that depends on its thickness. The well widths can be measured using the peak energy and a simple theoretical model. The peak energy is seen to be very sensitive to well width, and the peak width can give an indication of interface sharpness.

PL can be used as a sensitive probe of oxidative photodegradation in polymers.[9] After exposure to UV irradiation, materials such as polystyrene, polyethylene, polypropylene, and PTFE exhibit PL emission characteristic of oxidation products in these hosts. The effectiveness of stabilizer additives can be monitored by their effect on PL efficiency.

PL Excitation Spectroscopy

Instead of scanning the emission wavelength, the analyzing monochromator can be fixed and the wavelength of the incident exciting light scanned to give a PL excitation (PLE) spectrum. A tunable dye or Ti:Sapphire laser is typically used for solids, or if the signals are strong a xenon or quartz-halogen lamp in conjunction with a source monochromator is sufficient.

The resulting PL intensity depends on the absorption of the incident light and the mechanism of coupling between the initial excited states and the relaxed excited states that take part in emission. The spectrum is similar to an absorption spectrum and is useful because it includes higher excited levels that normally do not appear in the thermalized PL emission spectra. Some transitions are apparent in PLE spectra from thin layers that would only be seen in absorption data if the sample thickness were orders of magnitude greater.

This technique assists in the identification of compounds by distinguishing between substances that have the same emission energy but different absorption bands. In semiconductors, it can be valuable for identifying impurity PL peaks, especially donors, by enhancing certain PL transitions through resonant excitation. It is useful for determining the energy levels of thin-film quantum structures, which, when combined with appropriate models, are used to simultaneously determine well widths, interface band offsets, and effective masses. Information about higher energy transitions can also be obtained by Modulation Spectroscopy techniques such as photoreflectance and electroreflectance.

Time-Resolved PL

By monitoring the PL intensity at a chosen wavelength as a function of time delay after an exciting pulse, information can be obtained about the electron relaxation and recombination mechanisms, including nonradiative channels. The time scales involved may vary from two hundred femtoseconds to tens of seconds. A full emission spectrum may be measured also at successive points in time. Spectral analysis then yields, for example, the evolution of a carrier distribution as excitonic states form and as carriers are trapped by impurities. The progress of chemical reactions with time can be followed using time-dependent data. By monitoring the depolarization of luminescence with time of PL from polymer chains, rotational relaxation rates and segmental motion can be measured.

A useful application of time-dependent PL is the assessment of the quality of thin III-V semiconductor alloy layers and interfaces, such as those used in the fabrication of diode lasers. For example, at room temperature, a diode laser made with high-quality materials may show a slow decay of the active region PL over several ns, whereas in low-quality materials nonradiative centers (e.g., oxygen) at the cladding interface can rapidly deplete the free-carrier population, resulting in much shorter decay times. Measurements of lifetime are significantly less dependent on external conditions than is the PL intensity.

PL Mapping

Spatial information about a system can be obtained by analyzing the spatial distribution of PL intensity. Fluorescent tracers may be used to image chemical uptake in biological systems. Luminescence profiles have proven useful in the semiconductor industry for mapping impurity distributions, dislocations, or structural homogeneity in substrate wafers or epilayers. Similar spatial information over small regions is obtained by cathodoluminescence imaging.

For mapping, the sample (or the optical path) is translated, and at each position PL at a single wavelength or over an entire spectrum is measured. The image is formed from variations in intensity, peak energy, or peak line width. Lateral resolution of 1 μm is possible. Figure 4 shows an application of PL to identify imperfections in a 2-in InGaAsP epitaxial wafer.

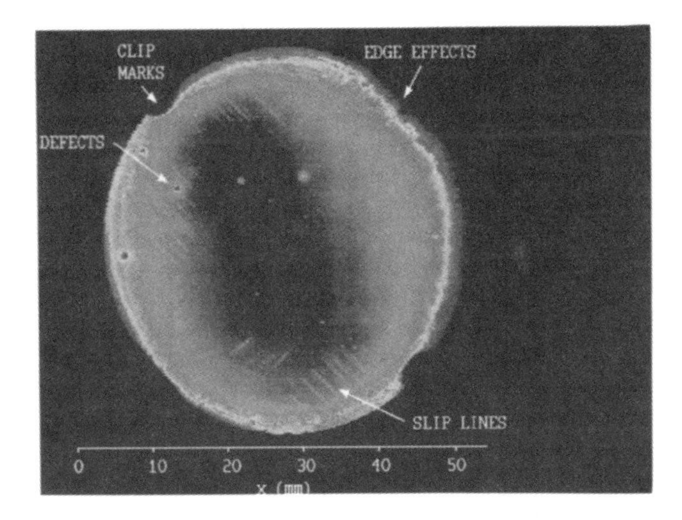

Figure 4 Spatial variation of PL intensity of an InGaAsP epitaxial layer on a 2-in InP substrate shows results of nonoptimal growth conditions. (Data from a Waterloo Scientific SPM-200 PL mapper, courtesy of Bell Northern Research)

Sample Requirements

PL measurements are generally nondestructive, and can be obtained in just about any configuration that allows some optically transparent access within several centimeters of the sample. This makes it adaptable as an *in situ* measurement tool. Little sample preparation is necessary other than to eliminate any contamination that may contribute its own luminescence. The sample may be in air, vacuum, or in any transparent, nonfluorescing medium.

Small probed regions down to 1–2 μm are possible using microscope lenses. Lasers can supply as much pump power as needed to compensate for weaker signals, but a limit is reached when sample heating or nonlinear optically induced processes become significant.

For semiconductor work, either whole wafers or small pieces are used, the latter often being necessary for insertion into a cryostat. Bulk solids may be analyzed in any form, but scattered light may be reduced and the signal increased if the emitting surface is specular.

Quantitative Abilities

Photoluminescence finds its greatest strengths as a qualitative and semiquantitative probe. Quantification based on absolute or relative intensities is difficult, although it is useful in applications where the sample and optical configurations may be carefully controlled. The necessary conditions are most easily met for analytical applications of molecular fluorescence, where samples may be reproducibly prepared in the form of controlled films or as dilute concentrations of material in a transparent liquid solvent, and where reference standards are available.[5]

PL intensities are strongly influenced by factors like surface conditions, heating, photochemical reactions, oxygen incorporation, and intensity, power density and the wavelength of the exciting light. If these factors are carefully controlled PL intensities can be used to study various aspects of the sample, but such control is not always possible. Other aspects that can cause intensity variations are the focal region of the incident and collection optics, the relationship of the sample's image to the monochromator entrance slit, and the spectral response of the detector and optical path.

Nevertheless, quantification is possible, a good example being the evaluation of the composition of chromatographic separations adsorbed onto glass, alumina, polyethylene, or paper. When compared with known standards, the presence of only a few nanograms of a strong fluorophore may be quantified to better than 10%.

As another example, PL from GaP:N at 77 K is a convenient way to assess nitrogen concentrations in the range 10^{17}–10^{19} cm^{-3} by observing the ratio of the peak intensity of the nitrogen-bound exciton transition to that of its LO phonon sideband, or to peaks involving nitrogen pairs. Similar ratio analysis allows estimates of EL2 defect concentration in GaAs wafers and has been used to quantify Mn concentrations in GaAs. Under carefully controlled conditions, PL intensity from layered-as-grown device structures can be correlated with device parameters (e.g., lasing threshold and transistor gain) and used to predict final device performance on other similar wafers.

Instrumentation

A variety of commercial instruments are available for PL measurements. These include spectrofluorometers intended primarily for use with liquids in a standard configuration, and simple filter-based systems for monitoring PL at a single wavelength. For use with opaque samples and surfaces, a few complete commercial systems are available or may be appropriately modified with special attachments, but due to the wide range of possible configuration requirements it is common to assemble a custom system from commercial optical components.

Four basic components make up a PL system:

1. A source of light for excitation. Surface studies generally require a continuous or pulsed laser. A dye or Ti:sapphire laser is used if tunability is needed.
2. A sample holder, including optics for focusing the incident light and collecting the luminescence. Efficient light collection is important, and the sample holder may need to allow for a cryostat, pressure cell, magnet, or electrical contacts.

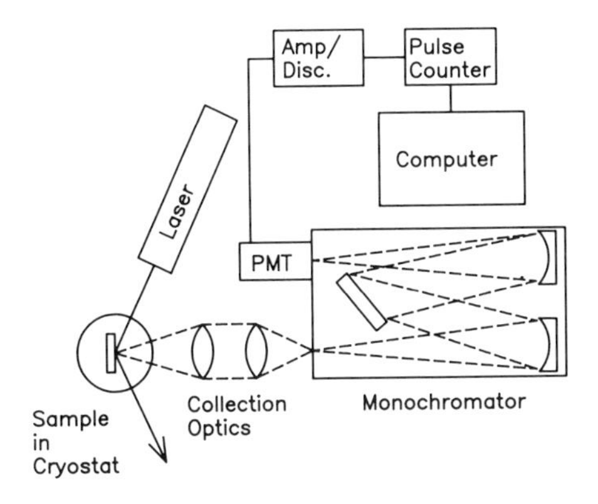

Figure 5 **Schematic layout of a high-sensitivity PL system incorporating a laser and photon-counting electronics.**

3 A dispersive element for spectral analysis of PL. This may be as simple as a filter, but it is usually a scanning grating monochromator. For excitation spectroscopy or in the presence of much scattered light, a double or triple monochromator (as used in Raman scattering) may be required.

4 An optical detector with appropriate electronics and readout. Photomultiplier tubes supply good sensitivity for wavelengths in the visible range, and Ge, Si, or other photodiodes can be used in the near infrared range. Multichannel detectors like CCD or photodiode arrays can reduce measurement times, and a streak camera or nonlinear optical techniques can be used to record ps or sub-ps transients.

A schematic of a PL system layout is shown in Figure 5. This optical system is very similar to that required for absorption, reflectance, modulated reflectance, and Raman scattering measurements. Many custom systems are designed to perform several of these techniques, simultaneously or with only small modifications.

Conclusions

Photoluminescence is a well-established and widely practiced tool for materials analysis. In the context of surface and microanalysis, PL is applied mostly qualitatively or semiquantitatively to exploit the correlation between the structure and composition of a material system and its electronic states and their lifetimes, and to identify the presence and type of trace chemicals, impurities, and defects.

Improvements in technology will shape developments in PL in the near future. PL will be essential for demonstrating the achievement of new low-dimensional quantum microstructures. Data collection will become easier and faster with the continuing development of advanced focusing holographic gratings, array and imaging detectors, sensitive near infrared detectors, and tunable laser sources.

Related Articles in the Encyclopedia

CL, Modulation Spectroscopy, Raman Spectroscopy, and FTIR

References

1 P. J. Dean. *Prog. Crystal Growth Charact.* **5,** 89, 1982. A review of PL as a diagnostic probe of impurities and defects in semiconductors by an important progenitor of the technique.

2 L. T. Canham, M. R. Dyball, and K. G. Barraclough. *J. Appl. Phys.* **66,** 920, 1989.

3 G. E. Stillman, B. Lee, M. H. Kim, and S. S. Bose. *Proc. Electrochem. Soc.* **88–20,** 56, 1988. Describes the use of PL for quantitative impurity analysis in semiconductors.

4 K. D. Mielenz, ed. *Measurement of Photoluminescence.* vol. 3 of *Optical Radiation Measurements.* (F. Grum and C. J. Bartleson, eds.) Academic Press, London, 1982. A thorough treatment of photoluminescence spectrometry for quantitative chemical analysis, oriented toward compounds in solution.

5 R. J. Hurtubise. *Solid Surface Luminescence Analysis.* Marcel Dekker, New York, 1981. Practical aspects of analysis for organics adsorbed onto solids.

6 H. B. Bebb and E. W. Williams. in *Semiconductors and Semimetals.* (R. K. Willardson and A .C. Beers, eds.) Academic Press, vol. 8, 1972. An extensive review of PL theory and technique, with emphasis on semiconductors. Some of the experimental aspects and examples are becoming outdated.

7 H. J. Queisser. *Appl. Phys.* **10,** 275, 1976. Describes PL measurements of a variety of semiconductor properties.

8 K. Mettler. *Appl. Phys.* **12,** 75, 1977. PL measurements of surface state densities and band bending in GaAs.

9 L. Zlatkevich, ed. *Luminescence Techniques in Solid-State Polymer Research.* Marcel Dekker, New York, 1989. Practical emphasis on polymers in the solid state rather than in solution.

7.2 Modulation Spectroscopy

FRED H. POLLAK

Contents

Introduction

Modulation Spectroscopy is an analog method for taking the derivative of an optical spectrum (reflectance or transmittance) of a material by modifying the measurement conditions in some manner.[1–6] This procedure results in a series of sharp, derivative-like spectral features in the photon energy region corresponding to electronic transitions between the filled and empty quantum levels of the atoms that constitute the bulk or surface of the material. Using Modulation Spectroscopy it is possible to meas-ure the photon energies of the interband transitions to a high degree of accuracy and precision. In semiconductors these band gap energies are typically 1 eV, and they can be determined to within a few meV, even at room temperature. The energies and line widths of the electronic transitions are characteristic of a particular material or surface. The energies are sensitive to a variety of internal and external parameters, such as chemical composition, temperature, strains, and electric and magnetic fields. The line widths are a function of the quality of the material, i.e., degree of crystallinity or dopant concentration.

The ability to measure the energy of electronic transitions and their line widths accurately, in a convenient manner, is one of the most important aspects of semiconductor characterization. The former can be used to evaluate alloy compositions (including topographical scans),[4] near-surface temperatures,[7] process- or growth-induced strains,[8] surface or interface electric fields associated with surface or interface states and metallization (Schottky barrier formation),[8] carrier types,[1–4] topographical variations in carrier concentrations,[4] and trap states.[8] The broadening parameter at a given temperature is a measure of crystal quality and hence can be used to evaluate the influence of various growth, processing and annealing procedures. These include ion implantation, reactive-ion etching, sputtering, and laser or rapid annealing.[7,8] In real device structures, such as heterojunction bipolar transistors, certain features of the Modulation Spectroscopy spectra have been correlated with actual device performance.[6] Thus, this method can be employed as an effective screening tool to select materials having the proper device characteristics before undertaking an expensive fabrication process. Various forms of Modulation Spectroscopy can be employed for *in-situ* monitoring of growth by molecular beam epitaxy (MBE), metal-organic chemical vapor deposition (MOCVD), or gas-phase MBE (GPMBE) at elevated temperatures.[7, 9–11] Modulation Spectroscopy has been used extensively to study semiconductors having diamond (Ge and Si), zincblende (GaAs, GaAlAs, InP, CdTe, and HgCdTe), and wurtzite (CdS) crystal structures. There also has been some work in the area of metals, including alloys.

The characteristic lines observed in the absorption (and emission) spectra of nearly isolated atoms and ions due to transitions between quantum levels are extremely sharp. As a result, their wavelengths (photon energies) can be determined with great accuracy. The lines are characteristic of a particular atom or ion and can be used for identification purposes. Molecular spectra, while usually less sharp than atomic spectra, are also relatively sharp. Positions of spectral lines can be determined with sufficient accuracy to verify the electronic structure of the molecules.

The high particle density of solids, however, makes their optical spectra rather broad, and often uninteresting from an experimental point of view. The large degeneracy of the atomic levels is split by interatomic interactions into quasicontinuous bands (valence and conduction bands). The energy difference between the highest lying valence and lowest lying conduction bands is designated as the fundamental band gap. Penetration depths for electromagnetic radiation are on the order of 500 Å through most of the optical spectrum. Such small penetration depths (except in the immediate vicinity of the fundamental gap), plus other considerations to be discussed later, make the reflection mode more convenient for characterization purposes, relative to absorption measurements.

These aspects of the optical spectra of solids are illustrated in the upper portion of Figure 1, which displays the reflectance curve (R) at room temperature for a typical semiconductor, GaAs. The fundamental absorption edge around 1.4 eV produces only a weak shoulder. Some structure is apparent in the two features around 3 eV and the large, broad peak near 5 eV. However, the dominant aspect of the line shape is the slowly varying background. The derivative nature of Modulation Spectroscopy suppresses the uninteresting background effects in favor of sharp, deriva-

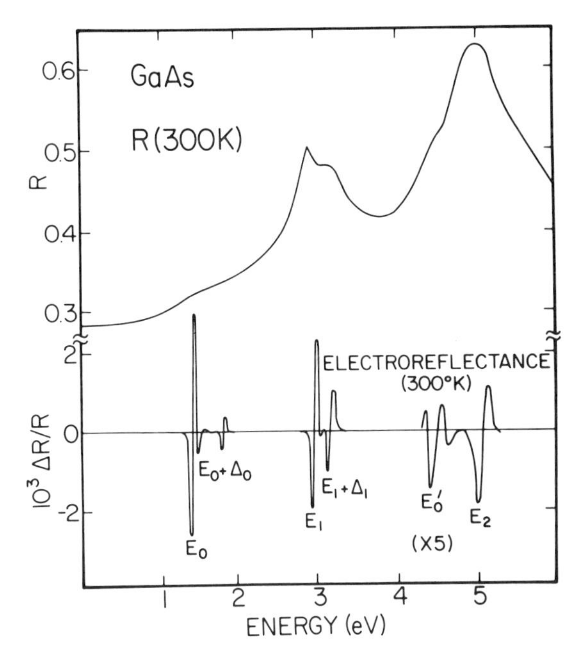

Figure 1 **Reflectance (R) and electroreflectance ($\Delta R/R$) spectra of GaAs at 300 K.**

tive-like lines corresponding to the shoulders and peaks in Figure 1. Also, weak structures that may go unseen in absolute spectra are enhanced.

Band gaps in semiconductors can be investigated by other optical methods, such as photoluminescence, cathodoluminescence, photoluminescence excitation spectroscopy, absorption, spectral ellipsometry, photocurrent spectroscopy, and resonant Raman spectroscopy. Photoluminescence and cathodoluminescence involve an emission process and hence can be used to evaluate only features near the fundamental band gap. The other methods are related to the absorption process or its derivative (resonant Raman scattering). Most of these methods require cryogenic temperatures.

For applied work, an optical characterization technique should be as simple, rapid, and informative as possible. Other valuable aspects are the ability to perform measurements in a contactless manner at (or even above) room temperature. Modulation Spectroscopy is one of the most useful techniques for studying the optical proponents of the bulk (semiconductors or metals) and surface (semiconductors) of technologically important materials. It is relatively simple, inexpensive, compact, and easy to use. Although photoluminescence is the most widely used technique for characterizing bulk and thin-film semiconductors, Modulation Spectroscopy is gaining in popularity as new applications are found and the database is increased. There are about 100 laboratories (university, industry, and government) around the world that use Modulation Spectroscopy for semiconductor characterization.

Basic Principles

The basic idea of Modulation Spectroscopy is a very general principle of experimental physics. Instead of measuring the optical reflectance (or transmittance) of a material, the derivative with respect to some parameter is evaluated. The spectral response of the material can be modified directly by applying a repetitive perturbation, such as an electric field (electromodulation), a heat pulse (thermomodulation), or stress (piezomodulation). This procedure is termed *external* modulation. The change may also occur in the measuring system itself, e.g., the wavelength or polarization conditions can be modulated or the sample reflectance (transmittance) can be compared to a reference sample. This mode has been labeled *internal* modulation. Because the changes in the optical spectra are typically small, in some cases 1 part in 10^6, phase-sensitive detection or some other signal-processing procedure is required.

To illustrate the power of Modulation Spectroscopy, displayed in the lower part of Figure 1 is the electromodulated reflectance spectra ($\Delta R/R$) of the semiconductor GaAs at 300 K in the range 0–6 eV. Although the fundamental direct absorption edge (E_0) at about 1.4 eV produces only a weak shoulder in R it is observed as a sharp, well-resolved line in $\Delta R/R$. There are also other spectral features, labeled $E_0 + \Delta_0$, E_1, $E_1 + \Delta_1$, E_0', and E_2, that correspond to transitions between other quantum levels in the semiconductor. In the region of the features at E_0 and $E_0 + \Delta_0$ the penetration depth of the light (the sampling depth) is typically several thousand Å, while for the peaks at E_1 and $E_1 + \Delta_1$ the light samples a depth of only a few hundred Å.

For characterization purposes of bulk or thin-film semiconductors the features at E_0 and E_1 are the most useful. In a number of technologically important semiconductors (e.g., $Hg_{1-x}Cd_xTe$, and $In_xGa_{1-x}As$) the value of E_0 is so small that it is not in a convenient spectral range for Modulation Spectroscopy, due to the limitations of light sources and detectors. In such cases the peak at E_1 can be used.[4] The features at E_0' and E_2 are not useful since they occur too far into the near-ultraviolet and are too broad.

Instrumentation

External Modulation

For characterization purposes the most useful form of external modulation is electromodulation, because it provides the sharpest structure (third derivative of R in bulk or thin films) and is sensitive to surface or interface electric fields.[1–5] The most widely used contactless mode of electromodulation is termed Photoreflectance (PR).[5, 7, 8]

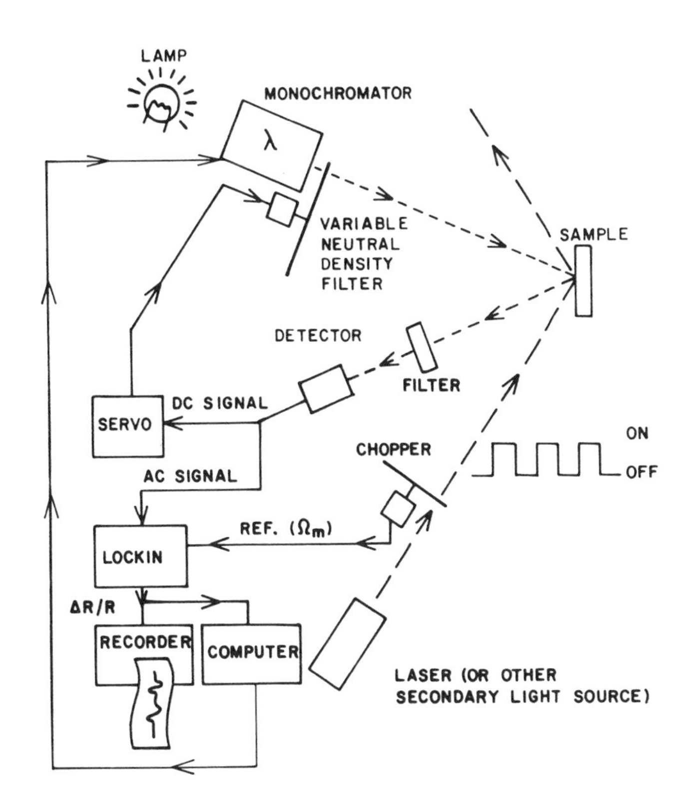

Figure 2 **Schematic representation of a photoreflectance apparatus.**

A schematic representation of a PR apparatus is shown in Figure 2.[5] In PR a pump beam (laser or other light source) chopped at frequency Ω_m creates photo-injected electron-hole pairs that modulate the built-in electric field of the semiconductor. The photon energy of the pump beam must be larger than the lowest energy gap of the material. A typical pump beam for measurements at or below room temperature is a 5-mW He–Ne laser. (At elevated temperatures a more powerful pump must be employed.)

Light from an appropriate light source (a xenon arc or a halogen or tungsten lamp) passes through a monochromator (probe monochromator). The exit intensity at wavelength λ, $I_0(\lambda)$, is focused onto the sample by means of a lens (or mirror). The reflected light is collected by a second lens (mirror) and focused onto an appropriate detector (photomultiplier, photodiode, etc.). For simplicity, the two lenses (mirrors) are not shown in Figure 2. For modulated transmission the detector is placed behind the sample.

The light striking the detector contains two signals: the dc (or average value) is given by $I_0(\lambda)R(\lambda)$, where $R(\lambda)$ is the dc reflectance of the material, while the modulated value (at frequency Ω_m) is $I_0(\lambda)\Delta R(\lambda)$, where $\Delta R(\lambda)$ is the change in reflectance produced by the modulation source. The ac signal from the detector, which is proportional to $I_0\Delta R$, is measured by a lock-in amplifier (or using another signal-averaging procedure). Typically $I_0\Delta R$ is 10^{-4}–10^{-6} I_0R.

To evaluate the quantity of interest, i.e., the relative change in reflectance, $\Delta R/R$, a normalization procedure must be used to eliminate the uninteresting common feature $I_0(\lambda)$. In Figure 2 the normalization is performed by the variable neutral density filter (VNDF) connected to a servo mechanism. The dc signal from the detector, which is proportional to $I_0(\lambda)R(\lambda)$, is introduced into the servo, which moves the VNDF in such a manner as to keep $I_0(\lambda)R(\lambda)$ constant, i.e., $I_0(\lambda)R(\lambda) = C$. Under these conditions the ac signal $I_0(\lambda)\Delta R(\lambda) = C\Delta R(\lambda)/R(\lambda)$.

Commercial versions of PR are available. Other contactless methods of electromodulation are Electron-Beam Electro-reflectance (EBER)[12] and Contactless Electroreflectance (CER)[13]. In EBER the pump beam of Figure 2 is replaced by a modulated low-energy electron beam (~ 200 eV) chopped at about 1 kHz. However, the sample and electron gun must be placed in an ultrahigh vacuum chamber. Contactless electroreflectance uses a capacitor-like arrangement.

An example of a contact mode of electromodulation would be the semiconductor–insulator–metal configuration, which consists of a semiconductor, about 200 Å of an insulator like Al_2O_3, and a semitransparent metal (about 50 Å of Ni or Au). Modulating (ac) and bias (dc) voltages are applied between the front semitransparent metal and a contact on the back of the sample. To employ this mode the sample must be conducting.

In temperature modulation, the sample may be mounted on a small heater attached to a heat sink and the temperature varied cyclically by passing current pulses through the heater.[1] If the sample is properly conducting, the current can be passed through the sample directly. Generally, for this method Ω_m must be kept below 10–20 Hz, and hence there are often problems with the $1/f$ noise of the detector.

In piezoreflectance (PzR), modulation is achieved by mounting the sample on a piezoelectric transducer that varies the lattice constant of the material, producing a band gap modulation.[14] Although PzR is contactless it requires special mounting of the sample, as does thermomodulation.

Internal Modulation

Differential Reflectivity

A commonly used form of internal modulation is differential reflectometry, in which the reflectance of the sample under investigation (or a portion of it) is compared to a standard material. This can be accomplished either by holding the sample stationary and scanning the probe beam between two regions[15] or by holding the light spot fixed and moving the sample.[16]

Reflection Difference Spectroscopy

In Reflection Difference Spectroscopy (RDS) the difference between the normal-incidence reflectance R of light polarized parallel and perpendicular to a principal crystallographic axis in the plane of the crystal is measured experimentally as a function of time, photon energy, or surface conditions.[9–11] Because of the cubic symmetry of zincblende semiconductors, the bulk is nearly isotropic (i.e., there is no distinction between parallel and perpendicular), while regions of lower symmetry, like the surface or interfaces can be anisotropic. In the case of (001) surfaces of zincblende semiconductors, the contribution from the bulk is expected to vanish. Thus, RDS is sensitive to both the chemical and structural state of the surface. Sensitivities to surface species of 0.01 monolayer have been demonstrated, with averaging times of 100 ms. Being an optical probe, RDS is well suited either to the reactive, relatively high-pressure sample environments in MOCVD reactors or to the ultrahigh-vacuum environment of MBE chambers. Moreover, the presence of a film deposited on the viewport can be overcome.

Line Shape Considerations

One of the great advantages of Modulation Spectroscopy is its ability to fit the line shapes of sharp, localized structures, as illustrated in the lower part of Figure 1. These fits yield important relevant parameters, such as the value of the energy gap and the broadening parameter.

Electromodulation

The most complicated form of Modulation Spectroscopy is electromodulation, since in certain cases it can accelerate the electron-hole pairs created by the light. If the electric field is not too large the quantity $\Delta R/R$ can be written as:[1–4]

$$\frac{\Delta R}{R} = \mathrm{Re}\left[Ae^{i\phi}(E - E_g + i\Gamma)^{-m}\right] \tag{1}$$

where A is the amplitude of the signal, ϕ is phase angle that mixes together the real and imaginary parts of dielectric function, E is the photon energy, E_g is the energy gap and Γ is a parameter that describes the broadening of the spectral line. The parameter $m = 2.5$ or 3.0 for the E_0 and E_1 optical features, respectively. Equation (1) is related to the third derivative of R.

At low temperatures the electron and hole created by the probe light beam can form a bound state (called an *exciton*) because of the Coulomb interaction between them. In this case the exponent m in Equation (1) becomes 2 and the line shape is only a first derivative.[1–5]

For sufficiently high built-in electric fields the electromodulation spectrum can display an oscillatory behavior above the band gap; these are called Franz-Keldysh oscillations (FK oscillations). In the presence of the field **F** the energy bands are tilted by an amount eFz, where e is the electronic charge and **z** is in the direction of **F**. Resonances appear whenever an integral number of de Broglie wavelengths fit into the triangular well formed by the electric field. The de Broglie wavelength is equal to $4\pi^2/hp$, where h is Planck's constant and p is the momentum of the electron (hole). The energy of the nth resonance E_n is proportional to $F^{2/3}$. Thus the periods of these resonances, or FK oscillations, are a direct measure of the built-in electric field.[3, 5]

Piezo- and Thermomodulation

These modulation methods do not accelerate the electron–hole pairs and hence produce only a first-derivative Modulation Spectroscopy. Their line shapes are given by Equation (1), with $m = 2$.

Applications and Examples

Alloy composition

Among the most important parameters for materials characterization are the compositions of binary $A_{1-x} B_x$ (e.g., $Ge_{1-x} Si_x$) alloys, ternary $A_{1-x} B_x C$ (e.g., $Ga_{1-x} Al_x As$, $Hg_{1-x} Cd_x Te$) alloys, and quaternary $A_{1-x} B_x C_y D_{1-y}$ (e.g., $In_{1-x} Ga_x As_y P_{1-y}$) alloys. The spectral features in Figure 1, e.g., E_0 and E_1 vary with alloy composition. Modulation Spectroscopy thus can be employed conveniently for this purpose even at 300 K.

Shown in Figure 3 is the variation of the fundamental direct band gap (E_0) of $Ga_{1-x}Al_xAs$ as a function of Al composition (x). These results were obtained at 300 K using electromodulation. Thus it would be possible to evaluate the Al composition of this alloy from the position of E_0.

The case of $Ga_{1-x}Al_xAs$ alloy determination is an example of the importance of the reflectance mode in relation to transmittance. In almost all cases the $Ga_{1-x}Al_xAs$ material is an epitaxial film (0.1–1μm) grown on a GaAs substrate (~0.5 mm thick). Since the band gap of GaAs is smaller than that of $Ga_{1-x}Al_xAs$, the reflectance mode must be used.

Some materials, such as $Hg_{1-x}Cd_xTe$, have a value of E_0 in certain composition regions that is too far into the infrared to be conveniently observed using Modulation Spectroscopy. In such circumstances other higher lying features, such as the peaks at E_1, can be used more readily.

The compositional variation of E_0 or higher lying features has been reported for a large number of alloys, including GeSi, GaAlAs, GaAlSb, GaAlP, InGaAs, InAsSb, InAsP, GaInSb, HgCdTe, HgMnTe, CdMnTe, CdZnTe, ZnMnTe, CdMnSe, InGaAsP lattice-matched to InP, GaAlInAs lattice-matched to InP, and

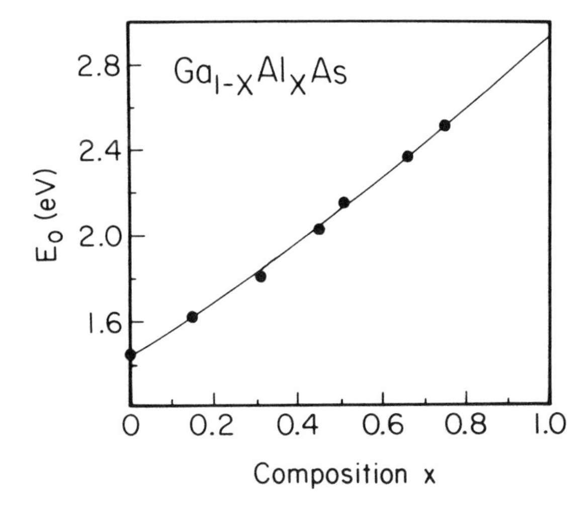

Figure 3 **Aluminum composition dependence of E_0 of $Ga_{1-x}Al_xAs$ at 300 K (solid line).**

GaAlInPAs lattice-matched to GaAs. The alloy composition x can be evaluated with a precision of $\Delta x = \pm 0.005$. By using a high-quality lens to focus the light from the probe monochromator onto the sample (see Figure 2) a spot size of about 100 μm can be achieved. By mounting the sample on an *x–y* stage it is possible to perform topographical scans with a spatial resolution of 100 μm.

Growth or Process-Induced Strain or Damage

Modulation Spectroscopy can be very useful in evaluating strains induced by growth (lattice-mismatched systems) or processing procedures, such as reactive-ion etching or oxide formation. The size and magnitude of the strain can be evaluated from the shifts and splittings of various spectral lines, such as E_0 or E_1.[6–8]

Device Structures

Certain features in the PR spectra at 300 K from $GaAs/Ga_{1-x}Al_xAs$ heterojunction bipolar transistor structures have been correlated with actual device performance; thus PR can be used as an effective screening tool.[6] From the observed FK oscillations it has been possible to evaluate the built-in dc electric fields F_{dc} in the $Ga_{1-x}Al_xAs$ emitter, as well as in the n–GaAs collector region. The behavior of F_{dc} (GaAlAs) has been found to have a direct relation to actual device performance, i.e., dc current gain. Shown in Figures 4a and 4b are the PR spectrum at 300 K for MBE and MOCVD fabricated samples, respectively. There are a number of FK oscillations in the vicinity of both the GaAs (~1.42 eV) and $Ga_{1-x}Al_xAs$

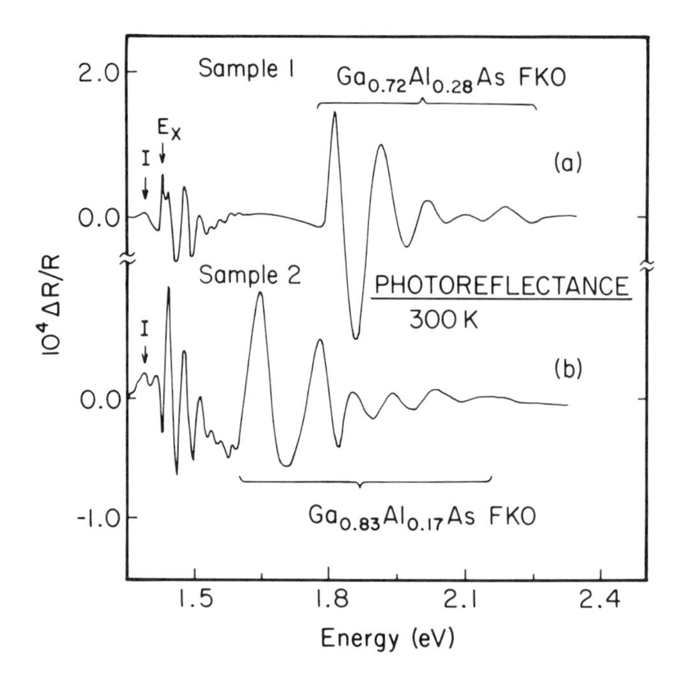

Figure 4 **Photoreflectance spectra for two GaAs/$Ga_{1-x}Al_xAs$ heterojunction bipolar transistor structures fabricated by MBE and MOCVD, respectively, at 300 K.**

band gaps. The $Ga_{1-x}Al_xAs$ portions of the two samples are 1.830 eV and 1.670 eV, which corresponds to $x = 0.28$ and 0.17, respectively, as shown in Figure 3. The most important aspects of Figure 4 are the FK oscillations associated with the $Ga_{1-x}Al_xAs$ band gap. From these features it is possible to evaluate F_{dc} in the emitter-base p–n junction. The electric fields, as deduced from the GaAlAs FK oscillations (F_{dc}, GaAlAs), were compared with fabricated heterojunction bipolar transistor MBE samples. Below electric field values of about 2×10^5 V/cm high current gains were obtained. Shown in Figure 5 is F_{dc} (in GaAlAs) as a function of dc current gain at 1 mA. Note that there is a sudden drop when F_{dc} (in GaAlAs) $> 2 \times 10^5$ V/cm. The explanation of this effect is the redistribution of the Be dopant in the p-region in these MBE samples. When the redistribution moves the p–n junction into the emitter, there is an increase in the electric field in this region; i.e., the value of F_{dc} becomes greater. The movement of the Be has been verified by Secondary Ion Mass Spectroscopy (SIMS). When the p–n junction and the GaAs/GaAlAs heterojunction are not coincident, carrier recombination occurs, reducing the current and the performance of fabricated heterojunction bipolar transistors.

These observations have made it possible to use PR as a contactless screening technique to eliminate wafers with unwanted characteristics before the costly fabrication step.

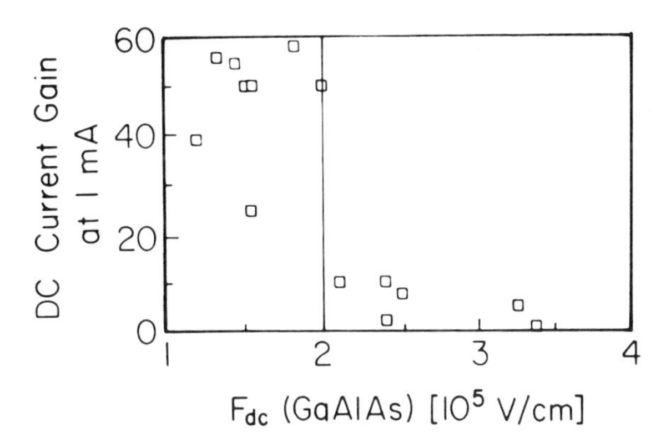

Figure 5 **Electric field F_{dc} (GaAlAs) in the p–n junction as evaluated from the GaAlAs FK oscillations as a function of the dc current gain of a fabricated heterojunction bipolar transistor.**

In-Situ Monitoring of Growth

RDS and PR are proving to be very useful methods for *in-situ* characterization of semiconductor thin-film growth by MBE, MOCVD, and GPMBE. RDS was first applied to study GaAs growth in an MBE environment. Results showed that the maximum surface anisotropy between (2 × 4) As-terminated and (4 × 2) Ga- and Al-terminated surfaces of GaAs and AlAs occur in the photon energy region between 2.0–2.5 eV and 3.5 eV, respectively. The strong dependence of this anisotropy on photon energy makes it possible to spectrally distinguish between Al–Al and Ga–Ga surface dimer bonds. The time dependence of RDS and simultaneously measured reflection high-energy electron diffraction (RHEED) signals for changes in surface conditions revealed that the RDS measurements follow surface structure. The RDS–RHEED correlation gives a valuable reference when RDS is applied to nonultrahigh-vacuum techniques, such as MOCVD, where RHEED cannot be used. A commercial model of an RDS system is available.

MBE Growth Studied by RDS

Figure 6 shows typical RDS (bottom) and RHEED (top) responses for an As-to-Ga-to-As surface stabilization sequence—from As-stabilized (2 × 4) to Ga-stabilized (4 × 2) (001) surface reconstructions and return—generated by interrupting and resuming the As flux at times t = 1 s and 10 s, respectively, during otherwise normal growth of GaAs at a rate of 1 GaAs monolayer per 4.6 s.[9, 10] The As growth-surface pressure of 6×10^{-6} torr provided 2.6 times the amount needed to consume the arriving Ga. The differences between the RDS data on the left and those on the right are due to the differences in energy of the photons used to obtain them. The differences in the RHEED data are due to small angle-of-incidence drifts of the electron beam in the time interval between the recording of successive sets of data.

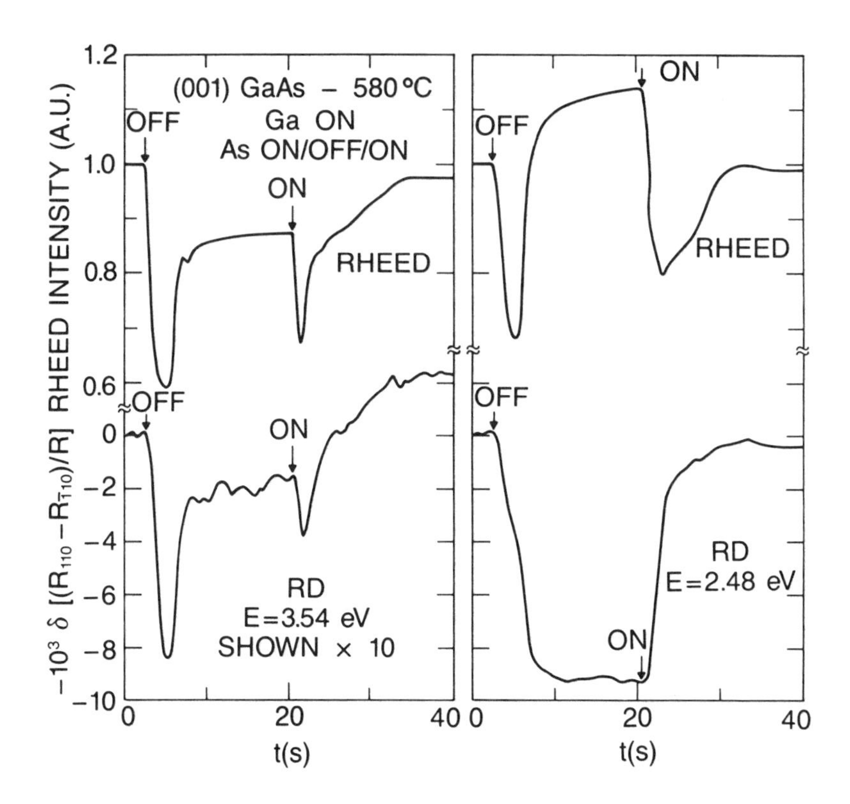

Figure 6 **RHEED (upper) and reflection anisotropy (lower) transients obtained by interrupting and resuming As flux during otherwise normal growth (001) GaAs at 1 semiconductor ML per 4.6 s. Data are shown for photon energies near the Ga RD peak at 2.5 eV (right) and minimum at 3.5 eV (left).**

The maximum change of the 2.48-eV RDS signal is nearly 1%, with R_{110} increasing relative to R_{110} as the surface becomes increasingly covered with Ga.

As soon as the As flux is terminated, the RDS signal begins to change nearly linearly in time, and saturates near $t = 5$ s; i.e., it tracks the amount of excess Ga accumulating on the surface up to one monolayer. Since RDS responds only to surface species that are in registry with the crystallographic axes of the substrate (i.e., have already reacted with it), and since it is insensitive to the presence of randomly oriented species, this time dependence implies that the excess Ga atoms are forming Ga–Ga dimer bonds instantly on arrival, with respect to laboratory time scales, and that the 2.48-eV RDS signal directly follows the chemistry of the (001) GaAs growth surface. It also implies that Ga diffusion lengths under Ga-stabilized surface conditions are large, in particular, hundreds of times greater than under As-stabilized conditions.

The 3.54-eV RDS response is completely different, exhibiting a striking similarity to the RHEED signal shown above it. Clearly, at this photon energy the RDS signal, as RHEED, is determined by surface structure. Thus RDS data either can complement or supplement RHEED data, depending on the measurement wavelength. As the saturation RDS signal at 3.54 eV is about an order of magnitude smaller than that at 2.48 eV, it follows that the small inflection in the otherwise linear initial 2.48-eV RD transient is due to the contribution of the structure-sensitive component, which is relatively minor at the lower photon energy.

Substrate Temperature and Alloy Composition by PR

It has been demonstrated that PR can be used to measure E_0 of technologically important materials, such as GaAs, InP, $Ga_{0.82}Al_{0.18}As$, and $In_xGa_{1-x}As$ ($x = 0.06$ and 0.15), to over 600° C.[6,7] Such temperatures correspond to growth conditions for thin-film methods like MBE, MOCVD, and gas-phase MBE. The value of E_0 can be evaluated to ± 5 meV at these elevated temperatures. Thus, the temperature of GaAs and InP substrates can be evaluated to ±10° C to within a depth of only several thousand Å from the growth surface. In addition, the alloy composition of epilayers of $Ga_{1-x}Al_xAs$ and $In_xGa_{1-x}As$ can be determined during actual growth. Measurements have been performed under actual growth conditions, including the case of a rotating substrate. Topographical scans can be performed to evaluate temperature or compositional homogeneity.

Figure 7 shows E_0 for GaAs and $Ga_{0.82}Al_{0.18}As$ as a function of temperature T to about 900 K. Additional measurements on samples having differing Al contents would generate a family of curves. The solid line is a least-squares fit to a semi-empirical relation that describes the temperature variation of semiconductor energy gaps:

$$E(T) = E(0) - \frac{\alpha T^2}{\beta + T} \qquad (2)$$

In Equation (2) $E(0)$ is the energy gap at $T = 0$, while α and β are materials parameters to be evaluated from experiment. Once the GaAs substrate temperature is measured from the position of E_0(GaAs), the Al composition of an epilayer can be determined readily from the position of E_0 (GaAlAs) at that temperature.

Conclusions

Modulation Spectroscopy has proven to be an important characterization method for semiconductors and semiconductor microstructures. The rich spectra contain a wealth of information about relevant materials, surfaces and interfaces, as well as device characteristics. In general, the apparatus is relatively simple, compact (except EBER), inexpensive (except EBER), and easy to use. One of the main advantages of Modulation Spectroscopy is its ability to perform relevant measurements at room

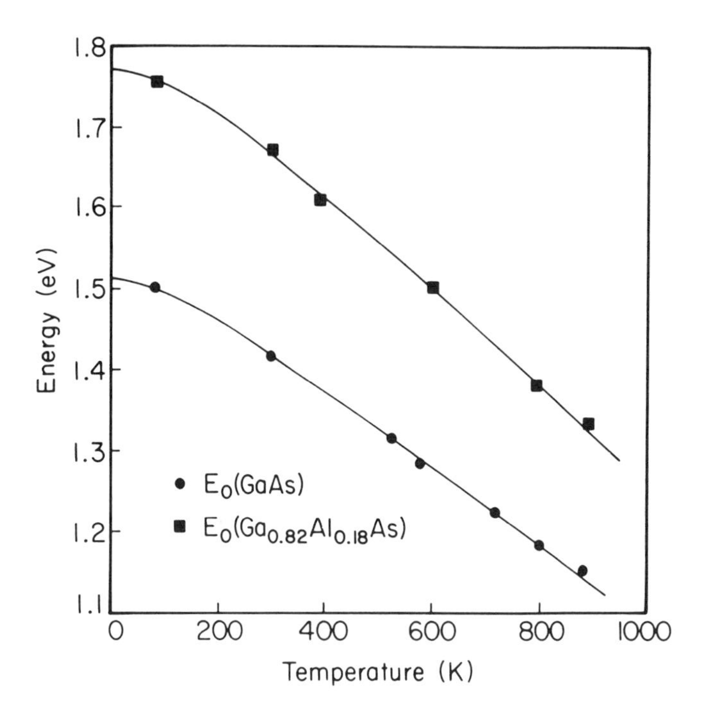

Figure 7 **Temperature dependence of E_0 of GaAs (circles) and $Ga_{0.82}Al_{0.18}As$ (squares). The solid lines are least-squares fits to Equation (2).**

temperature (or even above). Several modulation techniques, such as PR and RDS, not only are contactless but also require no special mounting of the sample and can be performed in any transparent ambient.

In bulk or thin films, material properties like the alloy composition (including topographical variations), the near-surface temperature, growth- or process-induced strain or damage, the influence of annealing procedures, surface or interface electric fields associated with Fermi level pinning, carrier types, topographical variations in carrier concentrations, and trap states, can be determined. Various contactless modulation methods, such as PR and RDS, can produce valuable information about surface and interface phenomena, including crystal growth at elevated temperatures.

In real device structures like heterojunction bipolar transistors, certain features in the PR spectrum can be correlated with actual device performance. Thus PR has been employed as an effective contactless screening technique to eliminate structures that have unwanted properties.

A major thrust in the future will be the use of contactless modulation methods like PR or RDS (together with scanning ellipsometry) for the *in-situ* monitoring and control of growth and processing, including real-time measurements. These methods can be used not only during actual growth at elevated temperatures but also for *in-situ* post growth or processing at room temperature before the sample is removed from the chamber. Such procedures should improve a material's quality and specifications, and also should serve to reduce the turn-around time for adjusting growth or processing parameters. The success of PR as a contactless screening tool for an industrial process, i.e., heterojunction bipolar transistor structures, certainly will lead to more work on real device configurations.

There also will be improvements in instrumentation and software to decrease data acquisition time. Changes can be made to improve lateral spatial resolution. For example, if the probe monochromator is replaced by a tunable dye laser spatial resolutions down to about 10 μm can be achieved.

Related Articles in the Encyclopedia

RHEED, VASE

References

1 *Semiconductors and Semimetals.* (R. K. Willardson and A. C. Beer, eds.) Academic, New York, 1972, Volume 9.

2 Proceedings of the First International Conference on Modulation Spectroscopy. *Surf. Sci.* **37,** 1973.

3 D. E. Aspnes. In: *Handbook on Semiconductors.* (T. S. Moss, ed.) North Holland, New York, 1980, Volume 2, p. 109.

4 F. H. Pollak. *Proc. Soc. Photo-Optical Instr. Eng.* **276,** 142, 1981.

5 F. H. Pollak and O. J. Glembocki. *Proc. Soc. Photo-Optical Instr. Eng.* **946,** 2, 1988.

6 D. E. Aspnes, R. Bhat, E. Coles, L. T. Florez, J. P. Harbison, M. K. Kelley, V. G. Keramidas, M. A. Koza, and A. A. Studna. *Proc. Soc. Photo-Optical Instr. Eng.* **1037,** 2, 1988.

7 D. E. Aspnes, J. P. Harbison, A. A. Studna, and L. T. Florez. *J. Vac. Sci. Technol.* **A6,** 1327, 1988.

8 F. H. Pollak and H. Shen. *J. Crystal Growth.* **98,** 53, 1989.

9 R. Tober, J. Pamulapati, R. K. Bhattacharya, and J. E. Oh. *J. Electronic Mater.* **18,** 379, 1989.

10 B. Drevillon. *Proc. Soc. Photo-Optical Instr. Eng.* **1186,** 110, 1989.

11 F. H. Pollak and H. Shen. *J. Electronic Mat.* **19,** 399, 1990.

12 Proceedings of the International Conference on Modulation Spectroscopy. *Proc. Soc. Photo-Optical Instr. Eng.* **1286,** 1990.

13 M. H. Herman. *Proc. Soc. Photo-Optical Instr. Eng.* **1286,** 39, 1990.

14 R. E. Hummel, W. Xi, and D. R. Hagmann. *J. Electrochem. Soc.* **137,** 3583, 1990.

7.3 VASE

Variable Angle Spectroscopic Ellipsometry

JOHN A. WOOLLAM AND PAUL G. SNYDER

Contents

Introduction

The technique of ellipsometry was introduced in the 1800s, but until computers became available, it was painfully slow to perform.[1] Rapid advances in small computer technology have made ellipsometric data acquisition rapid and accurate. Most important, fast personal computers make possible quick and convenient analysis of data from complex material structures.

Early work in ellipsometry focused on improving the technique, whereas attention now emphasizes applications to materials analysis. New uses continue to be found; however, ellipsometry traditionally has been used to determine film thicknesses (in the range 1–1000 nm), as well as optical constants.[1–6] Common systems are oxide and nitride films on silicon wafers, dielectric films deposited on optical surfaces, and multilayer semiconductor structures.

In ellipsometry a collimated polarized light beam is directed at the material under study, and the polarization state of the reflected light is determined using a second polarizer. To maximize sensitivity and accuracy, the angle that the light makes to the sample normal (the angle of incidence) and the wavelength are controlled.[4–6] The geometry of a typical ellipsometry set up is shown in Figure 1.

Ellipsometry is a very powerful, simple, and totally nondestructive technique for determining optical constants, film thicknesses in multilayered systems, surface and

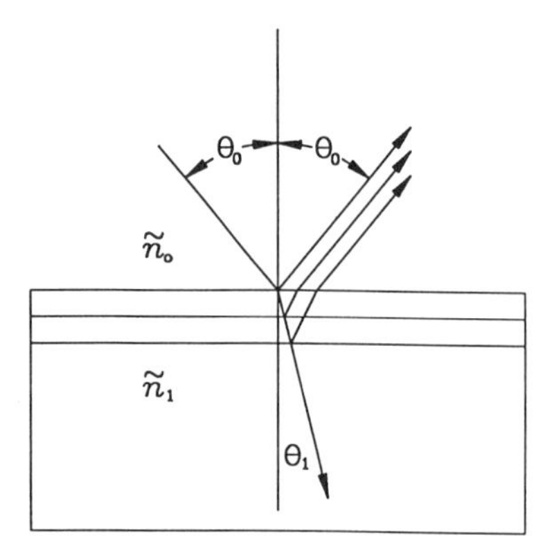

Figure 1 **Planar structure assumed for ellipsometric analysis: n_0 is the complex index of refraction for the ambient medium; n_1 is the complex index for the substrate medium; θ_0 is the value of the angles of incidence and reflection, which define the plane of incidence.**

interfacial roughness, and material microstructures. (An electron microscope may alter surfaces, as may Rutherford backscattering.) In contrast to a large class of surface techniques such as ESCA and AUGER, no vacuum chamber is necessary in ellipsometry. Measurements can be made in vacuum, air, or hostile environments like acids. The ability to study surfaces at the interface with liquids is a distinct advantage for many disciplines, including surface chemistry, biology and medicine, and corrosion engineering.

Ellipsometry can be sensitive to layers of matter only one atom thick. For example, oxidation of freshly cleaved single-crystal graphite can be monitored from the first monolayer and up. The best thicknesses for the ellipsometric study of thin films are between about 1 nm and 1000 nm. Although the spectra become complicated, films thicker than even 1 µm can be studied. Flat planar materials are optimum, but surface and interfacial roughness can be quantitatively determined if the roughness scale is smaller than about 100 nm. Thus ellipsometry is ideal for the investigation of interfacial surfaces in optical coatings and semiconductor structures.[2, 4, 7]

In some applications lateral homogeneity of a sample over large areas needs to be determined, and systems with stepper driven sample positioners have been built. Use of focused ellipsometer beams is then highly desirable. As normally practiced, the lateral resolution of ellipsometry is on the order of millimeters. However, the light beam can be focused to ~100 µm if the angle of incidence variation is not critical. For smaller focusing the beam contains components having a range of angles of incidence that may alter the validity of the data analysis.

Depth resolution depends on the (spectrally dependent) optical absorption coefficient of the material. Near-surface analysis (first 50 nm) frequently can be per-

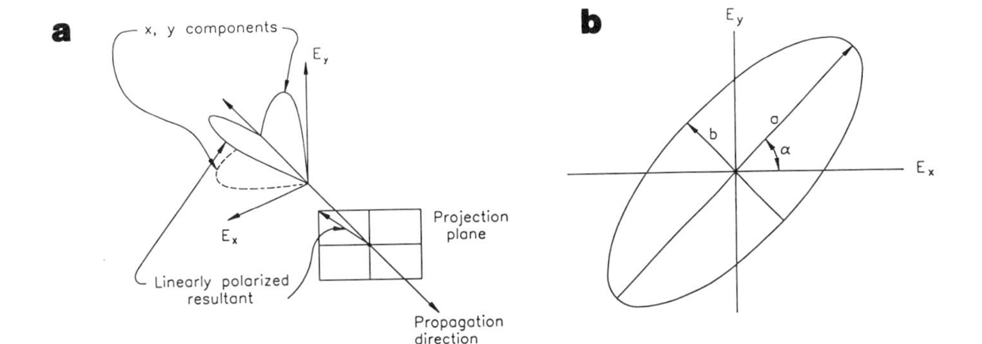

Figure 2 **(a) Representation of a linearly polarized beam in its *x*- and *y*- or (*p*- and *s*-) orthogonal component vectors. The projection plane is perpendicular to the propagation direction; (b) locus of projection of electric vector of light wave on the projection plane for elliptically polarized light—*a* and *b* are the major and minor axes of the ellipse, respectively, and α is the azimuthal angle relative to the *x*-axis.**

formed using short wavelength light (~300 nm) where absorption is strongest, and infrared radiation probes deeply (many µm) into many materials, including semiconductors.

Basic Principles

Light Waves and Polarization

Light is an electromagnetic wave with a wavelength ranging from 350 nm (blue) to 750 nm (red) for visible radiation.[8] These waves have associated electric (E) and magnetic (H) components that are related mathematically to each other, and thus the E component is normally treated alone. Figure 2a shows the electric field associated with linearly polarized light as it propagates in space and time, separated into its x- and y-vector components. In the figure the x- and y-components are exactly in phase with each other thus the electric vector oscillates in one plane, and a projection onto a plane perpendicular to the beam propagation direction traces out a straight line, as shown in Figure 2a.

When the vector components are not in phase with each other, the projection of the tip of the electric vector onto a plane perpendicular to the beam propagation direction traces out an ellipse, as shown in Figure 2b.

A complete description of the polarization state includes:[1]

1 The azimuthal angle of the electric field vector along the major axis of the ellipse (recall the angle α in Figure 2b) relative to a plane of reference

2. The ellipticity, which is defined by $e = b/a$
3. The handedness (righthanded rotation of the electric vector describes clockwise rotation when looking into the beam)
4. The amplitude, which is defined by $A = (a^2 + b^2)^{2/3}$
5. The absolute phase of the vector components of the electric field.

In ellipsometry only quantities 1 and 2 (and sometimes 3) are determined. The absolute intensity or phase of the light doesn't need to be measured, which simplifies the instrumentation enormously. The handedness information is normally not critical.

All electromagnetic phenomena are governed by Maxwell's equations, and one of the consequences is that certain mathematical relationships can be determined when light encounters boundaries between media.[1, 8] Three important conclusions that result for ellipsometry are:

1. The angle of incidence equals the angle of reflectance θ_0 (see Figure 1).
2. Snell's Law holds: $n_1 \sin \theta_1 = n_0 \sin \theta_0$ (Snell's Law), where n_1 and n_0 are the complex indexes of refraction in media 1 and media 0, and the angles θ_1 andθ_0 are shown in Figure 1.
3. The Fresnel reflection coefficients are:

$$r_p = \frac{E_{op}^{r}}{E_{op}^{i}} = \frac{n_1 \cos\theta_0 - n_0 \cos\theta_1}{n_1 \cos\theta_0 + n_0 \cos\theta_1} \qquad (1a)$$

$$r_p = \frac{E_{op}^{r}}{E_{op}^{i}} = \frac{n_1 \cos\theta_0 - n_0 \cos\theta_1}{n_1 \cos\theta_0 + n_0 \cos\theta_1} \qquad (1b)$$

where s refers to the light vector component perpendicular to the plane of incidence, p refers to the component parallel to the plane of incidence, and r and i refer to reflected and incoming light. The plane of incidence is defined by the incoming and outgoing beams and the normal to the sample. The complex indices of refraction for media 0 and 1 are given by n_0 and n_1. The relations r_s and r_p are the complex Fresnel reflection coefficients. Their ratio is measured in ellipsometry:

$$\rho = \frac{r_p}{r_s} = (\tan\Psi)\, e^{j\Delta} \qquad (2)$$

Since ρ is a complex number, it may be expressed in terms of the amplitude factor $\tan \Psi$, and the phase factor $\exp j\Delta$ or, more commonly, in terms of just Ψ and Δ. Thus measurements of Ψ and Δ are related to the properties of matter via Fresnel coefficients derived from the boundary conditions of electromagnetic theory.[1, 8]

There are several techniques for measuring Ψ and Δ, and a common one is discussed below.

Equations 1a and 1b are for a simple two-phase system such as the air–bulk solid interface. Real materials aren't so simple. They have natural oxides and surface roughness, and consist of deposited or grown multilayered structures in many cases. In these cases each layer and interface can be represented by a 2×2 matrix (for isotropic materials), and the overall reflection properties can be calculated by matrix multiplication.[1] The resulting algebraic equations are too complex to invert, and a major consequence is that regression analysis must be used to determine the system's physical parameters.[1, 2, 5, 9]

In a regression analysis Ψ_i^c and Δ_i^c are calculated from an assumed model for the structure using the Fresnel equations, where Ψ and Δ in Equation 2 are now indexed by *c*, to indicate that they are calculated, and by *i*, for each combination of wavelength and angle of incidence.

The unknown parameters of the model, such as film thicknesses, optical constants, or constituent material fractions, are varied until a best fit between the measured Ψ_i^m and Δ_i^m and the calculated Ψ_i^c and Δ_i^c is found, where *m* signifies a quantity that is measured. A mathematical function called the *mean squared error* (MSE) is used as a measure of the goodness of the fit:

$$\mathrm{MSE} = \frac{1}{N}\sum_{i=1}^{N} (\Psi_i^c - \Psi_i^m)^2 + (\Delta_i^c - \Delta_i^m)^2 \tag{3}$$

where N is the number of wavelength and angle of incidence combinations used. The MSE is low if the user's guess at the physical model for the system was correct and if starting parameters were reasonably close to correct values.

The model-dependent aspect of ellipsometric analysis makes it a difficult technique. Several different models fit to one set of data may produce equivalently low MSEs. The user must integrate and evaluate all available information about the sample to develop a physically realistic model. Another problem in applying ellipsometry is determining when the parameters of the model are mathematically correlated; for example, a thicker film but lower index of refraction might give the same MSE as some other combinations of index and thickness. That is, the answer is not always unique.

Access to the correlation matrix generated during the regression analysis is thus important[1, 9] to determine which, and to what degree, variables are correlated. It is common for the user of an ellipsometer mistakenly to make five wrong (correlated) measurements of an index of refraction and film thickness at, say, 632.8 nm and then to average these meaningless numbers. In reality all five measurements gave nonunique values, and averaging is not a valid procedure—the average of five bad numbers does not yield a correct number! The solution to the correlation problem

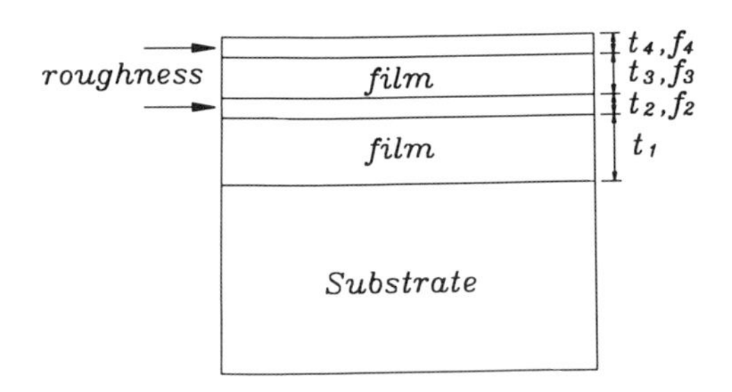

Figure 3 **Common structure assumed for ellipsometric data analysis: t_1 and t_3 are the thicknesses of the two deposited films, for example; t_2 and t_4 are interfacial and surface roughness regions; f_2 is the fraction of film t_1 mixed with film t_3 in an effective medium theory analysis of roughness—film f_3 could have void (with fraction 1–f_3) dispersed throughout; and f_4 is the fraction of t_4 mixed with the ambient medium to simulate surface roughness.**

is to make many measurements at optimum wavelength and angle combinations, and to keep the assumed model simple yet realistic. Even then, it is sometimes inherently not possible to avoid correlation. In this case especially it is important to know the degree of correlation. Predictive modeling can be performed prior to making any measurements to determine the optimum wavelength and angle combinations to use, and to determine when there are likely to be correlated variables and thus nonunique answers.[5, 6]

A typical structure capable of being analyzed is shown in Figure 3, consisting of a substrate, two films (thicknesses t_1 and t_3), two roughness regions (one is an interfacial region of thickness t_2, and the other is a surface region of thickness t_4). One of the films t_1 or t_3 may consist of microscopic (less than 100 nm size) mixtures of two materials, such as SiO_2 and Si_3N_4. The volume ratios of these two constituents can be determined by ellipsometry using effective medium theory.[10] This theory solves the electromagnetic equations for mixtures of constituent materials using simplifying approximations, resulting in the ability of the user to determine the fraction of any particular species in a mixed material. Likewise the roughness layers are modeled as mixtures of the neighboring media (air with medium 3 for the surface roughness, and medium 1 with medium 3 for interfacial roughness, as seen in Figure 3).

The example in Figure 3 is as complex as is usually possible to analyze. There are seven unknowns, if no indices of refraction are being solved for in the regression analysis. If correlation is a problem, then a less complex model must be assumed. For example, the assumption that f_2 and f_4 are each fixed at a value of 0.5 might reduce correlation. The five remaining unknowns in the regression analysis would then be t_1,t_2, t_3, t_4, and f_3. In practice one first assumes the simplest possible model, then makes it more complex until correlation sets in, or until the mean squared error fails to decrease significantly.

Polarization Measurement

Manual null ellipsometry is accurate but infrequently done, due to the length of time needed to acquire sufficient data for any meaningful materials analysis. Automated null ellipsometers are used, for example, in the infrared, but are still slow. Numerous versions of fast automated ellipsometers have been built.[1–3] Examples are:

1 Polarization modulation

2 Rotating analyzer

3 Rotating polarizer.

The most common versions are 2 and 3, and the rotating analyzer system will be briefly described here.[11] Such a system consists of a light source, monochromator, collimating optics, and polarizer preceding the sample of Figure 1, and a rotating polarizer (called the analyzer) and detector following the sample. The intensity of the light measured at the detector oscillates sinusoidally according to the relation

$$I = 1 + \alpha \cos 2A + \beta \sin 2A$$

where α and β are the Fourier coefficients, and A is the azimuthal angle between the analyzer "fast axis" and the plane of incidence. There is a direct mathematical relationship between the Fourier coefficients and the Ψ and Δ ellipsometric parameters. The actual experiment involves recording the relative light intensity versus A in a computer. The coefficients α and β, and thus Ψ and Δ, can then be determined. By changing the angle of incidence and wavelength, the user can determine N sets of Ψ_i and Δ_i values for the regression analysis used to derive the unknown physical properties of the sample.

The polarizer and analyzer azimuthal angles relative to the plane of incidence must be calibrated. A procedure for doing this is based on the minimum of signal that is observed when the fast axes of two polarizers are perpendicular to each other. For details the reader can consult the literature.[11]

Applications

In this section we will give some representative examples. Figure 4 shows the regression procedure for $\tan \Psi$ for the glass/TiO_2/Ag/TiO_2 system. The unknowns of the fit were the three thicknesses: TiO_2, Ag, and the top TiO_2. Initial guesses at the thicknesses were reasonable but not exact. The final thicknesses were 33.3 nm, 11.3 nm, and 26.9 nm, and the fits between measured Ψ_i^m and Δ_i^m and calculated (from Fresnel equations) Ψ_i^c and Δ_i^c were excellent. This means that the assumed optical constants and structure for the material were reasonable.

Because Ψ and Δ can be calculated for *any* structure (no matter how complex, as long as planar parallel interfaces are present), then the user can do predictive modeling. Figure 5 shows the expected Δ versus wavelength and angle of incidence for a

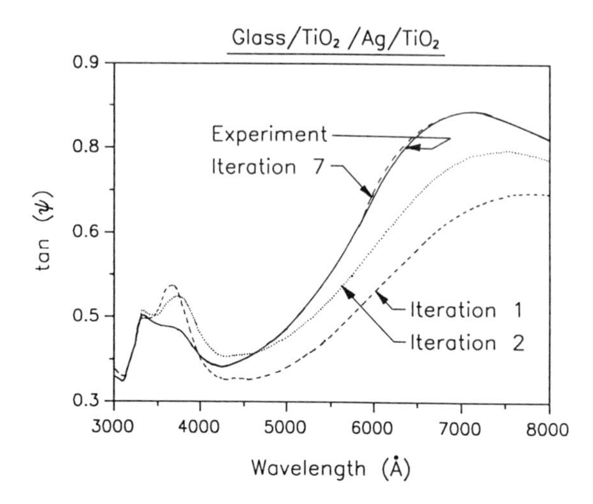

Figure 4 **Data plus iterations 1, 2, and 7 in regression analysis (data fit) for the optical coating glass / TiO_2 / Ag / TiO_2.**

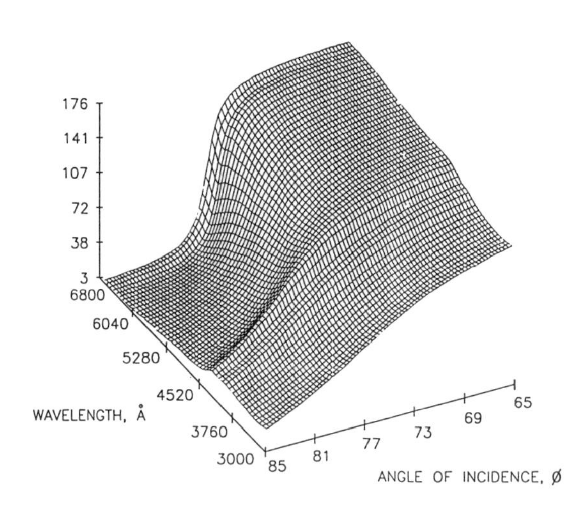

Figure 5 **Three-dimensional plot of predicted ellipsometric parameter data versus angle of incidence and wavelength.**

structure with a GaAs substrate/50 nm of $Al_{0.3}Ga_{0.7}As$/30 nm of GaAs/3 nm of oxide.[5] The best data are taken when Δ is near 90°, and generated surfaces such as Figure 5 help enormously in finding the proper wavelength and angle regions to take data.[4–6] Equally useful are contour plots made from the surfaces of Figure 5 which show quantitatively where the $90° \pm 20°$ regions of Δ will be found.[4, 6, 12]

Many materials have been studied; examples include:

1. Dielectrics and optical coatings: Si_3N_4, SiO_2, SiO_xN_y, Al_2O_3, a-C:H, ZnO, TiO_2, ZnO/Ag/ZnO, TiO_2/Ag/TiO_2, AgO, $In(Sn)_2O_3$, and organic dyes.
2. Semiconductors and heterostructures: Si, poly-Si, amorphous Si, GaAs, $Al_xG_{1-x}As$, $In_xGa_{1-x}As$, and numerous II-VI and III-V category compound semiconductors; ion implanted compound heterostructures, superlattices, and heterostructures exhibiting Franz-Keldysh oscillations. Work has been done on these materials at room temperature, as well as from cryogenic (4 K) to crystal growth temperatures (900 K).
3. Surface modifications and surface roughness: Cu, Mo, and Be laser mirrors; atomic oxygen modified (corroded) surfaces and films, and chemically etched surfaces.
4. Magneto-optic and magnetic disc materials: DyCo, TbFeCo, garnets, sputtered magnetic media (CoNiCr alloys and their carbon overcoats).
5. Electrochemical and biological and medical systems.
6. *In-situ* measurements into vacuum systems: In these experiments the light beams enter and leave via optical ports (usually at a 70° or 75° angle of incidence), and ψ and Δ are monitored in time. Example studies include the measurement of optical constants at high temperatures, surface oxide formation and sublimation, surface roughness, crystal growth, and film deposition. *In-situ* measurements were recently reviewed by Collins.[3]

Conclusions

Ellipsometry is a powerful technique for surface, thin-film, and interface analysis. It is totally nondestructive and rapid, and has monolayer resolution. It can be performed in any atmosphere including high-vacuum, air, and aqueous environments. Its principal uses are to determine thicknesses of thin films, optical constants of bulk and thin-film materials, constituent fractions (including void fractions) in deposited or grown materials, and surface and interfacial roughness. Recent trends in the relatively small community of scientists using ellipsometry in research have been towards *in-situ* measurements during crystal growth or material deposition or processing. Fast-acquisition automated ellipsometers have not been used widely in medical research, which represents an opportunity. Simple one-wavelength ellipsometers are in common use (and misuse due to correlated variables) in semiconductor processing. Use of a full spectroscopic ellipsometer is strongly advised.

The ellipsometer user will always get data; but unfortunately may not always know when the data or the results of analysis are correct. Improper optical alignment, bad calibration constants, reflection from the back surface of partially transparent materials, as well as correlation of variables are all potential problems to be aware of. Ellipsometry is a powerful technique when used properly.

The authors wish to recognize financial support under grants NAG 3-154 and NAG 3-95 from the NASA Lewis Research Center, Cleveland, Ohio.

Related Articles in the Encyclopedia

MOKE

References

1 R. M. A. Azzam and N.M. Bashara. *Ellipsometry and Polarized Light.* North Holland Press, New York, 1977. Classic book giving mathematical details of polarization in optics.

2 D. E. Aspnes. In: *Handbook of Optical Constants of Solids.* (E. Palik, ed.) Academic Press, Orlando, 1985. Description of use of ellipsometry to determine optical constants of solids.

3 R. E. Collins. *Rev. Sci. Insts.* **61,** 2029, 1990. Recent review of *in-situ* ellipsometry, in considerable depth.

4 P. G. Snyder, M. C. Rost, G. H. Bu-Abbud, J. A. Woollam, and S. A. Alterovitz. *J. of Appl. Phys.* **60,** 3293, 1986. First use of computer drawn three-dimensional surfaces (in wavelength and angle of incidence space) for ellipsometric parameters ψ and Δ and their sensitivities.

5 S. A. Alterovitz, J. A. Woollam, and P. G. Snyder. *Solid State Tech.* **31,** 99, 1988. Review of use of variable-angle spectroscopic ellipsometer (VASE) for semiconductors.

6 J. A. Woollam and P. G. Snyder. *Materials Sci. Eng.* **B5,** 279, 1990. Recent review of application of VASE in materials analysis.

7 K. G. Merkel, P. G. Snyder, J. A. Woollam, S. A. Alterovitz, and A. K. Rai. *Japanese J. App. Phys.* **28,** 1118, 1989. Application of VASE to complicated multilayer semiconductor transistor structures.

8 E. Hecht. *Optics.* Addison-Wesley, Reading, 1987. Well written and illustrated text on classical optics.

9 G. H. Bu-Abbud, N. M. Bashara, and J. A. Woollam. *Thin Solid Films.* **138,** 27, 1986. Description of Marquardt algorithm and parameter sensitivity correlation in ellipsometry.

10 D. E. Aspnes. *Thin Solid Films.* **89,** 249, 1982. A detailed review of effective medium theory and its use in studies of optical properties of solids.

11 D. E. Aspnes and A. A. Studna. *App. Optics.* **14,** 220, 1973. Details of a rotating analyzer ellipsometer design.
12 W. A. McGahan, and J. A. Woollam. *App. Phys. Commun.* **9,** 1, 1989. Well written and illustrated review of electromagnetic theory applied to a multilayer structure including magnetic and magneto-optic layers.

8

VIBRATIONAL SPECTROSCOPIES AND NMR

8.0 INTRODUCTION

In this chapter, three methods for measuring the frequencies of the vibrations of chemical bonds between atoms in solids are discussed. Two of them, Fourier Transform Infrared Spectroscopy, FTIR, and Raman Spectroscopy, use infrared (IR) radiation as the probe. The third, High-Resolution Electron Energy-Loss Spectroscopy, HREELS, uses electron impact. The fourth technique, Nuclear Magnetic Resonance, NMR, is physically unrelated to the other three, involving transitions between different spin states of the atomic nucleus instead of bond vibrational states, but is included here because it provides somewhat similar information on the local bonding arrangement around an atom.

The most commonly used of these methods, and the most inexpensive, is FTIR. In it a broad band source of IR radiation is reflected from the sample (or transmitted, for thin samples). The wavelengths at which absorption occurs are identified by measuring the change in intensity of the light after reflection (transmission) as a function of wavelength. These absorption wavelengths represent excitations of vibrations of the chemical bonds and are specific to the type of bond and the group of atoms involved in the vibration. IR spectroscopy as a method of quantitative chemical identification for species in solution, or liquids, has been commercially available for 50 years. The advent of fast Fourier transform methods in conjunction with interferometer wavelength detection schemes in the last 15 years has allowed drastic improvement in resolution, sensitivity, and reliable quantification. During this time the method has become regularly used also for solids. The sensitivity toward different bonds (chemical groups) is extremely variable, going from zero (no coupling of the IR radiation to vibrational excitations because of dipole selection rules) to high enough to detect submonolayer quantities. Intensities and line shapes are also sensitive to local solid state effects, such as stress, strain, and defects (which can therefore be characterized), so quantification is difficult, but with suitable standards 5–10% accuracy in concentrations are achievable. The depth probed depends strongly on the material (whether it is transparent or opaque to IR radiation) and can be as little as 100 Å or as much as 1 mm. The chemical nature of opaque interfaces beneath transparent overlayers can therefore be studied. Grazing angle measurements greatly reduce the probing depth, restricting it to a monolayer for molecules absorbed on metal surfaces. Often there is no spatial resolution (mm), but microfocus systems down to 20 μm exist. In Raman spectroscopy IR radiation of a single wavelength from a laser strikes the sample and the energy losses (gains) due to the Raman scattering process, which lead to some light being reemitted at lower (higher) frequencies, are determined. These loss (or gain) processes are again due to the coupling of the vibrational processes in the sample with the incident IR radiation. So, though the physics of the Raman process is quite different from that of IR spectroscopy (scattering instead of absorption), the information content is very similar. The selection rules defining which vibrational modes can be excited are different from IR, however, so Raman essentially provides complementary information. Cross sections for Raman scattering are extremely weak, resulting in Raman sensitivity being about a factor of 10 lower than for FTIR. However, better spatial resolution can be achieved (down to a few μm) because the single wavelength nature of the laser source allows an easy coupling to optical microscope elements. For the "fingerprinting" identification of chemical composition not nearly so extensive a library of data is available as for IR spectroscopy. Because of this, and because instrumentation is generally more expensive, Raman spectroscopy is less widely used, except where the microfocus capabilities are important or where differences in selection rules are critical.

Both IR and Raman have the great practical advantage of working in ambient atmosphere, and one can even study interfaces through liquids. The third vibrational technique discussed here, HREELS, requires ultrahigh vacuum conditions. A monochromatic, low-energy electron beam (a few eV) is reflected from a sample surface, losing energy by exciting vibrations (cf., Raman scattering) as it does so. Since the reflected part of the beam does not penetrate the surface, the vibrational information obtained relates only to the outermost layers. Actually two separate scattering mechanisms occur. Scattering in the specular direction is a long-range dipole process that has the same selection rules as for IR. Impact scattering is short range and nonspecular. It is an order of magnitude weaker than dipole scattering and has relaxed selection rules. Taking data in both the specular and off-specular directions therefore maximizes the amount of information obtainable. The wavelength range accessible is wider in HREELS than in IR spectroscopy, but the resolution is orders of magnitude poorer, leading to overlapped vibrational peaks and little detailed information on individual line shapes. The major uses of HREELS have been identifying chemical species, adsorption sites, and adsorption geometries (symmetry) for monolayer adsorption at single crystal surfaces. For non-single crystal surfaces the energy-loss intensities are drastically reduced, but the technique is still useful. It has been quite extensively used for characterizing polymer surfaces. For insulators charging can sometimes be a problem.

The last technique discussed here, NMR, involves immersing the sample in a strong magnetic field (1–12 Tesla), thereby splitting the degeneracy of the spin states of those nuclei that have either an odd mass or odd atomic number and hence possess a permanent magnetic moment. About half the elements in the periodic table have isotopes fulfilling these conditions. Excitation between these magnetic levels is then performed by absorption of radiofrequency (RF) radiation. By measuring the energy at which the absorptions occur (the "resonance" energies) the energy differences between the spin (magnetic) states are determined. For any given magnetic field the values are element specific, but the nuclear magnetic moments and electronic environment surrounding the target atoms also exert an influence, splitting the absorption resonances into multiple lines and shifting peak positions. From these effects the local environment of the atoms concerned—the coordination number, local symmetries, the nature of neighboring chemical groups, and bond distances—can be studied. H-atom NMR has been used as an analytical tool for molecules in liquids for about 40 years to identify chemical groupings, and the sequence of groupings containing H atoms. It is also, of course, the basis of Magnetic Resonance Imaging, MRI, which is used medically. In the solid state, crystalline phases can be identified, and quantitative analysis can be achieved directly in mixtures from the relative intensities of peaks and the use of well-defined model compound standards. In many cases the NMR spectra of solids are rather broad and unresolved due to strong anisotropic effects with respect to the applied magnetic field. There are a number of ways of removing these effects, the most popular being magic-angle spinning of the sample, which can collapse broad powder patterns into sharp resonances that can be easily assigned. NMR is intrinsically a bulk technique; the signal comes from the entire sample which is immersed in the magnetic field. At least 10 mg of material is required (powders, thin films, or crystals), and to get any information specific to surfaces or interfaces requires large surface areas (10–150 m^2/gm). Costs vary a lot ($200,000 to $1,200,000), depending on how wide a range of elements needs to be accessed, since this determines the range and magnitude of the magnetic fields and RF capabilities required.

8.1 FTIR

Fourier Transform Infrared Spectroscopy

J. NEAL COX

Contents

Introduction

The physical principles underlying infrared spectroscopy have been appreciated for more than a century. As one of the few techniques that can provide information about the chemical bonding in a material, it is particularly useful for the nondestructive analysis of solids and thin films, for which there are few alternative methods. Liquids and gases are also commonly studied, more often in conjunction with other techniques. Chemical bonds vary widely in their sensitivity to probing by infrared techniques. For example, carbon-sulfur bonds often give no infrared signal, and so cannot be detected at any concentration, while silicon-oxygen bonds can produce signals intense enough to be detected when probing submonolayer quantities, or on the order of 10^{13} bonds/cc. Thus, the potential utility of infrared spectrophotometry (IR) is a function of the chemical bond of interest, rather than being applicable as a generic probe. For quantitative analysis, modern instrumentation can provide a measurement repeatability of better than 0.1%. Accuracy and precision, however, are more commonly on the order of 5.0% (3σ), relative. The limitations arise from sample-to-sample variations that modify the optical quality of the material. This causes slight, complex distortions of the spectrum that are difficult to eliminate. Sensitivity of the sample to environmental influences that modify the chemical bonding and the need to calibrate the infrared spectral data to reference methods—such as neutron activation, gravimetry, and wet chemistry—also tend to degrade slightly the measurement for quantitative work.

The goal of the basic infrared experiment is to determine changes in the intensity of a beam of infrared radiation as a function of wavelength or frequency (2.5–50 μm or 4000–200 cm^{-1}, respectively) after it interacts with the sample. The centerpiece of most equipment configurations is the infrared spectrophotometer. Its function is to disperse the light from a broadband infrared source and to measure its intensity at each frequency. The ratio of the intensity before and after the light interacts with the sample is determined. The plot of this ratio versus frequency is the *infrared spectrum.*

As technology has progressed over the last 50 years, the infrared spectrophotometer has passed through two major stages of development. These phases have significantly impacted how infrared spectroscopy has been used to study materials. Driven in part by the needs of the petroleum industry, the first commercial infrared spectrophotometers became available in the 1940s. The instruments developed at that time are referred to as *spatially dispersive* (sometimes shortened to *dispersive*) instruments because ruled gratings were used to disperse spatially the broadband light into its spectral components. Many such instruments are still being built today. While somewhat limited in their ability to provide quantitative data, these dispersive instruments are valued for providing qualitative chemical identification of materials at a low cost. The 1970s witnessed the second phase of development. A new (albeit much more expensive) type of spectrophotometer, which incorporated a Michelson interferometer as the dispersing element, gained increasing acceptance. All frequencies emitted by the interferometer follow the same optical path, but differ in the time at which they are emitted. Thus these systems are referred to as being *temporally dispersive.* Since the intensity-time output of the interferometer must be subjected to a Fourier transform to convert it to the familiar infrared spectrum (intensity-frequency), these new units were termed Fourier Transform Infrared spectrophotometers, (FTIR). Signal-to-noise ratios that are higher by orders of magnitude, much better resolution, superior wavelength accuracy, and significantly shorter data acquisition times are gained by switching to an interferometer. This had been recognized for several decades, but commercialization of the equipment had to await the arrival of local computer systems with significant amounts of cheap memory, advances in equipment interfacing technology, and developments in fast Fourier-transform algorithms and circuitry.

Beyond the complexities of the dispersive element, the equipment requirements of infrared instrumentation are quite simple. The optical path is normally under a purge of dry nitrogen at atmospheric pressure; thus, no complicated vacuum pumps, chambers, or seals are needed. The infrared light source can be cooled by water. No high-voltage connections are required. A variety of detectors are available, with deuterated tri-glycene sulfate (DTGS) detectors offering a good signal-to-noise ratio and linearity when operated at room temperature. For more demanding applications, the mercury cadmium telluride (HgCdTe, or mer-cad telluride) detector, cooled by liquid nitrogen, can be used for a factor-of-ten gain in sensitivity.

With the advent of FTIR instrumentation, IR has experienced a dramatic increase in applications since the 1970s, especially in the area of quantitative analysis. FTIR spectrophotometry has grown to dominate the field of infrared spectroscopy. Experiments in microanalysis, surface chemistry, and ultra-thin films are now much more routine. The same is true for interfaces, if the infrared characteristics of the exterior layers are suitable. While infrared methods still are rarely used to profile composition as a function of depth, microprobing techniques available with FTIR technology permit the examination of microparticles and *xy*-profiling with a spatial resolution down to 20 μm. Concurrent with opening the field to new areas of research, the high level of computer integration, coupled with robust and nondestructive equipment configurations, has accelerated the move of the instrument out of the laboratory. Examples are in VLSI, computer-disk, and chemicals manufacturing, where it is used as a tool for thin-film, surface coating, and bulk monitors.

Unambiguous chemical identification usually requires the use of other techniques in conjunction with IR. For gases and liquids, Mass Spectrometry (MS) and Nuclear Magnetic Resonance Spectrometry (NMR) are routinely employed. The former, requiring only trace quantities of material, determines the masses of the molecule and of characteristic fragments, which can be used to deduce the most likely structure. MS data is sometimes supplemented with infrared results to distinguish certain chemical configurations that might produce similar fragment patterns. NMR generally requires a few milliliters of sample, more than needed by either the FTIR or MS techniques, and can identify chemical bonds that are associated with certain elements, bonds that are adjacent to each other, and their relative concentrations. Solids can also be studied by these methods. For MS, the sensitivity remains high, but the method is destructive because the solid must somehow be vaporized. While nondestructive, the sensitivity of NMR spectrometry is typically much lower for direct measurements on solids; otherwise, the solid may be taken into solution and analyzed. For thin films, both the MS and NMR methods are destructive. Complementary data for surfaces, interfaces, and thin films can be obtained by techniques like X-ray photoelectron spectroscopy, static secondary ion mass spectrometry, and electron energy loss spectrometry. These methods probe only the top few nanometers of the material. Depending upon the sample and the experimental configuration, IR may be used as either a surface or a bulk probe for thin films. For surface analysis, FTIR is about a factor of 10 less sensitive than these alternative methods. Raman spectroscopy is an optical technique that is complementary to infrared methods and also detects the vibrational motion of chemical bonds. While able to achieve submicron spatial resolution, the sensitivity of the

Raman technique is usually more than an order of magnitude less than that of FTIR.

As a surface probe, FTIR works best when the goal is to study a thin layer of material on a dissimilar substrate. Lubricating oil on a metal surface and thin oxide layers on semiconductor surfaces are examples. FTIR techniques become more difficult to apply when the goal is to examine a surface or layer on a similar substrate. An example would be the study of thin skins or surface layers formed during the curing cycles used for photoresist or other organic thin films deposited from the liquid phase. If the curing causes major changes in the bulk and the surface, FTIR methods usually cannot discriminate between them, because the beam probes deeply into the bulk material. The limitations as a surface probe often are dictated by the type of substrate being used. A metal or high refractive-index substrate will reflect enough light to permit sensitive probing of the surface region. A low refractive index substrate, in contrast, will permit the beam to probe deeply into the bulk, degrading the sensitivity to the surface.

The discussions presented in this article pertain to applications of FTIR, because most of the recent developments in the field have been attendant to FTIR technology. In many respects, FTIR is a "science of accessories". A myriad of sample holders, designed to permit the infrared light to interact with a given type of sample in an appropriate manner, are interfaced to the spectrophotometer. A large variety of "hyphenated" techniques, such as GC-FTIR (gas chromatography-FTIR) and TGA-FTIR (thermo-gravimetric analysis-FTIR), also are used. In these cases, the effluent emitted by the GC, TGA, or other unit is directed into the FTIR system for time-dependent study. Hyphenated methods will not be discussed further here. Still, common to all of these methods is the goal of obtaining and analyzing an infrared spectrum.

Basic Principles

Infrared Spectrum

Define I_0 to be the intensity of the light incident upon the sample and I to be the intensity of the beam after it has interacted with the sample. The goal of the basic infrared experiment is to determine the intensity ratio I/I_0 as a function of the frequency of the light (w). A plot of this ratio versus the frequency is the infrared spectrum. The infrared spectrum is commonly plotted in one of three formats: as transmittance, reflectance, or absorbance. If one is measuring the fraction of light transmitted through the sample, this ratio is defined as

$$T_w = \left(\frac{I_t}{I_0}\right)_w \tag{1}$$

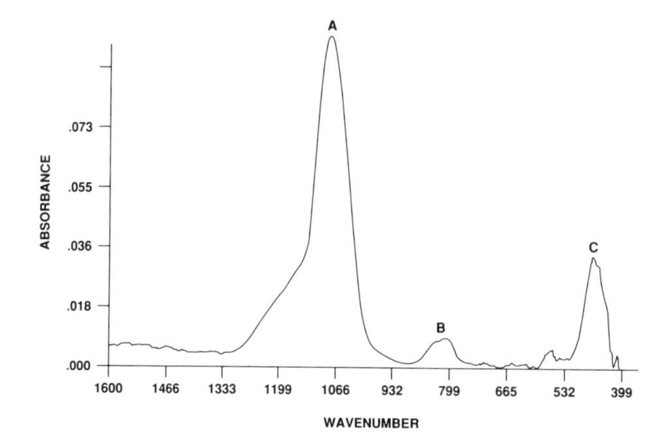

Figure 1 The FTIR spectrum of the oxide of silicon (thin film deposited by CVD). Primary features: (a), asymmetric stretching mode of vibration; (b), bending mode of vibration; (c), rocking mode of vibration.

where T_w is the transmittance of the sample at frequency w, and It is the intensity of the transmitted light. Similarly, if one is measuring the light reflected from the surface of the sample, then the ratio is equated to R_w, or the reflectance of the spectrum, with I_t being replaced with the intensity of the reflected light I_r. The third format, absorbance, is related to transmittance by the Beer-Lambert Law:

$$A_w = -\log T_w = (\varepsilon_w)(bc) \tag{2}$$

where c is the concentration of chemical bonds responsible for the absorption of infrared radiation, b is the sample thickness, and ε_w is the frequency-dependent absorptivity, a proportionality constant that must be experimentally determined at each w by measuring the absorbance of samples with known values of bc. As a first-order approximation the Beer-Lambert Law provides an simple foundation for quantitating FTIR spectra. For this reason, it is easier to obtain quantitative results if one collects an absorbance spectrum, as opposed to a reflectance spectrum. Prior to the introduction of FTIR spectrophotometers, infrared spectra were usually published in the transmittance format, because the goal of the experiment was to obtain qualitative information. With the growing use of FTIR technology, a quantitative result is more often the goal. Today the absorbance format dominates, because to first order it is a linear function of concentration.

Qualitative and Quantitative Analysis

Figure 1 shows a segment of the FTIR absorbance spectrum of a thin film of the oxide of silicon deposited by chemical vapor deposition techniques. In this film, sil-

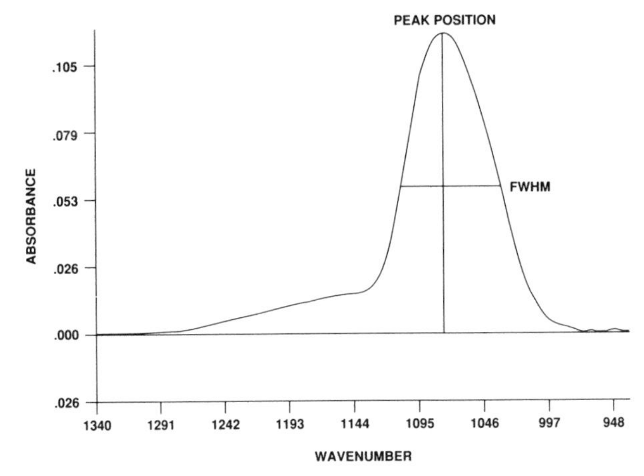

Figure 2 **Spectral parameters typically used in band shape analysis of an FTIR spectrum: peak position, integrated peak area, and FWHM.**

icon is tetrahedrally coordinated with four bridging oxygen atoms. Even though the bond angles are distorted slightly to produce the random glassy structure, this spectrum is quite similar to that obtained from crystalline quartz, because most features in the FTIR spectrum are the result of nearest neighbor interactions. In crystalline materials the many vibrational modes can be classified by the symmetry of their motions and, while not rigorous, these assignments can be applied to the glassy material, as well. Thus the peak near 1065 cm^{-1} arises from the asymmetric stretching motions of the Si and O atoms relative to each other. The band near 815 cm^{-1} arises from bending motions, while the one near 420 cm^{-1} comes from a collective rocking motion. These are not the only vibrational modes for the glass, but they are the only ones that generate electric dipoles that are effective in coupling with the electromagnetic field. For example, the glass also has a symmetric stretching mode, but since it generates no net dipole, no absorption band appears in the FTIR spectrum. For more quantitative work, the fundamental theory of infrared spectroscopy delineates a band shape analysis illustrated in Figure 2. Three characteristics are commonly examined: peak position, integrated peak intensity, and peak width.

Peak position is most commonly exploited for qualitative identification, because each chemical functional group displays peaks at a unique set of characteristic frequencies. The starting point for such a functional-group analysis is a table or computer database of peak positions and some relative intensity information. This provides a fingerprint that can be used to identify chemical groups. Thus the three peaks just described for the oxide of silicon can be used to identify that material. Typical of organic materials, C–H bonds have stretching modes around 3200 cm^{-1}; C = O, around 1700 cm^{-1}. Thus, the composition of oils may be qualitatively identified by classifying these and other peak positions observed in the spectrum. In addition, some bands have positions that are sensitive to physicomechanical properties. As a result, applied and internal pressures, stresses, and bond strain due to swelling can be studied.

The Beer-Lambert Law of Equation (2) is a simplification of the analysis of the second-band shape characteristic, the integrated peak intensity. If a band arises from a particular vibrational mode, then to the first order the integrated intensity is proportional to the concentration of absorbing bonds. When one assumes that the area is proportional to the peak intensity, Equation (2) applies.

In solids and liquids, peak width—the third characteristic—is a function of the homogeneity of the chemical bonding. For the most part, factors like defects and bond strain are the major sources of band width, usually expressed as the full width at half maximum (FWHM). This is due to the small changes these factors cause in the strengths of the chemical bonds. Small shifts in bond strengths cause small shifts in peak positions. The net result is a broadening of the absorption band. The effect of curing a material can be observed by peak-width analysis. As one anneals defects the bands become narrower and more intense (to conserve area, if no bonds are created or destroyed). Beyond the standard analysis, higher order band shape properties may also be examined, such as peak asymmetry. For example, the apparent shoulder on the high-frequency side of the band in Figure 2 may be due to a second band that overlaps the more prominent feature.

For many applications, quantitative band shape analysis is difficult to apply. Bands may be numerous or may overlap, the optical transmission properties of the film or host matrix may distort features, and features may be indistinct. If one can prepare samples of known properties and collect the FTIR spectra, then it is possible to produce a calibration matrix that can be used to assist in predicting these properties in unknown samples. Statistical, chemometric techniques, such as PLS (partial least-squares) and PCR (principle components of regression), may be applied to this matrix. Chemometric methods permit much larger segments of the spectra to be comprehended in developing an analysis model than is usually the case for simple band shape analyses.

Methodologies and Accessories

A large number of methods and accessories have been developed to permit the infrared source to interact with the sample in appropriate ways. Some of the more common approaches are listed below.

Single-pass transmission

The direct transmission experiment is the most elegant and yields the most quantifiable results. The beam makes a single pass through the sample before reaching the detector. The bands of interest in the absorbance spectrum should have peak absorbances in the range of 0.1–2.0 for routine work, although much weaker or stronger bands can be studied. Various holders, pellet presses, and liquid cells have been developed to permit samples to be prepared with the appropriate path length. *Diamond anvil cells* permit pliable samples to be squeezed to extremely thin path lengths or to be studied under applied pressures. Long path-length cells are used for samples in the gas phase. Thin films and prepared surfaces can be studied by transmission if the supporting substrate is transparent to the infrared. The highest sensitivity is obtained with double-beam or pseudo-double-beam experimental configurations. An example of the latter is the *interleaf* experiment, where a single beam is used, but a sample and reference are alternately shuttled into and out of the beam path.

Reflection

If the sample is inappropriate for a transmission experiment, for instance if the supporting substrate is opaque, a reflectance configuration will often be employed. Accessories to permit specular, diffuse, variable-angle, and grazing-incidence experiments are available. The angle of incidence can be adjusted to minimize multiple-reflection interferences by working at the Bragg angle for thin films, or to enhance the sensitivity of the probe to surface layers. A subset of this technique, Reflection Absorption (RA) spectroscopy, is capable of detecting submonolayer quantities of materials on metal surfaces. These grazing and RA techniques can enhance surface sensitivities by using geometries that optimize the coupling of the electromagnetic field at the metal surface. In some instances it has been possible to deduce preferred molecular orientations of ordered monolayers.

Attenuated Total Reflection

In this configuration an Attenuated Total Reflection (ATR) crystal is used, illustrated in Figure 3. The infrared beam is directed into the crystal. Exploiting the principles of a waveguide, the change in refractive index at the crystal surface causes the beam to be back-reflected several times as it propagates down the length of the crystal before it finally exits to the detector. If the sample is put into contact with the crystal surface, the beam will interact weakly with the sample at several points. For extremely thin samples, this is a means of increasing the effective path length. Since the propagating beam in the crystal barely penetrates through the surface of the sample adjacent to the crystal, signals at a sample surface can be enhanced, as well. This also helps in the study of opaque samples. Approximately fivefold amplifications in signals are typical over a direct transmission experiment. The quality of the crystal–sample interface is critical, and variability in that interface can make ATR results very difficult to quantify.

Emission

When samples are heated, they emit infrared radiation with a characteristic spectrum. The IR emission of ceramics, coals, and other complicated solids and thin films can be studied. Also, if conditions make it difficult to use an infrared source

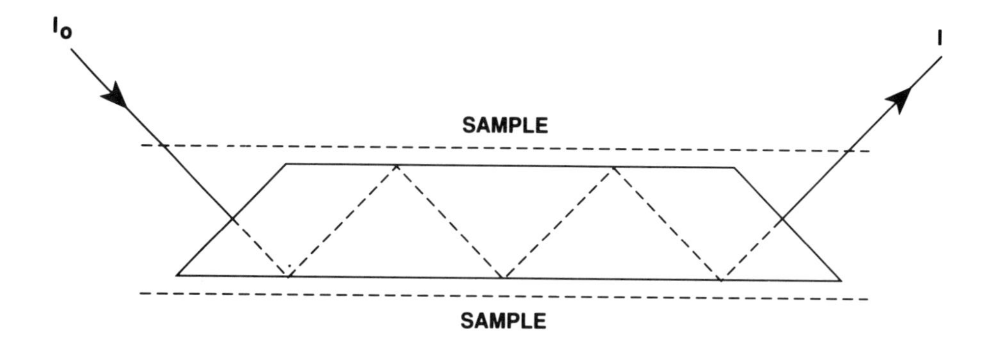

Figure 3 **Typical beam path configuration for collecting an FTIR spectrum using an attenuated total reflectance element: I_0 is the incident infrared beam, I is the exiting beam.**

(such as an *in situ* measurement of a thin film in the deposition chamber) but permit the controlled heating of a sample, then emission methods provide a means of examining these materials.

Microscopy

Infrared microscopes can focus the beam down to a 20-μm spot size for microprobing in either the transmission or reflection mode. Trace analysis, microparticle analysis, and spatial profiling can be performed routinely.

Interferences and Artifacts

Equipment

Equipment technology and processing software for FTIR are very robust and provide a high degree of reliability. Concerns arise for only the most demanding applications. For quantitative work on an isolated feature in the spectrum, the rule of thumb is that the spectrometer resolution be one-tenth the width of the band. FTIR instruments routinely meet that requirement for solids.

Short- and long-term drift in the spectral output can be caused by several factors: drift in the output of the infrared light source or of the electronics, aging of the beam splitter, and changes in the levels of contaminants (water, CO_2, etc.) in the optical path. These problems are normally eliminated by rapid, routine calibration procedures.

The two complicating factors that are encountered most frequently are the linearity of detector response and stray light scattering at low signal levels. DTGS detectors are quite linear and reliable, while MCT detectors can saturate at relatively low light levels. Stray light can make its way to the detector and be erroneously detected as signal, or it can be backscattered into the interferometer and degrade its output. A problem only at very low signal levels or with very reflective surfaces, proper procedures can minimize these effects.

Intrinsic or Matrix

Few cross sections for infrared absorption transitions have been published and typically they are not broadly applicable. The strength of the absorption depends upon changes in the dipole moment of the material during the vibrational motion of the constituent atoms. However, these moments are also very sensitive to environmental factors, such as nearest neighbor effects, that can cause marked changes in the infrared spectrum. For example, carbon–halogen bonds have a stretching mode that may be driven from a being very prominent feature to being an undetectable feature in the spectrum by adding electron-donating or -withdrawing substituents as nearest neighbors. For careful quantitative work, then, model compounds that are closely representative of the material in question are often needed for calibration.

Interface Optical Effects

For thin-film samples, abrupt changes in refractive indices at interfaces give rise to several complicated *multiple reflection* effects. Baselines become distorted into complex, sinusoidal, fringing patterns, and the intensities of absorption bands can be distorted by multiple reflections of the probe beam. These artifacts are difficult to model realistically and at present are probably the greatest limiters for quantitative work in thin films. Note, however, that these interferences are functions of the complex refractive index, thickness, and morphology of the layers. Thus, properly analyzed, useful information beyond that of chemical bonding potentially may be extracted from the FTIR spectra.

Many materials have grain boundaries or other microstructural features on the order of a micrometer or greater. This is on the same scale as the wavelength of the infrared radiation, and so artifacts due to the wavelength-dependent scattering of light at these boundaries can be introduced into the spectra. Thin films, powders, and solids with rough surfaces are the most affected. Again these artifacts are difficult to realistically model, but properly analyzed can provide additional information about the sample.

Conclusions

The principles of infrared spectroscopy can be exploited to extract information on the chemical bonding of an extremely wide variety of materials. The greatest strength of the technique is as a nondestructive, bulk probe of glassy and amorphous materials, where few alternate methods exist for obtaining chemical information. For other materials, FTIR is a valuable member in the arsenal of characterization tools. Other methods that are most likely to provide similar information include raman spectroscopy, X-ray photoelectron spectroscopy, NMR, MS, SIMS, and high-resolution electron energy-loss spectroscopy. The nondestructive, noninvasive potential of the infrared technique, and its ease of use, continues to distinguish it from these other methods, with the exception of Raman spectroscopy.

The trends begun with the general introduction of FTIR technology will undoubtedly continue. It is safe to say that the quality of the data being produced far exceeds our ability to analyze it. In fact, for many current applications, the principle limitations are not with the equipment, but rather with the quality of the samples. Thus, the shift from qualitative to quantitative work will proceed, reaching high levels of sophistication to address the optical and matrix interference problems discussed above.

With extensive computerization, the ease of use, and the robustness of equipment, movement of the instrumentation from the research laboratory to the manufacturing environment, for application as *in situ* and at-line monitors, will continue. *In situ* work in the research laboratory will also grow. New environments for application appear every day and improved computer-based data processing techniques make the rapid analysis of large sets of data more commonplace. These developments, coupled with rapid data acquisition times, are making possible the timely evaluation of the results of large-scale experiments. Most likely, much of the new physicochemical information developed by applying FTIR technology will come from trends observed in detailed studies of these large sample data sets.

Related Articles in the Encyclopedia

Raman Spectroscopy and HREELS

References

1 B. George and P. McIntyre. *Analytical Chemistry by Open Learning: Infrared Spectroscopy.* John Wiley & Sons, New York, 1987. A good primer on the basics of applied infrared spectroscopy.

2 P. R. Griffiths and J. A. de Haseth. *Fourier Transform Infrared Spectrometry.* John Wiley & Sons, New York, 1986. Chapters 1–8 review FTIR equipment in considerable detail. Chapters 9–19 describe applications, including surface techniques (Chapter 17).

3 *Practical Fourier Transform Infrared Spectroscopy: Industrial and Laboratory Chemical Analysis.* (J. R. Ferraro and K. Krishnan, eds.) Academic Press, New York, 1990. Chapters 3 (by K. Krishnan and S. L. Hill) and

Chapter 7 (by H. Ishida and A. Ishitani) review microscopic and surface analytical techniques. Chapter 8 (by D. M. Haaland) reviews developments in statistical chemometrics for data analysis.

4 O. S. Heavens. *Optical Properties of Thin Solid Films.* Butterworths, 1955. Chapter 4 presents a detailed mathematical description of the Fresnel fringing phenomenon for the transmission of light through thin films.

5 C. F. Bohren and D. R. Huffman. *Absorption and Scattering of Light by Small Particles.* John Wiley & Sons, New York, 1983. Parts 1 and 2 describe the theory of the scattering problem in some detail. Part 3 compares theory with experiment.

8.2 Raman Spectroscopy

WILLIAM B. WHITE

Contents

Introduction

To a surprisingly accurate approximation, molecules and crystals can be thought of as systems of balls (atoms) connected by springs (chemical bonds). These systems can be set into vibration, and vibrate with frequencies determined by the mass of the balls (atomic weights) and by the stiffness of the springs (bond force constants). Diatomic molecules (O_2, CO, HCl, etc.) having two balls connected by a single spring have only one fundamental vibrational frequency. The same is true for diatomic crystals, which have a single diatomic formula unit in the primitive unit cell (NaCl, ZnS, diamond, etc.), although the details are more complicated. The number of possible vibrational motions is $3n–6$ for nonlinear molecules and $3n–3$ for crystals, where n is the number of atoms in the molecule or in the primitive unit cell of the crystal. The mechanical molecular and crystal vibrations are at very high frequencies, in the range 10^{12}–10^{14} Hz (3–300 μm wavelength), which places them in the infrared (IR) region of the electromagnetic spectrum. Coupling between incident infrared radiation and the electronic structure of the chemical bond produces the infrared absorption spectrum as a direct means of observing molecular

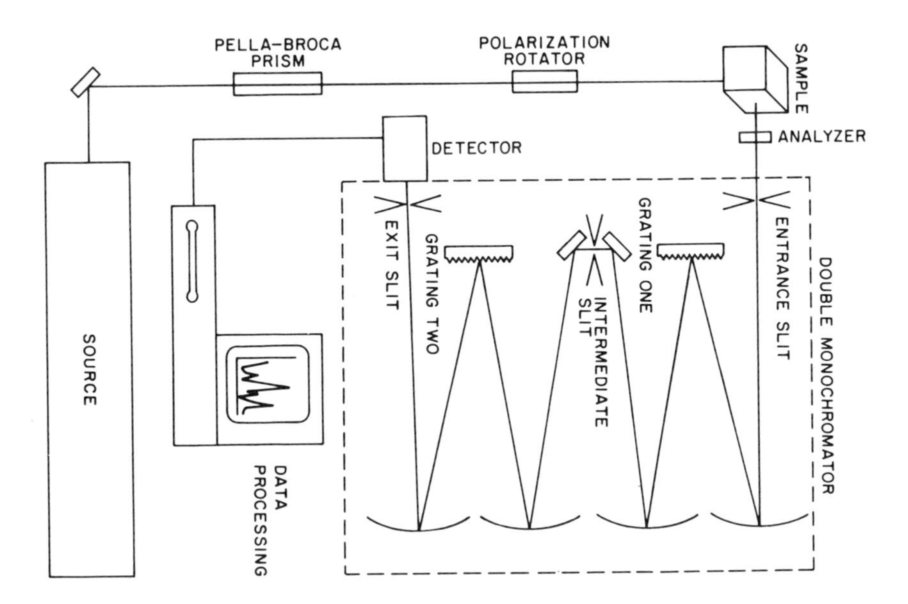

Figure 1 **Drawing of single-channel Raman spectrometer showing Czerny–Turner type double monochromator. Collecting optics for scattered beam are not shown.**

and crystal vibrations (see the article on FTIR). The Raman spectrum arises from an indirect coupling of high-frequency radiation (usually visible light, but also ultraviolet and near infrared) with the electron clouds that make up the chemical bonds. Thus, although both IR absorption and Raman spectroscopy measure the vibrational spectra of materials, the physical processes are different, the selection rules that determine which of the vibrational modes are excited are different, and the two spectroscopies must be considered to be complementary. Both, however, are characterization probes that are sensitive to the details of atomic arrangement and chemical bonding.

Raman spectroscopy is primarily a structural characterization tool. The spectrum is more sensitive to the lengths, strengths, and arrangement of bonds in a material than it is to the chemical composition. The Raman spectrum of crystals likewise responds more to details of defects and disorder than to trace impurities and related chemical imperfections.

Basic Principles

The essentials of the Raman scattering experiment are shown in Figure 1. An intense monochromatic light beam impinges on the sample. The electric field of the incident radiation distorts the electron clouds that make up the chemical bonds in the sample, storing some energy. When the field reverses as the wave passes, the

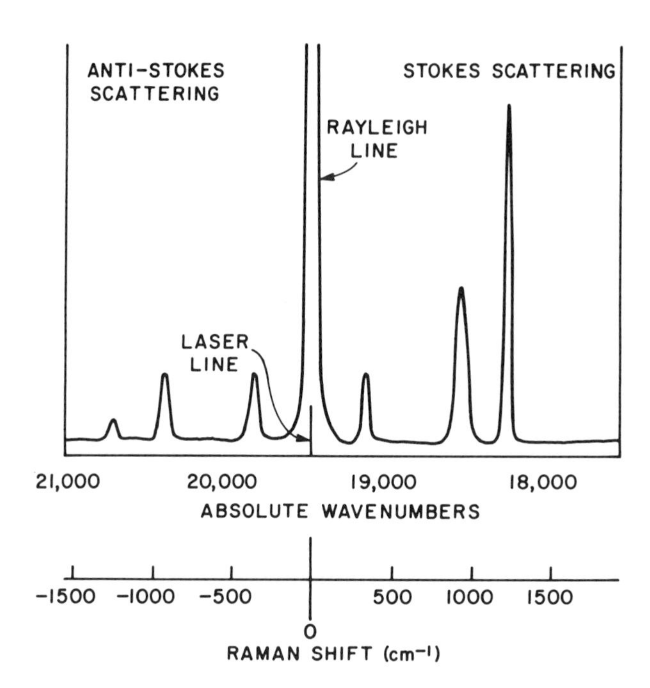

Figure 2 **Schematic Raman scattering spectrum showing Rayleigh line, Stokes Raman scattering and anti-Stokes Raman scattering. Note arrangements of wavenumber scales with Rayleigh line defining the zero on the Raman wavenumber scale.**

distorted electron clouds relax and the stored energy is reradiated. Although the incident beam may be polarized so that the electric field is oriented in a specific direction with respect to the sample, the scattered beam is reradiated in all directions, making possible a variety of scattering geometries. Most of the stored energy is reradiated at the same frequency as that of the incident exciting light. This component is known as the Rayleigh scattering and gives a strong central line in the scattering spectrum (Figure 2). However, a small portion of the stored energy is transferred to the sample itself, exciting the vibrational modes. The vibrational energies are deducted from the energy of the incident beam and weak side bands appear in the spectrum at frequencies less than that of the incident beam. These are the Raman lines. Their separation from the Rayleigh line is a direct measure of the vibrational frequencies of the sample.[1]

The reverse process also occurs. Existing vibrations that have been excited by thermal processes, can be annihilated by coupling with the incident beam and can add their energies to that of the source. These appear as side bands at higher wavenumbers. The Raman process that excites molecular and crystal vibrations is called Stokes scattering; the process that annihilates existing vibrations is called anti-Stokes scattering. The two spectra are mirror images on opposite sides of the Rayleigh line (Figure 2). However, because anti-Stokes scattering depends on the existence of thermally activated vibrations, the anti-Stokes intensities are strongly temperature dependent, whereas the Stokes intensities are only weakly temperature dependent. For this reason, anti-Stokes scattering is rarely measured, except for the specialized technique known as *coherent anti-Stokes Raman spectroscopy.*[2] Because the vibrational frequencies are measured by differences between the frequency of the Raman line and the Rayleigh line, most spectrometers are set up to display the difference frequency (wavenumber) directly, defining the exciting frequency as 0. The result sometimes causes confusion: as the displayed Raman wavenumber increases, the true wavenumber decreases. The Raman effect is extremely weak. Rayleigh scattering from optically transparent samples is on the order of 10^{-3}–10^{-5} of the intensity of the exciting line. Raman scattering is from 10^{-3} to 10^{-6} of the intensity of the Rayleigh line. For this reason Raman spectroscopy prior to the 1960s was a highly specialized measurement carried out in a few laboratories mainly with mercury discharge lamp sources, fast prism spectrographs, and photographic plate detectors. In such equipment, the most intense Raman lines are at the threshold of visibility to the darkness-adapted human eye. The invention of continuous gas lasers, which provided the needed intense monochromatic source, and the invention in 1965 of the double monochromator system for stray light discrimination combined to make the modern Raman spectrometer possible.

Instrumentation

Single-Channel Dispersive Instruments

The most widely used equipment for Raman spectroscopy follows the scheme shown in Figure 1. The source is usually a continuous gas laser. The 488-nm and 514.5-nm lines of the argon ion laser are commonly used, although argon has other, somewhat weaker lasing lines that can be used for special purposes. He–Ne lasers (632.8 nm) and the red line of the krypton laser (647.1 nm) also are used. The Raman scattering intensity decreases as the fourth power of the wavelength so that there is an advantage to using the shorter wavelength argon ion lines unless other circumstances, such as sample fluorescence, sample photodecomposition, or the location of the optical absorption edge in semiconductor materials, require a less energetic excitation source. The output from gas lasers is intrinsically polarized so that a polarization rotator is needed to control the polarization orientation with respect to the sample, especially with oriented single-crystal specimens.

The sample holder is indicated by a cube in Figure 1 to show the 90° scattering arrangement. The part of the sample actually measured is a volume, typically 200–500 µm on a side, where the laser beam and the scattered beam intersect. Omitted from the sketch are the focusing optics that collect the scattered light and bring it to a focus on the entrance slit. Many sample mounting arrangements are possible, depending on the physical state of the sample. An analyzer in front of the entrance slit permits the determination of the polarization of the scattered beam.

The commonly used monochromator is a Czerny–Turner grating type, usually of 1-m focal length, as sketched in Figure 1. This design features two collimating mirrors and a planar grating that is rotated to sweep the spectrum across the exit slit. Because Raman spectra are extremely weak, stray light within the monochromator must be effectively suppressed. The second monochromator, used as a stray light filter, reduces the stray light to the 10^{-12} range, providing a clean background for the Raman spectra.

Photomultipliers are used as detectors in the single-channel instruments. GaAs cathode tubes give a flat frequency response over the visible spectrum to 800 nm in the near IR. Contemporary Raman spectrometers use computers for instrument control, and data collection and storage, and permit versatile displays.

Diode Array and Charge-Coupled Detector Systems

A disadvantage of the single-channel scanning monochromator is that substantial times are needed to collect a spectrum, which is unsatisfactory for transient species, kinetic studies, and observation of unstable compounds. Linear diode array detectors allow the entire spectrum to be captured at one time, although resolution is limited by the number of pixels. Charge-coupled device detector arrays are now used on some commercial spectrometers. These are two-dimensional arrays of detector elements, 578×385 pixels being a common choice. Both diode array and charge-coupled detectors permit the use of pulsed laser sources and provide infrared sensitivity. Nd–YAG lasers operating at a wavelength of 1.06 μm can then be used to eliminate many problems of fluorescence and sample photodecomposition.

Fourier Transform Raman Spectrometers

A relatively recent development is the adaptation of Fourier transform spectrometers commonly used for infrared absorption spectroscopy to Raman spectroscopy. Fourier transform Raman spectroscopy has the same advantages as other applications of interferometric techniques. Elimination of the slits greatly increases energy throughput, an important advantage for weak Raman signals. The multiplex advantage permits more rapid accumulation of spectral data, although the alternative of diode array or charge-coupled detectors makes this advantage less important for Raman spectroscopy than for IR spectroscopy.

Sample Requirements

Liquids

Liquids and solutions can be measured in special cells that have optical windows at right angles, or they can be contained in capillary tubes or small vials. The latter are

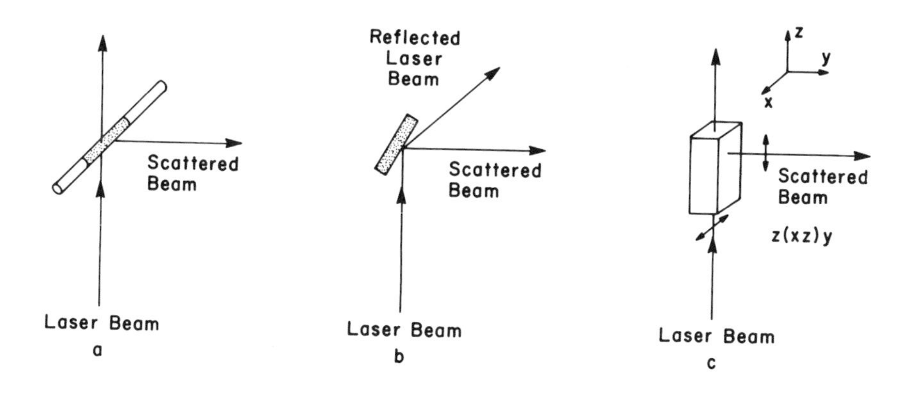

Figure 3 **Scattering geometries appropriate to (a) liquids in capillaries or glass fibers; (b) powder pellets, solid slabs of ceramic or rock, or films on substrates; and (c) oriented single crystals.**

mounted so that the laser passes along one diameter of the capillary while the scattered beam is along a perpendicular diameter (Figure 3).

Powders and Polycrystalline Solids

Usually, particle size has relatively little effect on Raman line shapes unless the particles are extremely small, less than 100 nm. For this reason, high-quality Raman spectra can be obtained from powders and from polycrystalline bulk specimens like ceramics and rocks by simply reflecting the laser beam from the specimen surface. Solid samples can be measured in the 90° scattering geometry by mounting a slab of the solid sample, or a pressed pellet of a powder sample so that the beam reflects from the surface but not into the entrance slit (Figure 3).

Single-Crystal Measurements

Maximum information is obtained by making Raman measurements on oriented, transparent single crystals. The essentials of the experiment are sketched in Figure 3. The crystal is aligned with the crystallographic axes parallel to a laboratory coordinate system defined by the directions of the laser beam and the scattered beam. A useful shorthand for describing the orientational relations (the Porto notation) is illustrated in Figure 3 as $z(xz)y$. The first symbol is the direction of the laser beam; the second symbol is the polarization direction of the laser beam; the third symbol is the polarization direction of the scattered beam; and the fourth symbol is the direction of the scattered beam, all with respect to the laboratory coordinate system.

Spectra From Thick and Thin Films

Raman spectroscopy is especially suited for structural characterization of films and coatings. Both oblique (Figure 3b) and backscattering geometries can be used. The effective penetration depth of the laser depends on film transparency and on the scattering geometry but in most cases the spectrum of the substrate will be superimposed on the spectrum of the film and must be accounted for. The minimum observable film thickness varies with the intrinsic scattering power of the material but is in the range of a few μm for most materials. Resonance enhancement may decrease the minimum thickness. Useful data usually can be obtained from both crystalline and noncrystalline films.

Surface-Enhanced Raman Spectroscopy

Intensity enhancement takes place on rough silver surfaces. Under such conditions, Raman scattering can be measured from monolayers of molecular substances adsorbed on the silver (pyridine was the original test case), a technique known as surface-enhanced Raman spectroscopy.[3] More recently it has been found that surface enhancement also occurs when a thin layer of silver is sputtered onto a solid sample and the Raman scattering is observed through the silver.

High-Temperature and High-Pressure Raman Spectroscopy

Because Raman spectroscopy requires one only to guide a laser beam to the sample and extract a scattered beam, the technique is easily adaptable to measurements as a function of temperature and pressure. High temperatures can be achieved by using a small furnace built into the sample compartment. Low temperatures, easily to 78 K (liquid nitrogen) and with some difficulty to 4.2 K (liquid helium), can be achieved with various commercially available cryostats. Chambers suitable for Raman spectroscopy to pressures of a few hundred MPa can be constructed using sapphire windows for the laser and scattered beams. However, Raman spectroscopy is the characterization tool of choice in diamond-anvil high-pressure cells, which produce pressures well in excess of 100 GPa.[4]

Fluorescence, Sample Heating and Sample Photodecomposition

Although Raman spectroscopy is very versatile with regard to sample form, particle size, and composition, certain interferences occur. Many transparent solid samples contain impurities that fluoresce under excitation by the laser beam. Fluorescence bands are usually extremely broad and are often of sufficient intensity to completely mask the weaker Raman scattering. Fluorescence can sometimes be avoided by using lower energy excitation, such as the red lines of He–Ne or Krypton lasers, or by using a Nd–YAG laser with a diode array detector. If solid samples of interest are being synthesized for measurement, one can sometimes eliminate fluorescence by deliberately doping the sample with luminescence poisons like ferrous iron. Narrow-band fluorescence can be distinguished from Raman scattering by measuring the spectra using two different wavelength exciting lines (the 488-nm and 514.5-nm lines of argon, for example). Because the instrument wavenumber scale is set to 0 at the laser line, Raman lines will appear at the same wavenumber position under both excitations whereas fluorescence lines will appear to shift by the wavenumber difference between the two exciting lines.

Because the laser beam is focused on the sample surface the laser power is dissipated in a very small area which may cause sample heating if the sample is absorbing and may cause break-down if the sample is susceptible to photodecomposition. This problem sometimes may be avoided simply by using the minimum laser power needed to observe the spectrum. If that fails, the sample can be mounted on a motor shaft and spun so that the power is dissipated over a larger area. Spinners must be adjusted carefully to avoid defocusing the laser or shifting the focal spot off the optic axis of the monochromator system.

Bulk Raman Spectroscopic Analysis

Fingerprinting

Raman spectra of molecules and crystals are composed of a pattern of relatively sharp lines. The wavenumber scale for most vibrations extends from 50 cm^{-1} to about 1800 cm^{-1} with some molecular vibrations extending to 3500 cm^{-1}. Line widths are on the order of 1–5 cm^{-1}. The Raman spectrum thus has a fairly high density of information and can be used as a fingerprint for the identification of unknown materials by direct comparison of the spectrum of the unknown with spectra in a reference catalog. An example for a series of strontium titanates is shown in Figure 4. This approach is widely used for NMR spectra, X-ray diffraction powder patterns, and infrared spectra as well. As with these other techniques, the success of fingerprinting depends on the availability of a complete, high-quality catalog of reference spectra. Catalogs of reference Raman spectra are much less complete than the corresponding catalogs for IR, ultraviolet, and nuclear magnetic resonance spectra. Users of Raman spectroscopy should compile their own reference spectra.

Crystal Spectra

From a knowledge of the crystal structure it is possible to calculate selection rules for each orientation position and thus gain considerable insight into the vibrational motions of the crystal. The interpretation of such spectra, which show a lot of detail, goes well beyond characterization applications.[5]

Ordered Structures and Phase Transitions

Raman spectroscopy is sensitive to ordering arrangements of crystal structures, the effect depending on the type of order. Ordering atoms onto specific lattice sites in

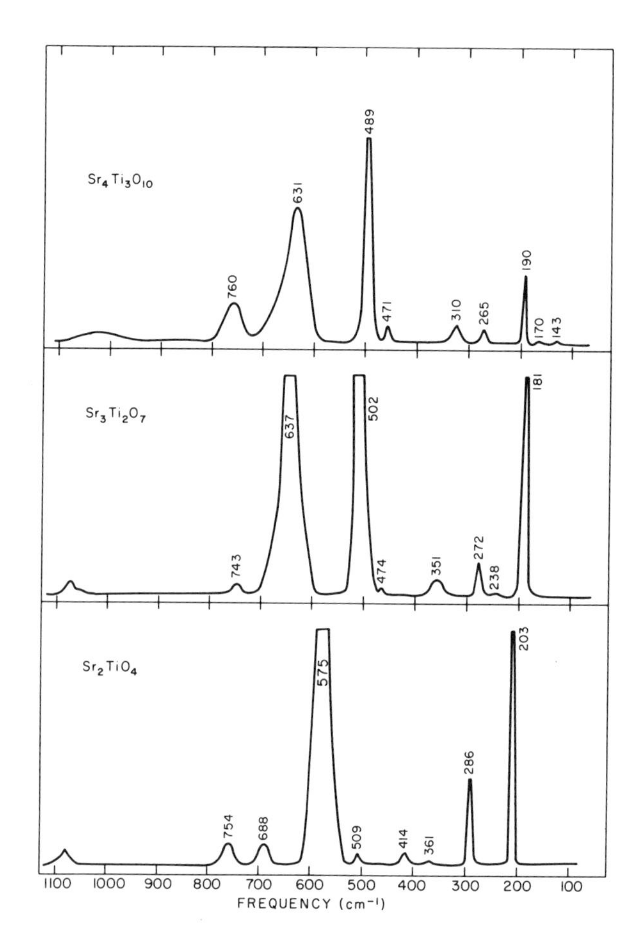

Figure 4 **Raman spectra of a series of strontium titanates showing typical line shapes and information available for fingerprinting. The broadening of the high-wavenumber lines is related to the polar character of the TiO_6 octahedra that occur in all of these structures.**

the parent structure often forms derivative structures having different space groups. The new symmetry leads to relaxed selection rules, which in turn lead to new Raman lines or to the splitting of the lines of the parent structure. Group theoretical calculations allow a good prediction of expected spectral behavior for possible ordering schemes, and Raman (IR) spectra can be used select the correct model.

Some materials undergo transitions from one crystal structure to another as a function of temperature and pressure. Sets of Raman spectra, collected at various temperatures or pressures through the transition often provide useful information on the mechanism of the phase change: first or second order, order/disorder, soft mode, etc.

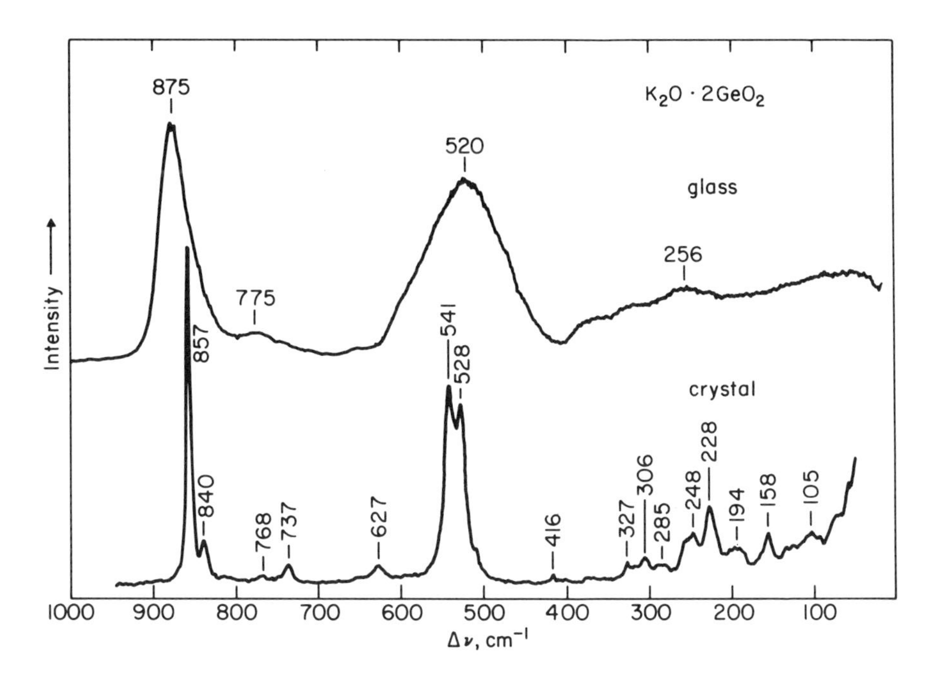

Figure 5 **Raman spectra of crystalline and glassy potassium digermanate showing comparison between crystal spectra and glass spectra.**

Defects and Structural Disorder

Raman spectra of solid solutions, crystals with lattice defects, and systems having other types of structural disorder usually exhibit a pronounced line broadening in comparison with ordered structures. Structural defects, such as lattice vacancies, produce line broadening with little temperature dependence. Orientational disorder, arising from alternative possible orientations of molecules in crystals, dipoles in highly polar crystals, and nonbonding lone pair electrons in ions like Pb^{2+} and As^{3+}, produce Raman lines that are broad at room temperature but become narrow as temperatures are lowered into the liquid nitrogen or liquid helium range. At high defect concentrations, greater than 10–20% mole, the broadened Raman lines give way to a scattering continuum having little structure.

Glasses and Gels

Raman spectroscopy is particularly useful for investigating the structure of noncrystalline solids. The vibrational spectra of noncrystalline solids exhibit broad bands centered at wavenumbers corresponding to the vibrational modes of the corresponding crystals (Figure 5). In silicate glasses shifts in the high-wavenumber bands is a measure of the degree of polymerization of the silicate network.[6] The processes of nucleation and crystallization in glasses can be readily followed by the Raman spectrum because the sharp crystal bands are easily detected against the much weaker and broader glass bands.

Gels are more disorganized than glasses and often have weaker Raman spectra. The processing steps from solution to sol, to gel, to glass in the sol–gel process of glass making can be followed in the Raman spectra. The loss of organic constituents of the gel can be followed as can the development of the bulk glass structure.

Microfocus Raman Spectroscopic Analysis

By successfully marrying a Raman spectrometer to an optical microscope it is possible to obtain spectra having the resolution of optical microscopy—a few μm. The essential feature is a beam splitter through which the laser beam passes on its way to the microscope objective. The beam emerges from the objective to strike a sample held on a standard microscope stage. Raman scattering is observed in the backscattering geometry. The scattered beam also passes through the beam splitter and reaches the entrance slit of the monochromator.

Grain-by-Grain Analysis and Crystal Zoning

The obvious application of microfocus Raman spectroscopy is the measurement of individual grains, inclusions, and grain boundary regions in polycrystalline materials. No special surface preparation is needed. Data can be obtained from fresh fracture surfaces, cut and polished surfaces, or natural surfaces. It is also possible to investigate growth zones and phase separated regions if these occur at a scale larger than the 1–2 μm optical focus limitation.

With a special optical system at the sample chamber, combined with an imaging system at the detector end, it is possible to construct two-dimensional images of the sample displayed in the emission of a selected Raman line.[7] By imaging from their characteristic Raman lines, it is possible to map individual phases in the multiphase sample; however, Raman images, unlike SEM and electron microprobe images, have not proved sufficiently useful to justify the substantial cost of imaging optical systems.

Precipitates and the Liquid/ Solid Interface

With the microfocus instrument it is possible to combine the weak Raman scattering of liquid water with a water-immersion lens on the microscope and to determine spectra on precipitates in equilibrium with the mother liquor. Unique among characterization tools, Raman spectroscopy will give structural information on solids that are otherwise unstable when removed from their solutions.

Fluid Inclusions

Natural crystals, synthetic crystals, and glasses often contain small bubbles that preserve samples of the fluid from which the crystals grew or of the atmosphere over the glass melt. Using a long focal length lens, the laser beam can be focused into inclusions at some depth below the crystal or glass surface. The Raman spectra then permit the identification of molecular species dissolved in the aqueous solutions or of components in the gas bubbles.[8]

Grain Boundaries, Cracks, and Stressed Materials

Stress in crystalline solids produces small shifts, typically a few wavenumbers, in the Raman lines that sometimes are accompanied by a small amount of line broadening. Measurement of a series of Raman spectra in high-pressure equipment under static or uniaxial pressure allows the line shifts to be calibrated in terms of stress level. This information can be used to characterize built-in stress in thin films, along grain boundaries, and in thermally stressed materials. Microfocus spectra can be obtained from crack tips in ceramic material; and by a careful spatial mapping along and across the crack estimates can be obtained of the stress fields around the crack.[9]

Thick and Thin Films

Because it is nondestructive, can be fine focused, and can be used in a backscattering geometry, Raman spectroscopy is a useful tool for characterizing films and layers either free standing or on various substrates. Semiconductors such as silicon, germanium and GaAs produce Raman spectra that give information on film crystallinity, the presence of impurity layers, and the presence of amorphous material. Films are ideal samples on the microscope stage of the microfocus instrument, and comparisons can easily be made between centers and margins of deposition zones. Raman spectroscopy has been applied to the characterization of chemical vapor deposition-grown diamond films, proving the formation of diamond and characterizing the presence of graphite and various amorphous carbons.[10]

The depth of film penetrated and sampled depends on the transparency of the sample to the laser radiation used. This is strongly dependent on the sample material and on the laser wavelength, and can vary from as little as a few μm to significant fractions of a mm. If a film or coating is transparent, the film / substrate interface and any other material deposited there is also accessible and may provide a detectable Raman signal. The absolute sensitivity of the Raman technique is routinely such that a film thickness of at least 0.1 μm is required to provide a workable signal. Special instrumentation, particularly high-sensitivity detectors, can reduce this to a few tens of nm and, as mentioned already, for surface-enhanced Raman scattering of molecules adsorbed on rough silver (and a few other) surfaces, submonolayer amounts are detectable.

Conclusions

Raman spectroscopy is a very convenient technique for the identification of crystalline or molecular phases, for obtaining structural information on noncrystalline solids, for identifying molecular species in aqueous solutions, and for characterizing solid–liquid interfaces. Backscattering geometries, especially with microfocus instruments, allow films, coatings, and surfaces to be easily measured. Ambient atmospheres can be used and no special sample preparation is needed.

In-situ Raman measurements will become more important in the future. In many types of vapor and liquid deposition systems, it is important to monitor the progress of reactions and changes in structure as films or crystals are grown. Fiber optic cables, sapphire light pipes and other optical systems are available for bringing the laser beam into the reaction chamber and for collecting the scattered light. Spectra can be collected without disturbing the deposition process.

Related Articles in the Encyclopedia

FTIR and HREELS

References

1 D. A. Long. *Raman Spectroscopy.* McGraw-Hill, New York, 1977. A standard reference work on Raman spectroscopy with much theoretical detail on the underlying physics. Most of the needed equations for any application of Raman spectroscopy can be found in this book.

2 W. M. Tolles, J. W. Nibler, J. R. McDonald, and A. B. Harvey. *Appl. Spectros.* **31,** 253, 1977.

3 R. K. Chang and T. E. Furtak. *Surface Enhanced Raman Scattering.* Plenum, New York, 1982. A collection of individually authored papers describing many different applications of SERS.

4 R. J. Hemley, P. M. Bell, and H. K. Mao. *Science.* **237,** 605, 1987.

5 G. Turrell. *Infrared and Raman Spectra of Crystals.* Academic, London, 1972. One of the best available texts describing the principles of Raman scattering from crystals. Includes factor group calculations, polarization measurements, force constant calculations, and many other aspects of crystal physics.

6 B. O. Mysen, D. Virgo, and F. A. Seifert. *Rev. Geophys. Space Phys.* **20,** 353, 1982.

7 P. Dhamelincourt, F. Wallart, M. Leclercq, A. T. N'Guyen, and D. O. Landon. *Analyt. Chem.* **51,** 414A, 1979.

8 J. D. Pasteris, B. Wopenka, and J. C. Seitz. *Geochim. Cosmochim. Acta.* **52,** 979, 1988.

9 R. H. Dauskardt, D. K. Veirs, and R. O. Ritchie. *J. Amer. Ceram.* Soc. 72, 1124, 1989.

10 D. S. Knight and W. B. White. *J. Mat. Res.* **4,** 385, 1989.

8.3 HREELS

High-Resolution Electron Energy Loss Spectroscopy

BRUCE E. KOEL

Contents

Introduction

High-resolution electron energy loss spectroscopy (HREELS) has emerged over the past decade as a sensitive, versatile, nondestructive surface analysis technique used to study the vibrations of atoms and molecules in and on solid surfaces.[1-4] Energy loss peaks in the spectra correspond to vibrations of atoms in the surface layers of a solid or in adsorbed molecules. The principal analytical use is to detect the presence of molecular groups on surfaces, often submonolayer amounts of adsorbed species, by observation of their vibrational spectra, and to chemically identify what these species are from the details (positions and intensities) of the vibrational peaks. For molecular materials, such as polymers, such information may be obtained also about the surface of the solid itself. The information obtained is of the same kind as that in infrared and Raman spectroscopy. The main reasons for the popularity of HREELS over these optical methods are its very small detection limit and its wide spectral range at high surface sensitivity. Most HREELS studies are of single-crystal surfaces, however polycrystalline and noncrystalline samples also can be studied with this technique, but with reduced effectiveness.

Fundamental information from vibrational spectra is important for understanding a wide range of chemical and physical properties of surfaces, e.g., chemical reactivity and forces involved in the atomic rearrangement (relaxation and reconstruction) of solid surfaces. Practical applications of HREELS include studies of:

1. Functional groups on polymer and polymer film surfaces
2. Phonon modes of metals and films, semiconductors, and insulators
3. Concentrations of free charge carriers in semiconductors
4. Adsorbed species (and even underlayer atoms) on single-crystal surfaces
5. Kinetics of surface processes when used in a time-resolved mode.

Of these, the most extensive use is to identify adsorbed molecules and molecular intermediates on metal single-crystal surfaces. On these well-defined surfaces, a wealth of information can be gained about adlayers, including the nature of the surface chemical bond, molecular structural determination and geometrical orientation, evidence for surface-site specificity, and lateral (adsorbate–adsorbate) interactions. Adsorption and reaction processes in model studies relevant to heterogeneous catalysis, materials science, electrochemistry, and microelectronics device failure and fabrication have been studied by this technique.

The first vibrational spectrum of adsorbed molecules obtained by inelastic scattering of low energy electrons was obtained by Propst and Piper in 1967. Over the next ten years, Ibach in Jülich achieved much higher resolution and, along with several other research groups, developed and used the new technique (HREELS) to study a wide range of surface vibrations, including polyatomic molecules adsorbed on surfaces. The enthusiasm existed mainly because HREELS could be used to study adsorbates on low surface-area, opaque, metal single-crystal samples, something that could not be done with infrared and Raman spectroscopy. It is now well-established as an important vibrational spectroscopy at surfaces, and its utility as an analytical tool with extreme surface sensitivity is rapidly being extended.

Basic Principles

Electron Scattering

At sufficiently high resolution, quasi-elastically scattered electrons have an inelastic scattering distribution from exciting surface vibrational modes such as surface phonons and adsorbate vibrations. (See, for example, Figure 1, the case of CO adsorbed on a Rh surface.) These modes have excitation energies below 0.5 eV (4000 cm^{-1}). The basis of the kinematic (single scattering) description of electron scattering from surfaces is conservation of energy and momentum parallel to the surface. These conservation laws define the scattering possibilities and, along with the vibrational energies, determine the peak energies in HREELS. The mechanisms

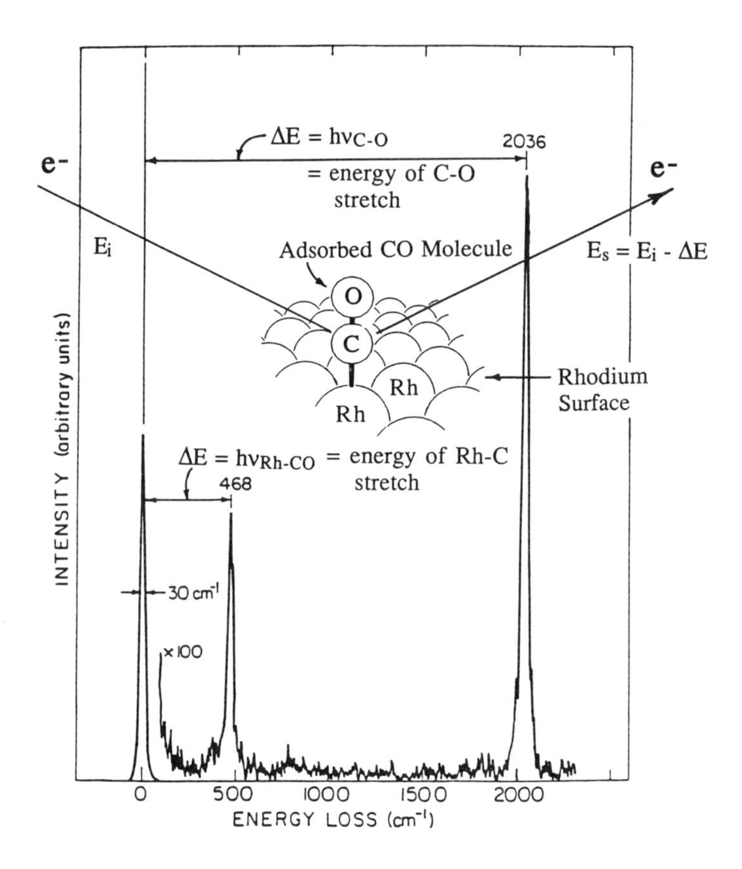

Figure 1 **Schematic of electron energy-loss scattering process for electrons of energy E_i striking a Rh single-crystal surface with adsorbed CO molecules present. The actual energy-loss spectrum, due to excitation of CO vibrations, is shown also.**

of electron inelastic scattering determine the scattering cross sections (probabilities), and therefore the intensities of vibrational energy loss peaks.

Two different electron inelastic scattering mechanisms are most important for explaining the intensities and angular distribution of peaks observed in HREELS: dipole and impact scattering. Dipole scattering is due to the long-range part of the electrostatic interaction between an incoming electron and the dipolar electric field of the vibrating (oscillating) atoms at the surface. Energy exchange takes place at a long distance (50 Å) from the surface. This leads to small angle scattering and so the scattered intensity is strongly peaked near the specular direction, with a half-angle $\Delta\theta \sim 0.1$–$1°$. The dominance of this mechanism for detection at the specular direction leads to the dipole selection rule in HREELS, that is identical to that associated with reflection infrared spectroscopy: Only vibrational modes which produce a dynamic (oscillating) dipole moment perpendicular to the metal surface are dipole-active, i.e., will produce energy loss peaks on-specular. The origin of this selection rule is the screening of a charge on a conducting surface by an image charge induced in the free electrons. Dynamic dipoles parallel to the surface generate no long-range dipole field, while dynamic dipoles perpendicular to the surface generate a long-range dipole field that is enhanced by a factor of 2.

At large scattering angles away from the specular direction, one enters the impact scattering regime. This mechanism is due to a short range electrostatic interaction between an incoming electron and the ion cores of the adsorbate and substrate lattice. Impact scattering is usually several orders of magnitude weaker than dipole scattering at the specular direction in the low-energy regime (below 10 eV). Broad angular scattering distributions having intensities proportional to vibrational amplitudes are characteristic. Selection rules are much less restrictive and are based on adsorbate site symmetry and vibrational mode polarization in relation to the scattering plane of incidence. The impact scattering regime extends to several hundred eV, and the theory of these interactions must include multiple scattering. Measurements of inelastic scattering cross sections in this regime have a great potential for structural analysis, but have been made experimentally only recently.

A type of molecular resonance scattering can also occur from the formation of short-lived negative ions due to electron capture by molecules on surfaces. While this is frequently observed for molecules in the gas phase, it is not so important for chemisorbed molecules on metal surfaces because of extremely rapid quenching (electron transfer to the substrate) of the negative ion. Observations have been made for this scattering mechanism in several chemisorbed systems and in physisorbed layers, with the effects usually observed as small deviations of the cross section for inelastic scattering from that predicted from dipole scattering theory.

While the underlying mechanisms of HREELS are pretty well understood, many important details relating to selection rules and scattering cross sections remain unknown.

Vibrations at Surfaces

Vibrations give molecular information by identifying which atoms are chemically bonded together. The frequency of a vibrational mode is related to the bond force constant and reduced mass of the vibrating atoms. Reasonable correlations exist between the number of bonds, bond energy, and force constant for vibrations within molecular species at surfaces and for adsorbate–substrate vibrations. It would be extremely useful if one could determine chemical bond strengths from vibrational spectra, but the accurate determination of this quantity is quite ambiguous. Frequency shifts occur as the concentration is changed in adsorbed layers as a result of local bonding variations due to changing sites or bond energies, dipole–dipole coupling leading to collective vibrations of the overlayer, or other factors.

The intensity of a vibrational mode in HREELS on-specular is given by the ratio of the inelastic to elastic intensities

$$S = 4\pi E_i^{-1} N P_p^2 F(\alpha, \theta) \arccos\theta \qquad (1)$$

where S is the dipole scattering cross section, E_i is the primary beam energy, N is the number of surface oscillators, P_p is the perpendicular component of the dynamic dipole moment, and $F(\alpha, \theta)$ is an instrument and geometry factor. This intensity (sensitivity) can be optimized by considering the detected inelastic current, I:

$$I = I_0 R(E_i)\, KS \qquad (2)$$

where I_0 is the incident beam current, $R(E_i)$ is the metal surface reflectivity (typically 1–10%) for the primary beam energy E_i, and K is the analyzer transmission constant. The function S is a smoothly decreasing function of energy and is proportional to E_i^{-1} for a layer of adsorbed dipoles, but R exhibits significant variation as a function of E_i. Operationally, the incident energy is tuned to give an intense elastic peak at a low energy (< 10 eV) where S is large.

The sensitivity of HREELS is largest for molecules, materials, or particular vibrations, that have large S. Normalizing for the other variables, such as concentration, leads to a general rule that vibrational modes that are observed as strong bands in infrared spectroscopy of gases or of condensed phases will give rise to intense loss peaks in HREELS. For example, carbonyl (C = O) stretching modes have larger intensities than hydrocarbon C–H stretching modes in IR of bulk phases and also in HREELS on surfaces. Orientational effects will affect this sensitivity in accord with the dipole selection rule.

Quantitative analysis and determination of the concentration of surface species requires measuring peak intensities and accounting for vibrational cross sections. This is a difficult task. Careful analysis of K and analyzer angular characteristics is required when determining $S(E_i)$. To obtain information on the charges on atoms at the surface, one could perform a calculation of dynamic dipole moments (or effective charges) from knowledge of S, in principle. In practice, one usually assumes a point dipole model and gas phase polarizabilities for surface species, and this has lead to anomalously low values in the case of adsorbed molecules on metals. Chemical bonding effects greatly enhance the electronic polarizability of chemisorbed molecules and enhance the dielectric screening by the adlayer, reducing the predicted vibrational (inelastic peak) intensity.

The width and shape of the energy loss peaks in HREELS are usually completely determined by the relatively poor instrumental resolution. This means that no information can be obtained from HREELS about such interesting chemical physics questions as vibrational energy transfer, since the influence of the time scale and mechanism of vibrational excitations at surfaces on the lifetimes, and therefore the line widths and shapes, is swamped. (Adsorbates on surfaces have intrinsic vibra-

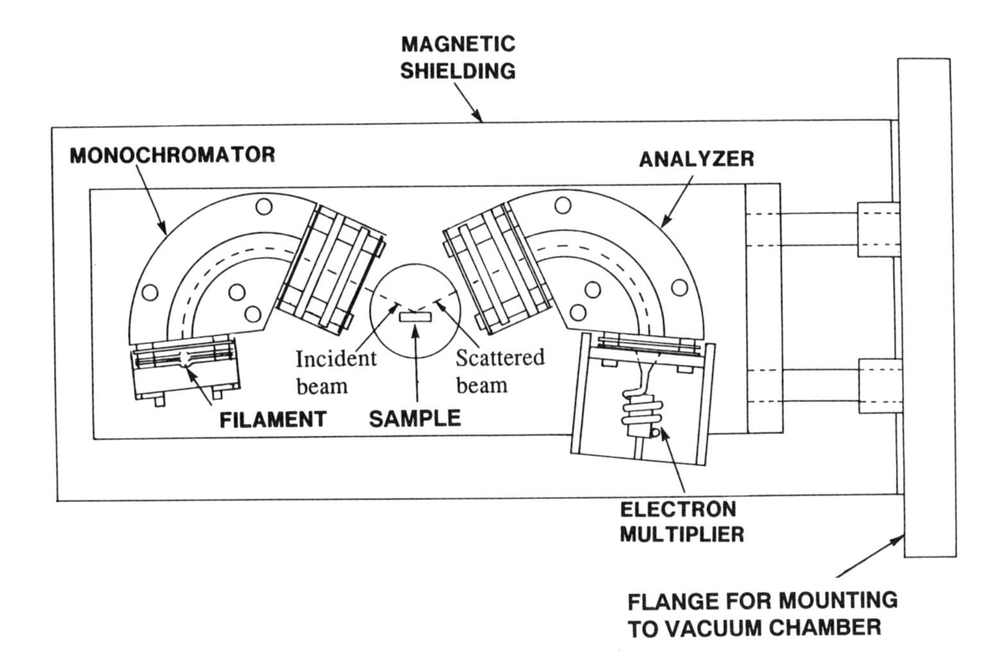

Figure 2 **Schematic of a 127° high-resolution electron energy-loss spectrometer mounted on an 8-in flange for studies of vibrations at surfaces.**

tional line widths of typically 5 cm^{-1} (0.6 meV) and Lorentzian line shapes.) A practical matter is that this poor resolution is insufficient to resolve closely overlapping vibrational frequencies.

Instrumentation

In HREELS, a monoenergetic beam of low energy electrons is focused onto the sample surface and the scattered electrons are analyzed with high resolution of the scattering energy (< 10 meV or 80 cm^{-1}) and angle ($\Delta\theta$ = 2–5°). This is achieved by using electrostatic spectrometers, typically with 127° cylindrical dispersive elements. Hemispherical and cylindrical mirror analyzers have been used also. Some typical analyzer parameters are 25-mm mean radius, 0.1-mm slit width, and 0.5-eV pass energies. Refinements also include the addition of tandem cylindrical sectors to the monochromator and analyzer. A number of commercial versions of spectrometers are capable of routine and dependable operation.

A simple spectrometer that we have used successfully is shown in Figure 2. Electrons from an electron microscope hairpin tungsten filament are focused with an Einzel lens onto the monochromator entrance slit, pass through the monochromator and exit slit, and are focused on the sample's surface by additional electrostatic lenses (in this case a double plate lens system). The incident beam energy is usually below 10 eV, where dipole scattering cross sections are strong, and the beam current to the sample is typically 0.1–1 nA. Electrons that are reflected from the sample's surface are focused on the analyzer entrance slit and energy analyzed to produce an electron energy loss (vibrational) spectrum. An electron multiplier and pulse counting electronics are used due to the small signals. Count rates in HREELS are typically 10^4–10^6 counts/sec for elastically scattered electrons and 10–10^3 counts/sec for inelastically scattered electrons. Scan times typically range from about five minutes to several hours.

The exact incident and scattering angles (~ 60° from the surface normal) are not critical, but a specular scattering geometry must be attainable. It is also very useful to be able to observe a nonspecular scattering angle either by rotating the crystal about an axis perpendicular to the scattering plane or by rotating one of the analyzers in the scattering plane. Magnetic shielding must surround the spectrometer because the magnetic field of the earth and any nearby ion pumps will distort the trajectories of the electrons within the spectrometer, because of their small kinetic energies. The spectrometer is mounted in a ultrahigh-vacuum chamber and analysis must be carried out at pressures below 10^{-4} torr. This requirement exists because of the sensitivity of the electron filament and slits to reactions with background gases. Higher pressure gas phase molecules also will cause inelastic scattering that obscures the surface spectra. The applicability of the HREELS technique can be greatly extended by combining it with a high-pressure reaction chamber and sample transfer mechanism. We have previously used this type of system to study hydrogen transfer in adsorbed hydrocarbon monolayers at atmospheric pressure.

Interpretation of Vibrational Spectra

In the following discussion, heavy emphasis is made of examples from studies of adsorbed layers on metal single-crystal samples. These illustrate the power of the HREELS technique and represent the main use of HREELS historically. Certainly HREELS has been used outside of the single-crystal world, and mention is made concerning its use on "practical" materials. This latter use of HREELS represents a true frontier.

Identification of Adsorbed Species

Determination of surface functional groups, e.g., –OH, –C ≡ C–, and >C = O, and identification of adsorbed molecules comes principally from comparison with vibrational spectra (infrared and Raman) of known molecules and compounds. Quick qualitative analysis is possible, e.g., stretching modes involving H appear for ν(C–H) at 3000 cm^{-1} and for ν(O–H) at 3400 cm^{-1}. In addition, the vibrational energy indicates the chemical state of the atoms involved, e.g., ν(C=C) ~ 1500 cm^{-1} and ν(C=O) ~ 1800 cm^{-1}. Further details concerning the structure of adsorbates

Mode assignment	$CH_3CCo_3(CO)_9$ [6]	$CH_3C–Rh$ (111) [5]
$\nu_{as}(CH_3)/\nu_{as}(CD_3)$	2930 (m)/2192 (w) e	2920 (vw)/2178 (vw) e
$\nu_s(CH_3)/\nu_s(CD_3)$	2888 (m)/— a_1	2880 (w)/2065 (vw) a_1
$\delta_{as}(CH_3)/\delta_{as}(CD_3)$	1420 (m)/1031 (w) e	1420 (vw)/— e
$\delta_s(CH_3)/\delta_s(CD_3)$	1356 (m)/1002 (vw) a_1	1337 (s)/988 (w) a_1
$\nu(CC)$	1163 (m)/1882 (ms) a_1	1121 (m)/1145 (m) a_1
$\rho(CH_3)/\rho(CD_3)$	1004 (s)/828 (s) e	972 (vw)/769 (vw) e
$\nu_s(M–C)$	401 (m)/393 (m) a_1	435 (w)/419 (w) a_1

* Intensities of the spectral bands are given in parentheses following the band frequencies using the following abbreviations: vs = very strong, s = strong, ms = medium strong, m = medium, w = weak, and vw = very weak. Symmetry assignments for each of the vibrational modes are also indicated after the band frequencies.

Table 1 **Comparison of the vibrational frequencies (cm^{-1}) of the ethylidyne surface species formed on Rh (111) with those of the ethylidyne cluster compound.***

comes from comparison to vibrational spectra of ligands in metal cluster compounds whose X-ray crystal structure is known. Isotopic substitution is extremely important in confirming vibrational assignments. Only H/D substitution can be carried out, due to the low resolution, but this is useful for an enormous range of adsorbed molecules, including hydrocarbons. Substituting D for H causes an isotopic shift of $\sqrt{2}$ for those modes with large-amplitude H motion.

As an example, Figure 3a shows the HREELS spectra after the adsorption of ethylene ($H_2C = CH_2$) on Rh(111) at 310 K.[5] Comparison with gas phase ethylene infrared spectra shows that large changes occurred during adsorption, e.g., ν(C–H) = 2880 cm^{-1}, indicative of aliphatic C–C–H bonds, rather than the expected ν(C-H) ~ 3000 cm^{-1} for olefinic C=C–H bonds. The complete agreement with frequencies, intensities, and H/D shifts observed in IR spectra (and normal mode analysis) of an organometallic complex, $CH_3CCo(CO)_9$,[6] allowed for the detailed assignment of the loss peaks to vibrational modes of a surface ethylidyne (CCH_3) species, as shown in Figure 3b and Table 1.

The lack of a well-defined specular direction for polycrystalline metal samples decreases the signal levels by 10^2 –10^3, and restricts the symmetry information on adsorbates, but many studies using these substrates have proven useful for identifying adsorbates. Charging, beam broadening, and the high probability for excitation of phonon modes of the substrate relative to modes of the adsorbate make it more difficult to carry out adsorption studies on nonmetallic materials. But, this has been done previously for a number of metal oxides and compounds, and also semicon-

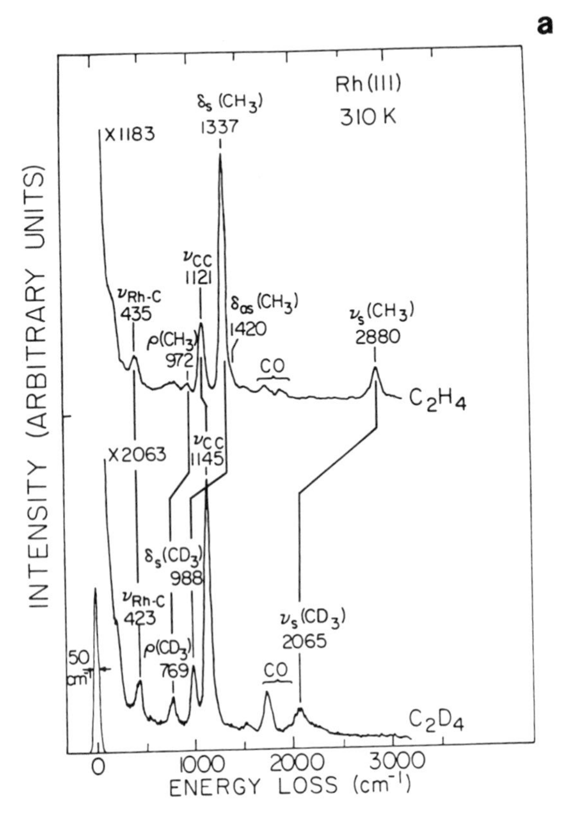

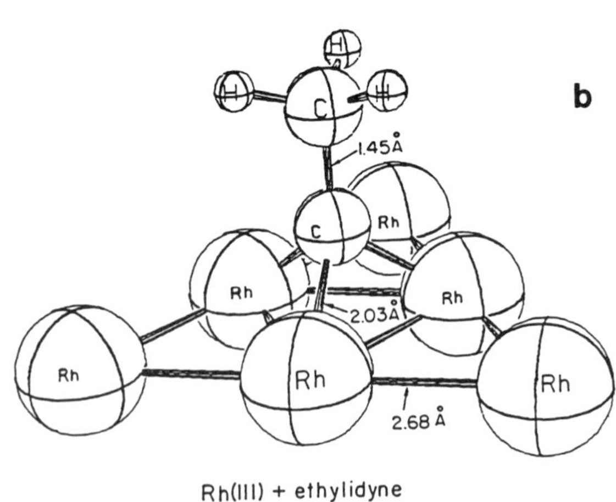

Figure 3 (a) Specular spectra in HREELS obtained following exposure of ethylene (C_2H_4 or C_2D_4) on Rh(111) at 310 K to form the ethylidyne (CCH_3) surface species.[2] (b) The atomic structure (bond distances and angles) of ethylidyne as determined by LEED crystallography.

ductors like Si, InP, and diamond. Dubois, Hansma, and Somorjai fabricated model supported metal catalysts by evaporating rhodium onto an oxidized aluminum substrate and studied CO adsorption by using HREELS. We also have used HREELS to characterize lubricating carbon films on small samples of actual magnetic recording disk heads.

Determination of Adsorption Geometry

The symmetry of an adsorbed molecule and its orientation relative to the surface plane can be established using group theory and the dipole selection rule for specular scattering. The angular variations of loss intensities determines the number and frequencies of the dipole active modes. Only those modes that belong to the totally symmetric representations of the point group which describes the symmetry of the adsorbed complex will be observed as fundamentals on-specular. The symmetry of the adsorbate-surface complex is then determined by comparing the intensity, number, and frequency of dipole-active modes with the correlation table of the point group of the gas phase molecule. In the previous example of adsorbed ethylidyne, observation of an intense symmetric C–H bending ($\delta_s CH_3$) loss peak and weak antisymmetric C–H bending ($\delta_{as} CH_3$) loss peak establishes C_{3v} symmetry for the surface complex.

The adsorption of nitrogen dioxide (NO_2) on metal surfaces is a beautiful example of linkage isomerism, as illustrated in Figure 4, that was discovered by using HREELS.[7] Gas phase NO_2 has C_{2v} symmetry which is retained in the top two binding geometries (Figures 4b and c) since the asymmetric ONO stretching (ν_{as}) is not observed on-specular. The symmetry is reduced to C_s when NO_2 is bonded as the bridging isomer (Figure 4a) and ν_{as} is dipole-active. Confirmation of these bonding geometries (and the correct assignment of the two C_{2v} isomers) comes from comparison with transition metal complexes containing the nitrite (NO_2^-) ligand.

Determination of Adsorption Site

While there is a general pattern of decreased metal-atom stretching frequency with increasing coordination of the adsorption site for the same metal–adatom combination, no site assignments can be made simply by observing the vibrational frequency. Surface chemical bonds clearly control the site dependent vibrational shifts of adsorbed species. Detailed studies that often involve impact scattering can determine the adatom adsorption site in some cases. For polyatomic molecules, bonding to one or more metal atoms at the surface can not be distinguished in general. Good correlation does exist between the C–O stretching frequency (ν_{CO}) for adsorbed CO and the adsorption site: $\nu_{CO} > 2000\ cm^{-1}$ indicates an *atop* site (bonding to a single metal atom); $1850\ cm^{-1} > \nu_{CO} > 2000\ cm^{-1}$ indicates a *bridge* site (bonding to two metal atoms); and $\nu_{CO} < 1850\ cm^{-1}$ indicates a threefold or fourfold bridge site (bonding in a region between three or four metal atoms). This correlation has

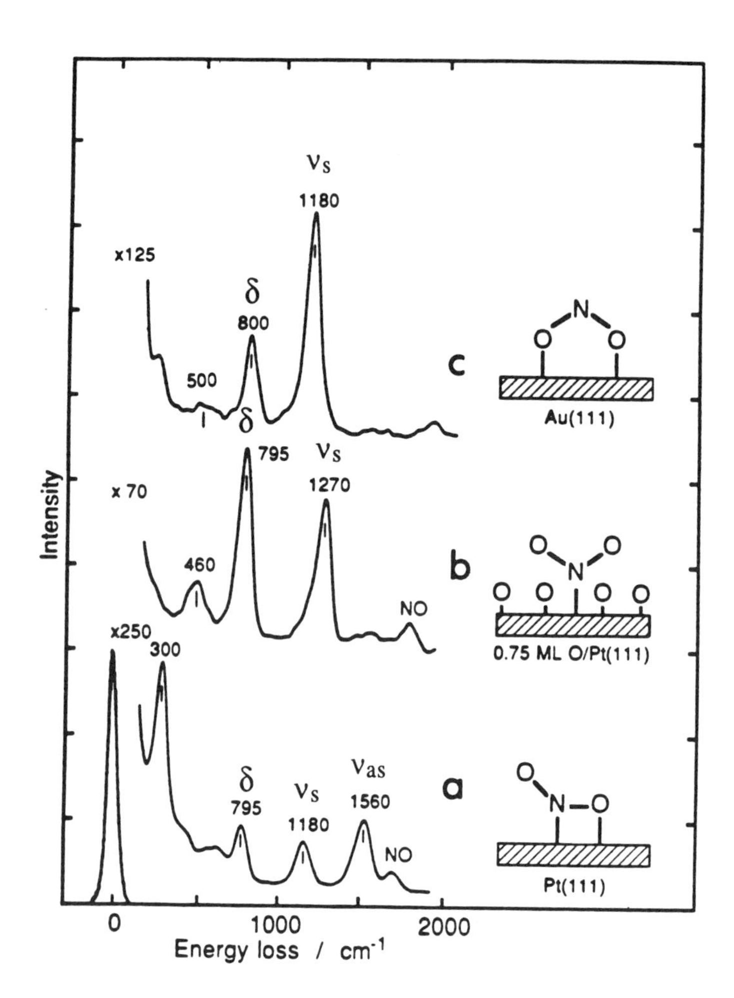

Figure 4 Specular spectra in HREELS of NO_2 adsorbed in three different bonding geometries.[7]

been established by measuring the vibrational spectrum in conjunction with determining the adsorption site for CO by LEED crystallography calculations, and also by examining the correlations for CO bonded in organometallic clusters. Similar correlations are likely to exist for other diatomic molecules bonded to surfaces, e.g., NO, based on correlations observed in organometallic clusters but this has not been investigated sufficiently.

Figure 5 shows the utility of HREELS in establishing the presence of both bridge-bonded and atop CO chemisorbed on Pt(111) and two SnPt alloy surfaces, and also serves to emphasize that HREELS is very useful in studies of metal alloys.[8] The ν_{CO} peaks for CO bonded in bridge sites appear at 1865, 1790, and 1845 cm^{-1} on the Pt(111), (2×2) and $\sqrt{3}$ surfaces, respectively. The ν_{CO} peaks for CO

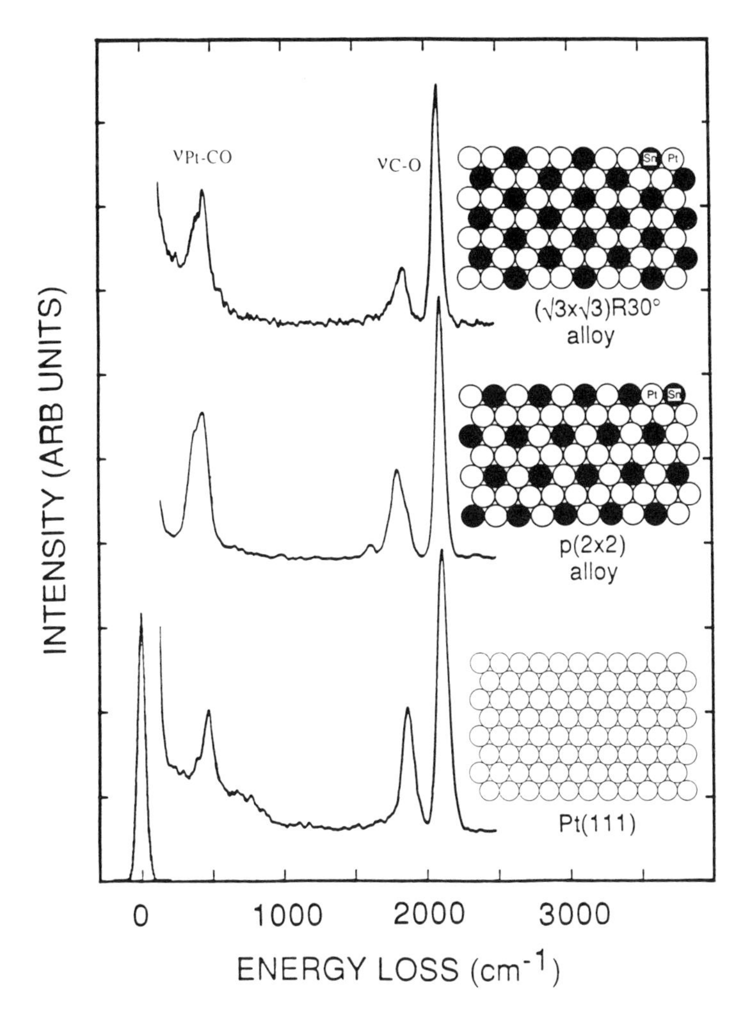

Figure 5 **HREELS of the saturation coverage of CO on Pt(111) and the (2 × 2) and ($\sqrt{3} \times \sqrt{3}$) R30° Sn/Pt surface alloys.**[8]

bonded in atop sites appear at 2105, 2090, and 2085 cm^{-1} on the Pt(111), (2 × 2) and $\sqrt{3}$ surfaces, respectively. Also, lower frequency $\nu_{Pt\text{-}CO}$ peaks accompany each of the ν_{CO} peaks. As discussed previously, the peak intensities are not necessarily proportional to the concentration of each type of CO species and the exact ν_{CO} frequency is determined by many factors.

Other Applications

Many other surfaces can be investigated by HREELS. As larger molecule and non-single-crystal examples, we briefly describe the use of HREELS in studies of polymer surfaces. The usefulness of HREELS specifically in polymer surface science

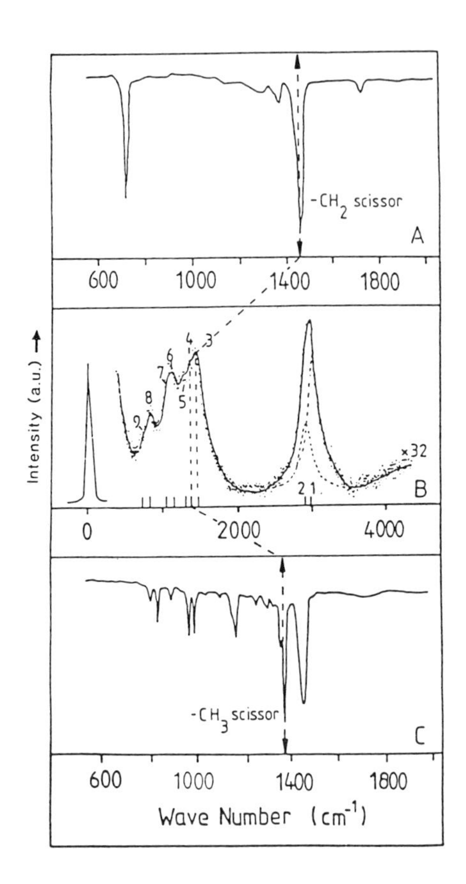

Figure 6 **Vibrational spectra of polymers. (a) Transmission infrared spectrum of polyethylene; (b) electron-induced loss spectrum of polyethylene; (c) transmission infrared spectrum of polypropylene.[10]**

applications has recently been reviewed by Gardella and Pireaux.[9] HREELS is absolutely nondestructive and can be used to obtain information on the chemical composition, morphology, structure, and phonon modes of the solid surface.

Many polymer surfaces have been studied, including simple materials like polyethylene, model compounds like Langmuir-Blodgett layers, and more complex systems like polymer physical mixtures. Figure 6 shows an HREELS spectrum from polyethylene [CH_3–$(CH_2)_n$–CH_3]. Assignment of the energy loss peaks to vibrational modes is done exactly as described for adsorbates in the preceding section. One observes a peak in the C–H stretching region near 2950 cm^{-1}, along with peaks due to C–C stretching and bending and C–H bending modes in the "fingerprint" region between 700–1500 cm^{-1} from both the –CH_3 (which terminate the chains) and –CH_2- groups in the polymer. Since the CH_3/CH_2 ratio is vanishingly small in the bulk of the polymer, the high intensity of the –CH_3 modes indicate

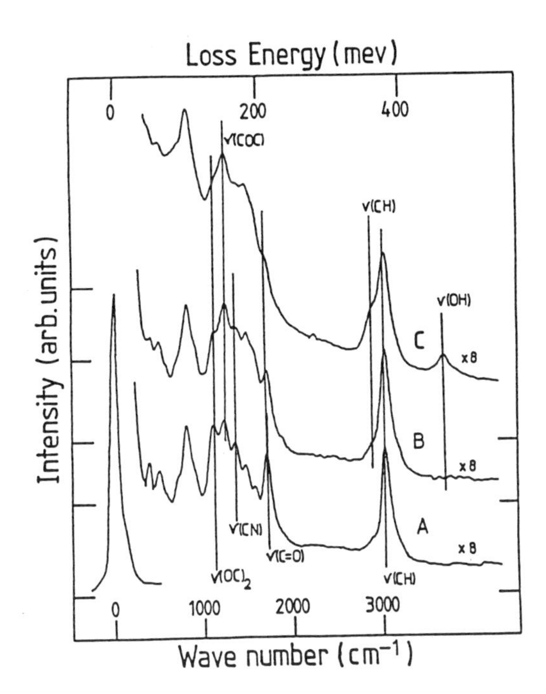

Figure 7 **HREELS vibrational spectra of the interface formation between a polyimide film and evaporated aluminum: (a) clean polyimide surface; (b) with 1/10 layer of Al; (c) with1/2 layer of Al.[11]**

that they are located preferentially in the extreme outer layers of the polymer surface.[10]

HREELS is useful in many interfacial problems requiring monolayer sensitivity. The incipient formation of the interface between a clean cured polyimide film and deposited aluminum has been studied using HREELS,[11] as shown in Figure 7. The film was PMDA-ODA [poly-N,N'–bis(phenoxyphenyl)pyromellitimide], shown schematically in Figure 8. At low Al coverage, the ν(C=O) peak at 1720 cm^{-1} is affected strongly, which indicates that Al reacts close to the carbonyl site. At higher Al coverage, new peaks at ~ 2950 and 3730 cm^{-1} appear which are due to aliphatic $–CH_x$ and –OH groups on the surface. This is evidence for bond scissions in the polymer skeleton.

In general, the main problems with the analysis of bulk polymers has been charging and rough surfaces. The latter characteristic makes the specular direction poorly defined, which causes diffuse and weak electron scattering. Preparation of the polymer as a thin film on a conducting substrate can overcome the charging problem. Even thick samples of insulating polymers can now be studied using a "flood gun" technique. Thiry and his coworkers[12] have shown that charging effects can be over-

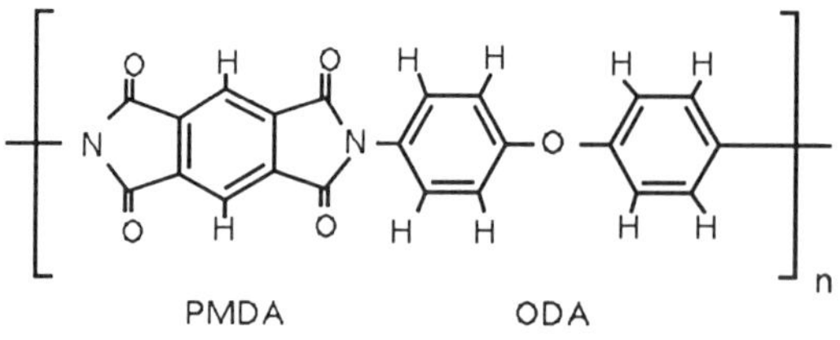

PMDA ODA

Figure 8 Structure of PMDA–ODA.

come by using an auxiliary defocused beam of high-energy electrons to give neutralization of even wide-gap insulators, including Al_2O_3, MgO, SiO_2, LiF, and NaCl.

Comparison to Other Techniques

Information on vibrations at surfaces is complementary to that provided on the compositional analysis by AES and SIMS, geometrical structure by LEED, and electronic structure by XPS and UPS. Vibrational spectroscopy is the most powerful method for the identification of molecular groups at surfaces, giving information directly about which atoms are chemically bonded together. These spectra are more directly interpreted to give chemical bonding information and are more sensitive to the chemical state of surface atoms than those in UPS or XPS. For example, the C(1s) binding energy shift in XPS between C=O and C–O species is 1.5 eV and that between C=C and C–C species is 0.7 eV, with an instrumental resolution of typically 1 eV. In contrast, the vibrational energy difference between C=O and C–O species is 1000 cm^{-1} and that between C=C and C–C species is 500 cm^{-1}, with an instrumental resolution of typically 60 cm^{-1}. Vibrational spectroscopy can handle the complications introduced by mixtures of many different surface species much better than UPS or XPS.

Many other techniques are capable of obtaining vibrational spectra of adsorbed species: infrared transmission-absorption (IR) and infrared reflection-absorption spectroscopy (IRAS), surface enhanced Raman spectroscopy (SERS), inelastic electron tunneling spectroscopy (IETS), neutron inelastic scattering (NIS), photoacoustic spectroscopy (PAS), and atom inelastic scattering (AIS). The analytical characteristics of these methods have been compared in several reviews previously. The principle reasons for the extensive use of the optical probes, e.g., IR, compared to HREELS in very practical nonsingle-crystal work are the higher resolution (0.2–8 cm^{-1}) and the possibility for use at ambient pressures. HREELS could be effectively used to provide high surface sensitivity and a much smaller sampling depth (< 2 nm) and wider spectral range (50–4000 cm^{-1}) than many of these other methods.

HREELS is used extensively in adsorption studies on metal single crystals, since its high sensitivity to small dynamic dipoles, such as those of C–C and C–H stretching modes, and its wide spectral range enable complete vibrational characterization of submonolayer coverages of adsorbed hydrocarbons.[13] The dipole selection rule constraint in IR, IRAS, and HREELS can be broken in HREELS by performing off-specular scans so that all vibrational modes can be observed. This is important in species identification, and critical in obtaining vibrational frequencies required to generate a molecular force field and in determining adsorption sites.

Conclusions

HREELS is one of the most important techniques for probing physical and chemical properties of surfaces. The future is bright, with new opportunities arising from continued fundamental advances in understanding electron scattering mechanisms and from improved instrumentation, particularly in the more quantitative aspects of the technique.[14] A better understanding of the scattering of electrons from surfaces means better structure determination and better probe of electronic properties. Improvements are coming in calculating HREELS cross sections and surface phonon properties and this means a better understanding of lattice dynamics. Extensions of dielectric theory of HREELS could lead to new applications concerning interface optical phonons and other properties of superlattice interfaces.

Novel applications of the HREELS technique include the use of spin-polarization of the incident or analyzed electrons and time-resolved studies on the ms and sub-ms time scale (sometimes coupled with pulsed molecular beams) of dynamical aspects of chemisorption and reaction. Studies of nontraditional surfaces, such as insulators, alloys, glasses, superconductors, model supported metal catalysts, and "technical" surfaces (samples of actual working devices) are currently being expanded. Many of these new studies are made possible through improved instrumentation. While the resolution seems to be limited practically at 10 cm^{-1}, higher intensity seems achievable. Advances have been made recently in the monochromator, analyzer, lenses, and signal detection (by using multichannel detection). New configurations, such as that utilized in the dispersion compensation approach, have improved signal levels by factors of 10^2–10^3.

Related Articles in the Encyclopedia

EELS, IR, FTIR, and Raman Spectroscopy

References

1 H. Ibach and D. L. Mills. *Electron Energy Loss Spectroscopy and Surface Vibrations.* Academic, New York, 1982. An excellent book covering all aspects of the theory and experiment in HREELS.

2 W. H. Weinberg. In: *Methods of Experimental Physics.* **22,** 23, 1985. Fundamentals of HREELS and comparisons to other vibrational spectroscopies.

3 *Vibrational Spectroscopy of Molecules on Surfaces.* (J. T. Yates, Jr. and T. E. Madey, eds.) Plenum, New York, 1987. Basic concepts and experimental methods used to measure vibrational spectra of surface species. Of particular interest is Chapter 6 by N. Avery on HREELS.

4 *Vibrations at Surfaces.* (R. Caudano, J. M. Gilles, and A. A. Lucas, eds.) Plenum, New York, 1982; *Vibrations at Surfaces.* (C. R. Brundle and H. Morawitz, eds.) Elsevier, Amsterdam, 1983; *Vibrations at Surfaces 1985.* (D. A. King, N. V. Richardson and S. Holloway, eds.) Elsevier, Amsterdam, 1986; and *Vibrations at Surfaces 1987.* (A. M. Bradshaw and H. Conrad, eds.) Elsevier, Amsterdam, 1988. Proceedings of the International Conferences on Vibrations at Surfaces.

5 B. E. Koel, B. E. Bent, and G. A. Somorjai. *Surface Sci.* **146,** 211, 1984. Hydrogenation and H, D exchange studies of $CCH_{3(a)}$ on Rh (111) at 1-atm pressure using HREELS in a high-pressure/low pressure system.

6 P. Skinner, M. W. Howard, I. A. Oxton, S. F. A. Kettle, D. B. Powell, and N. Sheppard. *J. Chem. Soc., Faraday Trans.* **2,** 1203, 1981. Vibrational spectroscopy (infrared) studies of an organometallic compound containing the ethylidyne ligand.

7 M. E. Bartram and B. E. Koel. *J. Vac. Sci. Technol.* **A 6,** 782, 1988. HREELS studies of nitrogen dioxide adsorbed on metal surfaces.

8 M. T. Paffett, S. C. Gebhard, R. G. Windham, and B. E. Koel. *J. Phys. Chem.* **94,** 6831, 1990. Chemisorption studies on well-characterized SnPt alloys.

9 J. A. Gardella, Jr, and J. J. Pireaux. *Anal. Chem.* **62,** 645, 1990. Analysis of polymer surfaces using HREELS.

10 J. J. Pireaux, C. Grégoire, M. Vermeersch, P. A. Thiry, and R. Caudano. *Surface Sci.* **189/190,** 903, 1987. Surface vibrational and structural properties of polymers by HREELS.

11 J. J. Pireaux, M. Vermeersch, N. Degosserie, C. Grégoire, Y. Novis, M. Chtaïb, and R. Caudano. In: *Adhesion and Friction.* (M. Grunze and H. J. Kreuzer, eds.) Springer-Verlag, Berlin, 1989, p. 53. Metallization of polymers as probed by HREELS.

12 P. A. Thiry, M. Liehr, J. J. Pireaux, and R. Caudano. *J. Electron Spectrosc. Relat. Phenom.* **39,** 69, 1986. HREELS of insulators.

13 B. E. Koel. *Scanning Electron Microscopy* **1985/IV**, 1421, 1985. The use of HREELS to determine molecular structure in adsorbed hydrocarbon monolayers.

14 J. L. Erskine. *CRC Crit. Rev. Solid State Mater. Sci.* **13,** 311, 1987. Recent review of scattering mechanisms, surface phonon properties, and improved instrumentation.

8.4 NMR

Solid State Nuclear Magnetic Resonance

HELLMUT ECKERT

Contents

Introduction

Solid state NMR is a relatively recent spectroscopic technique that can be used to uniquely identify and quantitate crystalline phases in bulk materials and at surfaces and interfaces. While NMR resembles X-ray diffraction in this capacity, it has the additional advantage of being element-selective and inherently quantitative. Since the signal observed is a direct reflection of the local environment of the element under study, NMR can also provide structural insights on a *molecular* level. Thus, information about coordination numbers, local symmetry, and internuclear bond distances is readily available. This feature is particularly useful in the structural analysis of highly disordered, amorphous, and compositionally complex systems, where diffraction techniques and other spectroscopies (IR, Raman, EXAFS) often fail.

Due to these virtues, solid state NMR is finding increasing use in the structural analysis of polymers, ceramics and glasses, composites, catalysts, and surfaces.

Examples of the unique insights obtained by solid state NMR applications to materials science include: the Si/Al distribution in zeolites,[1] the hydrogen microstructure in amorphous films of hydrogenated silicon,[2] and the mechanism for the zeolite-catalyzed oligomerization of olefins.[3]

Basic Principles

Nuclear Magnetism and Magnetic Resonance

NMR spectroscopy exploits the magnetism of certain nuclear isotopes.[4–6] Nuclei with odd mass, odd atomic number, or both possess a permanent magnetic moment, which can be detected by applying an external magnetic field (typical strength in NMR applications: 1–14 Tesla). Quantum mechanics states that the magnetic moments adopt only certain discrete orientations relative to the field's direction. The number of such discrete orientations is $2I+1$, where I, the nuclear spin quantum number, is a half-integral or integral constant. For the common case $I = ½$, two distinct orientations (states) result, with quantized components of the nuclear spin parallel and antiparallel to the field direction. Since the parallel orientations are energetically more favorable than the antiparallel ones, the populations of both states are unequal. As a consequence, a sample placed in a magnetic field develops a macroscopic magnetization M_0. This magnetization forms the source of the spectroscopic signal measured.

In NMR spectroscopy the precise energy differences between such nuclear magnetic states are of interest. To measure these differences, electromagnetic waves in the radiofrequency region (1–600 MHz) are applied, and the frequency at which transitions occur between the states, is measured. At resonance the condition

$$\omega = \gamma B_{loc} = \gamma(B_0 + B_{int}) \tag{1}$$

holds, where ω is the frequency of the electromagnetic radiation at which absorption occurs. The strength of the magnetic field present at the nuclei B_{loc} is generally very close to the strength of the externally applied magnetic field B_0 but differs slightly from it due to internal fields B_{int} arising from surrounding nuclear magnetic moments and electronic environments. The factor γ, the gyromagnetic ratio, is a characteristic constant for the nuclear isotope studied and ranges from 10^6 to 10^8 rad/Tesla-s. Thus, NMR experiments are always element-selective, since at a given field strength each nuclear isotope possesses a unique range of resonance frequencies.

Measurement and Observables

Figure 1 shows the detailed steps of the measurement, from the perspective of a coordinate system rotating with the applied radiofrequency $\omega_0 = \gamma B_0$. The sample is in the magnetic field, and is placed inside an inductor of a radiofrequency circuit

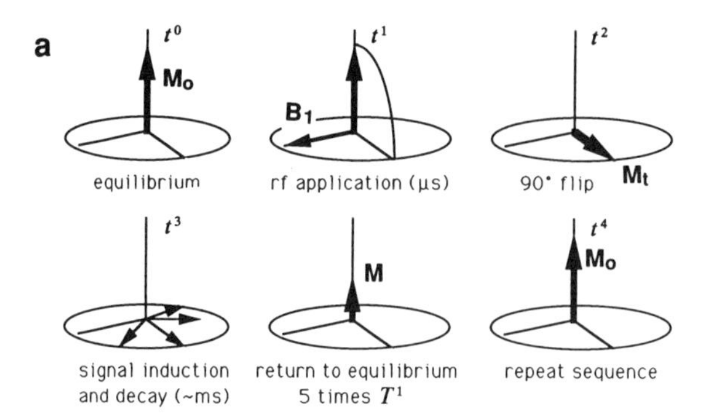

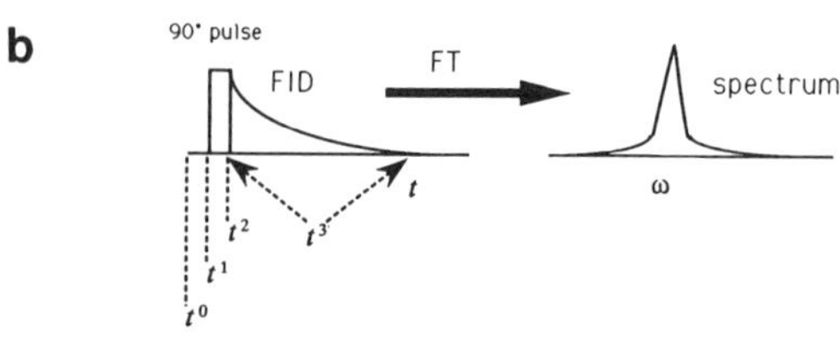

Figure 1 **Detection of NMR signals (a), shown in the rotating coordinate system associated with the oscillating magnetic field component B_1 at the applied radiofrequency ω_0 at various stages (t_0–t_4) of the experiment: t_0, spin system with magnetization (fat arrow) at equilibrium; t_1, irradiation of the B_1 field orthogonal to the magnetization direction tips the magnetization; t_2, the system after a 90° pulse resulting in transverse magnetization M_t; t_3, off-resonance precession and free induction decay in the signal acquisition period following the pulse; and t_4, return to spin equilibrium after spin–lattice relaxation; timing diagram of the experiment (b), followed by Fourier transformation.**

tuned to the resonance frequency of the nucleus under observation. The magnetization present at time t_0 is then detected by applying a short, intense (100–1000 W) radiofrequency pulse (typically 1–10 μs) in a direction perpendicular to B_0 (t_1). The oscillating magnetic component of the radiofrequency pulse stimulates transitions between the magnetic states and tips M_0 into the plane perpendicular to the direction of the magnetic field (90° pulse, t_2). Following this pulse, the magnetization oscillates in this plane at the transition frequency ω and also decays in time due to the various internal interactions present (t_3). It thereby induces an ac voltage signal in a coil, which is amplified, digitized, and acquired over a typical period of several ms (t_3). Fourier transformation of this free induction decay (FID) signal then results in the *NMR spectrum,* a plot of absorption intensity versus frequency. The position, width, and shape of the spectral peaks reflect the local fields present at the nuclei due to internal interactions and allow various chemical conclusions. The area under a spectral peak is directly proportional to the number of nuclei contributing to the resonance, and can be used for quantification purposes.

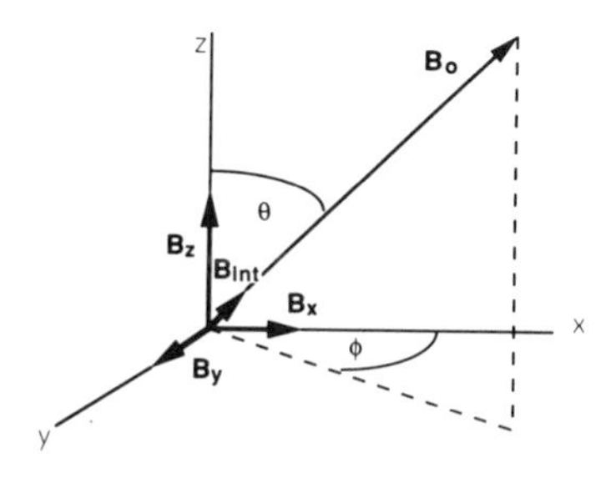

Figure 2 **Schematic illustration of the influence of chemical shift upon NMR spectra. See text for further explanation.**

Since typical NMR signals are quite weak, extensive signal averaging by repetitive scanning is generally necessary. The pulsing rate at which this can occur depends on the time it takes for the spin system to return into its initial state after the 90° pulse, with M_0 along the magnetic field direction (t_4). This process can generally be described by first-order kinetics. The associated time constant T_1, the spin-lattice relaxation time, can vary from a few ms to several hours in solids.

Structural and Chemical Information from Solid State NMR Line shapes

Internal Interactions

What makes NMR so useful for addressing structural questions in solids is the fact that B_{loc}, and hence the resonance frequency ω, are influenced by various types of internal interactions. These are a direct reflection of the local structural and chemical bonding environments of the nuclei studied, and hence are of central chemical interest. Generally, the observed nuclei experience three types of interactions:[6] *magnetic dipole–dipole interactions* with the magnetic moments from other, nearby nuclei; *chemical shift interactions* with the magnetic fields from the electron clouds that surround the nuclei; and (for nuclei with spin > ½) *electric quadrupole interactions* with electrostatic field gradients generated by the chemical bonding environment. Each of these interactions is characterized by a few spectroscopic parameters, which are listed in Table 1. Typically, these parameters are extracted from experimental spectra by computer-fitting methods or are measured by *selective averaging* techniques.

Due to the simultaneous presence of all three interactions, the resulting solid state NMR spectra can be quite complex. Fortunately, however, in many cases one interaction mechanism is dominant, resulting in spectra that yield highly specific information about local symmetry and bonding. In the following, we will discuss an application of the chemical shift anisotropy. Figure 2 illustrates that the anisotropic interaction between the molecule and the externally applied magnetic field

Interaction	Parameters	NMR measurement	Structural significance
Chemical shift (isotropic component)	δ_{iso}	Magic-angle spinning	Chemical bonding coordination number
Chemical shift anisotropy	$\delta_{xx}, \delta_{yy}, \delta_{zz}$	Line-shape analysis MAS-sidebands	Coordination symmetry
Dipole–dipole (homonuclear)	M_2(homo) (mean-squared local field)	Spin-echo NMR	Internuclear distances, number of surrounding nuclei
Dipole–dipole (heteronuclear)	M_2(hetero)	Spin-echo double resonance (SEDOR)	
Nuclear electric quadrupole	*QCC* (quadrupole coupling constant), (asymmetry parameter)	Line-shape analysis, nutation NMR	Coordination symmetry

Table 1 **Interactions in solid state NMR, parameters, their selective measurement, and their structural significance.**

induces local magnetic field components B_x, B_y, and B_z along the *x*-, *y*-, and *z*-directions of a molecular axis system. Quite generally, $B_x \neq B_y \neq B_z$. The vector sum of these components produces a resultant B_{int} along the direction of B_0, the axis of quantization, and hence affects the resonance condition. As seen in Figure 2, the magnitude of B_{int} (and hence the resonance frequency) will depend crucially on the orientation (θ, ϕ) of this molecular axis system relative to the magnetic field direction.

In a polycrystalline or amorphous material, the orientational statistics lead to a distribution of resonance conditions. Generally, we can distinguish three situations, illustrated in Figure 3a–c: The spectrum in Figure 3c is observed for compounds with asymmetric chemical environments. It shows three distinct features, which can be identified with the different Cartesian chemical shift components δ_{xx}, δ_{yy}, and δ_{zz} in the molecular axis system. Figure 3b corresponds to the case of cylindrical symmetry, where $\delta_{xx} = \delta_{yy} \neq \delta_{zz}$, and hence only two distinct line shape components appear. Finally, for chemical environments with spherical symmetry the chemical shift is the same in all three directions. Accordingly, the solid state NMR spectrum consists of only a single peak (see Figure 3a). The values of δ_{ii} extracted

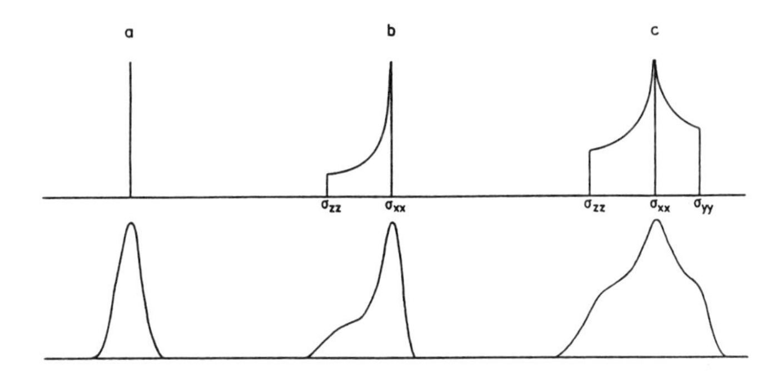

Figure 3 **Characteristic solid state NMR line shapes, dominated by the chemical shift anisotropy. The spatial distribution of the chemical shift is assumed to be spherically symmetric (a), axially symmetric (b), and completely asymmetric (c). The top trace shows theoretical line shapes, while the bottom trace shows "real" spectra influenced by broadening effects due to dipole–dipole couplings.**

from the spectra usually are reported in ppm relative to a standard reference compound. By definition,

$$\delta_{ii}[\mathrm{ppm}] = 10^6 \cdot \frac{\omega_{ii} - \omega_{ref}}{\omega_0} \tag{2}$$

An Example: Chemical Shift Anisotropy in Solid Vanadium Compounds

Figure 4 shows representative solid state ^{51}V NMR spectra of crystalline vanadates. Each model compound typifies a certain local vanadium environment with well-defined symmetry as shown. One can see from these representative data that the solid state ^{51}V chemical shift anisotropies are uniquely well suited for differentiating between the various site symmetries. VO_4^{-3} groups with approximate spherical symmetry yield single-peak spectra, dimeric $V_2O_7^{-4}$ groups (which possess a threefold axis and hence cylindrical symmetry) yield spectra resembling Figure 3b, while the spectra of the completely asymmetric $VO_{2/2}O_2^-$ groups are of the kind shown in Figure 3c. Highly diagnostic line shapes are also observed for vanadium in distorted octahedral environments (ZnV_2O_6) and in square-pyramidal environments (V_2O_5).

An Application: ^{51}V NMR of V oxide films on metal oxide supports

Investigations carried out within the past few years have revealed that multicomponent metal oxide systems may interact at interfaces by having one component form a two-dimensional metal oxide overlayer on the second metal oxide component. For example, vanadium oxide can be dispersed on TiO_2, ZrO_2, SiO_2, Al_2O_3, and

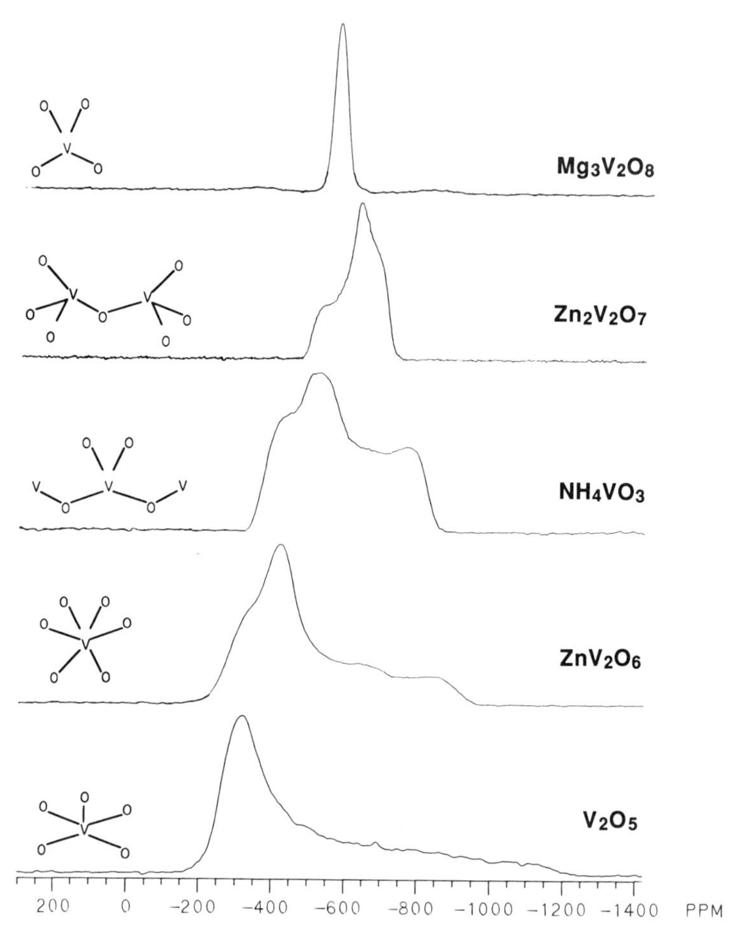

Figure 4 **Local microstructures and experimental solid state ^{51}V NMR spectra in crystalline vanadium oxide compounds.**

other oxide supports by impregnating the latter with a liquid molecular precursor and following with calcination. Many of these systems are potent oxidation catalysts, with significant inherent advantages to bulk V_2O_5. To explore a relationship between the catalytic activity and structural properties, extensive solid state ^{51}V NMR studies have been carried out on these phases.[8] These studies have benefited greatly from the chemical shift systematics discussed above. Figure 5 shows experimental spectra of V surface oxide on γ-Al_2O_3 support. In conjunction with the model compound work one can conclude that two distinctly different vanadia species are present at the surface: At low vanadia contents, a four-coordinated chain-type species dominates, whereas with increasing surface coverage a new site emerges whose spectroscopic parameters reveal the presence of a distorted octahedral vanadium environment. Similar trends have been seen with other metal oxide supports,

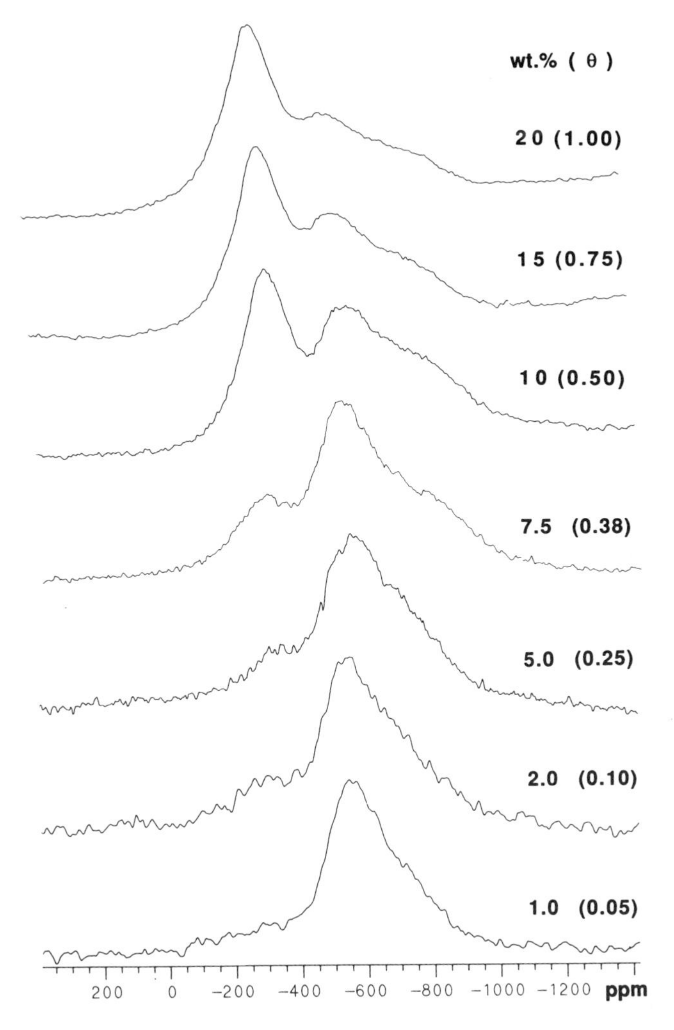

Figure 5 **Solid state ^{51}V NMR spectra of Vanadium oxide on γ-alumina as a function of vanadium loading (wt.%) and surface coverage θ. Note the gradual emergence of the six-coordinated vanadium site with increased loading.**

although the type of vanadium environment in the overlayer also depends strongly on the acidity of the surface.

Selective Averaging Techniques

In general, the specific information that can be obtained from a simple solid state NMR experiment depends on the "personality" of the nuclear isotope under study. In many cases, solid state NMR spectra are not as straightforwardly interpretable as in the preceding example. Furthermore, disordered materials, such as thin films,

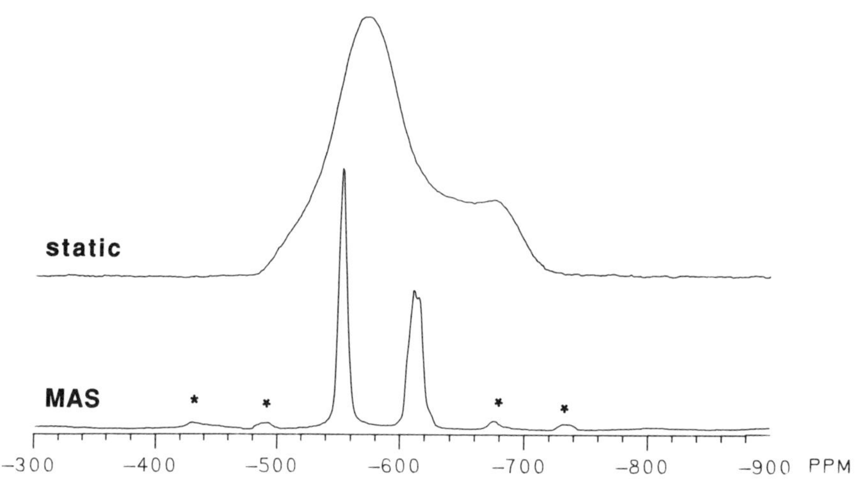

Figure 6 **Solid state ^{51}V static and magic-angle spinning NMR spectra of α-$Mg_2V_2O_7$. This compound has two crystallographically distinct vanadium sites. While the static spectrum is a superposition of two powder patterns of the kind shown in Figure 3, MAS leads to well-resolved sharp resonances. Weak peaks denoted by asterisks are spinning sidebands due to the quadrupolar interaction.**

glasses, and composites, often show only broad and unresolved spectra, because in such samples the spectroscopic parameters are subject to distribution effects. Here, the diagnostic character of solid state NMR can be enhanced dramatically by selective averaging techniques. The idea is to simplify the spectra by suppressing certain interactions while preserving others for analysis. The most popular and most widely applied experiment is to acquire the NMR spectrum while rotating the sample rapidly about an axis inclined by 54.7° (the "magic" angle) relative to the magnetic field direction. This technique, called Magic-Angle Spinning (MAS), results in an average molecular orientation of θ = 54.7° relative to the magnetic field over the rotation period, regardless of the initial molecular orientation. Theory predicts that at this specific angle the anisotropy of all internal interactions (which scale with the factor $3\cos^2\theta-1$) vanishes. Consequently, MAS converts broad powder patterns of the kind shown in Figure 3a–c into highly resolved sharp resonances that can be straightforwardly assigned to individual sites. For example, Figure 6 illustrates the superior ability of MAS to resolve the crystallographically distinct vanadium sites in the model compound α-$Mg_2V_2O_7$. The high resolution obtained by MAS and the simplicity of the spectra make solid state NMR a particularly useful technique for identifying crystalline phases in the bulk or at surfaces and interfaces.

A number of other, more sophisticated, selective averaging tools (including spin echo, double resonance and two-dimensional techniques) are available, both for spectral editing purposes and for obtaining quantitative information about interatomic distances.[7] However, among all these techniques, the conceptually simple MAS-NMR experiment has had by far the biggest impact in materials science applications.

Instrumentation

NMR instrumentation consists of three chief components: a magnet, a spectrometer console, and a probe. While in the past much solid state NMR research was conducted on home-built equipment, the current trend is toward the acquisition of commercial systems. The magnets used for solid state NMR applications generally are superconducting solenoids with a cylindrical bore of 89-mm diameter. The most common field strengths available, 4.7, 7.0, 9.4, and 11.7 Tesla, correspond to proton resonance frequencies near 200, 300, 400, and 500 MHz, respectively.

The spectrometer console comprises a radiofrequency part for the generation, amplification, mixing, and detection of radiofrequency and NMR signals, and a digital electronics part, consisting of a pulse programmer, a digitizer, and an on-line computer. Equipment normally used for pulsed liquid state NMR applications often can be modified for solid state experiments by adding high-power amplifiers (up to 1-kW output power) and fast digitizers (2 MHz or faster).

NMR probes are used to transfer the radiofrequency pulse to the sample and to detect the nuclear induction signal after the pulse. They contain radiofrequency circuitry, which is tunable to the nuclear resonance frequency via variable capacitors and which is based usually on a single solenoidal coil (diameter 4–25 mm). MAS-NMR experiments require special probes, enabling fast sample rotation within the magnet. Currently, MAS is done mostly on powdered samples packed within cylindrical containers (rotors) that are machined from single-crystal alumina, zirconia, or silicon nitride to precise dimensions. High-pressured gases (air, N_2, or Ar, at 40–60 lb/in^2) thrusting on turbine-shaped caps are used to accomplish fast rotation. For routine experiments, typical spinning speeds are 5–10 kHz; with suitable equipment up to 20 kHz can be reached.

Practical Aspects and Limitations

Sample preparation requirements in solid state NMR are strikingly simple because the measurement is carried out at ambient temperature and pressure. Wide-line NMR experiments can be carried out on solid samples in any form, as far as the sample dimensions fit those of the coil in the NMR probe. MAS experiments require the material to be uniformly distributed within the rotor.

Compared to other spectroscopic methods, NMR spectroscopy is a very insensitive technique. As a general rule of thumb, the sample studied must contain at least 10^{-5} moles of target nuclei. The required sample size thus depends on the percentage of the element present in the sample, as well as on the natural abundance of the

NMR isotope measured. For example, for the detection of phosphorus by ^{31}P NMR in a sample containing 3 wt.% phosphorus, approximately 10 mg of sample are required. By contrast, the corresponding detection limit for ^{29}Si in a similar situation is 22 times higher, due to the much lower natural abundance (4.7%) of the ^{29}Si isotope.

Naturally, the low sensitivity poses a particular obstacle to NMR studies of thin films and surfaces. Large surface areas are obviously favorable (the samples in Figure 5 have surface areas around 150 m^2/g), but good results can often be obtained on samples with surface areas as small as 10 m^2/g. Experimentally, the detection sensitivity can be increased by increasing the applied field strength; by increasing the sample size (although practical considerations often impose a maximum sample volume of several cm^3); and by using special NMR techniques (cross-polarization[4–6]) for sensitivity enhancement.

Additional limitations arise from the nuclear electric quadrupole interaction for nuclei with $I > ½$ and from the dipolar interaction of nuclei with localized electron spins in paramagnetic samples. Both interactions tend to interfere with the alignment of the nuclear spins in the external magnetic field, and to make the observation of NMR signals difficult. Due to these factors, less than half the elements in the periodic table are conducive to solid state NMR experiments. The following ranking holds with regard to detection sensitivity and general suitability in the solid state—highly favorable elements: H, Li, Be, B, F, Na, Al, P, V, Sn, Xe, Cs, Pt, and Tl; less well-suited elements, where NMR often suffers from sensitivity restrictions: C, N, Si, Se, Y, Rh, Ag, Cd, Te, W, Hg, and Pb; and elements whose suitability is often limited by quadrupolar interactions: N, O, Cl, Mn, Co, Cu, Ga, K, Rb, Nb, Mo, In, and Re. Elements not listed here can be considered generally unsuitable for solid state NMR.

Quantitative Analysis

In contrast to other spectroscopies, such as IR/Raman or VIS/UV, NMR spectroscopy is inherently quantitative. This means that for a given nucleus the proportionality factor relating the area of a signal to the number of nuclei giving rise to the signal is not at all sample-dependent. For this reason, NMR spectroscopy has been used extensively for absolute and relative quantitation experiments, using chemically well-defined model compounds as standards.

It is essential, however, to follow a rigorous experimental protocol for such applications. To maintain the quantitative character of NMR spectroscopy, the repetition rate of signal averaging experiments has to be at least five times the longest spin-lattice relaxation time present in the sample. This waiting period is necessary to ensure that the magnetization is probed in a reproducible state, corresponding to thermodynamic equilibrium.

Conclusions

To date, the simple one-pulse acquisition experiments (with or without MAS) reviewed here have been the mainstay for the majority of NMR applications in materials science. A current trend is the increasing use of NMR for *in situ* studies, using more sophisticated hardware arrangements.[3, 9] For the near future, a rapid diffusion of NMR know-how and methodology into many areas of solid state science can be foreseen, leading to the application of more complicated techniques that possess inherently greater informational content than MAS-NMR. Examples of this kind include multiple pulse techniques, such as one- and two-dimensional versions of spin-echo and double resonance methods, and experiments involving variable rotation angles.[7]

Also, new areas for applications are opening up. A most recent development has been the successful demonstration of three-dimensional imaging of ceramic and polymeric materials by solid state NMR techniques. This area is most likely to expand considerably.

Related Articles in the Encyclopedia

EXAFS, FTIR, XRD

References

1 J. Klinowski. *Prog. NMR Spectrosc.* **16,** 237, 1984. A summary of ^{29}Si MAS-NMR applications to zeolites.

2 J. Baum, K. K. Gleason, A. Pines, A. N. Garroway, and J. A. Reimer. *Phys. Rev. Lett.* **56,** 1377, 1986. Detection of hydrogen clustering in amorphous hydrogenated silicon by a special technique of dipolar spectroscopy, multiple-quantum NMR.

3 J. F. Haw, B. R. Richardson, I.S. Oshiro, N.D. Lazo, and J. A. Speed. *J. Am. Chem. Soc.* **111,** 2052, 1989. *In situ* NMR studies of catalytic properties.

4 T. M. Duncan and C. R. Dybowski. *Surf. Sci. Rep.* **1,** 157, 1981. An excellent review of relevant NMR theory, modern techniques, and applications to surfaces.

5 B. C. Gerstein and C. R. Dybowski. *Transient Techniques in NMR of Solids.* Academic Press, 1985. An in-depth treatment of the theoretical foundations of solid state NMR.

6 M. Mehring. *Principles of High Resolution NMR in Solids.* Springer Verlag, New York, 1983. An in-depth treatment of the theoretical foundations of solid state NMR.

7 H. Eckert. *Ber. Bunsenges. Phys. Chem.* **94,** 1062, 1990. A recent review of modern NMR techniques as applied to various Materials Science problems.

8 H. Eckert and I. E. Wachs. *J. Phys. Chem.* **93,** 6796, 1989. ^{51}V NMR studies of vanadia-based catalysts and model compounds.

9 J. F. Stebbins and I. Farnan. *Science.* **245,** 257, 1989. Highlights *in situ* NMR applications at ultrahigh temperatures.

9

ION SCATTERING TECHNIQUES

9.0 INTRODUCTION

In this chapter three ion-scattering methods for determining composition and geometric structure (for single crystal material) are discussed. They are Rutherford Backscattering Spectrometry, RBS, which typically utilizes high-energy He or H ions (usually 1–3.4 MeV energies), Medium-Energy Ion Scattering, MEIS (ion energies from 50 keV to 400 keV), and low-energy ion scattering (100 eV to 5 keV) which is more commonly known as Ion-Scattering Spectroscopy, ISS. A fourth technique, Elastic Recoil Spectrometry, ERS, is an auxiliary to these methods for the specific detection of hydrogen. All the techniques are performed in vacuum.

For the three ion-scattering techniques there are differences in information content that are a consequence of the different ion energy regimes involved, plus some differences in instrumentation. For RBS, the most widely used method, the high-energy ions penetrate well into the sample (up to 2 μm for He ions; 20 μm for H ions). On its way into the sample an individual ion loses energy in a continuous manner through a series of electronic scattering events. Occasionally an ion undergoes a billard ball-like collision with the nucleus of an atom in the sample material and is back scattered with a discrete, large energy loss, the value of which is characteristic of the atom struck (momentum transfer). Since this major energy loss is atom specific, whereas the small continuum energy loses depend on the depth traveled, the overall energy spectrum of the emerging back scattered ions reveals both the elemental composition and the depth distribution of those elements in a nondestructive manner. Since the scattering physics is quantitatively well understood at these high energies (Rutherford Scattering) a standardless depth profile is obtainable with a few percent accuracy. Other important factors are: the separation in backscattering energy of adjacent elements in the backscattered spectrum decreases with increasing mass such that Ni and Fe are not separable, whereas C and O are easily distinguished; the backscattering cross section is essentially proportional to $Z^2$and therefore heavy elements in light matrices have much better de- tection limits (by about a factor of 100) at 10–100 ppm than vice versa; the depth-resolution depends on ion energy, angle of incidence, and depth below the surface such that a resolution of 20 Å is achievable (low ion energy, grazing angle, analysis done right at the surface), but more typical values are several hundred angstroms.

For single crystal materials, aligning the ion beam with a crystallographic direction suppresses the signal from below the first few layers, since the atoms in these layers shadow bulk atoms below from the incoming ion beam. This technique, known as channeling, is used both to enhance the surface sensitivity and to determine the extent of crystalline defects, since if atoms are displaced from their correct positions the degree of shadowing in the channeling mode will be decreased.

MEIS is a more sophisticated form of RBS that uses lower energy ions (usually 100–400 keV) and a higher resolution ion energy analyzer. The lower energies restrict the probing depth. The better energy resolution improves the depth resolution down to a few angstroms. It also improves the ability to distinguish elements at high mass. When used for single crystal materials in conjunction with channelling of the incoming ions, and blocking of the outgoing backscattered ions, the method provides atomic positions at a surface, or an interface up to 4 or 5 layers below the surface, to an accuracy of a few hundredths of an angstrom. In addition it retains the standardless quantitation of the RBS method with sensitivities to submonolayer amounts. Both RBS and MEIS are extremely expensive, requiring an ion accelerator. The lower energy accelerator of MEIS is cheaper, but this is counteracted by the greater expense of the more sophisticated ion energy analysis. Both techniques typically cost around $1,000,000 and take up large laboratories. Beam diameters are usually millimeters in size, but microbeam systems with spatial resolution down to 1 μm exist. Ion-beam damage can be a problem, particularly for polymers. It can be mitigated by using low ion doses and by rastering the beam.

ISS involves the use of ions (usually He or Ar) in the 100–5000 eV range. At these energies essentially only backscattering from atoms in the outermost atomic layer produces peaks in the ion energy spectrum due to nearly complete neutralization of any ions scattered from below the surface. As with RBS and MEIS the ability to resolve adjacent elements becomes rapidly poorer with increasing Z. This can be mitigated, but not solved entirely, by changing the mass of the ion (eg Ar for He), the ion energy, and the angle of detection. All these variations significantly affect the scattering cross section and background, however, which complicates quantitative use. Quantitation is not standardless at these energies but requires suitable standards to determine relative cross sections for the set of scattering parameters used. Cross sections still depend roughly on Z^2, however, so the technique is much more sensitive to high-Z materials. Owing to its extreme surface sensitivity ISS is usually used in conjunction with sputter profiling over the top 50 Å or so. Spatial resolution down to about 150 µm is routinely obtained. The technique is not widely used owing to the lack of commercial equipment and its poor elemental resolution. Instrumentation is quite cheap, and simple, however, since an ordinary ion gun replaces the ion accelerator used in RBS and MEIS. It can be used as an auxiliary technique on XPS or AES spectrometers by reversing the voltage on the analyzer to pass ions instead of electrons.

In ERS, also known as Forward Recoil Spectrometry, FRS, Hydrogen Recoil Spectrometry, HRS, or Hydrogen Forward Scattering, HFS, hydrogen atoms present in a sample recoil from He ions striking the sample at grazing angle with sufficient forward momentum to be ejected. They are then separated from any He that also emerges by using a thin stopping foil that allows energetic H to pass but not He. In this way the hydrogen content can be quantitatively determined. The technique can be applied in RBS, MEIS, or ISS spectrometors and is used because a target atom that is lighter than the incident ion is only scattered in the forward direction; it is never backscattered. Therefore regular RBS cannot be used for H detection. The depths analyzed and depth-profiling capabilities are similar to those of the equivalent backscattering methods, but the depth resolution is poor (~500 Å at 1000-Å depths). NRA (Chapter 11), an alternative technique for detecting hydrogen, has greater sensitivity than ERS. SIMS (Chapter 10) has far greater sensitivity for hydrogen (down to trace amounts) than either technique and better depth resolution, but it is a destructive sputter-removal method and is difficult to quantify. Sample damage can also be a problem with ERS, particularly for polymers.

9.1 RBS

Rutherford Backscattering Spectrometry

SCOTT M. BAUMANN

Contents

Introduction

Rutherford Backscattering Spectrometry (RBS) is one of the more quantitative depth-profiling techniques available, with typical accuracies of a few percent. The depth profiling is done in a nondestructive manner, i.e., not by sputtering away the surface layers. Results obtained by RBS are insensitive to sample matrix and typically do not require the use of standards, which makes RBS the analysis of choice for depth profiling of major constituents in thin films. Detection limits range from a few parts per million (ppm) for heavy elements to a few percent for light elements. RBS depth resolution is on the order of 20–30 nm, but can be as low as 2–3 nm near the surface of a sample. Typical analysis depths are less than 2000 nm, but the use of protons, rather than helium, as the probe particle can increase the sampling depth by as much as an order of magnitude. Lateral resolution for most instruments is on the order of 1–2 millimeters, but some microbeam systems have a resolution on the order of 1–10 μm.

Three common uses of RBS analysis exist: quantitative depth profiling, areal concentration measurements (atoms/cm^2), and crystal quality and impurity lattice site analysis. Its primary application is quantitative depth profiling of semiconductor thin films and multilayered structures. It is also used to measure contaminants and to study crystal structures, also primarily in semiconductor materials. Other applications include depth profiling of polymers,[1] high-T_C superconductors, optical coatings, and catalyst particles.[2]

Recent advances in accelerator technology have reduced the cost and size of an RBS instrument to equal to or less than many other analytical instruments, and the development of dedicated RBS systems has resulted in increasing application of the technique, especially in industry, to areas of materials science, chemistry, geology, and biology, and also in the realm of particle physics. However, due to its historical segregation into physics rather than analytical chemistry, RBS still is not as readily available as some other techniques and is often overlooked as an analytical tool.

Basic Principles

RBS is based on collisions between atomic nuclei and derives its name from Lord Ernest Rutherford who first presented the concept of atoms having nuclei. When a sample is bombarded with a beam of high-energy particles, the vast majority of particles are implanted into the material and do not escape. This is because the diameter of an atomic nucleus is on the order of 10^{-4} Å while the spacing between nuclei is on the order of 1 Å. A small fraction of the incident particles do undergo a direct collision with a nucleus of one of the atoms in the upper few μm of the sample. This "collision" actually is due to the Coulombic force present between two nuclei in close proximity to each other, but can be modeled as an elastic collision using classical physics.

The energy of a backscattered particle detected at a given angle depends upon two processes: the loss of energy by the particle due to the transfer of momentum to the target atom during the backscattering event, and the loss of energy by the particle during transmission through the sample material (both before and after scattering). Figure 1 is a schematic showing backscattering events occurring at the surface of a sample and at a given depth *d* in the sample. For scattering at the sample's surface the only energy loss is due to momentum transfer to the target atom. The ratio of the projectile's energy after a collision to the its energy before a collision (E_1/E_0) is defined as the kinematic factor *K*:[3, 4]

$$K = \left(\frac{\sqrt{1 - ((M_1/M_2)\sin\theta)^2} + (M_1/M_2)\cos\theta}{1 + (M_1/M_2)} \right)^2 \qquad (1)$$

where M_1 is the mass of the incident particle (typically ^4He); M_2 is the mass of the target atom; and *R* is defined as the angle between the trajectory of the He particle before and after scattering.

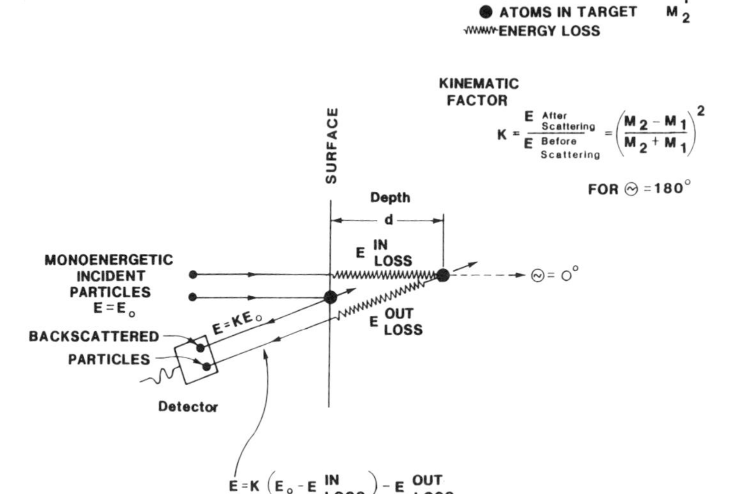

Figure 1 **A schematic showing the various energy loss processes for backscattering from a given depth in a sample. Energy is lost by momentum transfer between the probe particle and the target particle, and as the probing particle traverses the sample material both before and after scattering.**

As shown in Figure 1, when the probing particles penetrate to some depth in a sample, energy is lost in glancing collisions with the nuclei of the target atoms as well as in interactions with electrons. For a 2-MeV He atom, the energy loss is in the range of 100–800 eV/nm and depends upon the composition and density of the sample. This means that a particle that backscatters from some depth in a sample will have measurably less energy than a particle that backscatters from the same element on the sample's surface. This allows one to use RBS in determining the thickness of layers and in depth profiling.

The relative number of particles backscattered from a target atom into a given solid angle for a given number of incident particles is related to the differential scattering cross section:

$$\frac{d\sigma}{d\Omega} = \left(\frac{Z_1 Z_2 e^2}{4E}\right)^2 \frac{4\left(\sqrt{1-((M_1/M_2)\sin\theta)^2}+\cos\theta\right)^2}{(\sin\theta)^4\sqrt{1-((M_1/M_2)\sin\theta)^2}} \tag{2}$$

where Z_1 and Z_2 are the atomic numbers of the incident atom and the target atom, E is the energy of the incident atom immediately before scattering, and e is the electronic charge. A rule of thumb is that the scattering cross section is basically proportional to the square of the atomic number Z of the target species. This means that RBS is more than a hundred times more sensitive for heavy elements than for light

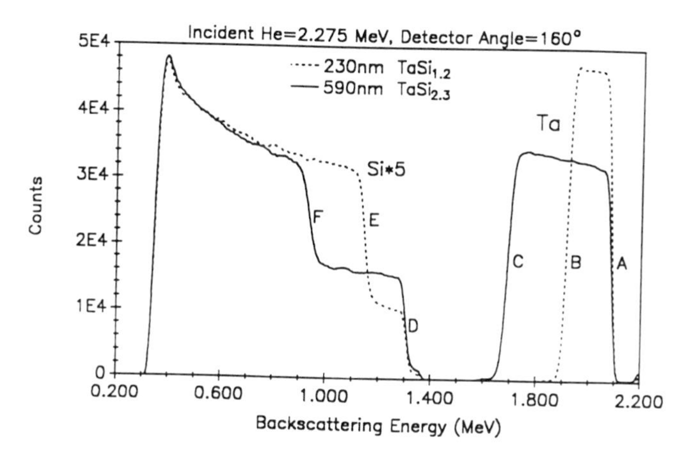

Figure 2 **RBS spectra from two $TaSi_x$ films with different Si / Ta ratios and layer thicknesses.**

elements, such as B or C. There is much greater separation between the energies of particles backscattered from light elements than from heavy elements, because a significant amount of momentum is transferred from an incident particle to a light target atom. As the mass of the target atom increases, less momentum is transferred to them and the energy of the backscattered particle asymptotically approaches the incident particle energy (see Equation 1). This means that RBS has good mass resolution for light elements, but poor mass resolution for heavy elements. For example, it is possible to resolve C from O or P from Si but it is not possible to resolve W from Ta, or Fe from Ni when these elements are present at the same depths in the sample, even though the difference in mass between the elements in each of these pairs is roughly 1 amu.

Figure 2 shows how the processes combine to create an RBS spectrum by displaying the spectra from two $TaSi_x$ films on Si substrates. Metal silicide films are commonly used as interconnects between semiconductor devices because they have lower resistivity than aluminum or polysilicon. The resistivity of the film depends upon the ratio of Si to metal and on the film thickness, both of which can be determined by RBS. The peak in each spectrum at high energy is due to scattering from Ta in the $TaSi_x$ layers while the peak at lower energy is from Si in the $TaSi_x$ layer and the Si substrate. The high-energy edge of the Ta peaks near 2.1 MeV (labeled *A*) corresponds to scattering from Ta at the surface of both samples, while the high-energy edge of the Si peaks (labeled *D*) near 1.3 MeV corresponds to backscattering from Si at the surface of the $TaSi_x$ layer. By measuring the energy width of the Ta peak or the Si step and dividing by the energy loss of He (the incident particle) per unit depth in a $TaSi_x$ matrix, the thickness of the $TaSi_x$ layer can be calculated. For example, the low-energy edge of the Ta peak corresponds to scattering from Ta at the $TaSi_x$–Si interface and the step in the Si peak corresponds to the increase in the

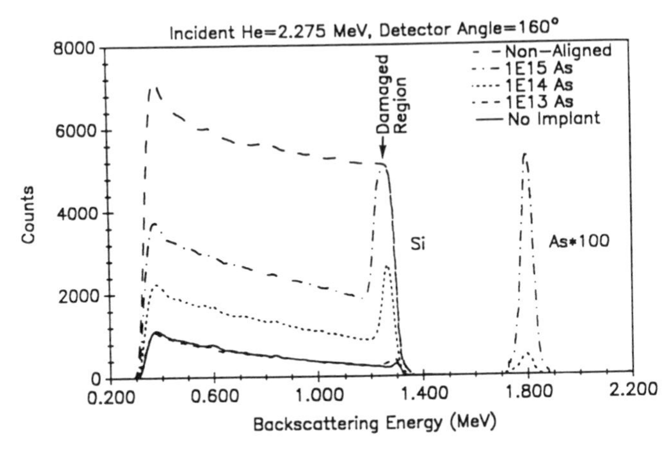

Figure 3 **Crystal channeled RBS spectra from Si samples implanted with 10^{13}, 10^{14}, and 10^{15} As atoms / cm^2. Also shown is a channeled spectrum from a nonimplanted Si sample and a nonaligned, or random, Si spectrum.**

Si concentration at the $TaSi_x$–Si interface. In this example, one of the films is 230 nm thick, while the other film is 590 nm thick. Particles scattered from Ta at the $TaSi_x$–Si interface of the 230-nm film have a final energy of about 1.9 MeV (labeled *B*) after escaping from the sample, while particles scattered from Ta at the $TaSi_x$–Si interface of the 590-nm film have a final energy of about 1.7 MeV (labeled *C*). Similarly, particles scattered from Si at the $TaSi_x$–Si interface of the 230-nm film have a final energy of about 1.1 MeV (labeled *E*) after escaping from the sample, while particles scattered from Si at the $TaSi_x$–Si interface of the 590-nm film have a final energy of about 0.9 MeV (labeled *F*). In these spectra the greater energy width of the Ta peak and the Si step for the 590-nm $TaSi_x$ film are directly related to the greater thickness of the film.

By measuring the height of the Ta and Si peaks and normalizing by the scattering cross section for the respective element, the ratio of Si to Ta can be obtained at any given depth in the film. Due to the smaller scattering cross section for Si, the Si peaks in Figure 3 have been multiplied by a factor of 5. The height of a backscattering peak for a given layer is inversely proportional to the stopping cross section for that layer, and in this case the stopping cross section of $TaSi_{2.3}$ is 1.37 times greater than that of Si. This explains why, even for the film with a Si/Ta ratio of 2.3, the height of the peak corresponding to Si in the $TaSi_x$ layer is less than ½ the height of the peak corresponding to Si in the sample substrate.

Channeling

In addition to elemental compositional information, RBS also can be used to study the structure of single-crystal samples.[5, 6] When a sample is *channeled*, the rows of atoms in the lattice are aligned parallel to the incident He ion beam. The bombarding He will backscatter from the first few monolayers of material at the same rate as a nonaligned sample, but backscattering from buried atoms in the lattice will be drastically reduced, since these atoms are shielded from the incident particles by the atoms in the surface layers. For example, the backscattering signal from a single-crystal Si sample that is in channeling alignment along the (100) axis will be approximately 3% of the backscattering signal from a nonaligned crystal, or amorphous or polycrystalline Si. By measuring the reduction in backscattering when a sample is channeled it is possible to quantitatively measure and profile the crystal perfection of a sample, or to determine its crystal orientation.

Figure 3 shows channeled spectra from a series of Si samples that were implanted with 10^{13}, 10^{14}, and 10^{15} arsenic atoms/cm^2. Only the As peaks for the two highest dose implants are shown, but with a longer data acquisition time the concentration 10^{13} As atoms/cm^2 could be detected. The damage caused to the Si crystal lattice by the As implants is reflected in the peaks near 1.25 MeV in the aligned spectra. In the case of the 10^{15}-atoms/cm^2 implant there is little or no single-crystal structure remaining in the damaged region of the Si, so the backscattering signal is the same height as for nonaligned Si. Measuring the energy width of the damage peak indicates that the damaged layer is approximately 200 nm thick. Integrating the damage peak and subtracting the backscattering signal obtained for the nonimplanted reference indicates that approximately 1.0×10^{18} Si atoms/cm^2 were displaced by the 10^{15}-atoms/cm^2 As implant, while 3.4×10^{17} and 1.7×10^{16} Si atoms/cm^2 were displaced by the 10^{14}-atoms/cm^2 and 10^{13}-atoms/cm^2 As implants, respectively. In this case RBS could be used to measure accurately the total concentration of arsenic atoms implanted in each sample, to profile the As implant, to determine the amount of As that is substitutional in the Si lattice and its lattice location, to measure the number of displaced Si atoms/cm^2, and to profile the damage in the Si crystal.

Quantification

As noted above, the calculation of elemental concentrations and thicknesses by RBS depends upon the scattering cross section of the element of interest and the stopping cross section of the sample matrix. The scattering and stopping cross sections for each element have been carefully measured and tabulated.[3, 4, 7] In general, scattering cross sections follow the Rutherford scattering model to within 5%. It is difficult to accurately describe the stopping cross sections for all elements with a single equation, so semiempirical values are employed. A polynomial equation with several terms is used so that the stopping cross sections for each element can be calculated over a range of energies. In general, the calculated stopping cross sections are accurate to 10% or better. The stopping cross section for a multi-elemental sample is calculated by normalizing the stopping cross section of each element to its concentration in the sample.

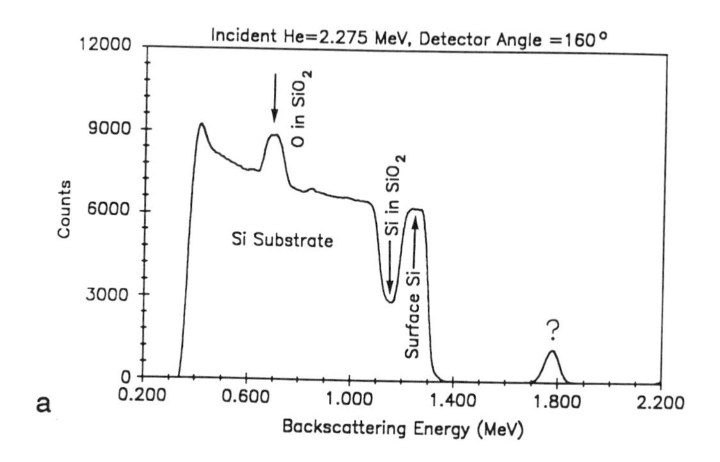

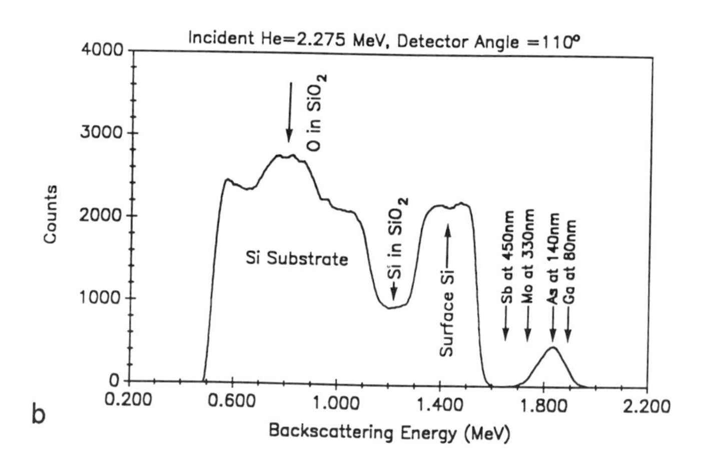

Figure 4 **RBS spectra from a sample consisting of 240 nm of Si on 170 nm of SiO_2 on a Si substrate. The spectrum in (a) was acquired using a scattering angle of 160° while the spectrum in (b) used a detector angle of 110°. This sample was implanted with 2.50×10^{16} As atoms/cm^2, but the As peak cannot be positively identified from either spectrum alone. Only As at a depth of 140 nm will produce the correct peak in both spectra.**

Due to the convoluted mass and depth scales present in an RBS spectrum, it may not be possible to accurately describe an unknown sample using a single RBS spectrum. For example, Figure 4a is an RBS spectrum acquired at a backscattering angle of 160° from a sample implanted with 2.50×10^{16} atoms/cm^2 of As at a depth of approximately 140 nm. If this were a totally unknown sample it would not be possible to determine positively the mass and depth of the implanted species from this spectrum alone, since the peak in the RBS spectrum also could have been caused by a heavier element at greater depth, such as Sb at 450 nm, or Mo at 330 nm, or by a lighter element at a shallower depth, such as Ga at 80 nm. If an additional spectrum is acquired at a glancing backscattering angle, the scattering kinematics will be changed and the backscattered particles will have a longer escape path through the sample material to the detector. As the detector angle approaches 90° (tangent to the sample surface) the backscattering peak for a buried element will be shifted to lower energies due to the greater loss of energy along the longer escape trajectory out of the sample. Figure 4b is the RBS spectrum acquired from the same sample but at a backscattering angle of 110°. Shown in this figure are the locations for the other possible elements and depths that would match the peak shown in Figure 4a. Only As at a depth of 140 nm will produce a peak at the correct energy in both spectra. By acquiring two backscattering spectra at different angles it is usually possible to determine the depth and mass of an unknown element. One should note also that the depth resolution for the surface Si layer and the SiO_2 layer are improved in the 110° spectrum due to greater energy loss per unit depth in the sample. This results in the wider peaks for the surface Si and SiO_2 layers in the 110° spectrum.

Artifacts

Although RBS does not suffer from matrix effects that are normally associated with profiling techniques using sputtering, such as SIMS, AES, or SNMS, there are other factors that do limit the application of the technique. The convoluted nature of the mass and depth information available in an RBS spectrum often results in a spectral interference between the peak for a light element and a buried heavier element. For example, in Figure 4a the He that backscatters from the oxygen in the SiO_2 produces the peak between 0.65–0.72 MeV, while backscattering from Si in the SiO_2 produces the peak between 1.2–1.3 MeV. Scattering from the Si substrate produces the peak between 0–1.2 MeV (the backscattering signal has been suppressed between 0–0.35 MeV). The peak from the Si substrate contributes noise to the oxygen peak and limits the accuracy to which the oxygen concentration can be measured. In cases where the matrix contains heavy elements it may not be possible to detect light elements at all, i.e., carbon in a bulk tungsten sample. Procedures have been developed to eliminate or minimize the effects of these spectral interferences. These include channeling crystalline substrates to reduce the backscattering signal from the substrate, using detectors at glancing angles to the sample's surface or orienting the sample at a glancing angle to the incident ion beam, and varying the energy of the incident ion. The repeatable nature of RBS allows the use of computer models to predict the RBS spectrum from a given sample structure, permitting the investigator to optimize the measurement parameters or the sample structure to maximize the accuracy and usefulness of the results.

Sample roughness also can produce problems in the interpretation of RBS spectra that are similar to problems encountered by sputtering techniques like AES,

SIMS, and SNMS; in rare cases, such as for $HgCd_xTe_{1-x}$ samples or some polymers, the sample structure can be modified by the incident ion beam. These effects can often be eliminated or minimized by limiting the total number of particles incident on the sample, increasing the analytical area, or by cooling the sample. Also, if channeling of the ion beam occurs in a crystal sample, this must be included in the data analysis or serious inaccuracies can result. To avoid unwanted channeling, samples are often manipulated during the analysis to present an average or "random" crystal orientation.

Finally, the fundamental unit of concentration obtained by RBS is in atoms/cm^2 or concentration in the sample-versus-backscattering energy loss. To convert the profile of a backscattering peak into a depth profile it is necessary to assume a density for the material being profiled. For single-element films, such as Si, Ti, and W, an elemental density can be assumed for the film and an accurate thickness is obtained. In the case of multi-elemental films with an unknown density, a density for the film is calculated by summing the density of each element, normalized to its concentration. The accuracy of this assumption is usually within 25%, but for some cases the actual density of the film may vary by as much as 50%–100% from the assumed density. It is useful to note that:

$$T_{RBS} \times D_{RBS} = (\text{atoms})/\text{cm}^2 = D_{real} \times T_{real} \tag{3}$$

where T_{RBS} and D_{RBS} are the thickness obtained by RBS and the density assumed to calculate this thickness; and T_{real} and D_{real} are the actual physical thickness and density of the film. If the physical thickness of a film can be measured by some other technique, such as SEM, TEM, or profilometry, then the actual film density can be accurately calculated.

Instrumentation

An RBS instrument can be divided into two basic components: the particle accelerator and the analysis chamber or end station. PIXE and ERS analyses employ similar instrumentation, but use different incident ion beams or detectors.

Particle Accelerators

Two types of particle accelerators are used to obtain the MeV energies used for RBS. Single-ended accelerators are similar to ion implanters used in the semiconductor industry but have an ion source located at the high-energy terminal of the accelerator. Ions are extracted from the source and are accelerated down the beam line to ground potential. Tandem accelerators use a source that is at ground potential and that emits a beam of negative ions that are accelerated toward the positively charged terminal of the accelerator, where their charge states are changed by passing the beam through a thin foil or a gas cell. The (now) positively charged particles are accelerated to higher energy as they are repelled from the positive terminal voltage and back to ground potential.

End Station

A multitude of analysis chambers exist that have been designed with specific measurements or sample sizes in mind. State-of-the-art systems have a multiple-axis goniometer, which allows positioning of many samples for analysis without breaking vacuum. High precision (on the order of ±0.01°) is required when orienting samples for channeling, which often makes the goniometers used for channeling both complicated and expensive. The minimum sample size is controlled by the dimensions of the incident ion beam. Typically, the ion beams used for RBS are about 1–2 mm in diameter and samples are between 0.1–1 cm^2 in area, however, some microbeam systems with beam diameters on the order of 10 μm have been built. Analysis chambers also have been made to accommodate large samples, such as entire silicon wafers. For the purposes of most standard RBS measurements the analysis chamber needs to be evacuated to at least 10^{-5} torr. Extremely good depth resolutions of less than 3 nm can be obtained by orienting either the incident ion beam or the detector at a glancing angle to the sample surface.

Applications

Listed below is a summary of some common applications of RBS.

Semiconductors:	Quantitative depth profiling of: Metal silicide films (WSi_x, $MoSi_x$, $TiSi_x$, etc.) Barrier metals (TiN_x, TiW_x, etc.) Insulating layers (SiO_x, SiN_x, and SiO_xP_y) Cu in Al interconnect III-V and II-VI materials ($Al_xGa_{1-x}As$, and $Hg_xCd_{1-x}Te$) Metal multilayer stacks Dose/lattice substitutionality of implanted species Crystal damage versus depth (Si, $SiGe_x$, $Al_xGa_{1-x}As$, and $Hg_xCd_{1-x}Te$)
High-T_C superconductors:	Quantitative depth profiling ($YBa_xCu_yO_z$, and $BiSr_wCu_xCa_yO_z$) Crystal orientation and damage versus depth
Optical/antireflective coatings:	Quantitative depth profiling of multilayered stacks (SiO_2, HfO_2, TiO_2, SnO_2, $InSn_xO_y$, etc.)
Polymers:	Depth profiling of halogens and impurities Metallization of surfaces
Catalysts:	Location of active ions on or in particles

Conclusions

RBS is a rapidly growing technique that has evolved from being used primarily in particle physics to being commonly applied and widely available. The size and cost of some RBS instruments are now equal to or less than that of other depth profiling techniques, such as SIMS, AES, and SNMS. RBS data analysis software allows most data to be rapidly and accurately analyzed and permits the automated acquisition of and (in some cases) analysis of data. Currently RBS is used primarily in the analysis of semiconductors, superconductors, optical coatings, and other thin films. Some applications have been developed for polymers and ceramics, and further growth is expected in these areas due to the technique's relatively lenient vacuum requirements and its insensitivity to charging problems for insulators. A few microbeam RBS systems are currently in service and the development of RBS imaging will certainly produce new applications for semiconductors and, possibly, even for biological samples, since the small size of cells that are typically analyzed has limited the use of RBS in the past.

Related Articles in the Encyclopedia

MEIS, ISS, PIXE, ERS, and NRA

References

1. S. J. Valenty, J. J. Chera, G. A. Smith, W. Katz, R. Argani, and H. Bakhru. *J. Polymer Sci. (Chem.)* **22,** 3367, 1984.
2. S. M. Baumann, M. D. Strathman, and S. L. Suib. *Analytical Chem.* **60,** 1046, 1988.
3. W. K. Chu, J. W. Mayer, and M. A. Nicolet. *Backscattering Spectrometry.* Academic Press, New York, 1978. This is a frequently used handbook that provides a thorough discussion of the technique.
4. J. R. Bird and J. S. Williams. *Ion Beams for Materials Analysis.* Academic Press, Australia, 1989. Chapter 3 provides an overview of RBS, while Chapter 6 reviews channeling techniques. This book also reviews NRA, PIXE, SIMS, and other related ion-beam analyses.
5. L. C. Feldman, J. W. Mayer, and S. T. Picraux. *Materials Analysis by Ion Channeling.* Academic Press, New York, 1982. This book provides an in-depth study of the principles and use of ion channeling for analyzing materials.
6. *Channeling.* (D. V. Morgan, ed.) John Wiley & Sons, London, 1973. Chapters 13–16 provide information regarding the use of channeling

measurements in the analysis of materials. The remainder of the book is a study of the physics of channeling.

7 J. F. Ziegler, J. P. Biersack, and U. Littmark. *Stopping Powers and Ranges in All Elements.* Pergamon Press, New York, 1977, vol.1-6.

8 D. K. Sadana, M. Strathman, J. Washburn, and G. R. Booker. *App. Phys. Letts.* **37**, 234, 1980.

9 *Ion Beam Handbook for Material Analysis.* (J. W. Mayer and E. Rimini, eds.) Academic Press, New York, 1977. This book provides useful tabular and graphic data for RBS, channeling, PIXE, and NRA.

9.2 ERS

Elastic Recoil Spectrometry

J.E.E. BAGLIN

Contents

Introduction

Overview

Elastic recoil spectrometry (ERS) is used for the specific detection of hydrogen (^1H, ^2H) in surface layers of thickness up to approximately 1 μm, and the determination of the concentration profile for each species as a function of depth below the sample's surface.[1, 2] When carefully used, the technique is nondestructive, absolute, fast, and independent of the host matrix and its chemical bonding structure. Although it requires an accelerator source of MeV helium ions, the instrumentation is simple and the data interpretation is straightforward.

The method may be contrasted to dynamic SIMS analysis, which, although capable of somewhat better depth resolution, is slower and matrix-dependent, and relies on ion milling (sputtering) for profiling. Nuclear Resonance Reaction Analysis (NRA) claims, in general, a better ability to identify trace (ppm) hydrogen levels, mainly because of enhanced (resonant) scattering cross sections. However, a depth profile determination by NRA is complex, requiring many sequential data runs, and it takes many times longer than ERS. Quantitative NRA data reduction is

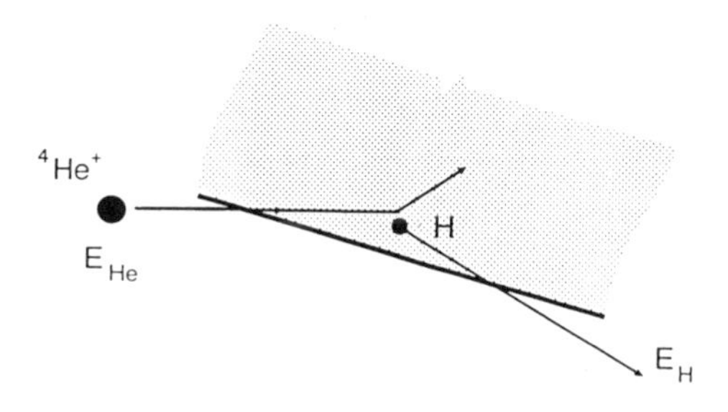

Figure 1 **The forward scattering concept of ERS.**

hampered by the difficulties of determining and deconvoluting the nuclear resonance shapes.

ERS may be regarded as an extension of Rutherford Backscattering Spectrometry (RBS).[3] It requires basically identical equipment, and it preserves many beneficial features of RBS: convenience, speed, precision, and simplicity. RBS is based on the simple point-charge scattering of ions (generally helium at 1–2 MeV) by the constituent atomic nuclei of the sample. The energy of ions scattered at a known angle is used to indicate both the mass of the scattering nucleus and the depth of penetration of the ion into the sample before the scattering collision occurred. To optimize mass resolution and sensitivity, those ions which are scattered backwards (near 180°) are frequently chosen for RBS spectrometry. This geometry does not work, however, when the projectile is heavier than the target nucleus. As illustrated in Figure 1, following the collision of a helium ion with a hydrogen (or deuterium) nucleus in the sample, both particles move in the forward direction, and it is therefore necessary to place detectors to receive forward scattered particles. Since most specimens will be too thick to allow either ^{4}He or H ions to escape in transmission geometry, a glancing angle arrangement is chosen. In this situation, it is advantageous to select the recoiling hydrogen ions themselves for energy spectrometry, rather than the lower energy, less-penetrating scattered helium ions. By covering the detector with a stopping foil of appropriate thickness, it is possible to admit H ions for analysis but exclude all scattered He ions, including the prolific He ion flux contributed by Rutherford scattering from other, heavier constituents of the sample.

The covered detector thus provides an energy spectrum of the forward-recoiling hydrogen ions. An important advantage of this technique is the uncomplicated relationship between this spectrum and the concentration-versus-depth profile for hydrogen in the sample. Derivation of the concentration profile is direct and unambiguous. This simplicity depends on the He–H scattering process being elastic (no residual excited states of scattered nuclei), and on the absence of nuclear reactions that might yield spurious detectable particles. The threshold energies for such reactions are 6.7 MeV (for $^4He + ^2H \rightarrow ^4He + n + p$), and tens of MeV for ^1H. Typically, ERS measurements are run with 1–2 MeV ^4He ions, where such problems do not arise.

Although most applications of ERS have used ^4He as the projectile ion, the principle clearly can be extended to recoiling ions from heavier projectiles. The depth resolution may be significantly improved in this way, e.g., in polystyrene, the resolution found[4] for ERS from 2.8-MeV He ions was 1000 Å, compared with 300 Å obtained by using 20-MeV Si ions. Also, the scattering cross section is larger for Si, leading to greater sensitivity. Severe radiation damage to samples can occur with heavy ions, however (factors of >100 worse than for He). In the interest of simplicity, this review will focus on the technique of hydrogen detection using helium ion beams.

Applications

ERS is an appropriate tool for a wide range of analytical applications. Some typical examples include the quantitation of hydrogen in glassy carbon films, the study of the dynamics of polymeric molecules,[5] studies of interface interactions between polymer films (using deuterium as a diffusion marker),[4] analysis for hydrogen in natural geological specimens,[6] a study of stress-induced redistribution of hydrogen in metal films,[7] and a study of the effects of hydrogen content upon the optical, mechanical, and structural properties of plasma-deposited amorphous silicon and silicon nitride films.[8, 9] Further applications will also be found in the specific examples cited in this article.

Principles of ERS

Apparatus

A typical experimental equipment layout for ERS is shown schematically in Figure 2. A beam of ^4He ions is produced at an energy of 1–2 MeV, generally using an electrostatic accelerator and a beam transport system including a mass- and energy-resolving bending magnet and knife edge collimators to define the beam spot size and position. A typical beam current required is 5–50 nA, with a beam spot dimension of 1–2 mm. A clean vacuum of $\leq 10^{-7}$ Torr is desirable for particle spectrometry, and specimens must be vacuum compatible. The specimen is tilted so that the incident beam makes an angle of approximately 15° with the plane of the surface, and a surface barrier detector is placed to receive particles scattered at a similar angle from the sample surface in the forward direction. The detector's aperture is set to limit the range of accepted scattering angles to ± 1° or less. A smooth foil of aluminum or Mylar is placed in front of the detector to stop scattered He ions, yet to transmit scattered ^1H (or ^2H) ions into the detector after they incur a small, well-defined energy loss. It is important for this foil to be uniform and free of

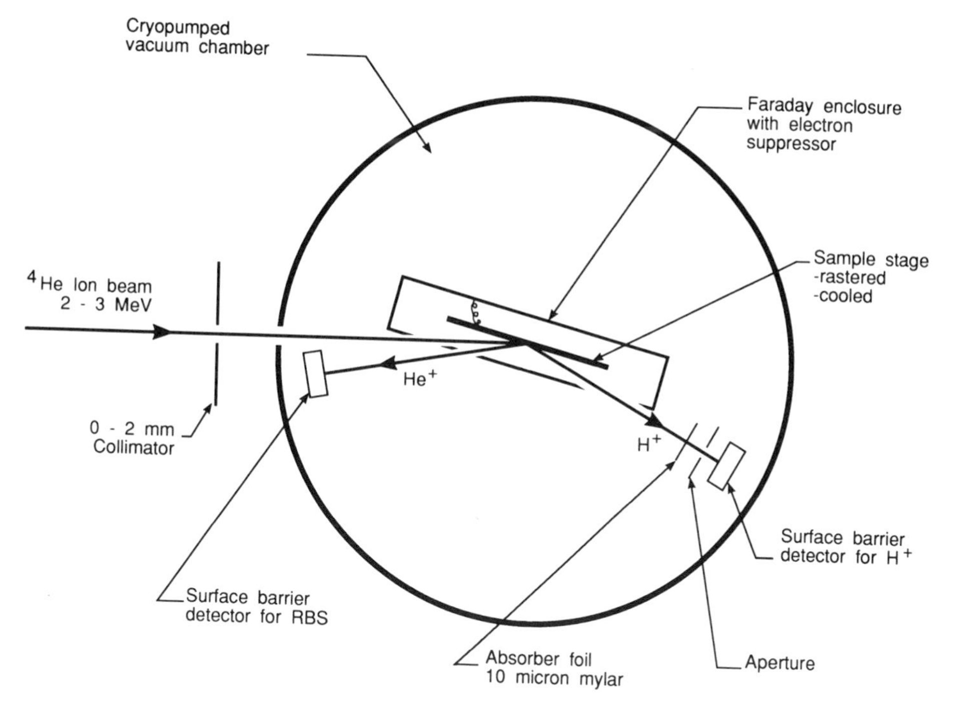

Figure 2 **Typical experimental layout for ERS (schematic).**

pinholes. A 10-μm Mylar foil will completely stop He ions of energy ≤ 2.3 MeV, while transmitting recoiling ^{1}H ions with an energy loss of 250 keV. An energy spread of ≤ 50 keV in the transmitted ions will be caused by straggling.[3] A second surface barrier detector normally is used in a backscattering position. Spectra accumulated simultaneously with this detector will provide RBS information on the sample's composition which must be combined with the ERS data for a complete sample analysis. Quantitative analysis requires a reliable measure of the incident ion fluence for each run. As in RBS, this may be obtained in various ways, either by direct collection of ion beam charge on the sample and in a surrounding Faraday enclosure (as illustrated in Figure 2), or by a well calibrated beam current sampling technique. In the former technique, complete collection of secondary electrons from the sample surface must be assured. A cooled sample stage mounted on a goniometer provides the means to adjust the sample's orientation and, ideally, to raster the sample, in order to limit local ion beam heating. The sample stage may be cooled with liquid nitrogen to reduce the effects of beam-induced damage on the analysis, especially for polymers. Signals from the two surface barrier detectors are sorted by a pulse height analyzer, and the resulting spectra are stored for subsequent display and quantitative analysis. Typical run times of 15–20 minutes are required to obtain good statistics. It should be noted that this technique responds exclusively

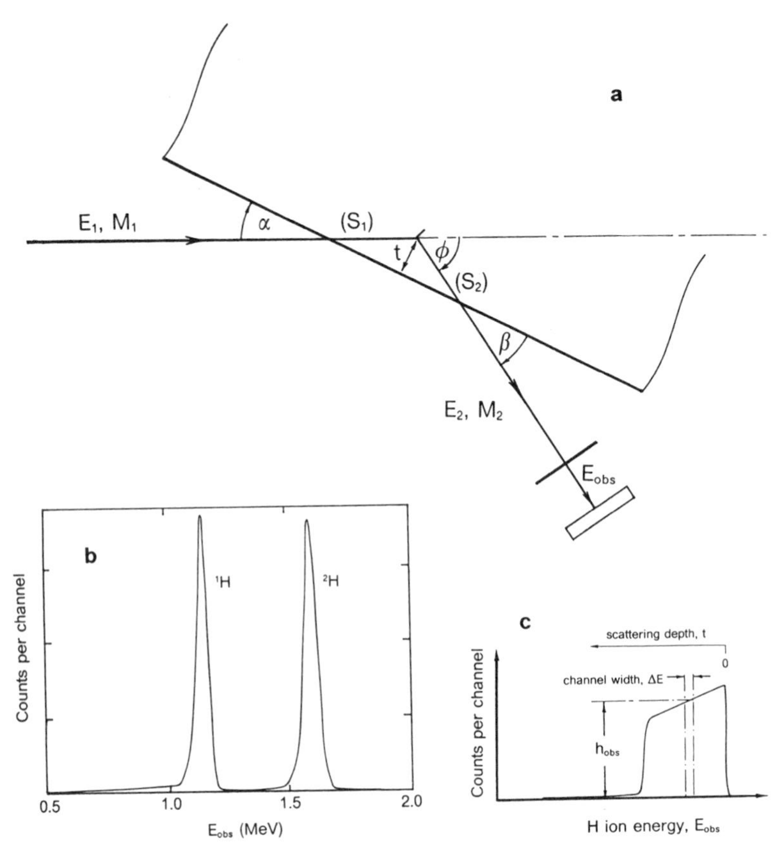

Figure 3 **(a) Scattering geometry for ERS; (b) ERS spectrum from 200-Å partially deuterated polystyrene on Si, E_1 = 3.0 MeV (adapted from ref. 10); and (c) schematic ERS depth profile spectrum.**

to the presence of ^{1}H or ^{2}H (or ^{3}H or ^{3}He) nuclei in a sample, and does not suffer from any "background" from other sources. The sensitivity of the technique to trace amounts of H is therefore largely determined by patience and counting statistics.

Energy Scale of ERS

The energy scale of an elastic recoil spectrum provides information about the mass of the recoiling species, and about the depth within a sample at which the scattering took place.

Figure 3a shows the ERS scattering geometry. To generalize the treatment, the following identities are adopted: ϕ is the angle of hydrogen recoil with respect to the incident beam; M_1 and E_1 are the mass and energy, respectively, of the incident ion; M_2 is the mass of the recoiling ion; E_2 is the energy of the recoiling particle at the point of the collision.

For the simple case of surface scattering (or scattering from a very thin layer), the ratio of E_2 to E_1 is given by

$$\frac{E_2}{E_1} = \frac{4M_1M_2\cos^2\phi}{(M_1+M_2)^2} \tag{1}$$

This ratio is known as the kinematic factor K'. The separation obtainable between ^1H and ^2H is evident ($K'(^1H) = 0.64\cos^2\phi$; $K'(^2H) = 0.88\cos^2\phi$), as seen in the energy spectrum of detected particles shown in Figure 3b, which was obtained from a 200-Å film of deuterated polystyrene on a silicon substrate.[10] In practice, ϕ is generally set in the range of 10–30°.

When a stopper foil of thickness δ_f(atom/ cm^2) is used, the hydrogen energy observed at the detector is not actually E_2, but a lower value, E_{obs}, where

$$E_{obs} = E_2 - \delta_f S_f \tag{2}$$

where S_f is the "stopping power" for hydrogen ions in the foil at the appropriate energy; i.e., the energy lost per unit thickness (atom/ cm^2) of foil material. Obviously, it is best to choose the smallest δ_f consistent with total stopping of unwanted ^4He ions.

In the more complex situation, a helium ion penetrates to a depth t (normal to the surface) into the sample before encountering a hydrogen nucleus, where scattering occurs. The helium ion loses energy in ionizing collisions before the scattering event, and in turn the recoiling H must undergo a similar energy loss before escaping the sample's surface. If the energy-averaged stopping powers for ^4He and recoiling H within the sample are S_1 and S_2, respectively, then the depth t of the recoiling collision may be derived from the expression[11]

$$t = (E_{obs} + \delta_f S_f - K'E_1)\left(\frac{K'S_1}{\sin\alpha} + \frac{S_2}{\sin\beta}\right)^{-1} \tag{3}$$

Generally $S_2 << S_1$. In this way, a depth scale may be associated with an ERS spectrum, as shown in the schematic spectrum of Figure 3c. That spectrum shows the H recoil rate (counts per energy channel of width ΔE) as a function of E_{obs} and hence as a function of depth below the sample surface (see Equation (3)).

Total Hydrogen Content

As with RBS, the simplest and most precise measurement that ERS can provide is the absolute measure of total hydrogen content (^1H and/or ^2H) within a surface layer or film of thickness less than a few thousand angstroms. When the layer is thin, this is determined from the total number of events, or *yield Y* summed over the elastic recoil spectral peak. The yield may be expressed as

$$Y(E_1, \phi) = \frac{d\sigma}{d\Omega}(E_1, \phi) \cdot Q \cdot \Omega \cdot N_H \qquad (4)$$

where $d\sigma/d\Omega\ (E_1, \phi)$ is the differential cross section for the nuclear scattering process $^{x}H(^{4}He,^{x}H)^{4}He$; Q is the total number of ^{4}He ions striking the sample (integral charge $\div\ q_{electron}$); Ω is the solid angle subtended by the recoil detector at the sample; and N_H is the density (atoms/cm^{2}) of ^{x}H in the layer. The quantities Q, Ω, and Y may be readily and precisely measured. Hence N_H may be derived provided that $d\sigma/d\Omega$ is precisely known.

Cross Section $d\sigma/d\Omega\ (E_1, \phi)$

In the analogous RBS analysis, $d\sigma/d\Omega$ is given precisely and analytically by the Rutherford scattering formula. Unfortunately, the case of (^{4}He, ^{x}H) scattering is not quite so simple. While the processes are indeed elastic, their cross sections are dominated by nuclear interaction components except at very low energies. (The $^{1}H(^{4}He, ^{1}H)^{4}He$ cross section approaches the Rutherford value for energies below 0.8 MeV.)

The absolute precision of ERS therefore depends on that of $d\sigma/d\Omega\ (E_1, \phi)$. Unfortunately, some disagreement prevails among measurements of the ^{1}H recoil cross section. One recent determination[6] is shown in Figure 4a for $\phi = 30°$ and 25°. The convergence of these data with the Rutherford cross section near 1 MeV lends support to their validity. The solid lines are least squares fits to the polynomial form used by Tirira et al.[6]. For $\phi = 30°$, the expression reads:

$$\ln \frac{d\sigma}{d\Omega} = 0.133\, E_1 + 4.383 + 2.196\, (1.6454)\ E_1^{-1} - 0.042\, E_1^{-2} \qquad (5)$$

where E_1 is expressed in MeV and the cross section in units of 10^{-27} cm^{2}/sr. Such an expression is of practical value for computer evaluation of measured spectra.

The measured cross section data for ^{2}H are shown[12] in Figure 4b. The dominant resonance at 2.13 MeV offers a powerful enhancement to sensitivity for deuterium detection, exceeding the Rutherford cross section by two orders of magnitude.

Concentration Profiles

As in RBS analysis, ERS can provide information on the atomic concentration of hydrogen as a function of depth (measured in atoms/cm^{2}). This is derived from the "height" h_{obs} of the ERS spectrum (counts per channel), at energies corresponding to particular depths within the sample (see Figure 3c). For a sample consisting of H and another material X, with composition $H_m X_{1-m}$, the spectrum height h_{obs} may be expressed

$$h_{obs} = \frac{m \cdot \frac{d\sigma}{d\Omega} \cdot Q \cdot \Omega \cdot \Delta E}{[S]^{H_m X_{1-m}} \cdot \sin\alpha} \left(1 - \delta \frac{dS_f}{dE}\right)^{-1} \qquad (6)$$

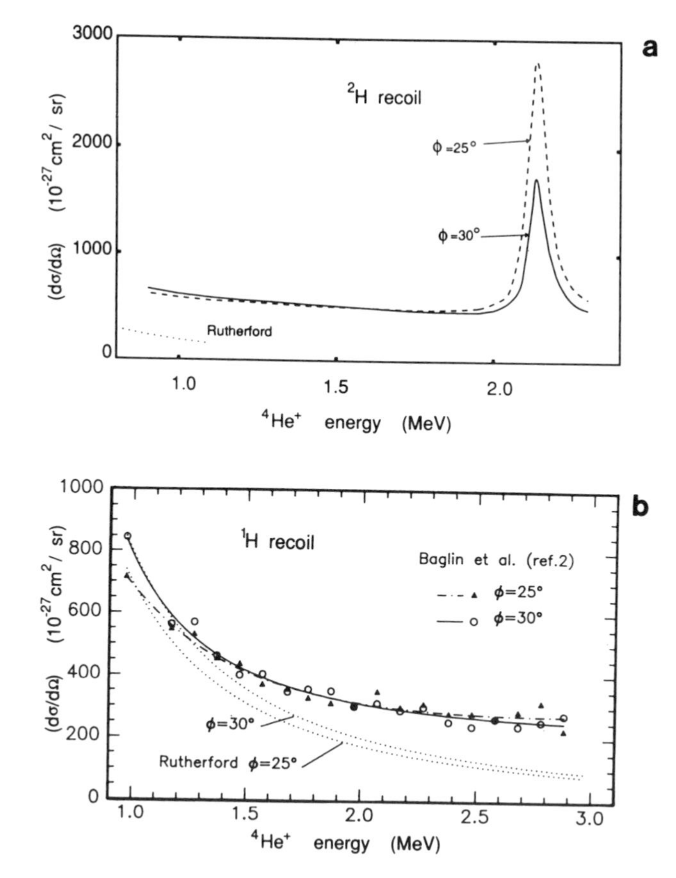

Figure 4 **(a) Differential cross section data for ^{1}H (^{4}He, ^{1}H) ^{4}He scattering at $\phi = 30°$ and 25°; and (b) differential cross section data for ^{2}H (^{4}He, ^{2}H) ^{4}He scattering measured at $\phi = 30°$ and 25° (adapted from ref. 12). The dotted lines locate the Rutherford scattering cross sections for comparison.**

where $[S]$ is a *reduced stopping power* serving to combine the effects of energy losses within the sample, given by the expression

$$[S]^{H_m X_{1-m}} = \frac{K'}{\sin\alpha} \cdot [S]_{He}^{H_m X_{1-m}} + \frac{1}{\sin\beta} \cdot [S]_{H}^{H_m X_{1-m}} \tag{7}$$

In turn, the stopping powers $[S]_{He}$ and $[S]_{H}$ are obtained by the application of Bragg's rule to combine the stopping powers in H and in X:

$$[S]_{He}^{H_m X_{1-m}} = m \cdot [S]_{He}^{H} + (1-m) \cdot [S]_{He}^{X} \tag{8}$$

The final term of Equation (6) corrects for the distortion of the ERS spectrum caused by velocity-dependent energy losses as the H ions pass through the stopper foil.

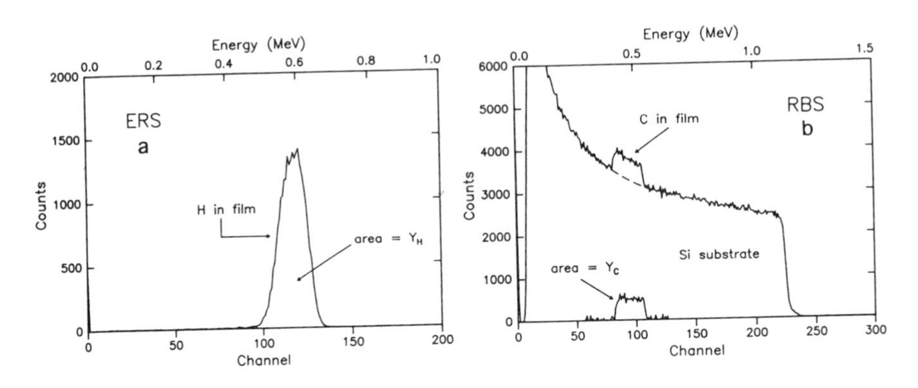

Figure 5 **ERS and RBS spectra for a 1000-Å sputter-deposited diamond-like carbon film on Si. Both spectra are required for complete analysis.**

The value of m may thus be derived from the spectrum height h_{obs} at an energy E_{obs}, whose value indicates the depth of the layer described. The precision of this measurement will be limited by the precision of $(d\sigma/d\Omega)$ and by the precision of stopping powers. Values of elemental stopping powers are available from compilations of experimental data that have been cross correlated for Z systematics.[13] Nevertheless, these stopping powers are still not well established, and their uncertainty may reach 10 % or more. The spectrum height therefore offers limited precision for direct measurement of m. However, the dependence of m upon depth can generally be accurately deduced from the energy dependence of h_{obs}.

Combined ERS and RBS Data

The use of RBS concurrently with ERS is necessary for the complete derivation of a hydrogen profile, and it offers some simplifications of analysis. For example, for thin-layer spectra that have been normalized for a common ion fluence Q and solid angle Ω, the total yields Y (the areas under the spectral peaks) may be compared in order to derive the layer composition. For $H_m X_{1-m}$,

$$\frac{m}{1-m} = Y_H \cdot \left(\frac{d\sigma}{d\Omega}\right)^{-1}_{ERS} \cdot \left(\frac{d\sigma}{d\Omega}\right)_{RBS} \cdot \frac{1}{Y_X} \tag{9}$$

where Y_H, Y_X are the areas for the H component in the ERS spectrum and for the X component in the RBS spectrum, respectively. Figure 5 illustrates such a measurement for a sputter-deposited diamond-like carbon film on a silicon substrate. Taking the areas of the peaks shown (normalized by the respective detector solid angles), and values of $d\sigma/d\Omega$ from Figure 4, and substituting in Equation (9) leads to the composition $H_{30}C_{70}$ in this case.

Similarly, it is possible to derive an expression relating the ratio of spectral heights h_{RBS} found for one or more elements in the RBS spectrum with that

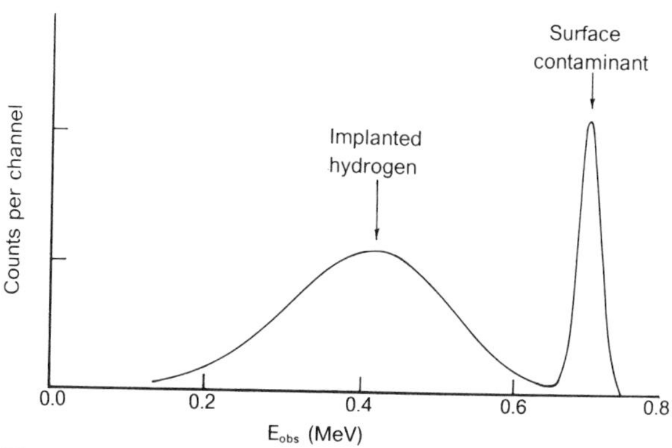

Figure 6 **Computer-simulated ERS spectrum (adapted from ref. 6) for the case of 1.6-MeV ^{4}He probing a silicon wafer implanted with 0.9×10^{16} H atoms/cm^2 at 10 keV (mean range of ions = 1750 Å, $\phi = 20°$, $\alpha = 10°$).**

obtained for hydrogen, h_{obs}, in the corresponding ERS spectrum. In this case, all data must correspond to a layer at a common depth in the sample. Since the RBS geometry differs from that of ERS, the RBS equivalents of Equation (6) must be derived. The resulting expression for h_{obs}/h_{RBS} may be used to calculate the sample composition at that depth.

Spectrum Simulation and Fitting

In the absence of a simple deconvolution procedure, an effective tool for determining a composition profile for a multilayered hydrogen-containing system is the simultaneous empirical fitting of observed ERS and RBS spectra with those calculated layer-by-layer for a trial sample structure. This requires straightforward software and data bases for $d\sigma/d\Omega$ and S, similar to those used for RBS fitting by RUMP and other standard packages. An example of such a computed ERS spectrum[6] is shown in Figure 6. The principal (broad) peak represents the expected distribution of ^{1}H (0.9×10^{16} atoms/cm^2) implanted in silicon at 10 keV (mean depth 1750 Å, range spread ± 700 Å), while the narrow peak corresponds to a surface contaminant layer. The simulated spectrum is an excellent fit to the corresponding experimental data.[6]

Heavy Ion Scattering

ERS can be applied to analysis for light elements other than hydrogen; the primary requirements are that the bombarding ion should be much heavier than the recoiling species being detected, and that inelastic nuclear reactions should not dominate; scattering cross sections must be known or determined; the spectrometer must be suitable for detecting recoiling species more massive than hydrogen; for depth profiling, greater ion energies are required, in general, increasing the risk of sample damage.

Practical Considerations

Sample Damage

High-energy He ions readily produce atomic displacements and broken chemical bonds in solids. While in many inorganic materials, the disordering is negligible during a 2-MeV He exposure for RBS, the situation for hydrogen is particularly sensitive. The probing He beam can readily liberate hydrogen from the near-surface regions of metal hydrides, and can cause substantial dissociation of polymers with concurrent loss of hydrogen from the sample. Special precautions must always be taken to minimize or quantify this effect, so that ERS will not alter the sample during analysis.

An example of the effect is the case of a 1000-Å polystyrene film deposited on a silicon wafer, which was subjected to successive exposures of 10 μC of helium ions, at room temperature (beam current 10 nA; sample area 4 mm^2). Each run produced ~1 % depletion in hydrogen content, while RBS showed no loss of carbon. The loss was shown to depend on cumulative ion fluence, rather than on beam current density. The sample analysis can be corrected for hydrogen loss by extrapolating to zero exposure. In general, for large beam current densities, thermal effects can become significant (depending on the nature of the polymer itself), increasing the loss rate of volatile hydrocarbons generated by radiation damage. In such cases, the use of a broadly rastered sample area and sample cooling may be helpful.

A further damage problem can be presented by thick samples of highly nonconducting material (e.g., ceramics), within which the incoming ion beam can build up electrostatic charge to the point of plasma discharge. Fracture or dissociation within the sample can result, and hydrogen (or other constituents) may well be liberated from the sample. The problem may be reduced with the use of conducting coatings and low beam current densities.

Reference Samples

Susceptibility to radiation damage must be considered seriously if reference samples are to be calibrated for use in place of absolute systems. For the measurement of absolute (^{4}He, ^{1}H) cross sections, films of polystyrene $(CH)_n$ (which is relatively radiation hard) have been used successfully, the RBS determination of carbon providing implied quantitation for the hydrogen present in the film. For a durable laboratory reference sample, however, there is much to recommend a known ion-implanted dose of H deep within Si or SiC, where the loss of hydrogen under room temperature irradiation will be negligible.

Depth Resolution

The depth resolution of ERS is typically found to be in the 300–600 Å region (at a depth of 1000 Å in silicon), which is not as good as one might wish for interface

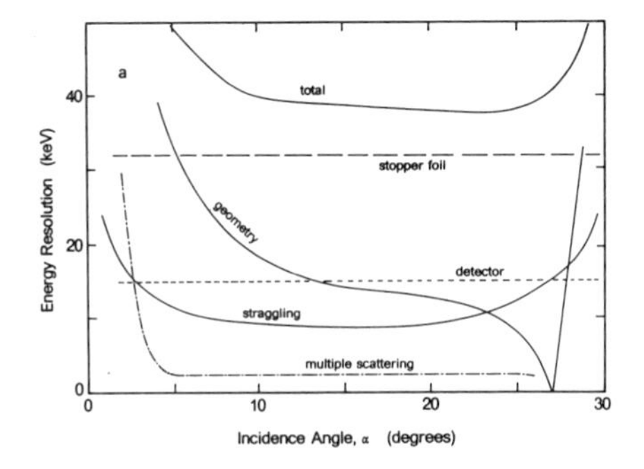

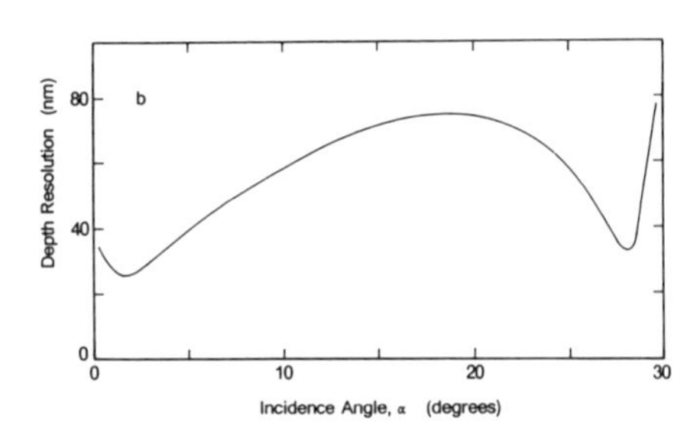

Figure 7 **(a) Major depth resolution contributions from experimental factors in ERS, as functions of sample tilt angle α, assuming $\phi = 30°$; and (b) net depth resolution calculated at a depth of 1000 Å in Si, as a function of α (adapted from ref. 14).**

and thin-film structure studies. The need to select analysis geometry and detectors that optimize resolution for a given situation must be stressed.

Figure 7a displays, for H at a depth of 1000 Å in silicon, the major contributions[14] to spectral energy resolution loss in a typical ERS configuration (recoil angle $\phi = 30°$), and their dependence on the sample tilt angle α.

Hydrogen ion energy straggling in the thick stopper foil is a large contribution. Care must be exercised to see that the foil does not contribute even more due to irregularities or pinholes. Otherwise, the only recourse is to replace the foil and detector with a cumbersome and costly electrostatic, magnetic, or time-of-flight spectrometer. This also could overcome some of the inherent limitations on the energy resolution of surface barrier detectors. It is worth noting that, due to the $\cos^2\phi$ dependence of K', the detector's aperture must limit its acceptance angle, $\Delta\phi$ to about 1°, to reduce geometrical degradation of resolution.

Figure 7 shows a contribution from ion energy straggling[3] in the sample. This, of course, is zero for near-surface layers and gets rapidly worse for layers several thousand Å deep, or for (α, ϕ) in grazing configurations.

The curve labeled *geometry* illustrates the kinematic energy spread due to the finite acceptance angle of the detector. The multiple scattering contribution arises from the spread in ion energies introduced by secondary scattering events.

Figure 7b shows the physical depth resolution resulting from the energy resolution of Figure 7a. Evidently, depth resolutions for H in silicon, using $\phi = 30°$, would be optimized by selecting a grazing geometry (α or $\beta \approx 2°$). Such a geometry presents practical problems, however. A more fruitful approach may be to accept the more convenient geometry ($\alpha = \beta$) and to control depth resolution by replacing the foil-covered detector with a magnetic spectrometer, thereby removing the largest contributor to resolution broadening.

Enhanced depth resolution could be obtained by using high-energy heavy ions (e.g., 20-MeV ^{28}Si) in place of ^{4}He, due to their higher stopping power in the sample.[4] Radiation damage problems then are much greater, however, and also a high-energy accelerator is required.

Conclusions

ERS constitutes a powerful, fast, and quantitative technique for depth profiling of hydrogen and deuterium in solids. The method is very appropriate, convenient, and fast for the quantitation of ^{1}H and ^{2}H in thin films or surface layers, in which total recoil counts lead directly to the hydrogen content in atoms/cm.[2] The result is free of solid-matrix effects, which can dominate lower energy techniques like SIMS. Its sensitivity to small amounts of hydrogen is high, and its natural identification of ^{1}H and ^{2}H makes deuterium tracer experiments easy. The depth resolution of ERS within a solid is not as good as that of SIMS, due mainly to the effect of the stopping foil and the energy resolution limits of surface barrier detectors. However, excellent concentration profiles have been obtained. While ERS is applicable in all materials that are vacuum compatible, radiation damage of samples can affect the results, especially in polymers or unstable hydrides; this problem usually can be overcome by sample rastering. ERS is easy to run concurrently with RBS, since identical apparatus is used. Such complementary runs provide a ready means for complete elemental analysis of hydrogen-containing thin film samples.

ERS is still being developed, refined, and enhanced to improve its depth resolution and absolute precision, to facilitate data reduction and to minimize sample damage. Excellent results can be achieved simply by implementing ERS with careful regard for the issues described in this article.

Related Articles in the Encyclopedia

RBS, NRA, and SIMS

References

1 J. L'Ecuyer, C. Brassard, C. Cardinal, and B. Terrault. *Nucl. Instr. and Methods.* **149,** 127, 1978.

2 B. L. Doyle and P. S. Peercy. *Appl. Phys. Lett.* **34,** 811, 1979.

3 W.-K. Chu, J. W. Mayer, and M. A. Nicolet. *Backscattering Spectrometry.* Academic Press, New York, 1978.

4 P. F. Green and B. L. Doyle. Hydrogen and Deuterium Profiling in Polymers using Light and Heavy Ion Elastic Recoil Detection. In *High Energy and Heavy Ion Beams in Materials Analysis.* (J. R. Tesmer, C. J. Maggiore, M. Nastasi, J. C. Barbour, and J. W. Mayer, eds.) Materials Research Society, Pittsburgh, 1990.

5 P. F. Green and E. J. Kramer. *J. Mat. Res.* **1,** 202, 1986.

6 J. E. E. Baglin, A. J. Kellock, M.A. Crockett and A. H. Shih, *Nucl. Instr. and Methods* **B64**, 469, 1992; see also J. Tirira, P. Trocellier, J. P. Frontier, and P. Trouslard. *Nucl. Instr. and Methods.* **B45,** 203, 1990.

7 L. S. Wielunski, R. E. Benenson, and W. A. Lanford. *Nucl. Instr. and Methods.* **218,** 120, 1983.

8 H. Cheng, Z.-Y. Zhou, F.-C. Yang, Z.-W. Xu, and Y.-H. Ren. *Nucl. Instr. and Methods.* **218,** 601, 1983.

9 M. F. C. Willemsen, A. E. T. Kuiper, L. J. van Ijzendoorn, and B. Faatz. ERD in IC Related Research. In *High Energy and Heavy Ion Beams in Materials Analysis. op cit.*

10 L. C. Feldman and J. W. Mayer. In *Fundamentals of Surface and Thin Film Analysis.* North Holland, New York, 1986, p. 61.

11 J. E. E. Baglin and J. S. Williams. High Energy Ion Scattering Spectrometry. In *Ion Beams for Materials Analysis.* (J. R. Bird and J. S. Williams, eds.) Academic Press, Sydney, 1989.

12 F. Besenbacher, I. Stensgaard, and P. Vase. *Nucl. Instr. and Methods.* **B15,** 459, 1986.

13 J. F. Ziegler. *The Stopping Power and Ranges of Ions in Matter.* Pergamon Press, New York, 1977, vols. 3 and 4.

14 A. Turos and O. Meyer. *Nucl. Instr. and Methods.* **B4,** 92, 1984.

9.3 MEIS

Medium-Energy Ion Scattering Spectrometry with Channeling and Blocking

T. GUSTAFSSON AND P. FENTER

Contents

Introduction

Medium-Energy Ion Scattering (MEIS) with channeling and blocking is a quantitative, real-space, nondestructive technique for studying the composition and structure of surfaces and buried interfaces. "Medium" energy is roughly the energy region between 50 keV and 300 keV. This region is sufficiently high so that the ion–surface interaction law is simple and well characterized—essentially only classical Rutherford scattering is involved—and sufficiently low so that the surface specificity is optimized. The basic quantities measured are the energy and angular distribution of backscattered ions. The technique derives elemental specificity from the fact that the energy of a backscattered ion is a strong function of the mass of the target atom. As the ions propagate through the sample, they also will lose smaller amounts of energy to the target electrons. As in Rutherford Backscattering (RBS), this will lead to depth sensitivity, provided that the backscattered ion energies are measured with sufficient resolution. The projectile ions are usually light—protons, alpha particles, or Li ions. In channeling, the incidence direction of the ion beam is fixed, and is aligned with a high symmetry direction of the substrate. In blocking, the angular distribution (as well as the energy distribution) of the backscattered flux is measured. This angular distribution is characterized by minima, whose positions are closely connected to the relative positions of atoms in the surface layer. The experiments are performed at high angular and energy resolution. The latter fact also makes it possible to perform very high resolution depth profiling (even on a layer-by-layer basis), an application that in coming years will find increased use in materials science. MEIS differs from low-energy ion scattering (ISS) in that the interaction law in the latter is much more complex and difficult to understand, and because ISS is essentially sensitive only to the top layer. Compared to high-energy ion scattering (conventional RBS), MEIS is more surface sensitive, and more complex instrumentally. The high depth resolution of the atomic composition in MEIS (resulting from the type of ion detector that is used) is useful for studies of all solids (crystalline, noncrystalline, metallic, semiconducting, and insulating), while the specific application of channeling and blocking is applicable only to single-crystal samples, and is used to extract highly accurate values for the structural parameters (atomic positions) of surfaces and interfaces. To date, the technique has been applied mainly to the study of metals, semiconductors, and overlayers on such surfaces (submonolayer adsorbate concentrations, thin films of silicides, etc.), but recently it has been applied to insulators as well.

The basic ideas behind channeling and blocking can be understood from Figure 1. A well collimated beam of ions is incident along a high symmetry (channeling) direction of the target (a single crystal). Most of the incident ions will propagate in the large channels between the nuclei, where they will lose energy quasicontinuously to the electrons in the target. These energy losses will be on the order of tens of eV and will be frequent, but will not lead to large angular deviations because of the enormous mismatch of the ion-electron masses. A few ions will collide with the first atom along a row of target nuclei. The energy loss is such a collision may be large ($\geq$ 1 keV) and a large angular deflection may also result. The angular distribution of the backscattered flux from the atoms in the first layer will be smooth. Due to geometrical distortions in the surface (contractions, reconstructions, etc.) or thermal vibrations, there may be a finite collision probability for deeper layers also. On their way back to the detector, electrons from the deeper layers cannot penetrate ions in layers closer to the surface. This means that the backscattered flux will be reduced in directions corresponding to a vector joining two atoms in different layers, and that the angular distributions will be marked with pronounced blocking dips, which contain direct information about the relative position of atoms in the first few layers of the crystal. Usually, the energy resolution of an experiment is not good enough to resolve the contributions from the different layers in the crystal, and the leading peak of the energy distribution (the surface peak) will contain contributions from several layers. The experimental parameters (the beam energy, the incidence direction, etc.) are usually set up such that the collision probabilities form a rapidly converging series; i.e., only three or four layers

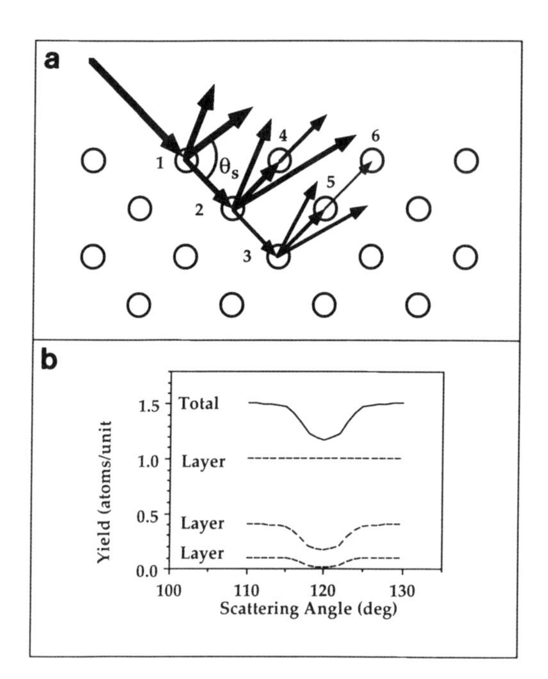

Figure 1 **Ion paths demonstrate MEIS and the effect of vibrations in a channeling and blocking experiment: (a) The widths of the arrows indicate the intensity of the ion flux at each point, and the atoms are numbered for reference to (b). (b) The angular distribution of the ions exiting the crystal, as well as the individual contributions from each layer of the crystal.**

contribute. The technique can also be used to study buried interfaces: One can easily separate the interface signal from the surface signal, say, for different lateral lattice parameters in the overlayer and the substrate (different channeling directions), and different atomic species at the interface can be distinguished through their signature backscattering energies.

We will review briefly the basic physics of ion scattering and will give a short overview of the experimental technology. We will conclude with some examples of the power of the technique. An exhaustive review of MEIS up to 1985 has been given by van der Veen; he covers both the basic principles and the results obtained.[1] More recently, Watson has made a comprehensive compilation of surface structural data obtained with MEIS and other ion scattering techniques.[2] More general introductions to ion scattering have been given by Feldman et al.[3, 4] The technique of medium-energy ion scattering originated at the FOM institute in Amsterdam, and the technical development is associated with the names of Frans Saris, Friso van der Veen, Ruud Tromp and their collaborators.

Basic Principles

Because of the high energy of the incident ions, the ion–target interaction is a series of binary events. The de Broglie wavelength of the ion is on the order of 10^{-3} Å, so that diffraction and other quantum mechanical effects are not important. By considering energy and momentum conservation, the energy loss in the collision may be calculated; it depends only upon scattering angle (θ_S) and the ratios of the ion and target masses ($\rho = m_1/m_2$). For an incident energy E_0 (and exit energy E_1) the fractional energy loss (or kinematic factor), K^2, is:

$$K^2 = \frac{E_1}{E_0} = \left[\frac{\rho\cos\theta_s + \sqrt{1-\rho^2\sin^2\theta_s}}{1+\rho}\right]^2 \tag{1}$$

The dependence on target mass makes ion scattering techniques ideal for the study of multielement systems. By increasing the incident ion mass, the energy separation between different elements becomes larger. On the other hand, radiation-induced damage becomes a more important consideration.

The amount of energy transferred to the target atom depends strongly upon the scattering angle and can be calculated directly from Equation (1). For a 100-keV beam, the energy losses are typically several keV. Such events will lead to sample damage. Due to the high velocity of the incident ions, the ion will have left the damaged region well before the recoiling target atom can cause any damage; damage avoidance therefore involves keeping the beam dose so low that the damaged region is not sampled by subsequent ions. This can be accomplished by efficient data collection (multidetection techniques), or by moving the ion beam to fresh spots on the sample and averaging the results. If the dose is not low enough beam damage, which leads to disorder, will become visible directly in the scattering spectra in the form of an increased background just behind the surface peak, and can be easily monitored in the spectra. The amount of damage has to be evaluated carefully in each experiment. Light substrates are more easily damaged than heavy ones. Many metals, fortunately, self anneal at room temperature, which facilitates analysis.

An important concept is the *shadow cone*, which is a region where no ions can penetrate due to the ion–nucleus repulsion (see Figure 2). This effect makes ion scattering surface sensitive. The size of the shadow cone R_s can be calculated for the classical Coulomb potential as:

$$R_s = \sqrt{\frac{4Z_1Z_2e^2d}{E}} \tag{2}$$

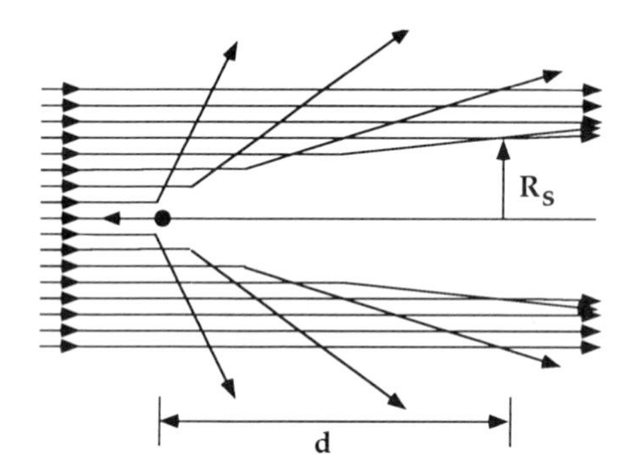

Figure 2 **Schematic of the shadow cone formed by the interaction of a parallel beam of ions with an atomic nucleus. For a static scatterer no ions can penetrate into the shadow cone.**

where R_s is measured a distance d behind the scattering center and Z_1 (Z_2) is the nuclear charge of the projectile (target atom). The shadow cone R_s is of a similar magnitude as a vibrational amplitude, and the visibility of the second layer atom will depend upon the ratio of these two numbers. In interpreting data, one also takes into account the fact that the surface vibrational amplitude is generally larger (by about a factor of 2) than the bulk amplitude, that it can be anisotropic, and that the vibrational amplitudes are correlated. Theoretical modeling of these effects rarely go beyond the Debye model.

The probability for a ion to scatter in a particular direction is determined by the ion–target interaction, and can be expressed in terms of a cross section σ_r. For a Coulomb potential, the differential cross section is the well-known Rutherford formula:

$$\sigma_r(\theta_s, E) = \left[\frac{Z_1 Z_2 e^2}{4E\sin^2\frac{\theta_s}{2}}\right]^2 g\left(\frac{M_1}{M_2}, \theta_s\right) \tag{3}$$

where $g(M_1/M_2, \theta_s)$ is due to the transformation from the center-of-mass frame to the lab frame, and is usually close to 1. Because the cross section is known on an absolute scale, one can predict the number of scattering events in a solid angle, given the number of incident ions. Conversely, one can invert this relationship and express the number of scattered ions in units of the number of atomic scatterers per surface unit cell. In practice, due to uncertainties in the exact angular acceptance and efficiency of the detector, as well as details due to the detector's geometry, one usually uses a calibrated standard, for example a known amount of Sb implanted in a Si wafer. The measured ion yield also has to be corrected for the fact that some scattered ions are neutralized at the surface.(The ion fraction P+, as it is known, is typically 0.8–0.9, which is very different from the energy range used in ISS, where P+ can be as small as 0.01.) In this way, it is possible to study surfaces in absolute units; i.e., in the number of visible atoms per surface unit cell. This is a valuable feature of the technique. As we shall see, this allows us to discriminate between different structural models, based only on a simple inspection of the data.

To quantify the interpretation of ion scattering yields, one performs Monte Carlo simulations of the scattering process. In these simulations, the ion scattering experiment is performed numerically on a computer (a SUN Sparcstation or similar machine is adequate). Since the scattering cross section is known, the simulated yields may be compared directly to the experimental yields. Other than the obvious input of charges and masses of the ion and substrate atoms, the only inputs to the simulation are the positions and vibrational amplitudes of each of the atoms in the crystal. To find the correct structure, one must vary the relevant structural and vibrational parameters until an optimal fit is found. Fortunately, there is a large amount of intuitive information in the blocking dips. Therefore, it is possible to determined the sign, as well as an estimate of the magnitude of a structural rearrangement, without doing any simulations.

To measure the goodness of fit, and to quantify the structural determination, a reliability (*R*-factor) comparison is used. In comparing the data and simulation of the experiment for many trial structures, a minimum *R* factor can be found corresponding to the optimal structure. In this way atomic positions can be determined in favorable cases to within a few hundredths of an Å, comparable to the accuracy achieved in Low-Energy Electron Diffraction (LEED).

Instrumentation

At present there are fewer than 10 laboratories worldwide using channeling and blocking for surface structural work, while the number of groups with the technical capability of doing high-resolution depth profiling is perhaps a factor of 3 larger. All of the necessary equipment is available commercially, but most groups have preferred to custom build at least a portion of it. The main drawback of MEIS is that the instrumentation is expensive even by surface science standards, and this has limited the number of workers in the field. The use of a ~100-keV ion beam implies that a small ion accelerator is needed. An ultrahigh vacuum compatible sample manipulator is needed to position the specimen to within ~0.02° along three orthogonal axes. To measure the angular distribution of the ions, it is necessary to have a detector that measures both the energy and the scattering angle of the ions with high precision. Multidetection schemes are useful to minimize data accumulation times and beam-induced sample damage. The ion energy analysis is usually done by a commercially available toroidal high-resolution ion energy analyzer that

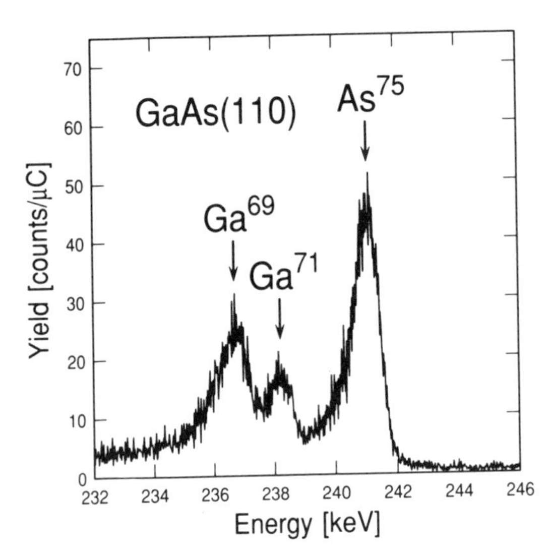

Figure 3 **Backscattering spectrum from GaAs (110), obtained with a 300-keV Li ion beam at a scattering angle of 85°.**

is free to move in the plane defined by the ion incidence direction and the surface normal. By determining the angular position at which the ion strikes the detector one recovers information about the angular distribution of the scattered ions; the radial position gives information about the energy. The energy resolution ΔE of the toroidal analyzer is determined primarily by the size of the beam spot on the sample and the size of the entrance slit. A total energy resolution (detector + ion beam width) ΔE of 150 eV at 100-keV primary energy is easily obtained. This is to be compared to the energy resolution of a conventional surface barrier detector (used in RBS), which can be ~10 keV at 1 MeV.

As an example, we show in Figure 3 a backscattering spectrum from GaAs (110), obtained with a 300-keV Li ion beam.[5] This is a well-chosen test example of energy resolution, as the atomic numbers of the two constituents are quite close (31 and 33 for Ga and As, respectively). Not only are these two species well resolved, but the two common isotopes of Ga are also well separated. Note that the peaks are asymmetric due to contributions from lower layers. Resolving power of this kind surely will find many new applications in materials science.

The main limitation to the accuracy of MEIS comes from systematic errors involving uncertainties in the vibrational modeling, the scattering cross section, and approximations in the ion scattering simulation code. All of these sources conspire to make a structural measurement of the complicated, highly distorted structures of heavy elements the most uncertain. On the other hand, the uncertainty in a measurement of the simple surface of a light element will approach 1% due to the angular resolution. It is difficult to estimate the magnitude of the systematic error

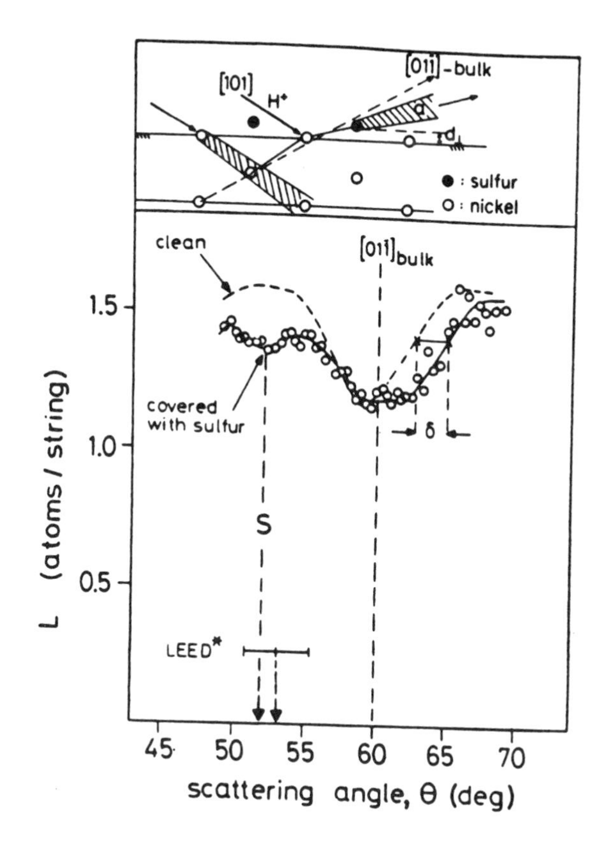

Figure 4 **Angular distribution of backscattered protons from clean and S-covered Ni(110). The top part of the figure shows the scattering geometry. The primary ion energy was 101 keV.**

in any given set of data. A typical MEIS experiment relies upon the analysis of many sets of data that overlap in sensitivity to a given structural parameter. The self-consistency of the analysis provides a direct measure of the magnitude of the systematic error. In several cases—for example, the (110) surfaces of Cu, Ni, Pt, Au, and III-V compound semiconductors like GaAs and InSb—both LEED and MEIS have been used to determine structural parameters with excellent mutual agreement and comparable accuracy.

Examples and Applications

We will illustrate the power of MEIS with three simple examples. In addition, we remind the reader of the existence of extensive reviews,[1, 2] and in particular would like to mention some quite recent, beautiful work on the melting of single-crystal surfaces.[6]

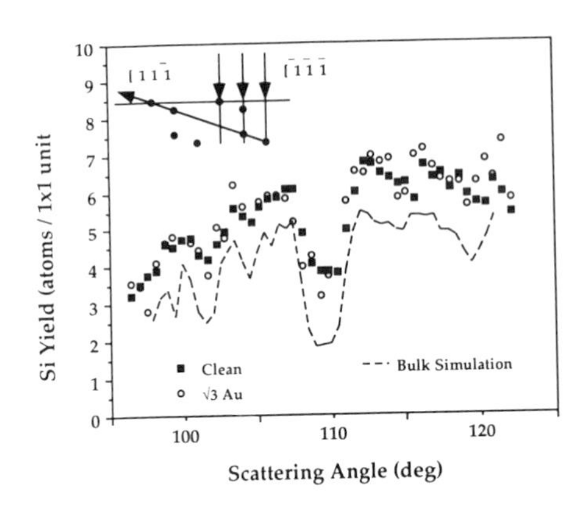

Figure 5 **Si backscattering yields (angular scans) for normal incidence on the Si (111) (7 × 7) surface (solid squares) and the Si (111) ($\sqrt{3} \times \sqrt{3}$) R30°-Au surface (open circles). The curve is the expected yield from a bulk terminated Si (111) surface. The scattering geometry is shown in the inset.**

S on Ni (110)

In Figure 4 we show MEIS data in the scattering geometry indicated for clean and sulphur-covered Ni (110).[7] For the clean surface, we observe a pronounced blocking dip at ~60°. The clean surface has a (1 × 1) LEED pattern, which means that the periodicity of the surface is that expected based on an extrapolation of the bulk geometry. However, the lattice spacing perpendicular to the surface may differ from that of the bulk. If the separation between the two outermost planes were unchanged, the blocking dip in Figure 4 would be observed at an angle of 60°. Clearly, the data are shifted towards smaller scattering angles, indicating a contraction of this spacing. By adsorbing 0.5 monolayers of sulphur on Ni (110), a (2 × 2) supercell is formed. The angular distribution of the Ni flux from this structure is also shown in the figure. One observes immediately that the dip is shifted to a larger scattering angle, indicating that the outermost Ni layer has now moved out past the bulk-like position and that the lattice is now expanded (a detailed numerical evaluation of the data show that the expansion is ~6%). In addition, a slight blocking dip is now observed around 53°. This dip corresponds to blocking of the outgoing Ni flux by the sulphur adatoms. The small size of this dip is due to the low concentration of the sulphur atoms and the fact that the light sulphur atoms are less efficient blockers than the heavier Ni atoms. The position of the dip allows us to determine the height of the sulphur atoms over the substrate (0.87 ± 0.03 Å).

Si (111) ($\sqrt{3} \times \sqrt{3}$) R30° Au

Gold is an example of a metal that does not form a silicide, and one may therefore expect the Au/Si interface to be abrupt. The $\sqrt{3}$ structures of Au on Si (111) are interesting in that the unit cell is much smaller than that of the well known (7 × 7)

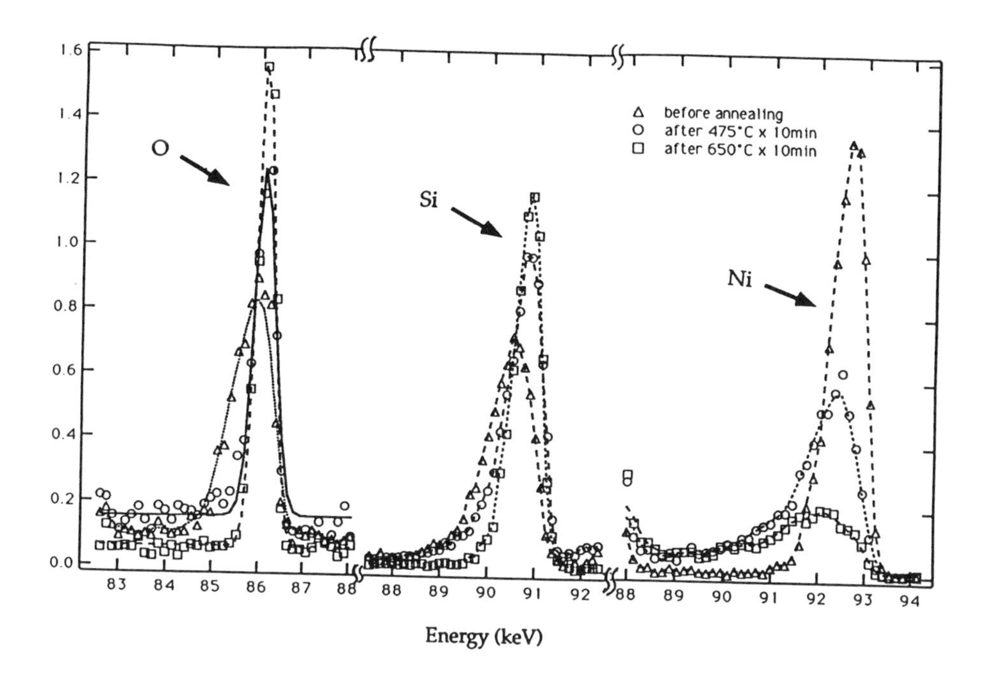

Figure 6 **Backscattering spectra for a thin film of Ni deposited on an amorphous SiO_2 film grown on top of Si (111) for three different annealing temperatures.**

structure of the clean surface. It might then seem plausible that this structure corresponds to a gentle modulation of an ideal (1 × 1) structure, with the Au atoms presumably close to Si lattice sites. Many drastically different models have been proposed for this structure; all are based implicitly on such assumptions. In Figure 5, we show MEIS data for the clean Si (111) (7 × 7) surface and for the ($\sqrt{3} \times \sqrt{3}$) R30° surface.[8] In addition, we show a computer simulation for what would be expected for an ideally terminated Si (111) (1 × 1) surface. Surprisingly, the two experimental spectra are rather similar and differ quite significantly from the calculated result. We find that Si atoms in more than one monolayer are displaced away from their lattice sites. The Au atoms do not block the outgoing Si flux. These conclusions, which are quite model independent, show that the Si lattice is severely distorted and that the Au atoms do not sit in Si lattice sites. The conclusions provide useful general constraints that more detailed models must obey. A more detailed analysis, based on Monte Carlo simulations for different trial structures, is necessary to establish a detailed structure. This shows that the structure most likely involves three Au atoms per unit cell, arranged in a trimer on a Si substrate, where the top half of the Si double layer is missing.

Ni/ SiO_2/ Si (111)

In Figure 6, we show MEIS energy spectra (for a fixed collection direction) for a thin film (initially some 6 monolayers) of Ni, deposited on top a thin film of SiO_2 grown on Si (111). As the three different atoms involved have widely differing masses, the signals from the three species are well resolved. The area under each peak is proportional to the concentration of each species. From the Ni peak, one can see that as the sample is annealed, the Ni starts diffusing into the bulk (the peak gets more and more asymmetric). The total concentration in the near surface region also decreases; evidently the diffusion into the bulk is quite rapid. The leading part of the Si peak falls initially at the energy of the clean Si (111) surface; the implication is that the surface is not completely covered by Ni, but that bare patches of SiO_2 remain. After annealing, the Si peak and the O peak move towards higher energies; this is consistent with less and less of the surface being covered by metal.

Conclusions

MEIS has proven to be a powerful and intuitive tool for the study of the composition and geometrical structure of surfaces and interfaces several layers below a surface. The fact that the technique is truly quantitative is all but unique in surface science. The use of very high resolution depth profiling, made possible by the high-resolution energy detectors in MEIS, will find increased applicability in many areas of materials science. With continued technical development, resulting in less costly instrumentation, the technique should become of even wider importance in the years to come.

This work was supported in part by National Science Foundation (NSF) Grant No. DMR-90-19868.

Related Articles in the Encyclopedia

RBS and ISS

References

1 J. F. van der Veen. *Surf. Sci. Rep.* **5,** 199, 1985. Basic principles of MEIS, with many results of structure determinations.

2 P. R. Watson. *J. Phys. Chem. Ref. Data.* **19,** 85, 1990. Compilation of structural data attained by MEIS and other ion scattering techniques.

3 L. C. Feldman, J. W. Mayer, and S. T. Picraux. *Materials Analysis by Ion Channeling.* Academic, New York, 1982. General introduction to ion scattering.

4 L. C. Feldman. *Crit. Rev. Solid State and Mat. Sci.* **10,** 143, 1981.

5 M. Copel and R. M. Tromp. Private communication.

6 J. F. van der Veen, B. Pluis, and A. W. van der Gon. In: *Chemistry and Physics of Solid Surfaces VII, Springer Series in Surface Sciences, Volume 10.* (R. Vanselow and R. F. Howe, eds.) Springer, Heidelberg, 1988, p. 455 and references therein.

7 J. F. van der Veen, R. M. Tromp, R. G. Smeenk, and F. Saris. *Surface Sci.* **82,** 468, 1979.

8 M. Chester and T. Gustafsson. *Phys. Rev. B.* **42,** 9233, 1990.

9.4 ISS

Ion Scattering Spectroscopy

GENE R. SPARROW

Contents

- Introduction
- Basic Principles
- Quantitation
- Advantages and Disadvantages
- Applications
- Conclusions

Introduction

Ion Scattering Spectroscopy (ISS) is one of the most powerful and practical methods of surface analysis available. However, it is underutilized due to a lack of understanding about its application and capabilities. This stems from its history, the limited number of high-performance instruments manufactured, and the small number of experienced surface scientists who have actually used ISS in extensive applications. Ironically, it is one of the easiest and most convenient surface analytical instruments to use and it provides useful information for almost any type of solid material.

The most useful application of ISS is in the detection and identification of surface contamination, which is one of the major causes of product failures and problems in product development. The surface composition of a solid material is almost always different than its bulk. Therefore, surface chemistry is usually the study of unknown surfaces of solid materials. To better understand the concept of "surface analysis," which is used very loosely among many scientists, we must first establish a definition for that term. This is particularly important when considering ISS because of its extreme sensitivity to the surface. In most applications *surface analysis* implies the analysis of a finite thickness or depth of the outermost layers of a material, generally from the outer few atomic layers to a depth of 100–200 Å. Techniques encompassing layers greater than that are better described as thin-film analyses, or as depth profiles directed at obtaining other information. Techniques like Energy-Dispersive X-Ray Spectroscopy (EDS) and FTIR with ATR (Attenuated Total Reflection) generally do not fit the description of surface analysis. Other techniques, such as Auger and ESCA, meet the definition by obtaining spectra that originate from a depth of up to approximately 50–80 Å.

ISS is the most surface sensitive technique known. It is routinely sensitive to the outermost layer of atoms. At this level of depth sensitivity, it can be shown by ISS that most practical solid materials have the same outer atomic layer, i.e., a layer of surface water molecules, or organic material, with the hydrogen oriented upward. Therefore in ISS, as in SIMS using low-energy ions, it is important to include spectra from several different sputtered depths into the surface or to specify the sputtered depth from which the spectrum was obtained. Usually a series of ISS spectra are obtained at successively greater depths into the surface and the resulting spectra are displayed to show the changing composition versus depth. Because of the extreme surface sensitivity of ISS, these depth profiles offer details about changes in surface composition in the outer 50 Å that are generally not obtainable by other techniques. These details are extremely important in many applications, such as the initiation of corrosion, adhesion, bonding, thin-film coatings, lubrication, and electrical contact resistance. Typical data and applications will be discussed.

History

Earlier studies of ion scattering were directed primarily at gas–ion interactions. As studies of ion–solid surfaces became common the energy of the scattered ions was eventually related mathematically to a simple binary elastic event involving a single atom on a surface element and a single probe ion.

The practical use of ion scattering was not developed until the late 1960s when David P. Smith of 3M Company first reported the use of low-energy inert ion scattering to analyze the composition of surfaces. This early pioneering work established ion scattering as a very useful and viable spectroscopy for studying surfaces. The first studies and instruments consisted of simple systems where the ion beam scattered through an angle of 90°; thus accepting only a small solid angle of the signal. Modern systems use ion beams that are coaxial with the detector and exhibit orders of magnitude higher sensitivity. These devices make use of a Cylindrical Mirror Analyzer (CMA) and include detection of ions scattered about a 360° solid angle. A typical device is shown in Figure 1. ISS has since become readily available commercially and is recognized as one of the four major surface techniques, generally including ESCA (XPS), Auger, and SIMS as well.

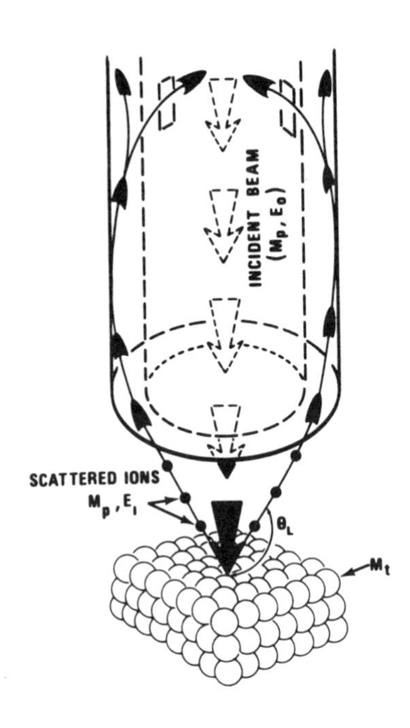

Figure 1 **Schematic of CMA ISS device showing primary ion beam, analyzer, and scattering at 138°.**

Basic Principles

ISS is relatively simple in principle and application. When a low-energy (100–5000 eV) beam of positive ions of some inert element, such as He, Ne, or Ar, strikes a surface, some of the ions are reflected back from the surface. This scattering process involves a single surface atom and a single incident ion. It is, therefore, a simple binary elastic collision that follows all the rules of classical physics. The incident ion scatters back with a loss of energy that depends only on the mass of the surface atom (element) with which the collision occurred. The heavier the surface atom, the smaller the change in energy of the scattered ion. Thus carbon, which is a light atom of mass 12, is readily displaced and the probe ion loses most of its energy, whereas a heavy atom like Pb, having mass 208, is not easily moved. An ion scattering from Pb retains most of its incoming energy. To obtain a spectrum, one merely records the number of scattered ions as their energy is scanned from near 0 eV to the energy of the primary incoming beam. Each element has a unique mass and therefore a unique energy at which the probe ion scatters. The energy of the scattered ion is mathematically related to the mass of the surface atom by the following equation:

$$\frac{E_1}{E_0} = E_r = \frac{M_1^2}{(M_1 + M_2)^2}\left(\cos\theta + \sqrt{\left(\frac{M_2}{M_1}\right)^2 - (\sin\theta)^2}\right)^2$$

where E_0 is the energy of the incident probe ion, E_1 is the energy of the ion scattered from surface atom, E_r is the ratio of the energies of the scattered and probe ions, M_1 is the mass of the primary ion, M_2 is the mass of the surface atom, and θ is the scattering angle measured from the direction of the ion beam.

Penetration of the incident beam below the very outermost atomic layer causes excessive and nondiscrete loss of energy such that the scattered ions do not yield sharp, discrete peaks. Only ions scattered from the outer atomic layer of a surface give rise to a sharp peak. ISS is therefore extremely sensitive to the surface and essentially detects only the outermost surface layer. To obtain more extensive surface information, it is therefore common to continuously monitor the ISS spectrum while sputtering into the surface. When the sputtering is done very slowly using a light atom, such as isotopically pure $^3He^+$, complete spectra can be obtained at successively greater depths into the surface. In routine practice, sputter rates on the order of about 1 to 5 Å per minute are used and approximately 15–20 ISS spectra are obtained throughout a sputtered depth of about 100 Å. Since the most important information is obtained near the surface, the majority of these spectra are obtained in the first few minutes of sputtering.

As the scattering angle θ is decreased to 90°, the physical size of the CMA must increase, until finally one cannot use a CMA but must resort to a sector analyzer. This decreases detection sensitivity by 2–3 orders of magnitude, increases multiple scattering at energies above the primary peaks, and requires much more precise positioning of the sample. Changing the mass of the primary ion beam gas controls not only the sputtering rate of the surface but also changes the spectral resolution and detection sensitivity. For example, using $^3He^+$ permits good detection of C, N, and O, whereas using $^4He^+$ does not. Using Ar^+ provides high sputtering rates for deeper profiles but does not permit the detection of elements having mass less than Ca. Argon also provides increased spectral resolution for higher elements not resolved by He. It is common to sometimes mix Ar and He to detect all elements while obtaining a high sputtering rate. Increasing the energy of the primary beam to above about 3000 eV dramatically increases the overall spectral background, thus decreasing sensitivity, but the spectral resolution increases. Decreasing the beam energy decreases this background and dramatically decreases the sputtering rate. It is possible to obtain useful ISS spectra at energies below 200 eV of He at less than a few nA. The sputtering rate under these conditions is extremely low.

During normal operation, the entire ISS spectrum, covering all elements, is scanned in about 1 second. A number of these scans are then added for signal enhancement and to control the predetermined depth to which sputtering is

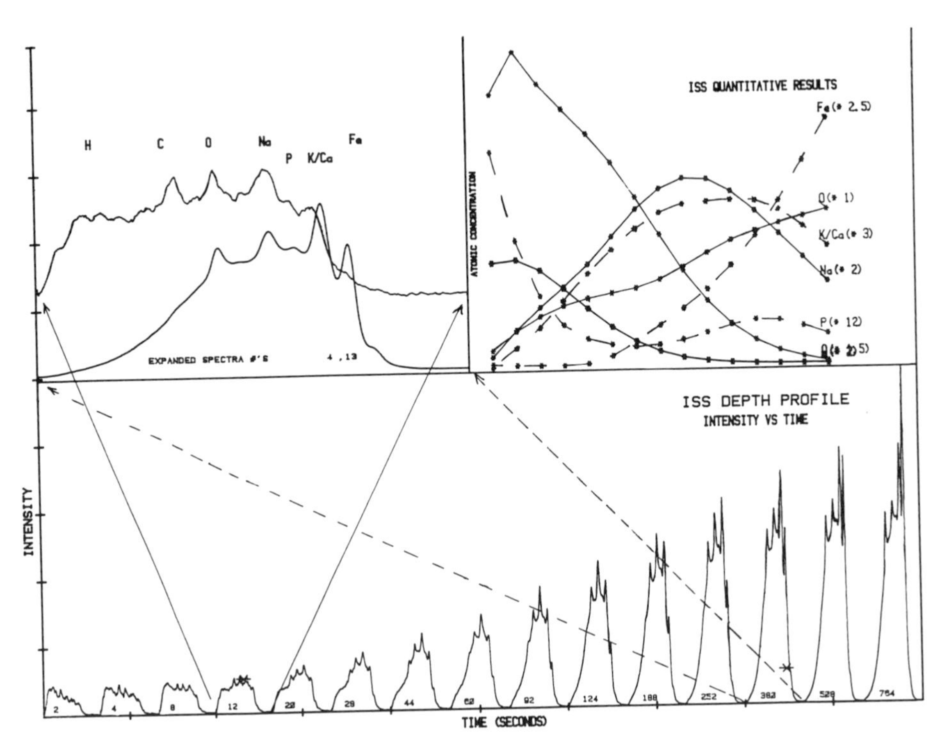

Figure 2 Typical ISS spectral data obtained from routine depth profile of cleaned washed steel. Left spectrum represents surface. Right spectrum represents about 50-Å depth. Expansions shown top left. Relative atomic compositions plotted top right.

desired. Since the greatest amount of surface information is usually obtained from the outermost layer, the initial spectra are obtained by restricting the accumulations to only 2 scans. At greater depths, many scans may be averaged to obtain one spectrum. When all of these spectra are normalized and plotted on a single figure, a depth profile is obtained that illustrates in great detail how the surface composition changes with depth. Figure 2 illustrates typical data obtained from such a plot. The spectrum on the extreme bottom left illustrates the composition at the outermost surface, whereas the spectrum on the extreme bottom right illustrates composition after sputtering to about 50 Å. In general, many of the compositional changes observed in the first few spectra are not detected in either AES or ESCA. In this particular sample, for example, changes in C, O, and Fe were detected in Auger profiles when sputtering to about 30 Å, but Ca was detected only in trace levels and H, Na, P, and Ba were not detected. The concentration of Fe at the surface was shown to be about 28% by AES (within the finite depth probed), whereas ISS (the first trace) clearly shows Fe is completely covered by contaminants. In the same figure, two spectra (numbers 4 and 13 in the profiled series) have been expanded in the

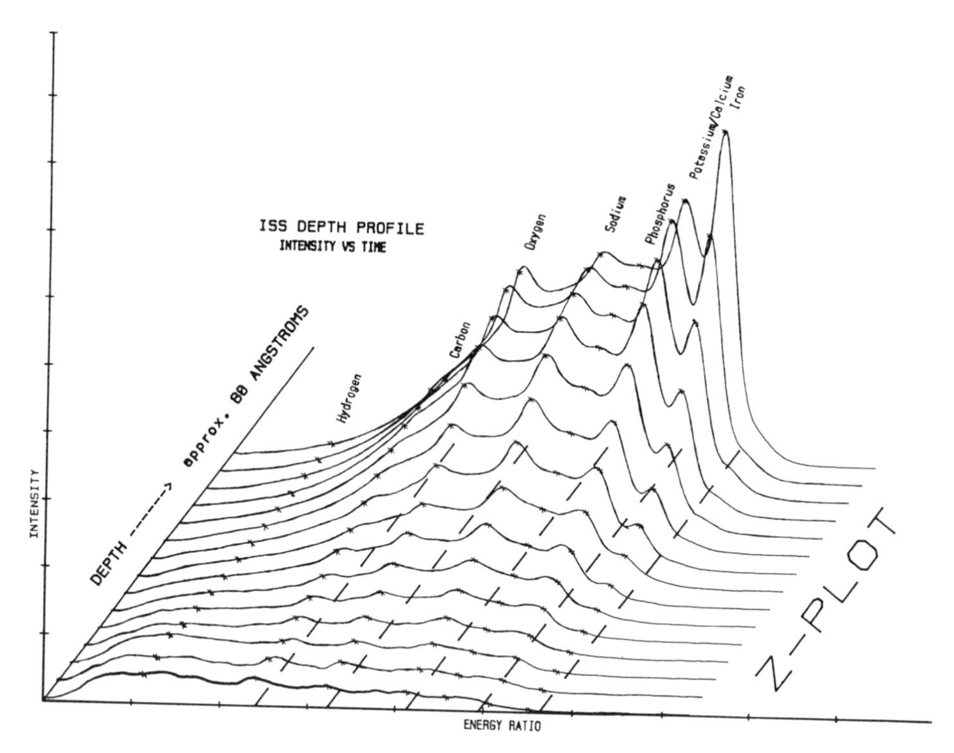

Figure 3 ***Z*-plot of ISS data shown in Figure 2. Each spectrum represents the composition of the surface at a different cross section in depth.**

upper lefthand corner to show the changes in detail, and actual atomic concentrations of the elements detected are shown in the upper righthand plot.

When all of the ISS spectra are plotted in a three-dimensional manner, such as the "*z*- plot" shown in Figure 3, the changes in surface composition with depth are much more obvious. In this figure, each spectrum represents the composition at a different cross section of the total depth sputtered, hence the spectra are plotted at different depths. Note that the spectra are not recorded at identical incremental depths.

Quantitation

ISS involves simple principles of classical physics and is one of the simplest spectroscopy for quantitative calculations. Under most standard instrumental operating conditions there is essentially no dependency on the chemical bonding or matrix of the sample. Several workers[1–6] have discussed quantitative aspects of ISS and elemental relative sensitivities. These have been compiled[7] with comparative measurements of sensitivity obtained from several different laboratories and are shown in

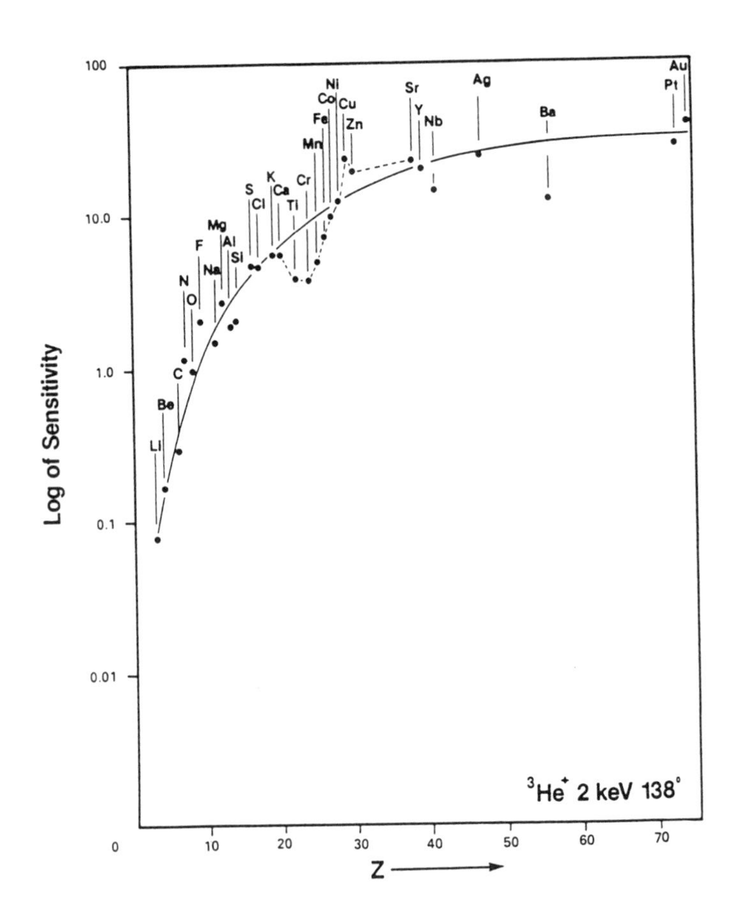

Figure 4 **Relative elemental sensitivities for ISS scattering using $^3He^+$ at 2000 eV.**

Figure 4 for $^3He^+$ scattering. In general, the precision of ISS is extremely high under routine conditions and can approach well under 1% relative for many measurements. When used with appropriate standards, it can provide very accurate results. This makes it extremely useful for comparisons of metal and oxygen levels, for example.

Several features of ISS quantitative analysis should be noted. First of all, the relative sensitivities for the elements increase monotonically with mass. Essentially none of the other surface spectroscopies exhibit this simplicity. Because of this simple relationship, it is possible to mathematically manipulate the entire ISS spectrum such that the signal intensity is a direct quantitative representation of the surface. This is illustrated in Figure 5, which shows a depth profile of clean electrical connector pins. Atomic concentration can be read roughly as atomic percent directly from the approximate scale at the left.

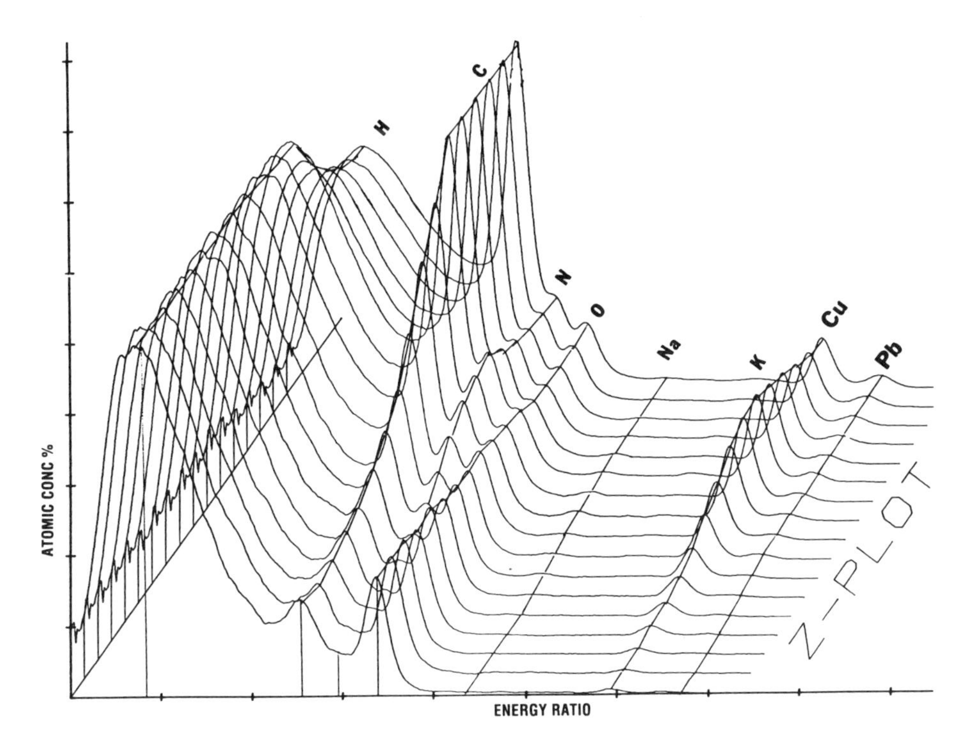

Figure 5 ***Z*-plot of polyamide delaminated from Cu metal film. Entire spectra have been mathematically treated to adjust for detector response versus energy. Each spectrum represents the composition at a different depth. Peak height can be read roughly as atomic concentrations at left.**

In addition to its precision and simplicity, ISS is also the only technique that can be used for quantitative analysis of hydrogen within the outer surface of material. Although SIMS can detect hydrogen, it is extremely difficult to quantitate it on the outer surface. Unlike detection of all other elements, the detection of hydrogen in ISS does not involve scattering from hydrogen but rather the detection of sputtered hydrogen, which passes through the detector and is detected at low energies in the spectrum. Through use of appropriate references, such as polymers, quantitative analysis has become possible. Even extremely small changes in hydrogen content, such as from differences in adsorbed water, are detectable. This makes ISS extremely valuable for the analysis of polymer surfaces.

There are two major drawbacks to ISS concerning quantitative analysis. First, it has very low spectral resolution. Thus it is very difficult either to identify or resolve many common adjacent elements, such as Al/SI, K/Ca, and Cu/Zn. If the elements of interest are sufficiently high in mass, this can be partially controlled by using a probe gas with a higher atomic mass, such as Ne or Ar. Second, ISS has an inherently high spectral background which often makes it difficult to determine true peak intensity. However, modern computer techniques provide significant ways to minimize these problems and quantitative results are obtained routinely. The relative detection sensitivity of ISS varies considerably depending on the type of sample and its composition. In general, the sensitivity can be as good as 20–50 ppm for a high-mass component, such as Pb in a low-mass substrate like Si, or as poor as a few percent, such as for C in a low-mass substrate like Al.

Advantages and Disadvantages

The most important features of ISS are its extreme speed—less than 0.5 s to obtain a single spectrum—and its extreme sensitivity to the outer surface. The speed is directly related to the high detection sensitivity of ISS, which can be well in excess of 10,000 counts of signal per nA (cps/nA) of ion beam signal for Ag. Other important features of ISS are that it is extremely simple in principle, operation, and instrumentation. The data presentation are extremely simple, exhibiting little noise and high precision and reproducibility. It is easily applied to nearly any material and is especially useful for the analysis of polymers or interfacial failures. ISS is normally very cost-effective, with pricing of instruments being very low and instrument size being small. Experimental set-up, data collection, and data manipulation are relatively simple.

Extreme sensitivity to the outer surface is the most useful advantage of ISS. It is unexcelled in this respect and has the unique capability to detect only the outermost atomic layer without signal dilution from many additional underlying layers. No other technique, including static SIMS or angle-resolved XPS, can detect only the outermost atomic layer. ISS is also very fast and sensitive, so that even very low level impurities within the outer few Å can be detected. Other very important advantages are the speed of depth profiling and the extreme detail one can obtain about the changes in chemical composition within the outer surface, especially the first 50–100 Å (i.e., the high depth resolution owing to sensitivity to the first atomic layer). The indirect detection of hydrogen also has proven extremely applicable to studies of polymers and other materials containing surface hydrogen in any form. This has been especially valuable in applications involving plasmas and corona treatments of polymers. ISS is routinely applicable to the analysis of insulators and irregularly shaped samples. In some research and development applications its ability to detect certain isotopes, such as O^{18}, are especially important. Quantitative analysis is also advantageous, since ISS does not miss elements that are often overlooked in other spectroscopies due to poor sensitivity (such as H, the alkalis, and the noble metals), and quantitative calculations are not affected by the matrix. In addition these relative sensitivities do not vary as dramatically as in some other spectroscopies and they are uniformly increasing with the mass of the elements.

One of the major disadvantages of ISS is its low spatial resolution. In most of the current systems, this is limited to about 120 μm because of limits on ion-beam diameter, although some work has been reported on ISS using an ion-beam diameter of about 5 μm. However, as the ion-beam diameter decreases, its energy normally increases, and this results in undesirable increases in the overall background of the spectrum. Another serious disadvantage of ISS is its low spectral resolution. Usually, this resolution is limited to about 4–5% of the mass of the detected element; hence it is very difficult to resolve unequivocally adjacent elements, especially at high mass. Although the spectral resolution can be improved to about 2% with instrument modifications or by computer deconvolution, this problem cannot be totally resolved. ISS also does not provide any information concerning the nature of chemical bonding, although a special technique called Resonance Charge Exchange (RCE)[8, 9] offers information about some elements. Ironically, the extreme surface sensitivity of ISS can become a disadvantage due to the "moving front" along which depth profiling can occur. For example, heavy surface atoms often are retained along this outer atomic layer during sputtering and are thus detected at levels far above what is representative of deeper layers in a thick film. Another key disadvantage is the technique's low sensitivity to certain important elements, such as N, P, S, and Cl, which are often more easily detected by AES or ESCA.

Typical Applications

Polymers and Adhesives

Applications of ISS to polymer analysis can provide some extremely useful and unique information that cannot be obtained by other means. This makes it extremely complementary to use ISS with other techniques, such as XPS and static SIMS. Some particularly important applications include the analysis of oxidation or degradation of polymers, adhesive failures, delaminations, silicone contamination, discolorations, and contamination by both organic or inorganic materials within the very outer layers of a sample. XPS and static SIMS are extremely complementary when used in these studies, although these contaminants often are undetected by XPS and too complex because of interferences in SIMS. The concentration, and especially the thickness, of these thin surface layers has been found to have profound affects on adhesion. Besides problems in adhesion, ISS has proven very useful in studies related to printing operations, which are extremely sensitive to surface chemistry in the very outer layers.

Metals

Perhaps the most useful application of ISS stems from its ability to monitor very precisely the concentration and thickness of contaminants on metals during development of optimum processing and cleaning operations. One particularly important application involves quantitatively monitoring total carbon on cleaned steels before paint coating. This has been useful in helping to develop optimum bond strength, as well as improved corrosion resistance. Other very common applications of ISS to metals include the detection of undesirable contaminants on electrical contacts or leads and accurate measurements of their oxide thickness. These factors can lead to disbonding, corrosion, tarnish, poor solderability, and electronic switch failures.

Ceramics

Two capabilities of ISS are important in applications to the analysis of ceramics. One of these is its surface sensitivity. Many catalyst systems use ceramics where the surface chemistry of the outer 50 Å or less is extremely important to performance. Comparing the ratio of H and O to Al or Si is equally important for many systems involving bonding operations, such as ceramic detectors, thin films, and hydroxyapatite for medical purposes.

Conclusions

ISS is too frequently thought of as being useful only for the analysis of the outer atomic layer. It is a powerful technique that should be considered strongly for nearly any application involving surface analysis. It is easy to use and displays results about the details of surface composition in a very simple, quantitative manner. It is relatively quick and inexpensive and extremely sensitive to changes and contamination in the outer surface, which is not as readily investigated by AES or ESCA. It has very high sensitivity to metals, especially in polymers or ceramics, and is applicable to virtually any solid, although its poor spectral resolution often make it difficult to distinguish adjacent masses. Future trends will most likely result in making ISS much more common than it presently is and instrumental developments will most likely include much improved spectral resolution and spatial resolution, as well as sensitivity. Computer software improvements will increase its speed and precision even further, and incorporate such things as peak deconvolution, database management, and sputtering rate corrections. Commercial instruments and analytical testing with excellent computer software and interfacing are readily available. As with all techniques, ISS is best used in conjunction with another technique, especially SIMS or ESCA. Further reading on the principles of ISS and some applications can be found in references 10 and 11.

Related Articles in the Encyclopedia

SIMS, XPS, AES, and RBS

References

1 H. Niehus and E. Bauer. *Surface Sci.* **47,** 222, 1975.
2 E.Tagluner and W. Heiland. *Surface Sci.* **47,** 234, 1975.

3 H. H.Brongersma and T. M. Buck. *Nucl. Instr. Meth.* **132,** 559, 1976.

4 M. A. Wheeler. *Anal. Chem.* **47,** 146, 1975.

5 E. N. Haussler. *Surf. Interface Anal.* 1979.

6 G. C. Nelson. *Anal. Chem.* **46,** (13) 2046, 1974.

7 G. R. Sparrow. *Relative Sensitivities for ISS.* Available from Advanced R & D, 245 E. 6th St., St. Paul, MN 55010.

8 T. W. Rusch and R. L. Erickson. Energy Dependence of Scattered Ion Yields in ISS. *J. Vac. Sci. Technol.* **13,** 374, 1976.

9 D. L. Christensen, V. G. Mossoti, T. W. Rusch, and R. L. Erickson. *Chem. Phys. Lett.* **44,** 8, 1976.

10 W. Heiland. *Electron Fisc. Applic.* 17, 1974. Covers further basic principles of ISS.

11 D.P. Smith. *Surface Sci.* 25, 171, 1971.

10

MASS AND OPTICAL SPECTROSCOPIES

10.0 INTRODUCTION

The analytical techniques covered in this chapter are typically used to measure trace-level elemental or molecular contaminants or dopants on surfaces, in thin films or bulk materials, or at interfaces. Several are also capable of providing quantitative measurements of major and minor components, though other analytical techniques, such as XRF, RBS, and EPMA, are more commonly used because of their better accuracy and reproducibility. Eight of the analytical techniques covered in this chapter use mass spectrometry to detect the trace-level components, while the ninth uses optical emission. All the techniques are destructive, involving the removal of some material from the sample, but many different methods are employed to remove material and introduce it into the analyzer.

In Dynamic Secondary Ion Mass Spectrometry (SIMS), a focused ion beam is used to sputter material from a specific location on a solid surface in the form of neutral and ionized atoms and molecules. The ions are then accelerated into a mass spectrometer and separated according to their mass-to-charge ratios. Several kinds of mass spectrometers and instrument configurations are used, depending upon the type of materials analyzed and the desired results.

The most common application of dynamic SIMS is depth profiling elemental dopants and contaminants in materials at trace levels in areas as small as 10 μm in diameter. SIMS provides little or no chemical or molecular information because of the violent sputtering process. SIMS provides a measurement of the elemental impurity as a function of depth with detection limits in the ppm–ppt range. Quantification requires the use of standards and is complicated by changes in the chemistry of the sample in surface and interface regions (matrix effects). Therefore, SIMS is almost never used to quantitatively analyze materials for which standards have not been carefully prepared. The depth resolution of SIMS is typically between 20 Å and 300 Å, and depends upon the analytical conditions and the sample type. SIMS is also used to measure bulk impurities (no depth resolution) in a variety of materials with detection limits in the ppb–ppt range.

By using a focused ion beam or an imaging mass spectrometer, SIMS can be used to image the lateral distribution of impurities in metal grain boundaries, biological materials (including individual cells), rocks and minerals, and semiconductors. The imaging resolution of SIMS is typically 1 μm, but can be as good as 10 nm.

Static SIMS is similar to dynamic SIMS but employs a much less intense primary ion beam to sputter the surface, such that material is removed from only the top monolayer of the sample. Because of the less violent sputtering process used during static SIMS, the chemical integrity of the surface is maintained during analysis such that whole molecules or characteristic fragment ions are removed from the surface and measured in the mass spectrometer. Measured molecular and fragment ions are used to provide a chemical rather than elemental characterization of the true surface. Static SIMS is often used in conjunction with X-Ray Photoelectron Spectroscopy (XPS), which provides chemical bonding information. The bonding information from XPS (see Chapter 5), combined with the mass spectrum from static SIMS, can often yield a complete picture of the molecular composition of the sample surface.

Static SIMS is labeled a trace analytical technique because of the very small volume of material (top monolayer) on which the analysis is performed. Static SIMS can also be used to perform chemical mapping by measuring characteristic molecules and fragment ions in imaging mode. Unlike dynamic SIMS, static SIMS is not used to depth profile or to measure elemental impurities at trace levels.

In Surface Analysis by Laser Ionization (SALI) ionized and neutral atoms are sputtered from the sample surface, typically using an ion beam (like SIMS) or a laser beam (like LIMS). However, SALI employs a high-intensity laser that passes parallel and close to the surface of the sample, interacting with the sputtered secondary ions and neutrals to enhance the ionization of the neutrals, which are then detected in the mass spectrometer. SALI, in the single-photon ionization mode (low-intensity laser), can be used to provide a chemical rather than elemental characterization of the true surface, like static SIMS. While in multiphoton ionization mode (intense laser), it can be used to provide enhanced sensitivity and improved quantification over dynamic SIMS in certain applications. Improved quantification is possible because ionization of the sputtered neutral atoms occurs above the surface, separate from the sputtering process, eliminating difficulties encountered during quantification of SIMS—especially at surfaces and interfaces. SALI can also be used in conjunction with other analytical techniques, such as LIMS, in which a laser is used to desorb material from the surface. Like static and dynamic SIMS, SALI can be used to map the distribution of molecular (organic) or elemental impurities.

In Sputtered Neutral Mass Spectrometry (SNMS), atoms are removed from the sample surface by energetic ion bombardment from an RF argon plasma (not an ion beam). Sputtered atoms are then ionized in the RF plasma and measured in a mass spectrometer. SNMS is used to provide accurate trace-level, major, and minor concentration depth profiles through chemically complex thin-film structures, including interfaces, with excellent depth resolution. Because ionization is separate from the sputtering process (unlike SIMS), semiquantitative analyses, through interfaces and multilayered samples, may be performed without standards; improved accuracy (±5–30%) is possible using standards. One of the primary advantages of SNMS over other depth profiling techniques is the extremely good depth resolution (as good as 10 Å) achievable in controlled cases. The detection limits of SNMS are limited to the 10 ppm–pph range. The analytical area of an SNMS depth profile is typically 5 mm in diameter, rendering analysis of small areas impossible, while providing a more "representative" sampling of inhomogeneous materials.

Laser Ionization Mass Spectrometry (LIMS) is similar to SIMS, except that a laser beam, rather than an ion beam, is used to remove and ionize material from a small area (1–5 μm in diameter) of the sample surface. By using a high-intensity laser pulse, the elemental composition of the area is measured, by using a low-intensity pulse, organic molecules and molecular fragments are desorbed from the surface, sometimes providing results similar to those of static SIMS. The elemental detection limits of LIMS are in the 1–100 ppm range which are not as good as those of SALI or SIMS but better than most electron-beam techniques, such as EDS and AES. LIMS is not usually used to acquire depth profiles because of the large depth (0.25–1 μm) to which the high-intensity laser penetrates during a single pulse and because of the irregularity of the crater shape. LIMS is used in failure analysis situations because it samples a relatively small volume of material (1 μm^3), is relatively independent of sample geometry (shape), and produces an entire mass spectrum from a single pulse of the laser (analysis time less than 10 minutes). LIMS mass spectra can be quantified using standards in certain cases, but LIMS data are usually qualitative only. Additionally, because LIMS employs a laser for desorption and ionization, it can be used to analyze nonconductors, such as optical components (glasses) and ceramics.

Spark-Source Mass Spectrometry (SSMS) is used to measure trace-level impurities in a wide variety of materials (both conducting and nonconducting specimens) at concentrations in the 10–50 ppb range. Elemental sensitivities are uniform to within a factor of 3, independent of the sample type, rendering SSMS an ideal tool for survey impurity measurements when standards are unavailable. SSMS is usually used to provide bulk analysis (no lateral or in-depth information) but can also be used to measure impurities on surfaces or in thin films with special sample configurations. In SSMS, a solid material, in the form of two conducting electrodes, is vaporized and ionized using a high-voltage RF spark. The ions from the sample are then simultaneously analyzed using a mass spectrometer.

Glow-Discharge Mass Spectrometry (GDMS) is used to measure trace level impurities in solid samples with detection limits in the ppb range and below, with little or no in-depth or lateral information. The sample, in the form of a pin measuring $2 \times 2 \times 20$ mm, forms the cathode of a noble gas DC glow discharge (plasma). Atoms sputtered from the surface of the sample are ionized in the plasma and analyzed in a high-resolution mass spectrometer. Depth profiles with a depth resolution poorer than 100 nm can be obtained from flat samples run in a special sample configuration.

GDMS is slowly replacing SSMS because of its increased quantitative accuracy and improved detection limits. Like SNMS and SALI, GDMS is semiquantitative without standards (± a factor of 3) and quantitative with standards (±20%) because sputtering and ionization are decoupled. GDMS is often used to measure impurities in metals and other materials which are eventually used to form thin films in other materials applications.

In Inductively Coupled Plasma Mass Spectrometry (ICPMS), ions are generated in an inductively coupled plasma and subsequently analyzed in a mass spectrometer. Detection of all elements is possible with the exception of a few because of mass interferences due to components of the plasma and the unit mass resolution of most ICPMS units. Typical samples are introduced into the plasma as liquids, but recent developments allow direct sampling by laser ablation. Solids and thin films (including interfaces) are usually digested into solution prior to analysis. Detection limits from solution are in the ppt–ppb range; with typical dilutions of 1000, the detection limits from solids are in the ppb–ppm range. ICPMS is fast and reproducible; survey mass spectra can be obtained from a solution in minutes. Quantitative analyses are performed with accuracies better than ±5% using standards, while semiquantitative analyses are accurate to ±20% or better. Surface and thin film analyses are performed by dissolving the surface or thin film into solution and analyzing the solution. This kind of methodology is often selected when the average composition of a surface or film over a large area must be measured, or when a thin film exceeds the thickness of the analytical depth of other analytical techniques.

ICP-OES is similar to ICPMS but uses an optical detection system rather than a mass spectrometer. Atoms and ions are excited in the plasma and emit light at characteristic wavelengths in the ultraviolet or visible region of the spectrum. A grating spectrometer is used for simultaneous measurement of preselected emission lines. Measurement of all elements is possible with the exception of a few blocked by spectral interferences. The intensity of each line is proportional to the concentration of the element from which it was emitted. Elemental sensitivities in the sub-ppb–100 ppb range are possible for solutions; dilutions of 1000 times yield detection limits in the ppm range. Direct sampling of solids is performed using spark, arc or laser ablation, yielding similar detection limits. By sampling a solid directly, the risk of introducing contamination into the sample is minimized. Like ICPMS, ICP-OES is quantified by comparison to standards. Quantitative analyses are performed with accuracies between 0.2 and 15% using standards (typically better than ±5%). ICP-OES is less sensitive than ICPMS (poorer detection limits) but is selected in certain applications because of its quantitative accuracy and accessibility. (There are thousands of ICP-OES systems in use worldwide and the cost of a new ICP-OES is half that of an ICPMS.)

10.1 Dynamic SIMS

Dynamic Secondary Ion Mass Spectrometry

PAUL K. CHU

Contents

Introduction

Dynamic SIMS, normally referred to as SIMS, is one of the most sensitive analytical techniques, with elemental detection limits in the ppm to sub-ppb range, depth resolution (z) as good as 2 nm and lateral (x, y) resolution between 50 nm and 2 μm, depending upon the application and mode of operation. SIMS can be used to measure any elemental impurity, from hydrogen to uranium and any isotope of any element. The detection limit of most impurities is typically between 10^{12} and 10^{16} atoms/cm^3, which is at least several orders of magnitude lower (better) than the detection limits of other analytical techniques capable of providing similar lateral and depth information. Therefore, SIMS (or the related technique, SALI) is almost always the analytical technique of choice when ultrahigh sensitivity with simultaneous depth or lateral information is required. Additionally, its ability to detect hydrogen is unique and not possible using most other non-mass spectrometry surface-sensitive analytical techniques.

Dynamic SIMS is used to measure elemental impurities in a wide variety of materials, but is almost new used to provide chemical bonding and molecular information because of the destructive nature of the technique. Molecular identification or measurement of the chemical bonds present in the sample is better performed using analytical techniques, such as X-Ray Photoelectron Spectrometry (XPS), Infrared (IR) Spectroscopy, or Static SIMS.

The accuracy of SIMS quantification ranges from ±2% in optimal cases to a factor of 2, depending upon the application and availability of good standards. However, it is generally not used for the measurement of major components, such as silicon and tungsten in tungsten silicide thin films, or aluminum and oxygen in alumina, where other analytical techniques, such as wet chemistry, X-Ray Fluorescence (XRF), Electron Probe (EPMA), or Rutherford Backscattering Spectrometry (RBS), to name only a few, may provide much better quantitative accuracy (±1% or better).

Because of its unique ability to measure the depth or lateral distributions of impurities or dopants at trace levels, SIMS is used in a great number of applications areas. In semiconductor applications, it is used to quantitatively measure the depth distributions of unwanted impurities or intentional dopants in single or multilayered structures. In metallurgical applications, it is used to measure surface contamination, impurities in grain boundaries, ultratrace level impurities in metal grains, and changes in composition caused by ion implantation for surface hardening. In polymers or other organic materials, SIMS is used to measure trace impurities on the surface or in the bulk of the material. In geological applications, SIMS is used to identify mineral phases, and to measure trace level impurities at grain boundaries and within individual phases. Isotope ratios and diffusion studies are used to date geological materials in cosmogeochemical and geochronological applications. In biology and pharmacology, SIMS is used to measure trace elements in localized areas, by taking advantage of its excellent lateral resolution, and in very small volumes, taking advantage of its extremely low detection limits.

Basic Principles

Sputtering

When heavy primary ions (oxygen or heavier) having energies between 1 and 20 keV impact a solid surface (the sample), energy is transferred to atoms in the surface through direct or indirect collisions. This creates a mixing zone consisting of primary ions and displaced atoms from the sample. The energy and momentum transfer process results in the ejection of neutral and charged particles (atomic ions and ionized clusters of atoms, called molecular ions) from the surface in a process called sputtering (Figure 1).

The depth (thickness) of the mixing zone, which limits the depth resolution of a SIMS analysis typically to 2–30 nm, is a function of the energy, angle of incidence,

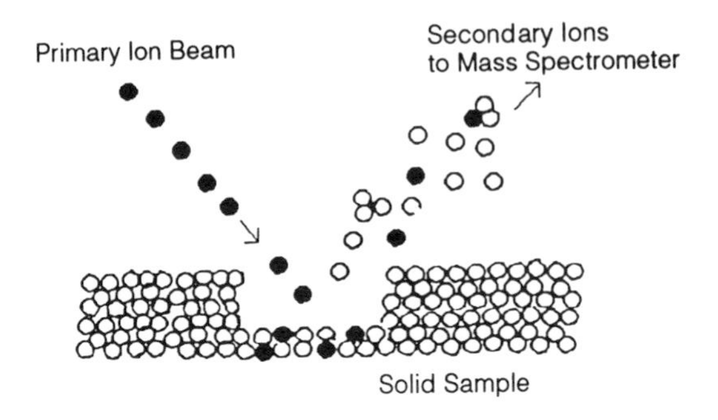

Figure 1 **Diagram of the SIMS sputtering process.**

and mass of the primary ions, as well as the sample material. Use of a higher mass primary ion beam, or a decrease in the primary ion energy or in the incoming angle with respect to the surface, will usually cause a decrease in the depth of the mixing zone and result in better depth resolution. Likewise, there is generally an inverse relationship between the depth (thickness) of the mixing zone and the average atomic number of the sample.

During a SIMS analysis, the primary ion beam continuously sputters the sample, advancing the mixing zone down and creating a sputtered crater. The rate at which the mixing zone is advanced is called the sputtering rate. The sputtering rate is usually increased by increasing the primary ion beam current density, using a higher atomic number primary ion or higher beam energy, or by decreasing the angle at which the primary ion beam impacts the surface. The primary ion beam currents used in typical SIMS analyses range from 10 nA to 15 μA—a range of more than three decades.

The depth resolution of a SIMS analysis is also affected by the flatness of the sputtered crater bottom over the analytical area; a nonuniform crater bottom will result in a loss in depth resolution. Because most ion beams have a Gaussian spatial distribution, flat-bottomed craters are best formed by rastering the ion beam over an extended area encompassing some multiples of beam diameters. Moreover, to reject stray ions emanating from the crater walls (other depths), secondary ions are collected only from the central, flat-bottomed region of the crater through the use of electronic gating or physical apertures in the mass spectrometer. For example, secondary ions are often collected from an area as small as 30 μm in diameter, while the primary ion beam sputters an area as large as 500 × 500 μm. Unfortunately, no matter what precautions and care are taken, the bottom of a sputtered crater becomes increasingly rough as the crater deepens, causing a continual degradation of depth resolution.

Detection Limits

The detection limit of each element depends upon the electron affinity or ionization potential of the element itself, the chemical nature of the sample in which it is contained, and the type and intensity of the primary ion beam used in the sputtering process.

Because SIMS can measure only ions created in the sputtering process and not neutral atoms or clusters, the detection limit of a particular element is affected by how efficiently it ionizes. The ionization efficiency of an element is referred to as its ion yield. The ion yield of a particular element *A* is simply the ratio of the number of *A* ions to the total number of *A* atoms sputtered from the mixing zone. For example, if element *A* has a 1:100 probability of being ionized in the sputtering process—that is, if 1 ion is formed from every 100 atoms of *A* sputtered from the sample—the ion yield of *A* would be 1/100. The higher the ion yield for a given element, the lower (better) the detection limit.

Many factors affect the ion yield of an element or molecule. The most obvious is its intrinsic tendency to be ionized, that is, its ionization potential (in the case of positive ions) or electron affinity (in the case of negative ions). Boron, which has an ionization potential of 8.3 eV, looses an electron much more easily than does oxygen, which has an ionization potential of 13.6 eV, and therefore has a higher positive ion yield. Conversely, oxygen possesses a higher electron affinity than boron (1.5 versus 0.3 eV) and therefore more easily gains an electron to form a negative ion. Figures 2a and 2b are semilogarithmic plots of observed elemental ion yields relative to the ion yield of iron (M^+/Fe^+ or M^-/Fe^-) versus ionization potential or electron affinity for some of the elements certified in an NBS 661 stainless steel reference material. From these plots, it is easy to see that an element like zirconium has a very high positive ion yield and, therefore, an excellent detection limit, compared to sulfur, which has a poor positive ion yield and a correspondingly poor detection limit. Likewise, selenium has an excellent negative ion yield and an excellent detection limit, while manganese has a poor negative ion yield and poor detection limit. The correlation of electron affinity and ionization potential with detection limits is consistent in most cases; exceptions due to the nature of the element itself or to the chemical nature of the sample material exist. For example, fluorine exhibits an anomalously high positive ion yield in almost any sample type.

One of three kinds of primary ion beams is typically used in dynamic SIMS analyses: oxygen (O_2^+ or O^-), cesium (Cs^+), or argon (Ar^+). The use of an oxygen beam can increase the ion yield of positive ions, while the use of a cesium beam can increase the ion yield of negative ions, by as much as four orders of magnitude. A simple model explains these phenomena qualitatively by postulating that *M*–O bonds are formed in an oxygen-rich mixing zone, created by oxygen ion bombardment. When these bonds break in the ion emission process, oxygen tends to become negatively charged due to its high ionization potential, and its counterpart

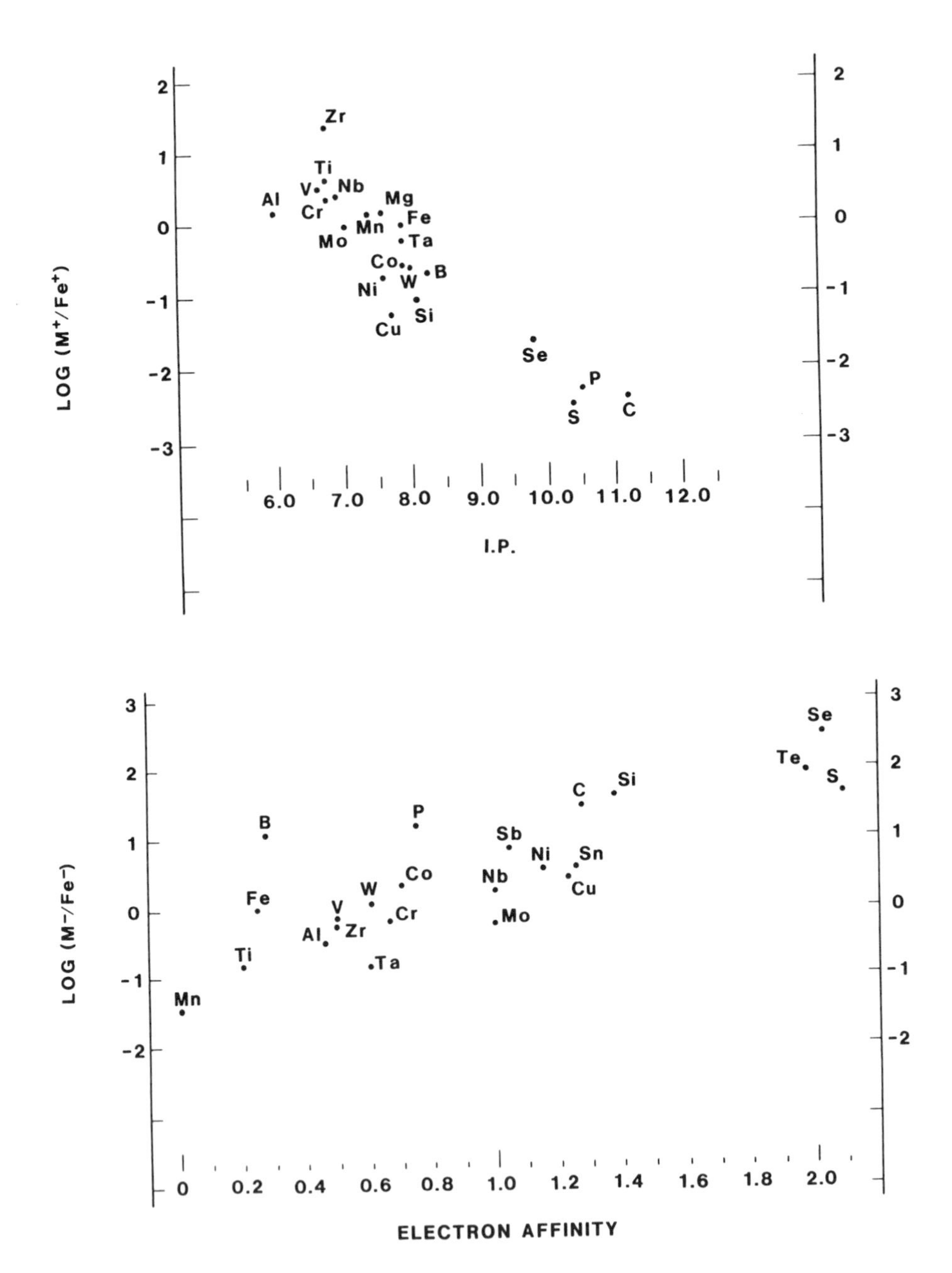

Figure 2 (a) Semilogarithmic plot of the positive relative ion yields of various certified elements (M^+/Fe^+) in NBS 661 stainless steel reference material versus ionization potential. (b) Semilogarithmic plot of the negative relative ion yields of various certified elements (M^-/Fe^-) in NBS 661 stainless steel reference material versus electron affinity.

M dissociates as a positive ion.[1] Conversely, the enhanced ion yields of the cesium ion beam can be explained using a work function model,[2] which postulates that because the work function of a cesiated surface is drastically reduced, there are more secondary electrons excited over the surface potential barrier to result in enhanced formation of negative ions. The use of an argon primary beam does not enhance the ion yields of either positive or negative ions, and is therefore, much less frequently used in SIMS analyses.

Like the chemical composition of the primary beam, the chemical nature of the sample affects the ion yield of elements contained within it. For example, the presence of a large amount of an electronegative element like oxygen in a sample enhances the positive secondary ion yields of impurities contained in it compared to a similar sample containing less oxygen.

Another factor affecting detection limits is the sputtering rate employed during the analysis. As a general rule, a higher sputtering rate yields a lower (better) detection limit because more ions are measured per unit time, improving the detection limits on a statistical basis alone. However, in circumstances when the detection limit of an element is limited by the presence of a spectral interference (see below), the detection limit may not get better with increased sputtering rate. Additionally and unfortunately, an increase in the sputtering rate nearly always results in some loss in depth resolution.

Common Modes of Analysis and Examples

SIMS can be operated in any of four basic modes to yield a wide variety of information:

1. The depth profiling mode, by far the most common, is used to measure the concentrations of specific preselected elements as a function of depth (z) from the surface.
2. The bulk analysis mode is used to achieve maximum sensitivity to trace-level components, while sacrificing both depth (z) and lateral (x and y) resolution.
3. The mass scan mode is used to survey the entire mass spectrum within a certain volume of the specimen.
4. The imaging mode is used to determine the lateral distribution (x and y) of specific preselected elements. In certain circumstances, an imaging depth profile is acquired, combining the use of both depth profiling and imaging.

Depth Profiling Mode

If the primary ion beam is used to continuously remove material from the surface of a specimen in a given area, the analytical zone is advanced into the sample as a function of the sputtering time. By monitoring the secondary ion count rates of selected

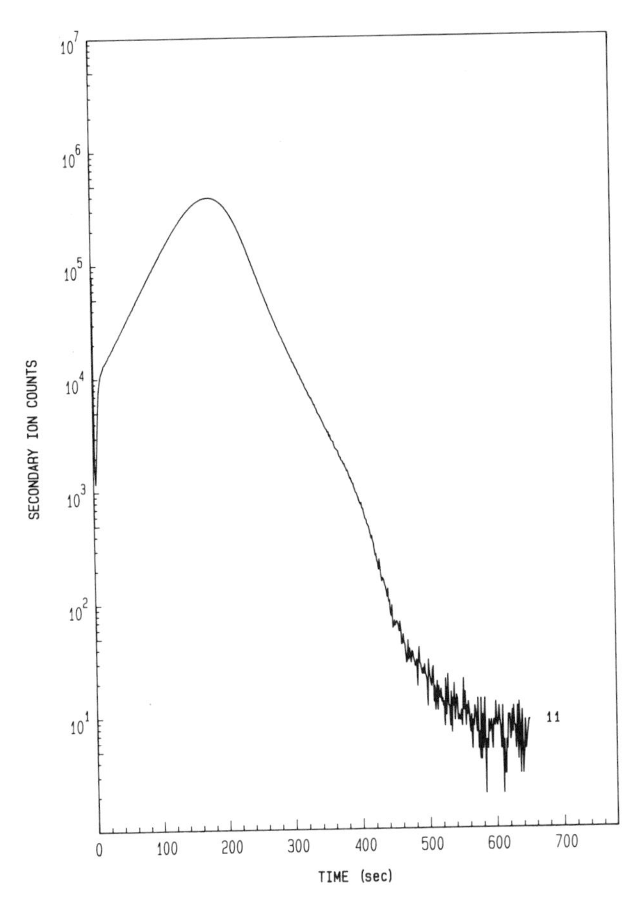

Figure 3a **Unprocessed depth profile (secondary ion intensity versus sputtering time) of a silicon sample containing a boron ion implant.**

elements as a function of time, a profile of the in-depth distribution of the elements is obtained. The depth scale of the profile is commonly determined by physically measuring the depth of the crater formed in the sputtering process and assigning that depth to the total sputtering time required to complete the depth profile. A depth scale assigned in this way will be accurate only if the sputtering rate is uniform throughout the entire profile. For samples composed of layers that sputter at different rates, an accurate depth scale can be assigned only if the relative sputtering rates of the different layers are known. A typical SIMS depth profile is collected as secondary ion counts per second versus sputtering time (typically one second per measurement) and converted to a plot of concentration versus depth by using the depth of the sputtering crater and comparing the data to standards. Figure 3a is an unprocessed depth profile of a silicon sample containing a boron ion implant.

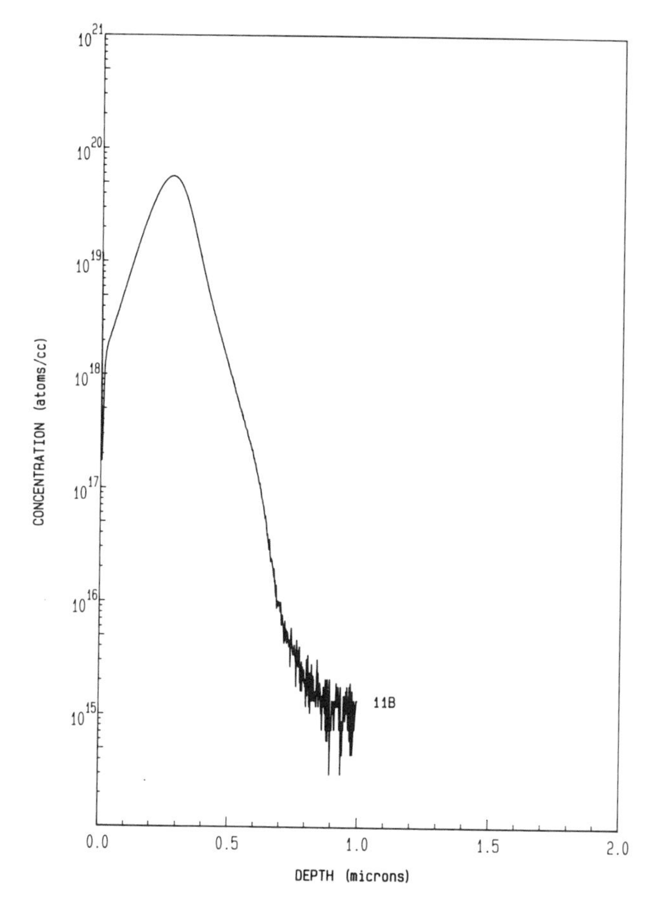

Figure 3b **Depth profile in (a), after converting the sputtering time to depth and the secondary ion intensities to concentrations.**

Figure 3b shows the same depth profile after converting to depth and concentration. Depth profiles can be performed to depths exceeding 100 μm and can take many hours to acquire; a more typical depth profile is several μm in depth and requires less than one hour to acquire.

Mass Scan Mode

A mass scan is acquired in cases when a survey of all impurities present in a volume of material is needed. Rather than measuring the secondary ion count rates of preselected elements as a function of sputtering time the count rates of all secondary ions are measured as a function of mass. Because a mass scan is continuously acquired over a mass range, no depth profiling or lateral information is available while operating in this mode. Figure 4 shows a mass scan acquired from a zirconia

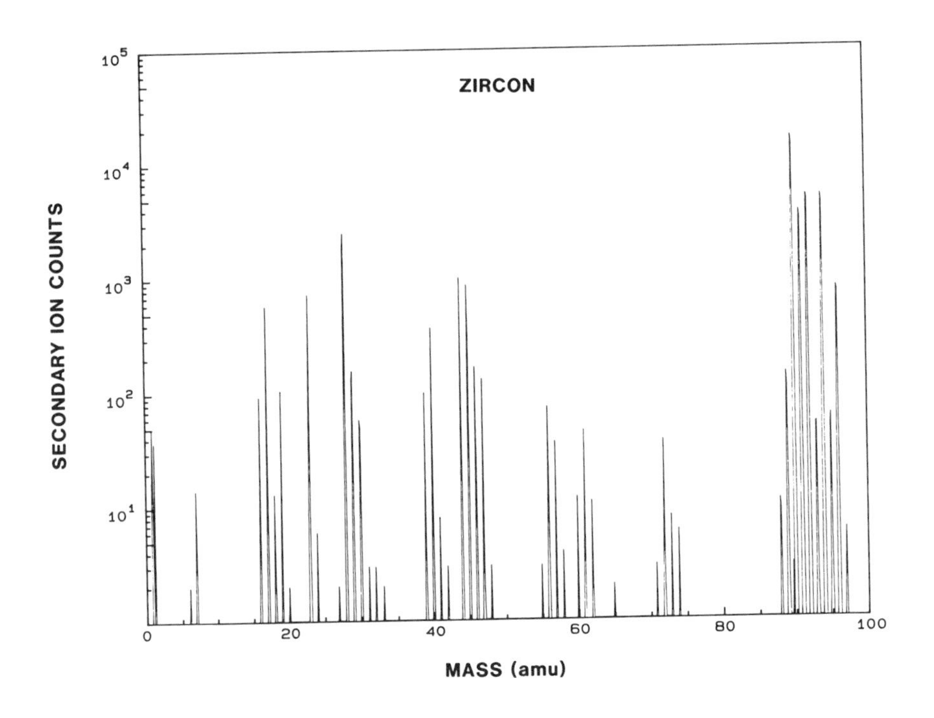

Figure 4 **Mass scan acquired from a zirconia crystal.**

crystal (geological sample). It shows peaks for many elements and molecules, but provides no information concerning the depth or lateral distribution of these impurities.

Bulk Analysis Mode

Bulk analysis mode is typically used to obtain the lowest possible detection limits of one or several elements in a uniform sample. This mode of operation is similar to a depth profile with the sputtering rate set to the maximum. This causes the crater bottom to lose its flatness and allows impurities from the crater walls to be measured, thereby sacrificing depth resolution. Therefore, accurate measurement of impurities is obtained only when they are uniformly distributed in the sample. This method of measurement usually results in at least a factor-of-10 improvement in detection limits over the depth profiling mode. As an example, the detection limit of boron in silicon using the bulk analysis mode is 5×10^{12} atoms/ cm^3, several orders of magnitude better than the boron background acquired using the depth profiling mode (6×10^{14} atoms/ cm^3), as shown in Figure 3b.

Imaging Mode

SIMS imaging is performed using one of two methods. The first, called ion microscopy or stigmatic imaging, is only possible using specially constructed mass spectrometers capable of maintaining the *x–y* spatial relationships of the secondary ions. These mass spectrometers are typically specially configured double-focusing magnetic-sector spectrometers and are actually better termed secondary ion *microscopes.* The lateral resolution of microscope imaging is typically no better than 1 μm. The second method, scanning imaging, is performed by measuring the secondary ion intensity as a function of the lateral position of a small spot scanning ion beam. The lateral resolution of this type of imaging is largely dependent on the diameter of the primary ion beam, which can be as small as 50 nm.

Figures 5a and 5b are mass-resolved secondary ion images of gold (Au) and sulfur (S) in a cross-sectioned and polished pyrite (gold ore) sample acquired using the microscope imaging method. The gray level is proportional to the secondary ion intensity measured at each location, i.e., more gold or sulfur is found in darker locations. These images show that the gold, the geologist's primary interest, is localized in the outer few μm of the sulfur-containing pyrite grain.

By acquiring mass-resolved images as a function of sputtering time, an imaging depth profile is obtained. This combined mode of operation provides simultaneous lateral and depth resolution to provide what is known as three-dimensional analysis.

Sample Requirements

Most SIMS instruments are configured to handle samples less than 2.5 cm in diameter and 1 cm in thickness. The surface of the sample must be as smooth as possible because surface roughness causes a significant loss in depth resolution; cross sections and other cut samples must be well polished before analysis. In SIMS instruments capable of stigmatic imaging, the sample should be planar, because it effectively is part of the secondary ion optics. Nonplanar samples are better analyzed using a quadrupole SIMS instrument (discussed below) in which the sample shape does not affect the results as strongly. Samples composed of materials that are dielectric (nonconducting) must be analyzed using special conditions (see below). Quadrupole SIMS instruments are also less affected by sample charging and are often used to analyze dielectric samples.

Artifacts

Although SIMS is one of the most powerful surface analysis techniques, its application is complicated by a variety of artifacts.

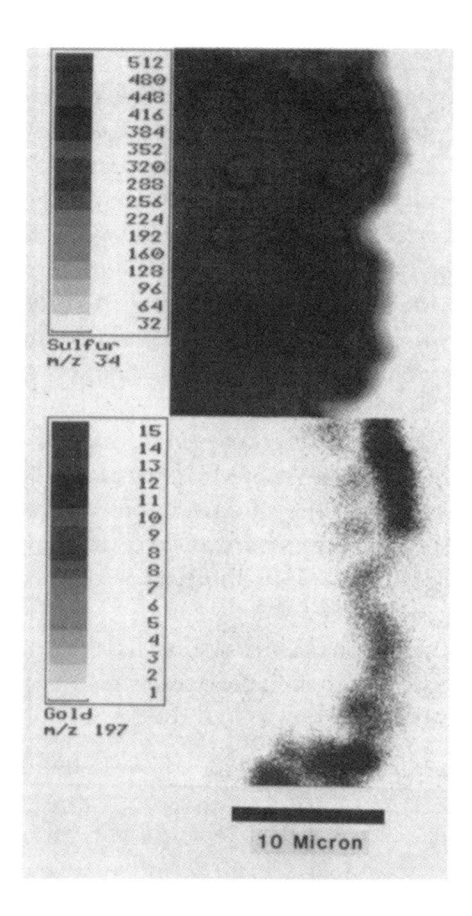

Figure 5 **Mass-resolved secondary ion images of sulfur and gold in a pyrite ore sample. A comparison of the two images clearly shows the gold is found in the outer several μm of the sulfur-containing pyrite grain. These images were acquired using a magnetic-sector mass spectrometer in the microscope-imaging mode.**

Mass Interferences

The most frequent artifacts arise from interferences in the mass spectrum, that is, ionized atomic clusters (molecular ions) or multiply charged ions whose nominal mass-to-charge ratio equals that of the elemental ions of interest. Such interferences can cause erroneous assignment of an element not present in the sample or simply can degrade the detection limit of the element of interest. Figure 6 is a mass spectrum obtained from high-purity silicon, using oxygen ion bombardment. In addition to the $^{28}Si^+$, $^{29}Si^+$, and $^{30}Si^+$ isotope peaks, there exist numerous other peaks of atomic and molecular ions typically composed of primary ion species (oxygen), ions

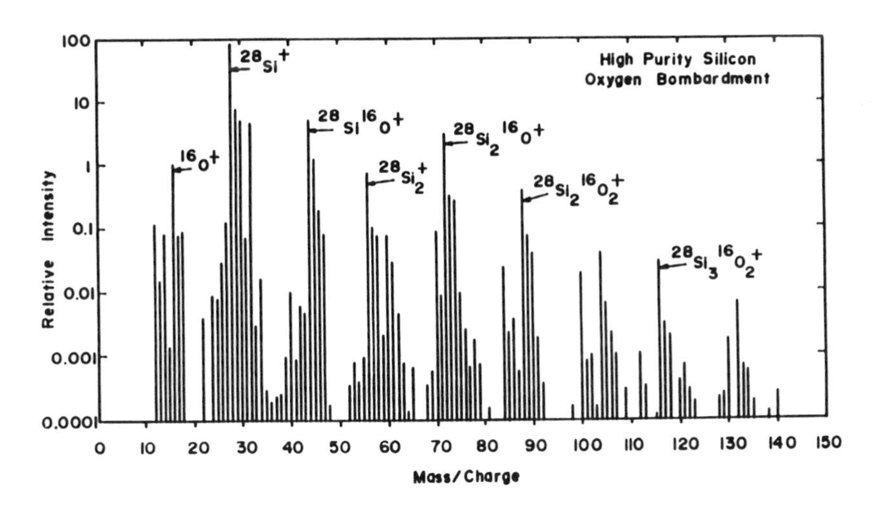

Figure 6 **Typical secondary ion mass spectrum obtained from high-purity silicon using an oxygen ion beam. Major ion peaks are identified in the spectrum.**

of the major components of the sample (silicon), or atmospheric species (hydrogen, carbon, oxygen, nitrogen, etc.) remaining in the high-vacuum sample chamber. Many of these peaks are sufficiently intense to produce a measurable background, which may preclude determination of a specific element (impurity), even in the ppm range.

Once identified, voltage offset and high mass resolution techniques may be used to reduce the detrimental effect of these interfering ions. In the voltage offset technique, the mass spectrometer is adjusted to accept only ions in a certain (usually higher) kinetic energy range. This technique is effective in discriminating against molecular ions because the energy distribution of atomic ions (typically the ions of interest) is broader than that of molecular ions at the same nominal mass. Figure 7 shows two SIMS depth profiles of the same silicon sample implanted with arsenic (^{75}As). These depth profiles were obtained under normal conditions (0-V offset) and under voltage offset conditions (50-V offset). The improvement in the detection limit of arsenic with the use of a 50-V offset results from discrimination of the $^{29}Si^{30}Si^{16}O$ molecular ion also at mass 75.

High mass resolution techniques are used to separate peaks at the same nominal mass by the very small mass differences between them. As an example, a combination of ^{30}Si and ^{1}H to form the molecular ion $^{30}Si^{1}H^{-}$, severely degrades the detection limit of phosphorous (^{31}P) in a silicon sample. The exact mass of phosphorous (^{31}P) is 31.9738 amu while the real masses of the interfering $^{30}Si^{1}H$ and $^{29}Si^{1}H_2$ molecules are 31.9816 amu and 31.9921 amu, respectively. Figure 8 shows a mass

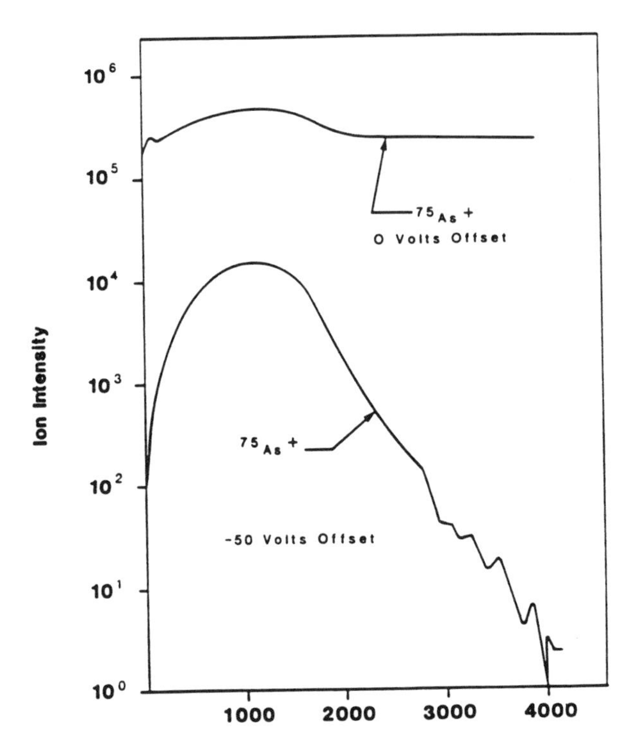

Figure 7 **Depth profile of an arsenic (^{75}As) ion implant in silicon with and without use of voltage offset techniques. Voltage offset provides an enhanced detection limit for As in Si.**

spectrum obtained from a phosphorus doped amorphous silicon thin film using high mass resolution techniques. The two mass interferences, $^{30}Si^{1}H^{-}$ and $^{29}Si^{1}H_2^{-}$, are completely separated from the $^{31}P^{-}$ peak. Quadrupole instruments are not usually capable of such high mass resolution.

Primary Ion Beam Sputtering Equilibrium

As explained above, the mixing zone contains a mixture of atoms from the primary ion beam and the solid sample. In the case of oxygen or cesium ion bombardment, these primary species become part of the material in the mixing zone and can significantly alter the ion yields of elements in the sample. However, when sputtering is first started (at the beginning of a depth profile), the mixing zone contains very few atoms from the primary ion beam, causing ions ejected from the mixing zone to be less affected by the enhancement process.

Crater Bottom Roughening

In polycrystalline solids or samples consisting of various phases, each grain may sputter at a different rate producing extensive roughness in the bottom of the cra-

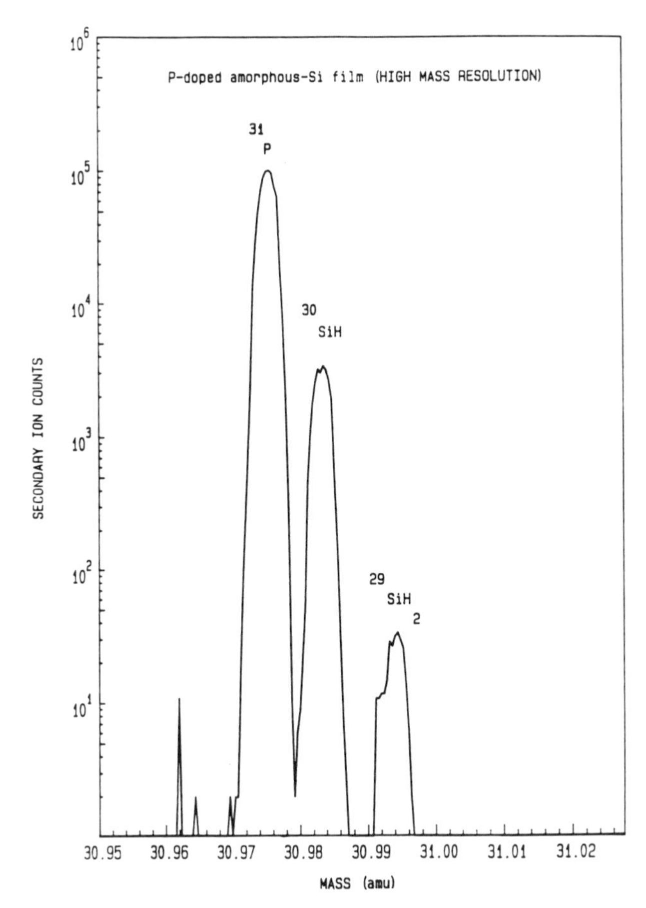

Figure 8 **High mass resolution mass spectrum obtained from a phosphorus-doped amorphous silicon hydride thin film using a magnetic sector ion microanalyzer. The ^{31}P peak is well separated from the hydride interferences.**

ter.[3] The roughness of the crater bottom will result in a loss in depth resolution and cause the depth profile to appear smeared in depth.

Surface Oxide

By enhancing the positive ion yield of most elements, the presence of an oxide on the surface of a sample can cause the first several points of a SIMS depth profile to be misleadingly high. Exceptionally large secondary ion signals for most elements are observed while profiling through the surface oxide, even though their concentrations are not higher. In these cases, the first several data points are corrected or simply disregarded.

Sample Charging

The charge carried by positive primary ions can accumulate on the surface of a non-conducting sample, causing the primary ion beam to be defocused or to move away from the analytical area of interest, thus preventing continued analysis. In addition, the accumulated charge can change the energy of the ejected ions, thereby affecting their transmission and detection in the mass spectrometer. This effect, called *sample charging*, is eliminated or reduced by flooding the sample surface with a low energy electron beam, providing compensation for the build-up of positive charge. As a general rule, samples with resistivities above 10^8 ohm/cm^2 require the use of electron flooding. In highly insulating samples, the use of a negative primary ion beam may also alleviate this charging problem.

Adsorption of Gaseous Species

During the sputtering process, residual atoms and molecules in the vacuum above the sample surface (typically containing hydrogen, carbon, nitrogen, and oxygen) are incorporated into the mixing zone by absorbing onto newly exposed and unsputtered reactive ions and molecules of the sample. The incorporated atmospheric species are eventually ejected from the mixing zone as elemental and molecular ions and detected as if they were originally present in the sample, complicating SIMS detection of these species and adding interfering molecular ions to the secondary ion mass spectrum. As an example, a mass interference between ^{31}P, and $^{30}Si^{1}H$ and $^{29}SiH_2$, all having mass 31, can be caused by hydrogen from the atmosphere in the sample chamber. The detrimental effects of these atmospheric species can be reduced by improving the vacuum in the sample chamber, but no matter how good the vacuum is, some adsorption will occur.

Impurity Mobility—Ion Beam-Induced Diffusion

Another difficulty is ion beam-induced diffusion of extremely mobile ions, such as lithium and sodium, in dielectric thin-film samples. This effect is normally observed when depth profiling a dielectric thin film on a conducting substrate with a positive primary ion beam. Diffusion occurs because the primary ion beam deposits a charge on the sample surface, creating a large electric field across the thin film, thereby driving the mobile ions away from the surface, to the interface between the thin film and substrate. In bulk insulators, this problem may be less severe because the electric field gradient is smaller. Nonetheless, the acquired depth profile no longer reflects the original composition of the sample. This effect is reduced or eliminated by flooding the sample surface with a low energy beam of electrons during sputtering. The current of electrons striking the sample surface must be carefully balanced against the build-up of charge due to the primary ion beam. Otherwise, distortion of the depth profile will still occur. As a general rule, quadrupole mass spectrometers have much less difficulty with impurity mobility artifacts than do magnetic sector spectrometers, and they are almost always used in these applications.

Quantification

Ion yields of different elements vary by several orders of magnitude and depend sensitively on the type of primary beam and sample. Accurate quantification requires comparison to standards or reference materials of similar or identical major element composition that must be measured using the same analytical conditions, especially using the same type of primary ion beam. For example, an aluminum sample with a known content of copper is not a good standard to use for quantification of copper in stainless steel. Similarly, a standard analyzed using a cesium primary ion beam must not be used as a standard for quantification of an unknown sample analyzed using an oxygen ion beam. In some cases, semiempirical ion yield systematics are successfully used to quantify certain analyses; this method of quantification is accurate only to within an order of magnitude.

Ion implantation is often used to produce reliable standards for quantification of SIMS analyses.[4] Ion implantation allows the introduction of a known amount of an element into a solid sample. A sample with a major component composition similar to that of the unknown sample may be implanted to produce an accurate standard. The accuracy of quantification using this implantation method can be as good as ±2%.

Instrumentation

SIMS instruments are generally distinguished by their primary ion beams, and the kinds of spectrometers they use to measure the secondary ions. Several types of primary ion beams—typically, oxygen, cesium, argon, or a liquid metal like gallium—are used in SIMS analyses, depending on the application. Nearly any SIMS instrument can be configured with one or more of these ion-beam types. The majority of SIMS mass spectrometers fall into three basic categories: double-focusing electrostatic or magnetic sector, quadrupole, and time-of-flight. Time-of-flight analyzers are primarily used for surface and organic analyses (especially for high molecular weight species) and are mentioned in the article on static SIMS.

A double-focusing, electrostatic or magnetic-sector mass spectrometer achieves mass separation using an electrostatic analyzer and magnet. Secondary ions of different mass are physically separated in the magnetic field, with light elements making a tight arc through the magnet and heavy elements making a broad arc. Ions of different charge-to-mass ratios are measured by changing the strength of the magnetic field in the magnet to align the ions of interest with a stationary detector. Magnetic-sector systems provide excellent detection limits because of their high transmission efficiency, and are capable of high mass resolution. Some of these spectrometers are capable of stigmatic imaging (also called ion microscopy) which is used to acquire mass-resolved ion images with a resolution as good as 1 μm.

In quadrupole-based SIMS instruments, mass separation is achieved by passing the secondary ions down a path surrounded by four rods excited with various AC and DC voltages. Different sets of AC and DC conditions are used to direct the flight path of the selected secondary ions into the detector. The primary advantage of this kind of spectrometer is the high speed at which they can switch from peak to peak and their ability to perform analysis of dielectric thin films and bulk insulators. The ability of the quadrupole to switch rapidly between mass peaks enables acquisition of depth profiles with more data points per depth, which improves depth resolution. Additionally, most quadrupole-based SIMS instruments are equipped with enhanced vacuum systems, reducing the detrimental contribution of residual atmospheric species to the mass spectrum.

The choice of mass spectrometer for a particular analysis depends on the nature of the sample and the desired results. For low detection limits, high mass resolution, or stigmatic imaging, a magnetic sector-based instrument should be used. The analysis of dielectric materials (in many cases) or a need for ultrahigh depth resolution requires the use of a quadrupole instrument.

Conclusions

SIMS is one of the most powerful surface and microanalytical techniques for materials characterization. It is primarily used in the analysis of semiconductors, as well as for metallurgical, and geological materials. The advent of a growing number of standards for SIMS has greatly enhanced the quantitative accuracy and reliability of the technique in these areas. Future development is expected in the area of small spot analysis, implementation of post-sputtering ionization to SIMS (see the articles on SALI and SNMS), and newer areas of application, such as ceramics, polymers, and biological and pharmaceutical materials.

Related Articles in the Encyclopedia

Static SIMS, SALI, SNMS, and Surface Roughness

References

1. G. Slodzian. *Surf. Sci.* **48,** 161, 1975.
2. C. A. Andersen. *Int. J. Mass. Spect. Io Phys.* **3,** 413, 1970.
3. E. Zinner, S. Dnst, J. Chaumont, and J. C. Dran. *Proceedings of the Ninth Lunar and Planetary Sciences Conference.* 1978, p. 1667.
4. P. Williams. *IEEE Trans. Nucl. Sci.* **26,** 1809, 1979.

10.2 Static SIMS

Static Secondary Ion Mass Spectrometry

BILL KATZ

Contents

Introduction

With today's technology, the definition of the surface as it effects a material's performance in many cases means the outer one or two monolayers. It is the specific chemistry of these immediate surface molecules that determines many of the chemical and physical properties. Therefore, it is important to have available a tool that is able to characterize the chemistry of these layers. One such method that has met with considerable success is Static Secondary Ion Mass Spectrometry (SIMS).

Static SIMS entails the bombardment of a sample surface with an energetic beam of particles, resulting in the emission of surface atoms and clusters. These ejected species subsequently become either positively or negatively charged and are referred to as secondary ions. The secondary ions are the actual analytical signal in SIMS. A mass spectrometer is used to separate the secondary ions with respect to their charge-to-mass ratios. The atomic ions give an elemental identification (see the article on dynamic SIMS), whereas the clusters can provide information on the chemical groups.

The mass spectrum of the clusters obtained represents a fingerprint of the compounds analyzed. The data show the various chemical functional groups as they fragment during analysis. The data may be acquired over relatively small areas (μm or less) for a localized analysis or larger areas (mm) for a macrocharacterization. Further, by monitoring a particular charge-to-mass ratio (i.e., a particular chemical group), one can obtain chemical maps depicting the lateral distribution of a specific fragment or compound.

Basic Principles

The analytical signal (he secondary ions) is generated by the interaction of an energetic particle with a sample surface. This interaction can be divided into two regimes by the total flux of primary particles used. In what is known as Dynamic SIMS, the incident flux of primary ions is large enough (above 5^{12} atoms/ cm^2) that, statistically, a given area has a high probability of being repetitively bombarded, causing a crater to be formed in the sample. This mode of SIMS provides an in-depth analysis of any element (including H) and its isotopes with excellent detection limits (ppm to ppb atomic). The disadvantage is the large primary fluxes tend to alter or totally obliterate the chemistry, limiting dynamic SIMS to elemental analysis.

When the total flux is kept below 5^{12} atoms/ cm^2, the probability of a primary ion striking a previously analyzed area is low, thus leaving the surface chemistry intact. This is the mode of operation used for Static SIMS. With respect to the primary beam, the total incident flux is the critical parameter for maintaining the chemical integrity of the material. Typically, static SIMS is performed using an inert gas beam operated at kinetic energies between 1 and 10 keV. The beam can be composed either of positively charged ions (Ar^+, Xe^+) or neutral particles (Ar, Xe) depending upon the type of ion source used.

In the SIMS process the energy of the incoming primary particles is dissipated by collisional cascades within the sample as the primary penetrates well into the material. The energy transferred to the sample is sufficient to cause atoms and molecules at the surface (within 1–3 atomic layers) to be ejected. A small fraction of these carry positive or negative charge, and are detected. In static SIMS, since one is primarily after chemical information, the identification of the cluster molecular ions is of particular interest as it allows one to identify the chemical compounds present at the surface. Unfortunately, in many-atom compounds severe fragmentation of the clusters takes place during the ejection and ionization process, with the parent ions rarely being observed. In most cases, therefore, identification of the parent species must rely on deductions made from the fragmentation patterns observed.

Sample Preparation

A major advantage of static SIMS over many other analytical methods is that usually no sample preparation is required. A solid sample is loaded directly into the instrument with the condition that it be compatible with an ultrahigh vacuum (10^{-9}–10^{-10} torr) environment. Other than this, the only constraint is one of sample size, which naturally varies from system to system. Most SIMS instruments can handle samples up to 1–2 inches in diameter.

In specialized cases, a treatment known as cationization sometimes is tried to improve the amount of molecular (chemical) information made available. If Ag or Na are deliberately introduced into the sample, they will often combine with the molecular species present to create Ag^+ or Na^+ molecular ions. These ions are more stable to fragmentation than the bare molecular ions, and can therefore be observed more easily in the mass spectrum. The identification of parent ion peaks in this manner aids in detailed chemical identification.

Static SIMS is also capable of analyzing liquids and fine particles or powders. A liquid is often prepared by putting down an extremely thin layer on a flat substrate, such as a silicon wafer. Particles are easily prepared by pressing them onto double-sided tape. No further sample preparation, such as gold- or carbon-coating, is required.

Because of the extreme surface sensitivity of static SIMS, care should always be exercised not to handle the samples. Clean tools and gloves should be used always to avoid the possibility of contaminating the surface. While it is possible to remove surface contamination with solvents like hexane, it is always desirable not to clean the surface to be analyzed.

Instrumentation

All commercially available SIMS systems have in common some type of computer automation, an ion source, a high-vacuum environment, and some type of mass spectrometer. While the specifics may vary from system to system, the basic requirements are the same. The hardware feature that tends to distinguish the various systems is the type of mass spectrometer used. These fall into three basic categories:[1]

- Quadrupole spectrometers. These are the least expensive mass spectrometers, and the easiest to operate. By applying AC and DC potentials to a set of four rods, ions are separated by mass as they pass through the quadrupole. The voltages can be changed quickly, allowing relatively rapid scanning of the mass range, which is usually limited to around 1000 amu. Because quadrupoles cannot effectively separate ions having a wide energy spectrum, an electrostatic filter is used between the sample and the quadrupole. Perhaps the major drawback to quadrupoles is that they have only moderate mass resolution and cannot separate peaks of the same nominal mass.

- Magnetic-sector spectrometers. These spectrometers use an electrostatic analyzer for energy filtering and a magnetic sector for mass separation. They are capable of achieving high mass resolution, with typical mass ranges of 250 amu.
- Time-of-flight spectrometers. *Time-of-flight* (TOF) analyzers are capable of both high mass resolution and extended mass range. Their design requires that the ion beam be pulsed (1–10 ns) to ensure high mass resolution. After the ion beam strikes the sample surface, the extracted secondary ions travel through a drift tube to a detector having a large area. Mass separation is accomplished by noting that ions having different masses take different times to reach the detector. For example, lighter secondary ions take less time to traverse the drift tube and reach the detector faster. Using this type of mass analyzer enables the entire mass spectrum to be collected in a few μs.

Each type of mass spectrometer has its associated advantages and disadvantages. Quadrupole-based systems offer a fairly simple ion optics design that provides a certain degree of flexibility with respect to instrument configuration. For example, quadrupole mass filters are often found in hybrid systems, that is, coupled with another surface analytical method, such as electron spectroscopy for chemical analysis or scanning Auger spectroscopy.

Contrasted with the quadrupole filter, both magnetic-sector and TOF analyzers provide the advantage of high mass resolution. This enables the separation of fragments having similar masses into distinct peaks. These instruments are useful when dealing with analytical situations requiring more exact mass determination. If there is a requirement to measure to high mass (thousands of amu), such as may be the case when studying polymers, TOF analyzers become the only choice.

Qualitative Analysis

One of the most common modes of characterization involves the determination of a material's surface chemistry. This is accomplished via interpretation of the fragmentation pattern in the static SIMS mass spectrum. This "fingerprint" yields a great deal of information about a sample's outer chemical nature, including the relative degree of unsaturation, the presence or absence of aromatic groups, and branching. In addition to the chemical information, the mass spectrum also provides data about any surface impurities or contaminants.

Figure 1 shows a positive static SIMS spectrum (obtained using a quadrupole) for polyethylene over the mass range 0–200 amu. The data are plotted as secondary ion intensity on a linear *y*-axis as a function of their charge-to-mass ratios (amu). This spectrum can be compared to a similar analysis from polystyrene seen in Figure 2. One can note easily the differences in fragmentation patterns between the

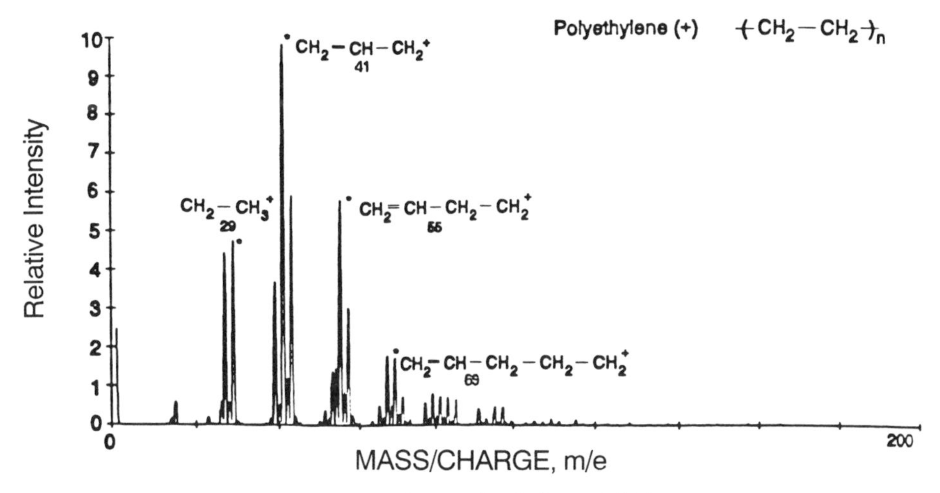

Figure 1 **Positive mass spectrum from polyethylene, 0–200 amu.**

two materials. Polystyrene is seen to have distinct fragment peaks characteristic of the aromatic nature of the compound. With the exception of the large peak at mass 91, most of the interesting information is found above 100 amu. Fragment peaks between 100 and 200 amu can be assigned to structures corresponding to rearrangements of the polystyrene backbone with one or more phenyl groups attached. This is seen to be dramatically different from the polyethylene spectrum, which is typical for a saturated short-chain hydrocarbon, with no significant peaks above 100 amu. This is a typical example of the current use of SIMS.

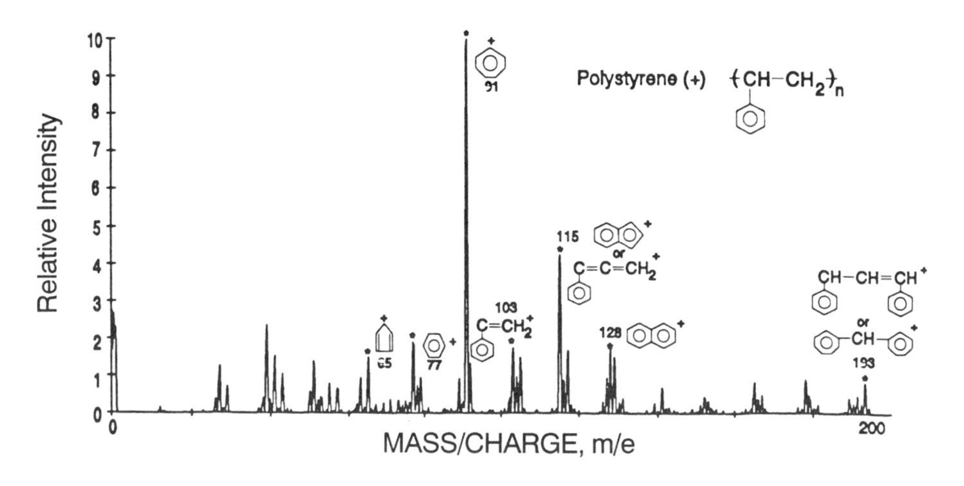

Figure 2 **Positive mass spectrum from polystyrene, 0–200 amu.**

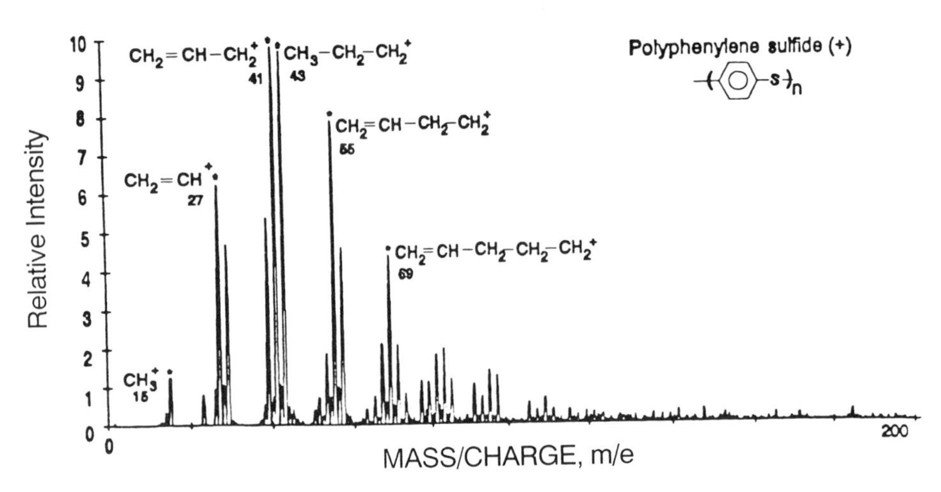

Figure 3 **Positive mass spectrum from polyphenylene sulfide, 0–200 amu.**

As mentioned earlier, the secondary ions ejected from a surface can be positively or negatively charged. Analytically, this is quite useful, as certain species are more readily ionized in one mode than another. The positive and negative mass spectra from polyphenylene sulfide shown in Figures 3 and 4, respectively, illustrate this point. The positive mass spectrum looks quite similar to that of polyethylene (Figure 1), with no indication of either the phenyl ring or the sulfur atom found in the polyphenylene sulfide. In marked contrast, the negative spectrum in Figure 4 clearly shows the presence of the phenyl group with the sulfur atom attached. This demonstrates how both positive and negative spectrometry may be needed to fully characterize a structure.

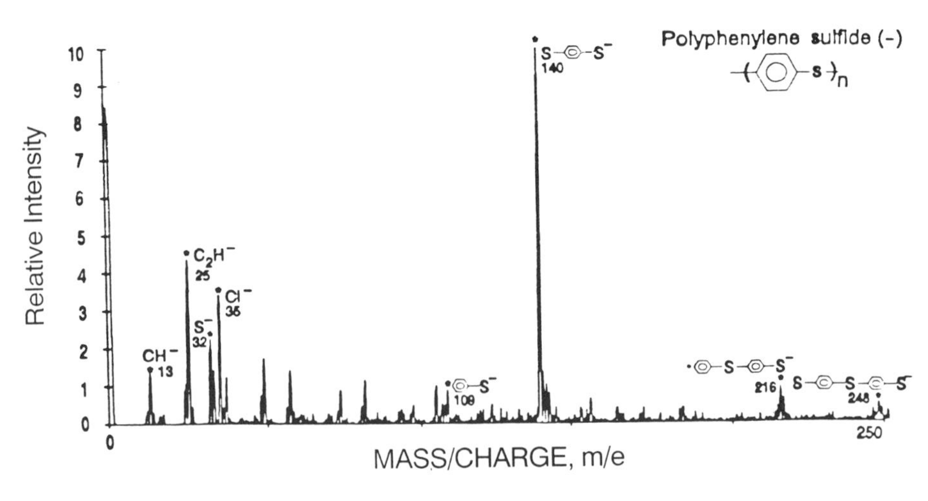

Figure 4 **Negative mass spectrum from polyphenylene sulfide, 0–250 amu.**

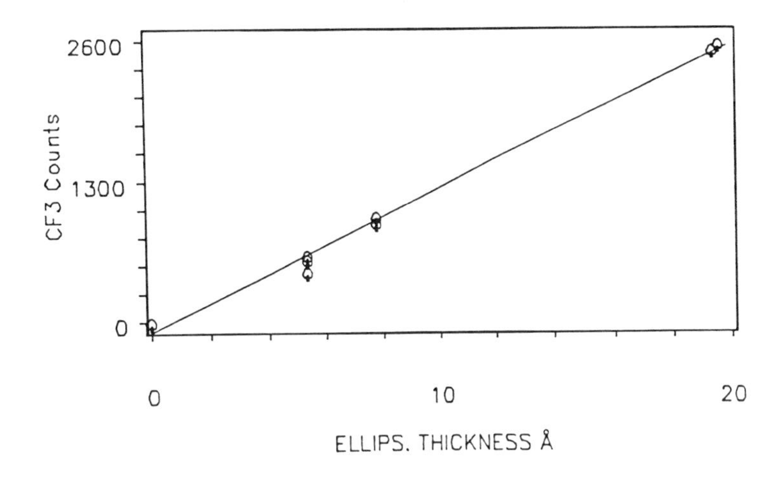

Figure 5 **Plot of positive CF_3 secondary ion intensity versus ellipsometric thickness from a set of perfluoropolyether standards.**

Quantitative Analysis

As with any analytical method, the ability to extract semiquantitative or quantitative information is the ultimate challenge. Generally, static SIMS is not used in this mode, but one application where static SIMS has been used successfully to provide quantitative data is in the accurate determination of the coverage of fluropolymer lubricants.[2] These compounds provide the lubrication for Winchester-type hard disks and are directly related to ultimate performance. If the lubricant is either too thick or too thin, catastrophic head crashes can occur.

Initially, a set of lubricant film standards of various thicknesses were prepared and their thickness measured by Fourier transform infrared spectroscopy and ellipsometry. Once good correlation between these two techniques was achieved, the same samples were analyzed using static SIMS. The CF_3 peak in the positive SIMS spectrum, which is characteristic of the fluoropolymer lubricant, was measured and its intensity plotted versus the thickness determined by ellipsometry. These results are presented in Figure 5, and show excellent correlation. It must be understood that since static SIMS analyzes the outer few monolayers, this measurement actually follows the coverage of the lubricant film. As more lubricant is put down, the coverage across the sample becomes more uniform, giving a higher secondary ion yield for the CF_3 fragment. Because of the excellent correlation between SIMS and other methods, one can conclude that all these techniques are actually measuring the effect of increasing lubricant coverage. However, static SIMS has been found to be more accurate for film thicknesses below 10 Å, owing to its extreme surface sensitivity. In addition, one also obtains an analysis of any contamination from the complete SIMS spectrum.

Another example of static SIMS used in a more quantitative role is in the analysis of extruded polymer blends. The morphology of blended polymers processed by extrusion or molding can be affected by the melt temperature, and pressure, etc. The surface morphology can have an effect on the properties of the molded polymer. Adhesion, mechanical properties, and physical appearance are just a few properties affected by processing conditions.

In a molded polymer blend, the surface morphology results from variations in composition between the surface and the bulk. Static SIMS was used to semiquantitatively provide information on the surface chemistry on a polycarbonate (PC)/polybutylene terephthalate (PBT) blend.[3] Samples of pure PC, pure PBT, and PC/PBT blends of known composition were prepared and analyzed using static SIMS. Fragment peaks characteristic of the PC and PBT materials were identified. By measuring the SIMS intensities of these characteristic peaks from the PC/PBT blends, a typical working curve between secondary ion intensity and polymer blend composition was determined. A static SIMS analysis of the extruded surface of a blended polymer was performed. The peak intensities could then be compared with the known samples in the working curve to provide information about the relative amounts of PC and PBT on the actual surface.

Chemical Mapping

In addition to data obtained using the spectral mode of analysis, it is often important to know the location of a particular chemical group or compound on the sample surface. Such information is achieved by static SIMS chemical mapping—a procedure in which a specific chemical functionality on the material is imaged, providing information as to its lateral distribution on the surface.

The use of chemical mapping is demonstrated in the following example involving the delamination of a silicone primer and polytetrafluoroethylene (PTFE) material. The positive mass spectrum acquired from the delaminated interface contains peaks known to be uniquely characteristic of PTFE (CF_3 at mass 69) and the silicone primer ($Si(CH_3)_3$ at mass 73). Figures 6 and 7 are secondary ion images of the CF_3 and $(Si(CH_3)_3$ fragments taken from a 1-mm^2 area of the delaminated interface. These maps clearly indicate that the PTFE and the silicone primer exist in well-defined and complementary areas.

Conclusions

Static SIMS has been demonstrated to be a valuable tool in the chemical characterization of surfaces. It is unique in its ability to provide chemical information with high surface sensitivity. The technique is capable of providing mass spectral data (both positive and negative spectrometry), as well as chemical mapping, thereby giving a complete microchemical analysis. The type of information provided by

Figure 6 Positive ion image of CF_3 taken from a 1-mm^2 area.

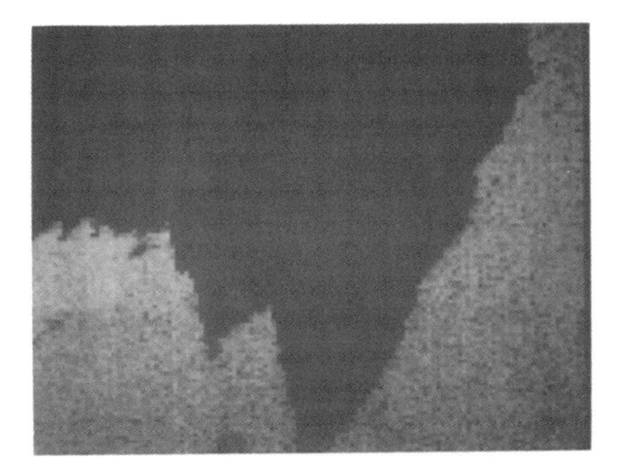

Figure 7 Positive ion image of $Si(CH_3)_3$ taken from a 1-mm^2 area.

static SIMS has been used to solve problems in a wide range of applications, including impurity analysis, the comparison of surface and bulk compositions, failure analysis, and the determination of adhesion or delamination mechanisms.[4–10]

With the increasing availability of TOF instruments, the field will see more applications involving the analysis of higher molecular weight fragments. This, coupled with the higher mass resolving power of TOF systems, will open up research in such fields as biomedical and pharmaceutical applications, in addition to all areas in high-technology materials where the identification of contaminants of high amu in trace amounts at surfaces is important. Residues from previous processing steps are prime examples, both in semiconductors and other thin-film technologies.

Related Articles in the Encyclopedia

Dynamic SIMS and SALI

References

1. F. A. White and G. M. Wood. *Mass Spectrometry.* John Wiley and Sons, New York, 1986.
2. J. G. Newman and K. V. Viswanathan. *J. Vac. Sci. Tech.* **A8** (3), 2388, 1990.
3. R. S. Michael, W. Katz, J. Newman, and J. Moulder. Proceedings of the seventh International SIMS Conference. 1989, p. 773.
4. W. Katz and J. G. Newman. *MRS Bulletin.* **12** (6), 40, 1987. Reviews the fundamentals of SIMS.
5. D. Briggs. *Polymer.* **25,** 1379, 1984. Review of static SIMS analysis.
6. A. Brown and J. C. Vickerman. *Surf. Interface Anal.* **6,** 1, 1984. Describes interpretation of fragmentation patterns in static SIMS.
7. W. J. van Ooij and R. H. G. Brinkhuis. *Surf. Interface Anal.* **11,** 430, 1988. Discusses fingerprint patterns characteristic of the molecular repeat unit of a polymer.
8. R. S. Michael and W. J. van Ooij. *Proc. ACS Div. Polymer Mater. Sci. Eng.* **59,** 734, 1988. Static SIMS analysis of plasma treated polymer surfaces.
9. D. Briggs. *Org. Mass Spectrom.* **22,** 91, 1987. Static SIMS analysis of copolymers.
10. M. J. Hearn, B. D. Ratner, and D. Briggs. *Macromol.* **21,** 2950, 1988. Use of peak intensities in static SIMS for quantification.

10.3 SALI

Surface Analysis by Laser Ionization

SUSAN G. MACKAY AND CHRISTOPHER H. BECKER

Contents

Introduction

In other articles in this section, a method of analysis is described called Secondary Ion Mass Spectrometry (SIMS), in which material is sputtered from a surface using an ion beam and the minor components that are ejected as positive or negative ions are analyzed by a mass spectrometer. Over the past few years, methods that *post-ionize* the major *neutral* components ejected from surfaces under ion-beam or laser bombardment have been introduced because of the improved quantitative aspects obtainable by analyzing the major ejected channel. These techniques include SALI, Sputter-Initiated Resonance Ionization Spectroscopy (SIRIS), and Sputtered Neutral Mass Spectrometry (SNMS) or electron-gas post-ionization. Post-ionization techniques for surface analysis have received widespread interest because of their increased sensitivity, compared to more traditional surface analysis techniques, such as X-Ray Photoelectron Spectroscopy (XPS) and Auger Electron Spectroscopy (AES), and their more reliable quantitation, compared to SIMS.

The advantages of SALI are seen most clearly when analyzing trace (ppm to ppb) amounts of material on surfaces or at interfaces. Typically, SALI analyzes the same samples as SIMS, with the added advantage of providing easily quantifiable data.

Technique	SALI	SIMS	XPS	AES	RBS
Analysis modes	Surface, depth-profiling, imaging	Surface, depth-profiling, imaging	Surface, depth-profiling, imaging	Surface, depth-profiling, imaging	Surface, depth-profiling
Common detection limit	ppm–ppb	ppm–ppb	0.05% at.	0.1% at.	2% at. –10 ppm
Probing depth	2–5 Å	2–5 Å	2–30 Å	2–30 Å	30–100 Å
Ultimate spatial resolution	60 nm	60 nm	10 μm	100 Å	1 mm
Quantitative?	Yes	Semi, with rigorous standards	Yes	Yes	Yes

Table 1 **Comparison of SALI to other surface spectroscopic techniques.**

Whereas SIMS provides highly matrix-dependent data, SALI can resolve problems associated with SIMS ion-yield transients. SALI has been applied to two basic groups of samples: inorganic and organic solid materials. For inorganic analysis or elemental analysis (e.g., semiconductors and metal alloys), ionization by absorption of more than one photon (multiphoton ionization) is generally used as the post-ionization source, while for organic analysis (e.g., polymers and biomaterials), a less destructive single-photon ionization probe is employed. In order to provide lateral and depth information, SALI can be operated in both mapping and depth profiling modes.

SALI compares favorably with other major surface analytical techniques in terms of sensitivity and spatial resolution. Its major advantage is the combination of analytical versatility, ease of quantification, and sensitivity. Table 1 compares the analytical characteristics of SALI to four major surface spectroscopic techniques.These techniques can also be categorized by the chemical information they provide. Both SALI and SIMS (static mode only) can provide molecular fingerprint information via mass spectra that give mass peaks corresponding to structural units of the molecule, while XPS provides only short-range chemical information. XPS and static SIMS are often used to complement each other since XPS chemical speciation information is semiquantitative; however, SALI molecular information can potentially be quantified directly without correlation with another surface spectroscopic technique. AES and Rutherford Backscattering (RBS) provide primarily elemental information, and therefore yield little structural information. The common detection limit refers to the sensitivity for nearly all elements that these techniques enjoy.

SALI, XPS, and AES have a nearly uniform sensitivity for all detectable elements with respect to the chemical composition of the sample matrix. SIMS and RBS, however, have sensitivities that vary greatly due to matrix effects. For RBS the matrix dependence is well understood, with the best sensitivities (10 ppm) found for heavy elements in light matrices while the worst sensitivities (3% at. to undetectable) are seen for light elements in heavy matrices. SIMS sensitivity varies for elements depending on both the chemical composition of the sample and the composition of the primary ion beam. Common primary ion beams are for SIMS O_2^+ and Cs^+ each of which enhances secondary ion yields. Again, an advantage of SALI is its relative insensitivity to the effects of changing matrix composition. Two drawbacks of SALI are that its maximum sensitivity is usually less than optimum case SIMS and, like all sputtering techniques, it is destructive.

A somewhat related technique is that of laser ionization mass spectrometry (LIMS), also known as LIMA and LAMMA, where a single pulsed laser beam ablates material and simultaneously causes some ionization, analogous to samples beyond the outer surface and therefore is more of a bulk analysis technique; it also has severe quantifiaction problems, often even more extreme than for SIMS.

Basic Principles

Figure 1 is a schematic of the SALI process. An energetic probe beam (ions, electrons, or light) is used to desorb material from the sample's surface. The neutral component is then intersected and ionized by a high-power, focused laser beam. The laser beam is passed parallel and close to the surface so that it intersects a large solid angle of the sputtered neutral species, thus improving the sensitivity. The positive ions created by the laser light (photoions) are then extracted by a high-voltage field (>3 kV) and pass into the mass analyzer. A time-of-flight tube is used to mass analyze the photoions created by each pulse of the laser. Lighter ions have a shorter flight time than heavier ions and therefore arrive at the ion detector (channel plate assembly) sooner. A typical SALI analysis looks at a mass range per laser pulse of approximately m/z = 12–300 amu (mass spectral fingerprint region) with a total acquisition time of 5 seconds (1000 pulses with a 200-Hz laser).

Analysis of Neutrals

Previous studies of the interaction of energetic particles with surfaces have made it clear that under nearly all conditions the majority of atoms or molecules removed from a surface are neutral, rather than charged. This means that the charged component can have large relative fluctuations (orders of magnitude) depending on the local chemical matrix. Calibration with standards for surfaces is difficult and for interfaces is nearly impossible. Therefore, for quantification ease, the majority neutral component of the departing flux must be sampled, and this requires some type of ionization above the sample, often referred to as post-ionization. SALI uses effi-

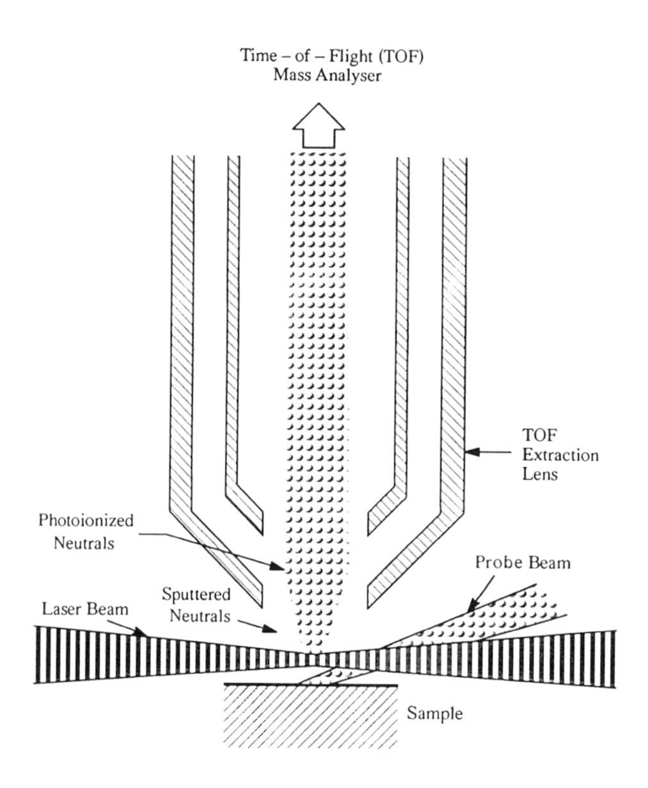

Figure 1 **Schematic representation of the fundamental SALI process.**

cient, nonresonant (not tuned to a specific energy level), and therefore nonselective photoionization by pulsed untuned laser radiation to accomplish this ionization and thus make this detection scheme a reality. The mass spectrometer, not the laser, performs the chemical differentiation. The commercial availability of intense laser radiation with high brightness makes this technique viable.

Photoionization

Two forms of nonselective photoionization have been used for SALI, one primarily for elemental analysis and the other primarily for molecular analysis. For elemental analysis, a powerful pulsed laser delivering focused power densities greater than 10^{10} W / cm^2 is used for multiphoton ionization (MPI). This typically ionizes all the species within the laser focus volume without the need for wavelength tuning and regardless of chemical type. This nonresonant photoionization yields the desired uniformity of detection probability essential for quantification. For molecular analysis, a soft (i.e., nonfragmenting) photoionization is needed so that the mass peaks in the mass spectrum correspond to larger chemical units. This is supplied by vacuum ultraviolet light with wavelengths in the range 115–120 nm (10–11 eV). Photoionization of this type is achieved by single-photon ionization (SPI).

Relative photoionization cross sections for molecules do not vary greatly between each other in this wavelength region, and therefore the peak intensities in the raw data approximately correspond to the relative abundances of the molecular species. Improvement in quantification for both photoionization methods is straightforward with calibration. Sampling the majority neutral channel means much less stringent requirements for calibrants than that for direct ion production from surfaces by energetic particles; this is especially important for the analysis of surfaces, interfaces, and unknown bulk materials.

Time-of-Flight Mass Spectrometry (TOFMS)

The advantage in using pulsed lasers is that they provide an excellent time marker for TOFMS. With TOFMS, a high mass resolution of several thousand can be achieved by energy focusing using a simple reflecting device, the instrument transmission is exceptional; and there is a multiplex advantage in mass. With the multiplex advantage, all masses are detected (parallel detection) within an extremely high mass range (up to 10,000 atomic mass units or more). The mass multiplex advantage has a dramatic impact on the instrument's sensitivity when numerous elemental or molecular species are present—a very common occurrence.

Surface Removal for Sampling

Surface removal for sampling involves removing atoms and molecules from the top surface layer into the vapor phase. The fact that the ionization step is decoupled from the surface removal step implies a great deal of flexibility and control in the types and conditions of the energetic beam of particles chosen to stimulate desorption. For elemental analysis of inorganic materials, typically a 50-μm Ar^+, or sub-μm diameter Ga^+ beam at several keV is used. Argon is used as an intense, high fluence ion beam that provides minimal chemical modification to the sample. Gallium is used as a liquid metal ion source that provides a highly focused, bright source for small area analysis (60–200 nm). Submonolayer or static analysis can be obtained by pulsing the beam and keeping the total dose extremely low ($< 10^{13}$ ions / cm^2). Depth profiling is accomplished by dc ion-beam milling and gating the pulsed photoionization to sample from the center of the sputter crater, which maintains state-of-the-art depth resolution. Ion-beam erosion is used to reveal buried interfaces during depth profiling, achieving a depth resolution often on the order of 20 Å after sputtering 1 μm in depth. The small-spot Ga^+ beam is well suited for quantitative chemical mapping with sub-μm spatial resolution. For other material types, such as bulk polymers, using energetic electrons, or another laser beam sometimes results in superior mass spectra; these sources often can remove clusters with less fragmentation, than pulsed ion beam sputtering and thus yield more characteristic mass peaks. For thermally sensitive samples, even thermal desorption can be used to investigate their temperature dependence.

Common Modes of Analysis and Examples

SALI applies two methods of post-ionization, MPI and SPI, each of which can be used in one of the three modes of analysis: survey analysis, depth profiling, and mapping:

1. Survey spectra using the MPI method are used primarily for quantification of surface components in inorganic materials, with a detection limit of ppm to ppb. The same mode coupled with SPI can be used for molecular characterization of polymer films.
2. Depth profiling by SALI provides quantitative information through interfaces and for extremely thin films, in the form of reliable chemical concentrations.
3. SALI mapping is a sensitive and quantitative method to characterize the spatial distribution of elements in both insulating and conductive materials.

Survey Mode

Surveys using MPI reveal the elemental composition of solid materials. Therefore this mode is employed most often in the analysis of inorganic materials like semiconductor devices and catalysts. Quantification can be achieved by using loosely matched standards and is accurate to within 10–20%. SPI has two advantages over MPI for the analysis of organic materials. First, it is a soft ionization method, so there is less fragmentation in addition to that of the primary beam, and second, the photoionization cross sections are nearly identical for molecules of similar size but different chemical type. This second characteristic enables SPI to provide semi-quantitative raw data for all classes of organic materials without rigorous standards. Figure 2 is an example of a SALI mass spectrum of polyethylene glycol using SPI. The dominance of the monomer peak is an example of the simple molecular identification using this technique.

Depth Profiling Mode

As stated above, SALI depth profiling is performed by gating the post-ionization beam by firing the laser only when the center of the crater is being sampled. This minimizes the contribution from the crater edge to the total signal at a specific depth, which increases the achievable depth resolution. Therefore, the depth resolution achieved by SALI easily equals that of SIMS which also employs gating. The major difference between these two depth profile techniques is that for SALI the sensitivity is nearly uniform for all elements, while for SIMS the sensitivity varies greatly. In selected cases this is an advantage for SIMS because the secondary ion yield for certain elements can be chemically enhanced, for example, by using a primary ion-beam composed of O_2^+ or Cs^+. However, it also severely limits the ability to quantify SIMS data because secondary ion yields can vary by orders of magnitude depending on the chemical composition of the matrix or probe beam. This is

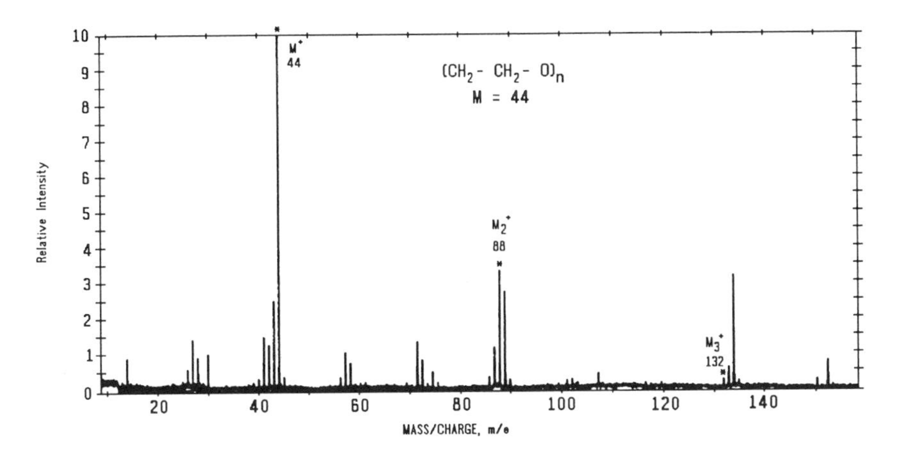

Figure 2 **SPI-SALI mass spectrum of a thin film of polyethylene glycol. The major peaks are identified on the spectrum. Analytical conditions: 7-keV Ar^+, 5-μs pulse length; 118-nm radiation.**

a problem when analyzing thin films and elemental distributions across chemically dissimilar interfaces because the changing ion yield causes changes in the ion signal intensity. In these cases SIMS ion yield transients can severely distort a depth profile and can be resolved only by using rigorous standards. An example[1] is a depth profile of a F implant (10^{15} F atoms/cm^2 at 93 keV) in a 2000 Å-thick polycrystalline Si sample on a thin SiO_2 layer on crystalline Si. Figure 3 is of an unannealed sample, where a smooth F distribution is expected. The SALI depth profile in Figure 3a shows the expected smooth distribution of the F implant. The SIMS data shown in Figure 3b, however, shows the common influence of matrix effects at an interface where the F positive ion yield is enhanced by the oxygen in the SiO_2 layer. The relative insensitivity of SALI to matrix effects is a tremendous advantage over SIMS in terms of quantitative depth profiling. Also, the useful yield (a measure of sensitivity) for the majority of elements falls into the 10^{-3} range when using SALI compared to the 10^{-2} to $< 10^{-7}$ range when using SIMS. Useful yield is defined as the number of ions detected versus the total material removed during analysis, and the efficiency of SALI can be equal to SIMS and orders of magnitude better than other nonselective post-ionization techniques (electron impact and radiofrequency low-pressure plasma).

Mapping Mode

The determination of the lateral distributions of chemical species on surfaces is of constantly increasing technological importance in many applications, such as integrated circuit manufacturing. The two major tools that have been available are

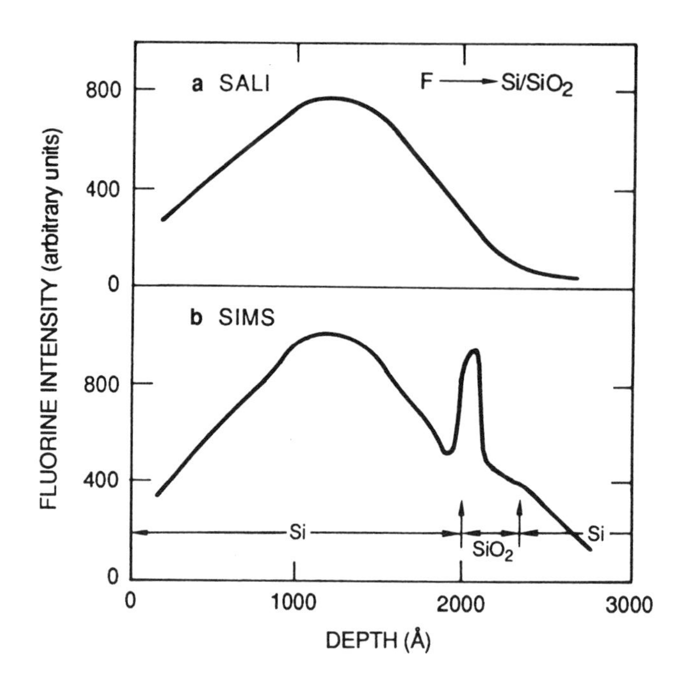

Figure 3 **Depth profiles of F implanted into 2000 Å Si on SiO_2: (a) SALI profile with Ar^+ sputtering and 248-nm photoionization; and (b) positive SIMS profile with O_2^+ sputtering. Analytical conditions: (SALI, SiF profile) 7-keV Ar^+, 248 nm; (SIMS, F profile) 7-keV O_2^+.**

Scanning Auger Electron Spectroscopy (SAM) and SIMS (in microprobe or microscope modes). SAM is the most widespread technique, but generally is considered to be of lesser sensitivity than SIMS, at least for spatial resolutions (defined by primary beam diameter *d*) of approximately ≥ 0.1 µm. However, with a field emission electron source, SAM can achieve sensitivities ranging from 0.3% at. to 3% at. for *d* ranging from 1000 Å to 300 Å, respectively, which is competitive with the best ion microprobes. Even with competitive sensitivity, though, SAM can be very problematic for insulators and electron-sensitive materials.

The sensitivities for SIMS are extremely variable, depending both on the species of interest and the local chemical matrix (so-called matrix effects). Quantification is very problematic for SIMS imaging because of matrix effects; on the very small scale associated with chemical imaging (sub-µm), it is not possible to generate closely matched reference materials because compositions change quickly and in an uncontrollable way. In the microscope mode, SIMS spatial resolutions are generally limited to about 1 µm. In the scanning mode, liquid-metal ion guns (notably Ga^+) have been used with better spatial resolution (sub-µm) but are somewhat unsatisfactory because Ga^+ is not effective for increasing secondary ion yields, unlike O_2^+

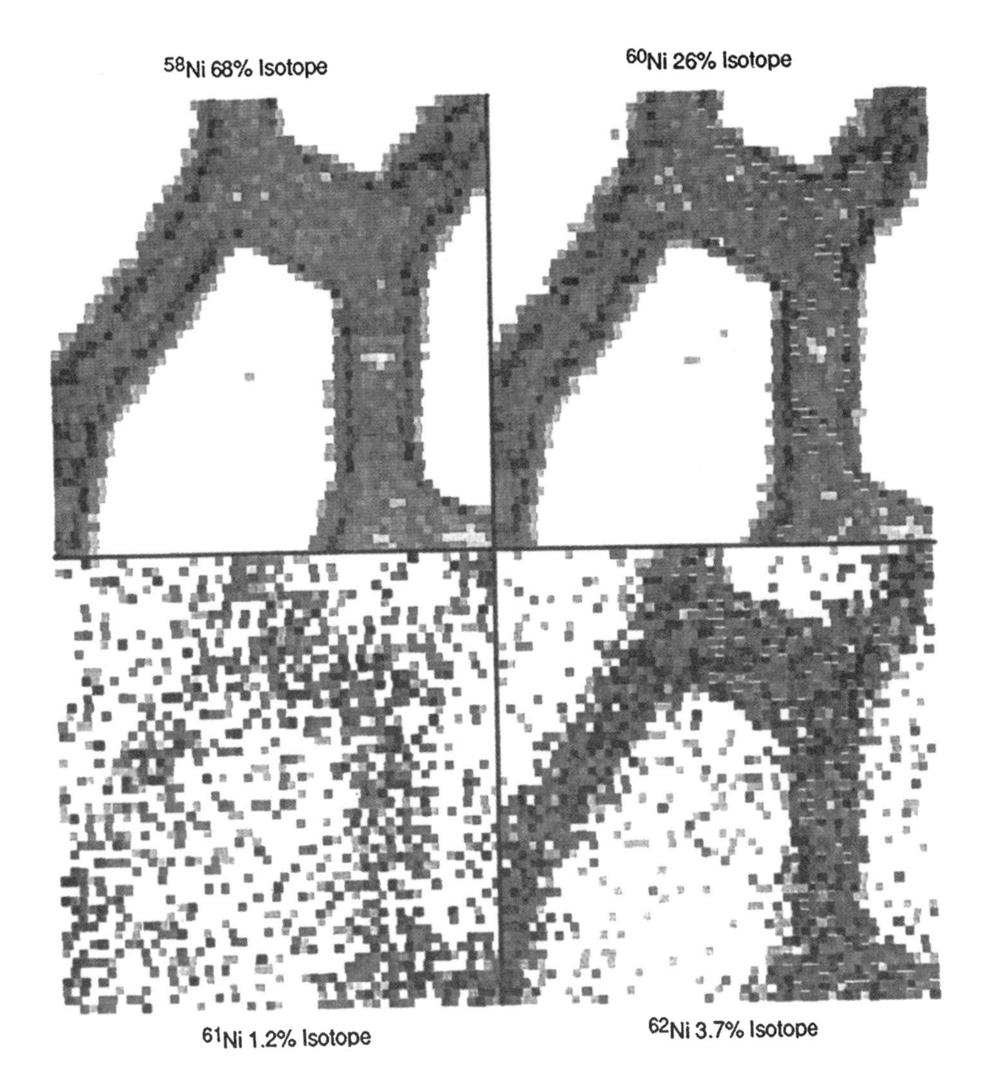

Figure 4 **Chemical images of a nickel TEM grid. Field of view is approximately 25 × 15 μm, 50 × 50 pixels. Analytical conditions: Ga^+ sputtering, spot size about 0.2 μm, 248-nm radiation, acquisition time 33 minutes.**

or Cs^+. The sensitivity for scanning SIMS can range, for example, from 0.01–10% at d = 1 μm (using O_2^+ or Cs^+ for ionization enhancement), to 1% to undetectable at d = 0.1 μm (using Ga^+).

By examining the sputtered neutral particles (the majority channel) using nonselective photoionization and TOFMS, SALI generates a relatively uniform sensitivity with semiquantitative raw data and overcomes many of the problems associated with SIMS. Estimates for sensitivities vary depending on the lateral spatial resolution for a commercial liquid-metal (Ga^+) ion gun. Calculated values[2] for SALI mapping show the sensitivity ranging from 0.2% to 3% at $d = 1$ to 0.1 μm. These sensitivities range as shown in Figure 4, which is a SALI image of a nickel TEM grid using Ga^+ sputtering and photoionization of the emitted neutrals at 248 nm (MPI, using KrF radiation). The pixel resolution achieved is < 0.5 μm, while the spot size d of the Ga^+ beam was 0.2 μm. As work in this area progresses and state-of-the-art liquid metal ion guns are used, the lateral resolutions achieved should approach the expected values. While the acquisition time for the sample image was somewhat long (33 minutes) this represents initial work. The acquisition time can be decreased readily by a factor of 10 with improvements in the computer system (factor of 2), and in off-the-shelf laser repetition rates (factor of 5). Since there exists a trade-off between analysis time an sensitivity, any decrease in acquisition time will make the application of SALI mapping more practical.

Instrumentation

A state-of-the-art SALI system combines both MPI and SPI capabilities. One commercial system[3] includes two laser sources: a Nd–YAG laser with a gas tube assembly used for frequency tripling to produce the coherent 118-nm light for SPI; and an excimer laser that produces both 248-nm (KrF) and 193-nm (ArF) wavelengths used for MPI. The system also includes two ion-beam sources: a duoplasmatron (Ar^+) or Cs^+ ion source, and a single or double lens liquid metal ion (Ga^+) source for SALI or TOF-SIMS mapping applications. Secondary Electron Detection (SED) images also can be obtained on this system, since it is equipped with an electron gun and the two ion guns. Each of these sources is compatible with the SED imaging system on the SALI instrument. The electron gun can also be used as an electron-stimulated desorption source. The instrument includes a TOF reflecting mass analyzer, a low-energy electron flood source for charge neutralization, a sample introduction system, a sample manipulator and a UHV chamber.

Conclusions

SALI is a relatively new surface technique that delivers a quantitative and sensitive measure of the chemical composition of solid surfaces. Its major advantage, compared to its "parent" technique SIMS, is that quantitative elemental and molecular information can be obtained. SPI offers exciting possibilities for the analytical characterization of the surfaces of polymers and biomaterials in which chemical differentiation could be based solely on the characteristic SALI spectra.

MPI is especially valuable for elemental analyses with typical useful yield of 10^{-3}. Because SALI is laser-based, expected improvements over the next few years, in particular for vacuum-ultraviolet laser technology, should have a significant impact. High repetition rate Nd–YAG systems with sufficient pulse energy are already available to 50 Hz, and probably can be extended to a few hundred Hz.

Ever brighter vacuum-ultraviolet sources are being developed that would further boost SPI sensitivity, which already is typically 10^{-5} useful yield; general, sensitive elemental analysis would then also be available using SPI, making possible a single laser arrangement for both elemental and molecular SALI.

Related Articles in the Encyclopedia

Static SIMS and SNMS

References

1 C. H. Becker. In: *Ion Spectroscopies for Surface Analysis.* (A. W. Czanderna and D. M. Hercules, eds.) Plenum Press, New York, 1991, Chapter 4, p. 273.

2 D. G. Welkie, S. M. Daiser, and C. H. Becker. *Vacuum.* **41,** 1665, (1991); S.P. Mouncey, L. Moro, and C.H. Becker, *Appl. Surf. Anal.* **52**, 39 (1991).

3 Perkin-Elmer Physical Electronics Division, Eden Prairie, MN, model 7000 SALI / TOF–SIMS instrument.

Bibliography

1 W. Reuter, in *Secondary Ion Mass Spectrometry SIMS V,* Springer-Verlag, Berlin, 1986, p. 94.

A comparison of the various post-ionization techniques: electron-gas bombardment, resonant and nonresonant laser ionization, etc. While some of the numbers are outdated, the relative capabilities of these methods have remained the same. This is a well-written review article that reiterates the specific areas where post-ionization has advantages over SIMS.

2 J. B. Pallix, C. H. Becker, and N. Newman, *MRS Bulletin*, **12**, no. 6, 52 (1987).

A discussion of the motivation behind doing sputtered neutral analysis versus SIMS, plus a description of the first prototype SALI instrument. A well written introduction for someone without previous surface analysis experience it also includes an historical overview of the various post-ionization techniques.

3 C. H. Becker, *J. Vac. Sci. Technol.*, **A5**, 1181 (1987).

This article discusses why one would choose nonresonant multiphoton ionization for mass spectrometry of solid surfaces. Examples are given for depth profiling by this method along with thermal desorption studies.

4 J. B. Pallix, C. H. Becker, and K. T. Gillen, *Appl. Surface Sci*, **32**, 1 (1988).

An applications oriented discussion of using MPI-SALI for depth profiling, interface analysis in inorganic material systems. Examples of SALI depth profiles are given of a B implant in Si and the fluorine implanted electronic test device which was referenced in this encyclopedia article.

5 J. B. Pallix, U. Schühle, C. H. Becker, and D. L. Huestis, *Anal. Chem.*, **61**, 805 (1989).

An introduction to the principles behind SPI-SALI, this article presents a theoretical discussion of why SPI-SALI is much less fragmenting than MPI-SALI. Examples are shown which describe the additional fragmentation induced by the desorption beam—in this case ESD is compared to ion sputtering. The main focus of the article is the advantages of SPI-SALI for surface analysis of bulk organic polymers.

10.4 SNMS

Sputtered Neutral Mass Spectrometry

JOHN C. HUNEKE

Contents

Introduction

The atom flux sputtered from a solid surface under energetic ion bombardment provides a representative sampling of the solid. Sputtered neutral mass spectrometry has been developed as method to quantitatively measure the composition of this atom flux and thus the composition of the sputtered material. The measurement of ionized sputtered neutrals has been a significant improvement over the use of sputtered ions as a measure of flux composition (the process called SIMS), since sputtered ion yields are seriously affected by matrix composition. Neutral particles are ionized by a separate process after sputter atomization, and SNMS quantitation is thus independent of the matrix. Also, since the sputtering and ionization processes are separate, an ionization process can be selected that provides relatively uniform yields for essentially all elements.

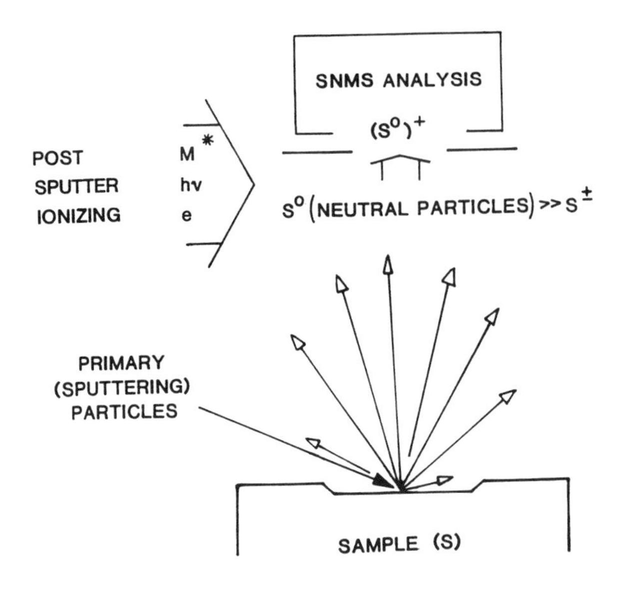

Figure 1 **Schematic of SNMS analysis. Neutral atoms and molecules sputtered from the sample surface by energetic ion bombardment are subsequently ionized for mass spectrometric analysis.**

The erosion of the surface by sputtering also provides a means to sample progressively deeper layers to determine concentration depth profiles. SNMS combines the features of sputter erosion, representative sampling, uniform ionization and matrix independence to provide a quantitative sputtering depth profile measurement with comparable sensitivity for all elements in complex thin-film structures comprising a variety of matrix compositions. This capability is used to good advantage, for example, in the compositional analysis of thin-film structures used for magnetic recording heads, optoelectronics, and semiconductor metallizations. The detection limits obtained in depth profiling by SNMS range from 100 to 1000 ppm, which are not as low as can be generally obtained using SIMS, but which are a significant improvement over the detection limits obtained using AES and ESCA.

Useful overviews of all SNMS modes have been provided by Oeschner[1] and Pallix and Becker,[2] and thorough reviews of electron impact SNMS in particular have been provided recently both by Ganschow[3] and by Jede.[4]

Basic Principles

The essentials of SNMS are illustrated in Figure 1. The surface of the solid sample is sputtered by energetic ion bombardment. Generally, at energies above a few hundred eV, several particles are ejected from the surface for each incident particle. A very small fraction of the particles are sputtered as ions, the so-called secondary ions measured by SIMS. By far the larger number of sputtered particles are neutral atoms and molecules, with atoms dominating. Coupling the nonselective sputter atomization with some method for nonselective ionization for mass spectrometric analysis, as illustrated in Figure 1, SNMS provides a technique in which the analytical signals directly reflect the sample's composition, unlike SIMS where the ionization can be very selective.

If the secondary ion component is indeed negligible, the measured SNMS ion currents will depend only on the ionizing mode, on the atomic properties of the sputtered atoms, and on the composition of the sputtered sample. Matrix characteristics will have no effect on the relative ion currents. SNMS analysis also provides essentially complete coverage, with almost all elements measured with equal facility. All elements in a chemically complex sample or thin-film structure will be measured, with no incompleteness due to insensitivity to an important constituent element. Properly implemented SNMS promises to be a near-universal analytical method for solids analysis.

The SIMS analytical ion signal of a specific element or isotope also can be enhanced by selective ionization of particular atoms, and the detection limit for that element thereby improved. This mode of SNMS is important to specific applications, but it lacks the generality inherent in nonselective SNMS methods. The focus of this article will be on the methods for obtaining complete, accurate, and matrix-independent compositions of chemically complex thin-film structures and materials.

SNMS Modes and Instrumentation

Three post-ionization (i.e., post-sputtering) mechanisms having relatively uniform ion yields appropriate for SNMS have been used to date (cf. Figure 1), and the combination of each of these with an appropriate mass spectrometer provides the basis for all present SNMS instruments. The postionization methods are: electron impact ionization, ionization by multiple photon interactions, and ionization by collision with metastable atoms in a plasma. The SNMS methods incorporating the latter two modes are nonresonant multiphoton ionization mass spectrometry, which is also referred to as Surface Analysis by Laser Ionization (SALI), and Glow-Discharge Mass Spectrometry (GDMS). Both are the subject of other articles in this encyclopedia but are mentioned here because of analytical affinities. This article will concentrate on the first mode, SNMS by electron impact ionization.

The various SNMS instruments using electron impact postionization differ both in the way that the sample surface is sputtered for analysis and in the way the ionizing electrons are generated (Figure 2). In all instruments, an ionizer of the electron-gun or electron-gas types is inserted between the sample surface and the mass spectrometer. In the case of an electron-gun ionizer, the sputtered neutrals are bombarded by electrons from a heated filament that have been accelerated to 80–

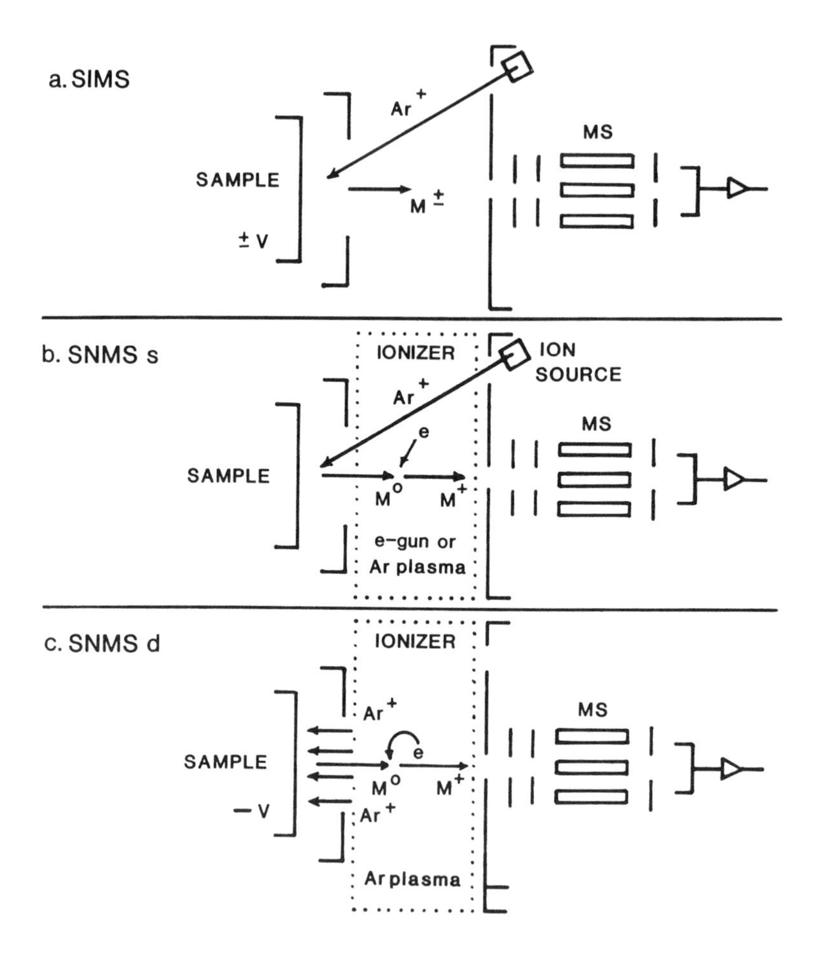

Figure 2 **Relationship of SIMS, separate bombardment SNMSs and direct bombardment SNMSd. (a) Materials for SIMS analysis are those ions formed in the sputtering with a focused primary ion beam. The largest fraction of the particles sputtered from the surface are neutral atoms. (b) Ions for SNMS analysis are formed by ionization of the sputtered neutrals. (c) When the plasma is used as an ionizer, plasma ions can also be used to sputter the sample surface at low energies.**

100 eV energy in the ionizer volume. Sputtering of the sample surface in electron-gun SNMS is accomplished by a focused ion beam, as provided for the SIMS instrumentation (Figures 2a and 2b). An electron-gun ionizer is commonly used to modify existing SIMS instrumentation to provide supplementary SNMS capability, and SNMS measurements using the separate ion-gun sputtering provided for SIMS will have the same imaging and thin-film depth resolution capabilities as for SIMS. In contrast to SIMS, a noble gas (e.g., Ar) or less reactive element (such as Ga) is most commonly used for ion-beam sputtering in SNMS to minimize

enhancement of secondary ion yields and thus improve quantitative accuracy. Following the initial development and use of electron-gun SNMS in the academic research environment, two such SNMS modules have been made commercially available.

In two other implementations of electron impact SNMS, a plasma is generated in the ionizer volume to provide an electron gas sufficiently dense and energetic for efficient postionization (Figure 2c). In one instrument, the electrons are a component of a low-pressure radiofrequency (RF) plasma in Ar, and in the second, the plasma is an electron beam excited plasma, also in Ar. The latter type of electron-gas SNMS is still in the developmental stages, while the former has been incorporated into commercial instrumentation.

The plasma electron-gas ionizer offers a distinct advantage for low-energy sputtering of the sample surface, since Ar ions are available in the overlying plasma and can be accelerated onto the surface by application of a sample bias voltage (Figure 2c). In neither of the electron-gas SNMS modes does the sample play an integral part in maintaining the plasma. Thus the bias voltage can be made arbitrarily small, enabling plasma-ion sputtering with minimal energy, ion-beam mixing, and thus optimal depth resolution (see below). The surface areas analyzed in the plasma mode of surface sputtering are large compared to SIMS or SNMS profiling by separate focused ion-beam bombardment, and lateral resolution is sacrificed.

The SNMS instrumentation that has been most extensively applied and evaluated has been of the electron-gas type, combining ion bombardment by a separate ion beam and by direct plasma-ion bombardment, coupled with postionization by a low-pressure RF plasma. The direct bombardment electron-gas SNMS (or SNMSd) adds a distinctly different capability to the arsenal of thin-film analytical techniques, providing not only matrix-independent quantitation, but also the excellent depth resolution available from low-energy sputtering. It is from the application of SNMSd that most of the illustrations below are selected. Little is lost in this restriction, since applications of SNMS using the separate bombardment option have been very limited to date.

Analysis and Quantitation

In the process of SNMS analysis, sputtered atoms are ionized while passing through the ionizer and are accelerated into the mass spectrometer for mass analysis. The ion currents of the analyzed ions are measured and recorded as a function of mass while stepping the mass spectrometer through the desired mass or element sequence. If the purpose of the analysis is to develop a depth profile to characterize the surface and subsurface regions of the sample, the selected sequence is repeated a number of times to record the variation in ion current of a selected elemental isotope as the sample surface is sputtered away.

Only the knowledge of relative useful ion yields and isotopic abundances is required to calculate elemental composition from the relative ion current measurements. The useful ion yield D_X is the number of ions x^+ detected relative to the number of atoms of element x sputtered. The measured relative ion current of two isotopes is

$$\frac{I_x}{I_s} = \left(\frac{i_x}{i_s}\right)\left(\frac{c_x}{c_s}\right)\left(\frac{D_x}{D_s}\right) \tag{1}$$

where c_X is the concentration and i_X is the isotopic fraction of the measured isotope of element x. Pragmatically, quantitation is accomplished by multiplying the ratios of the total ion currents for each element (summing over all isotopes of the element) by a multiplicative factor defined as the relative sensitivity factor or RSF.

$$\frac{c_x}{c_s} = \left(\frac{I_x}{I_s}\right)\text{RSF}\left(\frac{x}{s}\right) \tag{2}$$

The fraction of each element present in the material is then equal to the ratio of the RSF-corrected ion current for that element to the sum of the RSF-corrected ion currents for all elements.

It is important for quantitative SNMS that the fractions of element x forming molecules and sputtered ions be negligible, but such is not always the case.

Relative Sensitivity Factors

The relative sensitivity factors for most elements are comparable to within a factor of 25 for ionization with an energetic electron gas.[5] The RSFs for a number of elements determined from the analysis of NIST alloy samples vary by less than an order of magnitude for sputtering energies of 1250 V and more. The RSFs determined are reasonably independent of matrix. Nevertheless, there are differences of up to factors of 1.5 in RSFs of the same element determined from the analysis of several standards. Also, RSFs do change significantly with sputtering voltage. As a consequence, separate calibration is required when sputtering at the lower energies typical of depth profiling.

Similar detailed studies of RSFs have been carried out for GDMS, but not for electron-gun electron impact ionization or for SALI. The spread in elemental RSFs for electron-gas SNMS is comparable to that observed for Ar glow-discharge ionization of sputtered neutrals.[6]

Since elemental RSFs are reasonably similar for electron-gas SNMSd, a standardless analysis will result in compositions accurate to within a factor of 5 for matrices with major element RSFs close to the average, and to within a factor of 25 for matrices with major element RSFs at the extreme values. More importantly,

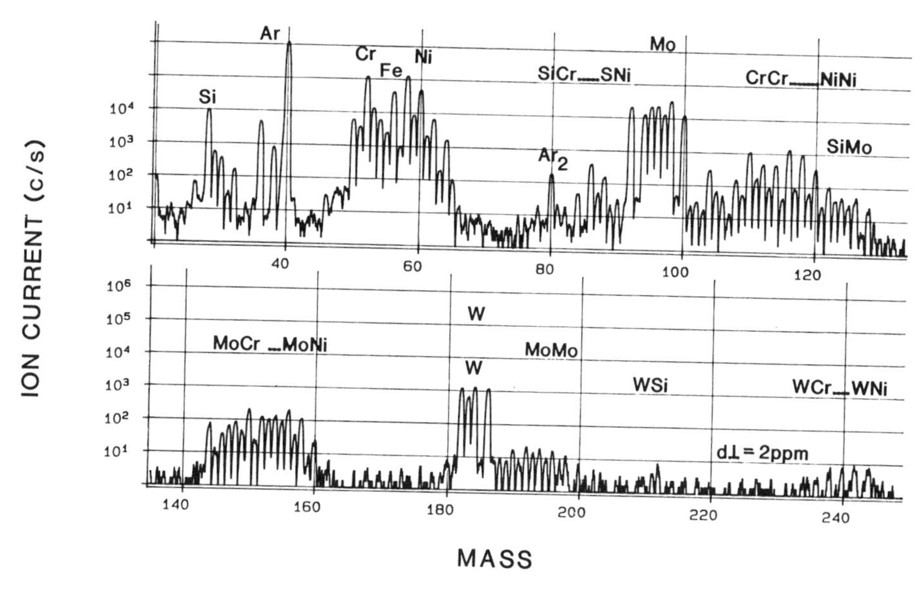

Figure 3 **Mass spectrum obtained from the NIST Hasteloy Ni-based standard alloy, using electron-gas SNMSd (Leybold INA-3). The sputtering energy was 1250 V, increasing the sputtered atom flux at the expense of depth resolution. Matrix ion currents were about 10^5 cps, yielding background limited detection at about 2 ppm.**

however, there will be no glaring gaps in the analytical results due to extreme insensitivity for a particular element. Every element present will be detected at roughly the same sensitivity. This characteristic of SNMS enables thorough materials characterization of complex samples in a single analysis and by one instrument.

Applications

Bulk Analysis

Independent of depth profiling considerations, SNMS provides a powerful bulk analysis method that is sensitive and accurate for all elements, from major to trace element levels. Since SNMS is universally sensitive, it offers obvious advantages over elementally selective optical methods.

As an example of a standardless bulk analysis by SNMS, a measurement of the complex Ni-based Hasteloy metal (NIST SRM 2402) is presented in Figure 3 and Table 1, in which the "composition" determined from ion-current ratios (not RSF corrected) is compared to the certified chemical composition.

It is very evident in Figure 3 that the chemical complexity of Hasteloy presents special problems for mass spectrometric analysis using a quadrupole mass spectrometer with low mass resolution. Molecular ions comprised of combinations of matrix and plasma atoms are formed in abundance and will obscure many elements

Element	Ion current ratio (%)	Certified content (% at.)
C		0.05
Si		1.90
P	0.005	0.014
S	0.02	0.035
V	0.6	0.27
Cr	26	19.5
Mn	0.6	0.73
Fe	9	8.2
Co	2	1.6
Ni	37	55
Cu	0.07	0.19
Mo	25	11.2
W	0.9	1.46

Table 1 **Semiquantitative bulk analysis by SNMSd of the NIST SRM 2402 Hasteloy metal standard.**

present at low concentrations. This is obviously a clear limitation to many applications, and the use of a high resolving power mass spectrometer is to be preferred in these instances.

Molecular ion mass interferences are not as prevalent for the simpler matrices, as is clear from the mass spectrum obtained for the Pechiney 11630 Al standard sample by electron-gas SNMSd (Figure 4). For metals like high-purity Al, the use of the quadrupole mass spectrometer can be quite satisfactory. The dopant elements are present in this standard at the level of several tens of ppm and are quite evident in the mass spectrum. While the detection limit on the order of one ppm is comparable to that obtained from optical techniques, the elemental coverage by SNMS is much more comprehensive.

Quantitative Depth Profiling

In addition to comprehensive elemental coverage, SNMS also provides for high-resolution depth profile measurements with the same quantitation capability

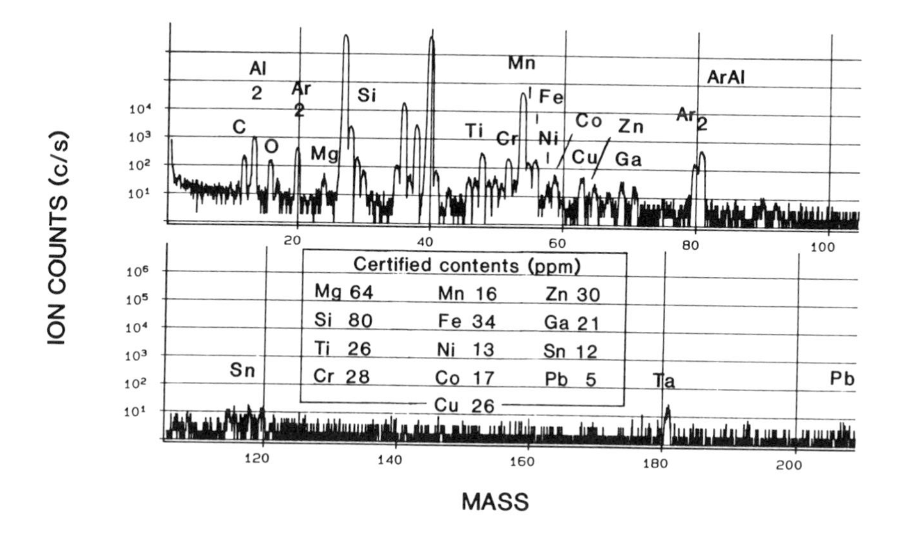

Figure 4 **Mass spectrum obtained from the Aluminium Pechiney standard Al 11630, using electron-gas SNMSd with a sputtering energy of 1250 V. The ^{27}Al matrix ion current was significantly greater than 10^6 cps, yielding a background count rate limit less than 1 ppm.**

throughout the depth profile regardless of film composition. This feature of SNMS is particularly useful for the measurement of elements located in and near interfaces, which are difficult regions for measurement by other thin-film analytical methods.

The advantage of SNMSd for high-resolution profiling derives from the sputtering of the sample surface at arbitrarily low energies, so that ion-beam mixing can be reduced and depth resolution enhanced. Excellent depth resolution by SNMSd depth profiling is well illustrated by the SNMSd depth profile of a laser diode test structure shown in Figure 5. Structures of this type are important in the manufacture of optoelectronics devices. The test structure is comprised of a GaAs cap overlying a sandwiched sequence of $Al_xGa_{1-x}As$ layers, where the intermediate Al-poor layer is on the order of 100 Å thick. The nominal compositions from growth parameters are noted in Figure 5. The layers are very well resolved to about a 30-Å depth resolution, with accurate composition measurement of each individual layer.

Every material sputters at a characteristic rate, which can lead to significant ambiguity in the presentation of depth profile measurements by sputtering. Before an accurate profile can be provided, the relative sputtering rates of the components of a material must be independently known and included, even though the total depth of the profile is normally determined (e.g., by stylus profilometer). To first order, SNMS offers a solution to this ambiguity, since a measure of the total number of atoms being sputtered from the surface is provided by summing all RSF- and

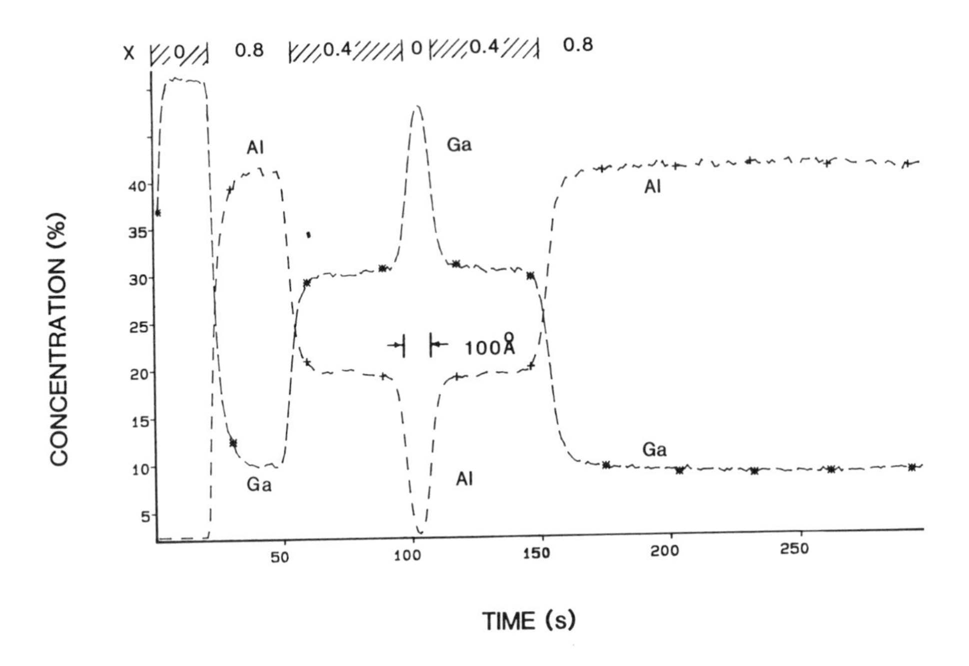

Figure 5 **Quantitative high depth resolution profile of a complex $Al_xGa_{1-x}As$ laser diode test structure obtained using electron-gas SNMS in the direct bombardment mode, with 600-V sputtering energy. The data have been corrected for relative ion yield variations and summed to Al + Ga = 50%. The 100-Å thick GaAs layer is very well resolved.**

isotope-corrected ion currents (assuming all major species have been identified and included in the measurements.) It is necessary only to scale the time required to profile through a layer by the total sputtered neutral current (allowing for atomic density variations) to have a measure of the relative layer thickness. The profiles illustrated in Figure 5 have not been corrected for this effect.

Al Metallization

The measurement of the concentration depth profiles of the minor alloying elements Si and Al in Al metallizations is also very important to semiconductor device manufacturing. The inclusion of Si prevents unwanted alloying of underlying Si into the Al. The Cu is included to prevent electromigration. These alloying elements are typically present at levels of 1% or less in the film, and the required accuracy of the measurement is several percent. Of the techniques that can be applied to this analysis, SNMS offers the combined advantages of sensitivity to both Si and Cu, good detection limits in the depth profiling (0.01–0.1%), and accuracy of analysis, as well as requiring measuring times on the order of only one-half hour.

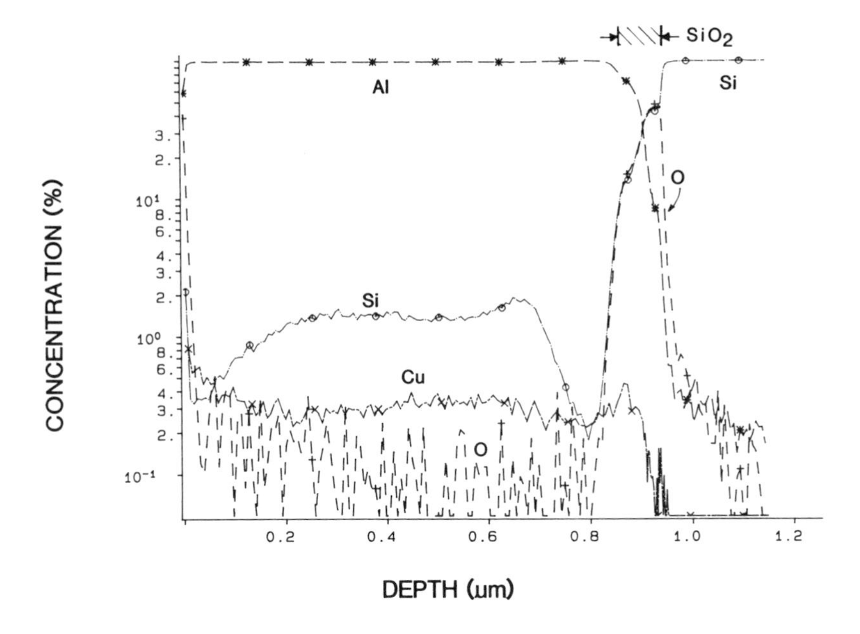

Figure 6 **Quantitative depth profile of the minor alloying elements Cu and Si in Al metallization on SiO_2/Si, using electron-gas SNMSd.**

A typical SNMSd profile of Al (1% Si, 0.5% Cu) metallization on SiO_2 is shown in Figure 6. The O signal is included as a marker for the Al/SiO_2 interface. The Al matrix signal is some 10^5 cps, yielding an ion count rate detection limit of 10 ppm for elements with similar RSF. The detection limit is degraded from this value by a general mass-independent background of 5 cps and by contamination by O and Si in the plasma. It does not help that in this instance the product (ion yield) × (isotopic abundance) for Cu is an order of magnitude lower than for Al. Nonetheless, the signals of both Si and Cu are quite adequate to the measurement. The Si exhibits a strongly varying composition with depth into the film, in contrast to the Cu distribution.

Diffusion Barriers

An important component of the complex metallizations for both semiconductor devices and magnetic media is the diffusion barrier, which is included to prevent interdiffusion between layers or diffusion from overlying layers into the substrate. A good example is placement of a TiN barrier under an Al metallization. Figure 7a illustrates the results of an SNMSd high-resolution depth profile measurement of a TiN diffusion barrier inserted between the Al metallization and the Si substrate. The profile clearly exhibits an uneven distribution of Si in the Al metallization and has provided a clear, accurate measurement of the composition of the underlying TiN layer. Both measurements are difficult to accomplish by other means and dem-

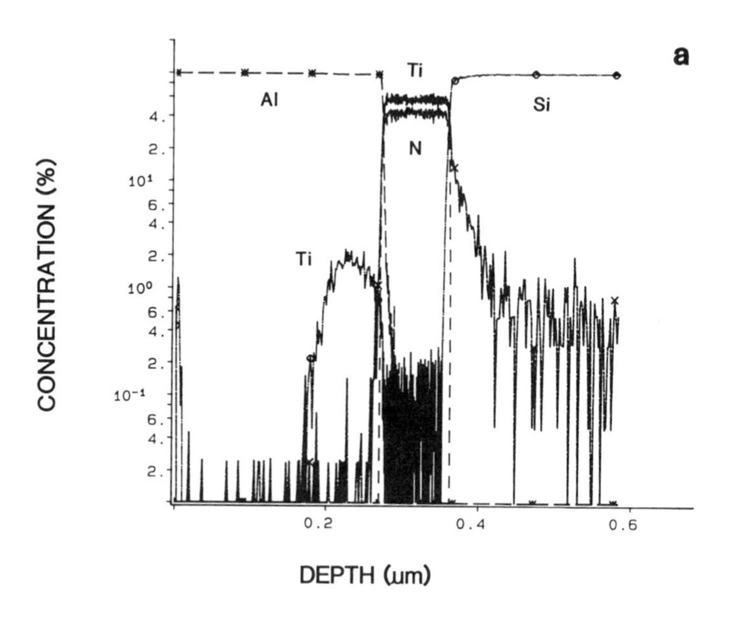

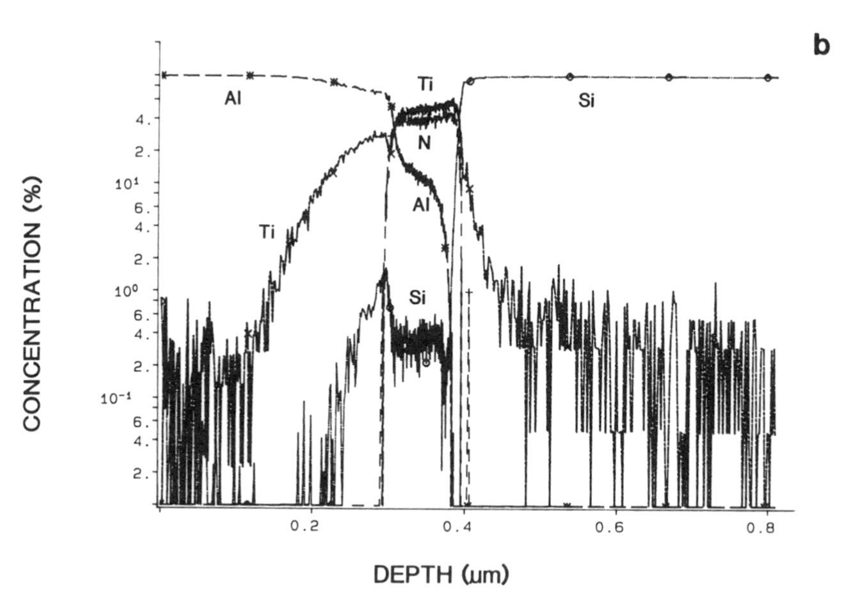

Figure 7 **Quantitative high depth resolution profile of the major elements in the thin-film structure of Al / TiN / Si, comparing the annealed and unannealed structures to determine the extent of interdiffusion of the layers. The depth profile of the unannealed sample shows excellent depth resolution (a). The small amount of Si in the Al is segregated toward the Al / TiN interface. After annealing, significant Ti has diffused into the Al layer and Al into the TiN layer, but essentially no Al has diffused into the Si (b). The Si has become very strongly localized at the Al / TiN interface.**

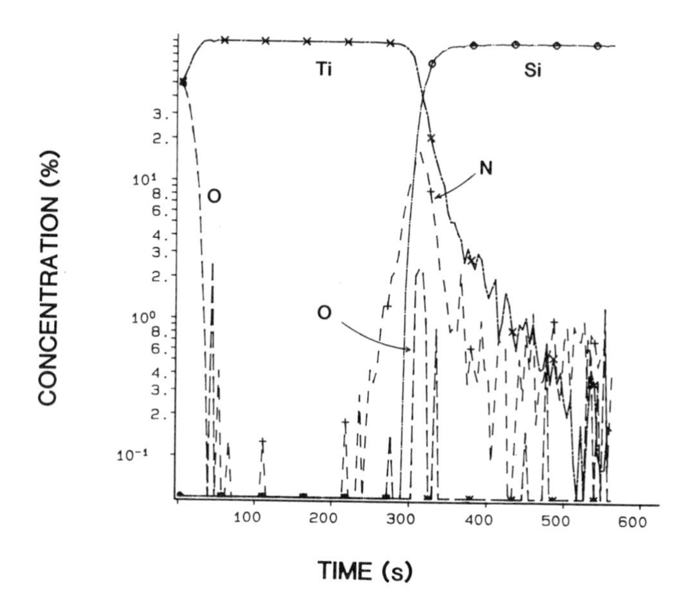

Figure 8 Quantitative high depth resolution profile of O and N in a Ti metal film on Si, using electron-gas SNMS in the direct bombardment mode. Both O and N are measured with reasonably good sensitivity and with good accuracy both at the heavily oxidized surface and at the Ti/Si interface.

onstrate the strength of SNMS for providing quantitative measurements in all components of a complex thin-film structure. The results of processing this structure of Al:Si/TiN/Si are shown in Figure 7b. The measurement identifies the redistribution of the Si to the interfaces, the diffusion of Al and Si into the TiN, and a strong diffusion out of Ti from the TiN into the overlying Al. However, no Al has diffused into the Si nor Si from the substrate into the Al, demonstrating the effectiveness of the TiN barrier.

Yet another strength of SNMS is the ability to measure elemental concentrations accurately at interfaces, as illustrated in Figure 8, which shows the results of the measurement of N and O in a Ti thin film on Si. A substantial oxide film has formed on the exposed Ti surface. The interior of the Ti film is free of N and O, but significant amounts of both are observed at the Ti/Si interface. SNMS is as sensitive to O as to N, and both the O and N contents are quantitatively measured in all regions of the structure, including the interface regions. Quantitation at the interface transition between two matrix types is difficult for SIMS due to the matrix dependence of ion yields.

Conclusions

The combination of sputter sampling and postsputtering ionization allows the atomization and ionization processes to be separated, eliminating matrix effects on elemental sensitivity and allowing the independent selection of an ionization process with uniform yields for essentially all elements. The coupling of such a uniform ionization method with the representative sampling by sputtering thus gives a "universal" method for solids analysis.

Electron impact SNMS has been combined most usefully with controlled surface sputtering to obtain accurate compositional depth profiles into surfaces and through thin-film structures, as for SIMS. In contrast to SIMS, however, SNMS provides accurate quantitation throughout the analyzed structure regardless of the chemical complexity, since elemental sensitivity is matrix independent. When sputtering with a separate focused ion beam, both image and depth resolutions obtained are similar to the those obtained by SIMS. However, using electron-gas SNMS, in which the surface can be sputtered by plasma ions at arbitrarily low bombarding energies, depth resolutions as low as 2 nm can be achieved, although lateral image resolution is sacrificed.

In summary, the forte of SNMS is the measurement of accurate compositional depth profiles with high depth resolution through chemically complex thin-film structures. Current examples of systems amenable to SNMS are complex III-IV laser diode structures, semiconductor device metallizations, and magnetic read–write devices, as well as storage media.

SNMS is still gaining industrial acceptance as an analytical tool, as more instruments become available and an appreciation of the unique analytical capabilities is developed. To date, SNMS has not become established as a routine analytical tool providing essential measurements to a significant segment of industry. The technique still remains largely in the domain of academic and research laboratories, where the full range of application is still being explored. The present stage of SNMS development is appropriate to this environment, and refinements in hardware and software can be expected, given a unique niche and the pressure of commercial or industrial use.

In addition to the analysis of complex thin-film structures typical of the semiconductor industry, for which several excellent examples have been provided, an application area that offers further promise for increased SNMS utilization is the accurate characterization of surfaces chemically modified in the outer several hundred-Å layers. Examples are surfaces altered in some way by ambient environments—a sheet steel surface intentionally altered to enhance paint bonding, or phosphor particles with surfaces altered to enhance fluorescence. A strength of SNMS that will also become more appreciated with time is its ability to provide, with good depth resolution, quantitative measurements of material trapped at interfaces, for example, contaminants underlying deposited thin films or migrating to interfacial regions during subsequent processing. As these and other application areas are explored more fully, the place of SNMS will become more evident and secure, and the evolution of SNMS instrumentation even more rapid.

Related Articles in the Encyclopedia

SIMS, SALI, and LIMS

References

1. H. Oechsner. *Scanning Microscopy.* **2,** 9, 1988.
2. J. B. Pallix and C. H. Becker. In *Advanced Characterization Techniques for Ceramics.* (G. L. McVay, G. E. Pike, and W. S. Young, Eds.) ACS, Westerville, 1989.
3. O. Ganschow. In *Analytical Techniques for Semiconductor Materials and Process Characterization.* The Electrochemical Society, Pennington, 1990, Vol. 90-11, p. 190.
4. R. Jede. In *Secondary Ion Mass Spectrometry.* (A. Benninghoven, C. A. Evans, K. D. McKeegan, H. A. Storms, and H. W. Werner, Eds.) J. Wiley and Sons, New York, 1989, p. 169.
5. A. Wucher, F. Novak, and W. Reuter. *J. Vac. Sci. Technol.* **A6,** 2265, 1988.
6. W. Vieth and J. C. Huneke. *Spectrochim. Acta.* **46B** (2), 137, 1991.

10.5 LIMS

Laser Ionization Mass Spectrometry

FILIPPO RADICATI DI BROZOLO

Contents

Introduction

Laser ionization mass spectrometry or laser microprobing (LIMS) is a microanalytical technique used to rapidly characterize the elemental and, sometimes, molecular composition of materials. It is based on the ability of short high-power laser pulses (~10 ns) to produce ions from solids. The ions formed in these brief pulses are analyzed using a time-of-flight mass spectrometer. The quasi-simultaneous collection of all ion masses allows the survey analysis of unknown materials. The main applications of LIMS are in failure analysis, where chemical differences between a contaminated sample and a control need to be rapidly assessed. The ability to focus the laser beam to a diameter of approximately 1 mm permits the application of this technique to the characterization of small features, for example, in integrated circuits. The LIMS detection limits for many elements are close to 10^{16} at/cm^3, which makes this technique considerably more sensitive than other survey microanalytical techniques, such as Auger Electron Spectroscopy (AES) or Electron Probe Microanalysis (EPMA). Additionally, LIMS can be used to analyze insulating samples, as well as samples of complex geometry. Another advantage of this technique is its ability sometimes to provide basic molecular information about inorganic as well as organic surface contaminants. A growing field of application is the characterization of organic polymers, and computerized pattern recognition techniques have been successfully applied to the classification of various types of mass spectra acquired from organic polymers.

The LIMS technique is rarely used for quantitative elemental analysis, since other techniques such as EPMA, AES or SIMS are usually more accurate. The limitations of LIMS in this respect can be ascribed to the lack of a generally valid model to describe ion production from solids under very brief laser irradiation. Dynamic range limitations in the LIMS detection systems are also present, and will be discussed below.

Basic Principles

LIMS uses a finely focused ultraviolet (UV) laser pulse (~10 ns) to vaporize and ionize a microvolume of material. The ions produced by the laser pulse are accelerated into a time-of-flight mass spectrometer, where they are analyzed according to mass and signal intensity. Each laser shot produces a complete mass spectrum, typically covering the range 0–250 amu. The interaction of laser radiation with solid matter depends significantly upon the duration of the pulse and the power density levels achieved during the pulse.[1] When the energy radiated into the material significantly exceeds its heat of vaporization, a plasma (ionized vapor) cloud forms above the region of impact. The interaction of the laser light with the plasma cloud further enhances the transfer of energy to the sample material. As a consequence, various types of ions are formed from the irradiated area, mainly through a process called nonresonant multiphoton ionization (NRMPI). The relative abundances of the ions are a function of the laser's power density and the optical properties and chemical state of the material. Typically, the ion species observed in LIMS include singly charged elemental ions, elemental cluster ions (for example, the abundant C_x^- negative ions observed in the analysis of organic substances), and organic fragment ions. Multiply charged ions are rarely observed, which sets an approximate upper limit on the energy that is effectively transferred to the material.[1, 5]

The material evaporated by the laser pulse is representative of the composition of the solid,[1] however the ion signals that are actually measured by the mass spectrometer must be interpreted in the light of different ionization efficiencies. A comprehensive model for ion formation from solids under typical LIMS conditions does not exist, but we are able to estimate that under high laser irradiance conditions ($>10^{10}$ W/cm^2) the detection limits vary from approximately 1 ppm atomic for easily ionized elements (such as the alkalis, in positive-ion spectroscopy, or the halogens, in negative-ion spectroscopy) to 100–200 ppm atomic for elements with poor ion yields (for example, Zn or As).

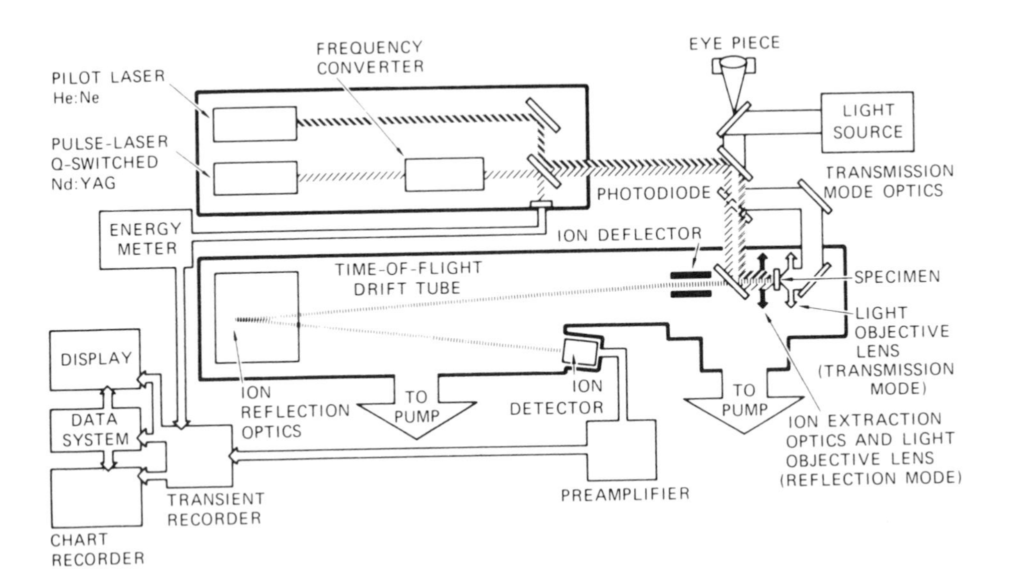

Figure 1 **Schematic diagram of a LIMS instrument, the LIMA 2A (Cambridge Mass Spectrometry, Ltd., Kratos Analytical, UK).**

The large variability in elemental ion yields which is typical of the single-laser LIMS technique, has motivated the development of alternative techniques, that are collectively labeled post-ablation ionization (PAI) techniques. These variants of LIMS are characterized by the use of a second laser to ionize the neutral species removed (ablated) from the sample surface by the primary (ablating) laser. One PAI technique uses a high-power, frequency-quadrupled Nd–YAG laser (λ = 266 nm) to produce elemental ions from the ablated neutrals, through nonresonant multiphoton ionization (NRMPI). Because of the high photon flux available, 100% ionization efficiency can be achieved for most elements, and this reduces the differences in elemental ion yields that are typical of single-laser LIMS. A typical analytical application is discussed below.

Instrumentation

The schematic diagram of a LIMS instrument is shown in Figure 1. The instrument's basic components include:

1 A *Q*-switched, frequency-quadrupled Nd–YAG laser (λ = 266 nm) and its accompanying optical components produce and focus the laser pulse onto the sample surface. The typical laser spot size in this instrument is approximately 2 µm. A He–Ne pilot laser, coaxial with the UV laser, enables the desired area to be located. A calibrated photodiode for the measurement of laser energy levels is also present

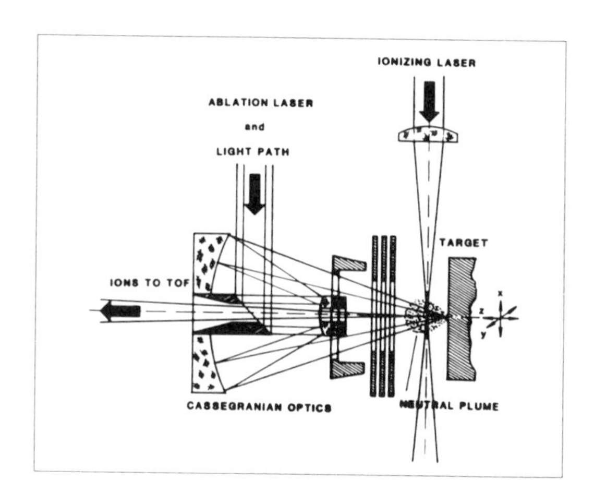

Figure 2 **Schematic view of the ion source region of the LIMS instrument in the PAI configuration.**

2 The time-of-flight mass spectrometer (under high vacuum) consists of a sample stage equipped with *xyz* motion, the ion extraction region, and the ion flight tube (approximately 2 m in length) with energy focusing capabilities

3 The ion detection system consists of a high-gain electron multiplier and the signal digitizing system, along with a computer for data acquisition and manipulation.

Figure 2 presents a schematic view of the ion source region in the PAI configuration. A second high-irradiance, frequency quadrupled pulsed Nd–YAG laser is focused parallel to and above the sample surface, where it intercepts the plume of neutral species that are produced by the ablating laser. Appropriate focusing optics and pulse time-delay circuitry are used in this configuration.

A typical LIMS analysis is performed by positioning the region of interest of the sample by means of the He–Ne laser beam, after which the Nd–YAG laser is fired. The UV laser pulse produces a burst of ions of different masses from the analytical crater. These ions are accelerated to almost constant kinetic energy and are injected into the spectrometer flight tube. As the ions travel through the flight tube and through the energy-focusing region, small differences in kinetic energy among ions of the same mass are compensated. Discrete packets of ions arrive at the detector and give rise to amplified voltage signals that are input to the transient recorder. The function of the transient recorder is to digitize the analog signal from the electron multiplier, providing a record of both the arrival time and the intensity of the signals associated with each mass. The data are then transferred to the computer for further manipulation, the transient recorder is cleared and rearmed, and the instrument is ready for the acquisition of another spectrum.

This sequence of events is quite rapid. If we take typical instrumental conditions of the LIMA 2A, where the UV laser pulse duration is 5–10 ns, the fight path is ~2 m, and the accelerating potential is 3 kV, then an H^+ ion arrives at the detector i n approximately 3 µs, and a U^+ ion arrives at the detector in approximately 40 µs. Since the time width of an individual signal can be as short as several tens of nanoseconds, a high speed detection and digitizing system must be employed.

Typical mass resolution values measured on the LIMA 2A range from 250 to 750 at a mass-to-charge ratio $M/Z = 100$. The parameter that appears to have the most influence on the measured mass resolving power is the duration of the ionization event, which may be longer than the duration of the laser pulse (5–10 ns), along with probable time broadening effects associated with the 16-ns time resolution of the transient recorder.[3]

The intensity of an ion signal recorded by the transient recorder is proportional to the number of detected ions. There are two limiting factors to this proportionality, one due to the nonlinear output of the electron multiplier at high-input ion signals, and the other due to the dynamic range of the digitizer. The dynamic range of typical venetian blind-type electron multipliers for linear response to fast transients is less than four orders of magnitude. Electron multipliers characterized by other geometries (mesh type) are currently being evaluated, and may provide a larger inherent dynamic range.[3]

The second limiting factor in the quantitative measurement of the ion signal intensity is associated with the digitization of the electron multiplier output signal by the transient recorder. For example, the Sony-Tektronix 390 AD transient recorder in the LIMA 2A is a 10-bit digitizer with an effective dynamic range of 6.5 bits for 10-MHz signals. This device provides approximately 90 discrete voltage output levels at input frequencies of typical ion signals.[4]

Limitations in the digitizer's dynamic range can be overcome by using multiple transient recorders operating at different sensitivities, or by adding logarithmic preamplifiers in the detection system. From the preceding discussion it appears, however, that quantitative analysis is not the primary area of application of LIMS. Semiquantitative and qualitative applications of LIMS have been developed and are discussed in the remainder of this article.

Applications

Most applications of LIMS are in failure analysis. A typical microanalytical failure analysis problem, for example, may involve determinating the cause of corrosion in a metallization line of an integrated circuit. One can achieve this by performing an elemental survey analysis of the corroded region. Since it is not always known which elements are normal constituents of the material in question and which are truly contaminants, the vast majority of these analyses are performed by comparing the elemental make-up of the defective region to that of a control region. The com-

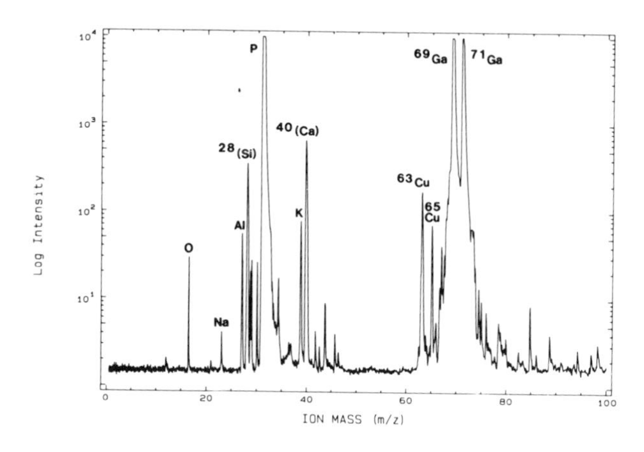

Figure 3 **Positive-ion mass spectrum acquired from defective sample. Intense copper ion signals are observed (*M*/*Z* = 63 and 65).**

parison of mass spectra of the two regions may reveal the presence of additional elements in the defective region. Those elements are often the cause or byproducts of the corrosion. In this type of analysis, the selection of a relevant control sample is obviously critical.[2]

LIMS analytical applications may be classified as elemental or molecular survey analyses. The former can be further subdivided into surface or bulk analyses, while molecular analyses are generally applicable only to surface contamination. In the following descriptions of applications, a comparison with other analytical techniques is presented, along with a discussion of their relative merits.

Bulk Analysis

One example of the application of LIMS to bulk contamination microanalysis is the analysis of low level contamination in GaP light emitting diodes (LEDs). The light emission characteristics of GaP LEDs can be severely affected by the presence of relatively low levels of transition elements. Although the nature of the poisoning species may be suspected or inferred from intentional contamination experiments, the determination of elemental contaminants in actual failures is a difficult analytical problem, in particular because of the small size and complex geometry of the parts. Figures 3 and 4 illustrate two positive-ion mass spectra that were acquired from cross sections of a defective and a nondefective GaP LED, respectively. The laser power density employed in this analysis was high to maximize the detection of low-level contaminants. The depth of sampling is estimated to be 1000–1500 Å. The two mass spectra exhibit intense signals for Ga^+, along with moderately intense signals for P^+. The defective LED also exhibits readily recognizable signals at $M/Z = 63$ and 65, matching in relative intensity the two Cu isotopes. The presence of Cu in the defective LED can explain its anomalous optical behavior. This

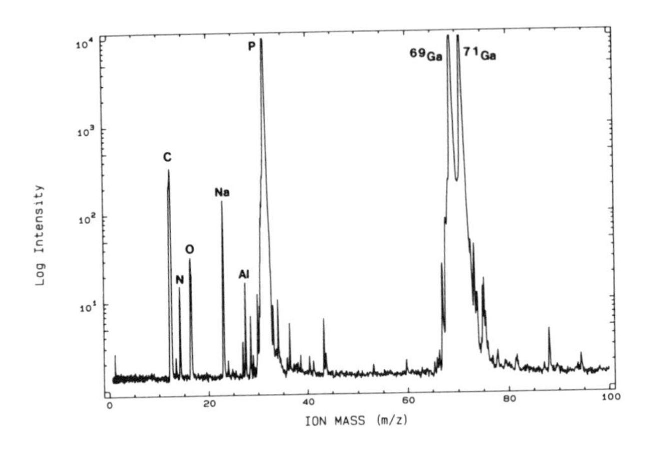

Figure 4 **Positive-ion mass spectrum acquired from the contact region of a control sample. Copper ion signals are absent.**

example is a good illustration of the unique advantages of LIMS over other analytical techniques. These include the ability to perform rapid survey analysis to detect unknown contaminants. Two other advantages of LIMS illustrated by this example are its ability to analyze a small sample having nonplanar geometry, without time-consuming sample preparation, and its sensitivity, which is superior to that of most electron beam techniques.

Surface Analysis

An example of elemental contamination surface microanalysis is shown inFigure 5. This is a negative-ion mass spectrum acquired from a small window (~4 µm) etched through a photoresist layer deposited onto a HgCdTe substrate. An Al film is then deposited in these windows to provide electrical contact with the substrate. Windows were found to be defective because of poor adhesion of the metallic layer. The spectrum shown in Figure 5 was acquired from a defective window, and reveals the presence of intense signals of Cl^- and Br^-, neither of which is observed in similar regions with good adhesion characteristics. In this case, the photoresist had been etched with solutions containing Cl and Br. The laser power density employed in this analysis was low, and the sampling depth was estimated to be < 500 Å. This analysis indicates that poor adhesion on the contaminated windows is due to incomplete rinsing of etching solutions. The ability of LIMS to operate on nonconductive materials is a major advantage in this case, since both the HgCdTe substrate and the surrounding photoresist are insulating. Techniques that use charged-particle beams (electrons, AES or EPMA; ions, SIMS) could probably not be applied in this case.

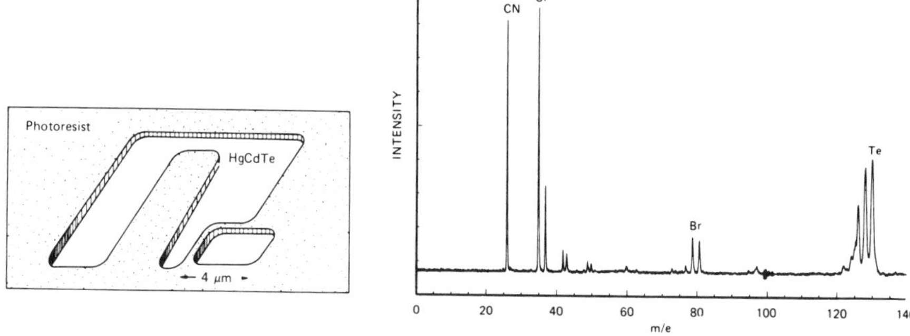

Figure 5 **Diagram of the windows cut in the photoresist on an HgCdTe substrate (top). Spectrum acquired from a defective window, and reveals the presence of intense signals of Cl^- and Br^- (bottom).**

Organic Surface Microanalysis

The laser irradiation of a material can produce molecular ions, in addition to elemental ions, if the power density of each pulse is sufficiently low. The analysis of such molecular species includes the study of organic materials ranging from polymers to biological specimens, as well as the analysis of known or suspected organic surface contaminants. LIMS organic spectroscopy is primarily a qualitative technique, which is used to identify a number of fragment ions in a spectrum that are diagnostic for a given class of organic species.

Organic contamination in the microelectronics industry is often related to the presence of organic polymer residues, for example, photoresist. These organic residues are a serious problem for surface adhesion, for example in the case of bond pads. Examples of LIMS organic surface analysis are shown in Figures 6 and 7: mass spectra acquired from a commercial photoresist, using positive- and negative-ion detection, respectively. The positive-ion mass spectrum in Figure 6 exhibits intense signals for alkalis (Na and K) along with a series of signals that are C-based fragment ion peaks. Some of these are undoubtedly aromatic fragment ions (M/Z = 77, 91, and 115, among others), and are diagnostic of this particular photoresist. Similarly, the negative-ion mass spectrum in Figure 7 exhibits an intense signal at M/Z = 107, which arises from Novolak resin, one of the constituents of the photoresist. Other, less intense signals of this spectrum include the species SO_2, SO_3, and HSO_3, which are also known to be present in the photoresist.

Laser Post-Ionization of Ablated Neutrals

A ZnSe-on-GaAs epitaxial layer required a sensitive survey of near-surface contamination. PAI was selected for ZnSe analysis because its major constituents and many of the expected impurities are elements that have poor ion yields in conventional LIMS. Figures 8 and 9 are two mass spectra acquired from the ZnSe epitaxial layer,

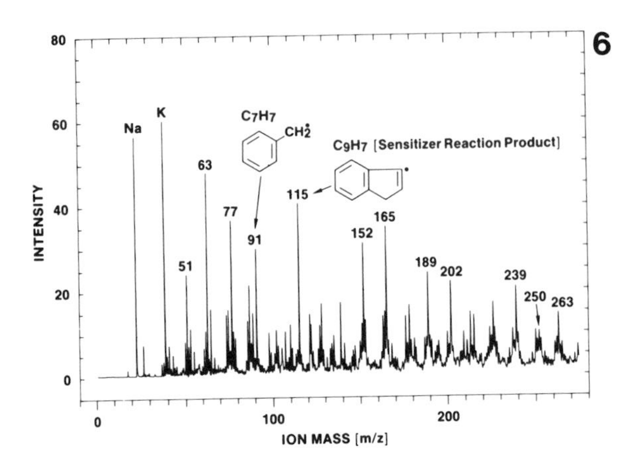

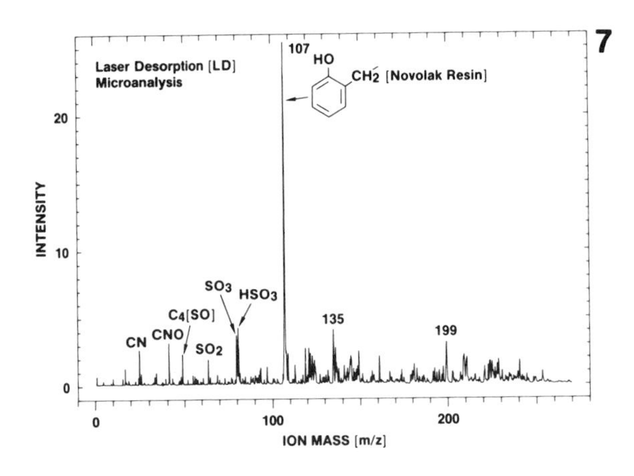

Figures 6 and 7 Mass spectra acquired from a commercial photoresist, using positive- and negative-ion detection, respectively.

using conventional single-laser LIMS and the PAI configuration, respectively. The single-laser spectrum in Figure 8 exhibits primarily the Zn^+ and Se^+ signals, and weak signals for Cr^+ and Fe^+. The high background signal level following the intense Se signal is related to detector saturation. The ablator laser irradiance for this spectrum was estimated to be > 10^{10} W/cm^2, hence the high background signal.

In contrast, the PAI mass spectrum in Figure 9 exhibits readily observable signals for Cd^+ and Te^+ in addition to the Zn and Se signals. Note also the low background in the region that follows the Se signal. The ablator laser irradiance in the PAI spectrum was approximately 10^9 W/cm^2, a factor of 10 lower than in the single-laser analysis. The lower ablator laser irradiance samples the top 100 Å of the sample, compared to 1000 Å or more in single-laser analysis, and hence provides better sur-

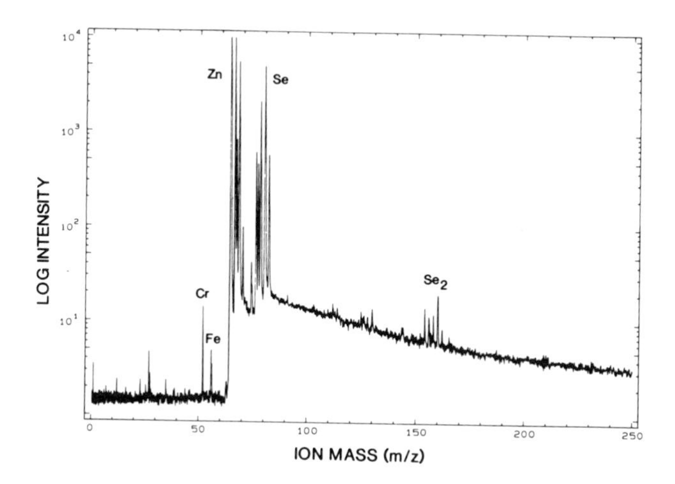

Figure 8 **Conventional, single-laser mass spectrum of ZnSe.**

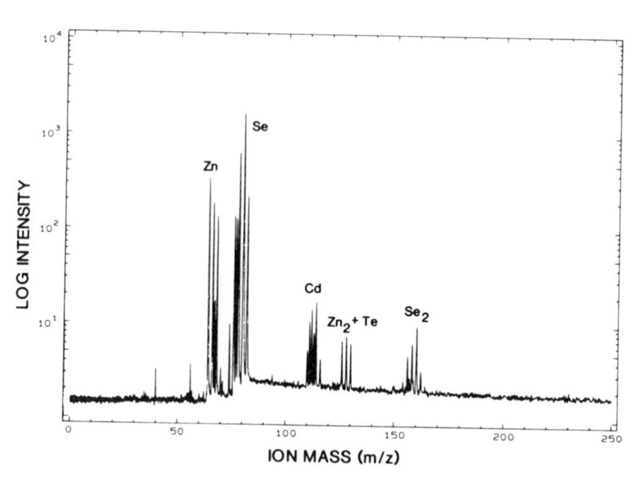

Figure 9 **Spectrum of ZnSe using the two-laser (PAI) instrumental configuration.**

face sensitivity. In conclusion, the PAI variant of LIMS is especially useful when the elements present have high ionization potentials that preclude efficient ion detection via conventional LIMS analysis, and in those cases when a higher surface sensitivity is desired.

Sample Requirements

A general requirement for LIMS analysis is that the material must be vacuum compatible and able to absorb UV laser radiation. With regard to the latter requirement, the absorption characteristics of UV-transparent materials can be improved with the use of thin UV-opaque coatings, such as Au or C. Care must be exercised, however, that the coating does not introduce excessive contamination, and practice is needed to determine the best coating for each sample.

A typical LIMS instrument accepts specimens up to 19 mm (0.75 in) in diameter and up to 6 mm in thickness. Custom designed instruments exist, with sample manipulation systems that accept much larger samples, up to a 6-in wafer. Although a flat sample is preferable and is easier to observe with the instrument's optical system, irregular samples are often analyzed. This is possible because ions are produced and extracted from μm-sized regions of the sample, without much influence from nearby topography. However, excessive sample relief is likely to result in reduced ion signal intensity.

The electrical conductivity of the sample is, to a first approximation, much less critical than in the case of charged-particle beam techniques (e.g., AES or SIMS), because the laser beam does not carry an electric charge, and is pulsed with a very low duty cycle. However, charging effects are sometimes observed in the negative-ion analysis of insulating samples, such as ceramics or silicon oxide. Charging probably arises from the acceleration of large numbers of electrons from the sample surface, along with the negative ions, which leaves behind a positively charged sample surface. Effects of this type may be alleviated with the use of conductive masks over the sample surface.

Conclusions

LIMS is primarily used in failure microanalysis applications, which make use of its survey capability, and its high sensitivity toward essentially all elements in the periodic table. The ability to provide organic molecular information on a microanalytical scale is another distinctive feature of LIMS, one that is likely to become more important in the future, with improved knowledge of laser desorption and ionization mechanisms.

Future trends for LIMS are likely to include hardware improvements, theoretical advances in the understanding of the basic mechanisms of laser–solid interactions, and improved methods for data handling and statistical analysis. Among the hardware improvements, one can count the advent of post-ionization techniques, which are briefly presented in this article and are discussed elsewhere in the Encyclopedia, and improvements in detection system dynamic ranges, through the use of different types of electron multipliers and improved transient recorders. These innovations are expected to result in improved quantification of the results. The introduction of faster pulsed lasers may also prove a significant improvement in mass resolution for LIMS, thus making it more suitable for organic analysis. Improvements in software may include compilations of computerized databases of LIMS organic mass spectra, the development of pattern recognition techniques, and the introduction of expert systems in the analysis of large bodies of LIMS data.[6]

Related Articles in the Encyclopedia

SALI, SIMS, SNMS, GDMS, and AES

References

1 I. D. Kovalev et al. *Int. J. Mass Spectrom. Ion Phys.* **27**, 101, 1978. Contains a discussion of laser–solid interactions and ion production under a variety of irradiation conditions.

2 T. Dingle and B. W. Griffiths.In *Microbeam Analysis-1985* (J. T. Armstrong, ed.) San Francisco Press, San Francisco, 315, 1985. Contains examples of quantitative analytical applications of LIMS.

3 R. W. Odom and B. Schueler. Laser Microprobe Mass Spectrometry: Ion and Neutral Analysis. in *Lasers and Mass Spectrometry* (D. M. Lubman, ed.) Oxford University Press, Oxford, 1990. Presents a useful discussion of LIMS instrumental issues, including the post-ablation ionization technique. Several analytical applications are presented.

4 D. S. Simons. *Int. J. Mass Spectrom. Ion Process.* **55**, 15, 1983. General discussion of the LIMS technique and its applications. Contains a discussion of detector dynamic range issues.

5 L. Van Vaeck and R. Gijbels. in *Microbeam Analysis-1989* (P. E. Russell, ed.) San Francisco Press, San Francisco, xvii, 1989. A synopsis of laser-based mass spectrometry analytical techniques.

6 P. B. Harrington, K. J. Voorhees, T. E. Street, F. Radicati di Brozolo, and R. W. Odom. *Anal. Chem.* **61**, 715, 1989. Presents a discussion of LIMS polymer analysis and pattern recognition techniques.

10.6 SSMS

Spark Source Mass Spectrometry

WILLIAM L. HARRINGTON

Contents

Introduction

No single trace elemental technique can provide a complete analysis of the many materials used in today's high technology applications. In the 30 years since its commercial introduction, SSMS has proven to be a versatile technique that can be applied to a wide variety of material types. The high-voltage radiofrequency (RF) spark source, which is used to volatilize and ionize the sample elements, has been shown to do so with relatively uniform probability across the entire periodic table. Although far from perfect in this respect, sensitivities for most elements in most matrices are therefore uniform and are constant within about a factor of 3, even at trace levels of less than 1 part per million atomic (ppma). Like most mass spectrometric techniques, SSMS is linear with respect to concentration over a wide range, achieving 8–9 orders of magnitude in cases where the signals are interference free. This relatively uniform, high sensitivity, combined with the ability to examine materials in a wide variety of forms, makes SSMS an excellent choice for trace elemental surveys of bulk and some thin-film specimens. The three most-used trace element techniques for survey analysis are Emission Spectrometry (ES), Glow Discharge Mass Spectrometry (GDMS), and SSMS. After a detailed discussion of SSMS, a comparison with these and other techniques will be made.

Bulk solids
Silicon (boules and wafers)
GaAs
Evaporation sources (e.g., Al, Au, and Ti)
Precious metals (Pt, Au, Rh wire, and melts)
Steel
Alloys (Ni-Co-Cr, Al-Si, Cu-Ni, and Inconel)
Ga metal (cooled to solid phase using liquid nitrogen)

Powders
Graphite
Rare earth oxides and phosphors
Ceramics (Al_2O_3) and glasses
Mining ores and rocks
Superconductors and precursor materials

Thin films
Silicon wafers and SiO_2 films
Si / sapphire (SOS), Si / insulators (SOI)
Epitaxial GaAs
Buried oxides (SIMOX)
Plated, sputtered, or evaporated metals

Table 1 **Typical materials analyzed.**

Table 1 lists some of the materials typically analyzed by SSMS and some of the forms in which these materials may exist. The basic requirement is that two conducting electrodes be formed of the material to be analyzed. Details of the analysis of each type of sample will be discussed in a later section.

Although a number of studies have been made concerning the basic properties of the RF vacuum spark used for excitation,[1–4] the discharge is typically erratic, producing a widely fluctuating signal for mass analysis. For this reason, the most widely used form of this instrumentation consists of a mass spectrometer of the

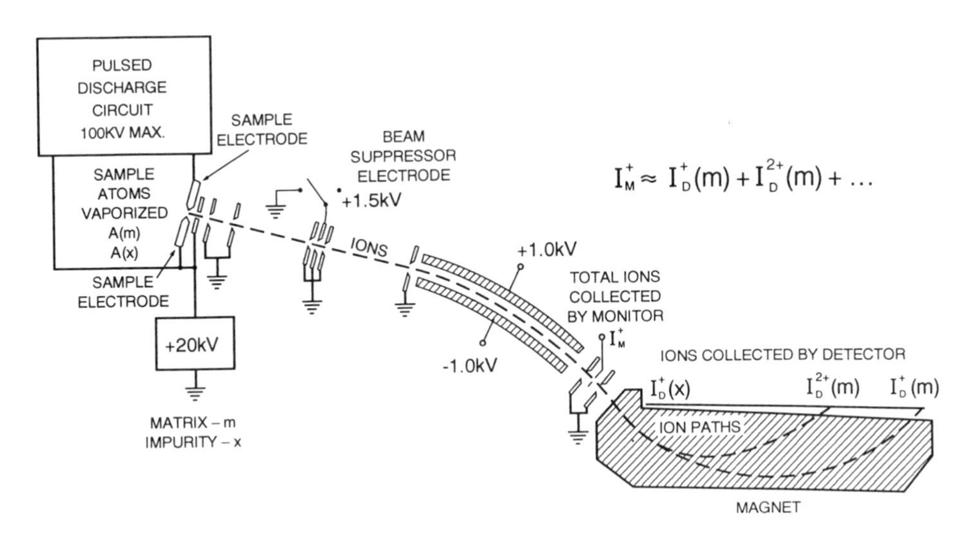

$$I_M^+ \approx I_D^+(m) + I_D^{2+}(m) + \ldots$$

Figure 1 **Schematic diagram of a Mattauch-Herzog geometry spark source mass spectrometer using an ion-sensitive plate detector.**

Mattauch-Herzog geometry,[5] which simultaneously focuses all resolved masses onto one plane, allowing the integrating properties of an ion-sensitive emulsion to be used as the detector. Although electrical detection with an electron multiplier can be applied,[6] the ion-sensitive emulsion-coated glass "photographic" plate is the most common method of detection and will be described in this article.

Basic Principles

General Technique Description

A schematic diagram of a spark source mass spectrometer is shown in Figure 1. The material to be analyzed forms the two electrodes separated by a spark gap. A pulsed 500-kHz high-voltage discharge across the gap volatilizes and ionizes the electrode material. The positive ions released are accelerated to 20–30 kV and passed into the mass spectrometer for energy and direction focusing. The electrostatic analyzer passes ions with an energy spread of about 600 eV, and focuses the beam onto a slit monitor that intercepts a constant fraction of the ions. This allows an accurate measurement of the number of ions (as Coulombic charge) entering the magnetic sector to be separated according to their mass-to-charge ratios and subsequently refocused and collected on the ion-sensitive plate detector. Figure 2 shows an example of SSMS data recorded on such a detector. The position of the collected ions (in the form of a line image of the source slit) provides qualitative identification of the isotopic masses (note the mass scale added to the plate to aid identification), and

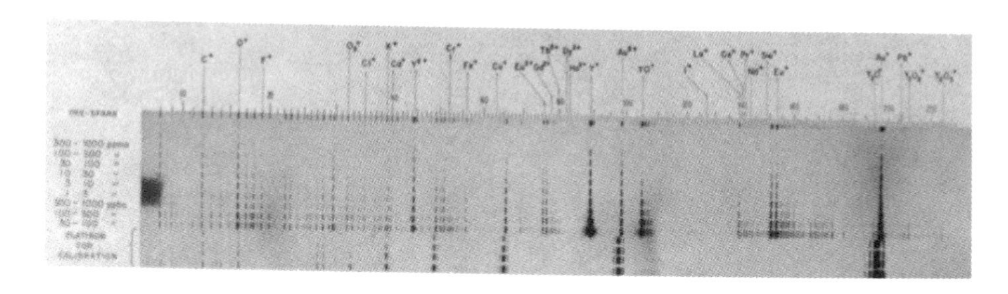

Figure 2 **Ion-sensitive plate detector showing the species produced by the SSMS analysis of a Y_2O_3:Eu_2O_3 mixture compacted with gold powder.**

the blackness of the lines can be related to the number of ions striking a position (i.e., the concentration).

Because the beam monitor allows accurate measurement of the total number of ions that are analyzed, a graded series of exposures (i.e., with varying numbers of ions impinging on the plate) is collected, resulting in the detection of a wide range of concentrations, from matrix elements to trace levels of impurities. In Figure 2, the values of the individual exposures have been replaced with the concentration range that can be expected for a mono-isotopic species just visible on that exposure. In this example, exposures from a known Pt sample have been added to determine the response curve of the emulsion.

Sample Requirements and Examples

For *bulk conductors* and *semiconductors,* sample preparation consists of breaking, cutting, or sawing the solid approximately into the dimensions 1/16 in. × 1/16 in. × 1/2 in. Large, irregular sample electrodes can be accommodated; however, they often shield the path of ions into the mass spectrometer, thereby reducing the beam current and increasing the analysis time. If the preparation, handling, or packaging contaminates the sample electrodes with elements that are not of interest, they can be degreased in high purity methanol, etched in an appropriate semiconductor-grade acid, and washed in several portions of methanol before analysis.

The final, and most critical cleaning is performed by presparking all electrode surfaces that will be consumed, before collecting the actual exposures for analysis. The presparking removes in vacuum the outer layers of the sample, which may contain trace levels of contamination due to handling or atmospheric exposure. This step also coats the surfaces in the analysis chamber to minimize memory effects, i.e., to minimize the chances of detecting elements from a previously analyzed sample. Cryosorption pumping[7] using an Al_2O_3 or charcoal-coated plate filled with liquid nitrogen is also used during presparking and analysis to maintain reproducible source pressures and to reduce hydrocarbon interferences.

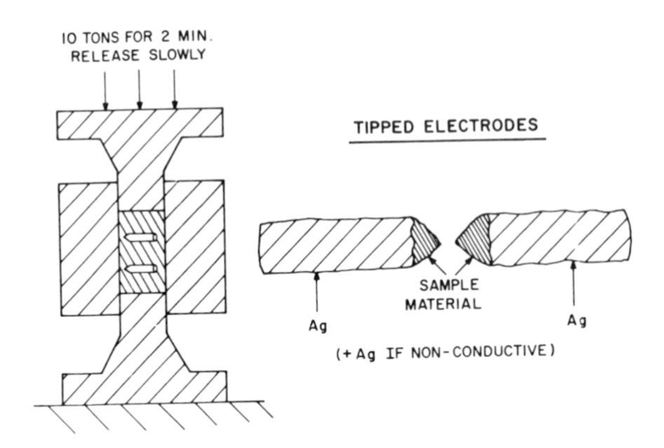

Figure 3 **Schematic of a polyethylene slug die used to compact powder samples. When the amount of material available for analysis is small, tipped electrodes can be formed.**

Table 2 is a typical example of SSMS analysis of a high-purity Pt wire that was simply broken in two and presparked. The elements from Be to U were determined in a single analysis requiring a total time of 1–2 hours. Hydrogen and Li must be measured in a separate set of exposures using a lower magnetic field and therefore generally are not included in a standard SSMS analysis. Because detection limits vary with plate sensitivity, background, isotopic abundance, and elemental mass, individual limits are listed for those elements not detected and are noted as less than (<). For practical reasons, a factor of about 3 better detection limits (3× longer analysis time) is generally the limit for this technique.

Powders or nonconductors represent important forms of materials that are well suited for SSMS analysis. Powders of conductive material generally can be prepared without a binder, but insulators first must be ground to a powder with a mortar and pestle, such as boron carbide and agate, then mixed with a high-purity conductive binder, such as Ag powder or graphite powder, and pressed to form solid, conductive electrodes. To prevent contamination from the metal die, a polyethylene cylinder is drilled to hold the powder such that the tips and sides of the electrodes touch only the polyethylene and not the steel parts of the die. If the sample material is limited in quantity, small portions (1–10 mg) can be tipped onto the end of high purity Ag.A die and tipped electrodes are shown diagrammatically in Figure 3.

Of course, this procedure for nonconducting powders dilutes the sample, causing poorer detection limits and limiting the purity that can be specified to that of the binder.

Although SSMS cannot be considered a surface technique due to the 1–5 μm penetration of the spark in most materials, few other techniques can provide a trace elemental survey analysis of surfaces consisting of films or having depths of interest

Element	Concentration (ppma)	Element	Concentration (ppma)
		Rh	3
		Pd	4
Be	≤ 0.003	Ag	0.1
B	0.003	Cd	< 0.04
C*	6	In	0.2
N*	0.1	Sn	< 0.04
O*	3	Sb	< 0.02
F	0.004	Te	< 0.04
Na	0.02	I	< 0.02
Mg	0.4	Cs	< 0.02
Al	0.2	Ba	< 0.02
Si	2	La	< 0.02
P	0.04	Ce	< 0.02
S	0.04	Pr	< 0.02
Cl	0.03	Nd	< 0.07
K	< 0.02	Sm	< 0.06
Ca	5	Eu	< 0.04
Sc	≤ 0.03	Gd	< 0.07
Ti	0.4	Tb	< 0.02
V	0.02	Dy	< 0.07
Cr	≤ 0.3	Ho	< 0.02
Mn	0.07	Er	< 0.06
Fe	4	Tm	< 0.03
Co	0.02	Yb	< 0.07
Ni	0.6	Lu	< 0.03
Cu	3	Hf	< 0.06
Zn	0.03	Ta**	4
Ga	0.02	W	< 0.08
Ge	< 0.02	Re	< 0.04
As	0.02	Os	< 0.2
Se	< 0.02	Ir	< 0.3
Br	< 0.02	Pt	Major
Rb	< 0.01	Au	≤ 3
Sr	≤ 0.02	Hg	< 0.2
Y	< 0.008	Tl	< 0.08
Zr	3	Pb	< 0.7
Nb	< 0.009	Bi	< 0.2
Mo	≤ 0.2	Th	< 0.04
Ru	0.2	U	< 0.04

* Upper limits; source not baked to reduce background.

Table 2 **SSMS analysis of high-purity Pt metal.**

of ~5 μm. Although the penetration depth can be somewhat controlled by the spark gap voltage, the electrode separation, and the speed of sample scanning, 1–5 μm is the typical range of penetration depths that can be achieved without punching through to deeper layers. Figure 3 shows a method for surface analysis using a high-purity metal probe (Au) as a counter electrode to spark an area of a sample's surface. The tip of the probe is positioned on the axis of the mass spectrometer, and the sample is scanned over the probe tip to erode tracks across the surface. By scanning over areas of about 1 cm^2, detection limits on the order of 1 ppma can be achieved. By combining this survey surface analysis with depth profiling techniques, such as Secondary-Ion Mass Spectrometry (SIMS) or Auger Electron Spectroscopy (AES), elements of interest can be identified by SSMS and then profiled in detail by the other methods.

The SSMS point-to-plane surface technique has been shown to be particularly useful in the survey analysis of epitaxial films, heavy metal implant contamination, diffusion furnace contamination, and deposited metal layers.

Data Evaluation

Qualitatively, the spark source mass spectrum is relatively simple and easy to interpret. Most instrumentation has been designed to operate with a mass resolution M/dM of about 1500. For example, at mass $M = 60$ a difference of 0.04 amu can be resolved. This is sufficient for the separation of most hydrocarbons from metals of the same nominal mass and for precise mass determinations to identify most species. Each exposure, as described earlier and shown in Figure 2, covers the mass range from Be to U, with the elemental isotopic patterns clearly resolved for positive identification.

The spark source is an energetic ionization process, producing a rich spectrum of multiply charged species ($M/2$, $M/3$, $M/4$, etc.). These masses, falling at halves, thirds, and fourths of the unit mass separation can aid in the positive identification of elements. In Figure 2, species like Au^{+2} and Y^{+2} are labeled. The most abundant species (matrix elements and major impurities) also form dimers and trimers (and so forth) at two and three times (and so forth) the mass of the monomer. Although these species can cause interference with certain trace elements, they also can aid in positively identifying a particular element. Finally, the spectrum generally contains mixed polyatomic species, such as MO^+, MO_2^+, MC^+ (in graphite), and MAg^+ (in silver). All such possibilities must be considered in the qualitative interpretation of a spark source mass spectrum.

Of course, the most reliable and accurate method of quantitative analysis is to calibrate each element with standards prepared in matrices similar to the unknown being analyzed. For a survey technique that is used to examine such a wide variety of materials, however, standards are not available in many cases. When the technique is used mainly in one application (typing steels, specifying the purity of alloys for a selected group of elements, or identifying impurities in silicon boules and wafers), such standards can be developed and should be applied. Because of the erratic nature of the spark (in terms of time) and variability in the response of the ion-sensitive emulsion detector, accuracy using standards to generate relative sensitivity factors is generally within 20–50%.

Due to the relative uniformity of ion formation by the RF spark (although its timing is erratic), the most widely used method of quantitation in SSMS is to assume equal sensitivity for all elements and to compare the signal for an individual element with that of the total number of ions recorded on the beam monitor. By empirically calibrating the number of ions necessary to produce a certain blackness on the plate detector, one can estimate the concentration. The signal detected must be corrected for isotopic abundance and the known mass response of the ion-sensitive plate. By this procedure to accuracies within a factor of 3 of the true value can be obtained without standards.

The optical density (blackness) of the lines recorded can be measured most accurately using a microdensitometer to scan each line and measure the transmission of light through it. A set of known relative exposures (from charge accumulated by the mass spectrometer beam monitor and known isotopic ratios) is used to establish the emulsion response curve relating transmission to exposure. The absolute position of this response curve on the exposure axis can be determined using standards or from isotopes of a pure element. For concentration determinations requiring the highest precision, the microdensitometer approach is recommended. This method, however, is time consuming; it can be considerably shortened by a well-established "visual" method.

If a graded series of exposures is made in relative steps of 1, 2, 5, 10, 20, 50, etc. (see the graded series of exposures in Figure 2), the exposure necessary to produce a barely detectable line for a particular isotope can be determined by simply observing a well-lighted plate with a 3–5× eyepiece. By determining the average exposures for which barely detectable lines appear for known concentrations of some elements in various matrices, a particular instrument can be calibrated to provide estimates of concentration without further analysis of standards, except to occasionally check the relationship between the beam monitor and the emulsion response. This visual method is surprisingly consistent when care is taken to provide accurate relative exposures, and it produces values that are generally accurate within a factor of 3. Several elements, such as Na, K, Ca, and Al, are best estimated using a multiply charged species (+2 in most cases). For the alkaline and alkaline earth elements in particular, the number of singly charged ions can be greatly enhanced by thermal excitation; a more accurate assessment is made by measuring the +2 species and applying an empirically determined correction factor. The accuracy for elemental concentrations determined in this manner is generally within a factor of 10.

Characteristic	ES	GDMS	SSMS
Detection limits	1–10 ppm	0.00001–0.01 ppm	0.003–0.03 ppm
Concentration	Minor, trace	Major, minor, Ultra-trace	Minor, trace
Elemental coverage	Metals	All elements	All elements
Accuracy			
without standards	± 10×	± 3×	± 3×
with standards	± 20%	± 10–20%	± 20–50%
Matrix effects	Strong	Weak	Weak/ medium
Bulk/ surface	Bulk	Bulk	Bulk and surface
Conductivity	Conductor or insulator	Conductor: run as is Insulator: + Ag	Conductor: run as is Insulator: + Ag or C
Sample Form	Solid/ powder	1 Conducting pin (shape important)	2 Conducting pins (can be irregular)

Table 3 SSMS—comparison with other techniques.

Comparison With Other Techniques

Although numerous analytical techniques have been developed for the quantitative determination of specific elements at trace levels in solids, the three most-used techniques providing multi-element surveys are Emission Spectroscopy (ES), Glow Discharge Mass Spectrometry (GDMS), and SSMS. GDMS is covered in detail elsewhere in this volume, but it is instructive to compare these techniques in tabular form. Table 3 provides this comparison for a number of characteristics that should be considered when choosing a technique. In most situations the required detection limits clearly define one's choice. Elemental coverage is important when nonmetals, such as As, P, Cl, F, C, and O, play a role as trace elements, making the mass spectrometric techniques a clear choice. Sample shape and form are also issues that must be considered. The versatility of SSMS in accommodating a wide variety of materials while maintaining high sensitivity for all elements is one of its prime features.

Inductively Coupled Plasma-Optical (ICP-optical) methods and ICPMS are extremely sensitive elemental survey techniques that also are described in this volume. ICP methods, however, require a solution for analysis, so that the direct

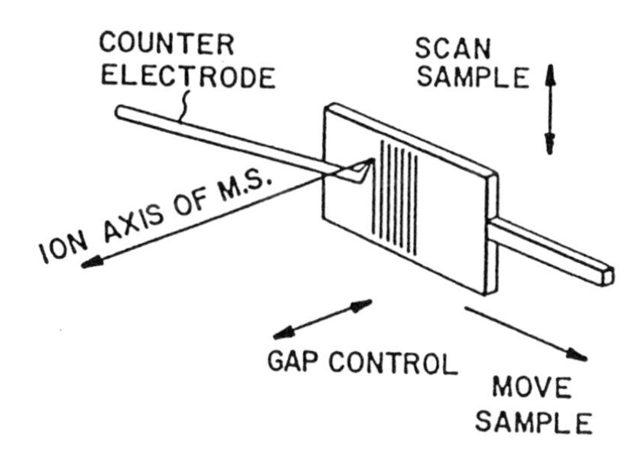

Figure 4 **SSMS surface analysis. The point-to-plane technique allows ppma elemental surveys over a depth of 1–5 μm.**

examination of solids is not possible. Because solution techniques offer relative ease of preparing standards, ICP-optical methods and ICPMS might be chosen in cases where accuracy is most important and the solids can be dissolved without contamination.

Conclusions

SSMS can provide a complete elemental survey with detection limits in the 10–50 ppba range and can deal with a wide variety of sample types and forms. Although GDMS offers higher sensitivity and accuracy of 20%, SSMS is still the technique of choice in many situations. Materials, such as carbon, that do not sputter rapidly enough for good GSMS detection limits, and insulators that cause erratic sputtering when combined with a conductive powder, are excellent candidates for SSMS analysis. In addition, the point-to-plane surface method is one of the few techniques available that can provide a complete elemental survey of 1–5 μm thick films with detection limits on the order of 1 ppma (see Figure 4).

Having described SSMS in some detail as a very useful technique for trace elemental survey analysis, one must note that the lack of manufacture of new instruments and the rising development of GDMS limit its future use. Industrial and service laboratories having SSMS instruments and experienced personnel will continue to use SSMS very effectively. Where there is the need for increased sensitivity, reaching detection limits of less than 1 ppba, and where there is sufficient justification to warrant the cost of GDMS ($600,000–$700,000 for magnetic sector instruments, and about $250,000 for quadrupole instruments), it is anticipated that SSMS gradually will be replaced. With progress being made in the instrumentation and methodology of GDMS, there are currently very few instances where

GDMS cannot be used instead of SSMS. As GDMS source designs are developed to allow clean, thin-film analyses, and some limitations are accepted for the analysis of insulators, GDMS instrumentation will replace more and more of the older SSMS installations. For the present, however, there are excellent laboratories having SSMS instrumentation and services, and SSMS should be used when it proves to be the technique of choice.

Related Articles in the Encyclopedia

GDMS, ICPMS, and ICP-Optical

References

1. J. Franzen and H. Hintenberger. *Zeit. fur Naturforschung.* **189,** 397, 1963.
2. J. R. Woolston and R. E. Honig. *Rev. Sci. Instr.* **35,** 69, 1964.
3. J. R. Woolston and R. E. Honig. In *Proceedings of the Eleventh Annual Conference on Mass Spectrometry and Allied Topics.* San Francisco, 1963.
4. J. R. Woolston and R. E. Honig. In *Proceedings of the Eleventh Annual Conference on Mass Spectrometry and Allied Topics.* Montreal, 1964.
5. J. Mattauch and R. F. K. Herzog. *Z. Physik.* **89,** 786, 1934. This is the original paper showing the double-focusing geometry necessary to focus onto a plane rather than at a single point.
6. C. W. Magee. *Critical Parameters Affecting Precision and Accuracy in Spark Source Mass Spectrometry with Electrical Detection.* PhD thesis, University of Virginia, University Microfilms, Ann Arbor, MI, 1973.
7. W. L. Harrington, R. K. Skogerboe, and G. H. Morrison. *Anal. Chem.* **37,** 1480–1484, 1965. The use of cryosorption pumping for SSMS is demonstrated.

10.7 GDMS

Glow-Discharge Mass Spectrometry

JOHN C. HUNEKE AND WOJCIECH VIETH

Contents

Introduction

Glow-Discharge Mass Spectrometry (GDMS) is a mass spectrometric analytical technique primarily used to measure trace level impurities in conducting or semiconducting solids. It can also be used, but less commonly, for elemental depth profile analyses. The primary advantages of GDMS are its sensitivity (ppt detection limits in some cases); its quantitative accuracy (20% on average), achieved without complicated standardization procedures; and its ability to detect essentially all elements in the periodic table from lithium to uranium at approximately the same sensitivity.

Because GDMS can provide ultratrace analysis with total elemental coverage, the technique fills a unique analytical niche, supplanting Spark-Source Mass Spectrometry (SSMS) by supplying the same analysis with an order-of-magnitude better accuracy and orders-of-magnitude improvement in detection limits. GDMS analysis has matured rapidly and has become more widely available since the recent introduction of commercial GDMS instrumentation.

Basic Principles

Ion Sources

In general, all GDMS ion sources use a noble gas glow-discharge plasma sustained at about 1 Torr pressure and a few Watts of discharge power. A conducting solid sample for GDMS analysis forms the cathode for a DC glow discharge (Figure 1). Atoms are sputtered nonselectively from the sample surface by ions accelerated from the plasma onto the surface by the cathode voltage. The sputtered atoms diffuse into the plasma and are mostly ionized by collision with metastable discharge gas atoms (so-called Penning ionization) but also in small part by electron impact. Penning ionization, more so than electron impact ionization, provides similar ion-production efficiencies for the majority of the elements. Plasma ions extracted through the exit aperture, including the analyte ions, are electrostatically accelerated into a mass analyzer for measurement.

Ion sources for GDMS have undergone significant evolution, ultimately resulting in discharge cells exposing only the sample (cathode) surface and the metal interior of the Ta discharge cell (anode) to the plasma, a design that minimizes contamination and enhances reliability. The glow-discharge cells accept a variety of shapes and surface conditions, requiring only sufficient length or diameter of the sample. Typically, pin- or wafer-shaped samples are required. It is very helpful to include cryocooling into the design of the discharge cell to enable the analysis of materials having low melting point and to reduce the density of molecular ions created from "atmospheric" gas related contaminants in the glow-discharge plasma.

At the current time, analytical glow-discharge sources incorporate a DC glow discharge. Efforts have been underway to develop glow-discharge sources appropriate for the analysis of electrically insulating materials (e.g., glass and ceramic), which comprise a very important class of materials for which few methods are currently available for complete, full-coverage analysis to trace levels. Two alternatives have been suggested as appropriate: RF-powered glow-discharge plasmas; and electron beam-assisted plasmas. While efforts are being made in both directions, no analytically viable sources have yet been made available.

The sputter sampling of the exposed surface also provides concentration depth profiling capability for GDMS. Depth resolution of some 0.1 μm has been demonstrated, with 1–2 orders of magnitude dynamic range due to geometric limitations and the high operating pressure of the glow-discharge source. With a rapid sputtering rate of about 1 μm/min, GDMS is particularly useful for thick-film (10–100 μm) depth profiling. GDMS can provide accurate, sensitive, and matrix indea

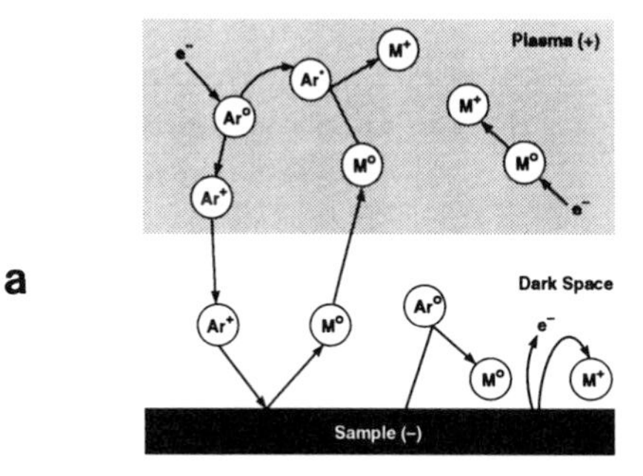

b

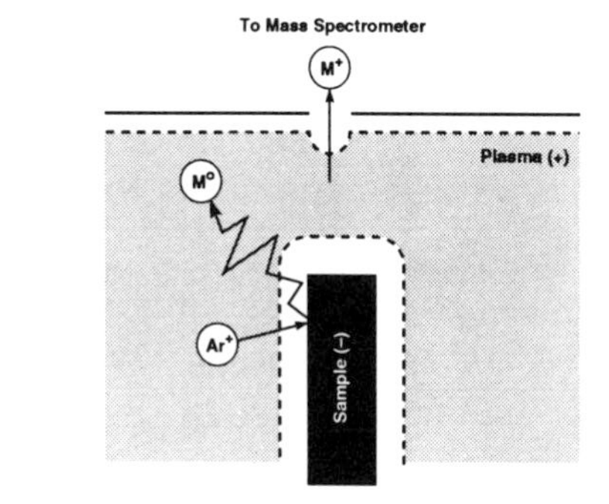

Figure 1 **Schematic of DC glow-discharge atomization and ionization processes. The sample is the cathode for a DC discharge in 1 Torr Ar. Ions accelerated across the cathode dark space onto the sample sputter surface atoms into the plasma (a). Atoms are ionized in collisions with metastable plasma atoms and with energetic plasma electrons. Atoms sputtered from the sample (cathode) diffuse through the plasma (b). Atoms ionized in the region of the cell exit aperture and passing through are taken into the mass spectrometer for analysis. The largest fraction condenses on the discharge cell (anode) wall.**

pendent concentration depth profiles throughout a complex film structure, including interfaces.

Mass Spectrometers

Demonstration of GDMS feasibility and research into glow-discharge processes has been carried out almost exclusively using the combination of a glow-discharge ion source with a quadrupole mass spectrometer (GDQMS). The combination is inexpensive, readily available and suitable for such purposes. In addition, the quadrupole provides the advantage of rapid mass spectrum scanning for data acquisition. Because ion transmission is limited and significant molecular ion mass interferences are unresolvable with the low mass resolution capabilities of the quadrupole, elemental detection limits by GDQMS are only slightly better than those provided by Optical Emission Spectroscopy (OES) methods, and the analytical usefulness of the GDQMS overlaps that of OES techniques (cf. Jakubowski, et al.[1]). Even so, the full elemental response of GDMS provides a substantial enhancement in analytical capability compared to the more selective OES. The introduction of GDMS instrumentation using high mass resolution, high-transmission magnetic-sector mass spectrometers has circumvented the major limitations of the quadrupole, providing an instrument with sub-ppb detection limits, albeit at the expense of analytical time. GDMS instrument and source descriptions have recently been the subject of an extensive review by Harrison and Bentz.[2]

The optimal analytical GDMS instrument for bulk trace element analysis is the one providing the largest analytical signal with the lowest background signal, the fewest problems with isobaric interferences in the mass spectrum (e.g., the interference of $^{40}Ar^{16}O^+$ with $^{56}Fe^+$), and the least contamination from instrument components or back contamination from preceding sample analyses. The first commercial GDMS instrument incorporated a high mass resolution magnetic-sector mass spectrometer to enable interfering isobaric masses to be eliminated, while at the same time providing high useful ion yields. The ion detection system of this instrument combined a Faraday cup collector, for the direct current measurement of the large ion beam associated with the matrix element, with a single-ion counting capability to measure the occasional trace element ion. The resulting ion current measuring system provides the necessary large dynamic range for matrix to ultratrace level measurements.

Instrument configurations other than a magnetic-sector mass spectrometer with a pin sample source are also suitable for analytical GDMS, but with some compromise in analytical performance. If analysis to ultratrace levels is not required, but only measurements to levels well above the background of isobaric mass spectral interferences, low-resolution quadrupole mass spectrometer based instruments can be configured. Such instruments have recently been made available by several instrument manufacturers. In these cases, the unique advantage of GDMS lies not with the ultratrace capability but with the full elemental coverage from matrix concentrations to levels of 0.01–0.1 ppm. Also, quadrupole MS mass spectral analysis requires significantly less time, enabling the more rapid analysis suitable for depth profiling of films.

Sample Preparation and Analytical Protocol

Accurate GDMS analysis has required the development of analytical procedures appropriate to the accuracy and detection limits required and specific to the material under analysis. Protocol particulars will differ from laboratory to laboratory. To use GDMS to advantage (i.e., to improve measurements to the ppb level) the surface exposed to the sampling plasma must be very clean. Common methods for surface cleaning are chemical etching and electropolishing using high-purity solutions. If such cleaning is not feasible (e.g., for pressed powders), presputtering of the surface in the glow-discharge source or with a separate sputtering unit to remove contaminants prior to analysis is generally required. (This procedure could not, of course, be used prior to concentration depth profiling measurements.) The risk of recontamination to ppb levels is high, and care must be taken in rinsing, handling, and transporting the cleaned sample.

The composition of the sample measured by GDMS reflects the surface composition, and the argument must be made that the measurement is representative of the bulk. This requires thorough sputter cleaning of residual contaminants and sputter equilibrium of the phases exposed at the surface. The pragmatic (and reasonably conservative) criterion that both goals have been accomplished is that the same composition has been obtained in consecutive measurements during the analysis (i.e., a "confirmed" analysis). Other than these requirements, the analysis protocol must be suitable for the instrument and the detection limits required, since in many instances the detection limit arises from lack of signal and not from backgrounds or interferences.

Accurate final results from GDMS are available very quickly. Samples for GDMS analysis requires little preparation other than shaping and cleaning, although the cutting of a pin or wafer from extremely hard materials can be time consuming. The actual GDMS analysis takes only on the order of one to two hours, depending on elemental coverage and detection limits required. Data reduction is on-line and essentially immediate.

GDMS Quantitation

Pragmatically, the relative concentrations of elements are determined from the measured ion beam ratios by the application of relative sensitivity factors, which are determined experimentally from standard samples:

$$\frac{M}{N} = \frac{I(M)}{I(N)} \mathrm{RSF}_N(M)$$

where M and N are the elements of concern, I is the measured ion current (including all isotopes of the element), and $\mathrm{RSF}_N(M)$ is the relative sensitivity factor for M relative to N.

Vieth and Huneke[3] have recently presented a thorough discussion of GDMS quantitation, including the measurement of relative GDMS sensitivity factors and a modeling of glow-discharge source processes to enable semiempirical estimates of

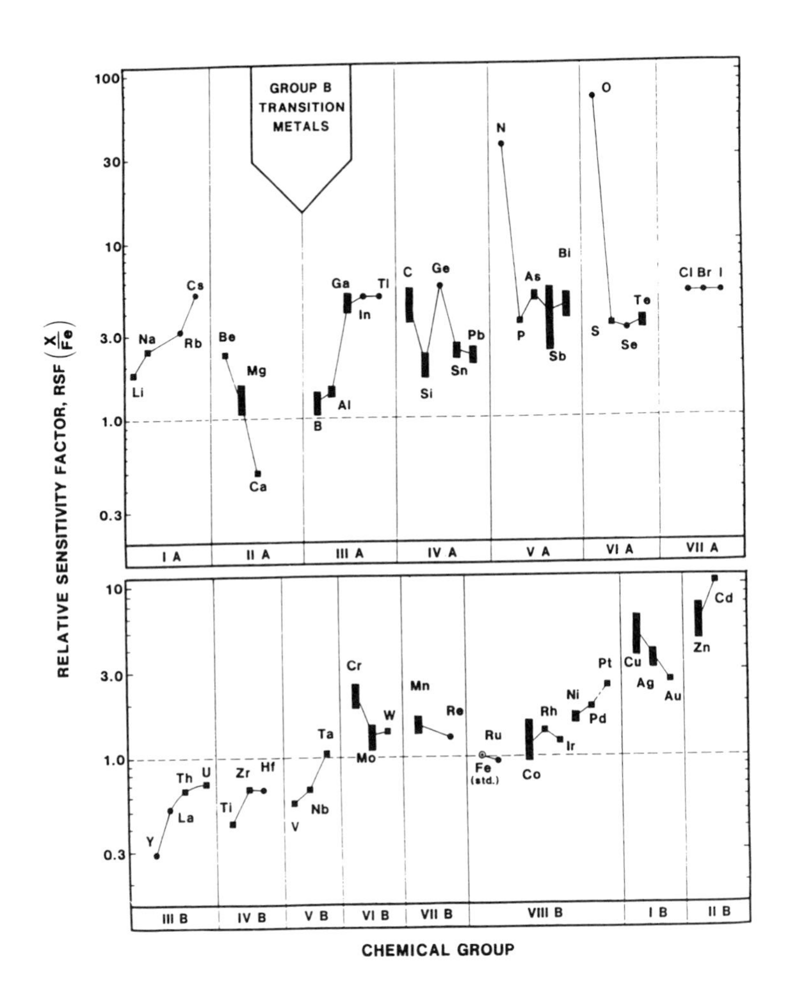

Figure 2 **Elemental relative sensitivity factors as a function of chemical grouping in the periodic table. The factors are determined from the measurement of 30 standard samples representing 6 different matrix elements. The factors are matrix-independent and similar within a factor of 10, except for C and O. There is a pronounced trend across the group-B elements, with a similar but separate trend across the group-A elements.**

RSFs. Figure 2 exhibits the average RSFs determined from analyses of 30 different standard samples of seven different matrices. The RSFs are matrix independent, with the spread in RSF determinations being due primarily to standard alloy inhomogeneities. It is remarkable that for all but N and O the factors range over only a decade.

The sensitivity factors summarized in Figure 2 are appropriate for analyses using a particular instrument (the VG 9000 GDMS) under specific glow-discharge conditions (3 mA and 1000 V in Ar, with cryocooling of the ion source) and with a well-controlled sample configuration in the source. RSFs depend on the sample–source configuration. In particular, they vary significantly with the spacing between the sample and the ion exit aperture from the cell. Use of the factors shown in Figure 2 under closely similar conditions will result in measurements with 20% accuracy. These factors can also be used to reduce the data obtained on other instruments, but the accuracy of the results will be reduced. In particular, the factors shown can be only approximately valid for results obtained using a quadrupole mass spectrometer, since ion-transmission characteristics differ significantly between the quadrupole mass spectrometer and the magnetic-sector spectrometer used to obtain the results of Figure 2.

It is clear from the RSF data shown in Figure 2 that even without the use of RSFs, a semiquantitative analysis accurate to within an order of magnitude is quite possible, and GDMS indeed will provide full coverage of the periodic table. The analysis of a material of unknown composition will be elementally complete to trace levels, with no glaring omissions that may eventually return to haunt the end user of the material.

Applications

The application of GDMS is strongest in the areas of:

1. Qualification of 5N–7N pure metals, since GDMS provides full elemental coverage to ultratrace levels
2. The analysis of specific elements for which GDMS is particularly well suited compared to other methods (e.g., measurement to ppb levels of S, Se, Te, Pb, Bi, Tl, in high-performance alloys; and measurement of U and Th in sputter targets)
3. The analysis of a material known to be impure, but with unspecified impurity elements
4. The analysis of a material in limited supply, and when too little is available for analysis by alternative methods.

A number of examples of the application of GDMS to various metals and alloys are exhibited below. All measurements were performed using the VG 9000 GDMS instrument with standard glow-discharge conditions of 3-mA discharge current and 1000-V discharge voltage except as noted. The standard pin dimensions were a diameter of 1.5–2.0 mm and a length of 18–22 mm.

Indium and Gallium Metal

Table 1 summarizes the results of an analysis to the 6N–7N total impurity level of very high purity In and Ga metals, such as would be used in the manufacture of III–

Element	Concentration (ppmw)		Element	Concentration (ppmw)	
	In	Ga		In	Ga
Li	<0.00006		Y	<0.00001	<0.00008
Be	<0.0001	<0.0005	Zr	<0.00006	<0.0003
B	0.002	<0.0003	Nb	<0.00009	<0.0001
C	(<20.)	(<0.9)	Mo	0.001	<0.001
N	(<200.)	(<20.)	Ru	<0.0002	<0.0006
O	(<60.)	(<8.)	Rh	<0.00006	<0.0003
F	(<0.2)	(<0.04)	Pd	<0.0003	<0.06
Na	0.005	0.05	Ag	0.001	<0.3
Mg	<0.00009	0.002	Cd	<0.004	<0.01
Al	0.002	0.005	In	Matrix	3.3
Si	0.005	0.03	Sn	<0.003	0.28
P	<0.0001	<0.0006	Sb	<0.002	<0.001
S	<0.0002	<0.0008	Te	<0.003	0.006
Cl	(<2.)	(<0.9)	I	0.001	<0.001
K	<0.001	<0.005	Cs	<0.02	<0.0005
Ca	<0.002	0.02	Ba	<0.0003	<0.04
Sc	<0.00001	<0.0005	La	<0.0002	<0.0003
Ti	0.001	0.001	Ce	0.00004	<0.005
V	<0.00002	<0.0001	Nd	<0.0001	<0.01
Cr	<0.0001	<0.0005	Hf	<0.0001	<0.0005
Mn	<0.00007	<0.0003	Ta	(<3.)	(<0.3)
Fe	0.0003	<0.0004	W	<0.0002	<0.0009
Co	<0.00004	<0.0002	Re	<0.00007	<0.0004
Ni	0.07	<0.001	Os	<0.0001	<0.0009
Cu	0.0007	0.04	Ir	<0.0001	<0.003
Zn	0.006	<0.003	Pt	<0.0003	<0.001
Ga	<0.0004	Matrix	Au	<0.002	<0.01
Ge	<0.002	<0.03	Hg	<0.0005	<0.005
As	<0.0009	<0.001	Tl	0.064	<0.001
Se	<0.02	<0.007	Pb	0.18	0.16
Br	<0.001	<0.04	Bi	0.004	0.008
Rb	<0.00005	<0.007	Th	<0.00003	<0.0001
Sr	<0.00002	<0.0001	U	<0.00003	<0.0001

Table 1 **Analysis of very high purity In and Ga metal by GDMS (VG 9000). Only three lanthanide elements have been measured as characteristic of all of the lanthanides. Concentrations preceded by a limit sign were not detected above the instrument background. The detection of elements included in parentheses were limited by instrumental or atmospheric contamination.**

V semiconductor material (e.g. InGaAs semiconductor). Very high purity Ga and In are required for the manufacture of semiconductor grade GaAs substrate material and in the deposition of the III–V alloy epilayer structures on these substrates, for example for the manufacture of laser diodes.

The analysis of Ga requires careful sample preparation to avoid altering the composition during solidification of the sample pin and preparation for analysis. The Ga pin was formed by drawing molten Ga into Teflon tubing and quickly freezing the Ga with liquid nitrogen. The low melting points of both metals requires analysis using low power and cryocooling of the discharge cell.

Some 70 elements are surveyed for their presence as impurities, and the detection limits must be on the order of 0.001–0.01 ppmw to qualify the material at the 6N–7N level. (The designation 6N is equivalent to specifying a total impurity content less than 1 ppmw; the metal is thus at least 99.9999% pure). The table shows mainly detection-limited values. The strict limit sign denotes the absence of an identifiable signal above the noise limit. Better limits result for elements with higher useful ion yields. Lower useful ion yields, or the need to measure a minor isotope (cf., Sn) results in a degradation of the detection limit. The detection of several elements is limited by contamination in the ion source. The gaseous atmospheric species are present at low, but not insignificant levels in the plasma gas. Ta and Au are obscured by Ta and TaO sputtered from source components. If all source components are not rigorously cleaned of material sputtered in previous analyses, residual material from these analyses will be observed at 0.1–10 ppm levels in the present analysis.

It is important to emphasize that GDMS provides an essentially complete elemental impurity survey to very low detection limits in a timely, cost effective manner. Although this level of analysis requires long signal integration times, the Ga measurement requires only a few hours to obtain an accurate, confirmed analysis to 7N levels. Except for the presence of rather high In in the Ga, the remaining impurities are at sub ppmw levels. Detection limits in the Ga are clearly adequate for 7N qualification.

The results of the analysis of 3N5 and 6N pure In metals for selected elements are summarized in Table 2. These measurements illustrate the precision possible when the impurity signals are given additional signal integration time at the cost of elemental coverage. The precision of GDMS elemental analysis for the homogeneously distributed impurities is much better than 5% to ppbw concentration levels. At lower levels the standard deviation increases due to detector noise and ion counting statistics, but the precision is still acceptable even at ppt levels. This data illustrate the trade-off between elemental coverage and improved detection limits. Very good detection limits can be obtained for almost all elements, but the time investment to achieve sub-ppb detection limits over the full elemental survey is substantial and not normally cost effective.

Element	Concentration (ppmw)	Standard deviation (%)	Element	Concentration (ppmw)	Standard deviation (%)
	3N5 (99.95%)			6N (99.9999%)	
Fe	0.038	4.7	Al	0.00056	33.
Ni	0.366	1.5	Fe	0.00025	25.
Cu	0.683	4.1	Ni	0.072	3.7
Cd	0.453	1.7	Cu	0.0069	9.6
Sn	20.5	0.6	Sn	0.0019	23.
Tl	0.698	1.9	Sb	0.0021	14.
Pb	61.3	1.3	Tl	0.044	5.4
Bi	0.202	1.7	Pb	0.066	3.4

Table 2 **Precision of trace elemental analysis of 3N5 (99.95%) and 6N (99.9999%) pure In metals by GDMS (VG 9000). The data are the average of five measurements. The integration time per isotope per measurement was 500 ms (3N5 In) and 1500 ms (6N In), respectively.**

Semiconductors

The results of GDMS analysis of several types of semiconductor substrates is shown in Table 3. Silicon is the most commonly used of these semiconductors. Gallium phosphide and ZnTe provide examples of III–V and II–VI semiconductors, respectively. The absence of the transition metals in particular is very important to the proper functioning of devices built on these substrates. Consequently, the detection limits for the full range of metals must be very good. GDMS provides detection limits at the ppb and lower levels. The detection limits in Si are significantly worse than the detection limits in the other two semiconductors, reflecting the fact that the sputter sampling rate, and thus the analytical signal of the Si, is significantly lower than for the other material. The Si results also provide an example for which the detection limit has been determined by a matrix-specific mass spectral interference (e.g., the $SiAr^{++}$ ions interfere with the measurement of the S^{+} ions and are not mass resolvable.) While such interferences may limit the measurement of a particular impurity in a particular matrix, they are not the general rule.

TiW Sputtering Target and W Metal Powder

The results of a GDMS analysis of high-purity TiW are summarized in Table 4. High-purity TiW is very commonly used as the metallization to provide the conducting links in the construction of semiconductor devices. The metallization is commonly deposited by sputtering from a high-purity alloy target onto the sub-

	Concentration (ppmw)				Concentration (ppmw)		
Element	Si	GaP	ZnTe	Element	Si	GaP	ZnTe
Li	<0.007	<0.0004	0.013	Y	<0.001	<0.00007	<0.00005
Be	<0.003	<0.0005	<0.0001	Zr	<0.003	<0.0003	<0.0003
B	<0.006	75.	0.002	Nb	<0.002	<0.002	<0.0004
C	(<5.)	(<5.)	(<3.)	Mo	<0.01	<0.001	<0.001
N	(<50.)	(<10.)	(<30.)	Ru	<0.009	<0.001	<0.001
O	(<60.)	(<10.)	(<30.)	Rh	<0.005	<0.001	<0.003
F	(<0.04)	(<0.8)	(<0.002)	Pd	<0.01	<0.002	<0.006
Na	<0.01	0.033	<0.019	Ag	<0.01	<0.05	<0.001
Mg	<0.006	<0.0004	0.004	Cd	<0.04	<0.004	<0.005
Al	<0.009	0.006	0.011	In	<0.007	<0.01	<0.1
Si	Matrix	0.008	0.024	Sn	<0.03	<0.004	<0.006
P	<0.01	Matrix	<0.002	Sb	<0.03	<0.002	<0.02
S	<0.1	0.9	<0.006	Te	<0.01	<0.007	Matrix
Cl	(<3.)	(<1.)	(<0.7)	I	<0.01	<0.0008	<0.04
K	<0.01	0.008	0.004	Cs	<0.003	<0.0002	<0.002
Ca	<0.03	<0.007	<0.04	Ba	<0.002	<0.0008	<0.003
Sc	<0.001	<0.00007	<0.0003	La	<0.001	<0.00009	<0.007
Ti	<0.002	0.0006	<0.0008	Ce	<0.009	<0.0006	<0.002
V	<0.002	<0.0002	<0.0001	Nd	<0.002	<0.0005	<0.01
Cr	<0.005	<0.0006	<0.002	Hf	<0.008	<0.0005	<0.0003
Mn	<0.005	0.002	<0.0007	Ta	(<7.)	(<7.)	(<3.)
Fe	<0.02	0.01	0.04	W	<0.008	<0.001	<0.0005
Co	<0.003	<0.0002	<0.0002	Re	<0.006	<0.0003	<0.0002
Ni	<0.01	<0.0005	0.003	Os	<0.005	<0.0007	<0.002
Cu	<0.015	<0.004	0.12	Ir	<0.01	<0.0007	<0.001
Zn	<0.03	<0.002	Matrix	Pt	<0.01	<0.002	<0.02
Ga	<0.02	Matrix	0.019	Au	<0.03	<0.009	<0.01
Ge	<0.06	<0.03	<0.02	Hg	<0.02	<0.003	<0.001
As	0.075	1.2	<0.0007	Tl	<0.01	<0.002	<0.0005
Se	<0.06	<0.007	0.4	Pb	<0.01	<0.001	<0.001
Br	<0.04	<0.005	<0.002	Bi	<0.008	<0.0006	<0.0002
Rb	<0.005	<0.0003	<0.0002	Th	<0.002	<0.0002	<0.00007
Sr	<0.002	<0.0001	<0.0001	U	<0.003	<0.0002	<0.00008

Table 3 **Results of GDMS analyses for impurities in three high-purity semiconductor substrates. Lower detection limits are achieved for materials with higher sputtering rates.**

	Concentration (ppmw)			Concentration (ppmw)	
Element	TiW	W	Element	TiW	W
Li	<0.0006	<0.005	Y	<2.	<0.00007
Be	<0.0007	<0.005	Zr	<0.04	≤0.005
B	0.03	0.014	Nb	<0.01	<0.003
C	(<12.)	(<9.)	Mo	0.68	0.47
N	(<14.)	(2.)	Ru	<0.0007	<0.0007
O	(<170.)	(50.)	Rh	<0.0003	<0.0003
F	(<0.2)	(<0.1)	Pd	<0.002	<0.002
Na	<0.01	0.02	Ag	<0.002	<0.004
Mg	<0.001	0.17	Cd	<0.008	<0.007
Al	0.27	≤0.08	In	<0.001	0.005
Si	1.2	0.34	Sn	0.04	0.07
P	2.5	1.0	Sb	<0.002	<0.002
S	0.3	0.01	Te	<0.01	<0.003
Cl	(<8.)	(<0.5)	I	<0.001	<0.001
K	<0.02	<0.06	Cs	<0.0006	<0.0006
Ca	<0.03	0.33	Ba	0.002	0.005
Sc	<0.005	0.00006	La	<0.0001	<0.0001
Ti	Matrix	0.35	Ce	<0.0001	<0.0001
V	0.19	0.004	Nd	<0.0006	<0.0006
Cr	3.7	0.03	Hf	0.004	<0.008
Mn	0.92	0.026	Ta	(<90.)	(<20.)
Fe	6.	0.26	W	Matrix	Matrix
Co	0.02	0.001	Re	<0.2	<0.06
Ni	0.87	0.11	Os	<0.009	<0.01
Cu	0.3	<0.01	Ir	<0.0008	<0.004
Zn	≤0.06	<0.01	Pt	<0.002	<0.05
Ga	<0.003	0.02	Au	<0.05	<0.02
Ge	<0.003	<0.003	Hg	<0.08	<0.07
As	0.40	0.14	Ti	<0.001	<0.002
Se	<0.005	<0.04	Pb	<0.001	≤0.005
Br	<0.01	<0.01	Bi	<0.0009	<0.001
Rb	<0.01	<0.0008	Th	<0.00004	0.0003
Sr	<40.	<0.001	U	<0.00003	0.00008

Table 4 **Results of GDMS analyses for impurities at the 4N–5N level in a TiW sputter target and a W metal powder. The W analysis required a pressed pin sample (1 mm × 1 mm x 10 mm). The dependence of RSFs on pin–aperture spacing require a different RSF suite for proper quantitation. Detection limits for U and Th have been improved using longer integration times.**

strate; the sputter target alloy must typically be at least 4N–5N pure. The impurity content of the transition metals in particular must be kept low. It is also a common requirement that the U and Th in these metallizations must be below 1 ppb total (U + Th). GDMS is particularly adept at combining measurements for total impurity qualification with the measurement of selected elements at ultratrace levels. In this instance it is also the case that the detection limits for the gaseous species, which are relatively high because of instrument background, are nonetheless adequate for the qualification of the metal for low C, N, O, F, and Cl contents. The accuracy of the elemental concentration determination is independent of the amount present, and GDMS can provide the contents of alloying elements as well as of impurity elements. The accuracy of GDMS analysis (better than 20%) is generally suitable for confirming alloy type.

The results of GDMS analysis for the qualification of W metal powder as incorporated, for example, into such TiW sputter targets is also presented in Table 4. The powder must be formed into a self-supporting pin, which can be done using an appropriate polypropylene die and suitably high pressures. The pressing procedure must be clean, so as to not compromise the analysis. As a further complication, the surface of the pressed pin cannot be cleaned in the "normal" way, by etching the surface. Instead, the surface must be presputtered in the GDMS source for some time to reduce surface contaminants to acceptable levels. Many impurity elements are present in the W powder at 0.1–1 ppmw levels. GDMS is capable of accurate measurement to much lower levels, as indicated by the ppbw detection limits of many of the elements. Very low detection limits are required if qualification analysis to 6N and lower impurity levels is to include essentially all elements. For many elements, detection limits depend on the specific matrix being analyzed. For example, the presence of ions of $TiAr^{++}$ interferes with the measurement of the Ca^{+} ions, and the corresponding detection limit on Ca is relatively high.

If the critical impurities are known, then only a selected list of elements need to be examined, with some improvement in the cost effectiveness of the analysis. However, the list of elements to be included in the qualification analysis is often historical and related to the limitations of the analytical methods previously used for qualification rather than for technological reasons related to the end use of the metal. As a result, problems in application can arise for no obvious reason. The time and cost of extending the impurity list for GDMS analysis to include essentially all elements is minimal, considering the additional information gained.

The analysis of nonconducting material using a DC glow-discharge source can be carried out in a manner similar to the analysis of the W powder. The sample must be ground to a fine powder, with care taken to minimize contamination, and then mixed with a high-purity, electrically conducting powder, such as Ag, to obtain an electrically conducting pin. The analysis of the nonconducting material by this method is limited mainly by the presence of corresponding impurities in the binding metal.

Conclusions

GDMS has now become a well-established analytical technique for direct multielemental analysis of conducting solids. Glow-discharge instruments incorporating high mass resolution magnetic mass spectrometers as well as quadrupole mass spectrometers are now commercially available. With high mass resolution GDMS, the quantitative measurement of essentially all elements in a single analysis with detection limits in the sub-ppbw concentration range is possible. With appropriate analytical protocol the time required for an elemental survey analysis to ppb limits is 1–2 hours, with an average accuracy not worse than 20%. Direct determinations in the pptw concentration range for single elements are also routinely possible for many elements, the primary consideration being analysis time and cost of analysis. GDMS has become a viable, and in many cases, preferable mode of analysis largely through the analytical capabilities brought by high mass resolution. With similar accuracy, glow-discharge quadrupole mass spectrometry with low mass resolving power provides a more rapid analysis compatible with thick-film depth profiling requirements. With nominal detection limits of 10–100 ppb and full elemental coverage, GDQMS offers a significant improvement on optical analytical methods.

Comparable mass spectrometric methods for solids analysis are spark-source (SSMS), Inductively Coupled Plasma (ICP-MS), and Secondary Ion Mass Spectrometry (SIMS). Of these, GDMS and SSMS are most similar in capability, but GDMS provides a 1–2 orders-of-magnitude better detection limit and an order-of-magnitude improvement in measurement accuracy. SIMS is generally hampered by extremely variable and matrix-dependent elemental sensitivities, and very limited sampling volume. However, SIMS does offer sub-ppb detection limits for selected elements, as well as microanalytical capability. ICP-MS was developed primarily for the accurate ppt level analysis of liquid samples, but can be used for solids analysis by sample dissolution or by laser-ablation sampling, yielding ppb detection limits. The advantage of ICP-MS lies in the sample homogenization resulting from dissolution.

Analytical GDMS instrumentation will continue to develop in response to market demands and as application areas are explored more thoroughly. GDMS is in a time of rapidly expanding industrial acceptance in the area of high-purity metals characterization, and the analytical niche for GDMS seems well assured. Serious efforts are underway to expand this niche to include ppb-level measurements on insulating solids. While the use of GDMS for the chemical characterization of steels and similar alloys has been limited, GDQMS may play a useful role in this area as well as providing unique capabilities for the accurate quantitative analysis of thick-film structures.

Related Articles in the Encyclopedia

SSMS, ICPMS, ICP-Optical, and SIMS

References

1 N. Jakubowski, D. Stuewer, and W. A. Vieth. *Fresnius Z. Anal. Chem.* **331,** 145, 1988.

2 W. W. Harrison and B. L. Bentz. *Prog. Analyt. Spect.* **11,** 53, 1988.

3 W. Vieth and J. C. Huneke. *Spectrochem. Acta.* **46B** (2), 137, 1991.

10.8 ICPMS

Inductively Coupled Plasma Mass Spectrometry

BARRY J. STREUSAND

Contents

Introduction

The importance of the electrical and physical properties of materials has strained the limits of characterization techniques in general, and elemental analysis techniques in particular. This includes not only the analysis of surfaces, films, and bulk materials, but also of the chemicals, gases, and equipment used to form them. Often properties of a material are affected by doping levels in the 10^{14} range, which means characterization at the parts-per-billion (ppb) to sub-ppb levels.Inductively Coupled Plasma Mass Spectrometry (ICPMS) is capable of this degree of sensitivity and has developed in the same time frame that high-purity materials have developed. These materials probably have played a large part in pushing instrument development quickly from a research stage to fairly common usage and wide ranging applications.

ICPMS can be considered a high-sensitivity extension of mass spectrometry, as well as an increased-sensitivity detector replacing optical ICP (ICP-OES) analysis. In fact, both viewpoints are accurate, and the wide application of ICPMS analysis in the world of materials science is evidence of that. ICPMS is an extremely sensitive technique. In high-purity water, for example, detection limits for many elements are under 100 parts per trillion (ppt). In higher sensitivity instruments, the limits are under 10 ppt.

The information derived from ICPMS analysis is, simply, a mass spectrum of the sample. This includes a wealth of information, however. In one sampling, which can take less than one minute, information on almost all elements in the periodic table can be derived to at least low ppb levels. This *multiplexing* advantage is extraordinarily valuable in materials analysis as it can give one a good look at a sample quickly and with surprisingly good quantitation results. (Semiquantitative analysis will be discussed later.) The mass spectrum contains not only elemental information but also isotopic information for each element. This is useful for giving a positive identification of most elements, for identifying interferences, and for providing alternative masses for characterization.

Other techniques that give elemental analysis information include the more established optical methods such as Atomic Absorption (AA), Graphite Furnace Atomic Absorption (GFAA), emission spectroscopy, Inductively Coupled Plasma–Optical Emission Spectroscopy (ICP-OES), and X-Ray Fluorescence (XRF). Newer mass spectrometry based techniques include Spark Source Mass Spectrometry (SSMS), Glow Discharge Mass Spectrometry (GDMS), and Secondary Ion Mass Spectrometry (SIMS). Elemental information may also be gained from other techniques such as Auger electron spectroscopy and X-Ray Photoelectron Spectroscopy (XPS). Of course there are other methods and new ones are being developed continually. Each of these techniques is useful for the purposes they were intended. Some, such as AA, have advantages of cost; others, such as XRF, can handle samples with minimal sample handling. ICPMS offers the detection limits of the most sensitive techniques (in many cases greater sensitivity) and easy sample handling for most samples.

Basic Principles and Instrumentation

An ICPMS spectrometer consists of:

1. An inductively coupled plasma for sample ionization
2. A mass spectrometer for detecting the ions
3. A sample introduction system.

All of these components are critical to the high sensitivity found in ICPMS instruments. Figure 1 shows their arrangement.

Mass Spectrometer

The mass spectrometer usually found on ICPMS instruments is a quadrupole mass spectrometer. This gives high throughput of ions and resolutions of 1 amu. Only a

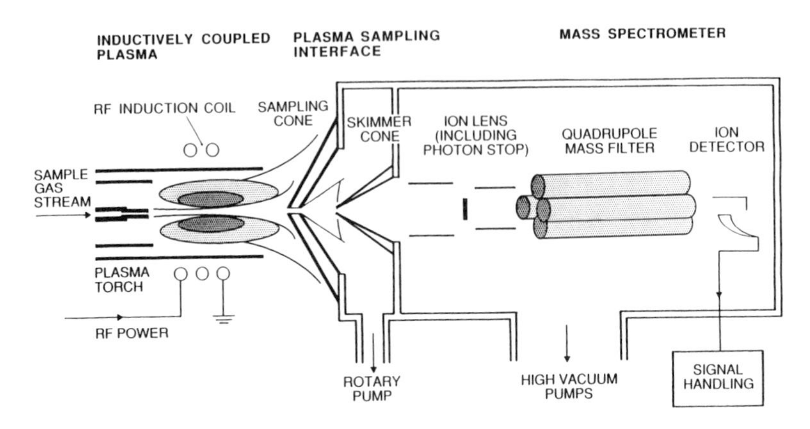

Figure 1 **Schematic of an ICPMS.**

relatively small mass range is required for analysis of materials broken down into their elemental composition as all atomic masses are below 300 amu.

A quadrupole mass spectrometer allows ions of a specific charge-to-mass ratio to pass through on a trajectory to reach the detector. This is accomplished by applying dc and rf potentials to four rods (hence the name quadrupole) that can be tuned to achieve different mass conductances through the spectrometer. The detector only counts ions, it is the quadrupole tuning that determines which ions are counted. The quadrupole can be tuned through a wide mass range quickly; a scan from 1 amu to 240 amu can take less than a second. An increased signal-to-noise ratio is accomplished by time averaging many scans.

Detectors used in ICPMS are usually electron multipliers operating in a pulse-counting mode. This gives a useful linear detector range of 10^6 (6 orders of magnitude). Some instruments can also use these detectors in an analog mode that is less sensitive. A combination of these modes allows an increase of operating range to over 10^8. This means that one can measure concentrations from 10 ppt to over a ppm in one sample. Another method of increasing the range is by using a Faraday detector in combination with the pulse counting, giving a 10^{10} range.

Two vacuum systems are used to provide both the high vacuum needed for the mass spectrometer and the differential pumping required for the interface region. Rotary pumps are used for the interface region. The high vacuum is obtained using diffusion pumps, cryogenic pumps, or turbo pumps.

Inductively Coupled Plasma

The inductively coupled plasma and the torch used in ICPMS are similar to that used in ICP-OES. In ICPMS, the torch is aimed horizontally at the mass spectrometer, rather than vertically, as in ICP-OES. In ICPMS the ions must be transported physically into the mass spectrometer for analysis, while in ICP-OES light is transmitted to the entrance slits of the monochromator. Most ICPMS instruments operate on 27.15 kHz.

Interface

The part that marries the plasma to the mass spectrometer in ICPMS is the interfacial region. This is where the 6000° C argon plasma couples to the mass spectrometer. The interface must transport ions from the atmospheric pressure of the plasma to the 10^{-6} bar pressures within the mass spectrometer. This is accomplished using an expansion chamber with an intermediate pressure. The expansion chamber consists of two cones, a sample cone upon which the plasma flame impinges and a skimmer cone. The region between these is continuously pumped.

The skimmer has a smaller aperture than the sample cone, which creates a pressure of 10^{-2} atmospheres in the intermediate region. The ions are conducted through the cones and focused into the quadrupole with a set of ion lenses. Much of the instrument's inherent sensitivity is due to good designs of these ion optics.

Sampling

Sample introduction into the ionizing plasma is normally carried out in the same manner as for ICP-OES. An aqueous solution is nebulized and swept into the plasma.

Obtaining the aqueous solution to analyze is often a challenge in materials analysis. Thin films usually can be dissolved by acids without dissolving the underlying substrate, however sometimes this is difficult. A film can also be oxidized and the oxide dissolved. Temperatures involved in this procedure are sometimes quite elevated so care must be taken to maintain sample integrity. The chemistry of the sample must be kept in mind so that the limits of the analysis are known.

By far the most simple acid to work with in ICPMS is nitric acid. This has minimal spectral interferences and in concentrations under 5% does not cause excessive wear to the sample cones. Other acids cause some spectral interferences that often must be minimized by dilution or removal. When HF is used, a resistant sampling system must be installed that does not contain quartz.

Organic polymer materials may be analyzed by ashing at relatively high temperatures. This involves oxidation of the carbon containing matrix, leaving an inorganic residue that is taken up in acid. An alternative in some cases is to dissolve the polymer in solvent and analyze the nonaqueous solution directly. Nonaqueous media will be discussed in a later section.

Solutions may typically be analyzed with up to 0.2% dissolved solids. This means a dilution factor of 1000. For example, an element that will give a 0.1 ppb detection limit in deionized water will give a detection limit of 100 ppb in a film dissolved in acid and diluted to 0.1% solids.

The role of the nebulizer in ICPMS is to transform the liquid sample into an aerosol. This is carried into the plasma by an argon flow after passing through a cooled spray chamber to remove excess vapor. Types of nebulizers in common use include Meinhard, DeGalen, and cross-flow nebulizers. A more novel nebulizer is the ultrasonic nebulizer.

Detection limits in ICPMS depend on several factors. Dilution of the sample has a large effect. The amount of sample that may be in solution is governed by suppression effects and tolerable levels of dissolved solids. The response curve of the mass spectrometer has a large effect. A typical response curve for an ICPMS instrument shows much greater sensitivity for elements in the middle of the mass range (around 120 amu). Isotopic distribution is an important factor. Elements with more abundant isotopes at useful masses for analysis show lower detection limits. Other factors that affect detection limits include interference (i.e., ambiguity in identification that arises because an elemental isotope has the same mass as a compound molecules that may be present in the system) and ionization potentials. Elements that are not efficiently ionized, such as arsenic, suffer from poorer detection limits.

There are fewer interferences in ICPMS, compared to other techniques. Because most elements have more than one isotope it is unusual to find an element that cannot be analyze: Several isotopes are almost always present. One of the most troublesome examples is the analysis of iron. Iron has three isotopes ^{54}Fe, ^{56}Fe, and ^{57}Fe; the most abundant by far is ^{56}Fe. These are all interfered with by argon molecules: ArN^+ at 54 amu, ArO^+ at 56 amu, and $ArOH^+$ at 57 amu. This gives detection limits of about 6–12 ppb for iron using ^{57}Fe rather than the < 0.1 ppb expected. Other interferences are almost always present, most involve molecular species formed by atmospheric constituents and argon. There are few interferences above 57 amu. The cone material, usually Ni, may also give a background peak. Matrix elements will give other interferences, for example, organic solvents give large interferences for ArC^+ at 52 amu, $Ar^{13}C^+$ st 53 amu, and CO_2^+ at 44 amu. A tungsten matrix will show tungsten isotope patterns for WO^+, WO_2^+, and WO_3^+.

Another type of interference in ICPMS is suppression of the formation of ions from trace constituents when a large amount of analyte is present. This effect depends on the mass of the analyte: The heavier the mass the worse the suppression.[1] This, in addition to orifice blockage from excessive dissolved solids, is usually the limiting factor in the analysis of dissolved materials.

Solvents

ICPMS offers a high-sensitivity method for the direct analysis of organic solvents. The large amount of carbon present introduces some problems unique to ICPMS. The need to transport ions directly from the plasma source into the mass spectrometer, and the small orifice needed to accomplish this, means that plugging is a problem. This is avoided by adding oxygen to the plasma, converting it from a reducing environment to an oxidizing one. Carbon dioxide is formed from the carbon. Other modifications include operating the spray chamber at a lower temperatures and increasing the RF power to the plasma. New interferences arise in organic solvent matrices.[2]

Solids

Direct sampling of solids may be carried out using laser ablation.[3] In this technique a high-power laser, usually a pulsed Nd–YAG laser, is used to vaporize the solid, which is then swept into the plasma for ionization. Besides not requiring dissolution or other chemistry to be performed on the sample, laser ablation ICPMS (LA-ICPMS) allows spatial resolution of 20–50 μm. Depth resolution is 1–10 μm per pulse. This aspect gives LA-ICPMS unique diagnostic capabilities for geologic samples, surface features, and other inhomogeneous samples. In addition minimal, or no, sample preparation is required.

Laser sampling is more a physical phenomenon than a chemical one. The energy of the laser is used to nonselectively ablate the sample. This insures homogeneous sampling of a physically defined area regardless of the nature of the components: Solubilities are not a factor. This technique shows much promise for ceramics, glasses, and geologic samples.

Another method devised for direct sampling of solids involves direct insertion of the sample into the plasma.[4, 5] In this procedure the sample is delivered through the central tube of the torch. The sample may be premixed with graphite powder.

Gases

Recently the high sensitivity of ICPMS has been applied to gas phase samples. This development has been driven mostly by new generation semiconductor processes, which use chemical vapor deposition techniques rather than the previously more common physical deposition techniques (i.e., sputtering). As geometries in devices shrink, more stringent purity is required for chemical precursors. Many of the gases and vapors are highly reactive, complicating the analysis.

One way to analyze gases is to simply add the gas or vapor to the plasma torch where the nebulized aqueous sample ordinarily would be introduced. This works for some gases but results in a dry plasma. It is difficult to know how the instrument is responding to the sample or how significant suppression effects are. For organometallic vapors the same problems arise as in sampling organic solvents. Carbon build-up on the sampling cone can plug the orifice into the mass spectrometer. Organometallic samples often react violently with oxygen or water and care must be taken when adding oxygen to the system to alleviate carbon deposition.

These problems are overcome through the use of a torch designed for both stable and reactive gases and vapors. The torch, which is shown in Figure 2, has an insertion tube to introduce the gas phase sample immediately preceding the plasma. It is mixed within the torch with an aqueous standard introduced through the nebulizer in the normal manner. The reactive gas or vapor will oxidize in the mixing region of the torch and be swept into the plasma for ionization and analysis. The standard

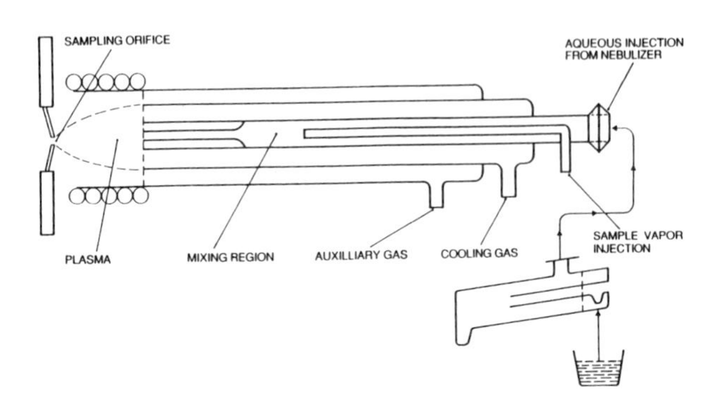

Figure 2 **Gas–vapor sampling torch.**

acts as an external measure of instrument performance and sensitivity.[6] Another innovation in the analysis of gases involves the use of a ceramic sample cone that maintains a higher temperature than metal cones during operation to minimize plugging, allowing a more concentrated sample to be used.[7]

Quantitation

One of the important advantages of ICPMS in problem solving is the ability to obtain a semiquantitative analysis of most elements in the periodic table in a few minutes. In addition, sub-ppb detection limits may be achieved using only a small amount of sample. This is possible because the response curve of the mass spectrometer over the relatively small mass range required for elemental analysis may be determined easily under a given set of matrix and instrument conditions. This curve can be used in conjunction with an internal or external standard to quantify within the sample. A recent study has found accuracies of 5–20% for this type of analysis.[8] The shape of the response curve is affected by several factors. These include matrix (particularly organic components), voltages within the ion optics, and the temperature of the interface.

Full quantitation is accomplished in the same manner as for most analytical instrumentation. This involves the preparation of standard solutions and matching of the matrix as much as possible. Since matrix interferences are usually minimized in ICPMS (relative to other techniques), the process is usually easier.

ICPMS is uniquely able to borrow a quantitation technique from molecular mass spectrometry. Use of the isotope dilution technique involves the addition of a spike having a different isotope ratio to the sample, which has a known isotope ratio. This is useful for determining the concentration of an element in a sample that must undergo some preparation before analysis, or for measuring an element with high precision and accuracy.[9]

Conclusions

ICPMS is a relatively new technique that became useful and commercially available early in its development. As a result, the field is continually changing and growing. The following is a summary of the directions of ICPMS instrumentation as described by three commercial instrument representatives.[10]

Trends in instrumentation are toward both lower and higher cost. Lower cost instruments may have limited capabilities, including less sensitivity than what is now typical of ICPMS. These instruments are used for the more routine types of analyses. Higher end instrumentation includes attaching the plasma source to a high-resolution magnetic sector mass spectrometer rather than a quadrupole. This avoids many mass-related interferences, such as occur for iron and calcium. Other instrument developments include improved ease of use, hardiness, and application specific software packages. Future improvements will include more extensive calculation software to correct for interferences by taking advantage of the large amount of isotopic information present. Combination instruments that offer a glow discharge source in addition to the ICP source have been introduced.

Like all techniques, ICPMS sampling is moving toward many hyphenated techniques. ICPMS instruments have been combined with flow injection analysis, electrothermal vaporization, ion chromatography, liquid chromatography, and chelation chromatography. Laser ablation-ICPMS has been discussed earlier. New lasers combined with frequency doubling and quadrupling crystals are being developed.

Gases for mixing with argon, such as N_2 and Xe, have been the subject of study for some time. Some new instrumentation will incorporate manifolds for making this process easier. Other plasma developments include microwave-induced plasmas with He to eliminate interferences from argon containing molecular species.

ICPMS, although a young technique, has become a powerful tool for the analysis of a variety of materials. New applications are continually being developed. Advantages include the ability to test for almost all elements in a very short time and the high sensitivity of the technique.

Related Articles in the Encyclopedia

ICP-OES, XRF, SSMS, and GDMS

References

1. Y. S. Kim, H. Kawaguchi, T. Tanaka, and A. Mizuike. *Spectrochim. Acta.* **45B,** 333, 1990.
2. R. C. Hutton. *J. Anal. Atom. Spec.* **1,** 259, 1986.
3. E. R. Denoyer, K. J. Fredeen, and J. W. Hager. *Anal. Chem.* **83,** 445A, 1991.

4 L. Blain, E. D. Salin, and D. W. Boomer. *J. Anal. Atom. Spec.* **4,** 721, 1989.

5 V. Karanassios and G. Horlick. *Spectrochim. Acta.* **44B,** 1345, 1989.

6 B. J. Streusand, R. H. Allen, D. E. Coons, and R. C. Hutton. US patent no. 4,926,021.

7 R. C. Hutton, M. Bridenne, E. Coffre, Y. Marot, and F. Simondet. *J. Anal. Atom. Spec.* **5,** 463, 1990.

8 D. Ekimoff, A. M. Van Nordstrand, and D. A. Mowers. *Appl. Spectrosc.* **43,** 1252, 1989.

9 A. A. van Heuzen, T. Hoekstra, and B. van Wingerden. *J. Anal. Atom. Spec.* **4,** 483, 1989.

10 P. Blair. Fison Instruments, private communication; J. Callaghan. Turner Scientific, private communication; and C. Fisher. Perkin-Elmer, private communication.

10.9 ICP-OES

Inductively Coupled Plasma-Optical Emission Spectroscopy

JOHN W. OLESIK

Contents

Introduction

The Inductively Coupled Plasma (ICP) has become the most popular source for multielement analysis via optical spectroscopy[1, 2] since the introduction of the first commercial instruments in 1974. About 6000 ICP-Optical Emission Spectrometry (ICP-OES) instruments are in operation throughout the world.

Approximately 70 different elements are routinely determined using ICP-OES. Detection limits are typically in the sub-part-per-billion (sub-ppb) to 0.1 part-per-million (ppm) range. ICP-OES is most commonly used for bulk analysis of liquid samples or solids dissolved in liquids. Special sample introduction techniques, such as spark discharge or laser ablation, allow the analysis of surfaces or thin films. Each element emits a characteristic spectrum in the ultraviolet and visible region. The light intensity at one of the characteristic wavelengths is proportional to the concentration of that element in the sample.

The strengths of ICP-OES are its speed, wide linear dynamic range, low detection limits, and relatively small interference effects. Automated instruments with multiple detectors can determine simultaneously 40 or more elements in a sample in less than one minute. The relationship between emission intensity and concentration is linear over 5–6 orders of magnitude. Therefore, trace and minor elements often can be measured simultaneously without prior separation or preconcentration. Detection limits are similar or better to those provided by Flame Atomic Absorption (FAA) which generally detects one element at a time. Detection limits are typically better for Graphite-Furnace Atomic Absorption (GFAA) or Inductively Coupled Plasma Mass Spectrometry (ICPMS) than for ICP-OES. However, commercial GFAA instruments do not provide the simultaneous multielement capabilities of ICP-OES. ICPMS can provide nearly simultaneously analysis via rapid scanning or hopping between mass-to-charge ratios. Detection limits are generally better for ICP-OES than for X-Ray Fluorescence Spectrometry (XFS), except for S, P, and the halogens.

ICP-OES is a destructive technique that provides only elemental composition. However, ICP-OES is relatively insensitive to sample matrix interference effects. Interference effects in ICP-OES are generally less severe than in GFAA, FAA, or ICPMS. Matrix effects are less severe when using the combination of laser ablation and ICP-OES than when a laser microprobe is used for both ablation and excitation.

The accuracy of ICP-OES ranges from 10% using simple, pure aqueous standards, to 0.5% using more elaborate calibration techniques. Precision is typically 0.2–0.5% for liquid samples or dissolved solids and 1–10% for direct solid analysis using electrothermal or laser vaporization. ICP-OES is used in a wide variety of applications because of its unique speed, multielement analysis capability, and applicability to samples having a wide range of compositions. Trace and minor elements have been determined in a variety of metal alloys. ICP-OES also has been applied to geological samples. Trace metals have been measured in petroleum samples, as have impurities in nuclear materials. ICP-OES has been used for elemental analysis of superconductors, ceramics, and other specialty materials. The technique also has been widely applied to measure impurities in the raw materials and acids used in semiconductor processing.

Basic Principles

An ICP-OES instrument consists of a sample introduction system, a plasma torch, a plasma power supply and impedance matcher, and an optical measurement system (Figure 1). The sample must be introduced into the plasma in a form that can be effectively vaporized and atomized (small droplets of solution, small particles of solid or vapor). The plasma torch confines the plasma to a diameter of about 18 mm. Atoms and ions produced in the plasma are excited and emit light. The intensity of light emitted at wavelengths characteristic of the particular elements of interest is measured and related to the concentration of each element via calibration curves.

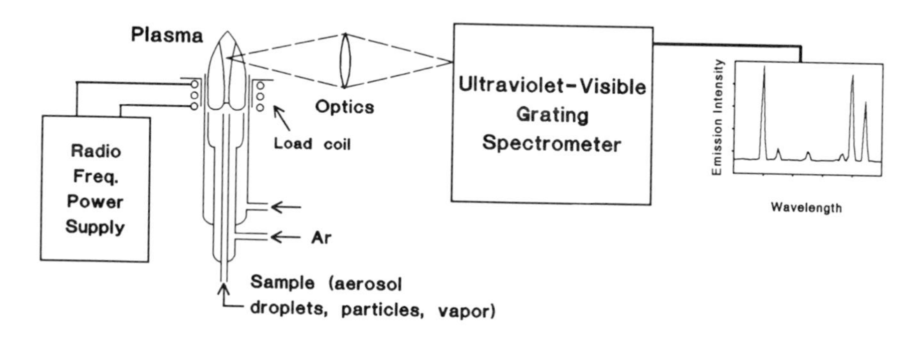

Figure 1 **Instrumentation for inductively coupled plasma-optical emission spectrometry.**

Plasma Generation and Sample Decomposition

The plasma is a high-temperature, atmospheric pressure, partially ionized gas. Argon is used most commonly as the plasma gas, although helium, nitrogen, oxygen, and mixed gas plasmas (including air) also have been used. The plasma is sustained in a quartz torch consisting of three concentric tubes (Figure 1). The inner diameter of the largest tube is about 18 mm. The outer and intermediate gases (typically, 10–16 L/min and 0–1 L/min Ar, respectively) are directed tangentially, producing a large swirl velocity resulting in efficient cooling of the quartz torch.[1, 2] The sample is carried into the center of the plasma through a third quartz or ceramic tube (with 0.7–1.0 L/min Ar), where it is introduced as a liquid aerosol (droplets less than 10 μm in diameter), fine powder, or vapor and particulates produced by laser or thermal vaporization.

The plasma is generated using a radiofrequency generator, typically at 27 or 40 MHz. Current is carried through a water cooled, three-to-five turn *load coil* surrounding the torch. Electrons in the plasma are accelerated by the resulting oscillating magnetic fields. Energy is transferred to other species, including the sample, through collisions.

In the plasma, the sample is vaporized and chemical bonds are effectively broken resulting in free atoms and ions. Temperatures of 5000–9000 K have been measured in the plasma compared to typical temperatures of 2000–3000 K in flames and graphite furnaces.

Generation of Emission Signals

Atoms and ions are excited via collisions, probably mainly with electrons, and then emit light. Most elements with ionization energies less than 8 eV exist mainly as singly charged ions in the plasma. Therefore, spectral lines from ions are most intense for these elements, whereas elements with high ionization energies (such as B, Si, Se and As), as well as the easily ionized alkalis (Li, Na, K, Rb, and Cs), emit most strongly as atoms.

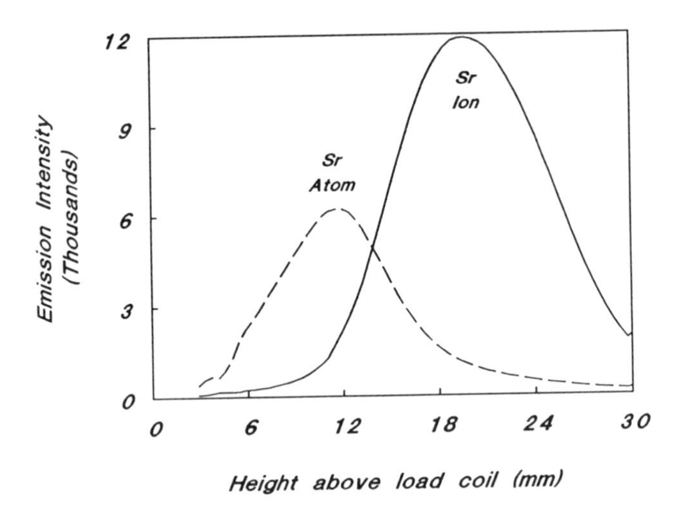

Figure 2 **Emission intensity for Sr atom (460 nm) and Sr ion (421 nm) as a function of height about the load coil in a 1-kW Ar plasma.**

Emission intensities depend on the observation height within the plasma Figure 2); the detailed behavior varies with the specific nature of the atom or ion. Emission from most ions peaks at nearly the same location, called the *normal analytical zone*, typically 10–20 mm above the top of the current-carrying induction coil. Similarly, atoms with high ionization energies (> 8 eV) or high excitation energies emit most intensely in the normal analytical zone. Emission usually is collected from a 3–5 mm section of the plasma near the peak emission intensity. Emission from atoms with low ionization energies and low excitation energies (Li, Na, K, Cs, and Rb) is most intense lower in the plasma. Unlike ion emission intensities, the atomic emission intensity peak location is a strong function of ionization and excitation energies.

Usually, the ultraviolet and visible regions of the spectrum are recorded. Many of the most intense emission lines lie between 200 nm and 400 nm. Some elements (the halogens, B, C, P, S, Se, As, Sn, N, and O) emit strong lines in the vacuum ultraviolet region (170–200 nm), requiring vacuum or purged spectrometers for optimum detection.

Quantitation

Calibration curves must be made using a series of standards to relate emission intensities to the concentration of each element of interest. Because ICP-OES is relatively insensitive to matrix effects, pure solutions containing the element of interest often are used for calibration. For thin films the amount of sample ablated by spark discharges or laser sources is often a strong function of the sample's composition. Therefore, either standards with a composition similar to the sample's must be used or an internal standard (a known concentration of one element) is needed.

0.1–1 ppb	1–10 ppb		10–50 ppb		50–100 ppb	100–500 ppb
Ba	Ag	Li	Al	Nb	As	Rb
Be	B	Lu	Au	Nd	K	U
Ca	Co	Mo	Bi	Ni	P	
Mn	Cu	Re	C	Pb	Pd	
Mg	Cd	Sc	Ce	Pr	S	
Sr	Cr	Ti	Cs	Pt	Se	
	Eu	V	Dy	Rh	Te	
	Fe	Y	Er	Ru	Tl	
	Ho	Yb	Ga	Sb		
	I	Zn	Ge	Si		
	La	Zr	Gd	Sm		
			Hf	Sn		
			Hg	Ta		
			In	Tb		
			Ir	Th		
			Na	W		

Table 1 **Typical detection limits (ppb) for ICP-OES (using a pneumatic nebulizer for sample introduction) of the most sensitive emission line between 175 nm and 850 nm for each element.**

Detection Limits

Typical elemental detection limits are listed in Table 1. The detection limit is the concentration that produces the smallest signal that can be distinguished from background emission fluctuations. The continuum background is produced via radiative recombination of electrons and ions ($M^+ + e^- \rightarrow M + h\nu$ or $M^+ + e^- + e^- \rightarrow M + e^- + h\nu$). The structured background is produced by partially or completely overlapping atomic, ionic, or in some cases, molecular emission. To obtain precision better than 10% the concentration of an element must be at least 5 times the detection limit.

Detection limits for a particular sample depend on a number of parameters,[1, 2] including observation height in the plasma, applied power, gas flow rates, spectrometer resolution,[3] integration time, the sample introduction system, and sample-induced background or spectral overlaps.[3]

Sample Introduction

Samples must be introduced into the plasma in an easily vaporized and atomized form. Typically this requires liquid aerosols with droplet diameters less than 10 μm, solid particles 1–5 μm in diameter, or vapors. The sample introduction method strongly influences precision, detection limits, and the sample size required.

Introduction of Liquids and Solutions of Dissolved Solids

Most often samples are introduced into the plasma as a liquid aerosol. Solid samples are dissolved using an appropriate procedure. Pneumatic nebulizers of various designs[1,2] generate aerosols by pumping or aspirating a flow of solution into a region of highly turbulent, high-speed gas flow. Concentric cross-flow nebulizers are used for solutions having less than 1% dissolved solids. V-groove (Babington) nebulizers can be used for highly viscous solutions having a high dissolved solid content. Most of the droplets produced by these nebulizers are too large to be vaporized effectively in the plasma; therefore, a spray chamber is used to remove large droplets via gravity and to cause them to impact onto the walls. The smallest droplets are able to follow the gas flow into the plasma. Typically only 1–2% of the sample reaches the plasma, and liquid sample volumes of 1 mL or more are required.

Flow injection techniques can be used to inject sample volumes as small as 10 μL into a flowing stream of water with little degradation of detection limits. Frit nebulizers[1, 2] have efficiencies as high as 94% and can be operated with as little as 2 μL of sample solution.

Electrothermal vaporization[1, 2] can be used for 5–100 μL sample solution volumes or for small amounts of some solids. A graphite furnace similar to those used for graphite-furnace atomic absorption spectrometry can be used to vaporize the sample. Other devices including boats, ribbons, rods, and filaments, also can be used. The chosen device is heated in a series of steps to temperatures as high as 3000 K to produce a dry vapor and an aerosol, which are transported into the center of the plasma. A transient signal is produced due to matrix and element-dependent volatilization, so the detection system must be capable of time resolution better than 0.25 s. Concentration detection limits are typically 1–2 orders of magnitude better than those obtained via nebulization. Mass detection limits are typically in the range of tens of pg to ng, with a precision of 10% to 15%.

Direct Introduction of Samples from Solids, Surfaces, or Thin Films

There are advantages to direct solid sampling. Sample preparation is less time consuming and less prone to contamination, and the analysis of microsamples is more straightforward. However, calibration may be more difficult than with solution samples, requiring standards that are matched more closely to the sample. Precision is typically 5% to 10% because of sample inhomogeneity and variations in the sample vaporization step.

In the direct insertion technique,[1, 2, 4] the sample (liquid or powder) is inserted into the plasma in a graphite, tantalum, or tungsten probe. If the sample is a liquid, the probe is raised to a location just below the bottom of the plasma, until it is dry. Then the probe is moved upward into the plasma. Emission intensities must be measured with time resolution because the signal is transient and its time dependence is element dependent, due to selective volatilization of the sample. The intensity–time behavior depends on the sample, probe material, and the shape and location of the probe. The main limitations of this technique are a time-dependent background and sample heterogeneity-limited precision. Currently, no commercial instruments using direct sample insertion are available, although both manual and highly automated systems have been described.[4]

Arc and spark discharges have been used to ablate material from a solid conducting sample surface.[1, 2] The dry aerosol is then transported to the plasma through a tube. Detection limits are typically in the low ppm range. The precision attainable with spark discharges that sample over a relatively large surface area (0.2–1 cm^2) is typically 0.5% to 5.0%. Calibration curves are linear over at least 3 orders of magnitude, and an accuracy of 5% or better is realized. Commercial instruments are available. In some cases it is possible to use pure aqueous standards to produce the calibration curves used for spark ablation ICP-OES. In general, calibration curves for spark or arc ablation followed by ICP-OES are more linear and less sample matrix-dependent than calibration curves in spark or arc emission spectrometry.

A vapor sample and dry aerosol also can be produced from surfaces via laser ablation.[1, 2] Typically, solid state pulsed Nd–YAG, Nd–glass, or ruby lasers have been used. The amount of material removed from the sample surface is a function of the sample matrix and the laser pulse energy, wavelength and focusing, but is usually in the μm range. Part-per-million detection limits are possible, and the technique is amenable to conducting and nonconducting samples. Precision is typically 3% to 15%. Shot-to-shot laser pulse energy reproducibility and sample heterogeneity are the two main sources of imprecision in this technique.

Instrumentation—Detection Systems

Three different types of grating spectrometer detection systems are used (Figure 3): sequential (slew-scan) monochromators, simultaneous direct-reading polychroma-

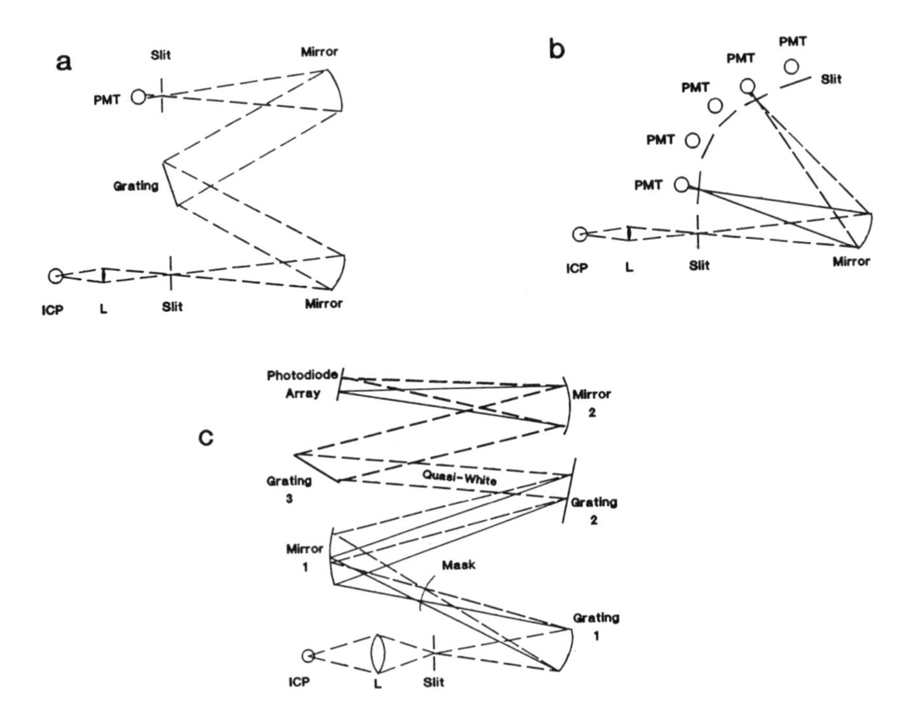

Figure 3 **Grating spectrometers commonly used for ICP-OES: (a) monochromator, in which wavelength is scanned by rotating the grating while using a single photomultiplier tube (PMT) detector; (b) polychromator, in which each photomultiplier observes emission from a different wavelength (40 or more exit slits and PMTs can be arranged along the focal plane); and (c) spectrally segmented diode-array spectrometer.**

tors, and segmented diode array-based spectrometers. The choice detection system depends on the number of samples to be analyzed per day, the number of elements of interest, whether the analysis will be of similar samples or of a wide range of sample types, and whether the chosen sample-introduction system will produce steady-state or transient signals.

Slew-scan spectrometers (Figure 3a) detect a single wavelength at a time with a single photomultiplier tube detector.[1, 2] The grating angle is rapidly slewed to observe a wavelength near an emission line from the element of interest. A spectrum is acquired in a series of 0.01–0.001 nm steps. The peak intensity is determined by a fitting routine. Background emission can be measured near the emission line of interest and subtracted from the peak intensity. The advantage of slew-scan spectrometers is that any emission line can be viewed, so that the best line for a particular sample can be chosen. Their main disadvantage is the sequential nature of the multielement analysis and the time required to slew from one wavelength to another (typically a few seconds).

Direct-reading polychromators[1, 2] (Figure 3b) have a number of exit slits and photomultiplier tube detectors, which allows one to view emission from many lines simultaneously. More than 40 elements can be determined in less than one minute. The choice of emission lines in the polychromator must be made before the instrument is purchased. The polychromator can be used to monitor transient signals (if the appropriate electronics and software are available) because unlike slew-scan systems it can be set stably to the peak emission wavelength. Background emission cannot be measured simultaneously at a wavelength close to the line for each element of interest. For maximum speed and flexibility both a direct-reading polychromator and a slew-scan monochromator can be used to view emission from the plasma simultaneously.

The spectrally segmented diode-array spectrometer[5] uses three gratings to produce a series of high-resolution spectra, each over a short range of wavelengths, at the focal plane (Figure 3c). A 1024-element diode array is used to detect the spectra simultaneously. By placing the appropriate interchangeable mask in the focal plane following the first grating, the short wavelength ranges to be viewed are selected. The light is recombined by a second grating, forming a quasi-white beam of light. A third grating is used to produce high-resolution spectra on the diode array. It is much easier to change masks in this spectrometer than to reposition exit slits in a direct-reading polychromator. The diode array-based system also provides simultaneous detection of the emission peak and nearby background. This capability is particularly advantageous when using a sample-introduction technique that generates a transient signal.

Limitations and Potential Analysis Errors

One of the major problems in ICP-OES can be spectral overlaps.[1, 2, 9] Some elements, particularly rare earth elements, emit light at thousands of different wavelengths between 180 nm and 600 nm. Spectral interferences can be minimized, but not eliminated, by using spectrometers with a resolving power ($\lambda / \Delta\lambda$) of 150,000 or higher.[1, 3] If a spectral overlap occurs, the operator can choose a different line for analysis; or identify the source of the interfering line, determine its magnitude, and subtract it from the measuring intensity. Tables of potential spectral line overlaps for many different emission lines are available.[6, 7] Some manufacturers provide computer database emission line lists. Most commercial direct-reading polychromators include software to subtract signals due to overlapping lines.[8] This is effective if the interferant line intensity is not large compared to the elemental line of interest and another line for the interferant element can be measured.

Although nonspectral interference effects are generally less severe in ICP-OES than in GFAA, FAA, or ICPMS, they can occur.[1, 2, 9] In most cases the effects produce less than a 20% error when the sample is introduced as a liquid aerosol. High concentrations (500 ppm or greater) of elements that are highly ionized in the

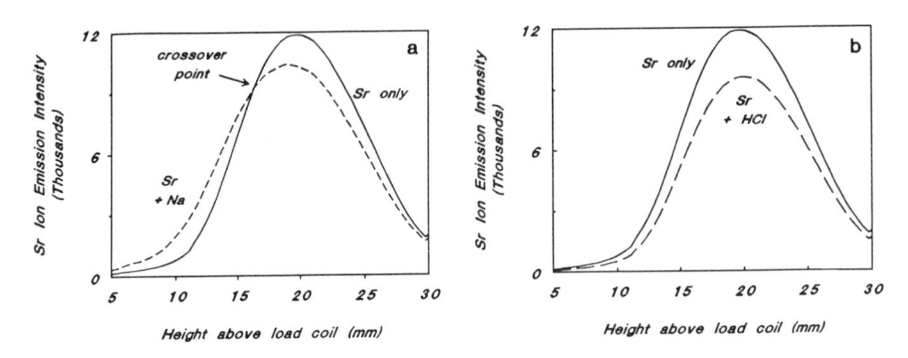

Figure 4 Effect of matrix on Sr ion emission at different heights in the plasma. Samples contained 50 ppm Sr in distilled, deionized water: (a) emission in the presence and absence of NaCl (solid line—no NaCl added; dashed line—0.05 M NaCl added); and (b) effect of the presence and absence of HCl (solid line—no HCl added; dashed line—0.6 M HCl added).

plasma can affect emission intensities. The magnitude and direction of the effect depends on experimental parameters including the observation height in the plasma, gas flow rates, power, and, to a lesser degree, the spectral line used for analysis and the identity of the matrix. A location generally can be found (called the *cross-over* point) where the effect is minimal (Figure 4a). If emission is collected from a region near the cross-over point, errors due to the presence of concomitant species will be small (generally less than 10% or 20%).

The presence of organic solvents (1% by volume or greater) or large differences in the concentration of acids used to dissolve solid samples can also affect the emission intensities (Figure 4b).[1, 2, 9] Direct solid-sampling techniques generally are more susceptible to nonspectral interference effects than techniques using solutions. The accuracy can be improved through internal standardization or by using standards that are as chemically and physically similar to the sample as possible.

Errors due to nonspectral interferences can be reduced via matrix matching, the method of standard additions (and its multivariant extensions), and the use of internal standards.[1, 2, 9]

Applications

ICP-OES has been applied to a wide range of sample types, with no single area or technology dominating. Elemental analysis can be performed on virtually any sample that can be introduced into the plasma as a liquid or dry aerosol. Metals and a wide variety of industrial materials are routinely analyzed. Environmental samples, including water, waste streams, airborne particles, and coal fly ash, are also amenable to ICP-OES. Biological and clinical samples, organic solvents, and acids used in semiconductor processing are widely analyzed.

Laser-ablation ICP-OES has been used to analyze metals, ceramics, and geological samples. This technique is amenable to a wide variety of samples, including surfaces and thin films (μm depths analyzed), similar to those analyzed by laser microprobe emission techniques (LIMS). However, interference effects are less severe using separate sampling and excitation steps, as in laser-ablation ICP-OES. Laser-ablation ICPMS is becoming more widely used than laser-ablation ICP-OES because the former's detection limits are up to 2 orders of magnitude. Spark discharge-ablation ICP-OES is used mainly to analyze conducting samples.

Conclusions

ICP-OES is one of the most successful multielement analysis techniques for materials characterization. While precision and interference effects are generally best when solutions are analyzed, a number of techniques allow the direct analysis of solids. The strengths of ICP-OES include speed, relatively small interference effects, low detection limits, and applicability to a wide variety of materials. Improvements are expected in sample-introduction techniques, spectrometers that detect simultaneously the entire ultraviolet–visible spectrum with high resolution, and in the development of intelligent instruments to further improve analysis reliability. ICPMS vigorously competes with ICP-OES, particularly when low detection limits are required.

Related Articles in the Encyclopedia

ICPMS, GDMS, SSMS, and LIMS

References

1 P. W. J. M. Boumans. *Inductively Coupled Plasma Emission Spectroscopy, Parts I and II.* John Wiley and Sons, New York, 1987. An excellent description of the fundamental concepts, instrumentation, use, and applications of ICP-OES.

2 A. Montaser and D. W. Golightly. *Inductively Coupled Plasmas in Analytical Atomic Spectrometry.* VCH Publishers, New York, 1987. Covers similar topics to Reference 1 but in a complementary manner.

3 P. W. J. M. Boumans and J. J. A. M. Vrakking. *Spect. Acta.* **42B,** 819, 1987. Describes how spectrometer resolution affects detection limits in the presence and absence of spectral overlaps.

4 W. E. Petit and G. Horlick. *Spect. Acta.* **41B,** 699, 1986. Describes an automated system for direct sample-insertion introduction of 10-μL liquid samples or small amounts (10 mg) of powder samples.

5 G. M. Levy, A. Quaglia, R. E. Lazure, and S. W. McGeorge. *Spect. Acta.* **42B,** 341, 1987. Describes the diode array-based spectrally segmented spectrometer for simultaneous multielement analysis.

6 P. W. J. M. Boumans. *Line Coincidence Tables for Inductively Coupled Plasma Atomic Emission Spectrometry.* Pergamon Press, Oxford, 1980, 1984. Lists of emission lines for analysis and potentially overlapping lines with relative intensities, using spectrometers with two different resolutions.

7 R. K. Winge, V. A. Fassel, V. J. Peterson, and M. A. Floyd. *Inductively Coupled Plasma Atomic Emission Spectroscopy: An Atlas of Spectral Information.* Elsevier, Amsterdam, 1985. ICP-OES spectral scans near emission lines useful for analysis.

8 R. I. Botto. In: *Developments in Atomic Plasma Spectrochemical Analysis.* (R. M. Barnes, ed.) Heyden, Philadelphia, 1981. Describes method for correction of overlapping spectral lines when using a polychromator for ICP-OES.

9 J. W. Olesik. *Analyt. Chem.* **63,** 12A, 1991. Evaluation of remaining limitations and potential sources of error in ICP-OES and ICPMS.

11

NEUTRON AND NUCLEAR TECHNIQUES

11.0 INTRODUCTION

All the techniques discussed here involve the atomic nucleus. Three use neutrons, generated either in nuclear reactors or very high energy proton accelerators (spallation sources), as the probe beam. They are Neutron Diffraction, Neutron Reflectivity, NR, and Neutron Activation Analysis, NAA. The fourth, Nuclear Reaction Analysis, NRA, uses charged particles from an ion accelerator to produce nuclear reactions. The nature and energy of the resulting products identify the atoms present. Since NRA is performed in RBS apparatus, it could have been included in Chapter 9. We include it here instead because nuclear reactions are involved.

Neutron diffraction uses neutrons of wavelengths 1–2 Å, similar to those used for X-rays in XRD (Chapter 4), to determine atomic structure in crystalline phases in an essentially similar manner. There are several differences that make the techniques somewhat complementary, though the need to go to a neutron source is a significant drawback. Because neutrons are diffracted by the nucleus, whereas X-ray diffraction is an electron density effect, the neutron probing depth is about 10^4 longer than X-ray. Thus neutron diffraction is an entirely bulk method, which can be used under ambient pressures, and to analyze the interiors of very large samples, or contained samples by passing the neutron flux through the containment walls. Along with this capability, however, goes the difficulty of neutron shielding and safety. Where X-ray scattering cross sections increase with the electron density of the atom, neutron scattering varies erratically across the periodic table and is approximately equal for many atoms. As a result, neutron diffraction "sees" light elements, such as oxygen atoms in oxide superconductors, much more effectively than X-ray diffraction. A further difference is that the neutron magnetic moment strongly interacts with the magnetic moment of the sample atoms, allowing determination of the spatial arrangements of magnetic moments in magnetic material. The equivalent interaction with X rays is a factor of 10^6 weaker. Neutron diffraction has proved useful in studying thin magnetic multilayers because, though it is a bulk technique, the magnetic scattering interactions are strong enough to enable usable data to be taken for as little as 500-Å thicknesses for metals.

In Neutron Reflectivity the neutron beam strikes the sample at grazing incidence. Below the critical angle (around 0.1°), total reflection occurs. Above it, reflection in the specular direction decreases rapidly with increasing angle in a manner depending on the neutron scattering cross sections of the elements present and their concentrations. On reaching a lower interface the transmitted part of the beam will undergo a similar process. H and D have one of the largest "mass contrasts" in neutron-scattering cross section. Thus, if there is an interface between a H-containing and a D-containing hydrocarbon, the reflection-versus-angle curve will depend strongly on the interface sharpness. Thus interdiffusion across hydrocarbon material interfaces can be studied by D labeling. For polymer interfaces the depth resolution obtained this way can be as good as 10 Å at buried interface depths of 100 nm, whereas the alternative techniques available for distinguishing D from H at interfaces, SIMS (Chapter 10) and ERS (Chapter 9), have much worse resolution. Also, neutron reflection is performed under ambient pressures, whereas SIMS and ERS require vacuum conditions. Labeling is not necessary if there is sufficient neutron "mass contract" already available—e.g., interfaces between fluorinated hydrocarbons and hydrocarbons. The technique has also been used for biological films and, magnetic thin films, using polarized neutron beam sources, where the magnetic gradient at an interface can be determined.

Though a powerful technique, Neutron Reflectivity has a number of drawbacks. Two are experimental: the necessity to go to a neutron source and, because of the extreme grazing angles, a requirement that the sample be optically flat over at least a 5-cm diameter. Two drawbacks are concerned with data interpretation: the reflectivity-versus-angle data does not directly give a a depth profile; this must be obtained by calculation for an assumed model where layer thickness and interface width are parameters (cf., XRF and VASE determination of film thicknesses, Chapters 6 and 7). The second problem is that roughness at an interface produces the same effect on specular reflection as true interdiffusion.

In NAA the sample is made radioactive by subjecting it to a high dose (days) of thermal neutrons in a reactor. The process is effective for about two-thirds of the elements in the periodic table. The sample is then removed in a lead-shielded container. The radioisotopes formed decay by B emission, γ-ray emission, or X-ray emission. The γ-ray or X-ray energies are measured by EDS (see Chapter 3) in special laboratories equipped to handle radioactive materials. The energies identify the elements present. Concentrations are determined from peak intensities, plus knowledge of neutron capture probabilities, irradiation dose, time from dose, and decay rates. The technique is entirely bulk and is most suitable for the simultaneous detection of trace amounts of heavy elements in non-γ-ray emitting hosts. Since decay lifetimes can be very variable it is sometimes possible to greatly improve detection limits by waiting for a host signal to decay before measuring that of the trace element. This is true for Au in Si where levels of 3×10^7 atoms/cc are achieved. An As- or Sb-doped Si, host would give much poorer limits for Au, however, because of interfering signals from the dopants.

In NRA a beam of charged particles (e.g., H, N, or F) from an ion accelerator at energies between a few hundred keV and several MeV (cf., RBS, Chapter 9) induces nuclear reactions for specific light elements (up to Ca). Various particles (protons, α particles, etc.) plus γ-rays are released by the process. The particles are detected as in RBS and, similarly their yield-versus-energy distribution identifies the element and its depth distribution. This can provide a rapid nondestructive, analysis for these elements, including H. The depth probed can be up to several μm with a re- solution varying from a few tens of nanometers at the surface to hundreds of nanometers at greater depths. Usually there is no lateral resolution, but a micro-beam systems with a few-micron capability exist. If particle detection is too inefficient (too low energies), γ-ray spectroscopy (cf., NAA) can yield elemental concentration, but not depth distributions. For some elements the nuclear reaction process has a maximum in its cross section at a specific beam energy, E_R (resonance energy). This provides an alternative method of depth profiling (resonance profiling), since if the incident beam energy, E_0, is above E_R, it will drop to E_R at a specific reaction distance below the surface (electronic energy losses, see RBS). By changing E_0 the depth at which E_R is achieved is changed, and so the depth at which the analyzed particles are produced is changed. Resonance profiling can have better sensitivity than nonresonance, but the depth resolution depends on the energy width of the resonance.

11.1 Neutron Diffraction

RAYMOND G. TELLER

Contents

Introduction

Since the recognition in 1936 of the wave nature of neutrons and the subsequent demonstration of the diffraction of neutrons by a crystalline material, the development of neutron diffraction as a useful analytical tool has been inevitable. The initial growth period of this field was slow due to the unavailability of neutron sources (nuclear reactors) and the low neutron flux available at existing reactors. Within the last decade, however, increases in the number and type of neutron sources, increased flux, and improved detection schemes have placed this technique firmly in the mainstream of materials analysis.

As with other diffraction techniques (X-ray and electron), neutron diffraction is a nondestructive technique that can be used to determine the positions of atoms in crystalline materials. Other uses are phase identification and quantitation, residual stress measurements, and average particle-size estimations for crystalline materials. Since neutrons possess a magnetic moment, neutron diffraction is sensitive to the ordering of magnetically active atoms. It differs from many site-specific analyses, such as nuclear magnetic resonance, vibrational, and X-ray absorption spectroscopies, in that neutron diffraction provides detailed structural information averaged over thousands of Å. It will be seen that the major differences between neutron diffraction and other diffraction techniques, namely the extraordinarily

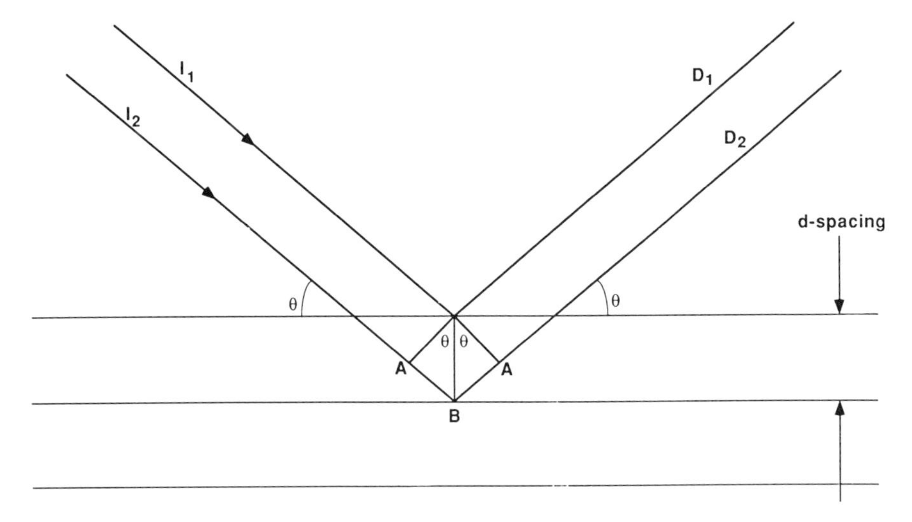

Figure 1 **Bragg diffraction. A reflected neutron wavefront (D_1, D_2) making an angle θ with planes of atoms will show constructive interference (a Bragg peak maxima) when the difference in path length between D_1 and D_2 ($2\overline{AB}$) equals an integral number of wavelengths λ. From the construction, $\overline{AB} = d \sin \theta$.**

greater penetrating nature of the neutron and its direct interaction with nuclei, naturally lead to its superior usage in experiments on materials requiring a penetration depth greater than about 50 µm. Neutron diffraction is especially well suited for structural analysis of materials containing atoms of widely varying atomic number, such as heavy metal oxides.

Basic Principles

Like X-ray and electron diffraction, neutron diffraction is a technique used primarily to characterize crystalline materials (defined here as materials possessing long-range order). The basic equation describing a diffraction experiment is the Bragg equation:

$$\lambda = 2d\sin\theta \tag{1}$$

where d represents the spacing between planes of atoms in the material in the neutron beam, λ is the wavelength of the impinging neutron wavefront, and 2θ is the diffracting angle. The diffraction geometry is illustrated in Figure 1. Inspection of the figure demonstrates that a diffraction maxima (a Bragg peak) is observed when there is constructive interference of the reflected neutron wavefront. The intensities of the Bragg peaks depend strongly upon the nature and number of atoms found lying in the planes responsible for the maxima. Consequently, a diffraction pattern can be obtained by fixing the wavelength of the neutron wavefront and scanning

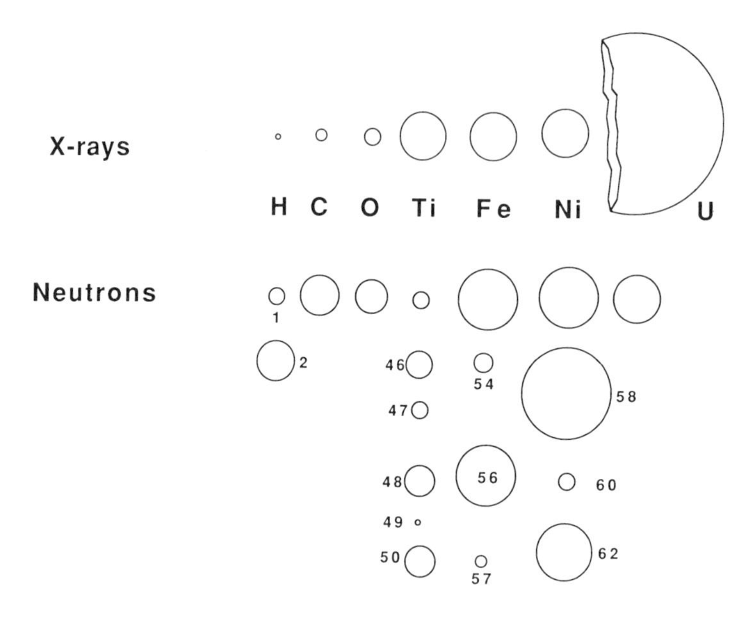

Figure 2 **Scattering physics of X rays and neutrons.**

the angle θ, or alternatively, by fixing θ and scanning a range of neutron wavelengths. It will be seen that both of these modes of operation are used at modern neutron sources.

It should be obvious from Figure 1 that if one wishes to probe *d*-spacings on the order of atomic spacings (Å) that wavelengths of the same length scale are required. Fortunately, X rays, electrons and "thermal" neutrons share the feature of possessing wavelengths of the appropriate size.

An important difference between neutron and X-ray diffraction is the way in which neutrons and X rays interact with matter. X rays are scattered primarily by electrons. Consequently, an X-ray diffraction pattern reflects the distribution of the electron density within a solid. Conversely, neutrons are scattered by nuclei, and the resultant diffraction pattern reflects nuclear distributions. Since the physics of the scattering differs significantly so does the sensitivity of each technique to various elements. While X-ray scattering from an element is roughly proportional to the local electron density (and the atomic number of the target atoms), neutron scattering is nucleus-dependent and can vary erratically as one proceeds through the periodic table. For this reason, X-ray diffraction analysis of heavy metal oxides (such as Bi_2O_3) provides information primarily about the metal atoms, whereas neutron diffraction analysis yields detailed positional information for all elements approximately equally.

One further important difference between neutron and X-ray diffraction is the former's sensitivity to magnetic structure. The magnetic moments of neutrons interact with the magnetic moments of target atoms, whereas this interaction is much weaker (~10^{-6}) for X rays. The interaction strength is proportional to the magnetic moments of the atoms in the material, and depends on their orientation relative to the neutron moment. These features make neutron diffraction the best technique for probing the spatial arrangement of magnetic moments in magnetic materials.

Experimental Considerations

Another major difference between the use of X rays and neutrons used as solid state probes is the difference in their penetration depths. This is illustrated by the thickness of materials required to reduce the intensity of a beam by 50%. For an aluminum absorber and wavelengths of about 1.5 Å (a common laboratory X-ray wavelength), the figures are 0.02 mm for X rays and 55 mm for neutrons. An obvious consequence of the difference in absorbance is the depth of analysis of bulk materials. X-ray diffraction analysis of materials thicker than 20–50 μm will yield results that are severely surface weighted unless special conditions are employed, whereas internal characteristics of physically large pieces are routinely probed with neutrons. The greater penetration of neutrons also allows one to use thick ancillary devices, such as furnaces or pressure cells, without seriously affecting the quality of diffraction data. Thick-walled devices will absorb most of the X-ray flux, while neutron fluxes hardly will be affected. For this reason, neutron diffraction is better suited than X-ray diffraction for *in-situ* studies.

A less obvious consequence of the difference in absorbance between X rays and neutrons is the large difference in the sizes of facilities using the two types of radiation (primarily for reasons of safety). While only a few millimeters of metal are required to assure the safety of workers near an X-ray source, several meters of absorbing material (usually steel, concrete, or boron-containing materials) are required around neutron sources. Because of the shielding requirement, neutron sources and instruments are orders of magnitude larger than the corresponding X-ray devices. While this leads to much greater expense for neutron sources, it also allows the analysis of larger samples. For example, railroad rails and large-circumference pipes have been analyzed for residual stress at the nuclear reactor at Chalk River, Ontario. This work could not have been done on a standard X-ray diffractometer.

Neutron Sources

Two types of sources are used. Originally developed in the 1940s, nuclear reactors provided the first neutrons for research. While reactors provide a continuous source of neutrons, recent developments in accelerator technology have made possible the construction of pulsed neutron sources, providing steady, intermittent neutron beams.

Name	Location	Type
HFBR	Brookhaven National Laboratory, USA	Reactor
HFIR	Oak Ridge National Laboratory, USA	Reactor
HFR	Institute Laue Langevin, France	Reactor
IPNS	Argonne National Laboratory, USA	Spallation
ISIS	Rutherford-Appleton Laboratory, UK	Spallation
LANSE	Los Alamos National Laboratory, USA	Spallation
NBS	National Institute of Standards and Technology, USA	Reactor
MURR	University of Missouri, USA	Reactor
OWR	Los Alamos National Laboratory, USA	Reactor
ORR	Oak Ridge National Laboratory, USA	Reactor

Table 1 **Some neutron sources.**

Within nuclear reactors, neutrons are a primary product of nuclear fission. By controlling the rate of the nuclear reactions, one controls the flux of neutrons and provides a steady supply of neutrons. For a diffraction analysis, a narrow band if neutron wavelengths is selected (fixing λ) and the angle 2θ is varied to scan the range of d values.

Pulsed sources use a process called *spallation.* If a high-energy pulsed beam of protons impinges upon a heavy metal target, a rather complex series of nuclear excitations and relaxations results in a burst of high-energy neutrons from the target. Since the spallation process occurs rapidly (in less than 1 μs), a pulsed source can be operated at 30–120 Hz (30–120 pulses per second), providing a steady, intermittent source of neutrons. Then, rather than select a narrow wavelength range for diffraction analysis, the full spectrum of neutron wavelengths are used in the diffraction experiment. Neutron wavelengths vary predictably with momentum according to the equation

$$\lambda = \frac{ht}{ml} \tag{2}$$

where λ is the neutron wavelength, h is Planck's constant, t is time, m is the neutron mass, and l is the neutron flight path. By noting the time of flight of a detached neutron, its wavelength can be deduced.

The difference between the neutron diffraction experiment performed at a steady-state (reactor) or a pulsed source are illustrated in Figure 3. Despite the

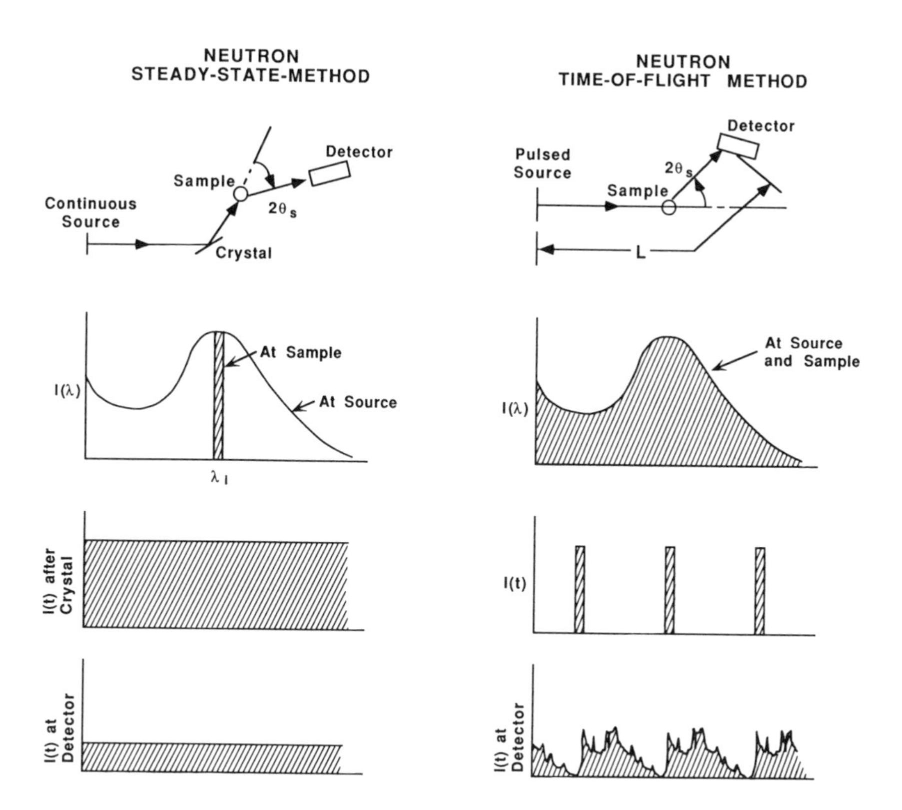

Figure 3 **Comparison of nuclear reactor and pulsed spallation sources. For reactor sources (steady-state method), a narrow band of wavelengths is selected with a monochromator crystal and the scattering angle ($2\theta_s$) is varied to scan *d* spacings. Pulsed sources (time-of-flight method) use almost the entire available neutron spectrum, fix the scattering angle ($2\theta_s$), and simultaneously detect a neutron while determining its time of flight.**

major differences in the design and instrumentation, the quality of data and usable neutron intensities for the two sources are comparable. Some currently available neutron sources are listed in Table 1.

Use of Neutron Diffraction

Neutron diffraction is particularly well suited for use in

1. Structural investigations of heavy metal oxides (particularly when oxygen positions or occupancies are important)
2. *In-situ* analysis
3. Determinations of magnetic ordering in crystalline materials
4. Bulk analysis of physically large pieces.

Specific examples of the first and third uses are given below.

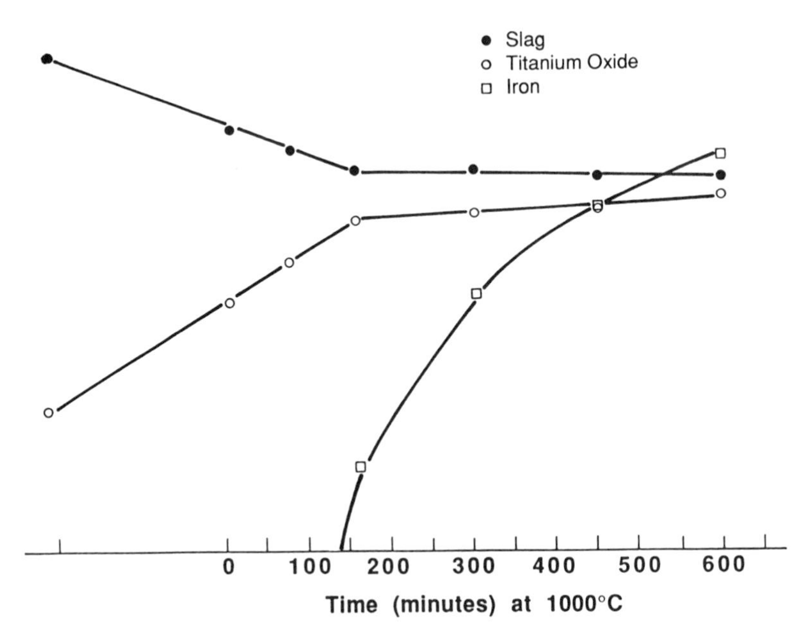

Figure 4 **Plot of time-resolved decomposition of titanium-enriched slags as extracted from neutron diffraction data collected at 1000° C.**

In-Situ Analysis

Figure 4 illustrates the results of neutron diffraction data collected at 1000° C as a function of time. The data were collected during the thermal degradation of commercial titanium slags.[1] The slags are produced by a smelting process used to enrich the titanium content of ilmenite ore. The purpose of the diffraction experiment was to determine the growth of particular undesirable phases as a function of time at the decomposition temperature of the slag. In addition to providing information about the type and number of phases that appeared during decomposition, it was also highly desirable to obtain detailed structural information about the newly appearing phases, as well as the changing nature of the decomposing phase.

Each set of data points for a particular time in Figure 4 represents a neutron diffraction pattern collected in a 15-minute period. Each of the data sets was analyzed to quantify the phases present, determine the identities and locations of metal atoms within each oxide phase, and accurately measure the unit cell parameters (an indication of the phase composition). Examples of the raw neutron diffraction data from which the useful data given in Figure 4 have been extracted are given in Figure 5. In Figure 5, peaks due to two of the decomposition products, titanium oxide (TiO_2) and iron metal, are marked; examination of the figure illustrates the utility of diffraction data. The appearance of Bragg peaks due to newly formed phases are clearly distinguished from those of the parent compound.

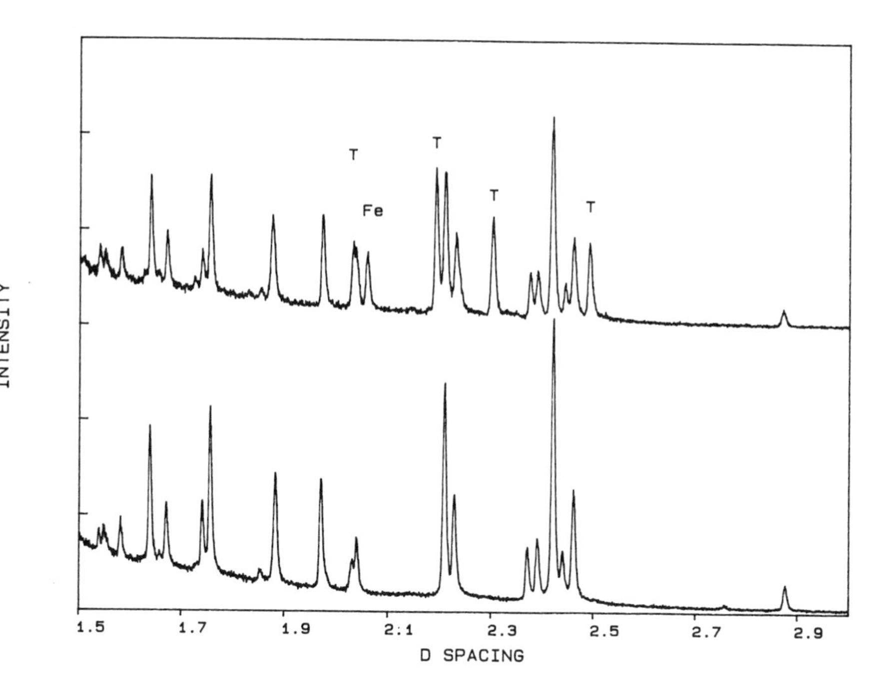

Figure 5 **Raw diffraction data at the start (bottom) and completion (top) of the *in-situ* decomposition of slag experiments. Most of the peaks in the pattern are due to the parent slag phase. Bragg peaks due to titanium oxide (T) and iron metal (Fe) are marked.**

It can be seen from Figure 4 that two distinct reaction pathways describe the decomposition of the slag. From the beginning of decomposition up to approximately 150 min, a pathway that results in the production of titanium oxide dominates the chemistry. After this initialization period, the decomposition rate diminishes greatly, and metallic iron formation becomes important. Additional features of the decomposition (not illustrated in Figures 4 and 5) also resulted from analysis of the neutron diffraction data. The most important of these was that a specific atomic arrangement of the titanium and iron atoms was required before decomposition could occur, and that a certain minimum temperature was required for this rearrangement. Knowledge of this atomic shifting and the temperature required for its occurrence led to an understanding of the maximum temperature above which slags would begin to decompose.

Superconducting Oxides

One of the most exciting and perhaps unexpected discoveries in science within the last decade has been the observation of superconductivity (the complete absence of resistivity to electric current) in metal oxides at temperature $\leq$ 90 K. This temperature range is particularly important because it can be reached with readily available liquid nitrogen (77 K). Important structural features of metal oxide superconductors have been revealed largely by the application of neutron diffraction.

In a sense, a superconductor is an insulator that has been doped (contains random defects in the metal oxide lattice).[2] Some of the defects observed via neutron diffraction experiments include metal site substitutions or vacancies, and oxygen vacancies or interstituals (atomic locations between normal atom positions). Neutron diffraction experiments have been an indispensable tool for probing the presence of vacancies, substitutions, or interstituals because of the approximately equal scattering power of all atoms.

Studies of the superconducting phase $YBa_2Cu_3O_{7-x}$ exemplify this point. *In-situ* neutron diffraction analysis revealed that during the synthesis of this material (above 900° C) oxygen vacancies occur and that the composition at this temperature is close to $YBa_2Cu_3O_6$ ($x = 1$). Upon cooling in an oxygen atmosphere, the oxygen vacancies are filled to form the superconducting material $YBa_2Cu_3O_{7-x}$, raising the average oxidation state of the copper above +2. These results are presented in Figure 6. The figure shows that as the temperature is reduced from about 600° C to room temperature, the occupancy of one oxygen site—O1, located at (0,½,0) in the unit cell—approaches 100%, while that of a second oxygen atom—O5, located at (½,0,0)—tends toward 0. It has been shown that the superconductivity of this metal oxide is a function of its oxygen content, and therefore a function of the partial occupancies of O1 and O5.

An advantage enjoyed by neutron diffraction over X-ray diffraction was outlined in the introduction. Since X rays are scattered by electrons, X-ray diffraction data from the $YBa_2Cu_3O_{7-x}$ system are mainly sensitive to the metal atom positions and occupancies, and much less sensitive to the oxygen atoms. Hence the key features of the $YBa_2Cu_3O_{7-x}$ superconductor—oxygen vacancies—would not be apparent from the analysis of X-ray diffraction data. However, since neutrons are scattered approximately equally by all atoms, a neutron diffraction experiment is very sensitive to the structural features of importance in this system. An additional advantage of neutrons over X rays in work of this type is the need for *in-situ* data. To understand the role of oxygen vacancies in the $YBa_2Cu_3O_{7-x}$ system, it was necessary to collect diffraction data over a wide range of temperatures and oxygen partial pressures. The greater penetrating nature of the neutron is well suited to the use of special equipment required for these types of experiments, since the neutrons can pass right through such equipment.

Another example of the use of neutron diffraction to understand the role of atomic vacancies in producing a superconducting metal oxide phase is work that has been performed on $Ba_{0.6}K_{0.4}BiO_3$. This work demonstrates that at the synthesis temperature (700° C), under the proper conditions, oxygen vacancies are created to allow the formation of the parent phase with bismuth largely in the +3 oxidation state. The presence of the vacancies allows the incorporation of potassium in the

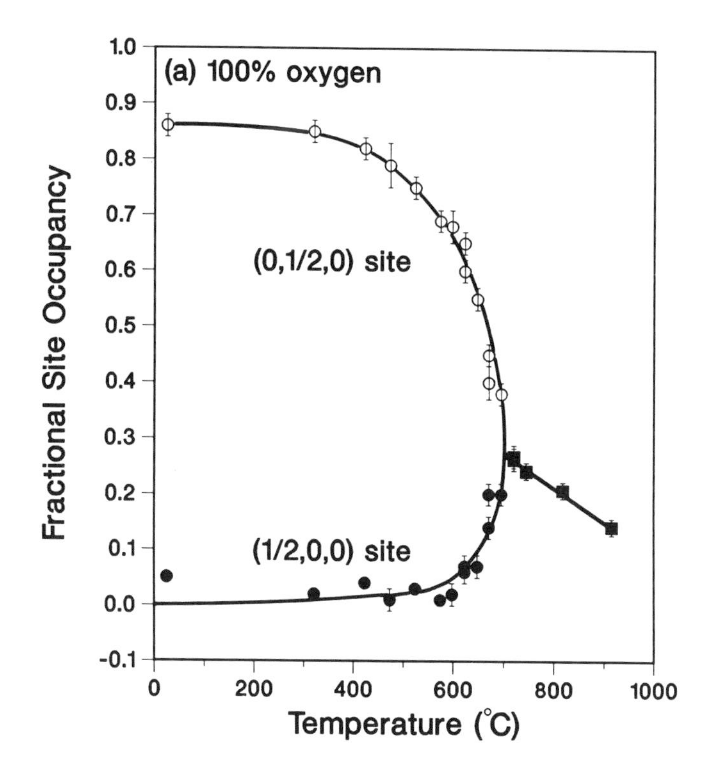

Figure 6 **Site occupancies for two of the oxygen atoms (O1 and O5) in the $YBa_2Cu_3O_{7-x}$ superconductor as a function of temperature. The site occupancies resulted from an analysis of *in-situ* neutron diffraction data. Reprinted by permission from Jorgensen and Hinks.[2]**

structure. As the temperature is reduced, the oxygen vacancies are filled, and because the potassium atoms lose their mobility at lower temperatures, the overall structure remains intact, producing a phase with bismuth in an unusually high oxidation state.

Magnetic Thin Films

Neutron diffraction is a powerful probe of the magnetic structure and ordering in magnetic thin films. Rare earth thin films and multilayers (materials having a repeating modulation in chemical composition) present an interesting class of materials, and neutron diffraction has been instrumental in elucidating their magnetic structure.[7] For multilayers of Dy, Er, and Gd alternating with Y, neutron diffraction has shown that the magnetic order is propagated through the intervening nonmagnetic Y layers. For DY-on-Y multilayers, it was found that the magnetically ordered state was an incommensurate helical antiferromagnetic state. That is, the magnetic moments in each basal plane are ferromagnetically aligned, but somewhat rotated between adjacent basal planes. Although this is similar to bulk Dy, the temperature dependence of the rotation, or turn, angle is different than in bulk Dy. It would be difficult or impossible to determine this microscopic information using a technique other than neutron diffraction. While the thickness of the magnetic films in these measurements was ~4000 Å, rare earth films as thin as ~500 Å and transition metal oxide films as thin as ~5000 Å can be analyzed. For multilayers, neutron measurements at very low angles are also useful in characterizing magnetic order; these are described in the article on neutron reflectivity.

Conclusions

Historically, due to the general unavailability of neutron sources, neutron diffraction has been a rather esoteric technique. Fortunately, the neutron users' community has expanded over the last decade, and concerted programs encouraging new users at many facilities have extended the use of the technique into the general scientific community. While neutron diffraction may never become a routine analytical tool, data collection times for studies requiring its use usually can be found

It has been shown that neutron diffraction offers the same kind of information that other diffraction techniques offer, namely atomically resolved structure determination and refinement, as well as phase identification and quantitation. Other uses not described herein include residual stress measurements, and determinations of average particle sizes for crystalline materials. The major advantage of neutrons with respect to the more readily available X rays lies in the greater penetrating power of the neutron, and the approximately equal scattering ability of nuclei. These features make neutron diffraction the proper choice when *in-situ* measurements, bulk penetration, or site occupancies of atoms are required.

Related Articles in the Encyclopedia

XRD and Neutron Reflectivity

References

1 Details of the thermal decomposition of commercial slags can be found in R. G. Teller, M. R. Antonio, A. Grau, M. Guegin, and E. Kostiner. *J. Solid State Chem.* 1990.

2 Discussion of neutron diffraction studies of superconductors was largely taken from J. D. Jorgensen and D. G. Hinks. *Neutron News.* **24,** 1, 1990.

3 For further discussion of neutron sources, see R. B. Von Dreete. *Reviews in Mineralogy. Volume 20: Modern Powder Diffraction.* **333,** 20, 1990.

4 For a detailed discussion of pulsed neutron sources, see J. D. Jorgensen and J. Faber. *ICANS-II, Proceedings of the Sixth International Collaboration*

on Advances in Neutron Sources. Argonne National Laboratory technical report ANL-82-80, 1983.

5 Important concepts in neutron diffraction can be found in G. E. Bacon. *Neutron Diffraction.* Clarendon Press, third edition, 1975.

6 The general principles of diffraction can be found in numerous books, for example, B. D. Cullity. *Elements of X-Ray Diffraction.* Addison-Wesley, New York, second edition, 1978.

7 For a review, see J. J. Rhyne, R. W. Erwin, J. Borchers, M. B. Salamon, R. Du, and C. P. Flynn. *Physica B.* **159,** 111, 1989.

11.2 Neutron Reflectivity

THOMAS P. RUSSELL

Contents

Introduction

Neutron reflectivity offers a means of determining the variation in concentration of a material's components as a function of depth from the surface or at an interface buried within the material, with a resolution of ~1 nm. Because of the large neutron contrast between hydrogen and deuterium, one may highlight a particular component through isotopic labeling with deuterium without substantially altering the thermodynamics of the system. This, however, normally means that reflectivity studies are relegated to the investigation of model systems that are designed to mimic the behavior of the system of interest.

Other technique—for example, dynamic secondary ion mass spectrometry or forward recoil spectrometry—that rely on mass differences can use the same type of substitution to provide contrast. However, for hydrocarbon materials these methods attain a depth resolution of approximately 13 nm and 80 nm, respectively. For many problems in complex fluids and in polymers this resolution is too poor to extract critical information. Consequently, neutron reflectivity substantially extends the depth resolution capabilities of these methods and has led, in recent years, to key information not accessible by the other techniques.

An additional advantage to neutron reflectivity is that high-vacuum conditions are not required. Thus, while studies on solid films can easily be pursued by several techniques, studies involving solvents or other volatile fluids are amenable only to reflectivity techniques. Neutrons penetrate deeply into a medium without substantial losses due to absorption. For example, a hydrocarbon film with a density of $1 g/cm^3$ having a thickness of 2 mm attenuates the neutron beam by only 50%. Consequently, films several μm in thickness can be studied by neutron reflectivity. Thus, one has the ability to probe concentration gradients at interfaces that are buried deep within a specimen while maintaining the high spatial resolution. Materials like quartz, sapphire, or aluminum are transparent to neutrons. Thus, concentration profiles at solid interfaces can be studied with neutrons, which simply is not possible with other techniques.

The single most severe drawback to reflectivity techniques in general is that the concentration profile in a specimen is not measured directly. Reflectivity is the optical transform of the concentration profile in the specimen. Since the reflectivity measured is an intensity of reflected neutrons, phase information is lost and one encounters the age-old inverse problem. However, the use of reflectivity with other techniques that place constraints on the concentration profiles circumvents this problem.

The high depth resolution, nondestructive nature of thermal neutrons, and availability of deuterium substituted materials has brought about a proliferation in the use of neutron reflectivity in material, polymer, and biological sciences. In response to this high demand, reflectivity equipment is now available at all major neutron facilities throughout the country, be they reactor or spallation sources.

Basic Principles

Considering Figure 1, radiation incident on a surface (light, X rays, or neutrons) will be reflected and refracted at the interface between the two media provided there is a difference in the index of refraction. In the case of neutrons and X rays, the refractive index of a specimen is slightly less than unity and, to within a good approximation, is given by

$$n = 1 - \delta + i\beta \tag{1}$$

The imaginary component of the refractive index is associated with absorption. In general, the absorption for thin films is not significant and, consequently, β can be ignored. However, for materials containing the elements Li, B, Cd, Sm, or Gd, where the absorption coefficient is large, β must be taken into account and the refractive index is imaginary.

The real component of the neutron refractive index δ is related to the wavelength λ of the incident neutrons, the neutron scattering length (a measure of the extent to which neutrons interact with different nuclei), the mass density and the atomic

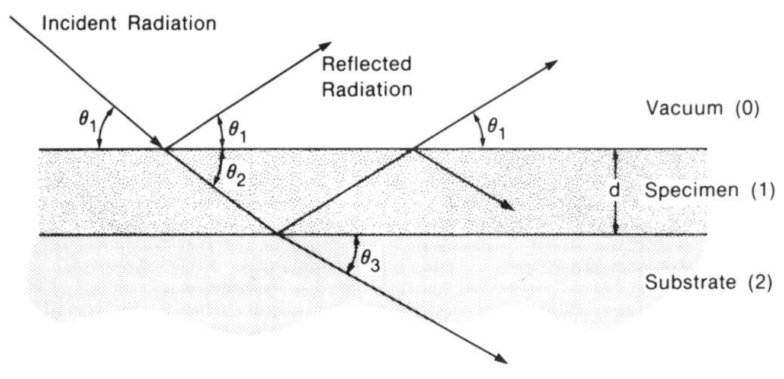

Figure 1 **Schematic diagram of the neutron reflectivity measurement with the neutrons incident on the surface and reflected at an angle θ_1 with respect to the surface. The angle θ_2 is the angle of refraction. The specimen in this case is a uniform film with thickness *d*, on a substrate.**

number of the components comprising the specimen. Values of the neutron scattering length for all the elements and their isotopes are tabulated.[1] The neutron scattering length does not vary systematically with the atomic number. This is shown in Figure 2. As can be seen, isotopes of a given element can have markedly different neutron scattering lengths while two different elements with vastly different atomic numbers can have similar scattering lengths. In fact, the difference between the proton and the deuteron provides one of the largest differences and offers the most convenient manner of labeling materials for neutron reflectivity studies.

In Table 1 values of δ are given for some common inorganic and organic compounds. First notice that δ is on the order of 10^{-6}. Therefore, the neutron refractive index differs from unity by only a small amount. With the exception of H_2O and

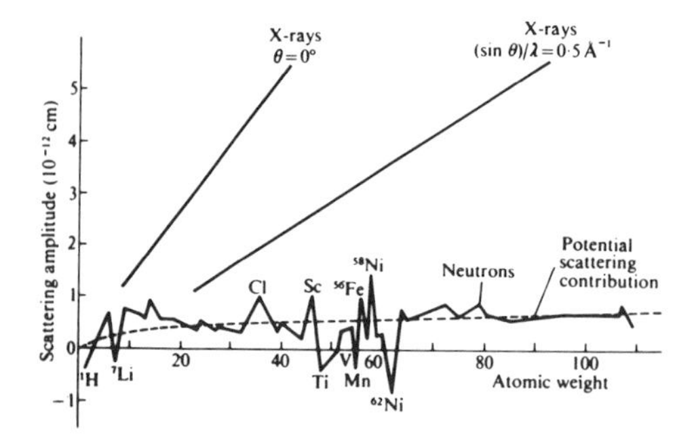

Figure 2 **Variations in the neutron scattering amplitude or scattering length as a function of the atomic weight. The irregularities arise from the superposition of resonance scattering on a slowly increasing potential scattering. For comparison the scattering amplitudes for X rays under two different conditions are shown. Unlike neutrons, the X-ray case exhibits a monotonic increase as a function of atomic weight.**

Substance	Chemical formula	ρ (g/cm³)	Σ b_i (10^{12} cm)	δ† (10^{12} cm)	θ_C† (10^3)	k_C† (Å^{-1})
Water	H_2O	1.00	–0.168	–0.212	—	—
	D_2O	1.11	1.914	2.41	2.19	0.0089
Benzene	C_6H_6	0.879	1.74	0.445	0.943	0.0038
		0.946	7.99	2.043	2.021	0.0082
Polyethylene	$(CH_2)_x$	0.95	–0.08	–0.128	—	—
	$(CD_2)_x$	1.08	1.99	3.067	2.477	0.0101
Polystyrene	$(C_8H_8)_x$	1.0	2.32	0.509	1.01	0.0041
	$(C_8D_8)_x$	1.08	10.656	2.336	2.16	0.0088
Silicon oxide	SiO_2	2.32	1.58	1.371	1.655	0.0067
Silicon	Si	2.32	0.42	0.791	1.26	0.0051

†Calculated for neutrons with $\lambda = 1.5$ Å.

In this list, ρ is the mass density, Σ b_i is the sum of scattering lengths of the atoms comprising the molecule, δ is the real part of the refractive index, θ_C is the critical angle, and k_C is the critical neutron momentum.

Table 1 Important neutron reflection parameters for some common materials.

polyethylene $(CH_2)_x$, $\delta > 0$ and therefore $n < 1$. The refractive index for air or vacuum is unity. From Snell's law it is easy to show that total external reflection occurs when the incidence angle, θ, is less than $(2\delta)^{1/2}$. Above this critical angle θ_C the reflectivity decreases and the manner of this decrease contains all the information pertinent to gradients in the concentration normal to the surface of the specimen.

Consider the simple case shown in Figure 1, where we have a uniform film on a substrate with a thickness of 50 nm. For demonstration purposes the film is a deuterated polystyrene on a silicon substrate. The reflectivity profile obtained from such a specimen is shown in Figure 3. The reflectivity is plotted as a function of the neutron momentum k_{z0}, which is proportional to $(\sin\theta)/\lambda$, where the subscript *z*0 indicates the *z*-direction (normal to the surface) in vacuum (0), λ is the wavelength and θ is the incidence angle. Using k_{z0} reduces data to a scale that is independent of λ and θ used in an experiment. Below θ_C total external reflection is seen, then the reflectivity is seen to decay with a series of oscillations. The oscillations, characterizing the film thickness, arise from interferences between neutrons reflected at the air and substrate interfaces. In general the higher the frequency of the oscillations, the thicker the specimen.

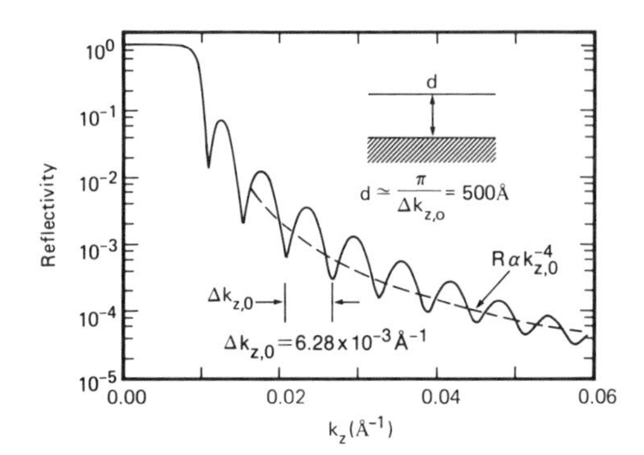

Figure 3 **A calculated reflectivity profile for a perdeuterated polystyrene film with a thickness of 50 nm on a silicon substrate. The calculation was for a specimen where the interfaces between the specimen and air and the specimen and the substrate were sharp. This causes the reflectivity on average (shown by the dashed line) to decrease in proportion to k_{z0}^4 or θ^{-4}. The separation distance between the minima of the oscillations directly yields the thickness of the specimen, as shown.**

On average, the reflectivity decays in proportion to θ^{-4} or k_{z0}^{-4} since both interfaces are sharp. However, if either surfaces is rough, then marked deviations are seen and the reflectivity is damped by a factor of $\exp\{-2k_{z0}^2\sigma^2\}$, where σ is the root-mean-square roughness. Thus, the reflectivity is very sensitive to surface roughness and to concentration gradients at interfaces.

The reflectivity for this simple case can be extended readily to more complex situations where there are concentration gradients in single films or multilayers comprised of different components. Basically the reflectivity can be calculated from a simple recursion relationship that effectively reduces any gradients in composition to a histogram representing small changes in the concentration as a function of depth. Details on this can be found in the literature cited.[2–4]

For specimens where gradients in the magnetic moment are of interest, similar arguments apply. Here, however, two separate reflectivity experiments are performed in which the incident neutrons are polarized parallel and perpendicular to the surface of the specimen. Combining reflectivity measurements under these two polarization conditions in a manner similar to that for the unpolarized case permits the determination of the variation in the magnetic moments of components parallel and perpendicular to the film surface. This is discussed in detail by Felcher et al.[5] and the interested reader is referred to the literature.

Instrumentation

The neutron's momentum can be varied by changing λ or θ. With a fixed wavelength the angle of incidence can be changed by rotating the sample. This, typically, is the situation one encounters at steady state nuclear reactors, where a specific wavelength of the neutrons emanating from the reactor core is selected with a monochromator. Such facilities are available at the Oak Ridge National Laboratory, the National Institute of Standards and Technology, and the Brookhaven National Laboratory in the United States, and at the Institut Laue Langevin (Grenoble) and the Kernforschungsanlage facility (Jülich) in Europe. Here, a collimated beam of monochromatic neutrons impinges on the surface of the specimen at an angle θ and a detector is placed behind a set of slits at an angle 2θ with respect to the incident beam (this is termed the specular condition). The geometry is shown in Figure 1. Both the specimen and detector are rotated synchronously at θ and 2θ, respectively, to measure the reflected neutrons as a function of θ. Since λ is known, θ is equivalent to k_{z0}. Normalization of the reflected intensity against the incident intensity yields the reflectivity directly.

Spallation sources are an alternative means of generating neutrons. As opposed to a reactor source, a target is bombarded with pulses of high-energy charged particles, for example, protons on a uranium target. Neutrons with a broad energy or wavelength distribution are generated at the target, passed through a cold moderator, and delivered into the experimental area. The velocity of a neutron is proportional to its energy and inversely proportional to its wavelength. Thus, knowing the time at which the pulse of protons hits the target and the distance to the source, we have that the time for a neutron to reach the detector is proportional to its wavelength. Hence the term *time of flight* is used. Argonne National Laboratory and the Los Alamos National Laboratory in the United States, and Rutherford Appleton Laboratories in Europe have reflectometers that are based on this principle.

At spallation sources, k_{z0} is varied by wavelength because pulsed streams of neutrons with a range of wavelengths are delivered onto the specimen surface at an angle θ. Knowing the incident distribution of wavelengths and measuring the distribution of wavelengths reflected at an angle θ with respect to the surface furnishes directly the reflectivity of the specimen. The beauty of the time-of-flight measurements is that there are no moving parts to the reflectometer. Unlike the fixed-wavelength spectrometers, with time-of-flight spectrometers exactly the same area of the specimen is measured for all values of k_{z0}. This is very important if the sample is not uniform across its surface.

One of the key advantages to time-of-flight reflectometers comes in the measurement of fluid surfaces. Simply delivering the neutrons onto the fluid surface at a fixed angle (without moving the specimen) and detecting the reflected neutrons yields the reflectivity profile.

Specimen Considerations

The measurements of concentration gradients at surfaces or in multilayer specimens by neutron reflectivity requires contrast in the reflectivity for the neutrons. Under most circumstances this means that one of the components must be labeled. Normally this is done is by isotopic substitution of protons with deuterons. This means that reflectivity studies are usually performed on model systems that are designed to behave identically to systems of more practical interest. In a few cases, however (for organic compounds containing fluorine, for example) sufficient contrast is present without labeling.

Neutron reflectivity measures the variation in concentration normal to the surface of the specimen. This concentration at any depth is averaged over the coherence length of the neutrons (on the order of 1 μm) parallel to the surface. Consequently, no information can be obtained on concentration variations parallel to the sample surface when measuring reflectivity under specular conditions. More importantly, however, this mandates that the specimens be as smooth as possible to avoid smearing the concentration profiles.

Typically specimens for reflectivity measurements are prepared on flat, smooth, rigid substrates. For example, these substrates can be polished fused silica, quartz, or silicon. It is important, however, that the substrates be thick to avoid distortions of the specimen when mounted in the reflectometer. Any curvature or bowing will increase the divergence of the incident beam and result in a deterioration of the resolution.

The substrates usually range from 5 cm to 10 cm in diameter. Such large specimens are required because experiments are performed at small angles of incidence and it is necessary to intercept the entire beam with the specimen. For example, consider the projection of a beam 0.1 mm in size onto a specimen at an angle of 2 mrad (~ 0.11°). Under such conditions, the incident beam will illuminate 5 cm of the surface. For specimens that are dominantly protonated, the critical angle is smaller and the projection is even larger. The large diameter of the specimens places stringent requirements on the sample's preparation, but it is not difficult to achieve uniformity for such large specimens and the preparation is routine.

Examples

The first studies on the use of neutron reflectivity appeared approximately 20 years ago and dealt with the investigation of Langmuir-Blodgett films. Surprisingly, the utility of the technique for the investigation of other materials was not fully realized and its use and availability virtually ceased. Only in the last four years has there been a resurgence of the use of neutron reflectivity, and now there are reflectometers available throughout the United States and abroad. This resurgence stems from the focus of the scientific community on surface and interfacial phenomena. In fact, interfacial behavior of polymeric materials has provided a rich area where the true power of neutron reflectivity has been brought to bear. There is no question, though, that surface and interfacial problems in other organic, inorganic and magnetic materials can be studied by neutron reflectivity. These areas have been discussed at length in the literature.[4–6] For the purposes of illustration, we shall focus on two specific problems in polymers that demonstrate the capabilities of reflectivity.

In numerous applications of polymeric materials multilayers of films are used. This practice is found in microelectronic, aeronautical, and biomedical applications to name a few. Developing good adhesion between these layers requires interdiffusion of the molecules at the interfaces between the layers over size scales comparable to the molecular diameter (tens of nm). In addition, these interfaces are buried within the specimen. Aside from this practical aspect, interdiffusion over short distances holds the key for critically evaluating current theories of polymer diffusion. Theories of polymer interdiffusion predict specific shapes for the concentration profile of segments across the interface as a function of time. Interdiffusion studies on bilayered specimen comprised of a layer of polystyrene (PS) on a layer of perdeuterated (PS) d-PS, can be used as a model system that will capture the fundamental physics of the problem. Initially, the bilayer will have a sharp interface, which upon annealing will broaden with time.

Neutron reflectivity is ideally suited to this problem, since concentration profiles can be resolved on the nanometer level and since, for an infinitely sharp interface, Rk_{z0}^4 will approach asymptotically a constant value. In addition, neutron reflectivity is nondestructive and multiple experiments can be performed on the same specimen. Figure 4 shows a plot of Rk_{z0}^4 as a function of k_{z0} for a bilayer of protonated polystyrene (h-PS) on a layer of, d-PS prepared as described. Near the critical angle, Rk_{z0}^4 reaches a maximum value then decreases, approaching an average constant value of 3×10^{-10} Å^{-4}. Thus, the initial interface is sharp. Evident in the reflectivity profile are a series of oscillations with frequencies characterizing the thicknesses of the h-PS and d-PS layers and the combined thickness. The solid line in the figure corresponds to a reflectivity profile calculated assuming a 67.5-nm layer of h-PS on an 185-nm layer of d-PS with an interfacial width of only 1 nm.

Upon heating the bilayer for two minutes at 105.5°C, where the glass transition temperature of PS is 100° C, interdiffusion occurs. As shown by the plot of Rk_{z0}^4 versus k_{z0} (offset by a factor of 10), the overall features of the reflectivity remained unchanged with the exception that the asymptotic limit decreased to 1.4×10^{-10} Å^{-4}. The solid line in the figure was calculated using the layer thicknesses mentioned above, but the interface was characterized now by an error function with a width of 3 nm. Combining these measurements with others at longer times and at different temperatures proved that reptation (the movement of the polymer along its own contour), coupled with rapid motions of the polymer chain between entanglements, quantitatively describes the interdiffusion process.

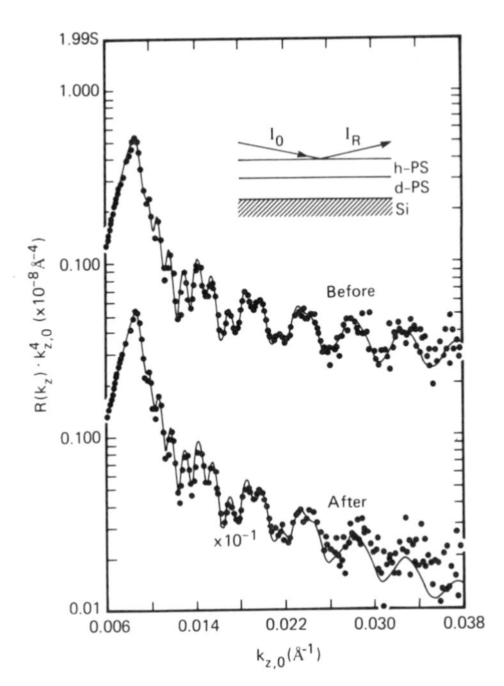

Figure 4 **The reflectivity multiplied by k_{z0}^4 as a function of k_{z0} for a bilayer of normal polystyrene on perdeuterated polystyrene before and after heating at 105.5° C for 2 minutes. The data is plotted in this manner since for sharp interfaces Rk_{z0}^4 is a constant at large k_{z0}.**

The second example deals with symmetric diblock copolymers. Diblock copolymers are finding widespread use as compatibilizing agents, adhesion promoters and surface-modifying agents. Diblock copolymers are comprised of two chemically different polymer chains covalently bonded together at one point. In cases where the length of the blocks are equal, lamellar domains of each component comparable to the size of the polymers are formed at the junction points between the two blocks at the interfaces. The chemical differences between the blocks causes a preferential segregation of one of the components to the air or substrate interfaces, in thin films, which leads to an orientation of the lamellar domains parallel to the film surface.

Figure 5 shows the neutron reflectivity profile of a diblock copolymer of PS and polymethylmethacrylate (PMMA), where the PS block is labeled with deuterium. The reflectivity profile exhibits five clearly resolved maxima characterizing the period of the lamellar microdomains. A high-frequency oscillation, characteristic of the total specimen thickness, is also seen. Note that below the critical angle ($k_{z0} \sim 0.006$ Å^{-1}) the reflectivity is not unity, due to the finite size of the specimen. The solid line in the figure was calculated using the scattering length density or concentration profile shown in the inset. This model is comprised of PS layers having a

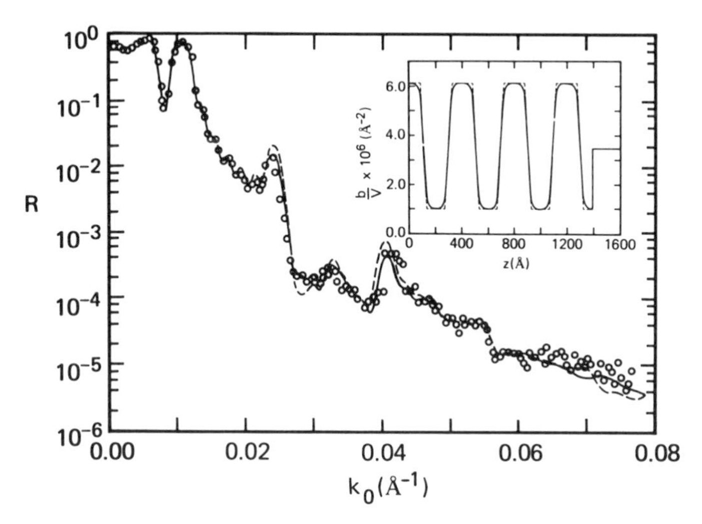

Figure 5 **Neutron reflectivity of a symmetric diblock copolymer of PS and PMMA annealed at 160° C for 24 hours under vacuum. The inset contains scattering length density profiles used to calculate the corresponding reflectivity profiles. The solid line represents the reflectivity profile calculated using the scattering length density (or concentration) profile as a function of distance *z* from the air surface where a hyperbolic tangent describes the interface between the layers. The dashed line was calculated using a linear gradient between the microdomains. This small change, as can be seen, produces a significant change in the reflectivity profile.**

thickness of 21 nm and PMMA layers having a thickness of 18.8 Å. The interface between the layers is characterized by a hyperbolic tangent function with an effective width of 5.0 ± 0.3 nm. At the air–copolymer and copolymer–substrate interfaces, the adjacent PS and PMMA layers, respectively, are one-half the thickness of those seen in the bulk. The major importance of these results comes in the precision to which the interface between the PS and PMMA layers can be determined. This far exceeds that possible by other techniques. To emphasize this, a second reflectivity profile (shown as the dashed line) was calculated assuming a linear gradient at the interface. While the differences in the segment density profiles are small, the differences in the reflectivity profile are pronounced, clearly demonstrating the details resolvable by neutron reflectivity.

Conclusions

Neutron reflectivity provides a depth resolution of ~1 nm and fills an important gap in the resolution between X-ray photoelectron spectroscopy and ion-beam techniques. In this regard, neutron reflectivity promises to play a decisive role in the investigation of solid materials. Equally important is the fact that reflectivity measurements can be used to study liquid specimens, which is not possible with other techniques requiring high vacuum. Many problems on the swelling and dissolution of materials with low molecular weight solvents, the adsorption of materials onto a surface or at the liquid–air interface or liquid–liquid interface, or transport through membranes in solution on specific interactions of materials with interfaces in the presence of a solvent can be addressed by neutron reflectivity.

Theoretical and experimental advances are being made on understanding reflectivity under nonspecular conditions. This permits the determination of correlations in composition parallel to the specimen's surface thereby adding a new dimension to existing capabilities. Studies under nonspecular conditions promise to be quite important in elucidating systems containing inhomogeneities parallel to the sample's surface. From this will stem a quantitative characterization of surface roughness, curvature of interfaces, capillary waves, and the configurations and conformations of molecules at surfaces and interfaces.

Related Articles in the Encyclopedia

Neutron Diffraction

References

1 G. E. Bacon. *Neutron Diffraction.* (3rd edition) Clarendon Press, Oxford, 1975.

2 O. S. Heavens. *Optical Properties of Thin Solid Films.* Butterworths, London, 1955.

3 J. Lekner. *Theory of Reflection.* Nighoff, Dordrecht, 1987.

4 T. P. Russell. *Mat. Sci. Rep.* **5,** 171, 1990.

5 G. P. Felcher, R. O. Hilleke, R. K. Crawford, J. Haumann, R. Kleb, and G. Ostrowski. *Rev. Sci. Instrum.* **58,** 609, 1987.

6 J. Penfold, R .C. Ward, and W. G. Williams. *J. Phys.* **E20,** 1411, 1987.

11.3 NAA

Neutron Activation Analysis

TIM Z. HOSSAIN

Contents

Introduction

Neutron activation analysis (NAA) is a sensitive analytical technique for measuring trace impurities in a wide variety of materials. NAA has been used for trace analysis in polymers, biological materials, and geological and environmental samples. It has been used successfully to study the role of trace elements in Alzheimer's Disease by measuring impurities in human brain tissue. Trace impurities in air particulates have been measured to correlate emissions from fossil fuel power plants. Since there are many excellent texts describing the applications of NAA in these fields, the present article will be devoted entirely to NAA of semiconductor materials. NAA is highly suited for elements of interest to semiconductor manufacturing. For example, NAA can detect Cu, Zn, and Au in Si with detection limits of 7×10^{12}, 1×10^{12}, and 3×10^{7} atoms/ cc, respectively.[1] Results obtained using this method usually provide bulk concentrations, however, surface-sensitive measurements can be made by combining chemical etching with NAA. The method is not suitable for depth profiling, such as can be done with SIMS. Neither is there a capability for

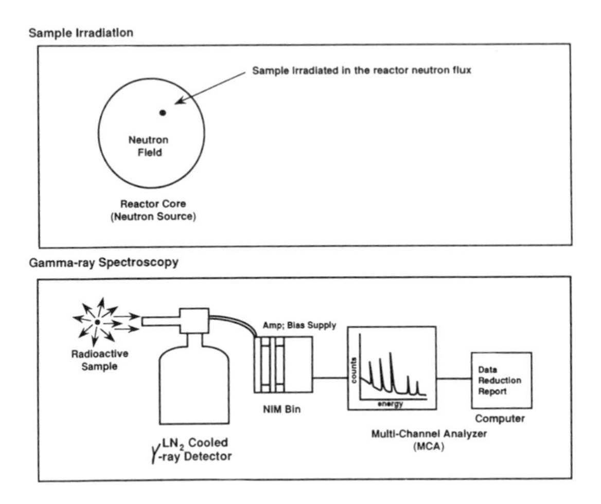

Figure 1 **Schematic of Neutron Activation Analysis.**

microspot analysis. The strength of NAA lies in its ability to perform quantitative bulk analysis at sub-ppb levels. Almost two-thirds of the elements in the periodic table can be analyzed easily. Light elements like H, N, O, or F are not suitable candidates for NAA, and Al and Cl can be analyzed only with high detection limits in a Si matrix.

Most of the transition elements that are of primary interest in the semiconductor industry such as Fe, Cr, Mn, Co, and Ni, can be analyzed with very low detection limits.[1] Second to its sensitivity, the most important advantage of NAA is the minimal sample preparation that is required, eliminating the likelihood of contamination due to handling. Quantitative values can be obtained and a precision of 1–5% relative is regularly achieved. Since the technique measures many elements simultaneously, NAA is used to scan for impurities conveniently.

It is applicable to plastic packaging materials, where purities with respect to mobile ions, such as Cl and Na, can be checked. In addition, α-particle precursors, such as U and Th, can be determined in solid plastics with sub-ppb detection limits.

Basic Principles

Irradiation

All NAA experiments are conducted in two steps: irradiation and counting as indicated in Figure 1. Samples are made radioactive by placing them in a neutron field. Typically a research nuclear reactor provides the necessary neutron flux. Elements present in the sample capture neutrons, and often become radioactive isotopes. This part of the experiment is known as irradiation. A typical irradiation in a reactor core position (where the neutron flux is high) may last about 1–7 days. The total integrated neutron dose received, in general, determines the detection limits obtained. Typical research reactors are capable of delivering $1–5 \times 10^{13}$ neutrons/cm^2 s.

It is important to note that the neutron capture probability, called the *cross section* σ, is vastly different for various elements. Excellent sensitivity for Au is due largely to its high cross section ($\sigma = 100$ barns; 1 barn = 1×10^{-24} cm^2). Other elements, such as Pb, have low cross sections and much poorer detection limits.

Elements with multiple stable isotopes may produce several radioisotopes that can be measured to assure the accuracy of the analysis. For example, Zn has five stable isotopes.[2] The isotope ^{64}Zn will produce the radioisotope ^{65}Zn, and ^{68}Zn will produce the radioisotope ^{69}Zn. Both of these radioisotopes can provide an independent measurement of the Zn concentration and therefore can be used to check the consistency and quality of the analysis. On the other hand, ^{66}Zn will produce ^{67}Zn, which is nonradioactive and therefore cannot be used in NAA.

At the end of the irradiation, the samples are withdrawn from the reactor and γ-ray spectroscopy is carried out. Most often the laboratory performing the γ-ray spectroscopy is located in a different city, in which case the samples are shipped and the reactor serves as a neutron source only. Many reactors also have γ-ray spectroscopy capability so that measurements can be made at the reactor site as well.

Counting, or γ-Ray Spectroscopy

All radioactive isotopes decay with a characteristic half-life. For example, ^{59}Fe decays with a half-life of 45 days, while ^{64}Cu decays with a half-life of 12.6 hours. As a result of the decay, signature high-energy photons or γ rays are emitted from a given radioisotope. Thus, ^{59}Fe emits two prominent γ rays at 1099 and 1292 keV, ^{24}Na emits at 1368 and 2754 keV, and ^{65}Zn emits at 1115 keV. Compilations of γ rays used in NAA can be found in γ-ray tables.

The emission of γ rays follows, in the majority of cases, what is known as β decay. In the β-decay process, a radionuclide undergoes transmutation and ejects an electron from inside the nucleus (i.e., not an orbital electron). For the purpose of simplicity, positron and electron capture modes are neglected. The resulting transmutated nucleus ends up in an excited nuclear state, which promptly relaxes by giving off γ rays. This is illustrated in Figure 2.

For the purpose of the activation analysis these high-energy γ rays are used or counted in a spectroscopy mode to obtain concentration values. Energy-dispersive spectroscopy of the γ rays is accomplished using a solid state detector constructed of high-purity single-crystal germanium. The detector is connected via a preamplifier and an amplifier to a multichannel analyzer (MCA). The sample holder and the detector are contained in a lead-shielded chamber. The radioactive sample serves as the γ-ray source and is placed in a convenient geometry in front of the detector. The response time of the detector must be fast enough so that each γ ray entering the

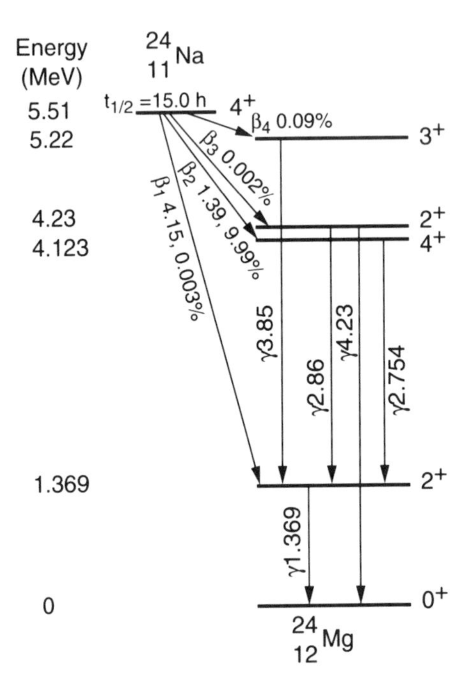

Figure 2 **Decay scheme of ^{24}Na. The transition energies are in MeV.[9]**

detector volume is processed as a single event. The γ rays are sorted by energy and counted in appropriate channels of the MCA. A spectrum is generated of intensity (counts) versus energy of the γ rays. While the energy of the γ ray is used to identify the element, the number of counts provides a measure of the concentration.

NAA can achieve ultralow detection limits for most impurities in a Si or SiO_2 matrix because Si itself has a low neutron capture cross section and produces radioisotopes having short half-lives. The only significant radioisotope, ^{31}Si, has a half-life of 2:6 hours. Radioactivity from this isotope, although initially intense, will therefore be reduced to a negligible intensity after 2–3 days. The impurities of interest, such as Cr and Fe, have long half-lives. By delaying the start of the count, a high signal-to-noise ratio is obtained. This indicates that for other matrices, such as HgCdTe and GaAs, in which both the cross sections and the half-lives are not as favorable, NAA is rendered less practical. On the other hand, organic matrices, such as polymers and biological samples, can be conveniently analyzed because the light elements C, H, O, and N have low neutron capture cross sections.

Sample Requirements

Size requirements are limited by packaging considerations for neutron irradiation. Typically, polyethylene or quartz containers are used to contain the sample in the reactor core. For example, Si wafers are cleaved into smaller pieces and flame sealed in high-purity quartz vials (e.g., suprasil quartz) for irradiation. The size of the irradiation port in the reactor core will typically limit the size of the polyethylene or quartz container. All handling is done wearing talc-free gloves and using white plastic tweezers. (Avoid the green-colored Teflon-coated stainless steel variety.) An intact full wafer (e.g., having a diameter of 4") usually cannot be inserted into an irradiation port of high neutron intensity ($> 1 \times 10^{13}$ neutrons/ (cm^2 s), since most ports have diameters of less than 4" It is desirable to perform NAA before any metallization steps (with Ti, W, or Cu, for example) have been carried out. This avoids a high intensity of radiation from the irradiated thin metal layer. Other important considerations are the dopants present in the wafer. Both As and Sb become readily activated (with half-lives 26.7 hours for As, and 64 hours and 60 days for the two radioisotopes of Sb), so that samples implanted with high doses of As and Sb ($> 1 \times 10^{14}$ atoms/ cm^2) are not suitable for NAA. However, B and P dopants do not pose problems during NAA.

Quantification

NAA is a quantitative method. Quantification can be performed by comparison to standards or by computation from basic principles (parametric analysis). A certified reference material specifically for trace impurities in silicon is not currently available. Since neutron and γ rays are penetrating radiations (free from absorption problems, such as those found in X-ray fluorescence), matrix matching between the sample and the comparator standard is not critical. Biological trace impurities standards (e.g., the *National Institute of Standards and Technology Standard Reference Material*, SRM 1572 Citrus Leaves) can be used as reference materials. For the parametric analysis many instrumental factors, such as the neutron flux density and the efficiency of the detector, must be well known. The activation equation can be used to determine concentrations:

$$A = \frac{\text{cps}}{\varepsilon_d \varepsilon_p} = n \phi f \sigma (1 - e^{-\lambda t_i}) e^{-\lambda t_d}$$

where A is the activity in disintegrations/ second; *cps* stands for counts/ second in a given photopeak (i.e., a specific γ ray); ε_d is the efficiency of the detector for the photopeak; ε_p is the photon fraction (i.e., the number of γ rays/ 100 decays); n is the number of atoms of the element of interest (unknown); ϕ is the isotopic abundance of the target; f is the neutron flux density; σ is the neutron cross section; λ is the decay constant for the radioisotope; t_i is the length of irradiation; and t_d is the time elapsed since the end of irradiation.

Applications

NAA has been used to determine trace impurities in polysilicon, single-crystal boules, silicon wafers, and processed silicon, as well as plastics used for packaging.[3–5]

Element	Polysilicon (ppba)	Single crystal (ppba)
Ag	< 0.00130	< 0.00110
As	0.02400	0.00810
Au	0.00007	0.00005
Ba	Not identified	Not identified
Co	< 0.00640	< 0.00490
Cr	0.0520	< 0.00600
Cs	< 0.00040	< 0.00020
Cu	0.00590	0.00730
Fe	< 0.44000	< 0.25000
Ga	Not identified	Not identified
Hf	< 0.00021	< 0.00015
In	< 0.01800	< 0.05900
Mo	< 0.01300	Not identified
Na	Not identified	0.62000
Rb	< 0.01100	< 0.01000
Sb	< 0.00040	< 0.00250
Sc	< 0.00005	< 0.00004
Ta	< 0.00008	< 0.00010
Zn	< 0.00150	< 0.00110
Zr	< 0.1800	< 0.13000

Table 1 **Comparison of NAA data for polysilicon and single-crystal CZ silicon.**[3]

An example of an analysis done on polysilicon and single-crystal Czochralski silicon (CZ) is shown in Table 1. As can be seen, polysilicon, which was used to grow the crystal, is "dirtier" than the CZ silicon. This is expected, since segregation coefficients limit the incorporation of each element into the crystal boule during the crystal growth process. All values shown in the table are from bulk analysis. Table 2 shows NAA data obtained in an experiment where surface analysis was accom-

Impurity surface (4 μ)	Concentration (atoms/cc)	Epitaxial film (> 4 μ)
Au	1×10^{13}	1×10^{12}
Cu	7×10^{14}	6×10^{13}
Na	3×10^{15}	7×10^{13}

Table 2 **Surface concentrations of impurities on a Si wafer, as measured by NAA.[4]**

plished by using chemical etching following irradiation. The wafer surface is usually etched with a HF/HNO_3 / glacial acetic acid mixture and a 2–3 μm layer is etched off by controlling the length of the etching time. Both the etched wafer and the etchant solution are then measured. Many etchants developed for wafer fabrication can be used for these measurements. Table 3 shows results obtained from NAA of plastic packaging materials. Low detection limits from U and Th are particularly noteworthy.

Although the majority of NAA applications have been in the area of bulk analysis, some specialized uses need to be mentioned. One such unique application is the measurement of phosphorus in thin films (about 5000 Å) of phosphosilicate (PSG) or borophosphosilicate (BPSG) glasses used in VLSI device fabrication. In this case,

Impurity element	PVDF (4 qualities) (ppbw)	PVI (5 lots) (ppbw)
Au	0.1–0.3	0.005–0.05
Cu	65–100	110–1800
Cr	< 10–70	50–300
Co	< 2–15	5–10
Ca	80,000–130,000	Not analyzed
Cd	Not analyzed	1000–3000
Mn	6–26	25–80
Na	1700–4200	300–1000
W	0.3–1	1–4
U	0.3–0.7	< 0.1–0.5
Th	< 0.05–0.3	< 0.06–0.7

Table 3 **Selected impurity concentrations in plastics, as measured by NAA.[5]**

a PSG or a BPSG wafer (usually a pilot wafer) is activated and ^{31}P (stable) is converted to ^{32}P (14.2-day half-life). After a week of decay, the major radioactivity remaining is due to ^{32}P, which can be counted with a gas flow counter.[6]

Another application involves the measurement of copper via the radioisotope ^{64}Cu (12.6-hour half-life). Since ^{64}Cu decays by electron capture to ^{64}Ni ($^{64}Cu \rightarrow ^{64}Ni$), a necessary consequence is the emission of X rays from Ni at 7.5 keV. By using X-ray spectrometry following irradiation, sensitive Cu analysis can be accomplished. Because of the short range of the low-energy X rays, near-surface analytical data are obtained without chemical etching. A combination of neutron activation with X-ray spectrometry also can be applied to other elements, such as Zn and Ge.

Neutron activation also has been combined with accelerator mass spectrometry and has been demonstrated to have part-per-billion sensitivities for bulk nitrogen analysis in silicon. This combination was also used to obtain depth profile of Cl in silicon semiconductors.[7]

Limitations

NAA cannot be used for some important elements, such as aluminum (in a Si or SiO_2 matrix) and boron. The radioactivity produced from silicon directly interferes with that from aluminum, while boron does not produce any radioisotope following neutron irradiation. (However, an in-beam neutron method known as *neutron depth profiling* can be used to obtain boron depth profiles in thin films.[8]) Another limitation of NAA is the long turn-around time necessary to complete the experiment. A typical survey measurement of all impurities in a sample may take 2–4 weeks.

For matrices other than silicon, such as GaAs, InSb, AlGaAs, and InP, it is difficult to measure trace elements because the activity from the matrix is intense and long-lived. In these cases, laborious radiochemical separation techniques are employed to measure impurities.

Conclusions

NAA is well suited for Si semiconductor impurities analysis. The sensitivity and the bulk mode of analysis make this an important tool for controlling trace impurities during crystal growth or for monitoring cleanliness of various processing operations for device manufacturing. It is expected that research reactors will serve as the central analytical facilities for NAA in the industry. Since reactors are already set up to handle radioactive materials and waste, this makes an attractive choice over installing individual facilities in industries.

Related Articles in the Encyclopedia

XRF, PIXE, Dynamic SIMS, and NRA

References

1 M. L. Verheijke, H. J. J. Jaspers, and J. M. G. Hanssen. *J. Radioanal. Nucl. Chem.* **131,** (1) 197, 1989.

2 F. W. Walker, J. R. Parrington, and F. Feiner. *Chart of the Nuclides, 14th Edition.* General Electric Company, 1989, pp. 26–27.

3 M. Domenici, P. Malinverni, and M. Pedrotti. *J. Cryst. Growth.* **75,** 80, 1986.

4 G. B. Larrabee and J. A. Keenan. *J. Electrochem. Soc.* **118,** (8) 1351, 1971.

5 E. W. Haas and R. Hofmann. *Solid State Electronics.* **30,** (3) 329, 1987.

6 P. M. Zeitzoff, T. Z. Hossain, D. M. Boisvert, and R. G. Downing. *J. Electrochem. Soc.* **137,** (12) 3917, 1990.

7 P. Sharma, P. W. Kubik, H. E. Gove, U. Fehn, R. T. D. Teng, S. Datar, and S. Tullai-Fitzpatrick. *Annual Report NSRL–360.* University of Rochester, Rochester, 1990.

8 R. G. Downing, J. P. Lavine, T. Z. Hossain, J. B. Russell, and G. P. Zenner. *J. Appl. Phys.* **67,** (8) 3652, 1990.

9 *Nuclear and Radiochemistry.* (Third edition) John Wiley & Sons, New York, 1981.

11.4 NRA

Nuclear Reaction Analysis

DANIELE J. CHERNIAK AND W.A. LANFORD

Contents

Introduction

Nuclear reaction analysis (NRA) is used to determine the concentration and depth distribution of light elements in the near surface (the first few μm) of solids. Because this method relies on nuclear reactions, it is insensitive to solid state matrix effects. Hence, it is easily made quantitative without reference to standard samples. NRA is isotope specific, making it ideal for isotopic tracer experiments. This characteristic also makes NRA less vulnerable than some other methods to interference effects that may overwhelm signals from low abundance elements. In addition, measurements are rapid and nondestructive.

NRA can be highly sensitive, with typical detection limits of 10^{19} atoms / cm^3, depending on the reaction involved. Depth resolutions typically range from a few nm to tens of nm, and lateral resolutions down to a few μm can be achieved with microbeams.

An especially significant application of NRA is the measurement of quantified hydrogen depth profiles, which is difficult using all but a few other analytical techniques. Hydrogen concentrations can be measured to a few tens or hundreds of parts per million (ppm) and with depth resolutions on the order of 10 nm.

As NRA is sensitive only to the nuclei present in the sample, it does not provide information on chemical bonding or microscopic structure. Hence, it is often used in conjunction with other techniques that do provide such information, such as ESCA, optical absorption, Auger, or electron microscopy. As NRA is used to detect mainly light nuclei, it complements another accelerator-based ion-beam technique, Rutherford backscattering (RBS), which is more sensitive to heavy nuclei than to light nuclei.

NRA has a wide range of applications, including use in investigations of metals, glasses, and semiconductor materials, and in such diverse fields as physics, archaeology, biology, and geology.

Basic Principles

General[1]

A beam of charged particles (an *ion* beam) with an energy from a few hundred keV to several MeV is produced in an accelerator and bombards a sample. Nuclear reactions with low-Z nuclei in the sample are induced by this ion beam. Products of these reactions (typically p, d, t, ^{3}He, α particles, and γ rays) are detected, producing a spectrum of particle yield versus energy. Many (p, α) reactions have energies that are too low for efficient detection. In these cases, the associated γ rays are detected instead. Important examples are:

$$^{19}F + p \rightarrow {}^{16}O + \alpha + \gamma$$

$$^{15}N + p \rightarrow {}^{12}C + \alpha + \gamma$$

These reactions may be used, respectively, to profile ^{19}F and ^{15}N, using incident proton beams, or to profile hydrogen, using incident beams of ^{19}F and ^{15}N.

NRA exploits the body of data accumulated through research in low-energy nuclear physics to determine concentrations and distributions of specific elements or isotopes in a material. Two parameters important in interpreting NRA spectra are reaction Q values and cross sections.

Q values are the energies released in specific nuclear reactions and are used to calculate the energies of particles resulting from the reaction. Reactions with large positive Q values are most suitable for NRA. Table 1 presents Q values for a number of nuclear reactions. More comprehensive compilations of these data exist.[2] Reaction cross sections have also been measured as functions of incident ion energy and beam–detector angle. As particle yields are directly proportional to reaction cross sections, this information permits the experimenter to select an incident beam energy and detector angle that will maximize sensitivity. In addition, concentrations can be calculated directly from particle yields without reference to standards if the cross sections are accurately known. Similarly, the yields and energies of γ rays

Isotope	Q (MeV)	Isotope	Q (MeV)	Isotope	Q (MeV)
		(p, α) reactions			
^{7}Li	17.347	^{6}Li	4.022	^{9}Be	2.125
^{11}Be	8.582	^{18}O	3.970	^{31}P	1.917
^{19}F	8.119	^{37}Cl	3.030	^{27}Al	1.594
^{15}N	4.964	^{23}Na	2.379	^{17}O	1.197
				^{10}B	1.147
		(d, α) reactions			
^{3}He	18.352	^{31}P	8.170	^{13}C	5.167
^{10}B	17.819	^{11}B	8.022	^{32}S	4.890
^{6}Li	22.360	^{15}N	7.683	^{18}O	4.237
^{7}Li	14.163	^{9}Be	7.152	^{30}Si	3.121
$^{14}N_{(\alpha 0)}$	13.579	^{25}Mg	7.047	^{16}O	3.116
^{19}F	10.038	^{23}Na	6.909	^{26}Mg	2.909
^{17}O	9.812	^{27}Al	6.701	^{24}Mg	1.964
$^{14}N_{(\alpha 1)}$	9.146	^{29}Si	6.012	^{28}Si	1.421
		(d, p) reactions			
^{3}He	18.352	^{17}O	5.842	^{30}Si	4.367
^{10}B	9.237	^{31}P	5.712	^{26}Mg	4.212
^{25}Mg	8.873	^{27}Al	5.499	^{12}C	2.719
$^{14}N_{(p0)}$	8.615	^{24}Mg	5.106	^{16}O	1.919
^{29}Si	8.390	^{6}Li	5.027	^{18}O	1.731
^{32}S	6.418	^{23}Na	4.734	$^{14}N_{(p5)}$	1.305
^{28}Si	6.253	^{9}Be	4.585	^{11}B	1.138
^{13}C	5.947	^{19}F	4.379	^{15}N	0.267

Table 1 ***Q* values for nuclear reactions induced by protons and deuterons on some light isotopes.**[14]

resulting from nuclear reactions have been determined. Mayer and Rimini[3] include reaction cross sections and γ yields as functions of the incident ion energy for a number of light elements, as well as a table of γ-ray energies as a function of ion energy for (p, γ) reactions involving many low- to medium-Z elements.

Resonant Profiling

Resonant profiling uses beam energies near narrow isolated resonances of relevant nuclear reactions to determine the depth distribution of elements in a sample. A good illustration of this technique is hydrogen profiling using the reaction ^{1}H (^{15}N, $\alpha\gamma$) ^{12}C, with a resonance energy of 6.385 MeV. The expression in the previous sentence is a shorthand form used to describe the nuclear reaction. The term before the parentheses is the nucleus of interest (the species to be profiled). The first term in parentheses is the incident particle. The remaining components are the products of the reaction. Those in parentheses (after comma) are the species that are detected. In this case, the target is bombarded with ^{15}N ions and the yield of characteristic γ rays resulting from the reaction of the ^{15}N with ^{1}H is measured. When the energy of the incident beam E_0 is equal to the resonance energy E_R, the γ yield is proportional to the hydrogen content on the sample surface. If the beam energy is raised, the resonance energy is reached at a depth x, where the energy lost by the incident ions in traversing a distance x in the target is $E_0 - E_R$. The γ yield is now proportional to the hydrogen concentration at x. This is illustrated schematically in Figure 1. Contributions to the γ yield due to H in the surface region are greatly diminished, as the nuclear reaction cross section is large near the resonance energy but drops by several orders of magnitude for energies more than a few keV away. By continuing to raise the incident beam energy, one can profile further into the sample.

In this case, converting the γ yield-versus-incident beam energy profile to a concentration-versus-depth profile is straightforward. This is because the energy loss rate of the ^{15}N ions with depth (dE/dx) is large with respect to variations in individual ion energies after they have traveled a distance x in the material. In practice, these energy *straggling* effects can be neglected.

The depth scale is determined simply by

$$x = \frac{E_0 - E_R}{dE/dx} \tag{1}$$

The detected yield is a function of the concentration of the element being profiled, the resonance cross section, the detector efficiency, and dE/dx. To be specific,

$$\rho = KY(dE/dx) \tag{2}$$

where ρ is the concentration of the element being profiled, Y is the reaction yield (e.g., number of γ rays per microcoulomb of ^{15}N beam), dE/dx is the energy loss

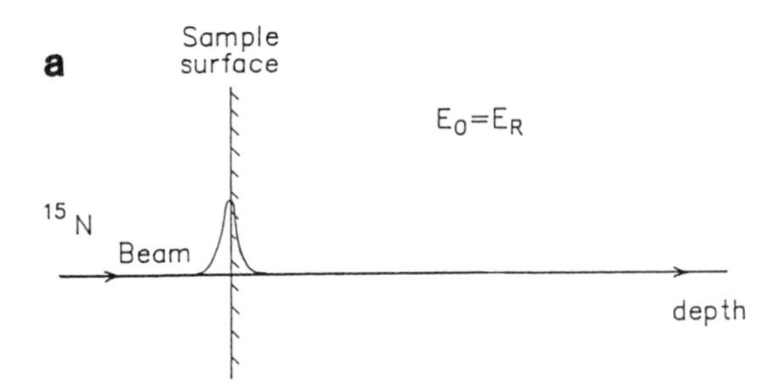

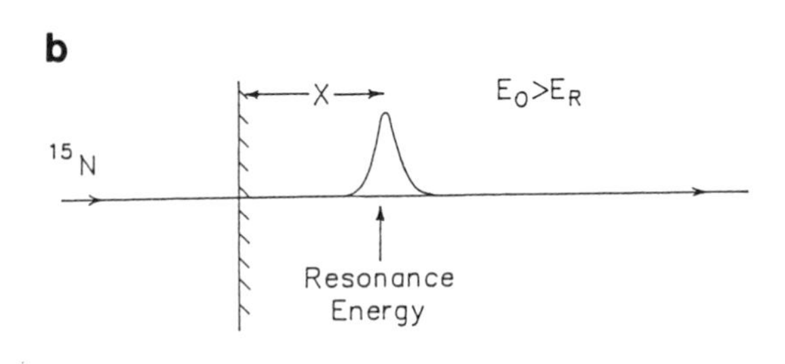

Figure 1 **Schematic illustration of resonant profiling technique. In (a), the incident ^{15}N beam is at resonance energy (E_R) and hydrogen on sample surface is detected. With higher beam energies (b), hydrogen is measured at depth x, where $x = (E_0 - E_R)/(dE/dx)$.**

rate of the incident ion in the target, and K is the *calibration* constant for the particular nuclear reaction and analysis chamber, a parameter independent of the material being analyzed.

Determination of concentration profiles from the raw data can be more complicated when protons are used as the incident particles. The energy loss (dE/dx) is smaller for protons and straggling effects are more important. The observed profile $N(E_0)$ is a convolution of the actual concentration profile $C(x)$ with a depth resolution function $q_0(x, E_0)$, which broadens with increasing x roughly as $\sqrt{x}$. Hence, resolution deteriorates with depth. However, near-surface resolution for resonant profiling may be on the order of tens of Å.

Nonresonant Profiling

When reaction cross sections are sufficiently large over an extended energy range, the entire depth profile may be obtained using a single incident beam energy. This is referred to as nonresonant profiling.

An example of this technique is the profiling of ^{18}O using the reaction ^{18}O (p, α) ^{15}N. Figure 2 shows the cross section of this reaction as a function of

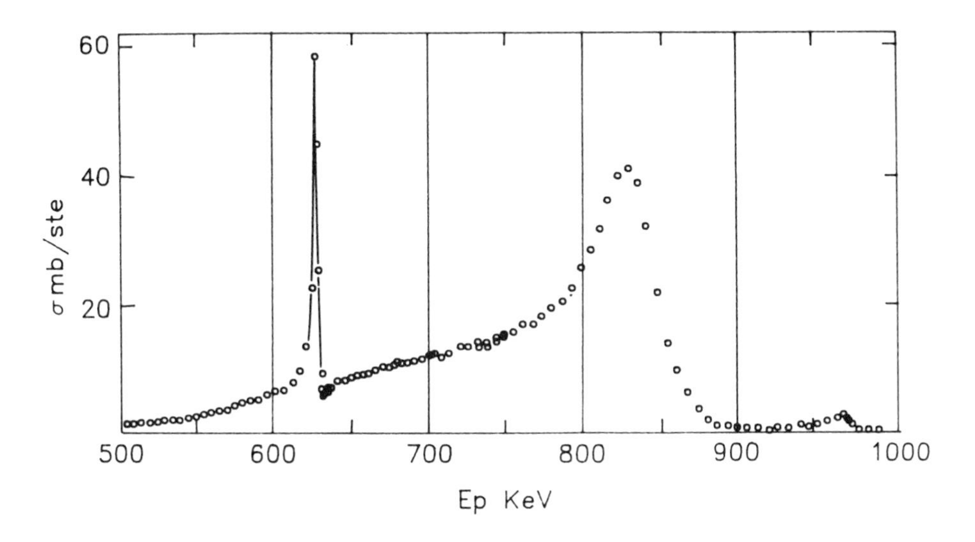

Figure 2 **Cross section versus incident proton energy for the ^{18}O (p, α) ^{15} N reaction, with a beam-detector angle of 165°.[14]**

incident proton energy, illustrating the large and smoothly varying cross section in the vicinity of 800 keV. It should be noted that ^{18}O also can be profiled using the resonant technique, employing the sharp resonance at 629 keV.

For nonresonant profiling, a sample is bombarded with protons at a suitable energy and the α particles resulting from the reaction of the protons with ^{18}O are detected. A spectrum of α particles over a range of energies is collected, representing contributions from ^{18}O at various depths in the material. The α spectra are converted to depth profiles in a manner analogous to that outlined above for H profiling. However, it must be noted that in this case not only the incoming protons, but also the outgoing α particles, lose energy traveling through the sample (unlike γ rays). The detected energy for ^{18}O on the sample surface can be calculated from the kinematics, using the incident proton energy, the angle between the incident beam and the particle detector, and the *Q* value for the reaction (3.97 MeV in this case). As they travel deeper into the sample, the protons lose energy. When these protons interact with the ^{18}O, the resultant α particles have a lower energy than those from the surface. The particles lose additional energy as they travel out of the material, so α particles from a certain depth will have a characteristic energy. This is illustrated schematically in Figure 3. To construct the depth scale from this information, the rate of energy loss for protons and α particles in the material must be known. This information is tabulated for most elements,[4] and values for compound targets can be calculated by weighting the elemental contributions according to their abundances in the material.

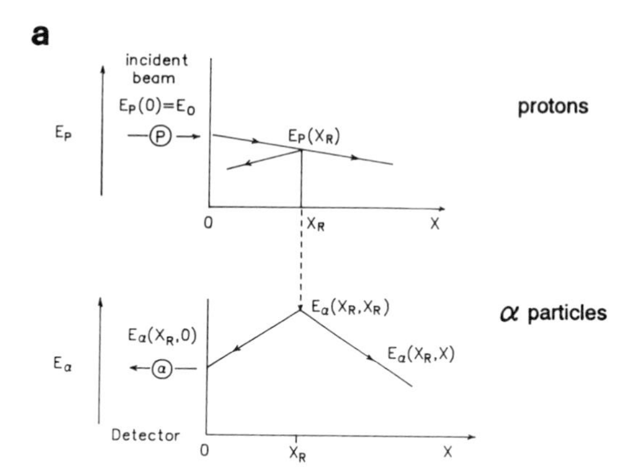

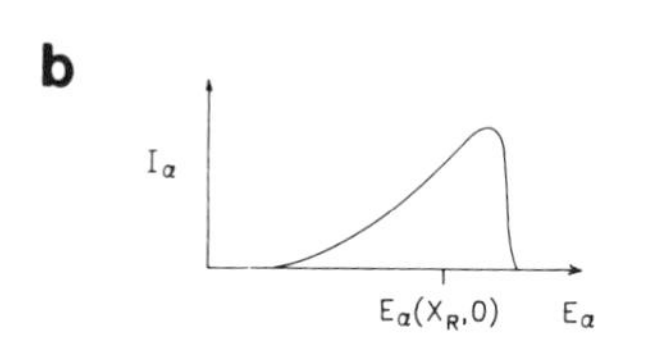

Figure 3 **Schematic illustration of nonresonant profiling technique. Incident protons lose energy with increasing depth x in the sample (a). At depth x_R, a proton with energy $E_P(x_R)$ induces a nuclear reaction with a target atom of ^{18}O, producing an α particle of energy $E_\alpha(x_R, x_R)$. The α particle loses energy as it travels out of the sample, resulting in the detected energy $E_\alpha(x_R, 0)$. The distribution of ^{18}O in the sample at various depths (b) results in a spectrum of α yield I_α versus detected alpha energy E_α.**

The number of detected α particles corresponding to a particular depth is a function of the detector solid angle, the total proton flux delivered to the sample, the reaction cross section at the appropriate energy, and the concentration of ^{18}O at that depth. Once again, the observed profile is a convolution of the actual concentration profile with a *spreading* or energy resolution function that takes into consideration such factors as the energy spread of the incident beam, proton and α-particle straggling in the sample, and detector resolution. Concentrations may be determined without reference to standards if these experimental parameters are known.

In nonresonant profiling, the silicon surface barrier detectors that detect the products of the nuclear reaction may also detect signals from incident ions that have been backscattered from the sample. Figure 4 shows an α particle spectrum from the reaction ^{18}O (p, α) ^{15}N, along with the signal produced by backscattered protons. The yield of backscattered particles (which is proportional to Z^2) may overwhelm the electronics in the detecting system, resulting in a pileup that greatly

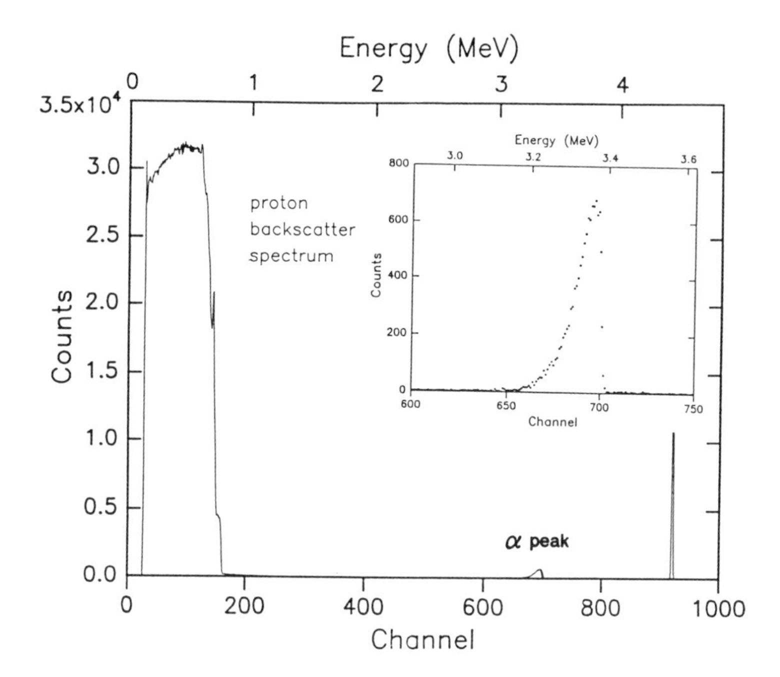

Figure 4 **Spectrum of ^{18}O diffusion in the mineral olivine ($(Mg, Fe)_2 SiO_4$) taken using nonresonant profiling technique with the reaction ^{18}O (p, α) ^{15}N. Both the α particles resulting from the nuclear reaction and backscattered protons are collected. Inset shows expanded region of the spectrum, where α yield indicates diffusion of ^{18}O into the material.**

reduces sensitivity. This difficulty is compounded by the fact that cross sections for nuclear reactions are generally much smaller than backscattering cross sections. A solution to this problem is to shield the detector with a thin-film absorber (such as Mylar). The absorber (of appropriate thickness) stops the backscattered incident ions, while permitting the higher energy ions from the nuclear reaction to pass into the detector. However, the presence of an absorbing material degrades depth resolution, since additional energy spreading occurs as the ions travel through the absorber to the detector. Because of this trade-off between depth resolution and sensitivity, the experimenter should weigh the usefulness of absorbers in each case. In ^{18}O profiling, for example, absorbers would be needed when profiling tantalum oxide, but may not be required during analysis of glasses and minerals having low-Z matrices, where ion backscattering is much reduced.

Characteristics

Selectivity and Quantification

Because of the nature of the technique, NRA is sensitive only to the nuclei present in the sample. While this characteristic prohibits obtaining direct information on chemical bonding in the material, it makes analyses generally insensitive to matrix effects. Therefore, NRA is easily made quantitative without reference to standard samples.

Since NRA focuses on inducing specific nuclear reactions, it permits selective observation of certain isotopes. This makes it ideal for tracer experiments using stable isotopes. Generally, there are no overlap or interference effects because reactions have very different *Q* values, and thus different resultant particle energies. This permits the observation of species present at relatively low concentrations. A good example is oxygen: ^{16}O and ^{18}O can be resolved unambiguously, as they are detected with completely different nuclear reactions; e.g., ^{16}O (d, p) ^{17}O, and ^{18}O (p, α) ^{15}N.

Resolution

Depth resolution in NRA is influenced by a number of factors. These include energy loss per unit depth in the material, straggling effects as the ions travel through the sample, and the energy resolution of the detection system.

As earlier discussed, the dominant factor in the near-surface region is the particle detection system. For a typical silicon surface barrier detector (15-keV FWHM resolution for ^{4}He ions), this translates to a few hundred Å for protons and 100–150 Å for ^{4}He in most targets. When γ rays induced by incident heavy ions are the detected species (as in H profiling), resolutions in the near-surface region may be on order of tens of Å. The exact value for depth resolution in a particular material depends on the rate of energy loss of incident ions in that material and therefore upon its composition and density.

In many cases, depth resolution in the near-surface region also can be improved by working at a grazing angle attained geometrically by tilting the sample. This increases the path length required to reach a given depth below the surface, which in turn produces an increase in effective depth resolution.

Straggling effects become more dominant further into the sample. They are most pronounced with proton beams, because the ratio of energy straggling to energy loss decreases with increasing ion mass. For protons, these effects may be quite substantial; for example, depth resolutions in excess of 1000 Å are typical for 1-MeV protons a few μm into a material.

With the use of a microbeam, lateral resolution with NRA on the order of several μm is possible. However, because of the small beam currents obtainable with microbeam systems, sensitivity is limited and reactions with relatively large cross sections are most useful. Only a few laboratories perform microbeam measurements.

Sensitivity

The sensitivity of NRA is affected by reaction cross sections, interfering reactions and other background effects. Hence, it is impossible to make general statements as to what sensitivities for NRA will be without considering the specific reactions and sample materials involved in each case. However, sensitivities on the order of 10–100 ppm are common.

Other Considerations

Sample Requirements

The maximum sample size is limited only by the design of the sample chamber. Typically, samples up to several cm in diameter can be accommodated. A diameter of a few mm is generally the lower limit because high-energy ion beams focused through standard beam optics are on the order of a few mm in diameter; however, microbeam setups permit the use of samples an order of magnitude smaller.

Nonconducting samples require special consideration. The incident ion beam causes a buildup of positive charge on the sample surface. Discharging of the sample may create noise in the spectrum collected by surface barrier detectors. In addition, the presence of accumulated positive charge on the sample may affect the accuracy of current integration systems, making it difficult to determine the exact beam dose delivered to the target. This problem may be obviated by flooding the sample surface with electrons to compensate for the buildup of positive charge or by depositing a thin layer of conducting material on the sample surface. If the latter option is chosen, the slowing down of ions in this layer must be considered when calculating depth scales. In addition, care must be taken to select a material that will not experience nuclear reactions that could interfere with those of the species of interest.

Accidental Channeling Effects

When analyzing single-crystal samples, the experimenter should be aware that accidental channeling may occur. This happens when the sample is oriented such that the ion beam is directed between rows or planes of atoms in the crystal, and generally results in reduced yields from reactions and scattering from lattice atoms. Such effects may be minimized by rotating the target in such a way to make the direction of the beam on the target more random. In some cases, the use of molecular ions (i.e. H_2^+ or H_3^+ instead of H^+) can also reduce the probability of accidental channeling. The molecular ions break up near the sample surface, producing atomic ions that repel and enter the material with more random trajectories, reducing the likelihood of channeling.

However, when deliberately employed, channeling is a powerful tool that may be used to determine the lattice positions of specific types of atoms or the number of specific atoms in interstitial positions (out of the lattice structure). Further information on this technique is available.[5]

Simulation Programs for NRA

There are a number of computer codes available[6, 7] to simulate and assist in the evaluation of NRA spectra. Most of these programs are similar to or compatible with the RBS simulation program RUMP. These programs require the input of reaction cross sections as a function of incident ion energy for the appropriate beam–detector geometry. The user interactively fits the simulation to the data by adjusting material parameters, such as the bulk composition and the depth distribution of the component being profiled. SPACES[6] is designed to deal specifically with narrow resonances (e.g., ^{27}Al (p, γ) ^{28}Si at 992 keV) and their associated difficulties, while SENRAS[7] is useful in many other cases.

Applications

In this section, a number of applications for NRA are presented. As this is not a review article, the following is only a sampling of the possible uses of this powerful technique. The reader interested in information on additional applications is directed to the proceedings of the Ion Beam Analysis Conferences[8] and those from the International Conferences on the Application of Accelerators in Research and Industry,[9] among other sources.

Hydration Studies of Glass

A combination of nuclear reactions have been used in studies of the processes involved in the hydration and dissolution of glass. Lanford et al.[10] investigated the hydration of soda-lime glass by measuring Na and H profiles. The profiles (Figure 5) indicate a depletion of sodium in the near-surface region of the glass and a complementary increase in hydrogen content. The ratio of maximum H concentration in the hydrated region and Na concentration in unhydrated glass is 3:1, suggesting that ionic exchange between H_3O^+ and Na^+ is occurring.

Residual Carbon in Ceramic Substrates

Multilayer ceramic substrates are used as multiple chip carriers in high-performance microelectronic packaging technologies. These substrates, however, may contain residual carbon which can adversely affect mechanical and electrical properties, even at ppm levels. Chou et al.[11] investigated the carbon contents of these ceramics with the reaction ^{12}C (d, p) ^{13}C. Carbon profiles for ceramic samples before and after surface cleaning are shown in Figure 6, and indicate significant reduction in the C content following the cleaning process.

Li Profiles in Leached Alloys

Schulte and collaborators[12] used the reaction ^{7}Li (^{3}He, p) ^{9}Be to measure the loss of Li from Al–Li alloys subjected to different environmental treatments. Figure 7 shows some of their results. Because they were interested in measuring how much

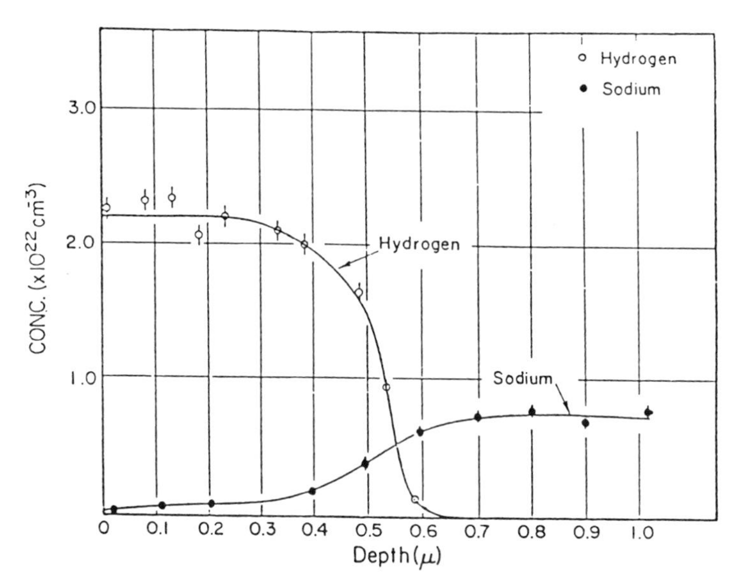

Figure 5 **Hydrogen and sodium profiles of a sample of soda-lime glass exposed to water at 90° C. The Na and H profiles were measured using ^{23}Na (p, γ) ^{24}Mg and ^{1}H (^{15}N, αγ) ^{12}C resonant nuclear reactions, respectively.**[10]

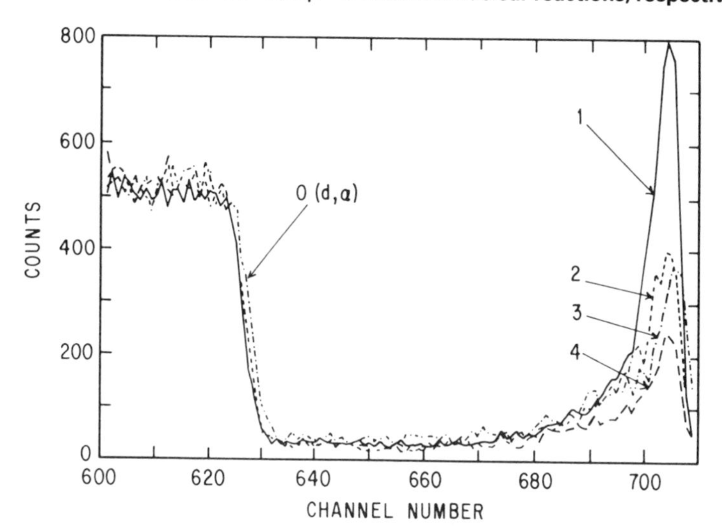

Figure 6 **Spectra of ceramic samples showing effects of surface cleaning on carbon content: (1) spectrum of specimen before cleaning; (2) spectrum of the same specimen after cleaning; (3) and (4) are spectra of two other surface-cleaned specimens.**[11]

Li was leached from a sample as a function of depth into the sample, they mounted the sample in epoxy and measured the Li as a function of distance from the alloy's surface using a finely collimated ^{3}He beam. To know when they were measuring in

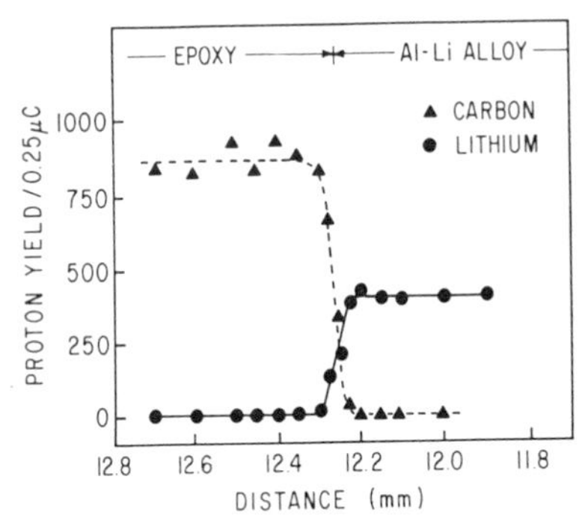

Figure 7 **Lateral profiles of carbon and lithium measured by nuclear reaction analysis. The sample was a lithium alloy mounted in epoxy. As the ion beam was scanned across the epoxy–metal interface, the C signal dropped and the Li signal increased.[12]**

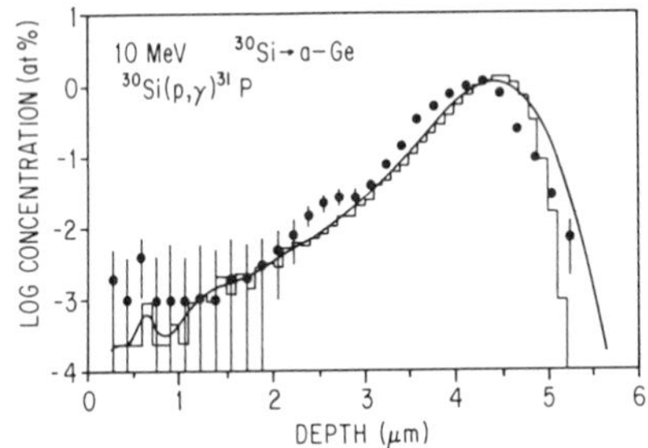

Figure 8 **Profiles of ^{30}Si implanted at 10 MeV into Ge measured by the ^{30}Si (p, γ) ^{31}P resonant nuclear reaction.[13]**

the metal and when in the epoxy, they also monitored the ^{12}C (^{3}He, p) ^{14}N reaction as a measure of the carbon content.

Si Profiles in Germanium

Kalbitzer and his colleagues[13] used the ^{30}Si (p, γ) resonant nuclear reaction to profile the range distribution of 10-MeV ^{30}Si implanted into Ge. Figure 8 shows their experimental results (data points), along with theoretical predictions (curves) of what is expected.

Conclusions

NRA is an effective technique for measuring depth profiles of light elements in solids. Its sensitivity and isotope-selective character make it ideal for isotopic tracer experiments. NRA is also capable of profiling hydrogen, which can be characterized by only a few other analytical techniques. Future prospects include further application of the technique in a wider range of fields, three-dimensional mapping with microbeams, and development of an easily accessible and comprehensive compilation of reaction cross sections.

Related Articles in the Encyclopedia

RBS and ERS

References

1 W. K. Chu, J. W. Mayer, and M. -A. Nicolet. *Backscattering Spectrometry.* Academic Press, New York, 1978, brief section on nuclear reaction analysis, discussions on energy loss of ions in materials, energy resolution, surface barrier detectors, and accelerators also applicable to NRA;

G. Amsel, J. P. Nadai, E. D'Artemare, D. David, E. Girard, and J. Moulin. *Nucl. Instr. Meth.* **92,** 481, 1971, classic paper on NRA, includes discussion of general principles, details on instrumentation, and applications to various fields; G.Amsel and W. A. Lanford. *Ann. Rev. Nucl. Part. Sci.* **34,** 435, 1984, comprehensive discussion of NRA and its characteristics, includes sections on the origin of the technique and applications; F. Xiong, F. Rauch, C. Shi, Z. Zhou, R. P. Livi, and T. A. Tombrello. *Nucl. Instr. Meth.* **B27,** 432, 1987, comparison of nuclear resonant reaction methods used for hydrogen depth profiling, includes tables comparing depth resolution, profiling ranges, and sensitivities.

2 E. Everling, L. A. Koenig, J. H. E. Mattauch, and A. H. Wapstra. *1960 Nuclear Data Tables.* National Academy of Sciences, Washington, 1961, Part I. Comprehensive listing of Q values for reactions involving atoms with $A < 66$.

3 J. W. Mayer, E. Rimini. *Ion Beam Handbook for Material Analysis.* Academic Press, New York, 1977. Useful compilation of information which includes Q values and cross sections of many nuclear reactions for low-Z nuclei. Also has selected γ yield spectra and γ-ray energies for (p, γ) reactions involving low to medium-Z nuclei.

4 J. F. Ziegler. *The Stopping and Range of Ions in Matter.* Pergamon Press, New York, 1980.

5 L. C. Feldman, J. W. Mayer, and S. T. Picraux. *Materials Analysis by Ion Channeling.* Academic Press, New York, 1982.

6 I. Vickridge and G. Amsel. *Nucl. Instr. Meth.* **B45,** 6, 1990. Presentation of the PC program SPACES, used in fitting spectra from narrow resonance profiling. A companion article includes further applications.

7 G. Vizkelethy. *Nucl. Instr. Meth.* **B45,** 1, 1990. Description of the program SENRAS, used in fitting NRA spectra; includes examples of data fitting.

8 Proceedings from Ion Beam Analysis Conferences, in *Nucl. Instr. Meth.* **B45,** 1990; **B35,** 1988; **B15,** 1986; **218,** 1983; **191,** 1981; **168,** 1980.

9 Proceedings from International Conferences on the Application of Accelerators in Research and Industry, in *Nucl. Instr. Meth.* **B40/41,** 1989; **B24/25,** 1987; **B10/11,** 1985.

10 W. A. Lanford, K. Davis, P. LaMarche, T. Laursen, R. Groleau, and R. H. Doremus. *J. Non-Cryst. Solids.* **33,** 249, 1979.

11 N. J. Chou, T. H. Zabel, J. Kim, and J. J. Ritsko. *Nucl. Instr. Meth.* **B45,** 86, 1990.

12 R. L. Shulte, J. M. Papazian, and P. N. Adler. *Nucl. Instr. Meth.* **B15,** 550, 1986.

13 P. Oberschachtsiek, V. Schule, R. Gunzler, M. Weiser, and S. Kalbitzer. *Nucl. Instr. Meth.* **B45,** 20, 1990.

14 G. Amsel and D. Samuel. *Anal. Chem.* **39,** 1689, 1967.

12

PHYSICAL AND MAGNETIC PROPERTIES

12.0 INTRODUCTION

In this last chapter we cover techniques for measuring surface areas, surface roughness, and surface and thin-film magnetism. In addition, the effects that sputter-induced surface roughness has on depth profiling methods are discussed.

Six methods for determining roughness are briefly explained and compared. They are mechanical profiling using a stylus; optical profiling by interferometry of reflected light with light from a flat reference surface; the use of SEM, AFM, and STM (see Chapter 2), and, finally, optical scatterometry, where light from a laser is reflected from a surface and the amount scattered out of the specular beam is measured as a function of scattering angle. All except optical scatterometry are scanning probe methods. A separate article is devoted to optical scatterometry. The different methods have their own strengths and weaknesses. Mechanical profiling is cheap and fast, but a tip is dragged in contact across the surface. The roughness "wavelength" has to be long compared to the stylus tip radius (typically 3 μm) and the amplitude small for the tip to follow the profile correctly. Depth resolution is about 5 Å. The optical profiler is a noncontact method, which can give a three-dimensional map, instead of a line scan, with a depth resolution of 1 Å. It cannot handle materials that are too rough (amplitudes larger than 1.5 μm) and if the surface is not completely reflective, reflection from the interior regions, or back interfaces, can cause problems. The lateral resolution depends on the light wavelength used, but is typically around 0.5 μm. The SEM operates in vacuum and requires a conducting surface, but is capable of 10-Å resolution in both vertical and lateral directions. AFM/STM measurements can provide surface topology maps with depth resolution down to a fraction of an angstrom and lateral resolution down to atomic dimensions. For practical surfaces, however, the instruments are usually operated in air at lower resolution. Optical Scatterometry is rather different in concept from the other methods in that it gives statistical information on the range of roughness, for flat reflective surfaces, within the area struck by the laser beam. Root-mean-squared (RMS) roughness values can be extracted from the data with a depth resolution of 1 Å. It can also be used to characterize the shapes and dimensions of periodic structures on a flat surface (e.g., patterned silicon wafers) with dimensions in the sub-μm range. To do this requires, however, calculation of the scattering behavior from an assumed model and a fit to the data. Optical scatterometry has been successfully used during on-line processing.

For many of the techniques discussed in this volume, composition depth profiling into a solid material is achieved by taking a measurement that is surface sensitive while sputtering away the material. Unfortunately, sputtering does not remove material uniformly layer by layer but introduces topography that depends on the material, the angle of sputtering, and the energy of the sputtering. This always degrades the depth resolution of the analysis technique with increasing depth. Specific examples are described here, as well as ways that the effect can be minimized.

In Magneto-Optic Kerr Rotation, MOKE, the rotation in polarization occurring when polarized laser light reflects from a magnetized material is measured. The rotation is due to the interaction of the light with the unpaired, oriented, valence electron spins of the magnetized sample. The degree of rotation is directly proportional to the magnetic moment, M, of the material, though absolute values of M are hard to obtain this way. This is because of the complex mathematical relationships between rotation and M, and the many artifacts that can occur in the experimental arrangement and also contribute to rotation. Usually, therefore, the method is used qualitatively to follow magnetic changes. These are either hysteresis loops in applied fields, or the use of a dynamic imaging mode to observe the movements and switching of magnetic domains in magnetic recording material. The lateral resolution capability is wavelength dependent and is about 0.5 μm for visible light. Sensitivity is enough to dynamically map domains at up to MHz switching frequencies. The depth of material probed depends on the light penetration depth; about 20–40 nm for magnetic material. Absolute sensitivity is high enough, though, to study monolayer amounts of magnetic material on a nonmagnetic substrate. Magnetic material buried under transparent overlayers can obviously be studied and this configuration is, in fact, the basis of magneto-optic data storage, which uses Kerr rotation to detect the magnetic bits. The technique is nondestructive and can be performed in ambient environments.

The final article of the volume deals with the use of adsorption isotherms to determine surface area. The amount of gas adsorbed at a surface can be determined volumetrically, or occasionally gravimetrically, as a function of applied gas pressure. Total surface areas are determined by physisorbing an inert gas (N_2 or Ar) at low temperature (77 K), measuring the adsorption isotherm (amount adsorbed versus pressure), and determining the monolayer volume (and hence number of molecules) from the Brunauer–Emmett–Teller equation. This value is then converted to an area by multiplying by the (known) area of a physisorbed molecule. The method is widely applied, particularly in the catalysis area, but requires a high surface area of material (at least 1 m^2 /gm): e.g., powders, porous materials, and large-area films. Selective surface areas of one material in the presence of another (e.g., metal particles on an oxide support) can sometimes be measured in a similar manner, but by using chemisorption where a strong chemical bond is formed between the adsorbed species and the substrate material of interest. Hydrogen is most commonly used for this, since by now it is known that for many metals it dissociates and forms one adsorbed H-atom per surface metal atom. From the measurement of the amount of hydrogen adsorbed and a knowledge of the spacing between metal atoms (i.e., a knowledge of the crystallographic surfaces exposed) the metal surface area can be determined.

12.1 Surface Roughness

Measurement, Formation by Sputtering, Impact on Depth Profiling

FRED A. STEVIE

Contents

Introduction

A surface property that has a direct impact on the results of many types of analysis is its texture or roughness. Roughness can also affect friction and other mechanical properties. A high percentage of surface analytical effort has been expended on samples that have very flat surfaces, such as polished silicon wafers, but there are many other materials of interest, for example, metals and ceramics, that can have roughness on the order of micrometers. Even a polished silicon surface has topographical variations that can be measured by very sensitive techniques, such as atomic force microscopy or scanning tunneling microscopy.

Two surface roughness terms are commonly used: average roughness *RA* and root-mean-square roughness *RMS*. For *N* measurements of height z and average height $\bar{z}$, the average roughness is the mean deviation of the height measurements

$$RA = \frac{1}{N}\sum_{i=1}^{N}\left|z_i - \bar{z}\right| \tag{1}$$

and the root-mean-square roughness is the standard deviation

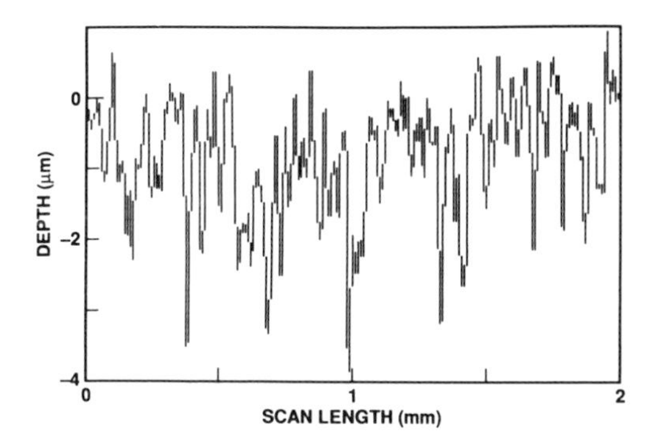

Figure 1 **Mechanical profiler trace of a region on the unpolished back of a silicon wafer.**

$$RMS = \left[\frac{1}{N}\sum_{i=1}^{N}(z_i - \bar{z})^2\right]^{1/2} \tag{2}$$

Several surface roughness measurement techniques are in common usage. The optimum method will depend upon the type and scale of roughness to be measured for a particular application.

Measurement Techniques

Mechanical Profiler

Mechanical profilers, also called profilometers, measure roughness by the mechanical movement of a diamond stylus over the sample of interest. No sample preparation is required and almost any sample that will not be deformed by the stylus can be measured very rapidly. The trace of the surface is typically digitized and stored in a computer for display on a cathode ray tube and for output to a printer. The stylus force can be adjusted to protect delicate surfaces from damage. Typical weight loading ranges from a few milligrams to tens of milligrams, but can be as low as one milligram. Small regions can be located with a microscope or camera mounted on the profiler. Lateral resolution depends upon the stylus radius. If the surface curvature exceeds the radius of curvature of the stylus, then the measurement will not provide a satisfactory reproduction of the surface. A typical stylus radius is about 3 μm, but smaller radii down to even submicron sizes are available. Arithmetic average or root-mean-square roughness can be calculated automatically from the stored array of measurement points.

As an example, consider the unpolished back of a silicon wafer. Figure 1 shows a mechanical profiler trace of a region on the wafer. The surface has variations that are generally 1–2 μm, but some of the largest changes in height exceed 3 μm. The average roughness is 0.66 μm.

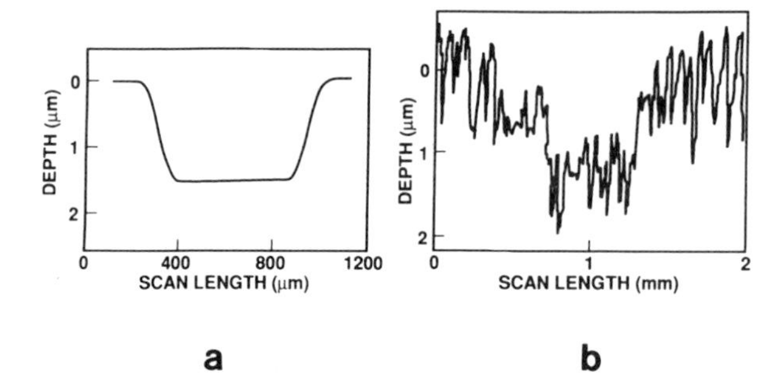

Figure 2 Mechanical profiler traces of craters sputtered with O_2^+ primary beam for an initially smooth surface of Si_3N_4/Si (a); and an initially rough SiC surface (b).

Mechanical profilers are the most common measurement tool for determining the depth of craters formed by rastered sputtering for analysis in techniques like Auger Electron Spectroscopy (AES) and Secondary Ion Mass Spectrometry (SIMS). Figure 2a shows an example of a 1.5-μm deep crater formed by a rastered oxygen beam used to bombard an initially smooth silicon nitride surface at 60° from normal incidence. The bottom of the crater has retained the smooth surface even though the 0.45-μm nitride layer has been penetrated. Depth resolution for an analytical measurement at the bottom of the crater should be good. Figure 2b shows a crater approximately 1 μm deep formed under similar conditions, but on a surface of silicon carbide that was initially rough. The bottom of the crater indicates that the roughness has not been removed by sputtering and that the depth resolution for a depth profile in this sample would be poor.

Even though the mechanical profiler provides somewhat limited two dimensional information, no sample preparation is necessary, and results can be obtained in seconds. Also, no restriction is imposed by the need to measure craters through several layers of different composition or material type.

Optical Profiler

Optical interferometry can be used to measure surface features without contact. Light reflected from the surface of interest interferes with light from an optically flat reference surface. Deviations in the fringe pattern produced by the interference are related to differences in surface height. The interferometer can be moved to quantify the deviations. Lateral resolution is determined by the resolution of the magnification optics. If an imaging array is used, three-dimensional (3D) information can be provided.

Figure 3 shows an optical profiler trace of the same portion of the wafer sample analyzed by the mechanical profiler. The resulting line scan in Figure 3a is similar to that for the mechanical system. The average and root-mean-square roughness are

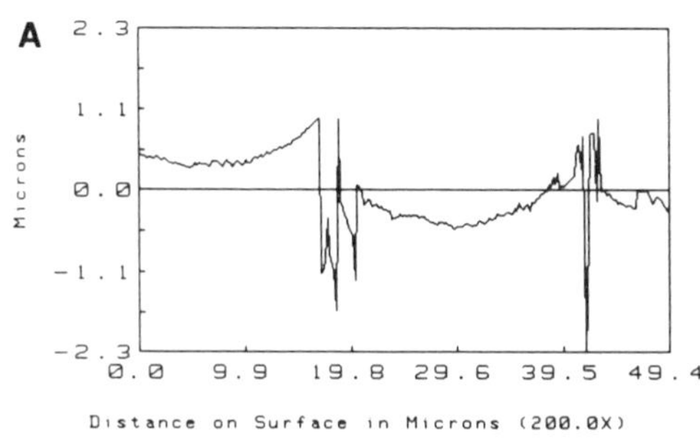

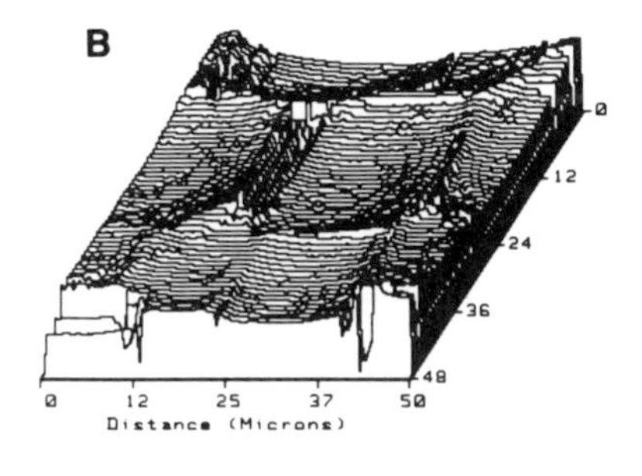

Figure 3 **Optical profiler measurements of a region on the unpolished back of a silicon wafer: line scan (a); and 3D display (b) (Courtesy of WYCO Corp.).**

determined by computer calculation using the stored data points for the line scan. A 3D representation, such as the one shown in Figure 3b, adds significantly to the information obtained about the surface from a line scan because crystallographic features can be identified.

In general, optical profilers have the same advantages as mechanical profilers: no sample preparation and short analysis time. However, the optical system also has some disadvantages. If the surface is too rough (roughness greater than 1.5 μm), the interference fringes can be scattered to the extent that topography cannot be determined. If more than one matrix is involved, for example, for multiple thin films on a substrate, or if the sample is partially or totally transparent to the wavelength of the measurement system, then measurement errors can be introduced. Software advances have improved the accuracy of measurements on a single film on a substrate. Even though a phase may be introduced because of a difference in indexes of refraction between the film and the substrate, a correction can be applied. Multiple matrix samples can be measured if coated with a layer that is not transparent to the wavelength of light used.

Scanning Electron Microscope (SEM)

SEM images are formed on a cathode ray tube with a raster synchronized with the raster of an electron beam moving over the sample of interest. Variations in the intensity of electrons scattered or emitted by the sample result in changes in the brightness on the corresponding points on the display. SEM measurements of the surface topography can be very accurate over the nanometer to millimeter range. Specific features can be measured best by cleaving the sample and taking a cross sectional view.

As an example, consider again the back surface of the silicon wafer used in the mechanical profiler example. Figure 4a, an SEM micrograph taken at 45° tilt, shows a surface covered with various sized square-shaped features that often overlap. This information cannot be discerned from the mechanical profiler trace, but can be obtained using a 3D optical profiler measurement. Figures 4b and 4c are also

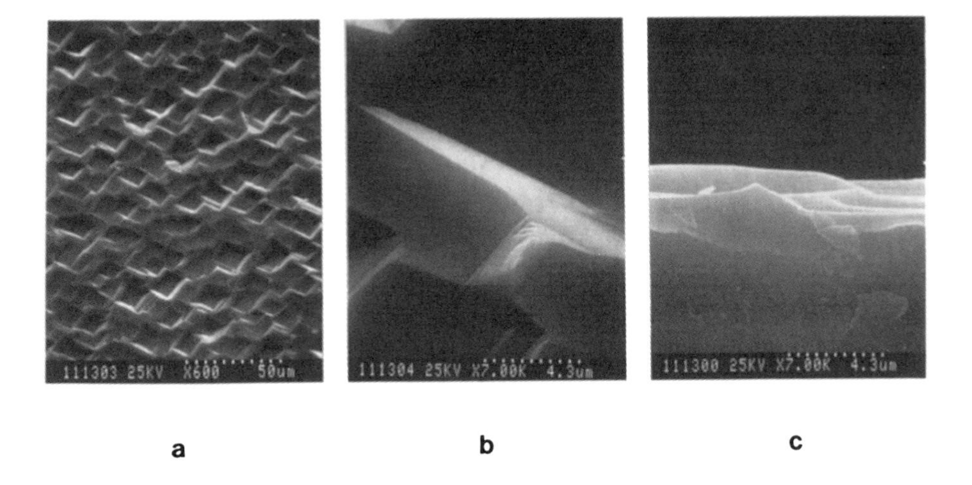

Figure 4 **SEM micrographs of a region on the back of a silicon wafer: (a) and (b) show the surface at different magnifications; (c) is a cross sectional view (Courtesy of P. M. Kahora, AT&T Bell Laboratories).**

SEM micrographs of the same sample. Figure 4b shows an area similar to that of Figure 4a, but at a higher magnification. Figure 4c is a cross sectional view that indicates the heights of several individual features. All three micrographs were taken at relatively low magnification for an SEM. Note that for many types of manufactured silicon wafers, the surface on the back of the wafer undergoes an acid etch after the lapping process and would exhibit a much more random surface roughness. The surface shown in the example results from a potassium hydroxide etch, which causes enhanced etching along certain crystallographic orientations.

Specific SEM techniques have been devised to optimize the topographical data that can be obtained. Stereo imaging consists of two images taken at different angles of incidence a few degrees from each other. Stereo images, in conjunction with computerized frame storage and image processing, can provide 3D images with the quality normally ascribed to optical microscopy. Another approach is confocal microscopy. This method improves resolution and contrast by eliminating scattered and reflected light from out-of-focus planes. Apertures are used to eliminate all light but that from the focused plane on the sample. Both single (confocal scanning laser microscope, CLSM) and multiple (tandem scanning reflected-light microscope, TSM or TSRLM) beam and aperture methods have been employed.

Some disadvantages for SEM measurements, compared with data from mechanical and optical profilers, are that the sample must be inserted into a vacuum system, and charging problems can make the analysis of insulators difficult. SEMs are also much more expensive than profilers.

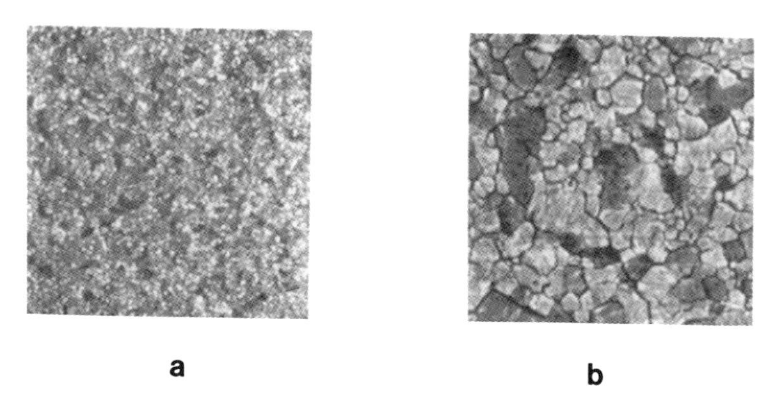

Figure 5 Atomic force microscope images of an aluminum film deposited on ambient (a) and heated (b) Si substrates. The scales are 15 μm × 15 μm (a) and 20 μm × 20 μm (b). The grain size can be clearly observed (Courtesy of M. Lawrence A. Dass, Intel Corporation).

Atomic Force Microscope

An Atomic Force Microscope (AFM), also called a Scanning Force Microscope (SFM), can measure the force between a sample surface and a very sharp probe tip mounted on a cantilever beam having a spring constant of about 0.1–1.0 N/m, which is more than an order of magnitude lower than the typical spring constant between two atoms. Raster scanning motion is controlled by piezoelectric tubes. If the force is determined as a function of the sample's position, then the surface topography can be obtained.[1, 2] Detection is most often made optically by interferometry or beam deflection. In AFM measurements, the tip is held in contact with the sample. Spatial resolution is a few nanometers for scans up to 130 μm, but can be at the atomic scale for smaller ranges. Both conducting and insulating materials can be analyzed without sample preparation.

Figure 5 shows AFM images of the surfaces of Al-0.5 % Cu thin films deposited on unheated (Figure 5a) and heated (Figure 5b) Si substrates. The aluminum grain size is smaller in the sample deposited at ambient temperature. Root-mean-square roughness was measured at 5.23 and 7.45 nm, respectively, for the ambient and heated samples. The depth of the grain boundaries can be determined from a 3D image. The roughness of the aluminum on the unheated substrate is dominated by the different grains, but the heated substrate sample roughness is determined by grain boundaries.

Scanning Tunneling Microscope (STM)

Electrons can penetrate the potential barrier between a sample and a probe tip, producing an electron tunneling current that varies exponentially with the distance.

The STM uses this effect to obtain a measurement of the surface by raster scanning over the sample in a manner similar to AFM while measuring the tunneling current. The probe tip is typically a few tenths of a nanometer from the sample. Individual atoms and atomic-scale surface structure can be measured in a field size that is usually less than 1 μm × 1 μm, but field sizes of 10 μm × 10 μm can also be imaged. STM can provide better resolution than AFM. Conductive samples are required, but insulators can be analyzed if coated with a conductive layer. No other sample preparation is required.

Examples of semiconductor applications include the imaging of surface coatings to determine uniformity and the imaging of submicron processed features.

Optical Scatterometry

An optical scatterometer can be used to measure angularly resolved light scatter. The light source for one of the systems in use is a linearly polarized He–Ne laser with the polarization plane perpendicular to the plane of incidence. Light scattered from the sample is focused onto an aperture in front of a photomultiplier. The multiplier is rotated in small increments ($< 0.5^\circ$) and the scattered light intensity is measured at each point. This method provides a noncontact measurement of roughness for reflecting samples and is capable of determining subsurface damage in silicon and gallium arsenide wafers.[3, 4] Root-mean-square roughness measurements as low as 0.1 nm can be obtained. No sample preparation is required for analysis.

If the sample is fully or partially transparent to the incident beam, light may be scattered from the back of the sample or from within the sample, and the surface measurement will be inaccurate.

Roughness Formed by Sputtering

The sputtering process is frequently used in both the processing (e.g., ion etching) and characterization of materials. Many materials develop nonuniformities, such as cones and ridges, under ion bombardment. Polycrystalline materials, in particular, have grains and grain boundaries that can sputter at different rates. Impurities can also influence the formation of surface topography.[5]

For several analytical techniques, depth profiles are obtained by sputtering the sample with a rastered ion beam to remove atoms from the surface and gradually form a crater. The most common elements used for primary beams are oxygen, argon, cesium, and gallium. For many materials, rastered or unrastered sputtering produces a rough surface. Even single-crystal materials are not immune to ion bombardment-induced topography formation. Ridges have been detected in Si, GaAs, and AlGaAs after O_2^+ bombardment. Figure 6 is a set of SEM micrographs that show the formation of a series of ridges in (100) Si after bombardment to increasing depth with a 6-keV O_2^+ primary beam at approximately 60° from normal inci-

	Mechanical profiler
Depth resolution	0.5 nm
Minimum step	2.5–5 nm
Maximum step	~150 μm
Lateral resolution	0.1–25 μm, depending on stylus radius
Maximum sample size	15-mm thickness, 200-mm diameter
Instrument cost	$30,000–$70,000
	Optical profiler
Depth resolution	0.1 nm
Minimum step	0.3 nm
Maximum step	15 μm
Lateral resolution	0.35–9 μm, depending on optical system
Maximum sample size	125-mm thickness, 100-mm diameter
Instrument cost	$80,000–$100,000
	SEM (see SEM article)
	Scanning force microscope (see STM/SFM article)
Depth resolution	0.01 nm
Lateral resolution	0.1 nm
Instrument cost	$75,000–$150,000
	Scanning tunneling microscope (see STM/SFM article)
Depth resolution	0.001 μm
Lateral resolution	0.1 nm
Instrument cost	$75,000–$150,000
	Optical scatterometer
Depth resolution	0.1 nm (root mean square)
Instrument cost	$50,000–$150,000

Table 1 **Comparison of the capabilities of several methods for determining surface roughness.**

dence.[6] The ridges that develop during this process are perpendicular to the direction of the ion beam. One explanation of the cause of this particular formation is based on the instability of a plane surface to periodic disturbances.[7] Topography

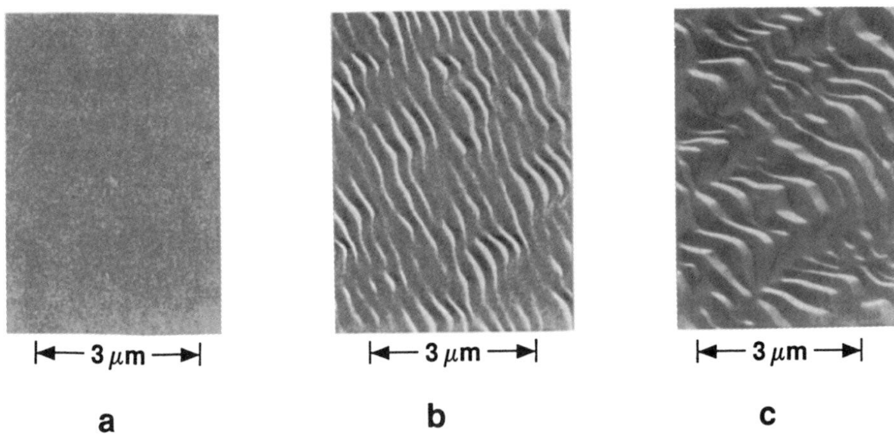

Figure 6 SEM micrographs of the bottoms of SIMS craters in (100) Si after 6 keV O_2^+ bombardment to 2.1 μm (a), 2.8 μm (b), and 4.3 μm (c). The angle of incidence is approximately 40° from normal.[6]

formation is different for different primary beams and for different angles of incidence. The ridges in Si do not form with Cs^+ bombardment or, at high angles of incidence from the normal, with O_2^+ bombardment.

Impact on Depth Profiling

Depth Resolution and Secondary Ion Yield

Roughness from sputtering causes loss of depth resolution in depth profiling for Auger Electron Spectroscopy (AES), X-Ray Photoelectron Spectroscopy (XPS), and SIMS.

Degraded depth resolution is especially apparent in the case of metals.[8] Figure 7 shows the analysis of a 1-μm film of aluminum on a silicon substrate. The interface between the layer and substrate is smeared out to the extent that only an approximate idea of the interface location can be obtained. The sputtering rates for aluminum and silicon under the conditions used differ by almost a factor of 2. Therefore, the sputtering rate varies significantly in the poorly resolved interface region and the depth axis cannot be accurately calibrated. The roughness at the bottom of the crater can be severe enough to affect the depth measurement of the crater.

For SIMS profiles, the secondary ion yield can also be affected by sputter-induced roughness. Figure 8 shows changes in secondary ion yield for silicon monomeric and polymeric species analyzed under the same conditions as the sample shown in the SEM micrographs from Figure 6. The micrographs correlate with the depths shown on the profile and prove that the change in ion yield is coincident with the topography formation.[6] The ion yield change (before and after topography formation) can vary for each secondary ion species. For the example, in

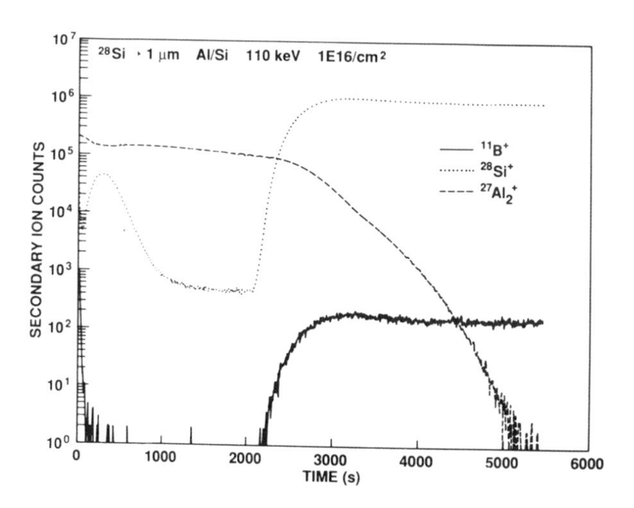

Figure 7 SIMS depth profile of Si implanted into a 1-μm layer of Al on a silicon substrate for 6-keV O_2^+ bombardment. The substrate is B doped.

Figures 6 and 8 the changes were approximately 65 % for $^{28}Si^+$ and over 250 % for $^{16}O^+$. Different ions can have yields affected in opposite directions, as shown by the two species in Figure 8. Other materials, such as GaAs, have also shown significant changes in ion yield that have been correlated with microtopography formation.

Sample Rotation During Sputtering

Corrective action for roughening induced by sputtering has taken several directions. The simultaneous use of two sputtering beams from different directions has been explored; however, rotation of the sample during ion bombardment appears to be the most promising. Attention to the angle of incidence is also important

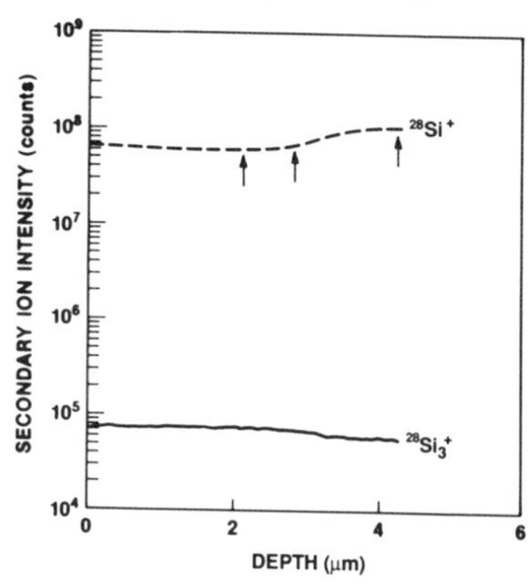

Figure 8 SIMS depth profile of (100) Si for 6-keV O_2^+ bombardment at approximately 40° from normal incidence. The arrows show the depths at which the SEM micrographs in Figure 6 were taken.[6]

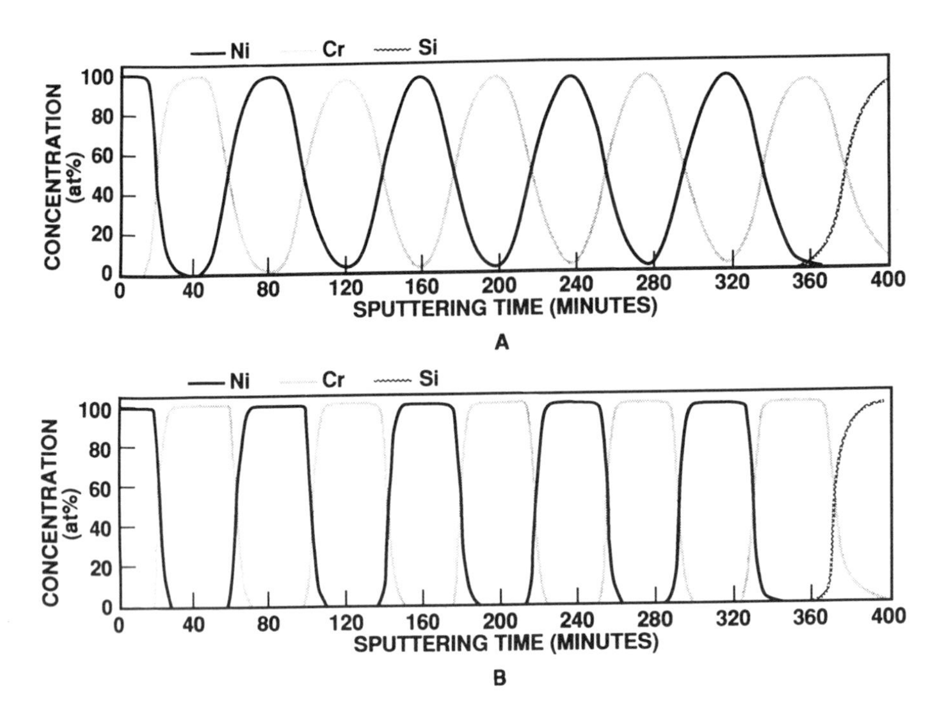

Figure 9 AES depth profiles of multilayer Cr/Ni thin film structures on a smooth substrate using a 5-keV Ar^+ primary beam: without rotation of the sample during bombardment (a), and with rotation (b).[9]

because topography formation can be reduced or eliminated for certain materials if the angle of incidence from the normal is 60° or higher.

If a sample of polycrystalline material is rotated during the sputtering process, the individual grains will be sputtered from multiple directions and nonuniform removal of material can be prevented. This technique has been successfully used in AES analysis to characterize several materials, including metal films. Figure 9 indicates the improvement in depth resolution obtained in an AES profile of five cycles of nickel and chromium layers on silicon.[9] Each layer is about 50 nm thick, except for a thinner nickel layer at the surface, and the total structure thickness is about 0.5 μm. There can be a problem if the surface is rough and the analysis area is small (less than 0.1-μm diameter), as is typical for AES. In this case the area of interest can rotate on and off of a specific feature and the profile will be jagged.

This technique has recently been sucessfully applied to SIMS depth profiling.[10] Figure 10 shows a profile of a GaAs/AlGaAs superlattice with and without sample rotation. The profile without rotation shows a severe loss of depth resolution for the aluminum and gallium signals after about 15 periods, whereas the profile with rotation shows no significant loss of depth resolution after almost 70 periods. The data

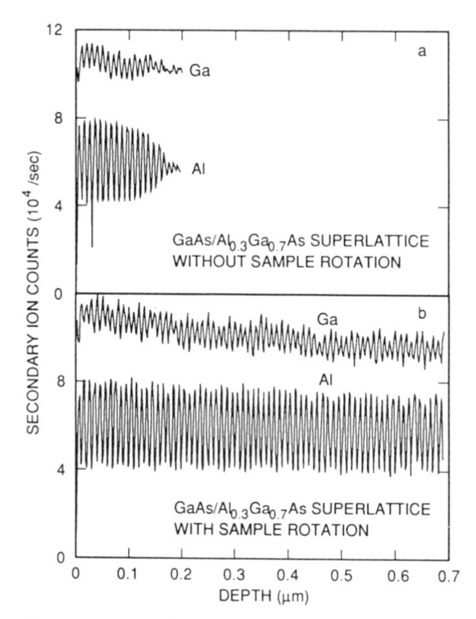

Figure 10 **SIMS depth profiles with and without sample rotation during bombardment by 3-keV O_2^+ at 40° from normal incidence.**[10]

were taken using a 3-keV oxygen primary beam rastered over a 1-mm × 1-mm area at 8 nm/ min. The rotation speed was approximately 0.6 cycles/ min. Additional work by the same group has shown that the secondary ion yield changes described above are removed also if the sample is rotated.

Related Articles in the Encyclopedia

Dynamic SIMS, AES, SEM, STM, and SFM

References

1 N. A. Burnham and R .J. Colton. *J. Vac. Sci. Technol.* **A7,** 2906, 1989.

2 N. A. Burnham and R. J. Colton. in *Scanning Tunneling Microscopy: Theory and Practice.* (D. A. Bonnell, ed.) V.C.H. Publishers, New York, 1991.

3 R. D. Jacobson, S. R. Wilson, G. A. Al-Jumaily, J. R. McNeil, J. M. Bennett, and L. Mattsson. *Applied Optics.* 1991.

4 J. R. McNeil, et al. *Optical Eng.* **26,** 953, 1987.

5 *Ion Bombardment Modification of Surfaces* (O. Auciello and R. Kelly, eds.) Elsevier, Amsterdam, 1984.

6 F. A. Stevie, P .M. Kahora, D .S. Simons, and P. Chi. *J. Vac. Sci. Technol.* **A6,** 76, 1988.

7 R. M. Bradley and J. M. E. Harper. *J. Vac. Sci. Technol.* **A6,** 2390, 1988.

8 R. G. Wilson, F. A. Stevie, and C. W. Magee. *Secondary Ion Mass Spectrometry: A Practical Handbook for Depth Profiling and Bulk Impurity Analysis.* Wiley, New York, 1989.

9 A. Zalar. *Thin Solid Films.* **124,** 223, 1983.

10 E.-H. Cirlin, J. J. Vajo, T. C. Hasenberg, and R. J. Hauenstein. *J. Vac. Sci. Technol.* **A8,** 4101, 1990.

12.2 Optical Scatterometry

JOHN R. MCNEIL, S.S.H. NAQVI, S.M. GASPAR,
K.C. HICKMAN, AND S.D. WILSON

Contents

Introduction

Many technologies involve the need to monitor the surface topology of materials. First the topology itself may be of direct interest. Second, topology is usually strongly influenced by the processing steps used to produce the surface; characterizing the topology therefore can serve as a process monitor. Angle-resolved characterization of light scattered from a surface, or scatterometry, is a very attractive diagnostic technique to characterize a sample's topology. It is noncontact, nondestructive, rapid, and often provides quantitative data. Scatterometry can be used as a diagnostic tool in the fabrication of microelectronics, optoelectronics, optical elements, storage media, and other, less glamorous areas such as the production of paper and rolled materials. Application of scatterometry in some cases eliminates the need for microscopic examination. The technique is amenable to automated processing, something which is not possible using microscopic examination.

Basic Principles and Applications

The arrangement illustrated in Figure 1 is commonly used for angular characterization of scattered light. The light source is usually a laser. The incident beam may be unpolarized, or it can be linearly polarized with provisions for rotating the plane of polarization. Typically the plane of polarization is perpendicular to the plane of

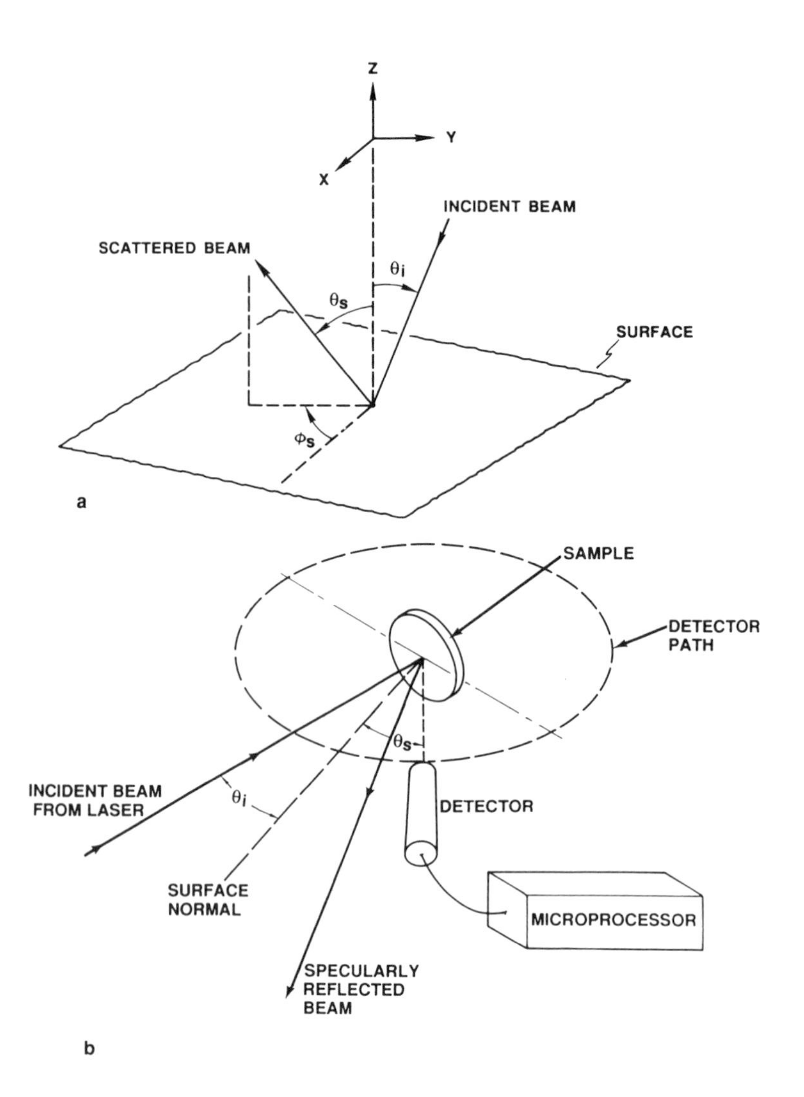

Figure 1 **Scatterometer arrangement, illustrating the geometry (a) and the experimental configuration (b).**

incidence (s-polarized light), as this avoids surface plasma wave coupling in conducting samples. The laser output is spatially filtered to provide a well-defined spot at the sample. This is critical for allowing measurements close to the specularly reflected beam or the directly transmitted beam (in the case of a sample which is transmitting at the wavelength of interest); the significance is described below. Sometimes the detector also has provisions for polarization discrimination. The detector is typically a photomultiplier or a Si photodiode. Other detection arrangements include multiple detectors or diode arrays. Arrangements that employ camera and screen configurations recently have shown utility for measuring scattering in two dimensions. Theoretical aspects of light scattering are reviewed below in connection with applications.

Applications of scatterometry can best be described by considering two general categories of surfaces that are examined: surfaces which are nominally smooth, and surfaces which are intentionally patterned. In the first category, scatterometry is used to measure surface roughness and other statistical properties of the sample's topology. Certain conditions of the surface are assumed, and these are discussed below. In addition, for some "smooth" surfaces, such as optical components, the scattered light intensity itself is the item of interest, and little or no additional interpretation is needed. This information might be sufficient to predict the performance of the sample, such as characterizing scattering losses from laser cavity elements.[1–4] Measuring light scattered from intentionally patterned surfaces is a very convenient process monitor in manufacturing areas like microelectronics and optoelectronics. This is an area of active research, with some results now appearing in manufacturing environments.[5–7]

Smooth Surfaces: Surface Topology Characterization

The relation between scattering of electromagnetic radiation and surface topography has been studied for many years, originally in connection with radar. In general this relationship is complicated. However, the relation is simple in the case of a clean, perfectly reflecting surface in which the heights of the surface irregularities are much smaller than the wavelength of the scattered light (i.e., the smooth-surface approximation). We present the results of Church's treatment.[2]

Vector scattering theories describe the differential light scatter dI_s as

$$\frac{1}{I_i}\left(\frac{dI_s}{d\omega_s}\right) = \frac{C}{\lambda^4} Q(\theta_i, \phi_i, \theta_s, \phi_s, N, \chi_i, \chi_s)\, P(p, q) \tag{1}$$

where C is a constant, I_i is the intensity of the incident light, and $d\omega_s$ is the solid angle of the detection system. The quantity Q in Equation (1), called the *optical factor,* is independent of the surface condition and is a function of the angles of incidence (θ_i, ϕ_i), the scattering angles (θ_s, ϕ_s), complex index of refraction N of the surface, and polarization states of the incident and scattered light, χ_i and χ_s, respectively. The *surface factor* $P(p,q)$ is the power spectral density of the surface roughness; it is the output of the scatterometer measurement and is the function which describes the surface structure.

If the surface (i.e., the best fit plane) is in the x–y plane, and $Z(x,y)$ is the surface height variation (surface roughness) relative to that plane, the power spectral density is given by

$$P(p, q) = \frac{1}{A}\left[\frac{1}{2\pi}\iint dxdye^{i(px+qy)}z(x, y)\right]^2 \quad (2)$$

where A is the area of the scatterer, and p and q are the surface spatial frequencies in the x- and y-directions, respectively. In other words, the power spectral density is the average squared magnitude of the two-dimensional Fourier transform of the surface roughness.

Although the power spectral density contains information about the surface roughness, it is often convenient to describe the surface roughness in terms of a single number or quantity. The most commonly used surface-finish parameter is the root-mean-squared (rms) roughness σ. The rms roughness is given in terms of the instrument's band width and modulation transfer function, $M(p, q)$ as

$$\sigma^2 = \int_{p_{\min}}^{p_{\max}} dp \int_{q_{\min}}^{q_{\max}} dqM(p, q)\,P(p, q) \quad (3)$$

Different values of σ^2 will result if the integral limits (i.e., band width) or modulation transfer function in the integral change. All surface characterization instruments have a band width and modulation transfer function. If rms roughness values for the same surface obtained using different instruments are to be compared, optimally the band widths and modulation transfer functions would be the same; they should at least be known. In the case of isotropic surface structure, the spatial frequencies p and q are identical, and a single spatial frequency (p) or spatial wavelength ($d = 1/p$) is used to describe the lateral dimension of structure of the sample.

An intuitive understanding of the power spectral density can be obtained by considering the surface to be composed of a number of surfaces, each having structure of a single spatial frequency that is sinusoidally varying in amplitude (height). Measuring the scattered light at a specific scattering angle corresponds to characterizing structure of a specific spatial frequency, and the intensity of the scattered light is proportional to the amplitude of the structure. The situation is analogous to Fourier analysis of an electrical signal. Figure 2 illustrates this idea by comparing polished Cu and Mo surfaces. The micrograph of the Cu surface in Figure 2a shows the dominance of fine texture and lines (high spatial frequency structure) which result from polishing the soft material. By contrast, the hard Mo surface has relatively little fine structure and is dominated by wide regions due to the large grains of the material. The two power spectral density characteristics shown in Figure 2b are a quantitative description of the information in the micrographs.

The modulation transfer function of the optical scatterometer is nearly unity.[4] The spatial frequency band width, using 0.633-nm photons from a He–Ne laser, is typically 0.014–1.6 μm^{-1}, corresponding to a spatial wavelength band width 70–0.633 μm. This corresponds to near normal sample illumination with a minimum

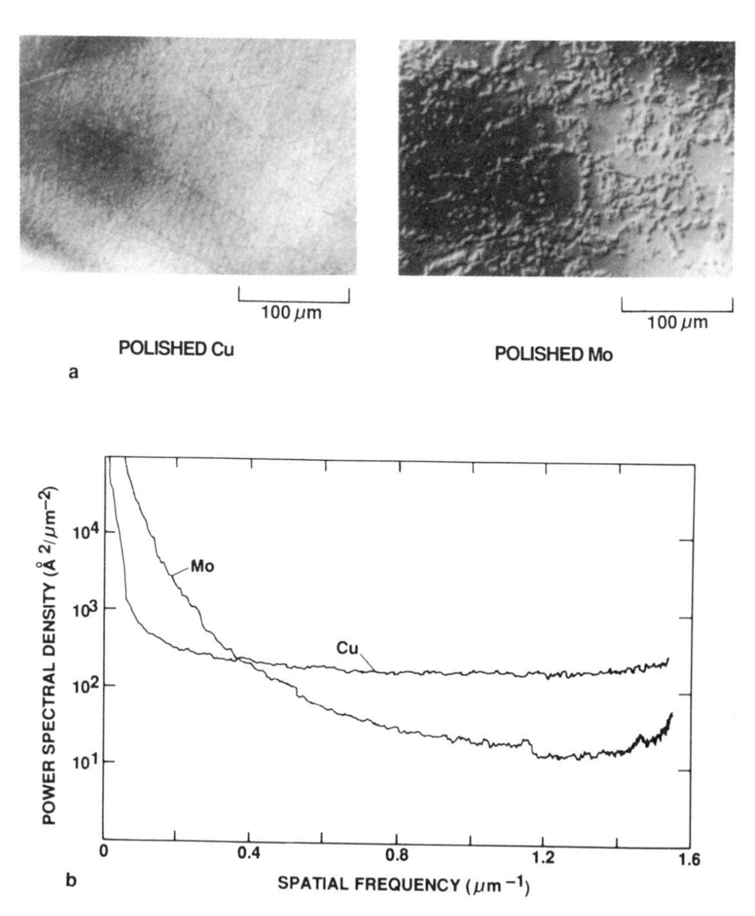

Figure 2 **Comparison of two polished metal surfaces: (a) photographs from Nomarski microscope examination; and (b) power spectral density characteristics of the same surfaces.**

scattering measurement angle of 0.5° from the specular beam. Measurements closer to the specular extend the 70-mm limit to larger values. For nonnormal angle of incidence these limits shift to smaller values. Similarly, a larger optical wavelength shifts these limits to larger values. This behavior is determined by the grating equation,

$$\sin\theta_s = \sin\theta_i + \frac{n\lambda}{d} \qquad (4)$$

where $n = \pm 1, \pm 2$, and so forth. For smooth surfaces, only the first order is considered.

The importance of instrument band width is illustrated by considering the rms roughness of the two samples of Figure 2. If the rms roughness is calculated over the band width 0.014–1.6 μm^{-1}, the roughness of the Mo sample is approximately 85 Å, and that of the Cu sample is 35 Å. The same power spectral density characteristics can be analyzed over a smaller band width, 0.06–1.6 μm^{-1}, and both surfaces have an rms roughness of approximately 30 Å. It is important to know these widths because all surface roughness measurement techniques have band widths.

The technique described above has been applied to characterize morphology of optical components[4] and microelectronics materials.[1] In the latter, an Al–Si (2%) alloy material was characterized that had been deposited at different substrate temperatures. The material grain size was determined using SEM inspection and was found to be highly correlated with the rms roughness results from the scatterometer characterization. This makes the scattering technique appear useful for grain-size analysis of this material, thus providing the analysis normally obtained using SEM inspection. Characterization of CVD W and WSi_2 is described in Gaspar et al.[1]

The preceding discussion relates only to a perfectly reflecting surface. If the surface transmits the incident light, either completely or only a short distance, the scattered light originates from the volume that is illuminated, as well as from the front surface and back surface (if illuminated). In this situation, the preceding treatment is not applicable, and analysis of the data is not straightforward. However, the technique still can provide very useful information on sample morphology. In general, the high spatial frequency structure of a surface or in the volume of the sample will scatter primarily at large angles. However, multiple scattering events in the material complicate the situation.

Smooth Surfaces: Characterization of Sample Bidirectional Scattering Distribution Function

The light scattered from a sample can fill the entire 4π steradians of space if the sample is transparent at the wavelength of interest, and 2π steradians if it is not. The angular distribution is a function of the optical properties (index, homogeneity, orientation, etc.), surface roughness and contamination of the sample, polarization of the source, and angles of incidence and detection. The bidirectional scattering distribution function (BSDF) is the term commonly applied to describe this pattern, and it is simply defined in radiometric terms as the ratio of the scattered surface radiance, measured at angle θ_s, to the incident surface irradiance. The former is the light flux or power P_s (Watts) scattered per unit surface area of the sample illuminated, per unit projected solid angle of the detection system. The incident surface irradiance is the light flux Pi (Watts) on the surface per unit of illuminated surface area. The projected solid angle is the solid angle $d\omega_s$ of the detection system times $\cos(\theta_s)$. BSDF is expressed as

$$\mathrm{BSDF} = \frac{dP_s / d\omega_s}{P_i \cos\theta_s} \tag{5}$$

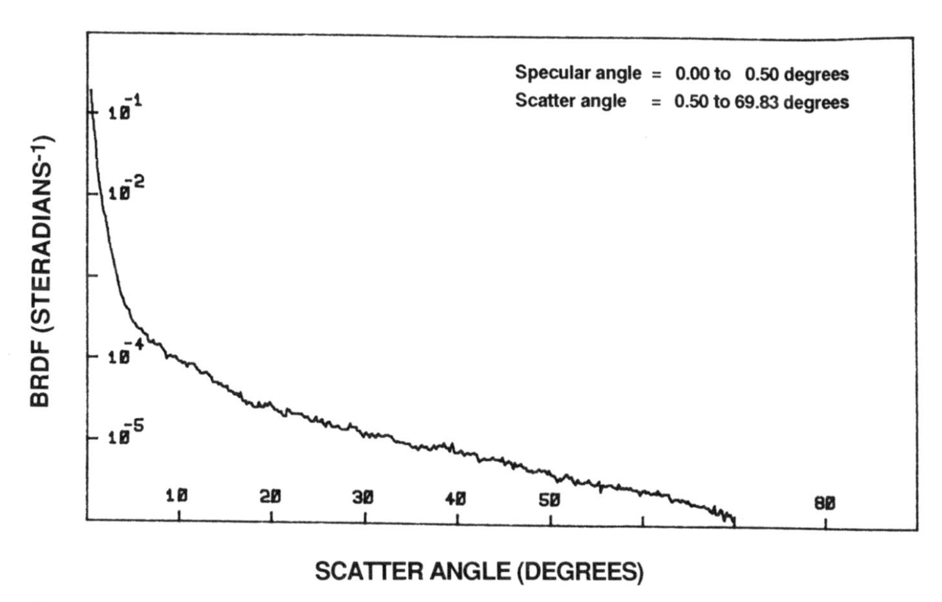

Figure 3 **Example of the BRDF characteristic of a polished optical surface. Note the rapid increase in scattering at small scatter angles.**

This expression for BSDF is appropriate for all angles of incidence and all angles of scatter, and it has units of (steradians^{-1}). Note that this expression was originally derived by Nicodemus, who assumed a plane wave input for the source beam. If BSDF is measured on the same side of the sample as the incident light source, the scattering characterization is referred to as bidirectional reflectance distribution function (BRDF); if scattering is measured on the other side of the sample, it is referred to as bidirectional transmittance distribution function (BTDF). Note that BRDF can have very large values. For example, when the specular reflection is measured, P_i/P_s is nearly unity, and BRDF is approximately $1/\omega_s$, which can be very large (e.g., 10^4). BRDF can be viewed as the directional reflectance of the sample per steradian. Note, too, that the factor $\cos\theta_s$ is sometimes not included in the expression above for BSDF, in which case it is said to be no longer "cosine corrected."

As discussed previously, BSDF characteristics of a sample depend strongly on the incident angles (θ_i, ϕ_i) and scattering measurement angles (θ_s, ϕ_s); in general they both increase very rapidly as scattered light measurements are made closer to the directly transmitted beam (BTDF) and the specularly reflected beam (BRDF). Figure 3 illustrates this by showing a typical BRDF characteristic of a polished optical surface.

Intentionally Patterned Surfaces

High technology surfaces (e.g., microelectronics) are often intentionally patterned, and light scattering may be used for subsequent characterization of the pattern. In particular, the pattern can be a structure that is periodic in one or more directions. The pattern may be a device of interest in a fabrication process, or it might be an imposed test pattern such as a diffraction grating. Light is diffracted (scattered) into several distinct orders as described by the grating equation. The intensity of the light in the different orders is a very sensitive function of the shape of the lines of the pattern. If the shape of the lines of the pattern is influenced by a processing step, scattering characterization can provide a simple way to monitor the process.

This technique has been used recently [5–7] to provide an attractive, simple monitor of processes involved in microelectronics fabrication. Sub-μm-wide lines of metal on glass used in photomasks have been accurately measured. The technique has been applied to characterize the depth and side wall angle of etched microelectronics structures. It also has been used to directly monitor the exposure level in photoresist during lithography for microelectronics processing. The technique has been used for noncontact temperature measurement of surfaces at room temperature and at elevated (700° C) temperatures; 1° C resolution has been achieved. Other applications to microelectronics and storage media are underway, and it is very realistic to expect this type of scattering characterization to develop into a valuable process monitor for use in many technological areas.

Theoretical modeling of the process consists first of predicting the fraction of incident power diffracted into the different orders by illuminating a known structure. The power distribution is a function of the shape of the lines in the periodic structure and is somewhat application specific. For example, in the case of trapezoidal shaped lines, the parameters of interest are the top line width, the side wall angle, and the height of the line structure. However, all problems involve application of Maxwell's equations in a rigorous vector diffraction approach to calculate this power distribution. A sufficient number of calculations are performed for different values of the line shape parameters of interest to facilitate addressing the inverse of this situation, namely identifying an unknown structure based upon its scattering characteristic. We have used neural network and linear statistical prediction techniques to compliment the theoretical calculations, so that an estimate of the line shape parameters can be obtained from experimentally measured diffracted intensities. This approach has been used successfully in the three applications given above.

Practical Considerations

Several practical issues of the scatterometer must be considered in the case of characterizing nominally smooth surfaces. The incident laser beam may be collimated, but more commonly it is brought to a focus at a distance defined by the arc in which the detector rotates. In addition, a deflection mirror or an optical fiber might be used to direct light to the detector element. These features permit measurements close to the specular and transmitted beams, and this is critical to fully characterize the scattered light. This is especially significant since the scattered light intensity can change several orders of magnitude within the first few degrees from the specular and transmitted beams. This is illustrated below in connection with sample data. It is important to perform a sufficient number of measurements to fully characterize the scattered light in this region of rapid change.

Another practical concern is the amount of light scattered by the optical elements of the scatterometer system. This instrument scatter, or *signature* can limit the scatterometer sensitivity (e.g., the minimum rms roughness that is measurable). Typically measurements can be performed at minimum scattering angles, θ_s = 0.5°–1.0°, from the specular and transmitted beams without the instrument signature being a concern. The instrument signature is a concern, however, when the intensity of light scattered by the instrument at a particular angle is comparable to the light scattered from a sample.

An additional concern involves how isotropically light is scattered from a sample. If the sample has nonisotropic structure (surface or volume), light will be scattered nonisotropically with respect to ϕ_s (see Figure 1). Examples of surfaces that have nonisotropic structure include fine diffraction gratings, machined parts, computer hard disks, and microelectronics circuits. In some instances it is important to fully characterize scattered light, such as in determining the roughness of machined parts. In this case, the scattering measurements might be performed in several planes. However, this involves complicated instrumentation. Alternatively, the scattered light might be measured in a single plane, as illustrated in Figure 1, and the sample can be rotated about a normal axis passing through the point of illumination. Allowance must be made for rotating the polarization of the input beam to maintain the appropriate geometry.

A final practical note involves instrument intensity measurement calibrations. The intensity measurement is self-calibrating relative to the incident beam from the source. However, measurements typically have a dynamic range of 10^8–10^{10}, and care must be taken to insure the detection system is linear. A method of calibrating the scatterometer is to characterize a diffuse reflector having a known scattering characteristic. For example, a surface coated with $BaSO_4$ makes a nearly Lambertian scatterer, which has a BRDF of $1/\pi$ at all angles.

Comparison to Other Techniques

Other methods available to characterize surface topology include optical and mechanical profilometers, and microscopy techniques. These techniques suffer from some combination of being contact, destructive, sensitive to vibration, qualitative, or slow to apply. Light scattering techniques avoid these. Another aspect of comparison has to do with the utility of a technique in advanced manufacturing environments. In particular, microscopy and profilometry techniques are not amenable to *in-situ* use, and this is a deterrent for their application as real-time, on-line process monitors. Scatterometers can be incorporated into many processing arrangements for *in-situ* use. This provides rapid feedback for process control. In addition, the scatterometer is available in rugged, user-friendly forms that can be operated by unskilled personnel.[8]

In the case of scatterometry applications mentioned above in connection with patterned surfaces, there are sometimes no alternative characterization techniques. For example, there currently is no direct monitor of photoresist exposure dose for lithography in microelectronics processing. This is very significant, as a Si wafer typically spends up to 50% of its processing time in lithography. There is no alternative to noncontact temperature measurements having the resolution and temperature range of the scattering technique described above; these are requirements for advanced processing of microelectronics.

The significance of instrument band width and modulation transfer function was discussed in connection with Equation (3) to characterize the roughness of nominally smooth surfaces. The mechanical (stylus) profilometer has a nonlinear response, and, strictly speaking, has no modulation transfer function because of this. The smallest spatial wavelength which the instrument can resolve, d_{min}, is given in terms of the stylus radius r and the amplitude a of the structure as

$$d_{min} = 2\pi\sqrt{ar} \tag{6}$$

This expression is applicable for a surface consisting of a single spatial frequency and is discussed in detail in Wilson et al.[9] Stylus curvatures of 12 μm are often used, and smaller curvatures are available. For a 12-μm curvature stylus to have a lateral resolution of 1 μm, the amplitude of the structure cannot exceed 20 Å, or the stylus will not follow the contour of the surface; to resolve 0.6 μm the amplitude cannot exceed 7.4 Å. A lateral resolution of 500 Å is quoted by some stylus instrument manufacturers. In this case the surface amplitude could not exceed 0.6 Å, using a 1-μm stylus, and 0.05 Å, using a 12-μm stylus. These surface amplitudes are clearly unrealistic. The presence of multiple spatial frequencies (i.e., realistic surfaces) causes harmonic distortion and other nonlinear effects. The long spatial wavelength limit of the band width is determined by the scan length of the stylus, with hundreds of μm being easily achievable. This limit is somewhat larger than that of the scatterometer. In general, using the stylus profilometer to profile a surface is valid only when the surface wavelengths are large compared to the stylus radius, and amplitudes are small compared to the radius. However these instruments are very useful for measuring step heights, the purpose for which they were originally designed.

Optical profilometers have nonlinear modulation transfer function characteristics resulting from an arrangement involving an incoherent imaging system.[4] Results from characterizing surfaces using an optical profilometer are compared to those from scatterometer and stylus measurements and discussed in Jacobson et al.[4] The optical profilometer equipped with a 20× objective lens has a short spatial wavelength resolution limit of approximately 2 μm and a long wavelength limit of approximately 160 μm. If the optical profilometer is used to profile a surface that has transparent thin films, the optical and mechanical "surfaces" will not necessarily be the same.

When using microscopy techniques to obtain topology information one must be aware of the spatial wavelength band width of the instrument, and this obviously depends on the instrument's magnification. In general the short spatial wavelength (resolution) limit of STM, AFM, SEM, and TEM techniques can be many times smaller than that of scatterometry. Because of this, applications of these techniques are sometimes very different from those of scatterometry, even though they involve characterizing topology or morphology. Instrument modulation transfer function can depend on a number of aspects of the instrument. For example, the STM and AFM probe characteristics strongly influence instrument response. Other microscopy techniques have less quantitative vertical resolution.

Conclusions

Light scattering techniques will play an increasingly significant role in materials processing, especially from surfaces that are intentionally patterned. Visible wavelength light has been used to easily characterize structures having a line width of 0.3 μm. Shorter wavelength laser output can be used to probe even smaller features. The technique is noncontact, nondestructive, noncontaminating, rapid, and it often yields quantitative measurements of surface structures. Future applications will include using the technique as an *in-situ* diagnostic tool.

Related Articles in the Encyclopedia

AFM, Optical Microscopy, STM, and Surface Roughness

References

1 S. M. Gaspar, K. C. Hickman, J. R. McNeil, R. D. Jacobson, Y. E. Strausser, and E. R. Krosche. Metal Surface Morphology Characterization Using Laser Scatterometry. In: *Proceedings of the Spring Meeting of the Materials Research Society.* MRS, 1990. Results are presented of scatterometer characterization of microelectronics materials, including Al-Si, CVD W, and WSi_2.

2 E. L. Church, H. A. Jenkinson, and J. M. Zavada. Relation Between the Angular Dependence of Scattering and Microtopographic Features. *Opt. Eng.* **18,** 125, 1979. This article presents an analysis of the relation between angle-resolved light scattering and surface topology.

3 J. C. Stover. *Optical Scattering: Measurement and Analysis.* McGraw-Hill, New York, 1990. This is a good presentation of angle-resolved optical scat-

tering technology, directed primarily toward characterization of optical surfaces.

4 R. D. Jacobson, S. R. Wilson, G. A. Al-Jumaily, J. R. McNeil, and J. M. Bennett. Microstructure Characterization by Angle-Resolved Scatter and Comparison to Measurements Made by Other Techniques. To be published in *Appl. Opt.* This work discusses the band width and modulation transfer function of the scatterometer, stylus profilometer, optical profilometer, and total integrated scattering systems, and gives results of measuring several surfaces using all techniques.

5 S. S. H. Naqvi, S. M. Gaspar, K. C. Hickman, and J. R. McNeil. A Simple Technique for Linewidth Measurement of Gratings on Photomasks. *Proc. SPIE.* **1261,** 495, 1990. K.P. Bishop, S.M. Gaspar, L.M. Milner, S.S.H. Naqvi, and J.R. McNeil. rasterization using Scatterometry. Proc. SPIE. **1545**, 64, 1991. These papers discusses a simple application of scattering from surfaces that are intentionally patterned.

6 K. C. Hickman, S. M. Gaspar, S. S. H. Naqvi, K. P. Bishop, J. R. McNeil, G. D. Tipton, B. R. Stallard, and B. L. Draper. Use of Diffraction From Latent Images to Improve Lithogrophy Control. Presented at the SPIE Technical Conference 1464: Symposium on I.C. Metrology, Inspection, and Process Control, San Jose, CA, 1991, Proc. SPIE. 1464, pp. 245-257, 1991. Another application is presented of scattering characterization and modeling from periodic structures for process control.

7 K. P. Giapis, R. A. Gottscho, L. A. Clark, J. B. Kruskal, D. Lambert, A. Kornblit, and D. Sinatore. Use of Light Scattering in Characterizing Reactively Ion Etched Profiles. To be published in *J. Vac. Sci. Technol.* 1991. This article gives a description of scattering measurements made to characterize line profiles of structures reactively etched in Si.

8 Sandia Systems Inc., Albuquerque, NM, and TMA Technologies, Inc., Bozeman, MT. Sandia systems specializes in systems for characterizing microelectronics and magnetic disk materials; TMA emphasizes optical materials characterization.

9 S. R. Wilson, G. A. Al-Jumaily, and J. R. McNeil. Nonlinear Characteristics of a Stylus Profilometer. *Opt. Eng.* **26,** 953, 1987. This describes modeling stylus profilometer response characteristics and explains their shortcomings.

12.3 MOKE

Magneto-optic Kerr Rotation

DAVID E. FOWLER

Contents

Introduction

The magneto-optic Kerr effect (MOKE), or Kerr rotation, provides a simple and straightforward optical method for magnetically characterizing the near-surface region of magnetic materials. Visible, linearly polarized light is reflected from a sample's surface and small rotations in the polarization and small changes in the ellipticity of the light are observed as schematically shown in Figure 1. Elliptical polarization of light results when the two orthogonal components of a light wave's electric field vector have a phase difference. These optical effects result from the interaction of the incident light with the conduction electrons in the magnetic solid. The magnitude of the rotation of the polarization is directly proportional to the net magnetization M of the material reflecting the light. Additionally, MOKE measurements can be used to determine the direction of magnetization in the domains of the material, i.e., for magnetic domain imaging, since the magnitude and sign of the rotation in polarization depends on the relative orientation of the plane of incidence, the incident angle of the light, and the orientation of M. While the amount of rotation is small, typically $\leq 0.5°$, it is well within the detection limits of simple optical hardware.

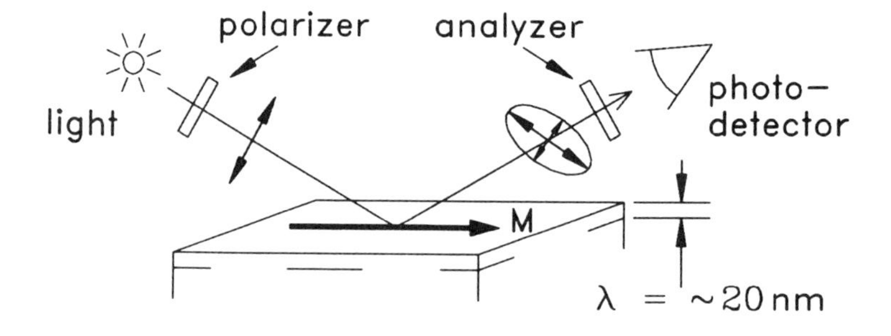

Figure 1 **Schematic diagram showing the basic elements of a MOKE experiment. The angle of incidence, the wavelength of the light, and the orientation of the magnetization, M, relative to the plane of incidence are variables in the experimental setup.**

Since MOKE is an optical probe, its lateral resolution is governed by the diffraction limit of the light source, from 0.3 µm to 0.5 µm for typical wavelengths. Its probing depth is determined by $I/I_0 = \exp(-t/\lambda)$, where the reflected light intensity from a given depth I_0 is attenuated to I for an optical path length t from the surface due to light absorption in the medium; the *absorption* is scaled by a characteristic attenuation length called the optical skin depth λ. For metals, which are good conductors, λ is of the order of 10–20 nm at visible frequencies. As a consequence of the fairly long probing depth of MOKE at optical wavelengths, it can be used to analyze ferromagnetic layers buried by 10 nm or so of an absorbing, nonmagnetic overlayer. Of course, there is no difficulty in obtaining Kerr-related signals from ferromagnetic layers that have been covered by transparent overlayers or vacuum isolated and examined through windows. Under some conditions these intervening transparent layers contribute to the ellipticity of the transmitted light. In fact, this effect can be used to an advantage by appropriately tuning the dielectric properties of the transparent layer to completely compensate the Kerr ellipticity. This results in an enhanced Kerr rotation,[1] which is used extensively in magneto-optic recording technology. For samples where the magnetic material has thickness $d \gg \lambda$ the technique is generally referred to as MOKE, whereas for $d \ll \lambda$ the acronym SMOKE (for *surface* MOKE) is sometimes used. This distinction highlights two important points for ultrathin-film Kerr analysis. First, while the Kerr effect is not intrinsically surface sensitive on the scale of many electron spectroscopies (which have signal attenuation lengths of 0.5–4 nm) it is, in effect, surface sensitive if the magnetic material is confined to the first few atomic layers of a sample, since the Kerr signal is only derived from the magnetic layers. Second, while the rotation of polarization is proportional to M of the magnetic material, it is also strongly influenced by the dielectric properties of the substrate,[2] as is described below.

The MOKE technique has a broad range of applications from the analysis of ultrathin films (less than about 2 nm) to the analysis of the near-surface region of bulk ferromagnets:

1 Hysteresis loops *M–H* have been determined for single atomic monolayers[3] of Fe and Ni that were prepared and measured in ultrahigh vacuum.
2 Maps of the remanent magnetic domain pattern in the near-surface region of magnetic material and thin films can be made routinely.
3 Dynamic domain imaging or Kerr microscopy of low coercivity thin films at MHz domain-switching frequencies allows one to examine domain wall motion in detail.[4]
4 The technology of magneto-optic recording is based on measuring the MOKE signal from remanently magnetized domain patterns in buried magnetic layers.
5 The electron–photon coupling that forms the microscopic basis of MOKE makes it possible, in principle, to determine the electron spin-dependent band structure of elements and alloys.[5] This is done by examining the dependence of the Kerr response on the wavelength of the incident light.

From a practical sense, MOKE is a versatile technique: it is an optical method; the polarization measurement is fairly easy to do; the necessary optical components are common and relatively inexpensive; and it has no intrinsic vacuum requirements.

History and Basic Principles

The first magneto-optic effects were discovered in transparent paramagnetic materials in the presence of a magnetic field by Faraday in 1846. He observed a rotation of polarized light transmitted through the material that depended on the magnitude of an axial magnetic field. This effect is generally quite small, on the order of 1°/cm of material in 10^4 gauss. Later, the polarization rotation of transmitted light through ferromagnetic films was measured and found to be large, on the order of 300,000°/cm in Fe, for example. In 1876, Kerr observed a rotation of the polarization of light reflected from a ferromagnetic surface. Kerr also discovered an electro-optic effect that bears his name, but it is unrelated to the magneto-optic effect of interest here. The Faraday and Kerr magneto-optic effects manifest themselves in a rotation of the polarization of the incident light and in a change in the ellipticity of the polarization upon interaction with a magnetic material. For the purposes of this discussion, it is convenient to describe linearly polarized incident light with components perpendicular to the plane of incidence, generally called *s*-light or *s*-polarized light, and parallel to the plane of incidence, generally called *p*-light or *p*-polarized light. The changes in polarization upon reflection are described schematically in Figure 2 for the case of incident *p*-polarized light. The signs of the rotation and the

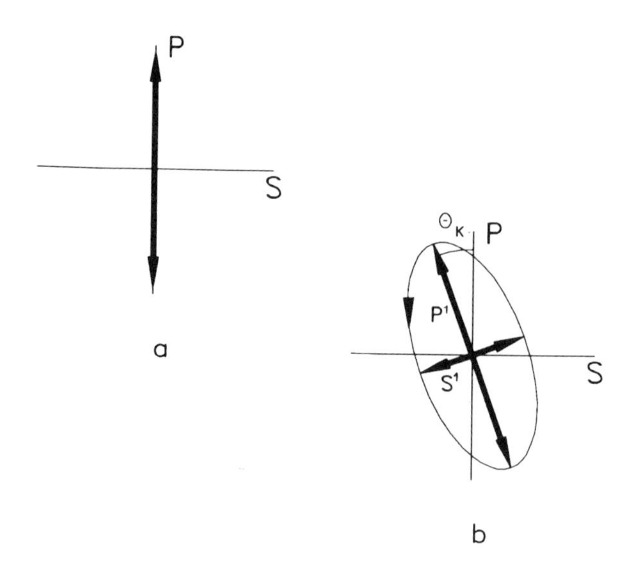

Figure 2 **Projection of the time-dependent *E*-field vector of light in the plane transverse to the direction of propagation for linearly polarized light (a), as for incident *p*-light in MOKE, and rotated and elliptically polarized light (b), which is the general case for light reflected from a Kerr-active surface.**

ellipticity changes invert when the magnetization of the sample is reversed. The measured rotation is given by θ_K and the ellipticity is $\varepsilon_K = \tan(S'/P')$, as defined in Figure 2.

The macroscopic optical analysis[6] of these effects requires the introduction of two complex indexes of refraction for the ferromagnetic material, one for left-circularly polarized light and another for right-circularly polarized light, which to first order, are given by

$$N_{(r,l)} \approx n \mp \frac{iQ}{2n} \tag{1}$$

Here, n is the complex index of refraction for the ferromagnet in the paramagnetic state, i.e., above the Curie temperature, and Q is the complex Kerr component. Since any polarization condition, including linearly polarized light, can be described as a linear combination of left- and right-circularly polarized light, the expected rotation and ellipticity of light reflected from a ferromagnet can be determined from standard solutions of the optics wave equations, using the appropriate boundary conditions for the structure of the sample and appropriate, known values for n and Q. In general, both n and Q must be measured for the material of interest. The result of this analysis for the highly symmetric case in which the light is nor-

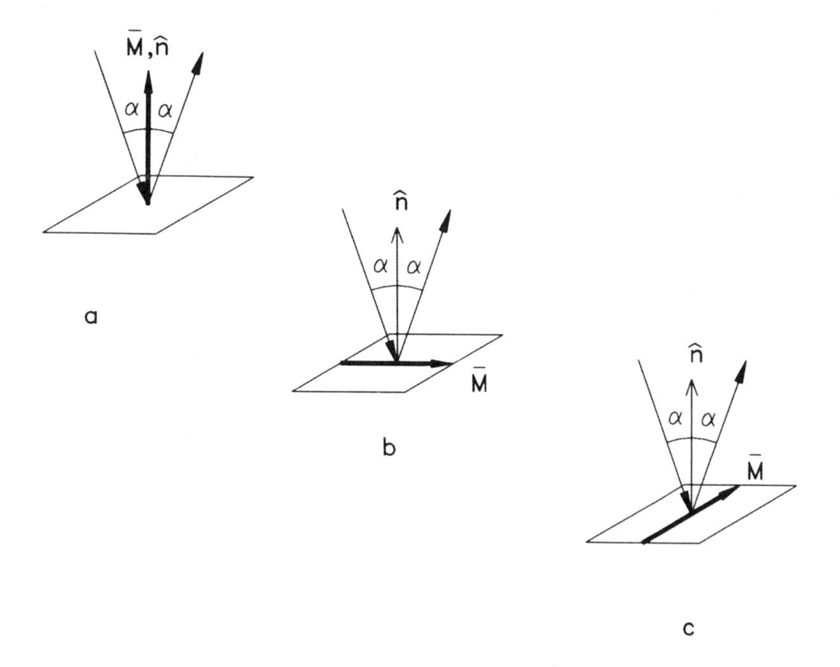

Figure 3 **MOKE geometries for light incident at an angle α with respect to the sample normal $\hat{n}$: polar geometry (a), longitudinal geometry (b), and transverse geometry (c).**

mally incident and the magnetic field is normal to the surface (polar geometry) gives

$$\Phi_K = \theta_K - i\varepsilon_K \approx \frac{-iQ}{n(n^2 - 1)} \tag{2}$$

for the Kerr effect response. Two important conclusions can be drawn from the these results. First, a rearrangement of the preceding relationships for θ_K and ε_K show that there is no MOKE activity when there is no absorption in the magnetic medium, i.e, when n and Q are real. Second, the magnitude of the Kerr rotation depends on the optical dielectric properties of the medium in the near-surface region, as well as on the Kerr activity.

Analyses and results for other geometries are more complicated and will be discussed only qualitatively. The general case of light reflected from a magnetized surface always can be reduced to combinations of the polar geometry, and two other special cases, the longitudinal (or meridional) geometry, and the transverse (or equatorial) geometry. The geometries of all three cases are defined in Figure 3.

From the solutions of the optical wave equations for these boundary conditions, the following statements can be verified;

1. There is no MOKE response for normally incident *p*- or *s*-light in either the longitudinal or transverse geometries.
2. There is no MOKE response for pure *s*-light for any angle of incidence in the transverse geometry.
3. There is no Kerr rotation or Kerr ellipticity for pure *p*-light or for mixed *s*- and *p*-light in the transverse geometry, but there is a magnetization induced change in the surface reflectivity.
4. There is a larger MOKE response at normal incidence for the polar geometry than for oblique angles.
5. There is a larger MOKE response as the angle of incidence becomes more oblique for the longitudinal geometry up to a maximum at an angle of 60° to 80°, depending on the specific material.

The longitudinal Kerr response versus angle is observed to be a slowly varying function with a broad maximum, and there is substantial variation in the angle for the maximum Kerr response reported in the experimental literature.[7] It is useful to note some other practical considerations regarding MOKE measurements. Since the reflectivity of *p*-light from metals is lower than that of *s*-light,[8] it is usually the case that the measured Kerr rotations are larger for incident *p*-light. The maximum MOKE response in the polar geometry is typically found to be 3–5 times greater than the maximum MOKE response for the other two geometries for a given material. The magnitude of the MOKE response has a wavelength dependence that results from the different possible intraband and interband transitions excited in a material's electronic band structure by a given wavelength of light. Some effort has been devoted to obtaining the magnetic band structure of a material through interpretation of the Kerr wavelength dependence. The information is complementary to magnetic band maps derived from spin-resolved photoemission.[9] Additionally, the wavelength dependence of MOKE may be of rudimentary use for distinguishing the Kerr signal of different ferromagnetic elements in an alloy or a layered structure.

The physical origin of Q lies in the microscopic quantum behavior of electron-photon interactions in a magnetic solid. It is not possible to discuss this subject in any detail here; however, a few general comments can be made. Many early attempts to explain MOKE[6] used classical concepts and incorrectly attributed MOKE to a Lorentz type force caused by interactions between a ferromagnet's internal field and the electrons in the solid, which are also in the presence of a photon field. It is now generally agreed that the Kerr effect is a consequence of coupling between the spin of the moving electron in the solid and its momentum due to the net atomic potentials of the medium, i.e., spin–orbit coupling, a purely quantum

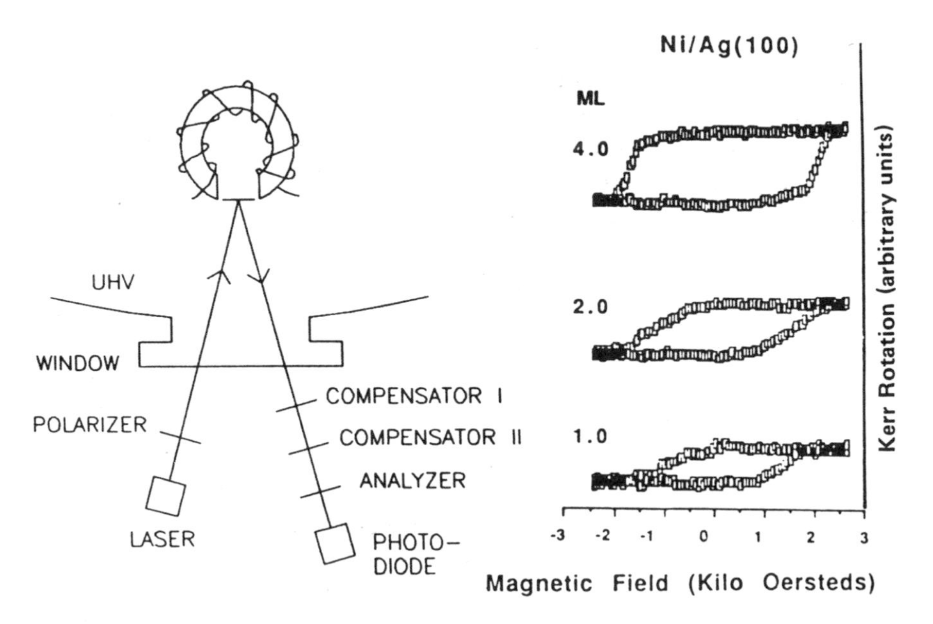

Figure 4 **Schematic of an ultrahigh-vacuum MOKE experiment using the longitudinal geometry (a). An electromagnet is used to magnetize the sample in vacuum. The polarized light from a laser source shines through a window onto the sample; the reflected light passes through the window, a compensator (not required when only θ_K is measured), and an analyzing polarizer onto the photodiode detector.[11] Experimental hysteresis loops (b) for different thicknesses, in monolayers (ML), of Ni on Ag (100) taken by MOKE using a similar apparatus to (a), with the sample at 110 K.[3]**

mechanical interaction. One consequence of the spin–orbit coupling is that the Kerr activity will be relatively large for materials with a large magnetic moment and with high atomic number. The first complete microscopic quantum theory of this effect was worked out by Argyres in 1955.[10] Nevertheless, it is still quite difficult, even today, to calculate the value of Q from first principles. Thus, absolute values of magnetization can, in general, be derived from MOKE only through careful calibration of appropriate known standards. Such calibrations require considerable care and often are unreliable, since the thickness and dielectric properties of the substrate, nonmagnetic overlayers, or windows in the optical path can have a considerable effect on the measured Kerr rotation in a real experiment.

Instrumentation

The instrumentation required to measure the hysteresis loop of a ferromagnetic surface with MOKE can be very simple. Figure 4a shows one such implementation of the experimental setup for analysis of ultrathin film samples maintained in a ultrahigh-vacuum environment.[11] The Kerr rotation θ_K and the ellipticity ε_K can be determined from a dc measurement in this setup, although ac detection schemes are straightforward to implement using polarization modulation. Often an even simpler setup is used that does not include one, or either, of the compensators. In this case, a single rotation angle is measured that is directly related to θ_K but which is modified by any ellipticity introduced by the vacuum window. Figure 4b is an example of hysteresis loops measured for Ni films of 1–4 atomic layers using this general type of apparatus.[3a] The acceptable *S/N* shown here for this extreme thin-film limit demonstrates the broad applicability of MOKE. Many variations of this simple setup are discussed in the literature.[3] Additional cautionary notes are warranted here. Some of the components of the setup, especially the vacuum windows, may distort slightly as a function of the external *H* field and cause artifacts in the *M–H* loops by adding undetermined ellipticity to the true MOKE signal. Another problem may arise if vacuum containment windows are included in the experimental light path (see Figure 4, for example) when changes in the sample temperature are required, e.g., when measuring the temperature dependence of *M*. While MOKE has no inherent temperature dependence, small distortions of the sample or the experimental apparatus during heating and cooling may change the light path. Vacuum windows often are under nonuniform states of stress and very slight changes in the light path can result in spurious rotations and ellipticity changes of the light that are comparable to the Kerr response. Clearly, when $M(T)$ is to be measured by MOKE, such polarization changes must be avoided or compensated. Temperature dependent MOKE data have been reported where critical parameters of the magnetic state in ultrathin films were derived from the resulting, measured $M(T)$.[3,12]

Imaging magnetic domains with MOKE or any other optical or electron imaging technique requires substantial magnification, since domains in interesting samples are typically small, on the order of 1–100 μm. Therefore, the instrumentation for Kerr microscopy requires the incorporation of basic polarization detection components with a high-quality optical microscope. In addition, much of the interest in imaging domains is in examining their dynamic behavior. This requires the addition of fast electronics capable of making MOKE measurements from ferromagnetic films switching at frequencies surpassing 1 MHz.[4] Figure 5 is an example of one such setup to image the dynamics of magnetic domains of magnetic recording heads using MOKE.[4b] Such microscopes generate a magnetization map of the head by spatially measuring the Kerr rotation in a point-by-point scan in steps of ~1 μm. Typically, the sample is moved under the polarized light, which is focused to about a 0.5-μm spot. Both the polar and longitudinal geometries are used in this setup. One useful kind of image obtained with this type of microscope shows just the changes in magnetization during a complete cycling of the head. (See Figure 6) The bright zones show the regions of the largest change in *M*, i.e., where the domain walls moved during a complete magnetization cycle. A variation on the

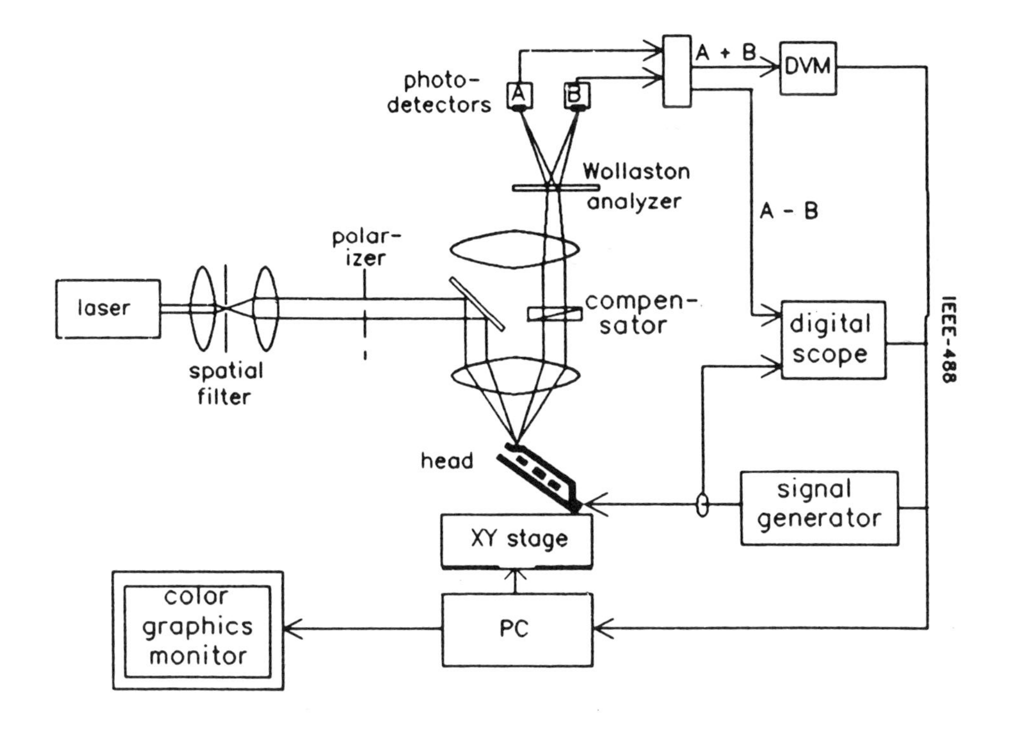

Figure 5 **Schematic diagram of a Kerr microscopy apparatus.[4]**

apparatus shown in Figure 5 provides a "parallel" method of generating the domain image.[13] A pulsed light source is defocused to illuminate a wide field of view in the microscope and the reflected light is collected by a time-synchronized video camera. A digital image for each time interval of interest can be stored and processed. The principle difference between the data provided by this approach and that of Figure 5 is that the former gives a domain image of a state of magnetization on a point-by-point basis, which is the average of many magnetization cycles, whereas the latter gives the instantaneous picture at all points in the field of view, at the resolution of the video digitizer, for a magnetization state from a single cycle. In the wide-field case the ratio S/N will be lower for a given spatial resolution. In either case, Kerr microscopy can be done using samples, with or without, thick transparent overlayers.

An interesting variation of the scanned sample, point-by-point experimental setup is the technological application of magneto-optic recording. A magneto-optic disk system is basically a miniaturized Kerr microscope using only the polar geometry. The magnetic sample or disk is rotated under the microscope, which moves radially from one track of magnetic domains or data bits to the next.

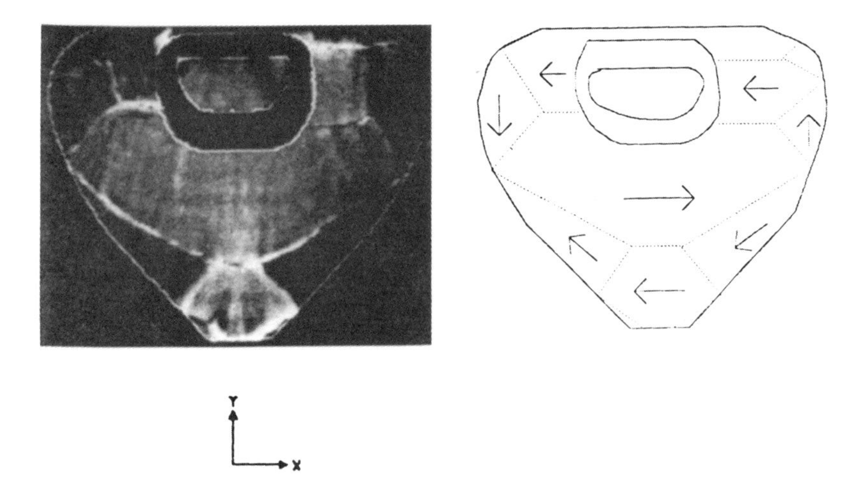

Figure 6 **Scanning Kerr image of the magnetization changes in the *y*-direction for a thin-film head having a 1-MHz, 5-mA p–p coil current, and the magnetic domain pattern deduced for this head from the observed domain wall motion.[4a]**

Comparison With Other Techniques

Even though MOKE is quite useful, it is not a panacea for all surface and interface magnetic characterization problems. It works best as a qualitative, diagnostic tool, as there is considerable difficulty in relating the Kerr rotation to the absolute value of M. This is the result of the complex coupling between the electronic band structure of the ferromagnet and incident light. Many other methods of characterizing ferromagnetic surfaces and interfaces exist, of course, and their use depends on which magnetic property, e.g., magnetization magnitude and orientation, magnetic anisotropy, or spin-dependent electronic band structure, is to be determined. Ferromagnetic resonance, FMR, is a technique that measures a quantity related to M and the anisotropy of the material. It is really a bulk technique, but can be applied to surfaces and interfaces in cases where these dominate the sample, as in ultrathin films and superlattices. Neutron reflectivity techniques, using polarized neutrons, are sensitive to the magnetization of a ferromagnet. These methods are rather specialized in view of the need for a neutron source, but they have the distinct advantage of being able to provide depth profiles of the magnetic moment. A large number of the standard electron spectroscopic techniques, including AES, UPS, LEED, and secondary electron emission, give signals related to the magnetic moment of a sample, when electron spin analysis is incorporated in the experiment. Any detailed comparison with these electron techniques is beyond our scope, but because of their widespread use it is worth noting some practical differences relative to MOKE analysis:

1 They all require substantially more sophisticated apparatus to make a measurement.
2 The application of external magnetic fields to the sample during analysis presents considerable problems and constraints for electron techniques, whereas external fields have no influence on MOKE.
3 Most electron probes have more elemental specificity.

Operationally, the electron techniques all require high-vacuum or, more likely, ultrahigh-vacuum environments, and the magnetic material of interest must be within a few atomic layers of the surface. MOKE analysis is not restricted by these constraints, although interesting samples may be.

Kerr microscopy is an excellent method for making magnetic domain images and maps. Other useful techniques include the Bitter colloid technique, secondary electron microscopy with polarization analysis (SEMPA), microscopy using other electron emission spectroscopies (like AES) and Lorentz transmission electron microscopy (LTEM). The Bitter method is simple to use, but compared to MOKE, it suffers in that it contaminates the sample's surface, it cannot be used in the presence of external fields and its intrinsic response time limits its ability to follow the dynamics of domain motion to very low frequencies. SEMPA and LTEM have all the advantages and disadvantages of the other spin-resolved electron techniques discussed. With regard to imaging, SEMPA can be used at a much higher magnification and lateral resolution (~50 nm) than MOKE, but cannot follow domain wall motion at relevant frequencies due to S/N considerations. LTEM currently has the highest magnetic spatial resolution available, when used in the differential phase contrast mode. One significant drawback to LTEM analysis is the special and tedious requirement that the entire sample to be analyzed be thin enough (0.1–0.3 μm) to transmit the high-energy incident electrons in TEM. Further details regarding SEM and TEM analysis are given elsewhere in this volume. None of the other magnetic domain imaging techniques has the inherent speed of MOKE microscopy, which obtains useful imaging signals from thin films at frequencies in the MHz range. Thus, it is really the only imaging technique with the requisite spatial resolution to examine the switching of high-speed recording heads used in the magnetic recording industry.

Conclusions

MOKE measurements can be made using relatively simple and inexpensive apparatus, compared to most other surface magnetic probes and surface analytical techniques. MOKE is useful for the magnetic characterization of films of one to several monolayers, thin films, or the near-surface regions of bulk materials. MOKE has been used most as a qualitative diagnostic of magnetic domain structure and domain dynamics. High spatial resolution can be obtained in this mode. Considerably less success has been achieved when attempting to make quantitative determinations of magnetic moment, with MOKE, because of the rather complex relationship of the Kerr rotation and ellipticity to the microscopic properties of the material, e.g., the magnetic moment and the electronic band structure.

MOKE will find continued application in analyses to determine the dynamics of domain motion in thin films and to determine values of the coercivity and information about magnetic anisotropies from hysteresis loop measurements. In addition, as the experimental method improves, reliable temperature-dependent MOKE measurements may become commonplace. There is considerable evidence that tuning the wavelength of the incident light to the dielectric properties of the magnetic film–substrate combination will offer somewhat enhanced Kerr rotations in the case of ultrathin films. As more wavelength-dependent studies are done and our knowledge of the detailed electronic band structure of alloys improves, one can expect progress in quantitative determinations of the magnetic moment of magnetic films and surfaces by MOKE. Nevertheless, a realistic assessment of MOKE in comparison with the spin-polarized electron spectroscopies suggests that the goal of obtaining moments at surfaces is more attainable using the electron spectroscopies. There is some potential for completely new methods and apparatus being developed in the field of MOKE analysis. For example, recent predictions[14] give rise to the possibility of developing intrinsically surface sensitive MOKE through the use of second harmonic detection of the reflected light. In such an experiment the intensity of the second harmonic light is quite low, but this light is emitted only from the surface or interfaces of the sample.

Related Articles in the Encyclopedia

SEM, TEM and VASE

References

1 K. Egashira and T. Yamada. *J. Appl. Phys.* **45,** 3643, 1974.

2 T. Katayama, Y. Suzuki, H. Awano, Y. Nishihara, and N. Koshizuka. *Phys. Rev. Lett.* **60**, 1426, 1988. A good demonstration of how the substrate influences MOKE in ultrathin films.

3 C. A. Ballentine, R. L. Fink, J. Araya-Pochet, and J. L. Erskine. *Appl. Phys. A.* **49,** 459, 1989; S. D. Bader, E. R. Moog, and P. Grünberg. *J. Magn. Magn. Mat.* **53,** **L**295, 1986.

4 P. Kasiraj, R. M. Shelby, J. S. Best, and D. E. Horne. *IEEE Trans. Mag.* **MAG-22,** 837, 1986; P. Kasiraj, D. E. Horne, and J. S. Best. *IEEE Trans. Mag.* **MAG-23,** 2161, 1987.

5 W. Reim. *J. Magn. Magn. Mat.* **58,** 1, 1986.

6 A. V. Sokolov. *Optical Properties of Metals.* Elsevier, New York, 1967, Chapters 10 and 11. A very detailed, mathematical description of solutions to the wave equations, with a nice historical perspective.

7 B. Thiel and H. Hoffmann. *J. Magn. Magn. Mat.* **6,** 309, 1977; C. C. Robinson. *J. Opt. Soc. Am.* **53,** 681, 1963. Dependence of MOKE on incidence angle.

8 P. Lorrain and D. Corson. *Electromagnetic Fields and Waves.* W. H. Freeman, San Francisco, 1970, p. 518. A general and clear text on electromagnetic wave phenomena at the undergraduate level.

9 T. E. Feuchtwang, P. H. Cutler, and J. Schmit. *Surf. Sci.* **75,** 401, 1978.

10 P. N. Argyres. *Phys. Rev.* **97,** 334, 1955. An excellent discussion of the quantum theoretical basis of MOKE.

11 E. R. Moog, C. Liu, S. D. Bader, and J. Zak. *Phys. Rev. B.* **39,** 6949, 1989.

12 C. Liu and S. D. Bader. in *Magnetic Properties of Low-Dimensional Systems* (L. M. Falicov, F. Mejia-Lira, and J. L. Moran-Lopez, eds.) Springer-Verlag, Berlin, 1990, vol. 2, p. 22.

13 D. A. Herman and B. E. Argyle. *IEEE Trans. Mag.* **MAG-22,** 772, 1986.

14 W. Hübner and K.-H. Bennemann. *Phys. Rev. B.* **40,** 5973, 1989.

12.4 Physical and Chemical Adsorption

Measurement of Solid Surface Areas

DAVID J. C. YATES

Contents

Introduction

When a gas comes in contact with a solid surface, under suitable conditions of temperature and pressure, the concentration of the gas (the adsorbate) is always found to be greater near the surface (the adsorbent) than in the bulk of the gas phase. This process is known as *adsorption.* In all solids, the surface atoms are influenced by unbalanced attractive forces normal to the surface plane; adsorption of gas molecules at the interface partially restores the balance of forces. Adsorption is spontaneous and is accompanied by a decrease in the free energy of the system. In the gas phase the adsorbate has three degrees of freedom; in the adsorbed phase it has only two. This decrease in entropy means that the adsorption process is always exothermic. Adsorption may be either physical or chemical in nature. In the former, the process is dominated by molecular interaction forces, e.g., van der Waals and dispersion forces.[1] The formation of the physically adsorbed layer is analogous to the condensation of a vapor into a liquid; in fact, the heat of adsorption for this process is similar to that of liquefaction.

Chemical adsorption (known as *chemisorption*) often, but not invariably, involves the formation of a chemical bond (i.e., the transfer of electrons) between gas and the solid. In other words, a specific chemical compound one layer thick

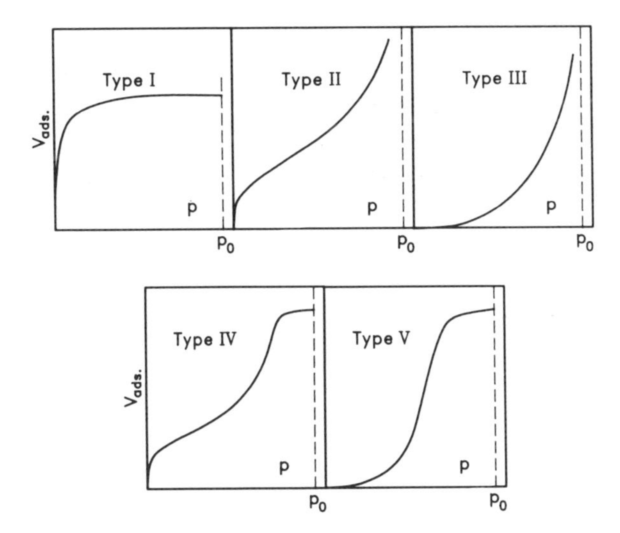

Figure 1 **The five types of physical adsorption isotherms.**[3]

(a *monolayer*) is formed between the adsorbate and the outer layer of the solid adsorbent The differences between physical and chemical adsorption have been discussed in detail,[1] but briefly they are as follows. Physically adsorbed gases can be removed readily from the adsorbent by evacuation at the same temperature at which adsorption took place. Chemisorbed gases can be removed isothermally in vacuum only rarely, especially on metal surfaces, where heating well above the adsorption temperature is usually required for complete desorption.

In this article, we will discuss the use of physical adsorption to determine the total surface areas of finely divided powders or solids, e.g., clay, carbon black, silica, inorganic pigments, polymers, alumina, and so forth. The use of chemisorption is confined to the measurements of metal surface areas of finely divided metals, such as powders, evaporated metal films, and those found in supported metal catalysts.

Surface Areas by the Brunauer, Emmett and Teller BET Method

The basic measurement of adsorption is the amount adsorbed v, which usually is given in units of cm^3 of gas adsorbed per gram of adsorbent. Usually this quantity is measured at constant temperature as a function of pressure p (in mm Hg), and hence is termed an isotherm. Isobars and isosteres also can be measured, but have little practical utility.[1] It has been found that isotherms of many types exist, but the five basic isotherm shapes are shown in Figure 1, where p_0 is the vapor pressure.

Type I isotherms occur only in systems where adsorption does not proceed beyond about one monolayer, and are found only in porous materials where the pore size is of molecular dimensions (e.g., zeolites and some carbons). Types IV and V are characteristic of multilayer adsorption on highly porous adsorbents, the flattening of isotherms near the saturation vapor pressure being due to the filling of capillaries. Type II isotherms are observed very commonly, being essentially the same as type IV isotherms except for the lack of capillary filling. Type III isotherms are but rarely found; they are usually a sign of a nonwetting system (e.g., the adsorption of water on a hydrophobic substance like pure charcoal). We will restrict our discussion in this article to types II and IV, as these isotherms are the only ones where the BET equation is applicable.

From an experimental point of view, the first step in determining an adsorption isotherm is to outgas the adsorbent. In the case of physical adsorption, this step serves to remove adsorbed water, and possibly other adsorbed gases, from the surface. This cannot be done efficiently by evacuation at room temperature, and is usually accelerated by heating the adsorbent. However, care must be taken to insure that the temperature maintained during the outgassing is not such[1] as to sinter the solid, and so decrease its surface area. In fact, it is only very recently that standardized methods for adsorption have been approved by the American Society for Testing and Materials (ASTM) Committee on Catalysts. In the formulation of the surface area standard,[2] the choice of degassing temperature occupied a considerable time, as no standard, or even recommendation for this temperature could be found in the earlier literature.[1] (This point, together with a detailed procedure for multipoint nitrogen isotherm determination for surface area by the BET method,[3] are discussed in the ASTM standard.) After outgassing, the next step is to determine the *dead space.* This is done by admitting a nonadsorbing gas, usually helium, when the cell is at the temperature at which the isotherm is to be measured. The initial pressure of the helium is measured by the pressure gauge in a system of accurately known volume. If a mercury manometer is used, the measuring side of the manometer must be kept at a fixed point in space when measuring the pressures. The author has found a commercially available quartz Bourdon gauge (of constant volume) to be much superior to any liquid manometer; the quartz gauge also has the advantage of keeping the system free of mercury or oil vapors. For example, a typical gauge I used[4] measured pressure changes as small as 0.004 cm, which is much better than can be realized with mercury manometers. In other words, helium at a known pressure and volume is admitted to the sample cell. From the pressure drop, the *effective* volume of the cell is determined. This procedure should be repeated several times. After this, the helium is evacuated, the cell shut and N_2 is admitted to the calibrated gauge volume. The gas pressure is measured, and the N_2 is admitted to the sample, which is held at 77 K. After allowing the pressure to come to equilibrium (which may take up to one hour) the final pressure is recorded. The same pro-[illegible] is followed for the second and subsequent doses. The amount adsorbed is

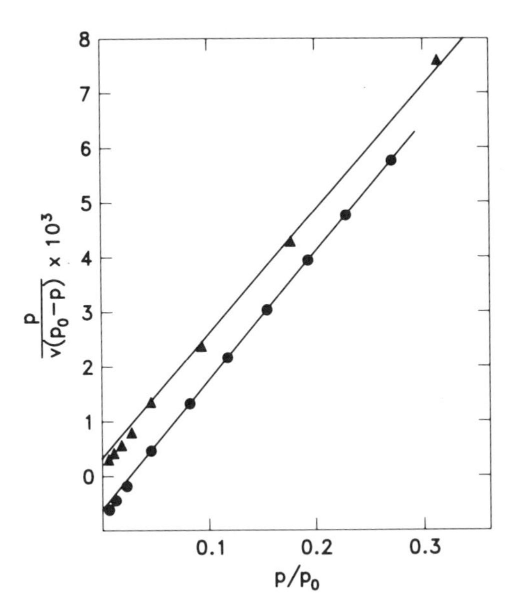

Figure 2 **BET plots for N_2 at 90 K (filled circles) and > 77 K (filled triangles) on porous glass.[5]**

calculated from the decrease in pressure beyond that expected from the increase in volume when the gas is admitted to the sample cell. Details of this procedure are given elsewhere.[2] Using the BET equation:

$$\frac{p}{v(p_0 - p)} = \frac{1}{v_m} + \left(\frac{c-1}{v_m c}\right)\frac{p}{p_0}$$

one plots $p/\,v(p_0-p)$ against p/p_0, as shown in Figure 2 for the adsorption of nitrogen at 77 K and 90 K on porous glass.[5]

In the BET equation, v is the volume of gas adsorbed in cm^3 at STP, per gram of solid at a pressure p. The constant c is related to the heat of adsorption[3] and v_m is the monolayer volume. From the plot one finds the slope $(1/\,v_m)$ and the intercept $1/\,cv_m$, so one can calculate v_m and c. At 90 K, v_m for the glass was 41.3 cm^3/gm, and assuming that nitrogen has a molecular area σ of 16.2 $Å^2$, one obtains[5] a surface area Σ of 179 m^2/gm. This value is obtained by using the equation $\Sigma = 0.269\,\sigma v_m$, where σ is in $Å^2$, v_m is in cm^3/gm, Σ is in m^2/gm, and the numerical factor comes from Loschmidt's number (molecules per cm^3 at STP). The values of σ usually are derived from the density of the liquid, and 16.2 $Å^2$ is the accepted value for nitrogen.[1, 2] Values for other gases are given in Young and Crowell,[1] and this parameter is a critical one in determining surface area from the monolayer capacity. Finally, it should be noted that when isotherms are measured close to the boiling point of the adsorbed gas, e.g., for nitrogen adsorbed at 77 K, the range of validity of the BET equation is in the relative pressure (p/p_0) range between 0.05 and 0.35, so that there is little point in measuring isotherms beyond those limits. In general, for best accuracy, at least four points should be measured in this region. Several other methods have been proposed for measuring surface areas by gas adsorption, (for example, the so-called *single-point* BET, in which the intercept value is ignored) and also by flow techniques (e.g., adsorption of N_2 from N_2:He mixtures) using a gas chromatography method, but an extensive review of these methods by the ASTM Committee on catalysts concluded that none were reliable, or as accurate, as the multipoint BET method, determined volumetrically.

While gravimetric methods that measure adsorption directly, for example, using helical quartz springs,[1] have been used in the past, the sensitivity in mass change per gram of adsorbent of early devices was very poor, and they were notoriously fragile and difficult to use. However, modern beam balances (e.g., the Cahn microbalance) have a much higher sensitivity, and their use has given accurate isotherms for ethylene adsorbed on a zeolite.[4] It also has been shown that the volumetric and gravimetric isotherms were in excellent agreement. This technique, however, is clearly more accurate for adsorbates of higher molecular weight; for instance, it has been rarely used to study the adsorption of hydrogen in chemisorption.

Chemisorption

This is a process that takes place via specific chemical forces, and the process is unique to the adsorbent or adsorbate used. In general, it is studied at temperatures much higher than those of the boiling point of the adsorbate; consequently, if supported metals are studied, little or no physical adsorption of the chemisorbing gas takes place on the high surface area support.

In particular, emphasis will be placed on the use of chemisorption to measure the metal dispersion, metal area, or particle size of catalytically active metals supported on nonreducible oxides such as the refractory oxides, silica, alumina, silica-alumina, and zeolites. In contrast to physical adsorption, there are no complete books devoted to this aspect of catalyst characterization; however, there is a chapter in Anderson[6] that discusses the subject.

As this field is very wide, we will discuss first the gases that can be used to study metal dispersion by selective chemisorption, and then some specific examples of their application. The choice of gases, is, of course, restricted to those that will strongly chemisorb on the metal, but will not physically adsorb on the support. Prior to determining the chemisorption isotherm, the metal must be reduced in flowing hydrogen; details are given elsewhere.[2] The isotherm measurement is identical to that used in physical adsorption.

The gases that have been used most often are hydrogen, carbon monoxide, and oxygen. Hydrogen is by far the most useful, and it has the best established adsorp- mechanism. It dissociates at room temperature on most clean metal surfaces of

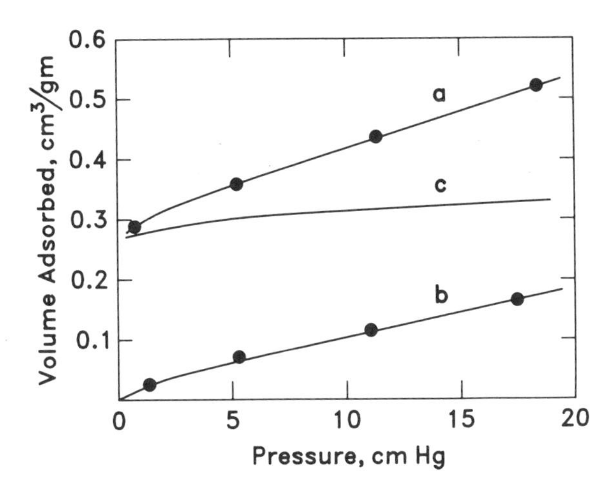

Figure 3 Adsorption isotherms[7] at room temperature on 0.1% Rh on silica: (a) CO isotherm; (b) CO isotherm after pumping out sample at end of isotherm (a) for 1 min; (c) difference between (a) and (b).

practical interest, one hydrogen atom being adsorbed per surface metal atom at saturation exposure. Due to its very low boiling point (20 K) there is no physical adsorption at room temperature on any support. Hydrogen can be used for all metals, with the exception of Pd, which forms bulk hydrides. If one knows the area occupied by one metal atom on the surface (readily obtained from crystallographic data) one can easily calculate the metal surface area from the monolayer capacity (v_m) of the adsorbed H_2.

The use of CO is complicated by the fact that two forms of adsorption—linear and bridged—have been shown by infrared (IR) spectroscopy to occur on most metal surfaces. For both forms, the molecule usually remains intact (i.e., no dissociation occurs). In the linear form the carbon end is attached to one metal atom, while in the bridged form it is attached to two metal atoms. Hence, if independent IR studies on an identical catalyst, identically reduced, show that *all* of the CO is either in the linear or the bridged form, then the measurement of CO isotherms can be used to determine metal dispersions. A metal for which CO cannot be used is nickel, due to the rapid formation of nickel carbonyl on clean nickel surfaces. Although CO has a relatively low boiling point, at very low metal concentrations (e.g., 0.1% Rh) the amount of CO adsorbed on the support can be as much as 25% of that on the metal; a procedure has been developed to accurately correct for this.[7] Also, CO dissociates on some metal surfaces (e.g., W and Mo), on which the method cannot be used.

Although considerable study has been devoted to oxygen chemisorption (mainly on platinum) there is considerable ambiguity in the surface stoichiometry of the reaction. In some cases Pt_2O is formed, in others PtO, the particular compound being a function of different lattice planes exposed.[8] Hence, even for a noble metal like Pt, O_2 chemisorption is unreliable as a means of measuring metal surface area.[8] It also follows that other, more complicated methods, e.g., those involving the titration of adsorbed O_2 with H_2 are unreliable, and should not be used.

In contrast to such procedures as the use of the BET equation, there are no general procedures that can be specified as standards for chemisorption. For example, ASTM has published a detailed, standard method specifically for measuring H_2 chemisorption for platinum on alumina catalysts.[2] Not only is this restricted to one metal, but also to one support. In other words, there are specific features of the Pt on Al_2O_3 system which may well not be the same for the Pt on SiO_2 system. In addition, it is well known that other noble metals on Al_2O_3 behave very differently from Pt. As an example, in the standard test method, before the reduction of the catalyst it is advised that the catalyst be calcined in air. This is done by flowing air (using a special flow through adsorption cell, see Figure 2 of *Standard Test Method for Hydrogen Chemisorption on Supported Platinum on Alumina Catalysts*[2]) over the catalyst, then heating to 450° C, and holding for 1 hr at this temperature. After evacuating the air at 425° C, the catalyst is cooled. Hydrogen is then passed through the cell, and the sample is heated again to 450° C. After a hold period, the hydrogen flow is stopped, and the cell evacuated at 425° C. After cooling again, the chemisorption of H_2 on the reduced catalyst is measured at room temperature. In the case of Pt on Al_2O_3 catalysts with Pt loading about 0.3%, the particle size of the platinum is very small, below 20 Å. In fact, one of the earliest[9] measurements of Pt dispersion on alumina, showed that nearly every Pt atom in the sample adsorbed one hydrogen atom, corresponding to a particle size of 10 Å. However, using the same support, alumina, but using 1% of iridium, the author has shown that a freshly prepared catalyst (air dried at 110° C) had an Ir particle size of 15 Å. If the procedure from *Standard Test Method for Surface Area of Catalysts* was followed (an *in situ* oxidation at 450° C), followed by a H_2 reduction at 450° C, then the Ir particle size was found to be 50 Å. In other words, air treatment of a platinum catalyst at 450° C has no deleterious effect on the metal dispersion, while drastically reducing the dispersion of iridium. Thus, following the "standard" procedure[2] for platinum would give entirely erroneous data for iridium. This is despite the general similarity of the two elements; it is due to the fact that after an oxidation treatment platinum remains as metallic particles, while the iridium forms 50A oxide particles on heating to 540° C,[10] as shown by X-ray diffraction.

To give an idea of the wide range of catalytic systems that have been investigated where chemisorption data were essential to interpret the results, some of the author's papers will be discussed. Measurements were reported on the surface areas of a very wide range of metals that catalyze the hydrogenation of ethane.[7, 11–14] In the earliest paper, on nickel, the specific catalytic activity of a supported metal was ··rately measured for the first time; it was shown also that the reaction rate was ··roportional to the nickel surface area.[11] Studies on the same reaction showed quite a marked variation in the nickel surface area when the support was changed—a factor of 2. However, a factor of 50 was found between the most active and least active catalyst, when expressed on a constant nickel surface area basis, due to effect of the support. Later work showed that the specific catalytic activity for 1% Ni on SiO_2 was about a hundredfold less than that of 10% Ni on the identical support. Using the most "inert" support, pure SiO_2, a series of metals were studied (Co, Pt, and Cu), all at 10% concentration and the specific catalytic activities of the metals were found to follow the order Ni > Co > Pt.[12] Other noble metals of known particle size were studied on the same silica,[12] and the order of activity was found to be Ru > Rh > Ir > Pd = Pt. For all group VIII metals the relative specific activity was not correlatable with any single property of the metal itself (e.g., % d character) but was dependent on both the % d character and the atomic radius.[13] The elements Ni and Co were on one line in the relation between the specific activity and % d character and the noble metals on another, very different, line (i.e., Ru, Rh, Pd, Ir and Pt).

Finally, the same reaction was studied with a series of unsupported Ni:Cu alloys.[14] In all cases, the total surface areas (equal to the metal area for an unsupported metal) of the pure catalysts and the eight alloys were measured by the BET method, using argon isotherms measured at 77 K. For all catalysts, hydrogen chemisorption was also measured. After the first H_2 isotherm, the catalysts were evacuated for 10 minutes, and H_2 readsorbed. The second isotherm measured the weakly adsorbed H_2. With pure copper, as expected, the total H_2 chemisorption was the same as the weakly adsorbed H_2; in other words, there was *no* strongly adsorbed H_2 on pure Cu. For pure Ni, in contrast, the weakly adsorbed H_2 was about 20% of the total H_2 adsorption. This technique enabled the *surface* composition of the alloys to be directly measured. It was shown for the first time that a catalyst containing 5% Cu overall had a surface composition in the range of 50% Cu. It should be noted here that although one can estimate the surface composition of such alloys by ESCA, Auger, or other spectroscopic techniques, they usually do not give selectively the composition of just the surface layer of metal atoms. The Auger electrons, for example, come from an escape depth equivalent to several atom layers. Auger spectroscopy is very difficult to use with insulating systems; it is less of a problem with ESCA. Some techniques do have very short probing depths, e.g., low-energy ion scattering spectroscopy (ISS), which gives essentially the top layer composition.

Conclusions

Both of the surface area techniques described in this article are well established. However, the determination of total surface area by physical adsorption using the BET equation is a very general method of wide applicability. The use of selective chemisorption to determine the surface area of metals is much newer, and has only been systematically applied in the last 20 years or so. It is expected that more metals than those discussed will be studied in the future. As mentioned, the method of measuring the surface area of supported metals is by no means the same for all metals. For example, Pt and Ir catalysts have to be prepared, calcined, and reduced in very different fashions, depending on the chemistry of the particular metal. In other words, very much depends on the particular knowledge of the individual investigator in studying supported metals.

Related Articles in the Encyclopedia

None

References

1. D. M. Young and A. D.Crowell. *Physical Adsorption of Gases.* Butterworths, London, 1962.
2. *Standard Test Method for Surface Area of Catalysts.* (D3663–78); *Standard Test Method for Hydrogen Chemisorption on Supported Platinum on Alumina Catalysts.* (D3908–80) American Society for Testing and Materials (ASTM), Philadelphia, PA.
3. S. Brunauer, P. H. Emmett, and E. Teller. *J. Amer. Chem. Soc.* **60,** 309, 1938.
4. D. J. C. Yates. *J. Phys. Chem.* **70,** 3693, 1966.
5. D. J. C. Yates. *Proc. Roy. Soc.* **A224,** 526, 1954.
6. J. R. Anderson. *Structure of Metallic Catalysts.* Academic Press, London, 1975.
7. D. J. C. Yates and J. H. Sinfelt. *J. Catal.* **8,** 348, 1967.
8. G. R. Wilson and W. K. Hall. *J. Catal.* **17,** 190, 1970.
9. L. Spenadel and M. Boudart. *J. Phys.Chem.* **64,** 604, 1960.
10. D. J. C. Yates and W. S. Kmak. (1979) US patent no. 4,172,817.
11. D. J. C Yates, W. F. Taylor, and J. H. Sinfelt. *J. Amer.Chem.Soc.* **86,** 2996, 1964.
12. W. F. Taylor, D. J. C. Yates, and J. H. Sinfelt. *J. Phys.Chem.* **69,** 95, 1965.
13. J. H. Sinfelt and D. J. C. Yates. *J. Catal.* **8,** 82, 1967.
14. J. H. Sinfelt, J. L. Carter, and D. J. C. Yates. *J. Catal.* **24,** 283, 1972.

Index

D

E

F

G

H

I

J

K

L

M

N

O

T

U

V

W

X

Z